L - logic

1 = LV - inactive

0 = HV - active

H - logic

1 = HV - active

0 = LV - inactive

Digital Engineering Design
A Modern Approach

RICHARD F. TINDER
Professor of Electrical Engineering and Computer Science
Washington State University

Prentice Hall
Englewood Cliffs, New Jersey 07632

Library of Congress Cataloging-in-Publication Data
Tinder, Richard F., [date]
    Digital engineering design: a modern approach / by Richard F.
Tinder.
        p.   cm.
    Includes bibliographical references.
    ISBN 0–13–211707–X
    1. Digital electronics 2. Logic design. I. Title.
TK7868.D5T56 1991
621.39′5—dc20   89–23137
                                                        CIP

Editorial/production supervision and
  interior design: Fred Dahl and Rose Kernan
Pre-press manufacturing buyer: L. Behrens
Print & bind manufacturing buyer: P. Fraccio

© 1991 by Prentice-Hall, Inc.
A Division of Simon & Schuster
Englewood Cliffs, New Jersey 07632

Printed in the United States of America
10 9 8 7 6 5 4 3 2

ISBN 0-13-211707-X

Prentice-Hall International (UK) Limited, *London*
Prentice-Hall of Australia Pty. Limited, *Sydney*
Prentice-Hall Canada Inc., *Toronto*
Prentice-Hall Hispanoamericana, S.A.., *Mexico*
Prentice-Hall of India Private Limited, *New Delhi*
Prentice-Hall of Japan, Inc., *Tokyo*
Simon & Schuster Asia Pte. Ltd., *Singapore*
Editora Prentice-Hall do Brasil, Ltda., *Rio de Janeiro*

Contents

Chapter 5   SYNCHRONOUS SEQUENTIAL MACHINES, 320

Chapter 6 ASYNCHRONOUS SEQUENTIAL MACHINES, 520

Preface

Sound Engineering Design Practices

Certainly the most important feature of the text is that it emphasizes the successful engineering design of digital devices and machines from first principles. A special effort has been made not to "throw" logic circuits at the reader so that questions remain as to how the circuits came about or whether or not they will function correctly. An understanding of the intricacies of digital circuit design, particularly in the area of sequential machines, is given the highest priority—the emphasis is on error-free operation. From an engineering point of view, the design of a digital device or machine is of little or no value unless it performs the intended operation(s) reliably.

Superior Design Tools

The fields of digital electronics and digital design are rapidly evolving, as they should, and an appropriate change in nomenclature, symbology, and design methods with this evolution is essential. We are convinced that the MIXED (positive and negative) LOGIC notation and symbology and the entered variable mapping method are superior design tools that should be taught beginning at the entry level. The MIXED LOGIC symbology is commonly used by practicing engineers and is found in various VLSI (very large scale integrated circuit) data books. Its use avoids the need for multiple applications of DeMorgan's laws and provides a very useful bridge between the logical and physical domains. The entered variable mapping method often greatly facilitates the function reduction process and permits the effortless

ix

transfer of information from fully documented state diagrams to Karnaugh maps. This text develops these tools in exhaustive detail and then utilizes them throughout.

Innovative Approaches

A significant contribution of the text is the use of innovative approaches to the presentation of some conventional subject matter. In providing these innovative approachees, we have also made every effort to fill important gaps left by previously published texts. One example is the use of Karnaugh map XOR patterns and multilevel minimization methods to design arithmetic and logic units. Another example is the heuristic development of memory which we use as an introduction to sequential machine design. Also, we introduce some useful new methods for recognizing and eliminating output race glitches and static hazards in synchronous sequential machines. In the area of asynchronous sequential machines, we introduce the use of the fully documented state diagram as a basis for diagnosing and eliminating (or preventing) timing defects such as endless cycles, critical races, and essential hazards. The nested machine approach to design is developed and applied to a variety of asynchronous sequential finite state machines including flip-flops. Other significant contributions include the use of the handshake interface to design asynchronous modular elements, such as the arbiter module, and to develop externally asynchronous internally clocked (delay insensitive) systems. The text also includes an introduction to system-level design which is illustrated by example and by specially chosen problems at the end of chapters.

Subject Area Continuity

Certain chapters in the text are unusually long and purposely so. It is the author's position that certain subject areas, such as synchronous sequential machines, should not be "sliced up" into shorter instructional segments. To do so, we believe, breaks the continuity and limits the opportunity to integrate subject matter for presentation and problem solving purposes. Fragmentation of a chapter might also imply (as it often does) that the various parts are not interrelated or that one particular part is more important than another.

Many Worked and Unworked Exercises and Problems

There are more than 500 examples, exercises, and problems (worked and unworked, single and multiple part) that are provided to enhance the learning process. They range in complexity from simple algebraic manipulations and calculations to multipart system-level designs; each has been carefully chosen with a specific purpose in mind. The exercises and problems fall into four categories:

1. *Exemplary worked exercises* appear in the body of the text and are unnumbered. These exercises are worked out in painstaking detail, leaving little to the imagination of the reader.
2. *Worked exercises,* presented in the end of the developing chapters (Chap-

ters 2 and 3), extend and amplify the coverage given in the body of these chapters. Such problems serve as a study guide and reference source for review of the developmental material.

3. *Unworked problems* take their traditional place at the end of each chapter. These problems, often used for homework assignments, call for an understanding of material covered to that point in the text. The solutions to these unworked problems are available to the instructor in the *Instructor's Manual*.

4. *Supplementary exercises* are provided in the *Instructor's Manual*. Their purpose is to help the instructor provoke the thinking of the reader beyond a particular point in the text discussion.

Better Placement of Number Systems

The subjects of number systems are removed from their usual place at the beginning of the text and are presented in a timely and effective way within the text so that they can be meaningfully applied to the logic design process.

An Extensive Glossary

For the convenience of the reader, an extensive glossary is provided at the back of the text. The glossary includes the terms and jargon coined in this text in addition to the standard terminology associated with the field.

Chapter Appendices

Important subject matter peripheral to the contents of a given chapter is provided in an appendix at the end of that chapter rather than in the traditional location at the end of the text. This is done primarily to establish connectivity of the appended material with related material in the chapter. However, it also serves as a convenient source of such information.

Annotated References

A few carefully selected references are given at the end of each chapter. These references are annotated to help guide the reader to the best sources of information that are known to be available.

Absence of Manufacturer Device Codes

Conspicuously absent in the chapter discussions is any mention of manufacturer device codes. Because such device codes are inconsistent among manufacturers and are constantly changing, we have purposely avoided them to eliminate possible sources of confusion. Data books, which are readily available from the various manufacturers, provide this information.

Practical Alternatives to the Traditional Quine-McCluskey Tabular Reduction Method

The treatment of the Quine-McCluskey (Q-M) tabular minimization method is purposely omitted in favor of the less tedious entered variable (EV) map method which is emphasized in this text. The (EV) map method yields absolute minimum results for most applications and yields near optimum results for most multioutput optimization problems. For complex multiinput/multioutput optimization problems, computer-aided solutions are necessary. For this purpose there exist modern minimization algorithms and associated software that are faster and more versatile than those which use the Quine-McCluskey algorithm.

CHAPTER SUMMARY

Background

Chapter 1 provides sufficient electronic background information for a beginning-level student to understand the nature and behavior of digital circuits. Various families of electronic switches are presented, and their performance characteristics are discussed. The subject matter contained in this chapter is considered to be useful but *not* critical to the developments which follow in the remaining chapters.

Fundamentals of Digital Design

Chapter 2 establishes the mixed logic terminology and basic logic functions which form the basis for digital design as we present it. Use is made of simple transistor switching circuits to distinguish between the physical gates and the logic interpretations they may assume. The reader is gradually lead into a detailed discussion of mixed logic symbology and its usage in implementing Boolean expressions. The laws of Boolean algebra, including XOR algebra, are presented in an informal manner, and use is made of truth tables and shaded logic maps to verify the laws. The information contained in this chapter is considered critical to an understanding of the treatment given in subsequent chapters.

Function Representation and Reduction

The subject matter of Chapter 3 begins with a discussion of the two basic forms of Boolean algebraic representation, called sum-of-products (SOP) and product-of-sums (POS), and their interrelationship. From there the reader is systematically lead through tabular minimization and shaded logic maps to a detailed discussion of Karnaugh maps which are used to minimize logic functions in both SOP and POS form. Considerable effort is made to develop the entered variable mapping method preparatory for combinational and sequential machine design presented in the remaining three chapters. A special section on XOR patterns and multilevel minimization highlights the EV mapping method. Like Chapter 2, the contents of this chapter are considered to be of fundamental importance to further developments in this text.

Combinational Logic Design

Chapter 4 deals mainly with the design of conventional combinational logic circuits and their use as building blocks in digital circuit systems. Included are arithmetic circuits such as ADDERs, SUBTRACTORs, MULTIPLIERs, DIVIDERs, COMPARATORs and error detection circuits, and nonarithmetic devices such as DECODERs, ENCODERs, CODE CONVERTERs, MULTIPLEXERs, and SHIFTERs. Number systems, binary arithmetic, and codes are integrated into the design process as they are needed. The design and application of ALUs (arithmetic and logic units) and array logic devices such as ROMs (read-only memories) and PLAs (programmable logic arrays) are discussed in great detail. A special section on hazards introduces the reader to the subject of logic noise in digital circuits. Mixed logic notation and the entered variable mapping method are used extensively throughout this chapter.

Synchronous Sequential Machines

Chapter 5 introduces the reader to an important class of machines that have true memory (i.e., ''feedback'' memory) and that can issue time-dependent sequences of logic signals controlled by present and past input information. The treatment begins with the concept of sequential logic operations, moves to the models for synchronous sequential machines, and then details the essential features of the state diagram. From there the concept of memory is heuristically developed, and the basic building blocks for memory are discussed in detail. This is followed by the design of various types of flip-flops and their use as memory elements in the design of more complex synchronous sequential machines. Detailed discussions of logic optimization, state code assignment rules, the glitch-free generation of outputs, the problem of asynchronous inputs, clock skew, the initialization and reset of the sequential machine, and debouncing circuits place heavy emphasis on the error-free operation of these machines. Modular bit-slice devices such as counters and shift registers are discussed at some length and are used as the memory devices in the design of programmable sequential machines. The chapter concludes with an introduction to system-level design.

Asynchronous Sequential Machines

The contents of Chapter 6 centers on the design and analysis of self-timed (clock-independent) sequential machines. The lumped path delay model, for fundamental mode operation, is developed and then applied to the design and analysis of a variety of asynchronous sequential machines. Timing defects such as endless cycles, critical races, static hazards, and essential hazards are considered in great detail. This is followed by a discussion of the setup and hold-time requirements for proper state branching. Emphasis is given to the rules for state diagram construction since the state diagram is central to the discussions of timing defects and to the design and analysis of asynchronous FSMs. The nested machine approach to asynchronous sequential machine design is considered, and some useful applications, such as flip-flop design, are provided. The chapter next turns to a discussion of asynchronous sequential machine design in the pulse mode. This is followed by a section that describes the use of the handshake interface to design modular elements such as the ARBITER MODULE and a module that is used for externally asynchronous/internally clocked (EAIC) systems.

READERSHIP AND COURSE PREREQUISITES

The text is designed to be used mainly by undergraduate engineering students or by advanced students and professionals in the field who may not be familiar with contemporary digital design methodology. Also, the text should provide a sound background for those planning advanced study in the area of digital design.

No advanced math or electronics is required for successful usage of this text. However, a minimum maturity level equivalent to a college sophomore is strongly advised. A semester or quarter course in elementary circuit concepts, used as a prerequisite, would be helpful but is not essential. High school physics may suffice for this purpose.

SUGGESTED TEXT USAGE

The contents of this text can be taught in a two-semester course sequence or in a three-quarter course sequence. We suggest the following usage for a semester system:

First semester
    Chapter 1 (optional)
    Chapters 2 and 3
    Chapter 4
    Chapter 5 (through Sect. 5.7)
Second semester
    Chapter 5
    Chapter 6

For the quarter system we suggest the following usage:

First quarter
    Chapter 1 (optional)
    Chapters 2 and 3
    Chapter 4 (through Sect. 4.8)
Second quarter
    Chapter 4 (Sect. 4.9 to end)
    Chapter 5
Third quarter
    Chapter 6

More information on text usage is provided in the *Instructor's Manual*.

Acknowledgments

First and foremost I must acknowledge, with love and appreciation, the patience, the understanding, and the support of my wife, Gloria. Special recognition and thanks are also owed to my colleague and friend Mark Manwaring for years of consultation and countless useful comments and suggestions. I extend gratitude to my other professional colleagues, Jack Meador and David Seamans for their critical eye and helpful suggestions. Thanks are due to Edwin Jones, Jr., of Iowa State University for his many helpful and constructive remarks during the review process and to the several unknown reviewers of the manuscript for numerous helpful suggestions. Also, I wish to thank Robert Sproull of Sutherland, Sproull, and Associates, Inc., for the preview of his and Sutherland's unpublished book *Asynchronous Systems,* which was of considerable help to me in developing many of the modular elements appearing in Chapter 6 of the text.

A debt of gratitude is owed to previous chairs of our department, Glen Hower and Harriett Rigas, and to Yacov Shamash, our present chair, for their support over the past several years during which the manuscript was in preparation. Since portions of the text have been used for text material in at least two courses over many semesters, I must acknowledge the many helpful comments made by the student users of this material.

I am unable to acknowledge secretarial help in the preparation of the manuscript since I typed the manuscript and created the drawings all on computers. However, I would like to express my appreciation to Mrs. Julia Davis, a retired English teacher and mother of Gloria Tinder, for her help regarding grammatical questions.

These acknowledgments would not be complete without recognizing the months of cooperative, thorough and professional editorial work by Rose

P. Kernan of Inkwell Publishing Services. Good production editors are indispensable to the creation of a quality text.

Finally, I acknowledge with gratitude the support and encouragement of Bernard Goodwin and Tim Bozik, former electrical engineering editors at Prentice Hall, and Peter Janzow, Senior Editor.

For the sake of persons of different types, scientific truth should be presented in different forms, and should be regarded as equally scientific, whether it appears in the robust form and the vivid coloring of a physical illustration, or in the tenuity and paleness of a symbolic expression.

JAMES CLERK MAXWELL
Address to the Mathematics and
Physical Section, British Association
of Sciences, 1870

CHAPTER 1
Background

1.1 INTRODUCTION

It would be easy to understate the importance of the role that digital device technology plays in our modern world. To emphasize the point, try to imagine how our lives would be affected by the absence of automatic washing machines, microwave ovens, calculators, computers, digital watches, microprocessors, compact disk players, and so on. Imagine also how businesses, industries, governments, and (of course) the defense establishments would fare without the use of high-speed, multitasking digital systems for computations, communications, data processing, and measurements. Thus, there is no doubt that the modern world has progressed well into the digital age. There is also no doubt that there are many more exciting, awe-inspiring things to come.

The word *digital* connotes discreteness or discrete quantities. A digital system receives, stores, manipulates, and communicates information in discrete form. The digital control systems for the appliances and devices mentioned in the preceding paragraph are assemblages of interacting subsystems that, in turn, consist of smaller, perhaps modular, digital devices. Equipped with some kind of human interface, these digital systems perform useful, highly complex tasks and can do so with an exactness and speed unattainable by any other means.

Digital engineering design involves both design and analysis of digital devices and systems—design and analysis cannot usually be considered separately. Also, the correct engineering design of a digital system can result only if the designer makes the proper choice of parts to be assembled. However, to do this the designer must have a complete understanding of the operation of each part and must know the precise manner in which the various parts interact within the system. All this requires considerable knowledge of the field, skill, experience, imagination, and, at times, even artistic ability. In learning any new field, knowledge and skill must usually come first and in that order. Experience, imagination, and artistic ability will follow. In this chapter we begin to build a knowledge base by considering the building blocks, that is, the simple electronic switches of which digital devices are constructed.

1.2 ELECTRONIC SWITCHES

We are all very familiar with push-button or lever-type mechanical switches that are used frequently in our daily lives to turn things on or off. But such operator-controlled switches are of no use in a digital system other than to activate or deactivate the system, or to initiate or terminate some process within the system. In digital circuits, nonlinear devices such as diodes and transistors are used as high-speed switching devices controlled by voltage, current, or light. Transistors can be driven between conducting and non-conducting (ON/OFF) modes extremely fast with a minimum consumption of electrical power. In this section we will discuss the various devices used in switching circuits with emphasis on the properties that make them suitable for such purposes.

1.2.1 The Diode Switch

A *diode*, that is, a p-n junction diode, is a doped semiconductor device that acts essentially as a voltage-controlled switch. It is either "ON" or "OFF." In Fig. 1-1 is shown the prototype of such a diode, its circuit symbol and its forward and reverse bias orientations. The term *bias* is used here to mean voltage "influence." Voltage is represented by the symbol V and is given in units of volts [V].

Under forward bias the diode is in the ON state. Here, majority charge carriers (holes in the p side and electrons on the n side) are injected across the p-n junction, and current flows in the direction indicated by the "ar-

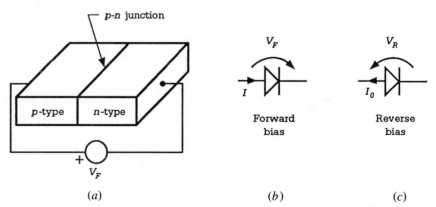

(a)	*(b)*	*(c)*

Fig. 1-1. *The* p-n *junction diode. (a) Physical prototype in forward bias. (b) Circuit symbol and current direction under forward bias. (c) Circuit symbol and current* I_0 *under reverse bias.*

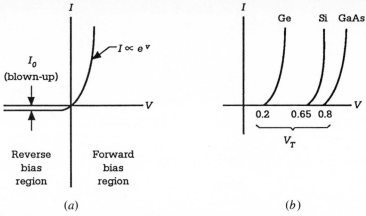

Fig. 1-2. I-V *characteristics for* p-n *junction diodes. (a) Typical* I-V *characteristic showing forward and reverse bias behavior. (b) Blown-up forward bias characteristics and threshold voltages for three different diodes.*

rowhead'' of the circuit symbol given in Fig. 1-1(b). A *hole* is defined as the absence of a valence electron. When the diode is under reverse bias, that is, OFF, the minority carriers (electrons on the p side and holes on the n side) are injected across the junction, permitting only a very small current I_0 (in amps [A]) to flow in the direction indicated in Fig. 1-1(c). The current I_0 is usually in the range of 10^{-12} A to 10^{-6} A and is called the *reverse saturation current.*

The *I-V* (current-voltage) characteristic for a typical diode is given in Fig. 1-2 together with the blown-up forward *I-V* characteristics for three diodes made of germanium (Ge), silicon (Si), and gallium arsenide (GaAs). Observe that the diodes of different semiconducting materials have different turn-on (threshold) voltages, V_T, which usually fall in the range of 0.2 V to 0.8 V. The threshold voltage is that voltage beyond which significant current flows in forward bias. Thus, its value for a given diode is somewhat arbitrary.

The current in a p-n junction diode varies exponentially with forward voltage across it. This exponential dependence is well known and can be derived from first principles by making a few simplifying assumptions. It is given by the equation

$$I = I_0[\exp(V/\eta V_{th}) - 1] \tag{1-1}$$

where I_0 is the reverse saturation current, V_{th} is the thermal voltage equal to about 26 mV (1 mV = 10^{-3} volts) at room temperature, and η is a fitting factor with values which lie in the range of about 1 and 2 for Si and GaAs, respectively.

Ideal Diode Models for Forward Bias. The ideal diode is a perfect voltage-controlled switch that has the *I-V* characteristic and equivalent circuits given in Fig. 1-3. For forward voltages greater than zero ($V_F > 0$), the ideal diode is a perfect conductor, and for reverse voltages ($V_F < 0$), it is a perfect nonconductor. So the ideal diode is turned ON or OFF depending on whether the forward voltage across the diode is greater than or less than zero, re-

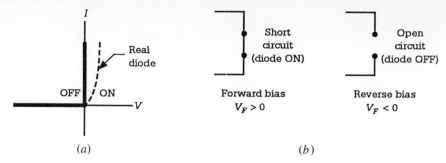

(a)   (b)

Fig. 1-3. *The ideal diode. (a)* I-V *characteristic. (b) Equivalent circuits under forward bias* ($V_F > 0$) *and reverse bias* ($V_F < 0$).

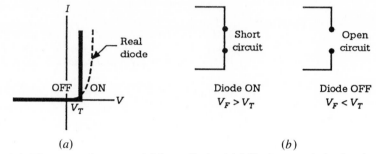

(a)   (b)

Fig. 1-4. *The approximate model for a diode. (a)* I-V *characteristic showing a threshold voltage,* V_T. *(b) The equivalent circuits for the ON and OFF conditions.*

spectively. The equivalent circuits for the ON and OFF modes are the short circuit and open circuit, respectively, as illustrated in Fig. 1-3(b).

The ideal diode has a threshold voltage of zero volts. The approximate diode model presented in Fig. 1-4(a) recognizes the existence of a turn-on voltage, $V_T > 0$, but retains the general features of the ideal model. Accordingly, for $V_F > V_T$ the equivalent circuit is a short circuit, while for $V_F < V_T$ it is an open circuit, as indicated in Fig. 1-4(b).

1.2.2 Classification of Transistor Switches

An active circuit element (device) is one that controls power flow from one part of a circuit to another. Active elements have at least three electrical terminals, and they may have power gain; that is, they may exhibit amplification. Transistors are by far the most common type of active element used as electronic switches in switching (digital) circuits. Transistor switches may be classified as indicated in Fig. 1-5. They may be further classified as npn or pnp bipolar junction transistor (BJT) switches; n- or p-channel junction field effect transistor (JFET) switches; and enhancement mode or depletion mode metal-oxide-semiconductor field effect transistor (MOSFET) switches of the n- or p-channel type. All this may seem complicated to the reader, and it is if a thorough analysis of the switching mechanisms of these devices is required. However, it is our intent only to discuss qualitatively the switching character of the BJT and MOSFET devices, thereby avoiding unnecessary detail that would add little or nothing to an entry-level understanding of these devices and their use in digital circuits.

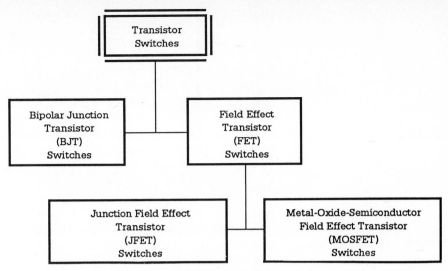

Fig. 1-5. *Classification of transistor switches.*

1.2.3 The BJT Switch

The prototype and circuit symbol for an npn bipolar junction transistor are given in Fig. 1-6(a), where the symbols B, C, and E represent the base, collector, and emitter, respectively, of the transistor. By convention, the currents I_B, I_C and I_E are all shown flowing into the device (I_E is a negative current), and the voltage drops across the terminals are V_{BE}, V_{CE}, and V_{BC} as indicated by the arrows. By applying Kirchhoff's current law (KCL: the algebraic sum of all currents into a node or circuit section is zero) and Kirchhoff's voltage law (KVL: the algebraic sum of all voltages around a closed path is zero) to Fig. 1-6, we obtain the results

$$I_B + I_C + I_E = 0 \tag{1-2}$$

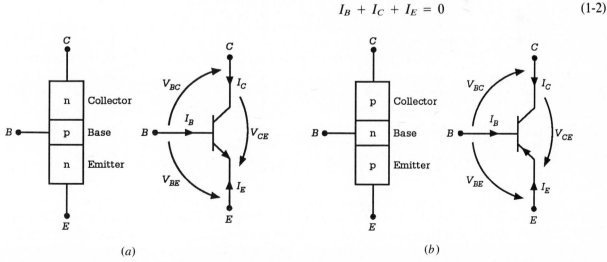

(a) (b)

Fig. 1-6. *Prototypes and circuit symbols for bipolar junction transistors showing directions of conventional currents and voltages. (a) npn BJT. (b) pnp BJT.*

and

$$V_{BE} - V_{CE} - V_{BC} = 0 \qquad (1\text{-}3)$$

where knowing any two currents or any two voltages permits calculation of the third current or voltage in the respective equation. Equation (1-3) results from the application of KVL to a closed path formed by the terminals of the transistor. The prototype and circuit symbol for a pnp BJT is presented in Fig. 1-6(b) where it is observed that the arrow in the BJT symbol is the reverse of that for the npn BJT. Also, we note that for a pnp BJT, Eqs. (1-2) and (1-3) are valid as written with the understanding that the current directions (signs) are the reverse of those for an npn BJT.

Shown in Fig. 1-7(a) is an npn BJT configured as a switch to connect or disconnect an output voltage, V_o, to ground. A switch of this type is commonly known as an RTL (resistor-BJT logic) switch. Here, we follow the shortened circuit form with V_{CC} as the supply voltage (relative to ground) and R_C as the collector load resistance.

The diode circuit in Fig. 1-7(b) and the mechanical switch analog in Fig. 1-7(c) are provided as an aid to understanding the switching character of the BJT. The two diodes in Fig. 1-7(b) are intended to represent the collector-base and base-emitter junctions of the BJT. However, since these two junctions are formed in the same crystal of semiconductor material and are very close together, a current flowing across one junction can have a strong influence on the properties of the other junction. Thus, if the base-emitter junction is forward biased, an emitter current I_E will be induced and will cause a collector current, I_C, to flow across the reverse-biased collector-base junction. The ratio $-I_C/I_E$ is called α and typically has values in the range of 0.980 to 0.995.

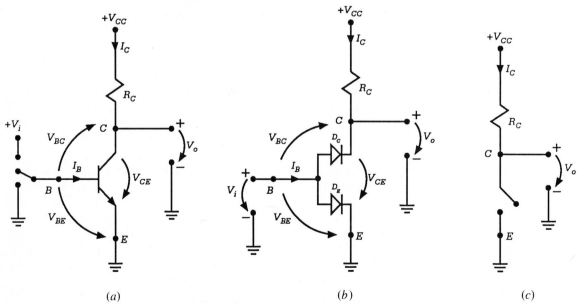

Fig. 1-7. *The* npn *BJT used as a switch. (a) Electrical circuit. (b) The diode model of the BJT switch in (a). (c) The mechanical switch analog.*

1 / Background

With reference to Fig. 1-7(c), we notice that no current is allowed to flow to ground when the switch is open and that V_{CC} falls across V_o ($V_o = V_{CC}$); and when the switch is closed, V_o is shorted to ground potential ($V_o = 0$). In the discussion that follows, we will demonstrate that the RTL switch of Fig. 1-7(a) obeys approximately this behavior.

By applying KVL and Ohm's law (Ohm's law: $V = IR$, where R is the resistance in ohms) to the I_C loop of Fig. 1-7(a), we obtain the result

$$V_{CE} = V_{CC} - I_C R_C \qquad (1\text{-}4)$$

which, when introduced into Eq. (1-3), gives a base-collector voltage of

$$V_{BC} = V_{BE} - V_{CC} + I_C R_C \qquad (1\text{-}5)$$

where V_{CC} is usually $+5$ V for most digital circuit applications. The emitter current I_E rises exponentially with V_{BE} according to Eq. (1-1), and the BJT enters the *active mode* (where the transistor can amplify) when V_{BE} just exceeds the turn-on voltage $V_{TBE} = 0.65$ V. Now, the collector current I_C and base current I_B are linearly related since $I_B = -I_C - I_E$ and $I_E = -I_C/\alpha$. Thus,

$$I_B = -I_C + I_C/\alpha$$

$$\alpha I_B = (1 - \alpha)I_C \quad \text{and}$$

$$I_C/I_B = \alpha/(1 - \alpha)$$

which permits us to write

$$V_{BC} = V_{BE} - V_{CC} + \beta I_B R_C \qquad (1\text{-}6)$$

where

$$\beta = I_C/I_B \qquad (1\text{-}7)$$

is the dc (direct current) gain in the active region with values typically ranging from 50 to 200 for most BJTs. So, just above the threshold, V_{TBE}, the forward voltage V_{BC} across the diode D_C is exponentially dependent on V_{BE}, a fact we will now make use of in qualitatively analyzing the switching properties of the BJT.

First, notice from Fig. 1-7(b) that V_o can be "shorted" to ground only if both diodes (D_C and D_E) are forward-biased beyond their threshold voltages, that is, turned ON. Beginning with zero input ($V_i = 0$), it is clear that Eq. (1-5) yields $V_{BC} = -V_{CC}$, obviously a reverse voltage bias. For very small values of input voltage, $V_i = V_{BE}$, the BJT is said to be in the *cut-off* mode with no significant collector current flowing ($I_C = 0$). In fact, in the cut-off region I_C is of the order of the reverse saturation current for diode D_C. As V_{BE} is increased beyond its threshold value ($V_{TBE} = 0.65$ V), the linearly related currents I_E, I_B, and I_C increase exponentially according to Eq. (1-1), and a point is rapidly reached in Eq. (1-6) when V_{BC} becomes positive and exceeds the threshold voltage for diode D_C. At that point the

BJT (actually the BJT circuit) enters the *saturation* mode, and V_o is virtually shorted to ground. Further increase in V_{BE} will have no significant effect on I_C whose value now is controlled by the external circuit.

The collector load resistance R_C in Fig. 1-7 has a special function. Without its presence, the BJT could not be driven into saturation, and "ON/OFF" switching would not be possible. Resistors used for this purpose are called *pull-up* resistors. The voltage drop developed across the pull-up resistor R_C permits the BC junction [diode D_C in Fig. 1-17(b)] to be forward biased, a necessary condition to reach saturation. Equation (1-6) demonstrates the function of R_C.

The behavior of the BJT we have just described also points to its inherent inverting character. As V_{BE} is increased from below its turn-on value (0.65 V for Si) to above the saturation value (0.75 V for Si), the output voltage V_o drops from its cut-off value of about $+V_{CC}$ to the saturation value near zero volts (about 0.2 V for Si BJTs). Conversely, when V_{BE} is decreased from above the saturation value to below the turn-on value, V_o rises from near ground potential to about $+V_{CC}$. All transistors exhibit this inverting property.

Shown in Fig. 1-8 is a summary of the approximate current and voltage conditions for the three modes of Si BJT operation. The quantity R_{CE} is the resistance across the collector-emitter terminals and varies from a very large value in the cut-off mode to only a few ohms in the saturation mode. Notice that only about 0.1 V separates the cut-off and saturation regions. This means that very little voltage change is required to toggle the BJT OFF and ON.

	V_{BE} (Volts)	V_{CE} (Volts)	I_C (Amps)	R_{CE} (Ohms)
Cutoff Mode	<0.65	$\approx V_{CC}$	$\sim 10^{-8}$	very large
Active Mode	0.65 to 0.75		βI_B	
Saturation Mode	>0.75	≈ 0.2	*	1–10 typically

* Determined by the external circuit

Fig. 1-8. *Typical Si* npn *transistor junction voltages, collector currents and collector-emitter resistances.*

Ideal Models for the BJT Switch. From Figs. 1-7 and 1-8, we see that in cut-off, $I_C \cong 0$ with $V_{CE} \cong V_{CC}$, and in saturation, V_{CE}(sat) $\cong 0$ (actually about 0.2 V for Si BJTs). From this information two very useful (ideal) models may be deduced and are presented in Fig. 1-9. These models will be of

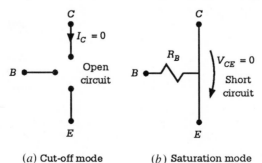

(a) **Cut-off mode**   (b) **Saturation mode**

Fig. 1-9. *Ideal models for BJTs operated in the cut-off and saturation modes.*

1 / Background

considerable value in reading the TTL (transistor-transistor logic, that is, BJT-BJT logic) circuits in Chapter 2, since in these circuits the BJTs operate strictly in either the OFF or ON states. We will also find that these models are quite similar to those for the MOS transistor switches, which are discussed next in Sect. 1.2.4.

1.2.4 The MOSFET Switch

There are two types of field effect transistors: the junction field effect transistor and the metal-oxide-semiconductor field effect transistor. The JFET is used rather extensively in linear amplifier circuits and in a number of nonlinear circuits used to process analog signals. For digital (switching) circuit applications, the MOSFETs are by far the most important members of the FET family. In particular, *complementary* configured MOSFETs (CMOSs) are used extensively in IC (integrated circuit) switching circuit systems due to their extremely low power consumption and small size. Therefore, it is our intention to concentrate mainly on the switching character of MOSFETs.

In Fig. 1-10 we illustrate the physical structure for an n-channel MOSFET (NMOS) before and after the application of a potential $+|V_G|$ to the in-

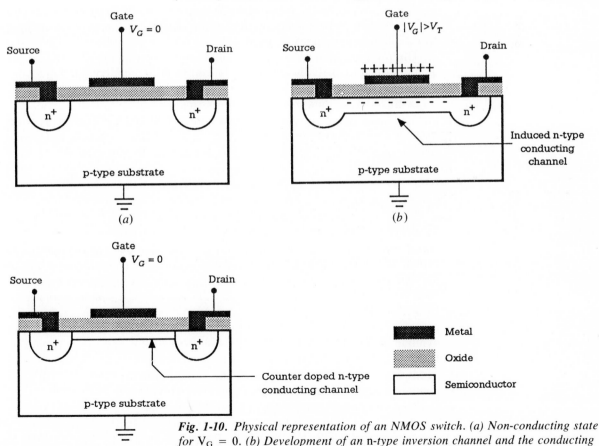

Fig. 1-10. *Physical representation of an NMOS switch. (a) Non-conducting state for* $V_G = 0$. *(b) Development of an* n-*type inversion channel and the conducting state for* $+|V_G| > V_T$. *(c) A depletion-mode NMOS indicating a conducting state for* $V_G = 0$.

sulated gate terminal. This NMOS is said to operate in the *enhancement* mode. Observe that the MOSFET is divided into four sections: two heavily doped n^+ regions onto which the *source* and *drain* leads are attached, a p-type *substrate* between the n^+ regions, and a *gate* terminal which is electrically insulated by silicon oxide from the substrate. The symbol n^+ signifies a heavily doped (highly conductive) n region. Now, when $V_G = 0$, as in Fig. 1-10(a), the drain-to-source path is nonconducting. That is, any voltage V_{DS} (less in magnitude than the breakdown value for a p-n junction) applied to the drain-source terminals will cause negligible current to flow since one of the p-n^+ junctions will always be reverse-biased. However, as $+|V_G|$ is increased to the n^+-p junction threshold voltage V_T and beyond, electrons are drawn out of the heavily doped n^+ regions to form an n-type *inversion* channel between the drain and source as illustrated in Fig. 1-10(b); hence the name n-channel MOSFET or simply NMOS. This channel is now a low-resistance (enhanced) path allowing current I_D to flow from source to drain when a voltage V_{DS} is impressed across these terminals. Therefore, under the condition $V_G > V_T$, the NMOS is said to be in the "ON" condition. The "OFF" or cut-off condition of the NMOS is established when V_G falls below V_T. In this region there are too few "free" electrons to support a significant conductivity, hence the enhancement mode NMOS is normally OFF. For most such NMOS, V_T lies in the range of 1 V to 4 V, depending on the relative doping levels used.

The p-type region between the source and drain in Fig. 1-10(a) can be counterdoped (usually by ion implantation) to produce a conducting n-type inversion channel in the absence of a voltage $+|V_G|$. Figure 1-10(c) illustrates this. Then, by applying an increasingly negative voltage $-|V_G|$ to the gate terminal, the n-type channel becomes depleted and eventually loses its ability to support conduction when V_T is reached or exceeded (negatively). Conversely, when positive gate voltage $+|V_G|$ is applied, the drain current increases. MOSFETs of this type are said to operate in the *depletion* mode and have threshold voltages V_T typically in the range of -1 V to -6 V.

Three circuit symbols commonly used for NMOS are given in Fig. 1-11(a). They consist of a detailed symbol (left), a simplified enhancement-mode symbol (middle), and a simplified depletion-mode symbol (right). The broken line connection between drain and source that is used in the detailed symbol serves as a reminder that an n-type conducting inversion channel must be established before the device will conduct. The source is identified by its location nearest the L-shaped gate symbol and the direction of the arrow is indicative of the p-type substrate. The two simplified NMOS circuit symbols in Fig. 1-11(a) do not have these distinguishing features. However, it is because of this simplicity that these two circuit symbols are made useful in MOS circuit construction. When the depletion mode NMOS symbol appears in a MOS switching circuit, it is understood that it will conduct when the gate-source voltage V_{GS} is zero, hence normally ON, and that the application of a negative voltage V_{GS} is necessary to prevent conduction.

A p-channel MOSFET, or simply PMOS, is formed by reversing the n- and p-type regions in Fig. 1-10 as well as the voltages. Now, the drain and source regions are heavily doped p^+-type semiconductor, the substrate is

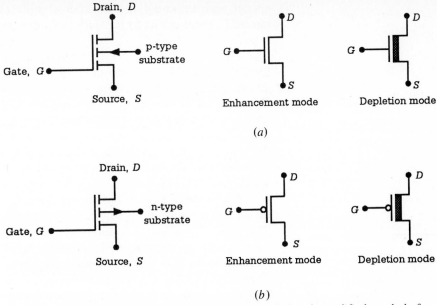

Fig. 1-11. *Circuit symbols for MOSFETs. (a) Detailed and simplified symbols for NMOS. (b) Detailed and simplified symbols for PMOS.*

n-type, and the formation of a p-type inversion channel establishes the conducting state. Therefore, since the voltages are now reversed, a voltage $+|V_G| > V_T$ that turns an NMOS "ON" will turn a PMOS "OFF," and vice versa for $V_G = 0$. The detailed and simplified circuit symbols for PMOS are given in Fig. 1-11(b). Note the direction of the arrow in the detailed symbol which is indicative of an n-type substrate. Of the three symbols, only the enhancement-mode symbol will be used in our treatment of switching circuits. Depletion-mode PMOS transistors have little use in switching circuit design.

Ideal Models for MOSFETs. As a help in reading MOSFET switching circuits, we offer the ideal models for the OFF and ON conditions of NMOS and PMOS given in Fig. 1-12. Notice that, except for the terminal names, the models are basically the same as those in Fig. 1-9. For the ON condition

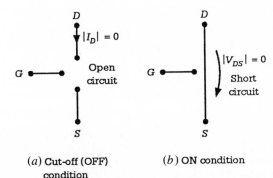

(a) Cut-off (OFF) condition

(b) ON condition

Fig. 1-12. *Idealized equivalent circuits for N and P MOSFETs operated in the OFF (a) and ON (b) conditions.*

1.2 / Electronic Switches

model, an open circuit in the gate lead replaces the base resistance used for the BJT model owing to the very high resistance of the insulated gate.

The NMOS Switch. An NMOS switch is formed when an enhancement-mode (switching) NMOS is joined in series with a depletion-mode (load) NMOS, as shown in Fig. 1-13(a). Here, the drain-source resistance, R_{T1}, for NMOS T1 becomes, in effect, a variable resistor. This is demonstrated by the *I-V* characteristic in Fig. 1-13(b), where R_{T1} is shown to be the inverse of the slope of the *I-V* characteristic for NMOS T1. When T2 is ON (V_i = HV), V_{DS} is equal to V_{DD}, and the resistance R_{T1} becomes very large (typically a few hundred thousand ohms). The high resistance occurs because the conducting n-channel is "pinched off" at the source end of T1, thereby creating a high-resistance path. Thus, when V_i turns T2 ON, V_o is "shorted" to ground, and the supply voltage V_{DD} drains to ground via a large resistance R_{T1}, as indicated in Fig. 1-13(c). Conversely, when V_i turns T2 OFF (V_i = LV), a voltage equal to $+V_{DD}$ falls across V_o to ground causing R_{T1} to become small. This occurs because the drain and source of T1 are now at approximately the same potential (V_{DD}) which maintains the n-channel open and conducting. That is, the n-channel is not pinched off as is the case when

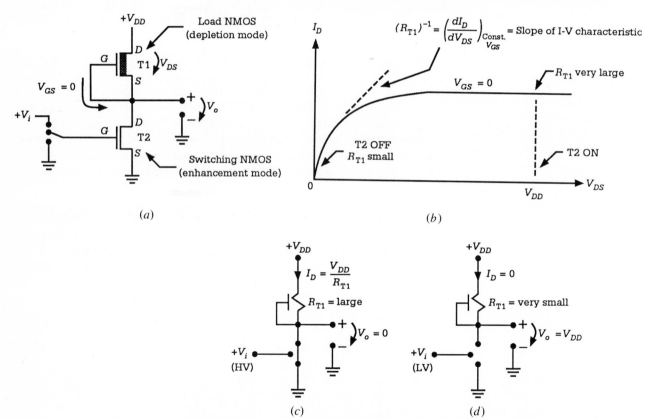

Fig. 1-13. Configuration and operation of an NMOS switch. (a) Complete circuit. (b) I-V characteristic for the load (depletion mode) NMOS T1 at V_{GS} = 0. (c) Idealized equivalent circuit for V_i = HV (T2 ON). (d) Idealized equivalent circuit for V_i = LV (T2 OFF).

1 / Background

T2 is turned ON. The low-resistance condition is illustrated in Fig. 1-13(d). Notice that the inverting character of the basic NMOS switch is established, since when V_i is at high voltage (HV), V_o is at low voltage (LV) and vice versa.

The CMOS Switch. Another important type of MOSFET switch results when a PMOS and NMOS are joined at their drains to form the series connection shown in Fig. 1-14. This is the physical picture of a basic p-well CMOS switch that is widely used in switching circuits and that requires very small power consumption. The formation of the p-well permits both types of MOS-FETs to exist in the n-type substrate. Notice that a p-type conducting channel is induced for LV input whereas an n-type conducting channel is formed for HV input. Proper operation of the CMOS switch requires that the two types of conducting channels not be formed simultaneously.

The circuit symbol and operation of the CMOS switch is illustrated in Fig. 1-15. When high voltage is applied to the input V_i, T1 turns OFF and

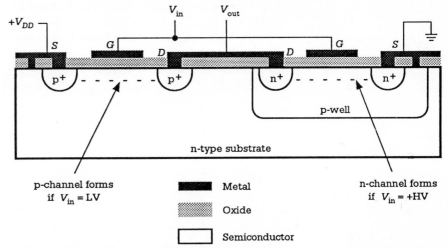

Fig. 1-14. Physical picture of a p-well CMOS switch showing locations of the p and n channels and the requirements for their formation.

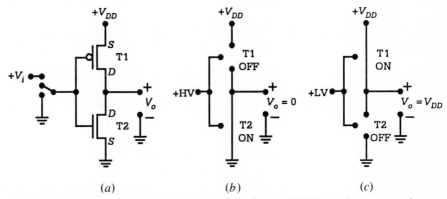

Fig. 1-15. Configuration and operation of the basic CMOS switch. (a) Complete circuit. (b) Idealized equivalent circuit for high voltage (HV) input. (c) Idealized equivalent circuit for low voltage (LV) input.

1.2 / Electronic Switches
13

T2 turns ON, essentially shorting the output V_o to ground. Or, when low voltage is applied to the input, T1 turns ON while T2 goes OFF, causing V_{DD} to fall across the output terminal relative to ground. Notice that in either case one of the two MOSFETs is always OFF, a feature that minimizes "leakage" current drain and, hence, minimizes the power dissipation. Power dissipation (defined as VI) is typically in the low nW (nanowatt) range for simple CMOS switching devices.

1.2.5 The Pass Transistor Switch

All the transistor switches we have discussed so far are classified as *active* or *restoring* switching devices, since they are capable of delivering more electric power than they consume. The supply voltage (V_{CC} or V_{DD}) makes possible current (hence power) amplification, a characteristic of all active switching devices. A switching device that cannot provide power gain is said to be *passive*. Diodes are good examples of passive switching devices since they can only dissipate (consume) power. In this section we will introduce another type of passive switch, one that has become important in IC design.

A MOS transistor switch that functions as a passive (nonrestoring) switching device and that does not invert a voltage signal is called a *pass transistor* or *transmission gate*. In Fig. 1-16(a) is shown the circuit symbol and idealized equivalent circuits for an NMOS pass transistor which is operated in the enhancement mode. Simply put, current can flow in either direction when a voltage $V_G > V_T$ (HV), supplied to the gate terminal, turns the NMOS ON. But for $V_G \cong 0\ V$ (LV), the NMOS is turned OFF, creating a very high resistance (say, $10^9\ \Omega$) between input and output terminals, thereby permitting no significant current to flow. This device generates no output current in the ON state other than that which is supplied to the input terminals. Furthermore, any "ON" current flowing through the switch creates a small voltage drop due to the channel resistance of the switch.

The circuit symbol and ideal equivalent circuits for the CMOS pass transistor switch are given in Fig. 1-16(b). It operates in a manner similar to the NMOS switch just described. Here, however, antiphase gate (ENABLE) voltages, EN and $\overline{EN}$ (meaning LV enabled), are supplied simultaneously so that both MOSFETs are either turned ON or OFF at the same time. The ideal equivalent circuits for the transfer and disconnect modes are shown in Fig. 1-16(b). Notice that the NMOS is turned ON by HV and OFF by LV, whereas the reverse is true for the PMOS.

The CMOS pass transistor (PT) has the advantage of a lower resistance in the ON state than the NMOS switch. This is so because the channel resistance for the NMOS portion of the CMOS switch rises rapidly with increasing source voltage (V_i) while the channel resistance of the PMOS portion falls off rapidly with increasing V_i. As a result, the two-channel CMOS switch has a relatively low, nearly constant overall resistance, as shown in Fig. 1-16(c). Thus, the NMOS switch passes LV well and the PMOS switch passes HV well, but only the CMOS device transfers both the high voltage level and the low voltage level without significant signal degradation—an important advantage. It is for this reason that CMOS pass transistor switches can operate over a larger range of source voltage, typically 1.5 V

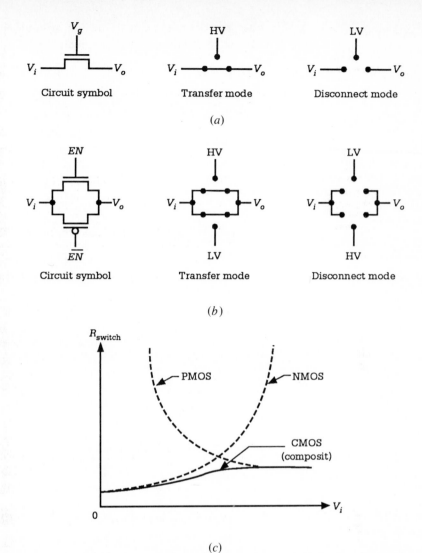

Fig. 1-16. *The MOS pass transistor switches. (a) Circuit symbol and idealized equivalent circuits for the NMOS pass transistor switch. (b) Circuit symbol and ideal equivalent circuits for the CMOS pass transistor switch. (c) Channel resistance, R_{switch}, verses input voltage, V_i, for NMOS, PMOS, and CMOS pass transistor switches.*

to 15 V. NMOS pass transistor switches are usually held to an upper limit of about 5 V.

Pass transistor switches are used mainly as *steering* logic devices. Their main function is to conditionally steer digital signals and connect circuit nodes together. It is important to remember that they can neither boost current nor invert a signal as can the active transistor switches considered previously. Thus, PT switches dissipate power but do not restore it. Also, PT switches permit the passage of signals in either direction, at times making the input and output terminals indistinguishable. It is these properties that often restrict the use of PT switching devices in logic circuits. Examples of PT switch application will be given in Chapter 2.

1.2.6 The Tristate Driver Switches

An active device that operates in either an inverting mode or a disconnect mode is called an *inverting tristate driver*. As the name implies, the tristate driver is used as a current boosting device in the inverting mode. But when the driver is in the disconnect mode, the output is electrically isolated from the input due to an extremely large channel resistance. Shown in Fig. 1-17 is a simple CMOS inverting tristate driver with antiphase ENABLE inputs required for its operation. The idealized equivalent circuits for the inverting and disconnect modes are presented in Figs. 1-17(b) and (c). Thus, when EN is at HV with $\overline{EN}$ at LV, the ENABLE MOSFETs are ON and the tristate driver functions as a simple inverting CMOS switch (as in Fig. 1-15). However, when EN is at LV with $\overline{EN}$ at HV, the ENABLE MOSFETs are turned OFF. This condition disconnects the output from the input and disables the driver.

Noninverting tristate drivers can be designed simply by placing an inverting switch in front of an inverting tristate driver. In Fig. 1-18 is shown a CMOS tristate driver which operates either in a transfer (noninverting) mode or a disconnect mode, where again use must be made of antiphase gate ENABLE inputs. So, when EN is at HV ($\overline{EN}$ is at LV), the two ENABLE MOSFETs are turned ON, thereby creating two inverting switches in series as is indicated in Fig. 1-18(b). This is the transfer mode in which the driver can be used to boost current. Or for opposite ENABLE voltages, the ENABLE MOSFETs are turned OFF resulting in the disconnect mode of Fig. 1-18(c). Interchanging the ENABLE MOSFETs of Fig. 1-18(a) pro-

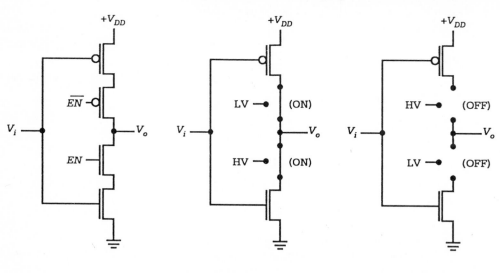

(a)   (b) Inverting mode   (c) Disconnect mode

Fig. 1-17. The CMOS inverting tristate driver. (a) Complete circuit showing antiphase ENABLE inputs. (b) and (c) Idealized equivalent circuits for the inverting and disconnect modes.

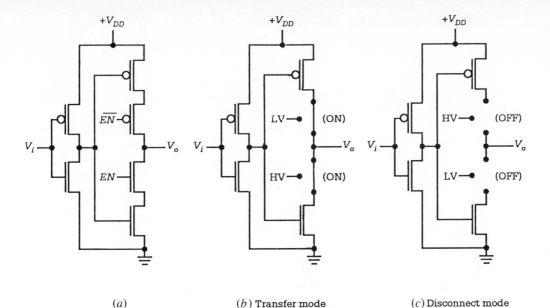

(a)   (b) Transfer mode   (c) Disconnect mode

Fig. 1-18. *CMOS tristate driver (noninverting). (a) Complete circuit. (b) and (c) Idealized equivalent circuits for the transfer and disconnect modes.*

duces a noninverting tristate driver that will transfer when *EN* is at LV and disconnect when at HV.

Tristate drivers are used to interface various IC devices to a common data bus so that the devices will not interfere with each other. That is, all such devices which are not communicating information to a data bus at any point in time can be disconnected by disabling their respective tristate drivers leaving those devices enabled that are communicating. In Sect. 4.11.5 we will have the opportunity to demonstrate one such application of these drivers. There, as usual, the antiphase ENABLE inputs are connected together through an inverting switch so that a single input voltage change enables or disables the driver.

1.3   PERFORMANCE CHARACTERISTICS OF ELECTRONIC SWITCHES

To begin, let us list the most desirable features a designer would want in a switch or switching device, say, for integrated circuit applications. They are

1. Fast switching speed.
2. Low power dissipation.
3. Wide noise margins.
4. High fan-out capability.

While no single family of electronic switches has all these features, some may come close. In this section we will discuss these performance characteristics and conclude with a summary of characteristics for four commonly used IC logic families.

1.3.1 Propagation Time Delay and Switching Speed

When an input to an electronic switch changes to a value that exceeds the switching threshold of the switch, the output of the switch does not respond instantaneously. The time required for a change in the input of a switch to cause a change in the output is called the *propagation delay time*, which is illustrated in Fig. 1-19. Here, t_{PHL} measures the time interval between the input and output signals in passing the 50% points when the output changes from V_{OH} to V_{OL}, hence the extended subscript *HL*. Similarly, t_{PLH} measures the propagation delay when the output level changes from V_{OL} to V_{OH}. Since t_{PHL} and t_{PLH} are not usually the same for a given electronic switch, an average propagation delay time can be obtained from the expression

$$\tau_{p(\text{avg})} = \frac{t_{PHL} + t_{PLH}}{2} \tag{1-8}$$

The voltage levels V_{IH}, V_{IL}, V_{OH}, and V_{OL} are defined by Fig. 1-19. Also shown are the rise and fall times, which are measured between the 10% and 90% voltage marks, and the pulse width, which is measured between the 50% voltage marks.

Propagation delay time of a switch or switching device is a measure of the *switching speed* of the switch or device and, therefore, is of great concern to the designer. These periods of time limit the rate at which information can be processed in a digital system and, in some cases, determine whether or not a given digital design will function properly.

Propagation delays in integrated circuit systems are caused mainly by capacitances inherent in the electronic switches used and by stray (or parasitic) capacitances in the signal paths. These delays result from the fact that it takes time to charge or discharge a capacitor. Thus, the greater the capacitance, the longer it will take the switch to reach the switching threshold in response to an input signal, and the greater will be τ_p. For any given

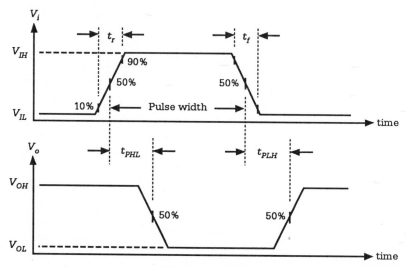

Fig. 1-19. *Propagation delay and rise and fall times for an electronic switch.*

1 / Background

family of switches, the manufactures will give *worst case* propagation delays, which should be used by the designer when conservative estimates are needed for design purposes.

1.3.2 Power Dissipation

When current flows through a conducting medium of finite resistance, power is dissipated (consumed); that is, heat is generated. Consequently, we want the power dissipation to be minimal, but not just to conserve on power consumption, which may be desirable. The greater the amount of circuitry packed within a given volume of IC chip, the greater will be the heat generated in that volume and the greater will be the chance of overheating and malfunction of the switching devices. Obviously, it is desirable to minimize the power dissipation and, if necessary, provide special heat conduction paths to external heat sinks.

Power is consumed due to the momentary current required for the transition from HIGH level to LOW level, or vice versa. Power is also consumed due to the continuous current required to maintain a particular switching level or state. The continuous average power (in watts [W]) consumed by a BJT-type switch is given as the product of the supply voltage and supply current

$$P_{dis} = V_{CC}I_{CC} \tag{1-9}$$

where the supply current is usually taken as the average

$$I_{CC} = \frac{I_{CCH} + I_{CCL}}{2} \tag{1-10}$$

Here, I_{CCH} and I_{CCL} are defined as the supply currents for HV and LV output, respectively, and are usually different. Expressions similar to Eqs. (1-9) and (1-10) can be written for MOS-type switches by replacing the supply designation CC with DD. Note that the unit equation for power is [W] = [V][A].

NMOS and CMOS switches have an extremely low continuous power dissipation, but it is not zero. In these devices a very small leakage current occurs during maintenance of the OFF or ON state, and transient currents occur during the state-to-state transitions, all contributing to the total power dissipation. Thus, MOS switches claim one of the lowest power consumption rates of any of the switching families.

A useful figure of merit for switching devices is called the *power-delay product* (*PDP*) given by

$$PDP = P_{dis}\tau_p \tag{1-11}$$

In this expression P_{dis} is the average power dissipation of the device and τ_p is the propagation delay. The unit of PDP is [W][s] = [J], or joule, usually expressed as pJ (10^{-12} joules). Since it is desirable to achieve both a low-power dissipation and small propagation time, a low PDP is also desirable.

The maximum voltage fluctuation that can be tolerated in a digital signal without crossing the switching threshold of the switching devices used in a digital system is called the *noise margin* or *noise immunity* of that system. Noise can originate from external sources such as line transients and radio frequency signals, or it can be generated within the digital system by electromagnetic coupling (cross-talk) or by asymmetric path delays in complex digital circuits. In any case, if the noise level is insufficient to cross the switching thresholds of the devices in the digital system, it (the noise) will be sharply attenuated in passing through the various switching stages of the system and no problem will result—noise does not accumulate from one switching stage to another as in some analog circuits, and this gives digital systems a distinct advantage. On the other hand, if the noise margins are exceeded, unwanted signals are likely to appear in the outputs of the switching devices, possibly causing system failure.

The role that noise margins play in a digital system is illustrated in Fig. 1-20. Here, a source switch delivers a digital signal with noise to a load switch as indicated in Fig. 1-20(a). To help understand the cause and effect switching process involved here, the following quantities are defined:

$V_{IL(\text{max})}$ = maximum voltage that the load switch will accept as LOW level

$V_{IH(\text{min})}$ = minimum voltage that the load switch will accept as HIGH level

$V_{OL(\text{max})}$ = maximum LOW-level output voltage of the source switch

$V_{OH(\text{min})}$ = minimum HIGH-level output voltage of the source switch

From the foregoing definitions it follows that a voltage level from the source switch that exceeds $V_{IL(\text{max})}$ could appear as a possible HIGH to the load switch and that a voltage level from the source that falls short of $V_{IH(\text{min})}$ could appear as a possible LOW to the load switch. Thus, when either of these two worst case limits is exceeded in the appropriate direction, the load switch may not operate predictably. With this information in mind we define the LOW and HIGH noise margins to be

$$NML = V_{IL(\text{max})} - V_{OL(\text{max})}$$

and

$$NMH = V_{OH(\text{min})} - V_{IH(\text{min})}$$

(1-12)

respectively, as illustrated in Figs. 1-20(b) and (c). The noise margins can be regarded as margins of safety within which digital systems must be operated if their behavior is to be predictable.

Returning to Fig. 1-20, we see that only when the inner boundaries of the noise margins, $V_{IH(\text{min})}$ and $V_{IL(\text{max})}$, are exceeded will unpredictable switch behavior result. While we show error transitions taking place for each inner boundary crossing, this need not necessarily be so. These boundaries are simply the dividing lines between predictable and unpredictable behavior and, accordingly, represent worst case values. The external limits of the

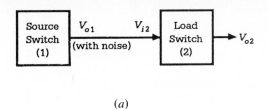

(a)

Fig. 1-20. *Effect of noise margins. (a) Source-to-load switch configuration. (b) Voltage pulse with low noise condition. (c) Voltage pulse with high noise condition showing possible error transitions in output.*

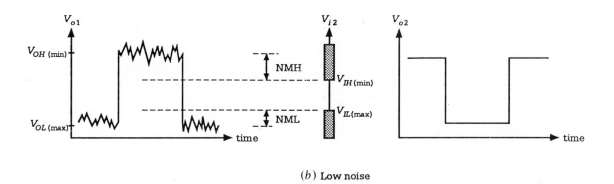

(b) **Low noise**

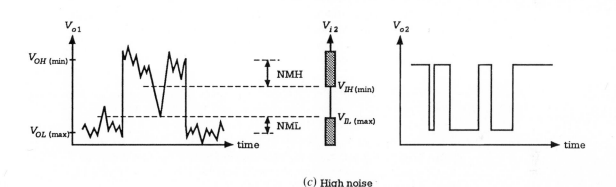

(c) **High noise**

shaded signal range bars shown in Figs. 1-20(b) and (c) are of no consequence as long as they lie outside of valid output levels from the source.

So it is obvious that wide noise margins are a desirable feature in a switch. Narrow noise margins make switches more difficult to work with since even small fluctuations in the input signal may cause error transitions.

1.3.4   Fan-out and Fan-in

Since an output to a switch has a definite limit to the amount of current it can supply (or sink), it is not surprising that there is a definite limit to the number of parallel switches that can be driven by a single switch output. This limit (number) is called the *fan-out* of the device and is illustrated in Fig. 1-21 for the case of CMOS switches, although the concept is perfectly general and applicable to other types of switches or switching devices.

The fan-out of a restoring device can be measured by comparing the

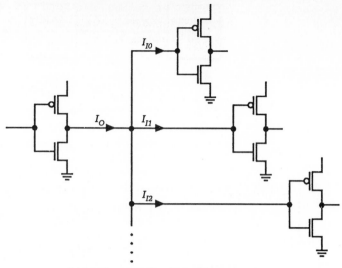

Fig. 1-21. Fan-out of CMOS switches.

output current available from the driver device with the input current requirements of the load device to be driven. Thus, $|I_{OH}|/|I_{IH}|$ or $|I_{OL}|/|I_{IL}|$, whichever is the smaller, is a measure of the fan-out capacity of the device. Here, I_{OH} is the driver output current at V_{OH}, I_{IH} is the load device input current requirement at V_{IH}, and so on. If either of these ratios is less than unity, the driver-load device combination is not compatible. The fan-out requirements may be generalized as follows:

$$I_{OH} \geq \sum_i (I_{IH})_i$$

and

$$I_{OL} \geq \sum_i (I_{IL})_i$$

(1-13)

Figure 1-21 illustrates these requirements. The output current I_O of the driver device must be at least equal to the sum of all input currents of the load devices for both the high-voltage and low-voltage states. Thus, both requirements expressed by Eqs. (1-13) must be met before one device can predictably drive other devices.

Loading the output of a switching device beyond the fan-out limit set by the designer or manufacturer can seriously reduce the noise margins and switching speeds of the system and can result ultimately in failure due to insufficient current input. Thus, high fan-out capability is a desirable feature in a switch, especially if that switch is to be used in the design of a complex digital circuit.

The maximum number of inputs permitted to control the operation of a digital circuit is called *fan-in*. For example, an eight-input device has a fan-in of 8. There are circuits called *extenders* that are designed to be connected to a given digital circuit to increase fan-in capability. However, the resulting

1 / Background

fan-in also increases the input capacitance, which, in turn, increases the propagation delay of the circuit. Thus, extenders should be avoided for high-speed applications.

1.3.5 Other Factors

In IC design it is usually desirable to pack as many switches as possible into a given chip area without jeopardizing the operation of the digital device. The practical limit to which switches of the same family can be packed is called the *packing density* and is usually given in numbers of switches (or switching devices) per square millimeter. Densely packing of switches in ICs can improve the switching speed of the system by reducing lead capacitances, but there is a price to be paid. The more densely packed the switches or switching devices, the more difficult it is to remove the joule heat that is generated and the greater chance there is for cross-talk problems. A low-PDP factor favors high packing densities.

To one extent or another all the performance characteristics previously discussed, including packing density, affect the *cost* of an IC device or system. But the subject of cost is highly complex and one that is continuously changing with time. Usually the cost factor depends most sensitively on the total number of IC devices to be made and on the number of processing steps involved such as ion implantation and photomasking, to name just two. As manufacturers gain more familiarity with new and improved processing techniques and as the failure rate decreases, costs generally are reduced substantially. Research and design costs must be taken into account in arriving at a bottom-line figure, but these costs can also be highly variable.

1.3.6 Summary of IC Logic Family Characteristics

Any summary of switching (logic) family characteristics is risky because the relative assessments change, sometimes rapidly, with technological developments and because these assessments often have to be qualified to be valid. However, in spite of the risks, we will attempt to summarize the characteristics in Fig. 1-22 for four commonly used IC logic families, TTL, ECL, NMOS, and CMOS. We have not mentioned the ECL family, which stands for emitter-coupled logic. Because of its high switching speed, the ECL family is currently the fourth most important logic family next to CMOS, NMOS, and TTL. Other logic families, important for certain applications, are not represented in Fig. 1-22. These include integrated Schottky transistor logic (STL), integrated injection logic (IIL), low-voltage

Parameter	TTL	ECL	NMOS	CMOS
Switching speed	Good	Very good	Fair	Fair
Power dissipation*	Medium	High	Low	Low
Noise immunity	Very good	Fair	Good	Very good
Fan-out	Fair	Fair	Very good	Very good
Packing density	Medium	Low	High	High

* The power dissipation of CMOS ICs increases rapidly at high frequencies.

Fig. 1-22. Summary of characteristics for four commonly used IC logic families.

injection logic (LVIL), gallium arsenide logic, and optical and superconducting logic families. In addition, there are subgroups of logic families, such as silicon-on-sapphire CMOS (CMOS/SOS) logic, that are of some importance but not included in the summary.

ANNOTATED REFERENCES

Good reviews of the integrated logic families are provided in the texts by Tocci (Chapter 8) and Jones (Chapter 5). These reviews are electronics oriented and include performance characteristics. Pertinent references are provided by Jones at the end of the chapter.

Tocci, R. J., *Digital Systems, Principles and Applications*, 4th ed., Prentice Hall, Englewood Cliffs, N.J., 1988.

Jones, L. D., *Principles and Applications of Digital Electronics*, Macmillan, New York, 1986.

PROBLEMS

1.1 In Fig. P1-1(a) is shown an NMOS switch and its approximate equivalent circuit, where R_{T1} and R_{T2} represent variable drain-source resistances. In the table of Fig. P1-1(b) typical NMOS drain-source resistances are given for two input voltage levels, 0 V and +5 V. Use this information to determine the output voltage V_o, the current I_D, and the power dissipated for each of the input voltages given. Present the results in a table. [Hint: Use Ohm's law ($V = IR$) and the expression for power $P = IV$.]

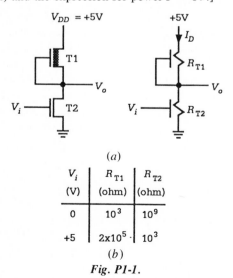

(*a*)

V_i	R_{T1}	R_{T2}
(V)	(ohm)	(ohm)
0	10^3	10^9
+5	2×10^5	10^3

(*b*)

Fig. P1-1.

1.2 **(a)** Apply KVL and Ohm's law to the approximate equivalent circuit of Fig. P1-1(a) to determine the output voltage V_o in terms of the two resistances and the supply voltage V_{DD}, independent of I_D. The result is the well-known *voltage divider equation*.

(b) Use the voltage divider concept to determine V_{RT1}, the voltage drop across R_{T1}.

1.3 The circuit in Fig. P1-2 shows a BJT switch driving a light-emitting diode (LED). Assume that this BJT operates such that the output V_o changes from 4.1 V to 0.2 V when the input V_i changes from 0 V to +5 V and that the LED is rated at 18 mA and 2.5 V for full brightness.

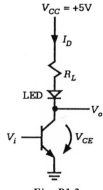

Fig. P1-2.

(a) Calculate the value (in ohms) of the current limiting resistor, R_L, required to prevent damage to the LED at full brightness, and determine the IR voltage drop across R_L.

(b) For the full brightness condition, determine the value R_{CE} of the BJT and R_{LED}. Use the value of R_L taken from part (a).

(c) Calculate the total average power dissipated by the system at full brightness of the LED.

1.4 Two diodes are connected in series opposition with a 5 V supply as shown in Fig. P1-3.

(a) Assuming that these diodes are real diodes and that Eq. (1-1) applies to each with $n = 1$, calculate $V_o = V_2$ relative to ground.

(b) Repeat part (a) if the diodes are assumed to be ideal according to Fig. 1-3.

(c) Repeat part (a) if the diodes are assumed to obey the model in Fig. 1-4.

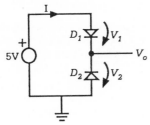

Fig. P1-3.

1.5 In Fig. P1-4(a) is shown the common-emitter connection for an npn BJT. It consists of two loops: an input loop in which the base current I_B flows and an output loop in which the collector current I_C flows.

(a) Use the short form representation of Fig. P1-4(b) and the definition of β given by Eq. (1-7) to express the output voltage $V_o = V_{CE}$ as a function of input voltage V_i and the resistances R_C and R_B. On a sketch of V_o versus V_i, called the *transfer characteristic*, indicate the three regions of operation: the cut-off region, the active region, and the saturation region. Give the slope of the trans-

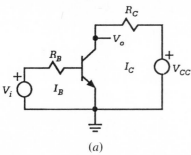

(a)

Fig. P1-4.

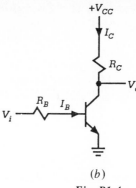

(b)

Fig. P1-4.

fer characteristic in terms of the resistances R_C and R_B. (Hint: Begin by writing the KVL equations for the input and output loops. Then solve these two equations simultaneously for $V_o = V_{CE}$.)

(b) Determine the voltage gain V_o/V_i in the active range, discuss its meaning, and determine the limits over which it is valid for a supply voltage of $V_{CC} = +5$ V. Let $R_C = 2$ kohm (2×10^3 ohm), $R_B = 10$ kohm, and β = 100, and let $V_{TBE} = 0.6$ V, $V_{BE,sat} = 0.8$ V and $V_{CE,sat} = 0.2$ V (see Fig. 1-8).

(c) Discuss the use of a BJT in a digital circuit, given the results of parts (b) and (c).

1.6 A logic switch draws a supply current of $I_{CCH} = 6$ mA when the output is at HV = 4 V, but draws $I_{CCL} = 18$ mA when the output is at LV = 0.5 V. Find the average power dissipated for a 0.5 V to 4 V square waveform output operated as follows:

(a) A 50% duty cycle

(b) A 20% duty cycle

Take the supply voltage to be $V_{CC} = 5$ V. Duty cycle is the ratio of the time spent in the high-voltage state to the period of the waveform. The period is the time required for a periodic waveform to repeat itself.

1.7 For the two systems represented in Fig. P1-5, determine whether or not the input specifications of system II are compatible with the output specifications of system I, if system I drives system II as shown.

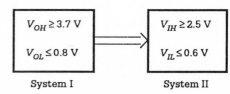

$V_{OH} \geq 3.7$ V	$V_{IH} \geq 2.5$ V
$V_{OL} \leq 0.8$ V	$V_{IL} \leq 0.6$ V
System I	System II

Fig. P1-5.

1.8 The fan-out capability of a driver switch relative to the load switches can be determined from the ratios $|I_{OL}|/|I_{IL}|$ and $|I_{OH}|/|I_{IH}|$ as suggested by the requirements of Eqs. (1-13). If these current parameters are given in Fig. P1-6 for switch systems I and II and if system I is to drive system II, what is the predicted fan-out and what does it mean? Can system II drive system I? Explain.

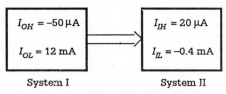

| $I_{OH} = -50\,\mu A$ | | $I_{IH} = 20\,\mu A$ |
| $I_{OL} = 12\,mA$ | | $I_{IL} = -0.4\,mA$ |

System I   System II

Fig. P1-6.

1.9 Listed in the table of Fig. P1-7 are the operating specifications for two switches, A and B. Determine the following:

(a) The switch with the wider HV noise margin.

(b) The better combination (A driving B switches or vice versa) for fan-out capability.

(c) The switch that can be operated at the higher frequency.

(d) The switch that will draw the lesser supply current.

(e) The switch with the more favorable PDP factor.

Parameter	Switch A	Switch B	Parameter	Switch A	Switch B
V_{supply}	5 V	5 V	t_{PHL}	5 ns	12 ns
$V_{IH(min)}$	2.0 V	1.8 V	I_{IH}	0.04 mA	0.1 mA
$V_{IL(max)}$	0.8 V	0.6 V	I_{IL}	2 mA	3 mA
$V_{OH(min)}$	2.4 V	2.6 V	I_{OH}	0.4 mA	0.6 mA
$V_{OL(max)}$	0.4 V	0.5 V	I_{OL}	17 mA	21 mA
t_{PLH}	7 ns	18 ns	P_{dis}	25 mW	20 mW

Fig. P1-7.

CHAPTER 2
Fundamentals of Digital Design

2.1 INTRODUCTION

The contents of this chapter are considered to be vitally important to the reader's understanding of the remainder of this text and, hence, to an understanding of digital design as we present it. In this chapter the reader will learn some of the vocabulary and jargon, the fundamental concepts, and the algebra and symbols that are common to the field of digital design. Furthermore, the design of simple digital (logic) circuits begins in this chapter.

In Chapter 1 we introduced the electronic switch by giving several examples and by discussing the operation of each. Now, we will extend that coverage by using at least one of several logic families to illustrate the electrical analog (equivalent) of each logic circuit symbol we introduce. We do this as a continuing reminder that the logic symbolism has a physical realization which must ultimately be dealt with to complete a given logic design task. The logic families to be included are CMOS and NMOS logic, BJT-BJT logic called TTL (for transistor-transistor logic), and diode-BJT logic called DTL.

2.2 BINARY STATE TERMINOLOGY

Digital systems use switching devices that operate in only one of two possible states at any given time but that can be switched back and forth from one state to another at very high speed (millions of times per second). The two states are high voltage (HV) and low voltage (LV). To design a useful digital device, meaningful names must be assigned to the various input and output voltage levels in agreement with some algorithm on which the device is

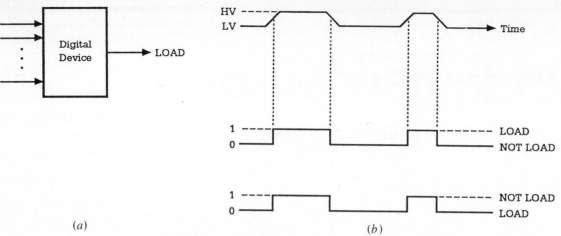

Fig. 2-1. *Positive logic interpretation of a physical waveform. (a) The digital device and its voltage output called LOAD. (b) Ambiguous logic interpretation of the waveform LOAD in (a).*

based. Typical of the names that could be assigned to HV or LV are quit, 1, set, load, false, true, 0, or any other name the user wishes to use. For example, HV and LV for a given signal could be assigned the names true and false, respectively. But as we shall soon see, it is not enough simply to assign names to an input or output signal. More information is needed to avoid ambiguous interpretation.

Shown in Fig. 2-1(a) is a digital (switching) device having an output which has been assigned the name LOAD. The problem is, there are two possible interpretations that can be given to the physical waveform representing the output LOAD, and these are given in Fig. 2-1(b). Notice that LOAD is represented by HV in one interpretation of the waveform and by LV in the other. So how many times has the signal LOAD been issued for the time interval shown? The answer is: two or three depending on which interpretation is chosen. Thus, as it stands, the nomenclature used in Fig. 2-1 is ambiguous since the name LOAD conveys no information as to which waveform interpretation is the valid one.

2.2.1 ACTIVE and INACTIVE States

Two important terms, which will be used throughout the remainder of this text and which correspond closely to the IEEE standards and to the nomenclature used in data books of leading IC manufacturers, are defined as follows:

> *ACTIVE—a descriptor denoting an action condition.*
> *INACTIVE—a descriptor denoting a condition which is not ACTIVE.*

Other words that can be used in place of ACTIVE include AFFIRMATIVE, ASSERT, UP, EXCITED, ON, and so on, with their opposites used for INACTIVE.

A state is said to be ACTIVE if it is the condition for causing something to happen. And for every ACTIVE state there must exist one that is IN-ACTIVE. In the binary (base 2) system of 1's and 0's, these descriptors take the logic values

$$\text{ACTIVE} = \text{logic 1}$$
$$\text{INACTIVE} = \text{logic 0}$$

(2-1)

which assigns logic 1 as the ACTIVE condition and logic 0 as the INACTIVE condition. Thus, we may state:

> *In the logic domain, "1" is the ACTIVE state while "0" is the IN-ACTIVE state.*

This will always be so. As we shall see later, use will be made of the descriptor ACTIVE in defining the binary logic operators which underlie all of digital electronics.

2.2.2 ACTIVATION LEVEL INDICATORS

A symbol that is attached to the name of a signal and that establishes which physical state, HV or LV, is to be the ACTIVE state for that signal, is called an ACTIVATION LEVEL INDICATOR. The ACTIVATION LEVEL IN-DICATORs we use, together with their meanings are

> *(H)   meaning ACTIVE HIGH.*
> *(L)   meaning ACTIVE LOW.*

Thus, a line signal representing LOAD(H) is one for which the ACTIVE state occurs at high voltage (HV), or if LOAD(L) one for which the ACTIVE state occurs at low voltage (LV). For example, when the waveform representing LOAD(H) crosses the HV threshold, the action LOAD occurs. To illustrate, we show in Fig. 2-2 the results of applying this symbology to the same digital device represented in Fig. 2-1. Notice that all ambiguity has been removed and that now we can indicate at which voltage level we want the action LOAD to take place.

There are alternatives to using the symbols (H) and (L) as ACTIVATION LEVEL INDICATORs. One alternative is to attach the symbols "+" and "−" to the signal name with the meaning (H) and (L), respectively. We have avoided this alternative symbolism because the symbol "+" is also used as a binary operator, a possible source of confusion. Even less appealing is another alternative symbolism which replaces (H) and (L) with the numeric symbols "1" and "0." The problem is that these symbols often appear like the literals "1" and "O" leading to confusion and possible error.

2.2.3 Positive and Negative Logic

The positive and negative logic systems are defined by Fig. 2-3(a), and the implications of Fig. 2-2 are presented in Fig. 2-3(b). In these figures the

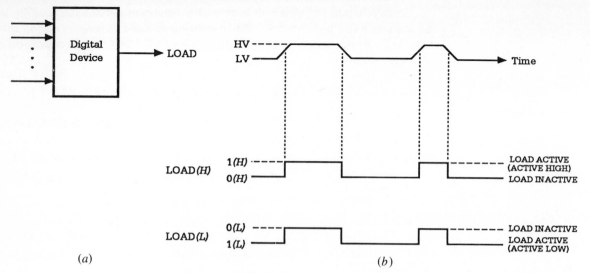

(a)

Fig. 2-2. MIXED LOGIC interpretation of a physical waveform. (a) The digital device and its voltage output called LOAD. (b) Positive and negative logic interpretations of the waveform in (a) showing the use of ACTIVATION LEVEL indicator symbols.

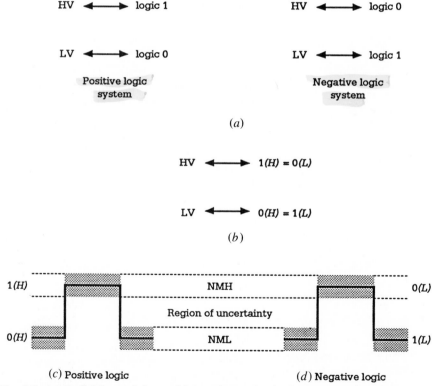

Fig. 2-3. (a) Definitions of the positive and negative logic systems. (b) Implications of Fig. 2-2. Noise margins shown superimposed on (c) the positive logic and (d) the negative logic interpretations of a voltage waveform.

symbol ↔ is read in either direction as "corresponds to". Thus, HV corresponds to the logic 1 state in the positive logic system but corresponds to the logic 0 state in the negative logic system. Conversely, LV corresponds to logic 0 in the positive logic system but corresponds to logic 1 in the negative logic system. So, in the logic domain, logic 1 is always the ACTIVE state while logic 0 is always the INACTIVE state. To bridge the gap between the physical domain and the logic domain, Fig. 2-2 requires that HV corresponds to $1(H) = 0(L)$ and that LV corresponds to $0(H) = 1(L)$. This is summarized in Figs. 2-3(b), (c), and (d). The noise margins, NMH and NML, are included in Figs. 2-3(c) and (d) as a reminder that their inner boundaries are also the inner limits of the logic states (1 and 0) as well as the outer limits of the uncertainty region.

2.2.4 POLARIZED MNEMONICs

A contracted signal name onto which is attached an ACTIVATION LEVEL INDICATOR is called a POLARIZED MNEMONIC. A signal representing a logic operation or action can be ACTIVE HIGH or ACTIVE LOW but never both—hence, it is polarized. MNEMONICs can be used in algebraic formulations whereas use of their corresponding full names would be unwieldy and even ambiguous in cases where the operation or action is represented by more than one word. Listed in Fig. 2-4 are several examples of POLARIZED MNEMONICs together with their original meanings and ACTIVATION LEVELs. Notice that names consisting of more than one word can be contracted to MNEMONIC form in more than one way, but that *all* MNEMONICs are composed of a *single* group of symbols short enough to be easily used yet long enough to convey their meaning.

At the risk of sounding repetitive, let us make certain the reader understands the meaning of the polarized MNEMONICs given in Fig. 2-4. Take for example CL(H). As the label for an input or output signal to a digital device, CL(H) would be read as CLEAR ACTIVE HIGH with the understanding that when the voltage waveform for CLEAR crosses the logic 1 threshold in the upgoing direction, the action CLEAR occurs. Similarly, the line labeled PR(L), meaning PRESET ACTIVE LOW, must provide the action PRESET when its voltage waveform crosses the logic 0 threshold in the downgoing direction.

Action Name	Activation level	Polarized Mnemonic
LOAD	LOW	LD(L)
SYSTEM CLOCK	HIGH	SYSCK(H)
PRESET	LOW	PRE(L) or PR(L)
CLEAR	HIGH	CLR(H) or CL(H)
DATA READY	HIGH	DRDY(H) or DATRDY(H)
REQUEST TO SEND	LOW	RTS(L) or REQSND(L)
COUNT 8	HIGH	CNT(H)

Fig. 2-4. Examples of POLARIZED MNEMONIC with their original meanings and ACTIVATION LEVELs.

2.2.5 MIXED LOGIC Notation

A notational scheme which permits both positive and negative logic systems to exist within the same digital environment is called MIXED LOGIC notation. This notational scheme results quite naturally from the use of PO-LARIZED MNEMONICs which carry the following logic connotations:

> X(H) represents an action signal called X from a positive logic source (ACTIVE HIGH), X(L) represents an action signal called X from a negative logic source (ACTIVE LOW),

where X is any MNEMONIC. That the foregoing connotations are valid is easily demonstrated. The POLARIZATION LEVEL INDICATOR (H) signifies ACTIVE HIGH, which, in turn, implies that the logic 1 state occurs at HV, the definition of POSITIVE LOGIC. Similarly, (L) signifies ACTIVE LOW, which implies that the logic 1 state occurs at LV, the definition of negative logic. Thus, if the outputs, shown in Fig. 2-2, are present in the same digital system, MIXED LOGIC notation is said to be in use.

It is not a requirement that both (H) and (L) INDICATORs be used within the same digital system to say that MIXED LOGIC notation is employed. The use of POLARIZED MNEMONICs is itself a sufficient requirement, even if only one ACTIVATION LEVEL exists. However, if for such a system all ACTIVATION LEVEL INDICATORs, say, (H), are dropped, a homologic or monologic system is implied. It is our recommendation that the removal of the ACTIVATION LEVEL INDICATORs *never* be practiced even if all signal leads are of the same ACTIVATION LEVEL. Removal of these symbols means loss of logic system information which might later result in design errors.

From all the information we have gathered so far, we may summarize with the following observation:

> Use of the MIXED LOGIC notation can be viewed as the necessary vehicle required to bridge the gap between the logic domain and the physical world.

The use of the MIXED LOGIC notation conveys both logical and physical meaning to the designer, since it is the medium through which logic abstraction and physical reality join.

2.3 LOGIC LEVEL CONVERSION—THE INVERTER

When a positive logic source is converted to a negative logic source, or vice versa, the LOGIC LEVEL is said to be *converted* or *transformed*. The physical device that performs the electrical analog of LOGIC LEVEL CONVERSION is called an INVERTER. Except for the pass transistor switch, all switches discussed in Sect. 1-2 are INVERTERs, each capable of performing the electrical analog of LOGIC LEVEL CONVERSION. Shown in

2 / Fundamentals of Digital Design

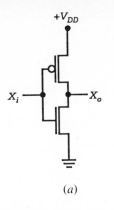

(a)

X_i (Volts)	X_0 (Volts)
0	$+V_{DD}$
$+V_{DD}$	0

(b)

X_i	X_o
LV	HV
HV	LV

(c)

Fig. 2-5. *(a) Transistor circuit for the CMOS INVERTER. (b) I/O specification table for the CMOS INVERTER. (c) I/O specification table for any INVERTER.*

Fig. 2-5(a) is the CMOS version of the INVERTER, and in Figs. 2-5(b) and (c) are two equivalent forms of its input/output (I/O) specification table or physical truth table, the latter I/O table being applicable to any INVERTER. Clearly, when X_i is at low voltage (LV) X_o is high (HV), and vice versa, as explained in Sect. 1.2.4.

The LOGIC LEVEL CONVERSION function of an INVERTER can be interpreted in two ways: ACTIVE HIGH (positive logic) to ACTIVE LOW (negative logic), and vice versa. These two interpretations of the physical truth table are presented in Figs. 2-6(b) and (c), respectively, together with the appropriate circuit symbol for each. Here, the variable X with its ACTIVATION LEVEL INDICATOR represents any POLARIZED MNEMONIC. Notice that a "bubble" is placed on the symbol to indicate the ACTIVE LOW logic level and that its absence indicates ACTIVE HIGH.

What is important to remember is that the *only* logic function of an INVERTER in a logic (digital) circuit is to perform the electrical analog of one of the two LOGIC LEVEL CONVERSION interpretations given in Fig. 2-6, and that is all. This point cannot be overemphasized. Therefore, the presence of either logic symbol in a logic circuit represents an INVERTER in the physical sense, but is otherwise taken to be a LOGIC LEVEL CONVERSION symbol.

Two versions of the INVERTER which have not been discussed previously are the TTL (BJT-BJT logic) INVERTER of Fig. 2-7(a), and the DTL (diode-BJT logic) INVERTER of Fig. 2-7(b). The TTL INVERTER represents a family of switching devices commonly used in IC technology. The logic family represented by the DTL INVERTER is rarely used except in power switching circuits where large currents must be accommodated.

The operation of the TTL INVERTER as an electronic switch is best understood by discussing the operation of the DTL INVERTER. The BE and BC junctions of the transistor T1 in Fig. 2-7(a) function as simple diode

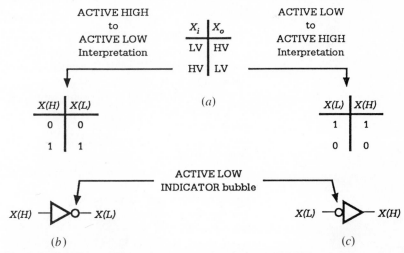

Fig. 2-6. *LOGIC LEVEL CONVERSION of an INVERTER. (a) Physical truth table for an INVERTER. (b) ACTIVE HIGH to ACTIVE LOW logic interpretation and circuit symbol. (c) ACTIVE LOW to ACTIVE HIGH logic interpretation and circuit symbol.*

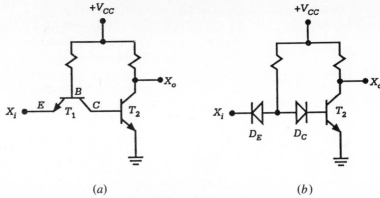

Fig. 2-7. Two types of BJT logic INVERTERs. (a) TTL INVERTER. (b) DTL INVERTER used to explain the operation of the TTL INVERTER.

junctions and can be represented by the diodes D_E and D_C, respectively, in the DTL circuit of Fig. 2-7(b). Recall that this was done earlier in Fig. 1-7. Thus, when input X_i is low (LV), diode D_E is ON, diode D_C is OFF, and transistor T2 is in cut-off, bringing V_o to about $+V_{CC}$ relative to ground. Turned about, when X_i is high (HV), diode D_E is OFF, diode D_C is ON, and the transistor T2 is forced into saturation virtually shorting V_o to ground.

2.4 AND and OR OPERATORS and THEIR PHYSICAL REALIZATION

Remarkable as it may seem, there are just *two* binary logic operations which underlie all digital logic design-the AND and OR operations. These operations are purely logic concepts, or abstractions, and as such have no physical significance. The following are the operator symbols (or connectives) which are used for AND and OR:

$$(\cdot) \rightarrow \text{AND}$$

$$(+) \rightarrow \text{OR}$$

So, if one writes $X \cdot Y$ it is read as X AND Y, and $LD + CNT$ is read as LOAD OR COUNT. The connective $(\cdot)$ is commonly omitted as in XY or $(X)(Y)$, but such expressions are still read as X AND Y.

By using the AND and OR operator symbols, we can formulate a mathematical logic expression representing a conditional statement. For example, we may state

the device will run if switches A and B are ON or if switch C is ON.

Then, taking the mnemonic SW to represent a switch in the ON position and RUN to represent the run condition of the device, we write and read the following logic expression:

$$\text{RUN} = \text{SWA} \cdot \text{SWB} + \text{SWC}$$

RUN IF SWITCH *A* AND SWITCH *B* OR SWITCH *C*

2 / Fundamentals of Digital Design

Notice that the equal sign is properly read as "if" or "whenever" in a logic expression, although it is common practice to say "equals" due to the association with conventional algebra. In this expression the AND operation takes precedence over the OR operation; similar to the multiplication and addition operations in decimal arithmetic.

As we shall soon learn, it is possible to construct a physical device that will perform the electrical analog of the conditional statement just given, or any other mathematical logic expression for that matter. To do this, however, requires more information than we now have. Therefore, our next task is an extremely important one: *identify a symbolism that can be used to transform any mathematical logic "statement" into its physical realization.* It is the primary purpose of this chapter to develop such a symbolism. Once this is accomplished, the reader will be able to take a conditional statement of reasonable complexity, formulate it into a mathematical logic expression, and then construct a logic circuit that will perform the electrical analog of the original statement.

2.4.1 Logic Circuit Symbology for AND and OR

For our purposes the meanings of the AND and OR operators are best understood in terms of their logic circuit symbols. Shown in Fig. 2-8 are the distinctively shaped logic circuit symbols commonly used to represent the AND and OR operators, but which have inputs and an output for each. The functions of these symbols are stated as follows:

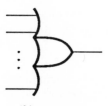

(a) **AND circuit** · **symbol**

(b) **OR circuit** **symbol**

Fig. 2-8. Logic circuit symbols for the binary operators. (a) AND circuit symbol. (b) OR circuit symbol.

> *The output of a logic AND circuit symbol is ACTIVE if, and only if, all inputs are ACTIVE.*
>
> *The output of a logic OR circuit symbol is ACTIVE if one or more of the inputs are ACTIVE.*

Remember that the term ACTIVE implies logic 1. Thus, the output of a logic AND symbol is logic 1 if, and only if, all inputs are logic 1, and so on.

The distinctively shaped symbols of Fig. 2-8 together with their functional description constitute the special symbology (mentioned earlier) that connects a logic concept with its physical realization called the *gate*. The following definition of gate applies:

> *A gate is the interconnection of electronic switches and other circuit elements required to produce the electrical analog of a logic operation.*

We are about to discover that under the proper I/O conditions, any one of four different gates is capable of producing the electrical analog of either binary logic operator, AND or OR.

2.4.2 NAND Gate Realization of AND and OR

The physical device presented in Fig. 2-9(a) is called a NOT-AND or NAND gate for short, and because this particular gate is composed of both NMOS-

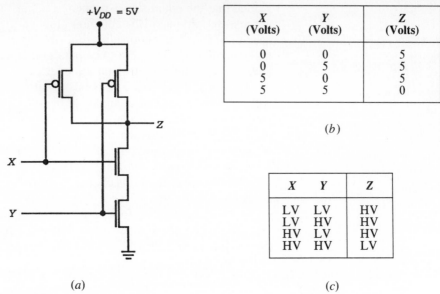

X (Volts)	Y (Volts)	Z (Volts)
0	0	5
0	5	5
5	0	5
5	5	0

(b)

X	Y	Z
LV	LV	HV
LV	HV	HV
HV	LV	HV
HV	HV	LV

(a)

(c)

Fig. 2-9. *(a) CMOS NAND gate circuit. (b) The I/O specification table for the CMOS NAND gate. (c) Physical truth table for any NAND gate.*

FETs and PMOSFETs in a complementary configuration, it is called a CMOS NAND gate. The I/O specification table (or physical truth table) for this gate is given in Fig. 2-9(b) and for any two-input NAND gate in Fig. 2-9(c). The physical truth table is obtained by "reading" the output voltage for all possible input voltage combinations. In Fig. 2-9(a) we read a high voltage (HV) across output Z to ground for all input voltage combinations except when both the X and Y inputs are HV (relative to ground), in which case the output is virtually shorted to ground (LV). Notice that when the PMOSFETs are ON the NMOSFETs are OFF, and vice versa, which accounts for the fact that essentially no static power is dissipated by a CMOS NAND gate.

Now that we have characterized the physical device called a NAND gate by a physical truth table, it is necessary to determine what meaningful logic interpretations it possesses. By using the functional descriptions of the AND and OR logic circuit symbols given in Sect. 2.4.1 together with the binary state terminology presented in Sect. 2.2, we arrive at the results presented in Fig. 2-10. Here, it is seen that a logic AND interpretation of the physical truth table results if the inputs are ACTIVE HIGH (from positive logic sources) and the output is ACTIVE LOW (a negative logic source). This gives rise to the logic truth table for AND and the characteristic AND circuit symbol, both shown in Fig. 2-10(b). The ACTIVE LOW INDICATOR bubble has been placed on the output of the circuit symbol to account for the ACTIVE LOW output which is read as $Z(L) = (X \cdot Y)(L)$. Hence, this symbol and its corresponding truth table carry the meaning that "the output of an AND circuit symbol is ACTIVE if, and only if, all inputs are ACTIVE." Again, we remind the reader that the descriptor ACTIVE implies logic 1.

In Fig. 2-10(c) is given the logic OR interpretation of the NAND gate. This results when both inputs are ACTIVE LOW and the output is ACTIVE

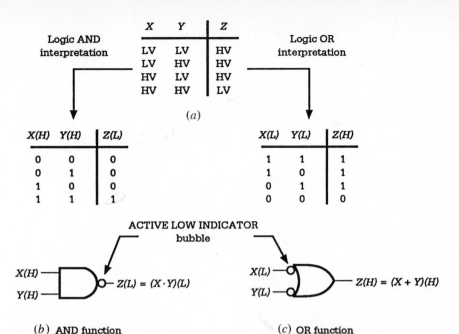

	X	Y	Z
	LV	LV	HV
	LV	HV	HV
	HV	LV	HV
	HV	HV	LV

Logic AND interpretation

Logic OR interpretation

(a)

X(H)	Y(H)	Z(L)
0	0	0
0	1	0
1	0	0
1	1	1

X(L)	Y(L)	Z(H)
1	1	1
1	0	1
0	1	1
0	0	0

ACTIVE LOW INDICATOR
bubble

$X(H)$ —
$Y(H)$ — $Z(L) = (X \cdot Y)(L)$

$X(L)$ —
$Y(L)$ — $Z(H) = (X + Y)(H)$

(*b*) AND function   (*c*) OR function

Fig. 2-10. AND and OR functions of the NAND gate. (a) The physical truth table for the NAND gate. (b) Logic AND interpretation and circuit symbol for the NAND gate. (c) Logic OR interpretation and circuit symbol for the NAND gate.

HIGH, as indicated by the logic truth table. Then, from the functional description of the logic OR circuit symbol and use of the terminology presented in Sect. 2.2, there results the logic truth table and characteristic OR circuit symbol shown in Fig. 2-10(c). Both are read as $Z(H) = (X + Y)(H)$ and convey the meaning that "the output of a logic OR circuit symbol is ACTIVE if one or more of the inputs are ACTIVE." ACTIVE LOW INDICATOR bubbles are attached to the inputs to account for the ACTIVE LOW input requirements.

The AND and OR circuit symbols given in Figs. 2-10(b) and (c) both represent the physical NAND gate. For this reason they can be regarded as *conjugate NAND gate forms*, even though, strictly speaking, they are only logic circuit symbols. Being midway between that which is pure logic and that which is the physical world, these symbols have a dual personality. So, when either conjugate form appears in a logic circuit, it may be interpreted as the physical NAND gate for the purpose of implementation. Or putting it another way, the NAND gate performs the electrical analog of either logic circuit symbol.

More inputs can be added to the CMOS NAND gate shown in Fig. 2-9(a) by connecting one PMOSFET in parallel and one NMOSFET in series for each input. This is illustrated in Fig. 2-11, together with the conjugate logic circuit symbols representing the AND and OR logic interpretations of the NAND gate. Thus, an n-input CMOS NAND gate of the type shown in Fig. 2-11 must have 2*n* MOSFETs, n-PMOS in parallel and n-NMOS in series. The number of inputs is limited due to the fact that the discharge path of the PMOSFETs to ground must be through the series connection of

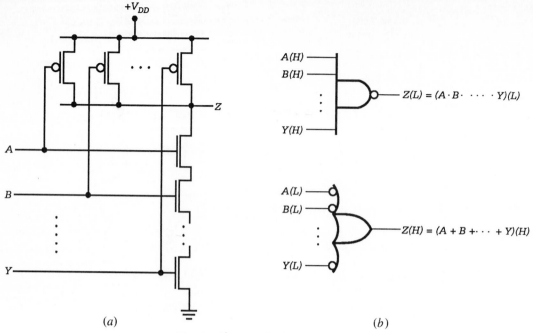

Fig. 2-11. (a) Multiple input CMOS NAND gate. (b) Logic circuit symbols for the logic AND and OR interpretations of a multiple input NAND gate.

NMOSFETs, each of which has a small ON resistance associated with it. Therefore, the addition of too many inputs can seriously degrade the output signal waveform.

There are many different NAND gate circuit configurations capable of performing the electrical analog of the AND and OR operations. The circuit given in Fig. 2-9(a) is one type belonging to the CMOS family. Three other NAND gate circuits, representing three different transistor logic families, are presented in Fig. 2-12. We will briefly discuss their operation.

The NMOS NAND gate in Fig. 2-12(a) operates in a manner similar to the CMOS version in Fig. 2-9(a), except that now a depletion mode NMOS-FET (T1) must function like a pull-up resistor, as discussed in Sect. 1.2.4. Here, NMOSFETs T2 and T3 are enhancement mode devices. Also, additional inputs are possible by connecting more NMOSFETs in series, but the number that may be added suffers from the same discharge path limitation as the CMOS version discussed earlier. In any case an NMOS NAND gate requires $n + 1$ MOSFETs for n inputs, which is considerably less than the $2n$ MOSFETs required by the CMOS version. Obviously NMOS NAND gates can be packed more densely in an IC chip than can CMOS NAND gates having the same number of inputs.

The operation of the TTL NAND gate in Fig. 2-12(b) requires some explanation. This device uses a two-emitter BJT (T1), the operation of which can be qualitatively understood by analyzing the diode-equivalent DTL NAND gate circuit in Fig. 2-12(c). Here, the diodes D_{EX} and D_{EY} represent the *BE* junctions connected to inputs X and Y, respectively, and the diode D_C represents the *BC* junction. Now, if either X or Y is at LV (assumed to

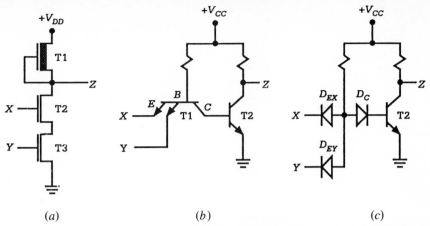

(a)   *(b)*   *(c)*

Fig. 2-12. *Other NAND gates. (a) NMOS circuit. (b) TTL circuit. (c) DTL circuit used to explain the operation of the TTL NAND gate of (b).*

be ground), or both are at LV, $+V_{CC}$ has a "direct" path to ground through D_{EX} or D_{EY}, or both. When this happens, diode D_C is OFF, T2 is in cut-off, and $+V_{CC}$ (approximately) falls across Z to ground. However, if both X and Y are at HV, D_C is turned ON, T2 is forced into saturation, and Z is virtually shorted to ground (LV).

Taken collectively, the operation of the DTL just described, when applied to the junctions of T1 in Fig. 2-12(b), also describes qualitatively the operation of the TTL NAND gate with one qualification. When both X and Y are at HV, the BJT T1 in the TTL NAND gate circuit is said to be operated in *reverse* mode, since the emitter and collector roles are interchanged. This is a mode of BJT operation that was not mentioned in Chapter 1.

The number of inputs to a TTL NAND gate can be increased by increasing the number of emitter input connections. However, this number is usually limited to eight or fewer for technological reasons.

2.4.3   NOR Gate Realization of AND and OR

Just as there are several families of switching devices called NAND gates, so also are there several families of switching devices called NOT-OR or NOR gates. A CMOS version of the NOR gate is presented in Fig. 2-13(a). In comparison with the CMOS NAND gate of Fig. 2-9(a), we see that there is a pattern reversal. PMOSFETs that form a parallel pattern in the NAND gate form a series pattern in the NOR gate, and vice versa for the NMOS-FETs.

The I/O specification tables for the CMOS NOR gate and for any NOR gate are presented in Figs. 2-13(b) and (c), respectively. A cursory inspection of the gate circuit confirms that if either or both of the inputs are at HV, Z is "shorted" to ground (LV), and only when both inputs are at LV will the output be at HV.

The logic AND interpretation of the physical truth table for the NOR gate in Fig. 2-14(a) is represented by the logic truth table and circuit symbol shown in Fig. 2-14(b). These results are valid for ACTIVE LOW inputs (those from negative logic sources) and for an ACTIVE HIGH output (a

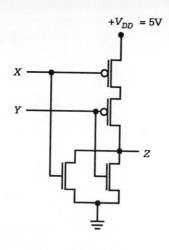

$+V_{DD} = 5V$

X (Volts)	Y (Volts)	Z (Volts)
0	0	5
0	5	0
5	0	0
5	5	0

(b)

X	Y	Z
LV	LV	HV
LV	HV	LV
HV	LV	LV
HV	HV	LV

(a)

(c)

Fig. 2-13. *The NOR gate. (a) CMOS version. (b) I/O specification table for the CMOS NOR gate. (c) Physical truth for any NOR gate.*

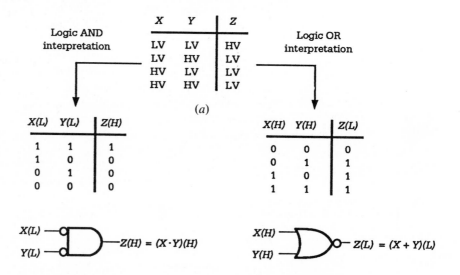

X	Y	Z
LV	LV	HV
LV	HV	LV
HV	LV	LV
HV	HV	LV

Logic AND interpretation

Logic OR interpretation

(a)

X(L)	Y(L)	Z(H)
1	1	1
1	0	0
0	1	0
0	0	0

X(H)	Y(H)	Z(L)
0	0	0
0	1	1
1	0	1
1	1	1

X(L) — Y(L) — $Z(H) = (X \cdot Y)(H)$

X(H) — Y(H) — $Z(L) = (X + Y)(L)$

(b) AND function

(c) OR function

Fig. 2-14. *AND and OR functions of the NOR gate. (a) The physical truth table for the NOR gate. (b) Logic AND interpretation and circuit symbol for the NOR gate. (c) Logic OR interpretation and circuit symbol for the NOR gate.*

positive logic source). An inspection of the truth table in Fig. 2-14(b) indicates that the requirements set forth in Sect. 2.4.1 are met, namely, that "the output of a logic AND circuit symbol is ACTIVE if, and only if, all inputs are ACTIVE." Notice that ACTIVE LOW INDICATOR bubbles are used to indicate ACTIVE LOW inputs.

In a similar manner the logic OR interpretation of the NOR gate is represented by the truth table and logic circuit symbol shown in Fig. 2-14(c), but this time for ACTIVE HIGH inputs and ACTIVE LOW output. As can

2 / Fundamentals of Digital Design

be seen, these results are in agreement with the requirements for the logic OR circuit symbol: "The output of a logic OR circuit symbol is ACTIVE if one or more of the inputs are ACTIVE."

The logic circuit symbols of Figs. 2-14(b) and (c) are conjugate forms representing the NOR gate. This means that the appearance of either of these symbols in a logic circuit can be interpreted as the physical NOR gate for purposes of implementation of the circuit. That is, the NOR gate performs the electrical analog of the AND and OR operations represented by these two symbols.

A CMOS NOR gate with multiple inputs is shown in Fig. 2-15(a), and the conjugate gate symbols representing its AND and OR interpretations are given in Fig. 2-15(b). The number of inputs that can be added to a NOR gate is nearly unlimited since the discharge path of the PMOSFETs to ground is through a parallel combination of NMOSFETs that collectively have a smaller ON resistance than a single MOSFET. This makes CMOS NOR gates more attractive than CMOS NAND gates for IC manufacturing purposes. Consequently, one would expect to see more logic circuits composed of CMOS NOR gates than CMOS NAND.

Other logic families can be used to produce NOR gates, and three examples are shown in Fig. 2-16. The NMOS version in Fig. 2-16(a) operates with a depletion mode pull-up resistor and two parallel NMOSFETs, one for each input. Notice the pattern differences between NAND an NOR gate configurations. The NMOSFETs of the NOR gate are configured in parallel, while those for the NMOS NAND gate are shown to be in series, a pattern difference similar to that shown earlier for the CMOS NAND and NOR

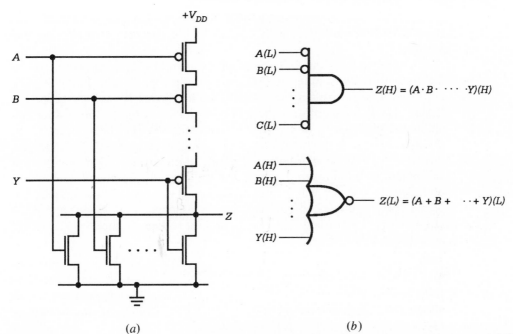

(a)   (b)

Fig. 2-15. *(a) Multiple input CMOS NOR gate (b) Logic circuit symbols representing the logic AND and OR interpretations of a multiple input NOR gate.*

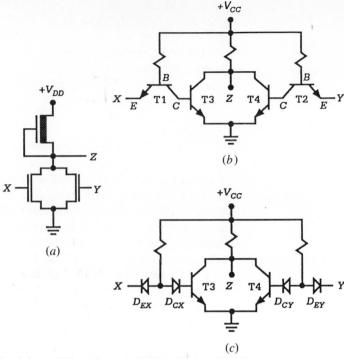

Fig. 2-16. *Other NOR gates. (a) NMOS circuit. (b) TTL circuit. (c) DTL circuit used to explain the operation of the TTL NOR gate circuit.*

gates. Thus, both NMOS and CMOS NOR gates have an advantage over their NAND counterparts with respect to the number of inputs that are permitted.

The operation of the TTL NOR gate shown in Fig. 2-16(b) can best be understood by explaining the DTL nor gate in Fig. 2-16(c). Following the approach used to explain Fig. 2-12(b), we let diodes D_E and D_C represent the *BE* and *BC* junctions for the BJTs T1 and T2 in Fig. 2-16(b), where an additional subscript (X or Y) is added to identify the associated input. Then, if either or both of the inputs are at HV, one or both of the D_C diodes are turned ON, forcing one or both of the parallel transistors (T3 and/or T4) into saturation, with the result that output Z is "shorted" to ground. Only when both inputs are at LV will the output Z be at HV. This is so because for LV inputs both D_E diodes are ON causing the associated D_C diodes and BJTs (T3 and T4) to be turned OFF, resulting in a voltage of about $+V_{CC}$ to fall across Z to ground.

2.4.4 NAND and NOR Gate Realization of LOGIC LEVEL CONVERSION

It may come as a surprise to the reader, but NAND and NOR gates can also perform the electrical analog of LOGIC LEVEL CONVERSION—that is, they can also function as INVERTERs. To see how this is accomplished one need only study the AND and OR logic interpretations of these two gates. From the truth tables in Figs. 2-10(b) and (c), for example, we see that a NAND gate symbol functions as a LOGIC LEVEL CONVERTER

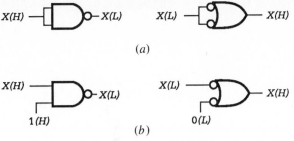

(a)

$$X(H) \quad \text{—▷○—} \quad X(L)$$
$$1\,(H)$$

$$X(L) \quad \text{—▷○—} \quad X(H)$$
$$0\,(L)$$

(b)

Fig. 2-17. *NAND gate realizations of LOGIC LEVEL CONVERSION.*

if its inputs (two in this case) are connected together. Thus, for the logic AND interpretation, $Z(L)$ is $0(L)$ or $1(L)$ if $X(H)[= Y(H)]$ is $0(H)$ or $1(H)$, respectively. Similarly, for the logic OR interpretation $Z(H)$ is $0(H)$ or $1(H)$ if $X(L)[= Y(L)]$ is $0(L)$ or $1(L)$, respectively. The resulting symbols derived from these observations are given in Fig. 2-17(a), one for the AND interpretation and one for the OR interpretation.

But the two-input NAND gate can also function as an INVERTER if one of the input leads is connected to HV. Referring again to the truth tables of Figs. 2-10(b) and (c), notice that for the AND interpretation if either input is held at $1(H)$, then $Z(L)$ is $0(L)$ or $1(L)$ if the other input is $0(H)$ or $1(H)$, respectively. Similarly, for the OR interpretation, if either input is held at $0(L)$, then $Z(H)$ is $0(H)$ or $1(H)$ if the other input is $0(L)$ or $1(L)$. The logic symbols for these two cases are given in Fig. 2-17(b).

In a manner similar to that used for the NAND gate, we find that the NOR gate can also function as an INVERTER. Shown in Fig. 2-18 are the NOR gate realizations of LOGIC LEVEL CONVERSION, which are obtained from the logic AND and OR interpretations of the NOR gate physical truth table given in Fig. 2-14. The LOGIC CONVERTER circuit symbols of Fig. 2-18(a) result from a comparison of the first and fourth rows of the truth table in Figs. 2-14(b) and (c). So, for the AND interpretation, $Z(H)$ is $0(H)$ or $1(H)$, if $X(L)[= Y(L)]$ is $0(L)$ or $1(L)$, respectively, and for the OR interpretation, $Z(L)$ is $0(L)$ or $1(L)$, if $X(H)$ $[= Y(H)]$ is $0(H)$ or $1(H)$, respectively. Also, the two-input NOR gate becomes an INVERTER if one of its input leads is connected to LV. Then, for the AND interpretation, if either of the inputs is held at $1(L)$, $Z(H)$ is $0(H)$ or $1(H)$ if the other input is $0(L)$ or $1(L)$, respectively. Similarly, for the OR interpretation, if either input is held at $0(H)$, $Z(L)$ is $0(L)$ or $1(L)$ if the other input is $0(H)$ or $1(H)$, respectively.

$$X(L) \quad \text{—◁○▷—} \quad X(H) \qquad X(H) \quad \text{—▷○—} \quad X(L)$$

(a)

$$X(L) \quad \text{—▷○—} \quad X(H)$$
$$1\,(L)$$

$$X(H) \quad \text{—▷○—} \quad X(L)$$
$$0\,(H)$$

(b)

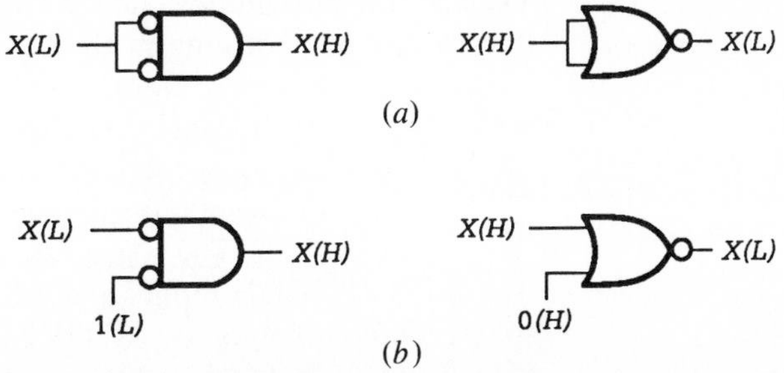

Fig. 2-18. *NOR gate realizations of LOGIC LEVEL CONVERSION.*

The reader should note that implicit in our development are the interpretations

$$1(H) = 0(L) \rightarrow HV$$
$$1(L) = 0(H) \rightarrow LV$$

(2-2)

which follow from Fig. 2-3 and the logic interpretations of the physical truth tables used to explain the logic symbolism presented in this chapter. These physical interpretations of logic levels are perfectly general and will be used throughout this text.

Some other observations are in order at this time. The subject of NAND and NOR gate INVERTERs seems to suggest that an INVERTER is not a unique device but rather a degenerate form of NAND or NOR gate. We take this position and assert that any single input NAND or NOR gate is an INVERTER. Furthermore, we recognize that the LOGIC CONVERTER forms of Figs. 2-17(b) and 2-18(b) are the more efficient of the two types, since only one input need be driven (compared to two for the other type), requiring less power and permitting a faster response time.

2.4.5 AND Gate Realization of AND and OR

NAND and NOR gates are what can be termed *natural* electrical realizations of the AND and OR logic operators, but the AND and OR gates are not. This statement can be understood when it is recalled that an active transistor switch is, by its nature, an INVERTER. Consequently, one might expect the physical realizations of NOT-AND and NOT-OR to be simpler (by transistor count) than the equivalent physical realizations of AND and OR. Let us look into this further.

Shown in Fig. 2-19 are two CMOS versions of the AND gate together with their physical truth table. In Fig. 2-19(a) the AND gate is recognized as a NAND gate with an INVERTER on its output and, therefore, issues HV only when both inputs are at HV, the opposite of the NAND gate behavior given in Fig. 2-9. But there is a simpler way of configuring a CMOS AND gate. By interchanging the NMOS and PMOS transistors in the NOR gate of Fig. 2-13(a), there results the AND gate of Fig. 2-19(b). This configuration has the disadvantage of two pull-up NMOS in series which do not transfer HV well and two PMOS in parallel which do not transfer LV well. This results in the distortion of the output waveform. The discussion in Sect. 1.2.5 together with the use of Fig. 1-16(c) explains why this is so.

The pass transistor realization of the AND gate is a series combination of PTL MOS switches, one for each input. However, the PTL switch is passive or nonrestoring in the sense that it cannot supply power but only dissipates it. For this reason every four or five PTL switches used in a logic circuit must be followed by an INVERTER or BUFFER (noninverting active device) so as to boost the current level.

The logic interpretations of the physical truth table for an AND gate are given in Fig. 2-20. The logic AND interpretation occurs (not surprisingly) for ACTIVE HIGH inputs and ACTIVE HIGH output, while the logic OR interpretation results for ACTIVE LOW inputs and ACTIVE LOW output. The logic truth tables and logic circuit symbols reflect these I/O ACTI-

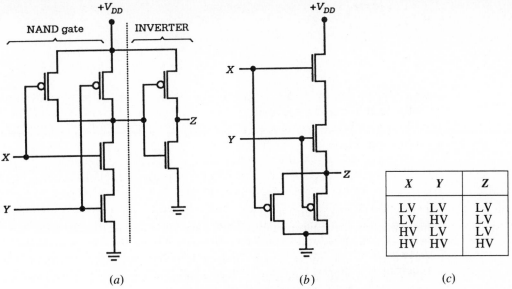

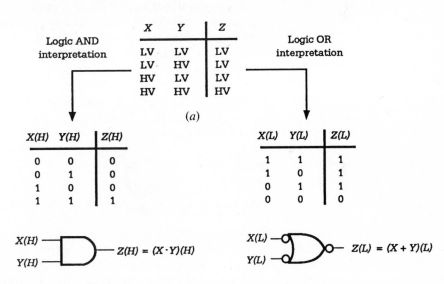

Fig. 2-19. (a) A CMOS AND gate composed of a NAND gate and an INVERTER. (b) Direct configuration of a CMOS AND gate. (c) Physical truth table for an AND gate.

X	Y	Z
LV	LV	LV
LV	HV	LV
HV	LV	LV
HV	HV	HV

Logic AND interpretation

Logic OR interpretation

(a)

X(H)	Y(H)	Z(H)
0	0	0
0	1	0
1	0	0
1	1	1

X(L)	Y(L)	Z(L)
1	1	1
1	0	1
0	1	1
0	0	0

$$X(H) \longrightarrow Z(H) = (X \cdot Y)(H)$$
$$Y(H) \longrightarrow$$

$$X(L) \longrightarrow Z(L) = (X + Y)(L)$$
$$Y(L) \longrightarrow$$

(b) **AND function**

(c) **OR function**

Fig. 2-20. AND and OR functions of the AND gate. (a) Physical truth table for the AND gate. (b) Logic AND interpretation and circuit symbol. (c) Logic OR interpretation and circuit symbol.

VATION LEVEL requirements, including the use of ACTIVE LOW INDICATOR bubbles on the OR symbol.

A multiple-input AND gate can be produced by adding an INVERTER to the multiple-input NAND gate. For CMOS logic one would add the INVERTER of Fig. 2-5(a) to the output of the NAND gate shown in Fig. 2-11(a). This requires $2n + 2$ MOSFETs for n inputs, a significant factor in

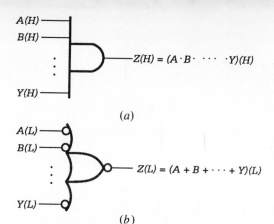

$$Z(H) = (A \cdot B \cdot \cdots \cdot Y)(H)$$

(a)

$$Z(L) = (A + B + \cdots + Y)(L)$$

(b)

Fig. 2-21. *Logic circuit symbols representing (a) the logic AND interpretation and (b) the logic OR interpretation of a multiple input AND gate.*

IC design. The logic circuit symbols representing the AND and OR interpretations for any multiple-input AND gate are given in Fig. 2-21.

Various logic families other than CMOS can be used to implement an AND gate. To do so, it is only necessary to add an INVERTER of that family to the NAND gate output. For example, a TTL AND gate can be produced by adding the INVERTER of Fig. 2-7(a) to the output of the NAND gate shown in Fig. 2-12(b). There are, however, other methods of producing an AND gate beyond those just described. A simple approach, which does not require the use of an INVERTER on the output of a NAND gate, involves the use of a *pull-down resistor* placed just above the ground symbol on a NAND gate of the type shown in Fig. 2-12. Then, with the reposition of the output lead Z immediately above the pull-down resistor, the NAND gate is converted to an AND gate. The function of the pull-down resistor is to produce a voltage drop (about equal to the supply voltage) when supply current flows through it, a result of the HV input condition. See Example 2-17 in Sect. 2-10 for a typical application of this latter approach to AND gate implementation.

2.4.6 OR Gate Realization of AND and OR

Just as the AND gate can perform the electrical analog of the AND and OR logic operators, we should expect that the OR gate can also perform the electrical analog of the two binary operators. This we will demonstrate after we have discussed the physical realizations of the OR gate.

Presented in Fig. 2-22 are two CMOS versions of the OR gate. Again we notice that in Fig. 2-22(a) an INVERTER is placed on the output of a gate already familiar to us, the NOR gate. But direct configuration of a CMOS OR gate can also be accomplished. By interchanging the NMOS and PMOS transistors in the NAND gate of Fig. 2-9(a), there results a simpler CMOS realization of the OR gate as shown in Fig. 2-22(b). In either case the OR gate circuit produces a HV for all input voltage combinations except when both inputs are at LV, a condition which yields a LV output. However, the direct configuration in Fig. 2-22(b) will cause the output waveform to be distorted because NMOS do not pass HV well and PMOS do not pass LV well as explained in Sect. 2.4.5.

OR gates of other logic families can be produced by adding INVERTERs to NOR gates of the same logic family. Alternatively, a pull-down resistor

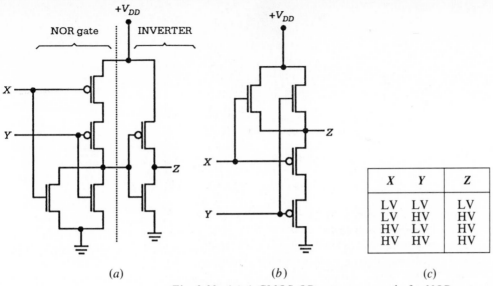

Fig. 2-22. (a) A CMOS OR gate composed of a NOR gate and an INVERTER. (b) Direct configuration of an OR gate. (c) The physical truth table for an OR gate.

X	Y	Z
LV	LV	LV
LV	HV	HV
HV	LV	HV
HV	HV	HV

can be added to each NOR gate of Fig. 2-16 just above the ground symbol. Then, with a relocation of the output lead, Z, just above the pull-down resistor, an OR gate is formed. This converts the NOR to an OR gate since when supply current flows through the pull-down resistor a voltage drop, equal approximately to the supply voltage, falls across Z relative to ground. This method is the same as that described for converting a NAND gate to an AND.

The logic AND and OR interpretations of an OR gate are given in Fig. 2-23 together with the appropriate logic circuit symbol for each. Observe

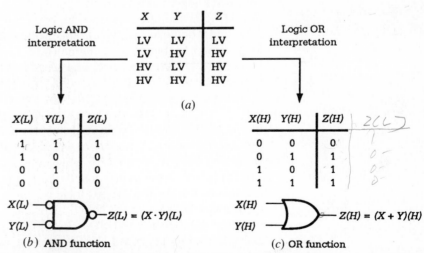

Fig. 2-23. AND and OR functions of the OR gate. (a) The physical truth table for the OR gate. (b) Logic AND interpretation and circuit symbol for the OR gate. (c) Logic OR interpretation and circuit symbol for the OR gate.

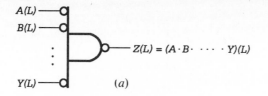

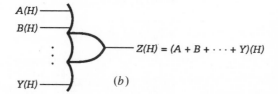

$$Z(L) = (A \cdot B \cdot \cdots \cdot Y)(L)$$

(a)

$$Z(H) = (A + B + \cdots + Y)(H)$$

(b)

Fig. 2-24. *Logic circuit symbols representing (a) the logic AND interpretation and (b) the logic OR interpretation of a multiple input OR gate.*

that the AND interpretation of the OR gate results when both inputs and the output are ACTIVE LOW and that the OR interpretation results if both inputs and output are ACTIVE HIGH. The presence of the ACTIVE LOW INDICATOR bubbles in the AND symbol or their absence in the OR symbol is consistent with these interpretations.

Presented in Fig. 2-24 are the logic circuit symbols representing a multiple-input OR gate. They should be compared with those of the multiple-input AND gate given in Fig. 2-21.

2.5   SUMMARY OF LOGIC CIRCUIT SYMBOLS

Now that we have labored through a discussion of the logic symbols used to represent LOGIC LEVEL CONVERSION, and AND and OR operations, it is desirable to summarize these symbols for future reference purposes. We present this summary from a logic function point of view since this is how one uses these symbols in designing a logic circuit. In Fig. 2-25 are shown the logic circuit symbols for LOGIC LEVEL CONVERSION and for the AND and OR operators, all presented with polarized mnemonics— X, Y, and Z being used to represent any mnemonic. Also included with each logic symbol is the name of the physical device (in parentheses) that must be used to perform the electrical analog of the operation implied by the symbol.

An interesting pattern emerges in Fig. 2-25. By comparing diagonally placed logic symbols for the AND and OR logic operations, we observe that the ACTIVE LOW INDICATOR bubbles are placed on one symbol where they are absent on the other. For example, compare the NAND gate symbol with the NOR gate symbol for either the AND operation or OR operation. The reader may wish to use this "bubble" pattern as an aid to remembering the gate names for the various logic circuit symbols given in Fig. 2-25.

We have shown that the NAND gate and the NOR gate can each perform the electrical analog of LOGIC LEVEL CONVERSION as well as the AND and OR operations, given the appropriate input conditions. It is for this reason that the NAND and NOR gates are often called *universal* gates. Furthermore, because an active transistor switch is a natural INVERTER, the NAND and NOR gates are physically simpler than are their AND and

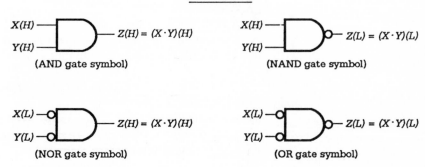

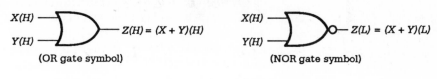

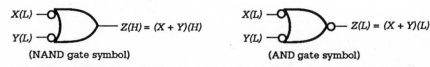

Fig. 2-25. *Summary of logic circuit symbols for LOGIC LEVEL CONVERSION and for the AND and OR logic operations. The gate name associated with each symbol is included in parentheses.*

OR counterparts. Thus, it is easy to understand why NAND and NOR gates are the most frequently used gates in IC technology.

It is appropriate to mention at this time that another gate symbology is in use. It is the new ANSI/IEEE Std. 91-1984, which includes SSI, MSI, and LSI device symbols. The gate (small-scale) symbology embraced by this new standard is rectangular with an identifying operator symbol located within. A summary of the new gate symbology is provided in Appendix 2.2, where a comparison is made with the distinctive gate symbols adopted in this text. We have purposely avoided the use of the rectangular gate symbols in favor of the distinctive shapes for instructional purposes. In a complex logic circuit, the rectangular symbols all look much the same—hence, a possible source of confusion and error. Sources of information regarding the new ANSI/IEEE Standard are given in the annotated references at the end of this chapter.

2.6 LOGIC LEVEL INCOMPATIBILITY, COMPLEMENTATION, AND CONJUGATED SYMBOL FORMS

We now reach a critical point in our development of digital design fundamentals. What follows is a natural extension of the material presented previously—and, more important, it is basic to an understanding of the digital design methodology presented in this text. This section is mainly about a new concept, complementation, and the subject of conjugation of logic circuit symbols. Mastery of the subject matter in this section will permit the reader to operate efficiently between the extremes of logic abstraction and physical reality with the least amount of effort and a minimum probability of error.

A MIXED LOGIC notation has been established by using the term ACTIVE to bridge the gap between that which is pure logic and that which is physical realization. The ACTIVE LOW INDICATOR bubble is a necessary part of the circuit symbolism for the MIXED LOGIC notation. Referring to Fig. 2-25, we understand, for example, that the presence of an ACTIVE LOW INDICATOR bubble on the input to a gate symbol requires that that input signal arrive from a negative logic source if logic level compatibility is to exist. Likewise, the absence of ACTIVE LOW INDICATOR bubble on the input to a gate requires that the input signal be from a positive logic source so as to satisfy logic level compatibility. Clearly, the logic compatibility condition is satisfied for all logic gate symbols shown in Fig. 2-25.

But suppose a signal arrives at the input to a logic symbol from a source whose ACTIVATION LEVEL differs from that required for logic level compatibility by the symbol. When this happens a condition of *logic incompatibility* exists, which requires that each occurrence of the signal name in the output be complemented. The operation of *complementation* is defined by the important relations

$$\alpha(L) = \bar{\alpha}(H) \quad \text{and} \quad \bar{\alpha}(L) = \alpha(H) \tag{2-3}$$

such that

$$(\alpha \cdot \bar{\alpha})(H) = 0(H)$$
$$(\alpha \cdot \bar{\alpha})(L) = 0(L)$$

and $\qquad\qquad\qquad\qquad\qquad\qquad\qquad\qquad\qquad\qquad$ (2-4)

$$(\alpha + \bar{\alpha})(H) = 1(H)$$
$$(\alpha + \bar{\alpha})(L) = 1(L)$$

where the overscore (or bar) is read as "the complement of." Thus, in a given logic system, a mnemonic α ANDed with its complement $\bar{\alpha}$ is logic 0, or a mnemonic ORed with its complement is logic 1. Now let us apply Eqs. (2-3) to actual logic circuit symbols.

In Fig. 2-26 are given four typical examples of logic incompatibility, where Eqs. (2-3) have been applied so as to be consistent with the requirement of logic incompatibility. To serve as a visual reminder that an incompatible polarized mnemonic must be complemented in the output expression,

2 / Fundamentals of Digital Design

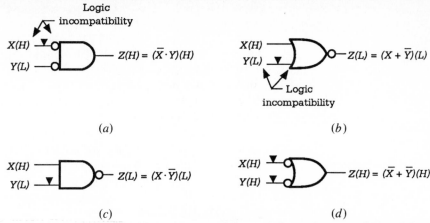

(a)   *(b)*

Fig. 2-26. *Examples of logic incompatibility and the application of Eqs. (2-3).*

an *incompatibility indicator flag*, ▼, is placed on the appropriate input line—this flag is never placed on the output of a circuit symbol.

Earlier we stated that the only logic function of the INVERTER is to perform the electrical analog of LOGIC LEVEL CONVERSION, and this is true. But to what purpose? The answer is simply stated:

> *The logic function of the INVERTER is to create or remove a logic incompatibility depending on the output requirements of the particular logic symbol.*

Consider, for example, that we want the output $Z(H) = (\overline{X} \cdot Y)(H)$ as in Fig. 2-26(a) but for inputs that are presented to the gate as $X(L)$ and $Y(L)$, that is, both from negative logic sources. If we are told that a NOR gate symbol must be used to create the output $(\overline{X} \cdot Y)(H)$ given these input constraints, then how are we to do this? As indicated in Fig. 2-27(a), we would

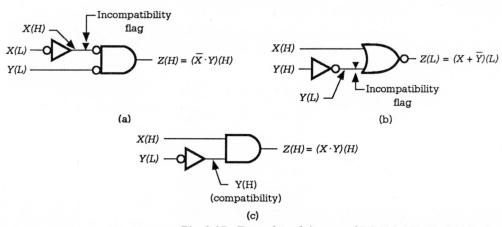

Fig. 2-27. *Examples of the use of LOGIC LEVEL CONVERSION to create or remove logic incompatibilities. (a), (b) Creation of a logic incompatibility. (c) Removal of a logic incompatibility.*

use a LOGIC LEVEL CONVERTER (an INVERTER) to create a logic incompatibility on the $X(L)$ input so that the output would read $(\overline{\overline{X}} \cdot Y)(H)$. Notice that the ACTIVE LOW INDICATOR bubble is situated on the $X(L)$ side of the INVERTER symbol, thereby creating the incompatibility at the gate input.

Two other examples are given in Fig. 2-27. In Fig. 2-27(b) the INVERTER symbol is used to create a logic incompatibility for ACTIVE HIGH inputs so that the output of a NOR gate (OR symbol) reads $(X + \overline{Y})(L)$. In Fig. 2-27(c) the INVERTER symbol is used to remove a logic incompatibility for mixed inputs so that the output of the AND gate (AND symbol) reads $(X \cdot Y)(H)$. Observe that in each case of Fig. 2-27 the INVERTER symbol input requirement is compatible with the signal ACTIVATION LEVEL so that the creation or removal of an incompatibility exists at the gate input. We conclude then that *an incompatibility must never be produced at the input of an INVERTER*, since the INVERTER's only logic function is to create or remove an incompatibility at the input of a logic device.

Logic circuit symbols which derive from different logic interpretations of the same gate are called *conjugate gate symbols*. In Fig. 2-28 we summarize the four pairs of conjugate gate symbols discussed previously in Sect. 2.4; the double arrow signifies their conjugate relationship. Because conjugate gate symbol pairs are derived from the same gate, their outputs must be identical for identical inputs. The following identities, called the *DeMorgan relations*, are presented in the order that the conjugate symbol pairs appear in Fig. 2-28:

$$
\begin{array}{ll}
\text{(a)} & (X \cdot Y)(L) = (\overline{X} + \overline{Y})(H) \\[6pt]
\text{(b)} & (\overline{X} \cdot \overline{Y})(H) = (X + Y)(L)
\end{array}
\left.\begin{array}{c}\\[6pt]\\\end{array}\right\} \text{Dual relations}
$$

$$
\begin{array}{ll}
\text{(c)} & (X \cdot Y)(H) = (\overline{X} + \overline{Y})(L) \\[6pt]
\text{(d)} & (\overline{X} \cdot \overline{Y})(L) = (X + Y)(H)
\end{array}
\left.\begin{array}{c}\\[6pt]\\\end{array}\right\} \text{Dual relations}
$$

$$(2\text{-}5)$$

The DeMorgan relations were obtained by using ACTIVE HIGH inputs, as demonstrated in Fig. 2-29. An equivalent set of relations could have been generated by using ACTIVE LOW inputs, and still other such relations would result from the use of mixed inputs. However, these other equivalent

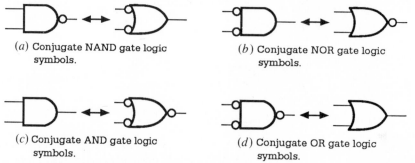

(*a*) Conjugate NAND gate logic symbols.

(*b*) Conjugate NOR gate logic symbols.

(*c*) Conjugate AND gate logic symbols.

(*d*) Conjugate OR gate logic symbols.

Fig. 2-28. *Conjugate pairs of logic symbols and the gates they represent.*

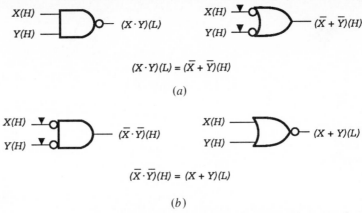

$$(X \cdot Y)(L) = (\overline{X} + \overline{Y})(H)$$

(a)

$$(\overline{X} \cdot \overline{Y})(H) = (X + Y)(L)$$

(b)

Fig. 2-29. *Conjugate pairs of logic gate symbols and their DeMorgan relations for ACTIVE HIGH inputs. (a) NAND gate logic symbols. (b) NOR gate logic symbols.*

relations would contribute nothing new and so will not be presented. Notice that relations (a) and (b) of Eqs. (2-5) are called duals as are relations (c) and (d). A *dual* relationship exists when the operators ($\cdot$) and ($+$) are interchanged simultaneously with the interchange of logic 1 and logic 0. Obviously, there are no 1's and 0's to be interchanged in Eqs. (2-5).

The DeMorgan relations of Eqs. (2-5) attest to the interchangeability of the conjugate logic symbols given in Fig. 2-28, and knowing this can be very useful in digital logic design. The DeMorgan relations establish a definite break with the homologic approach which has been a burden to designers from the start. As will be clear by chapter's end, symbol interchangeability represented by the DeMorgan relations eliminates the need for extensive algebraic manipulation of logic variables as would be required in a homologic system.

2.7 READING AND CONSTRUCTION OF LOGIC CIRCUITS

We are but a small step away from useful logic circuit construction. All that remains is to use correctly the information that we have gathered to this point. In this section we will demonstrate by example the very simple procedures that are necessary to construct and read any logic circuit regardless of its complexity.

Consider the logic function

$$X = (\overline{A} \cdot B) + (A \cdot \overline{B})$$

OR output stage

AND input stage

and notice that it is formed by ORing together two ANDed terms. There is an AND stage input and an OR stage output. Suppose it is required that this function be ACTIVE HIGH, hence, $X(H) = [(\overline{A} \cdot B) + (A \cdot \overline{B})](H)$, and

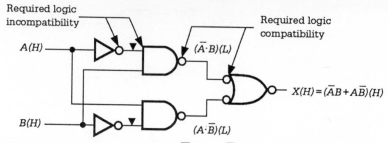

Fig. 2-30. *Implementation of* $X(H) = (\overline{A}B + A\overline{B})(H)$ *by using NAND gate logic symbols with ACTIVE HIGH inputs.*

that all inputs are from positive logic sources, that is, also ACTIVE HIGH. At this point we must decide from Fig. 2-25 or from Fig. 2-28 which logic symbols should be used to implement this function. Let us construct this logic circuit in NAND logic, meaning that we use NAND gates. By choosing the conjugate logic symbols representing the NAND gate and by implementing *output to input*, we arrive at the logic circuit given in Fig. 2-30. Notice that two INVERTERs were required to produce the necessary logic incompatibilities in the terms $(\overline{A} \cdot B)(L)$ and $(A \cdot \overline{B})(L)$, which are the outputs from the two NAND gate AND symbols. These two ANDed terms when ORed by the NAND gate OR symbol produce the required function.

The physical realization of the function given in Fig. 2-30 can be altered without changing the original function by complementing between the AND and OR stages so as to preserve logic compatibility. This is indicated within the area enclosed by the dashed lines in Fig. 2-31. In this way a NAND gate realization is replaced by an equivalent AND/OR gate realization, or vice versa. Observe that the AND stage in Fig. 2-31(b) yields $(\overline{A} \cdot B)(H)$ and $(A \cdot \overline{B})(H)$ instead of the ACTIVE LOW terms of Fig. 2-30. Nevertheless, the output remains the same if the ACTIVATION LEVELs for the inputs remain the same.

Next, let us implement the function

$$Y(H) = [(A + B) \cdot (\overline{A \cdot B})](H)$$

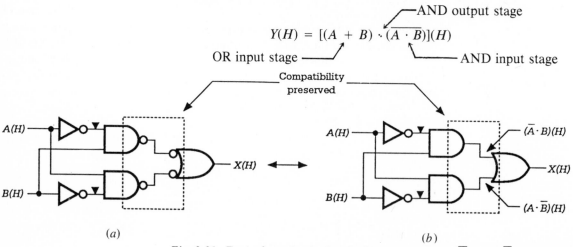

(a)   *(b)*

Fig. 2-31. *Equivalent circuits for the function* $X(H) = (\overline{A}B + A\overline{B})(H)$ *produced by complementing between stages. (a) NAND gate version. (b) AND/OR gate version.*

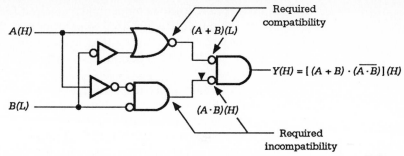

Fig. 2-32. *NOR gate realization of the function* Y(H) = [(A + B) · ($\overline{A \cdot B}$)](H).

in NOR logic (all NOR gates) given that *A* is from a positive logic source and *B* is from a negative logic source. Noting the position of the AND and OR operators, we choose the appropriate NOR gate symbols from Fig. 2-28(b). Then, from output to input, the logic circuit is constructed as shown in Fig. 2-32. Clearly, the presence of the logic incompatibility between the input and output stages is required to produce the $\overline{A \cdot B}$ term in the output, and the use of two INVERTERs removes the incompatibilities at the input stage.

As a final example, consider the somewhat more complex function

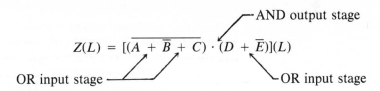

Let us implement this function in "free style" logic (i.e., by using any gate type we wish), given that the input signals *A*, *B*, and $\overline{C}$ are from positive logic sources, and *D* and *E* are from negative logic sources. Since there are several gate combinations to choose from in a free-style design, we will choose one. Beginning from output to input, we notice that two terms, $(\overline{A + \overline{B} + C})$ and $(D + \overline{E})$, are ANDed together and that the former $(\overline{A + \overline{B} + C})$ requires a logic incompatibility at the input of the AND output stage. Shown in Fig. 2-33 is our choice of logic circuit symbols for this function. Notice that one NAND gate performs the AND operation of the output stage, while another NAND gate and an AND gate perform the two OR operations of the input stage. The output of the AND gate OR symbol

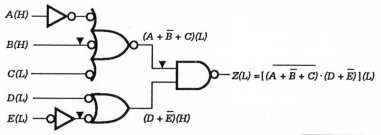

Fig. 2-33. *Free-style gate realization of the function* Z(L) = [($\overline{A + \overline{B} + C}$) · (D + $\overline{E}$)] (L).

2.7 / Reading and Construction of Logic Circuits

creates the logic incompatibility required for the $(\overline{A + \overline{B} + C})$ term. Also, two INVERTERs are used, one to remove the incompatibility on the $A(H)$ input and the other to create an incompatibility on the the $E(L)$ input. Finally, note that Eqs. (2-3) permit us to write $\overline{C}(H) = C(L)$, as indicated in Fig. 2-33.

Alternative physical realizations of the logic function $Z(L)$ can be produced by complementing between stages as was done earlier in Fig. 2-31. For example, by complementing between the input OR stage and the output AND stage, the NAND/AND realization of Fig. 2-33 is changed to one consisting of an OR gate (AND) output with a NAND gate and AND gate forming the input stage as before but switched around.

There are still other ways to alter the gate selection for a given function. Shown in Fig. 2-34 is a three-input NAND/NOR gate realization of the logic function $Z(L)$, this one produced by complementing between the input stage and previous stage of the AND gate OR symbol of Fig. 2-33. Here, it is seen that relocation of one INVERTER and the addition of a second one is required to preserve the input conditions $A(H)$, $B(H)$, and $C(L)$. Also, observe that the unused inputs on two of the three gates have been tied to HV, a requirement deduced from Eqs. (2-2) and Figs. 2-15 and 2-17. Later, in Sect. 2-9, additional information will be given in support of the fact that all unused inputs must be tied to either HV or LV, but never left "dangling."

Summary of Important Points to Remember. This section represents a culmination of all previous development in this chapter. For this reason it seems desirable to provide the following summary of important points to remember:

1. Before a logic function can be expressed in the form of a logic circuit, its input and output specifications must be given. This includes the name (mnemonic) of each input signal and its ACTIVATION LEVEL.
2. Implementation of a logic function must *always* be from output to input.
3. Reading a logic circuit must *always* be from input to output and, hence, the opposite of implementation.
4. Alternative physical realizations of a logic function can be obtained by the appropriate choice of logic symbols for the AND and OR operations. If a logic circuit already exists, an alternative logic circuit can be obtained by complementation between stages. Remember that complementation

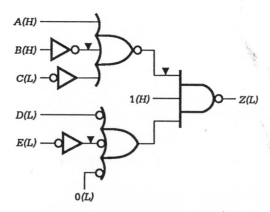

Fig. 2-34. Three input NAND/NOR realization of the function Z(L)
$$= [\overline{(A + \overline{B} + C) \cdot (D + \overline{E})}]$$
(L).

2 / Fundamentals of Digital Design

of only one input to a gate is *not* permitted since the resulting symbol would have no known physical identity.

5. Unused inputs must *never* be left "dangling" but must be tied either to HV or LV in accordance with the relations of Eqs. (2-2) given in Sect. 2.4.4.

We consider these points to be important for successful construction and reading of logic circuits. Failure to follow them will likely lead to serious design error.

2.8 XOR and EQV FUNCTIONS and THEIR PHYSICAL REALIZATIONS

Certain functions consisting of the AND and OR operators occur so often in digital logic design that special names and operator symbols have been assigned to them. By far the most common of these are the EXCLUSIVE OR (XOR) and EQUIVALENCE (EQV) functions represented by the following symbols:

$$\oplus \rightarrow \text{XOR}$$

$$\odot \rightarrow \text{EQV}$$

Thus, if we write $X \oplus Y$, it is read as X XOR Y, or if $X \odot Y$, it is read as X EQV Y. As we shall see in Chapter 4, these functions form the basis of arithmetic-type circuits, an important part of computer design. Like the AND and OR operators, the XOR and EQV functions are best understood in terms of logic circuit symbols representing them. In Fig. 2-35 are given the distinctively shaped circuit symbols for which the following functional descriptions apply:

(a) **XOR circuit symbol**

(b) **EQV circuit symbol**

Fig. 2-35. *Distinctively shaped logic circuit symbols representing (a) the XOR function and (b) the EQV function.*

> *The output of a logic XOR circuit symbol is ACTIVE if one or the other of two inputs is ACTIVE but not both ACTIVE or INACTIVE.*
>
> *The output of a logic EQV circuit symbol is ACTIVE if, and only if, both inputs are ACTIVE or if both inputs are INACTIVE.*

The reader should note that the functional description of an XOR or EQV circuit symbol is valid only for two inputs.

That the logic circuits of Figs. 2-30 and 2-32 represent the XOR function is evident from the functional description for XOR just given. Thus, we write

$$X = (\overline{A} \cdot B) + (A \cdot \overline{B}) = A \oplus B \qquad (2\text{-}6)$$

and read as

X is ACTIVE if A or B is ACTIVE but not both ACTIVE; that is, X is ACTIVE if A and B are not equivalent.

So, X is logic 1 only if A or B is logic 1 but not both logic 1 (or logic 0).

Alternatively, we write

$$Y = (A + B) \cdot (\overline{A \cdot B}) = A \oplus B \qquad (2\text{-}7)$$

and read as

Y if A OR B AND NOT BOTH

Similarly, for the EQV function, we write

$$F = (\overline{A} \cdot \overline{B}) + (A \cdot B) = A \odot B \qquad (2\text{-}8)$$

and read as

F is ACTIVE only if A and B are both ACTIVE or both INACTIVE,

which agrees with the functional description given earlier.

Equations (2-6) and (2-8) are usually considered to be the defining expressions for the XOR and EQV functions, respectively. Thus, whenever these functions are encountered, it is appropriate, and even necessary, to think of them in terms of their defining expressions. We do so in this text, often without specific reference to Eqs. (2-6) and (2-8).

2.8.1 The XOR and EQV Functions of the XOR Gate

Shown in Fig. 2-36 is a CMOS version of the XOR gate and its physical truth table. Although somewhat more complicated than the previous gates, its function is readily understood by inspection. First, notice it is constructed of two CMOS INVERTERs with a CMOS pass transistor switch separating

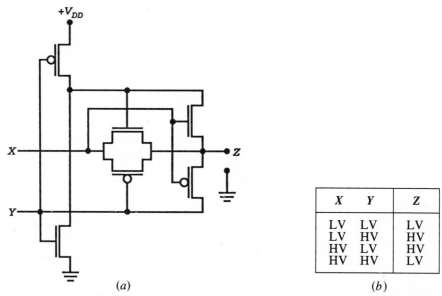

X	Y	Z
LV	LV	LV
LV	HV	HV
HV	LV	HV
HV	HV	LV

(a)   (b)

Fig. 2-36. *(a) A CMOS circuit for an XOR gate. (b) The physical truth table for the XOR gate.*

them. Let normal operation of the INVERTER refer to HV MOS gate supply so that the NMOS is ON and the PMOS is OFF, and vice versa for the inverse operation. Now, if both X and Y inputs are at HV, both INVERTERs are in normal operation, the PTL switch is OFF, and the output is shorted to ground (LV). Or if both inputs are at LV, both INVERTERs are in inverse operation, the PT switch is ON, and the output is again at LV via both inputs. However, if X is at HV and Y at LV, the output INVERTER switches to the normal mode, thereby connecting the output to $+V_{DD}$ (HV). Finally, if X is at LV and Y is at HV, the input INVERTER is in normal operation, the PT switch is OFF, and the output INVERTER is in inverse operation again connecting the output to HV but via the Y input.

That the physical device shown in Fig. 2-36 can perform the electrical analog of the XOR and EQV functions is easily demonstrated. Presented in Fig. 2-37 are the two logic interpretations of the XOR gate together with their distinctively shaped circuit symbols. The logic XOR interpretation of the physical truth table [Fig. 2-37(a)] results for ACTIVE HIGH inputs and output as indicated in Fig. 2-37(b). The logic EQV interpretation of the physical truth table is produced for ACTIVE HIGH inputs and ACTIVE LOW output. Notice the distinctively shaped circuit symbols used to represent these functions and notice that each logic truth table agrees with the functional description of the respective circuit symbol given earlier.

An interesting property of the logic interpretation for an XOR gate is the fact that changing both inputs to ACTIVE LOW does not alter the output function, as indicated in Figs. 2-38(a) and (b). The same is true if both inputs create logic incompatibilities [see Fig. 2-38(c)]. However, if any single incompatibility is produced, as in Fig. 2-38(d), the output function is complemented.

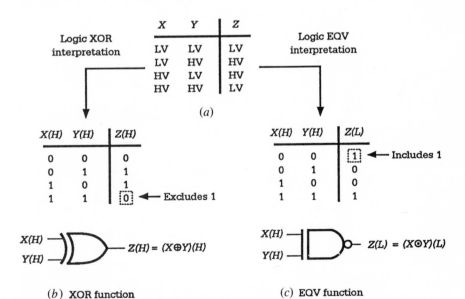

(a)

(b) XOR function   *(c)* EQV function

Fig. 2-37. *XOR and EQV functions of the XOR gate. (a) Physical truth table for the XOR gate. (b) Logic XOR interpretation and circuit symbol for the XOR gate. (c) Logic EQV interpretation and circuit symbol for the XOR gate.*

$X(L)$ $Y(L)$ — $Z(H) = (X \oplus Y)(H)$

(a)

$X(L)$ $Y(L)$ — $Z(L) = (X \odot Y)(L)$

(b)

$X(L)$ $Y(L)$ — $Z(H) = (X \oplus Y)(H)$

(c)

$X(L)$ $Y(H)$ — $Z(H) = (X \odot Y)(H)$

(d)

Fig. 2-38. *The effect of ACTIVE LOW inputs on the output functions of circuit symbols representing the XOR gate.*

2.8.2 The XOR and EQV Functions of the EQV Gate

A simple MOS version of the EQV gate, together with its physical truth table, is given in Fig. 2-39. In operation, if inputs X and Y are both at HV or both at LV, there is no path from output Z to ground, and the supply $+V_{DD}$ falls across Z, hence HV. However, if any one input is at HV and the other at LV, the output connects to ground by one path or the other and Z is at LV. Note that an XOR gate results if an INVERTER is placed on the output of an EQV gate.

The logic interpretations of the EQV gate are given in Fig. 2-40. Here, the logic EQV interpretation of the physical truth table is produced if both inputs and output are ACTIVE HIGH, while the XOR interpretation results for ACTIVE HIGH inputs and ACTIVE LOW output. Notice again the distinctively shaped logic circuit symbols which are used to represent this function. Just as with the XOR gate, the logic interpretations of the EQV

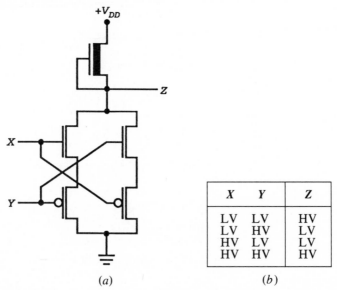

X	Y	Z
LV	LV	HV
LV	HV	LV
HV	LV	LV
HV	HV	HV

(a)　　　　　(b)

Fig. 2-39. *The EQV gate. (a) A simple MOS version. (b) The physical truth table for the EQV gate.*

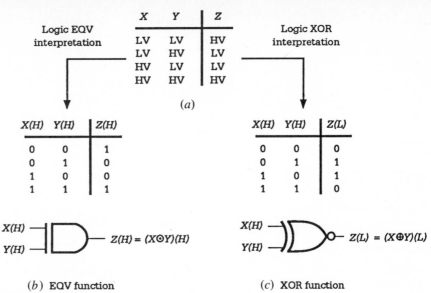

X	Y	Z
LV	LV	HV
LV	HV	LV
HV	LV	LV
HV	HV	HV

Logic EQV interpretation

Logic XOR interpretation

(a)

X(H)	Y(H)	Z(H)
0	0	1
0	1	0
1	0	0
1	1	1

X(H)	Y(H)	Z(L)
0	0	0
0	1	1
1	0	1
1	1	0

$X(H)$
$Y(H)$ — $Z(H) = (X \odot Y)(H)$

$X(H)$
$Y(H)$ — $Z(L) = (X \oplus Y)(L)$

(b) EQV function

(c) XOR function

Fig. 2-40. *XOR and EQV functions of the EQV gate. (a) Physical truth table for the EQV gate. (b) Logic EQV interpretation and circuit symbol for the EQV gate. (c) Logic XOR interpretation and circuit symbol for the EQV gate.*

$X(L)$
$Y(L)$ — $Z(H) = (X \odot Y)(H)$

$X(L)$
$Y(L)$ — $Z(L) = (X \oplus Y)(L)$

(a)

(b)

$X(L)$
$Y(L)$ — $Z(H) = (X \odot Y)(H)$

$X(H)$
$Y(L)$ — $Z(H) = (X \oplus Y)(H)$

(c)

(d)

Fig. 2-41. *The effect of ACTIVE LOW inputs on the output functions of circuit symbols representing the EQV gate.*

gate are unaltered if both inputs are ACTIVE LOW or if both inputs create logic incompatibilities. However, as with the XOR gate, if only one incompatibility is produced, the output function is complemented. These results are illustrated in Fig. 2-41 and may be proven by altering the appropriate logic truth tables of Fig. 2-40 to agree with the input ACTIVATION LEVELs indicated for each circuit symbol.

Let us now comment on the names and circuit symbol shapes associated with the XOR and EQV functions. The XOR function truth table in Fig. 2-37(b) would be that of an OR function had the 1 not been "excluded" (see box), hence, EXCLUSIVE OR and the OR-shaped symbol. By this reasoning the OR function could be labeled INCLUSIVE OR and is occasionally. The second curved line at the input of the XOR symbol is used to distinguish the XOR circuit symbol from that for the OR.

As suggested, the use of descriptive words such as INCLUSIVE or EX-

CLUSIVE are often used to identify these logic functions. However, use of such terms leads to ambiguity since it is never certain what is being included or excluded. The application of these terms to the EQV function is a good example. Thus, the EQV function truth table in Fig. 2-37(c) would appear as that of the AND function had the 1 (see box) not been "included," hence INCLUSIVE AND and the AND-shaped symbol with a double bar. But we could have "excluded" the 1 in the box to achieve the same result, so EXCLUSIVE AND (XAND). Worse yet is the use of the term XNOR to represent the EQV function. In this case the 1 at the bottom of the EQV truth table in Fig. 2-40(b) must be excluded to yield the "NOR" function, which is not even recognized as a fundamental operation—AND and OR are the only fundamental binary operations.

To avoid unnecessary confusion and in keeping with popular usage, we will take the contractions XOR and EQV to represent the logic interpretations of the physical truth tables given in Figs. 2-37 and 2-39. While the use of XOR seems unrivaled, the reader should be aware that other terms such as EXCLUSIVE NOR, COINCIDENCE, IDENTITY, and EQUALITY have also been used in place of EQUIVALENCE.

2.8.3 Controlled LOGIC LEVEL CONVERSION

An XOR or EQV gate has the unique property that it can be operated either to be transparent to an input signal, or it can be operated as an INVERTER, depending on how the second input is connected. This means that a signal is either transferred from input to output (transparent mode) or its logic level is converted (INVERTER mode). Shown in Figs. 2-42(b) and (c) are examples of controlled LOGIC LEVEL CONVERSION interpretations for an XOR gate. The reader can easily verify these examples by using the appropriate logic truth table provided in Fig. 2-42(a) and by using Eqs. (2-3). Thus,

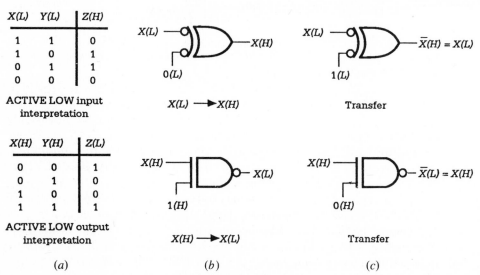

X(L)	Y(L)	Z(H)
1	1	0
1	0	1
0	1	1
0	0	0

ACTIVE LOW input
interpretation

$X(L) \longrightarrow X(H)$

Transfer

X(H)	Y(H)	Z(L)
0	0	1
0	1	0
1	0	0
1	1	1

ACTIVE LOW output
interpretation

$X(H) \longrightarrow X(L)$

Transfer

(a)   (b)   (c)

Fig. 2-42. Controlled LOGIC LEVEL CONVERSION interpretations of an XOR gate. (a) Logic truth tables. (b) LOGIC LEVEL CONVERSION mode. (c) Transfer mode.

if one input to the XOR gate is connected to HV, the INVERTER mode results, or if connected to LV, the transfer mode results.

A set of circuit symbols similar to those of Fig. 2-42 can be established for the EQV gate. In this case the physical requirements for INVERSION and transparency are the reverse of those for the XOR gate.

2.9 LAWS OF BOOLEAN ALGEBRA

To design a digital circuit that will perform a prescribed function, one must be able to manipulate and combine the various switching variables (those that are an inherent part of the design) in certain ways which are in agreement with mathematical logic. Use of the laws of Boolean algebra makes these manipulations and combinations relatively simple. This special algebra of mathematical logic, named in recognition of the English logician George Boole, can be rigorously and eloquently presented by using sets of theorems and postulates. However, for our purposes there is no need for such a formal approach. The laws of Boolean algebra are relatively few in number and can ultimately be deduced from the truth tables for the three basic logic operators NOT, AND, and OR. A reference regarding a brief history of Boolean algebra is provided in the Annotated References at the end of this chapter.

In this section the reader can expect that the development will be nearly exclusively in the logic domain. We do this so as to provide the background necessary for what is to come in Chapter 3 and beyond.

2.9.1 NOT, AND, and OR Laws

Recall that we have already introduced the binary operators, AND and OR, in Sect. 2.4, and that the logic AND and logic OR interpretations of their respective physical truth tables were given in Figs. 2-20 and 2-23. The AND and OR laws are inherent in these logic truth tables as we shall soon see. Also, without specifically saying so, the NOT laws were implied in Sect. 2.6, where the concepts of LOGIC LEVEL INCOMPATIBILITY and COMPLEMENTATION were developed by using MIXED LOGIC notation.

NOT Laws. The unary operator NOT is the logic equivalent of COMPLEMENTATION, which, in turn, is intimately associated with the concept of LOGIC LEVEL INCOMPATIBILITY, discussed in Sect. 2.6. In effect, COMPLEMENTATION can be viewed as the MIXED LOGIC (hence quasi-physical) equivalent of the unary logic operation NOT, and NOT connotes inversion in the sense of supplying the lack of something. Although NOT is a purely logic concept while COMPLEMENTATION arose more from a physical standpoint, the two terms, NOT and COMPLEMENTATION, will be used interchangeably following established practice.

Shown in Fig. 2-43(a) is the NOT truth table, which is the positive logic form of the MIXED LOGIC truth table given in Fig. 2-6(b). The NOT laws are derived from the positive logic truth table and are given in Fig. 2-43(b). The NOT operation, like COMPLEMENTATION, is designated by the overscore (or bar). A double COMPLEMENTATION (or double bar) of a function, sometimes called *involution*, is the function itself.

X	$\overline{X}$
0	1
1	0

(a)

NOT Laws
$\overline{0} = 1$
$\overline{1} = 0$
$\overline{\overline{X}} = X$

(b)

Fig. 2-43. (a) Truth table for NOT. (b) The NOT laws.

X	Y	$X \cdot Y$
0	0	0
0	1	0
1	0	0
1	1	1

(a)

AND Laws
$X \cdot 0 = 0$
$X \cdot 1 = X$
$X \cdot X = X$
$X \cdot \overline{X} = 0$

(b)

Fig. 2-44. *(a) Truth table for AND. (b) The AND laws.*

X	Y	$X + Y$
0	0	0
0	1	1
1	0	1
1	1	1

(a)

OR Laws
$X + 0 = X$
$X + 1 = 1$
$X + X = X$
$X + \overline{X} = 1$

(b)

Fig. 2-45. *(a) Truth table for OR. (b) The OR laws.*

In the following examples, we will use X to represent a multivariable function $X = X(A,B,C,...)$. Suppose $X = A \cdot \overline{B}$ (read as A AND NOT B), then $\overline{X} = \overline{A \cdot \overline{B}}$ (read as A AND NOT B the quantity complemented), and $\overline{\overline{X}} = \overline{\overline{A \cdot \overline{B}}} = A \cdot \overline{B}$. Or suppose $X = 0$, then $\overline{X} = \overline{0} = \overline{1} = 0$, and so on. Finally, notice that Eqs. (2-3), given in Sect. 2.6, can be generated one from the other by involution—even in mixed-rail notation. Thus, $\overline{\alpha}(L) = \overline{\overline{\alpha}}(H) = \alpha(H)$, and so on.

AND Laws. The AND laws are easily deduced from the truth table for AND in Fig. 2-44(a), which is the positive logic form of the MIXED LOGIC truth table given in Fig. 2-20(b). Thus, by taking Y equal to logic values 0, 1, X, and $\overline{X}$, the AND laws of Fig. 2-44(b) result. For example, if $Y = 1$ then we deduce that $X \cdot 1 = X$. This follows since if $X = 0$, there results $0 \cdot 1 = 0$, or if $X = 1$, then $1 \cdot 1 = 1$. Or if $Y = \overline{X}$, then we deduce that $X \cdot \overline{X} = 0$, since if $X = 0$, there results $0 \cdot \overline{0} = 0 \cdot 1 = 0$, and if $X = 1$ then $1 \cdot \overline{1} = 1 \cdot 0 = 0$, and so on.

To illustrate the application of the AND laws, we will again take X as representing a multivariable function $X = X(A,B,C,...)$. Then, let $X = A + \overline{B}$ (read as A OR NOT B) so that $(A + \overline{B}) \cdot 0 = 0$, $(A + \overline{B}) \cdot 1 = (A + \overline{B})$, $(A + \overline{B}) \cdot (A + \overline{B}) = (A + \overline{B})$, and $(A + \overline{B}) \cdot \overline{(A + \overline{B})} = 0$. Thus, these laws are valid regardless of the complexity of the function X.

OR Laws. Presented in Fig. 2-45(b) are the OR laws which are deduced from the positive logic truth table for OR in Fig. 2-45(a) by assigning to Y the values 0, 1, X, and $\overline{X}$. [Note that the positive logic truth table for OR is taken directly from Fig. 2-23(c).] So, if $Y = 1$ then $X + 1 = 1$, since $1 + 1 = 1$ and $0 + 1 = 1$. Or when $Y = \overline{X}$ we deduce that $X + \overline{X} = 1$, since if $X = 0$ it follows that $0 + 1 = 1$, and if $X = 1$ then $1 + 0 = 1$.

We illustrate the application of the OR laws by letting the function be $X = \overline{B}C$ (read as B NOT AND C). Now, according to the OR laws $\overline{B}C + 0 = \overline{B}C$, $\overline{B}C + 1 = 1$, $\overline{B}C + \overline{B}C = \overline{B}C$, and $\overline{B}C + \overline{\overline{B}C} = 1$. Here again we have taken the symbol X to represent a multivariable function BC attesting the general nature of the OR laws.

Notice that the AND laws and OR laws are easily verified by setting $X = 0, 1$ in the AND and OR laws, and then comparing the results with the AND and OR truth tables in Figs. 2-44(a) and 2-45(a), respectively.

2.9.2 Concept of Duality

We interrupt our development of the laws of Boolean algebra to reflect briefly on what we have established so far. An inspection of the AND and OR laws reveals an interesting relationship which may not be obvious to the reader. If the 1's and 0's are interchanged while at the same time interchanging the $(\cdot)$ and $(+)$ logic operators, the AND laws generate the OR laws, and vice versa. This relationship is called logic *duality*; that is, the AND and OR laws are the *dual* of each other as indicated by the double arrow.

2 / Fundamentals of Digital Design

AND Laws		OR Laws
$X \cdot 0 = 0$		$X + 1 = 1$
$X \cdot 1 = X$	By	$X + 0 = X$
$X \cdot X = X$	$\longleftrightarrow$	$X + X = X$
$X \cdot \overline{X} = 0$	duality	$X + \overline{X} = 1$

$$(2\text{-}9)$$

As we proceed through the remainder of this text, we will find that the concept of duality pervades the entire field of digital design. But this is really not surprising, since duality is an inherent property of Boolean algebra, which, in turn, is the foundation of digital design.

2.9.3   Associative, Commutative, Distributive, and Absorptive Laws

Associative Laws. The associative laws follow conventional algebra in appearance and are given here in terms of three variables, X, Y, and Z, any or all of which could be multivariable functions.

$$(X \cdot Y) \cdot Z = X \cdot (Y \cdot Z) = X \cdot Y \cdot Z \qquad \text{AND form}$$
$$(X + Y) + Z = X + (Y + Z) = X + Y + Z \qquad \text{OR form}$$

$$(2\text{-}10)$$

The dual relationship between the AND and OR associative laws is obvious—one generates the other by interchanging the logic operators ($\cdot$) and ($+$).

Formal proof of the associative laws is no trivial matter, but their verification is straightforward by using truth tables. In Fig. 2-46 we construct the truth table for the AND form of Eqs. (2-10) by assigning all possible combinations of logic levels to X, Y, and Z in a binary sequence equivalent to decimals 0 through 7, as indicated. Notice that repeated application of the AND laws was used to construct the truth table, which clearly verifies the AND associative law.

Verification of the OR associative law can also be easily accomplished by using a similar truth table. To demonstrate the power of duality, the reader should verify the OR form of Eqs. (2-10) by constructing a dual truth table with repeated applications of the OR laws given earlier.

Commutative Laws. Like the associative laws, the commutative laws follow conventional algebra in appearance and are given by the following equations

Decimal $X\ Y\ Z$	$X \cdot Y$	$Y \cdot Z$	$(X \cdot Y) \cdot Z$	$X \cdot (Y \cdot Z)$	$X \cdot Y \cdot Z$
0 0 0 0	0	0	0	0	0
1 0 0 1	0	0	0	0	0
2 0 1 0	0	0	0	0	0
3 0 1 1	0	1	0	0	0
4 1 0 0	0	0	0	0	0
5 1 0 1	0	0	0	0	0
6 1 1 0	1	0	0	0	0
7 1 1 1	1	1	1	1	1

Fig. 2-46. Truth table for the AND form of the associative laws.

in terms of the three multivariable functions, X, Y, and Z, where the dual relationship between the AND and OR forms is easily seen:

$$X \cdot Y \cdot Z = X \cdot Z \cdot Y = Z \cdot X \cdot Y = \cdots \qquad \text{AND form}$$
$$X + Y + Z = X + Z + Y = Z + X + Y = \cdots \qquad \text{OR form}$$

$$(2\text{-}11)$$

Verification of the commutative laws is accomplished simply by assigning logic 0 and logic 1 to the X's and Y's in the two variable forms of these laws and then comparing the results with the AND and OR truth tables given in Figs. 2-44(a) and 2-45(a). Thus, we see that $0 \cdot 1 = 1 \cdot 0 = 0$ and $1 + 0 = 0 + 1 = 1$ are verified by the AND and OR truth table, respectively. The commutative properties of $1 \cdot 1$ and $0 \cdot 0$ are indistinguishable.

Distributive and Absorptive Laws. There are four laws in this category: two distributive laws and two absorptive laws. The two *distributive laws* are given by Eqs. (2-12) and (2-13) and are named the *factoring* and *distributive laws*, respectively, primarily for future reference purposes. The two *absorptive laws* are represented by Eqs. (2-14) and (2-15) and are not given separate names. These laws are presented in terms of the multivariable functions X, Y, and Z so as to dramatize their generality.

Factoring law $\qquad (X \cdot Y) + (X \cdot Z) = X \cdot (Y + Z) \qquad (2\text{-}12)$

Distributive law $\qquad (X + Y) \cdot (X + Z) = X + (Y \cdot Z) \qquad (2\text{-}13)$

Absorptive laws $\begin{cases} X \cdot (\overline{X} + Y) = X \cdot Y & (2\text{-}14) \\ X + (\overline{X} \cdot Y) = X + Y & (2\text{-}15) \end{cases}$

The first thing to notice about these laws is the dual relationship that exists between the factoring and distributive laws and between the two absorptive laws. By interchanging logic operators, $(\cdot)$ and $(+)$, we see that one law is generated from the other. Second, one notices that the factoring law follows conventional algebra in appearance, while the remaining three laws do not.

Verification of the distributive and absorptive laws is accomplished by constructing a truth table similar to that used to verify the AND form of the associative law in Fig. 2-46. Shown in Fig. 2-47 is such a truth table for the factoring law [Eq. (2-12)], where again the logic values for X, Y, and Z are unfolded in a sequence equivalent to decimal 0 through 7. The factoring law

Decimal	X Y Z	$X \cdot Y$	$X \cdot Z$	$Y + Z$	$(X \cdot Y) + (X \cdot Z)$	$X \cdot (Y + Z)$
0	0 0 0	0	0	0	0	0
1	0 0 1	0	0	1	0	0
2	0 1 0	0	0	1	0	0
3	0 1 1	0	0	1	0	0
4	1 0 0	0	0	0	0	0
5	1 0 1	0	1	1	1	1
6	1 1 0	1	0	1	1	1
7	1 1 1	1	1	1	1	1

Fig. 2-47. Truth table for the factoring law.

is verified by repeated application of the AND and OR laws and by comparing the two columns on the right.

The distributive law can also be verified by using a truth table. However, it is simpler to prove this law by algebraic manipulation with the AND and OR and factoring laws which have now been verified. We do this in the following sequence of steps by using the square brackets to draw the reader's attention to those portions where the laws indicated on the right are applied:

$$[(X + Y) \cdot (X + Z)] = [X \cdot (X + Z)] + [Y \cdot (X + Z)] \qquad \text{Factoring law}$$
$$= [X \cdot X] + (X \cdot Z) + (Y \cdot X) + (Y \cdot Z) \qquad \text{AND law}$$
$$= [X + (X \cdot Z) + (Y \cdot X)] + (Y \cdot Z) \qquad \text{Factoring law}$$
$$= X \cdot [1 + Z + Y] + (Y \cdot Z) \qquad \text{OR laws}$$
$$= X + (Y \cdot Z)$$

The absorptive law given by Eq. (2-14) is proved in a similar manner:

$$[X \cdot (\overline{X} + Y)] = [X \cdot \overline{X}] + X \cdot Y \qquad \text{Factoring and AND laws}$$
$$= [0 + X \cdot Y] \qquad \text{OR laws}$$
$$= X \cdot Y$$

Finally, it is left as an exercise for the reader to verify the remaining absorptive law by algebraic means using the laws which have already been verified. The use of the concept of duality or the use truth tables are also adequate means of verification. However, it is our belief that more is to be gained by using the algebraic methods demonstrated earlier.

2.9.4 DeMorgan's Laws

In the latter half of the nineteenth century, the English logician and mathematician Augustus DeMorgan proposed two theorems of mathematical logic that have since become known as DeMorgan's theorems. The Boolean algebraic representations of these theorems are known as DeMorgan's laws and, in terms of the two multivariable functions X and Y, are given by

$$\overline{X \cdot Y} = \overline{X} + \overline{Y} \qquad (2\text{-}16)$$
$$\overline{X + Y} = \overline{X} \cdot \overline{Y} \qquad (2\text{-}17)$$

Observe that the two laws are the dual of each other.

More generally, for any number of multivariable functions, the DeMorgan laws take the following form:

$$\overline{X \cdot Y \cdot Z \cdot \cdots \cdot N} = \overline{X} + \overline{Y} + \overline{Z} + \cdots + \overline{N} \qquad (2\text{-}18)$$
$$\overline{X + X + Z + \cdots + N} = \overline{X} \cdot \overline{Y} \cdot \overline{Z} \cdot \cdots \cdot \overline{N} \qquad (2\text{-}19)$$

Verification of DeMorgan's laws is easily accomplished by using truth tables and logic maps in two variables. Shown in Fig. 2-48(a) is the truth table representing Eq. (2-16). This law is verified by comparing the truth

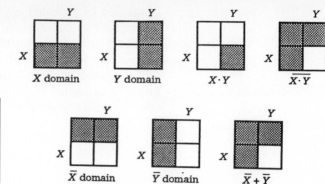

X Y	X · Y	$\overline{X \cdot Y}$	$\overline{X}$	$\overline{Y}$	$\overline{X} + \overline{Y}$
0 0	0	1	1	1	1
0 1	0	1	1	0	1
1 0	0	1	0	1	1
1 1	1	0	0	0	0

(a)

(b)

Fig. 2-48. *(a) Truth table and (b) logic maps for DeMorgan's law* $\overline{X \cdot Y} = \overline{X} + \overline{Y}$.

table columns for $\overline{X \cdot Y}$ and $\overline{X} + \overline{Y}$ following repeated application of the NOT, AND, and OR laws.

Some explanation of the logic maps presented in Fig. 2-48(b) is needed. The shaded areas shown in the logic maps are called *domains*. For example, the X domain represents "all that is X," and the $\overline{Y}$ domain represents "all that is NOT Y," and so on. In logic terminology the AND operation is properly called the *intersection*, while the OR operation is appropriately termed the *union*. Applied to logic maps, $X \cdot Y$ represents the intersection of the X domain with the Y domain. Or the function $X + Y$ represents the union (combination) of the X domain with the Y domain. So, by referring to the logic maps of Fig. 2-48(b), it is clear to us that these maps express vividly the equivalence between "all that is NOT" the intersection of X and Y, or $\overline{(X \cdot Y)}$, and the union of NOT X and NOT Y, or $(\overline{X} + \overline{Y})$, as required by Eq. (2-13).

The verification of DeMorgan's second law, Eq. (2-17), follows in a similar fashion as shown in Fig. 2-49. A comparison of the last two columns of the truth table in Fig. 2-49(a) shows us that Eq. (2-17) is verified. However, notice how clearly the shaded logic maps express the equivalence of "all that is NOT" the union of X and Y, or $\overline{(X + Y)}$, and the intersection of NOT X and NOT Y, or $(\overline{X} \cdot \overline{Y})$. The dual relationship between Eqs. (2-16) and (2-17) is evident when one compares the last logic maps in Figs. 2-48(b) and 2-49(b).

To illustrate the use of DeMorgan's laws, let us prove the second absorptive law [Eq. (2-15)] in variables A and B by using what could be termed the "double-bar" method of simplification as follows:

$$A + \overline{A} \cdot B = \overline{[\overline{A + \overline{A} \cdot B}]} \qquad \text{Involution and DeMorgan's law [Eq. (2-17)]}$$

$$= \overline{\overline{A} \cdot [\overline{\overline{A} \cdot B}]} \qquad \text{DeMorgan's law [Eq. (2-16)]} \qquad (2\text{-}20)$$

$$= \overline{[\overline{A} \cdot (A + \overline{B})]} \qquad \text{Factoring, AND and OR laws}$$

$$= \overline{\overline{A} \cdot \overline{B}} \qquad \text{DeMorgan's law [Eq. (2-16)]}$$

$$= A + B$$

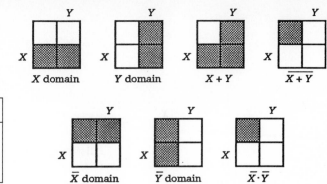

X Y	X + Y	$\overline{X + Y}$	$\overline{X}$	$\overline{Y}$	$\overline{X} \cdot \overline{Y}$
0 0	0	1	1	1	1
0 1	1	0	1	0	0
1 0	1	0	0	1	0
1 1	1	0	0	0	0

(a)

(b)

Fig. 2-49. *(a) Truth table and (b) logic maps for DeMorgan's law $\overline{X + Y} = \overline{X} \cdot \overline{Y}$.*

Here, again, the square brackets are used to draw the reader's attention to those portions of the expressions where the laws indicated on the right are applied. Notice that we simplify under the complementation bar before actually applying it, that is, by working *inside outward*. This procedure often avoids multiple applications of DeMorgan's laws.

This brings us to a final point which is noteworthy. Working within the homologic (single logic) system as we have been doing in this section, often requires us to repeat application of DeMorgan's laws as we have just demonstrated. However, recall that in Sect. 2.6 we developed from the conjugate gate forms of Fig. 2-28 the DeMorgan relations given by the Eqs. (2-5). These relations attest to the fact that in the MIXED LOGIC system the need to apply DeMorgan's laws is minimized if not eliminated. DeMorgan's laws mixed-are still important, certainly, but their application is implicit in the MIXED LOGIC symbology that is used. There is little or no need to manipulate Boolean expressions by using DeMorgan's laws in the MIXED LOGIC system and, as a result, the task of creating an error-free design is greatly simplified.

2.9.5 Summary of the Laws of Boolean Algebra

For the convenience of the reader, we present a summary of the laws of conventional Boolean algebra. Remember that the symbols X, Y, and Z may each represent a single variable or a complex multivariable function.

NOT Laws	AND Laws	OR Laws	
$\overline{0} = 1$	$X \cdot 0 = 0$	$X + 0 = X$	
$\overline{1} = 0$	$X \cdot 1 = X$	$X + 1 = 1$	(2-9)
$\overline{\overline{X}} = X$	$X \cdot X = X$	$X + X = X$	
	$X \cdot \overline{X} = 0$	$X + \overline{X} = 1$	

Associative laws

$$(X \cdot Y) \cdot Z = X \cdot (Y \cdot Z) = X \cdot Y \cdot Z \qquad \text{AND form}$$
$$(X + Y) + Z = X + (Y + Z) = X + Y + Z \qquad \text{OR form}$$

(2-10)

Commutative laws

$$X \cdot Y \cdot Z = X \cdot Z \cdot Y = Z \cdot X \cdot Y = \cdots \qquad \text{AND form}$$
$$X + Y + Z = X + Z + Y = Z + X + Y = \cdots \qquad \text{OR form} \tag{2-11}$$

Distributive and Absorptive Laws

Factoring law	$XY + XZ = X(Y + Z)$	(2-12)
Distributive law	$(X + Y)(X + Z) = X + YZ$	(2-13)
Absorptive laws	$X(\overline{X} + Y) = XY$	(2-14)
	$X + \overline{X}Y = X + Y$	(2-15)

DeMorgan's Laws

$$\overline{X \cdot Y} = \overline{X} + \overline{Y} \tag{2-16}$$
$$\overline{X + Y} = \overline{X} \cdot \overline{Y} \tag{2-17}$$

2.9.6 The XOR Laws

The laws of XOR algebra share many similarities with those of conventional Boolean algebra discussed in the previous sections and can be viewed as a natural extension of the conventional laws. In this section we will examine these lesser known XOR laws and provide some examples of their application.

Just as we deduced the AND and OR laws from their respective truth tables, we can also deduce the XOR and EQV laws from their respective truth tables. Shown in Fig. 2-50 are the positive logic truth tables for the XOR and EQV functions which follow directly from Figs. 2-37(b) and 2-40(b). Then, by letting Y take on values 0, 1, X, and $\overline{X}$, we deduce from Fig. 2-50 the following laws:

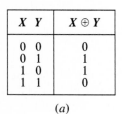

$X\ Y$	$X \oplus Y$
0 0	0
0 1	1
1 0	1
1 1	0

(a)

$X\ Y$	$X \odot Y$
0 0	1
0 1	0
1 0	0
1 1	1

(b)

Fig. 2-50. *Positive logic truth table for (a) the XOR function and (b) the EQV function.*

XOR Laws		EQV Laws	
$X \oplus 0 = X$		$X \odot 1 = X$	
$X \oplus 1 = \overline{X}$	Duals	$X \odot 0 = \overline{X}$	(2-21)
$X \oplus X = 0$	$\longleftrightarrow$	$X \odot X = 1$	
$X \oplus \overline{X} = 1$		$X \odot \overline{X} = 0$	

Here, the dual relationship is established by interchanging the 1's and 0's while simultaneously interchanging the logic operators $\oplus$ and $\odot$.

The associative and commutative laws for XOR and EQV follow from the conventional associative and commutative laws given in Sect. 2.9.3 by exchanging operator symbols: $\oplus$ for $(+)$ and $\odot$ for $(\cdot)$. The distributive and absorptive laws for the XOR and EQV operations also follow from the conventional forms by replacing the appropriate $(+)$ operator symbols with the $\oplus$ operator symbol, and by replacing the appropriate $(\cdot)$ symbols with the $\odot$ symbol, but not both in any given expression. These laws are stated as follows:

2 / Fundamentals of Digital Design

Associative Laws

$$(X \odot Y) \odot Z = X \odot (Y \odot Z) = X \odot Y \odot Z \qquad \text{EQV form} \qquad (2\text{-}22)$$

$$(X \oplus Y) \oplus Z = X \oplus (Y \oplus Z) = X \oplus Y \oplus Z \qquad \text{XOR form} \qquad (2\text{-}23)$$

Commutative Laws

$$X \odot Y \odot Z = X \odot Z \odot Y = Z \odot X \odot Y = \cdots \qquad \text{EQV form} \qquad (2\text{-}24)$$

$$X \oplus Y \oplus Z = X \oplus Z \oplus Y = Z \oplus X \oplus Y = \cdots \qquad \text{XOR form} \qquad (2\text{-}25)$$

Distributive and Absorptive Laws

Factoring law $\qquad (X \cdot Y) \oplus (X \cdot Z) = X \cdot (Y \oplus Z) \qquad (2\text{-}26)$

Distributive law $\qquad (X + Y) \odot (X + Z) = X + (Y \odot Z) \qquad (2\text{-}27)$

Absorptive laws $\qquad \begin{cases} X \cdot (\overline{X} \oplus Y) = X \cdot Y & (2\text{-}28) \\ X + (\overline{X} \odot Y) = X + Y & (2\text{-}29) \end{cases}$

DeMorgan's Laws

$$\overline{X \odot Y} = \overline{X} \oplus \overline{Y} = X \oplus Y \qquad\qquad (2\text{-}30)$$

$$\overline{X \oplus Y} = \overline{X} \odot \overline{Y} = X \odot Y \qquad\qquad (2\text{-}31)$$

Verification of the associative, commutative, and distributive laws is easily accomplished by using truth tables, as was done in Sect. 2.9.3. For example, the truth table in Fig. 2-51 verifies the distributive law of Eq. (2-27). Here, Eq. (2-8) together with the OR laws have been used to show the identity of the terms $(X + Y) \odot (X + Z)$ and $X + (Y \odot Z)$.

The absorptive laws are most easily verified by Boolean manipulation. Beginning with Eq. (2-28), we write

$$X \cdot [(\overline{X} \oplus Y)] = X \cdot (\overline{X}\overline{Y} + XY) \qquad \text{Eq. (2-6)}$$

$$= [X \cdot (\overline{X}\overline{Y} + XY)] \qquad \text{Factoring law and}$$
$$\qquad\qquad\qquad\qquad\qquad \text{AND and OR laws}$$
$$= XY$$

where the square brackets [ ] are used to draw the reader's attention to those portions where the laws or equations indicated on the right are applied. Similarly, Eq. (2-29) is verified by the following sequence of steps:

$X\ Y\ Z$	$X + Y$	$X + Z$	$Y \odot Z$	$(X + Y) \odot (X + Z)$	$X + (Y \odot Z)$
0 0 0	0	0	1	1	1
0 0 1	0	1	0	0	0
0 1 0	1	0	0	0	0
0 1 1	1	1	1	1	1
1 0 0	1	1	1	1	1
1 0 1	1	1	0	1	1
1 1 0	1	1	0	1	1
1 1 1	1	1	1	1	1

Fig. 2-51. *Verficiation of the distributive law, Eq. (2-27).*

$$X + [(\overline{X} \odot Y)] = X + (X\overline{Y}) + (\overline{X}Y) \qquad \text{Eq. (2-6)}$$
$$= [X + (X\overline{Y})] + \overline{X}Y \qquad \text{Factoring and OR laws}$$
$$= [X + (\overline{X}Y)] \qquad \text{Eq. (2-15)}$$
$$= X + Y$$

Notice that in the foregoing verifications use is tacitly made of the relations

$$\overline{X} \oplus Y = X \oplus \overline{Y} = \overline{X \oplus Y} = X \odot Y$$

and

$$\overline{X} \odot Y = X \odot \overline{Y} = \overline{X \odot Y} = X \oplus Y \qquad (2\text{-}32)$$

These relations are easily verified by replacing the variable (X or Y) by its complement ($\overline{X}$ or $\overline{Y}$) in the appropriate defining expression, Eq. (2-6) or Eq. (2-8), or by using DeMorgan's XOR laws, Eqs. (2-30) and (2-31), which are yet to be verified. Identities of the following type, which have been produced by repeated application of Eqs. (2-32), are presented for future reference purposes:

$$\begin{aligned}
F &= X \oplus Y \oplus Z & \overline{F} &= X \oplus \overline{Y} \oplus Z \\
&= X \odot Y \odot Z & &= X \odot \overline{Y} \odot Z \\
&= X \odot \overline{Y} \oplus Z & &= X \oplus Y \odot Z \\
&= \overline{X} \oplus Y \oplus \overline{Z} & &= X \odot Y \oplus Z \\
&\qquad \vdots & &\qquad \vdots
\end{aligned} \qquad (2\text{-}33)$$

Clearly, similar sets of equations having three or more XOR and/or EQV operators in each equation can be produced by repeated application of Eqs. (2-32).

The DeMorgan laws for the XOR and EQV functions tell us what we should already know—that the complement of one function yields the other, a fact that is evident from the truth tables given in Fig. 2-50. To gain further experience in the use of logic maps, let us use these maps to verify this fact for one of these laws. Shown in Fig. 2-52 is the logic map development of Eq. (2-31), where use is made of the defining relations given by Eqs. (2-6) and (2-8). Observe how vividly the functional descriptions of XOR and EQV are portrayed by the shaded logic maps.

A brief inspection of the XOR laws reveals that they are divided into sets of two, one the dual of the other, as was true for the conventional forms. Duality is established in the associative, commutative, and DeMorgan laws by interchanging the operators $\oplus$ and $\odot$. Duality is established in the distributive and absorptive laws by interchanging the logic operators $\oplus$ and $\odot$ simultaneously with an interchange of the operators ($+$) and ($\cdot$).

The parallelism between the XOR algebra and conventional Boolean algebra is made complete by recognizing the following two corollaries which follow directly from the definitions of the XOR and EQV operations:

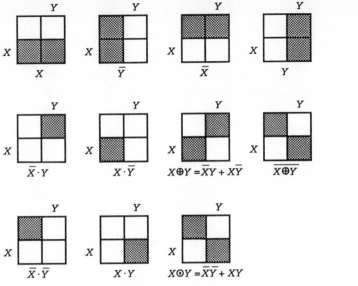

Fig. 2-52. *Logic map verification of Eq. (2-31)*

COROLLARY I

If two functions, α and β, never take the logic 1 value simultaneously, then

$$\alpha \cdot \beta = 0 \quad \text{and} \quad \alpha + \beta = \alpha \oplus \beta \qquad (2\text{-}34)$$

and the logic operators $(+)$ and $(\oplus)$ are interchangeable.

COROLLARY II

If two functions, α and β, never take the logic 0 value simultaneously, then

$$\alpha + \beta = 1 \quad \text{and} \quad \alpha \cdot \beta = \alpha \odot \beta \qquad (2\text{-}35)$$

and the logic operators $(\cdot)$ and $\odot$ are interchangeable.

Although the implications of the corollaries given by Eqs. (2-34) and (2-35) are quite extensive, their most obvious application is in operator exchange. Take, for example, the function $(AB + \overline{B}C)$. If $\alpha = AB$ and $\beta = \overline{B}C$, then by corollary I it follows that $(AB + \overline{B}C) = (AB) \oplus (\overline{B}C)$. Or consider the function $(A + \overline{B}) \cdot (B + D)$ as a second example. Now, if $\alpha = (A + \overline{B})$ and $\beta = (B + D)$, corollary II permits $(A + \overline{B}) \cdot (B + D) = (A + \overline{B}) \odot (B + D)$. Notice that the defining expressions for XOR and EQV, given by Eqs. (2-6) and (2-8), satisfy corollaries I and II, respectively.

This concludes our treatment of Boolean algebra. While not intended to be an exhaustive treatment, we believe it to be adequate for the needs of digital design.

2.10 WORKED EXERCISES

In the development of this chapter we have purposely omitted a number of subtle points and extensions of the coverage so as not to break up the continuity of the subject matter. It is the aim of this section to fill in some of this detail and, at the same time, provide the reader with a study guide for the chapter.

EXAMPLE 2-1 Presented in Fig. 2-53(a) is an NMOS INVERTER. Describe how it works and find the output voltage waveform for the set of random voltage input pulses given in Fig. 2-53(b). Finally, draw the input and output logic waveforms representing the voltage waveform of Fig. 2-53(b) by assuming the logic interpretations given in Fig. 2-6. Neglect propagation delay times.

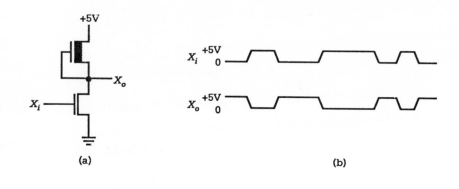

(a)   (b)

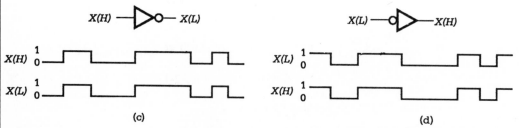

(c)   (d)

Fig. 2-53. (a) An NMOS INVERTER. (b) The input and output voltage waveforms for the INVERTER. (c) ACTIVE HIGH to ACTIVE LOW CONVERTER symbol and its logic waveforms corresponding to the physical waveforms in (b). (d) ACTIVE LOW to ACTIVE HIGH CONVERTER symbol and its logic waveforms corresponding to the physical waveforms in (b).

SOLUTION When X_i is at HV, the NMOS switch is ON, which causes X_o to be nearly shorted to ground (LV). Conversely, when X_i is at LV, the NMOS switch is OFF, causing +5 V (HV) to fall across X_o relative to ground. The output voltage waveform, resulting from the input waveform of Fig. 2-53(b), is also shown in Fig. 2-53(b). Notice that the input and output voltage waveforms are the inverse of one another, a fact that is *always* true for an INVERTER.

The two LOGIC LEVEL CONVERTER symbols of Figs. 2-53(c) and (d) are the only possible logic interpretations of an INVERTER. The logic waveforms, consistent with the waveforms of Fig. 2-53(b), are shown in Figs. 2-53(c) and (d). Observe that the input and output waveforms are identical, except that one represents positive logic while the other represents negative logic, the result of LOGIC LEVEL CONVERSION. The logic waveform for $\overline{X}(H)$ would be the inverse of that for $X(H)$.

2 / Fundamentals of Digital Design

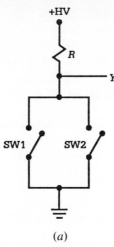

SW1	SW2	Y
OPEN	OPEN	HV
OPEN	CLOSED	LV
CLOSED	OPEN	LV
CLOSED	CLOSED	LV

(a)   (b)

Fig. 2-54. *(a) Switch analog circuit for a logic gate. (b) The physical truth table for a NOR gate.*

EXAMPLE 2-2   In Fig. 2-54(a) is given the electrical switch analog for a logic gate. Construct a physical truth table for this switch circuit, indicate what logic gate this circuit represents, and give two logic interpretations for it.

SOLUTION   Let the switches be designed as OPEN or CLOSED and the output as LV or HV. So, if one switch is CLOSED (or both are CLOSED), Y is nearly shorted to ground (LV), and only if both switches are OPEN will the output Y be at HV. The result of all possible switch combinations is the physical truth table shown in Fig. 2-54(b), which indicates that the circuit is the switch analog of a NOR gate. As indicated in Fig. 2-14, the NOR gate has both AND and OR logic interpretations depending on the ACTIVATION LEVELS chosen for the two switches.

EXAMPLE 2-3   The electrical analog of a logic gate is shown in Fig. 2-55(a). Explain how this circuit operates, construct its physical truth table, and indicate what logic functions it performs. To what logic family does this gate belong?

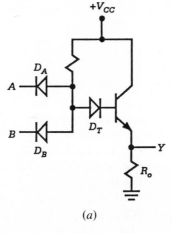

A	B	Y_0
LV	LV	LV
LV	HV	LV
HV	LV	LV
HV	HV	HV

(b)

Fig. 2-55. *(a) Electrical circuit analog for a logic gate. (b) Physical truth table for an AND gate.*

(a)

2.10 / Worked Exercises

75

SOLUTION   If either input, A or B (or both), is connected to ground (LV), the input diodes D_A and D_B are turned ON, the transistor diode D_T is turned OFF, and the BJT is in the cut-off mode electrically isolating the output resistor R_o from the supply voltage $+V_{CC}$. Under this condition the output Y must be at LV since there is no IR_o drop across R_o. Only when both inputs are at HV will the diode D_T be turned ON forcing the BJT into saturation electrically connecting $+V_{CC}$ to the output resistor R_o. When this happens the output is at HV since now there is a current I flowing through R_o which produces an IR_o drop across it to ground. The physical truth table that results from all possible LV and HV combinations of the two inputs is given in Fig. 2-55(b). This truth table is recognized to be that of an AND gate, which, according to Fig. 2-20, can be interpreted as either a logic AND or logic OR function, depending on the ACTIVATION LEVELs chosen. The circuit of Fig. 2-55(a) belongs to the diode-BJT logic (DTL) family.

EXAMPLE 2-4   The MOS circuit analog of a logic gate is provided in Fig. 2-56(a). Describe how it operates, construct its physical truth table, and indicate what logic functions it performs.

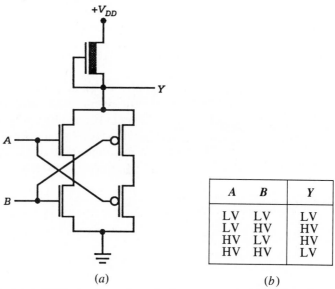

A	B	Y
LV	LV	LV
LV	HV	HV
HV	LV	HV
HV	HV	LV

(a)   *(b)*

Fig. 2-56. (a) A MOS circuit analog of a logic gate. (b) The physical truth table for an XOR gate.

SOLUTION   Since a HV input turns an NMOS transistor ON while turning a PMOS transistor OFF, and vice versa for LV, output Y in Fig. 2-56 will be virtually shorted to ground by one path or the other if the inputs are either both at HV or both at LV. Any other combination of input voltages "opens up" the path from output to ground causing $+V_{DD}$ (HV) to fall across the output Y relative to ground. The resulting physical truth table is given in Fig. 2-56(b) and is recognized as that of an XOR gate. Referring to Fig. 2-37, it is clear that this gate can perform the XOR or EQV operations depending on the ACTIVATION LEVEL assumed for the output.

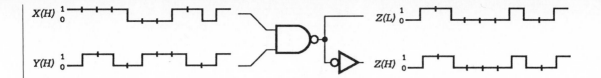

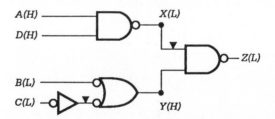

(a)

(b)

Fig. 2-57. (a) *The NAND gate symbol with rectangular logic waveform inputs.* (b) *Outputs* Z(L) *and* Z(H) *corresponding to the inputs given in* (a).

EXAMPLE 2-5

In Fig. 2-57(a) is shown the logic symbol for a NAND gate with ACTIVE HIGH logic waveform inputs. Determine the output waveform for $Z(L)$. Also, determine the waveform for $Z(H)$ taken from the *INVERTER symbol.* Neglect propagation delay times.

SOLUTION

The $Z(L)$ waveform is found by ANDing the waveforms for $X(H)$ and $Y(H)$, that is, $Z(L) = X \cdot Y(L)$, as given in Fig. 2-57(b). Also shown in Fig. 2-57(b) is the $Z(H)$ waveform. Note that while $Z(L)$ and $Z(H)$ are identical logic waveforms they represent inverse voltage waveforms.

EXAMPLE 2-6

Read the NAND/INV logic circuit of Fig. 2-58 in MIXED LOGIC notation and find $X(L)$, $Y(H)$, and $Z(L)$.

Fig. 2-58. *NAND/INV logic circuit for Example 2-6.*

SOLUTION

Reading a logic circuit begins with the input and ends with the output. Proceeding in this manner for the circuit of Fig. 2-58 gives us the following results:

$$X(L) = (A \cdot D)(L)$$

$$Y(H) = (B + \overline{C})(H)$$

$$Z(L) = [(\overline{A \cdot D}) \cdot (B + \overline{C})](L)$$

Notice that the output expression $Z(L)$ is written exactly as required by the AND and OR operations, taking into account the two logic incompatibilities indicated by the flags. Unless there is reason to do otherwise, terms such as $\overline{A \cdot D}$ are best left in their original form without applying DeMorgan's laws. Remember that the reason for reading a circuit in MIXED LOGIC notation is to avoid the unnecessary application of DeMorgan's laws.

EXAMPLE 2-7

The logic circuit shown in Fig. 2-59 is a redundant circuit, meaning that it uses more gates than are required to represent the function $F(H)$. Read this

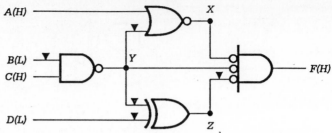

Fig. 2-59. *A redundant logic circuit for Example 2-7.*

circuit in MIXED LOGIC notation at points X, Y, and Z, and then give the output $F(H)$ in its simplest form, all without applying DeMorgan's laws.

SOLUTION   Reading the circuit in Fig. 2-59 from input to output, we obtain

$$X(L) = (A + \overline{\overline{BC}})(L)$$
$$Y(L) = (\overline{BC})(L)$$
$$Z(H) = [(\overline{\overline{BC}}) \oplus \overline{D}](H)$$
$$F(H) = (XY\overline{Z})(H)$$
$$= (A\overline{B}CD)(H)$$

Here, we recognize by inspection that

$$XY = AY$$

and that

$$AY(\overline{Y \oplus D}) = AY(Y \odot D)$$
$$= AY(\overline{Y}\overline{D} + YD)$$
$$= AYD$$

where $Y = \overline{B}C$. Obviously, a four-input AND gate and an INVERTER would suffice for the function $F(H)$.

EXAMPLE 2-8   The logic circuit in Fig. 2-60(a) is given in the positive logic system and represents a well-known logic function. Redraft this circuit in MIXED LOGIC notation and identify its logic function: use the appropriate conjugate gate forms (see Fig. 2-28) where needed to issue $F(H)$ and eliminate *all* operations which require the application of DeMorgan's laws. Remember that the integrity of each gate must be preserved in the redraft so as not to alter its logic content.

SOLUTION   In redrafting a circuit, as in any logic design, it is necessary to begin at the output and work back to the input (top-down). Replacing the output NAND gate and the OR gate with their conjugate gate forms yields the results presented in Fig. 2-60(b), where the function $F(H)$ is easily recognized as the EQV function $F(H) = (\overline{A}\overline{B} + AB)(H) = A \odot B(H)$ according to Eq. (2-8) in Sect. 2.8. Notice that the positive logic circuit would be read as $F =$

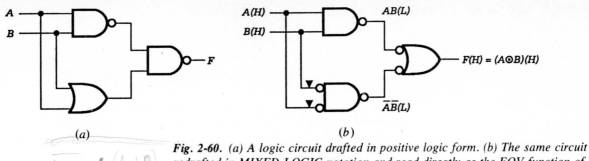

(a)

(b)

Fig. 2-60. *(a) A logic circuit drafted in positive logic form. (b) The same circuit redrafted in MIXED LOGIC notation and read directly as the EQV function of inputs ACTIVE HIGH A and B.*

$\overline{AB} \cdot (A + B)$, which gives the desired results but only after two applications of DeMorgan's laws.

EXAMPLE 2-9

Unused inputs must not be left dangling. Instead, they must be tied to other inputs, or be connected to HV or LV depending on the logic operation involved. The four gates in Fig. 2-61 have specific input conditions which demonstrate this. Write the output expression for each gate symbol; then name the gate and indicate what logic function it performs. Remember that $1(H)$ is interpreted as HV and $0(H)$ as LV.

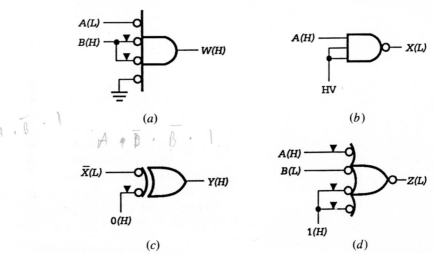

(a)

(b)

(c)

(d)

Fig. 2-61. *Four gates with specific input conditions demonstrating how unused inputs can be connected.*

SOLUTION

a. $W(H) = (A\overline{B})(H)$ — A four-input NOR gate performing the AND operation of two variables.

b. $X(L) = A(L)$ — A three-input NAND gate performing the $A(H)$ to $A(L)$ LOGIC CONVERSION operation.

c. $Y(H) = \overline{X}(H)$ — An XOR gate performing the $\overline{X}(L)$ to $\overline{X}(H)$ LOGIC CONVERSION operation.

d. $Z(L) = (\overline{A} + B)(L)$ — A four-input AND gate performing the OR operation of two variables.

EXAMPLE 2-10

By using logic maps, verify the expression

$$\overline{X} + XY = \overline{X} + Y$$

which is seen to be a form of the absorptive law given by Eq. (2-28).

SOLUTION

Verification of this form of the absorptive law begins with the logic maps for the individual variables and continues as shown in Fig. 2-62.

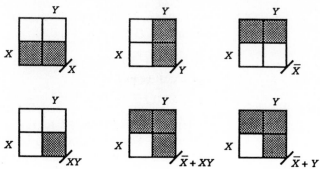

Fig. 2-62. *Logic map verification of the expression* $\overline{X} + XY = \overline{X} + Y$ *given in Example 2-10.*

EXAMPLE 2-11

By using AND/OR logic implement the expression

$$F(L) = [A\overline{D} + (\overline{\overline{B} + C})](L)$$

exactly as written with the fewest number of gates if the inputs *A*, *C*, and *D* arrive from positive logic sources and if input *B* arrives from a negative logic source.

SOLUTION

Implementation of $[A\overline{D} + (\overline{\overline{B} + C})](L)$ must begin at the output and end at the inputs, a top-down procedure which should always be followed when implementing a function. Then, by using the appropriate conjugate gate forms from Fig. 2-28, there results the AND/OR logic diagram of Fig. 2-63. Notice that it was necessary to use an INVERTER on the *A(H)* input line to remove the logic incompatibility.

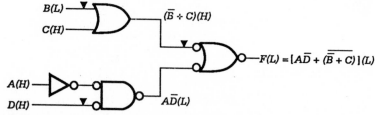

Fig. 2-63. *The AND/OR implementation of the expression given in Example 2-11.*

EXAMPLE 2-12

Use NOR/XOR logic to implement the function

$$Z(H) = \{[\overline{X} \odot (\overline{A} + Y)] \cdot \overline{B}\}(H)$$

exactly as written with the fewest number of gates assuming that inputs A and B arrive ACTIVE LOW and inputs X and Y are ACTIVE HIGH.

SOLUTION By following a top-down procedure and by inserting the appropriate conjugate gate forms, we obtain the logic circuit shown in Fig. 2-64. Here, use is made of the ACTIVE LOW input XOR symbol so as to produce the term $\overline{X} \oplus (\overline{A} + Y)$. Removal of the ACTIVE LOW indicator bubbles would yield the $X \oplus (\overline{\overline{A}} + Y)$, which is logically identical to $\overline{X} \oplus (\overline{A} + Y)$ by Eqs. (2-32) in Sect. 2.9.6.

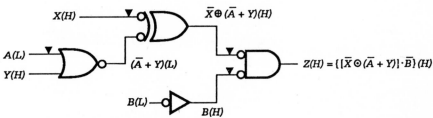

Fig. 2-64. *NOR/XOR implementation of the expression given in Example 2-12.*

EXAMPLE 2-13 Reduce the following expressions to their simplest form and name the law(s) to be used for each step. Application of the associative and commutative laws need not be indicated. Refer to Sect. 2.9.5 for a review of the laws of Boolean algebra.

a. $Y = \overline{A} + A \cdot B \cdot \overline{C} + \overline{A \cdot \overline{C}}$

b. $Z = \overline{(b \oplus \overline{c}) + (a \cdot b)(\overline{\overline{a} + c})}$

c. $F = (X + Y)(X + \overline{Z})[Y \cdot (X + \overline{Z}) + \overline{Y}]$

SOLUTION **a.** $Y = \overline{A} + (A \cdot B \cdot \overline{C}) + \overline{A} \cdot \overline{\overline{C}}$ DeMorgan's law [Eq. (2-16)]

$\quad\quad = \overline{A} + (A \cdot B \cdot \overline{C}) + \overline{A} + C$ OR laws and Absorptive law [Eq. (2-15)]

$\quad\quad = \overline{A} + (B \cdot \overline{C}) + C$ Absorptive law [Eq. (2-15)]

$\quad\quad = \overline{A} + B + C$

b. Reduction of this expression is best accomplished by an inside-out procedure to avoid multiple applications of DeMorgan's laws:

$Z = \overline{(b \oplus \overline{c}) + (\overline{a \cdot b})(\overline{\overline{a} + c})}$ DeMorgan's laws and Eqs. (2-32)

$\quad = \overline{(b \odot c) + (\overline{a} + \overline{b})(a\overline{c})}$ Factoring law; AND and OR laws Eqs. (2-8)

$\quad = \overline{(b \odot c) + \overline{a} \, b \, \overline{c}}$ Factoring and OR laws

$\quad = \overline{bc + \overline{b}\,\overline{c}}$ Eqs. (2-8) and (2-32)

$\quad = b \oplus c$

c. This expression is most easily reduced to minimum form by first applying the distributive and absorptive laws:

$$F = (X + Y)(X + \overline{Z})\,[Y \cdot (X + \overline{Z}) + \overline{Y}] \quad \text{Distributive, absorptive law [Eq. (2-15)]}$$
$$= (X + Y\overline{Z})\,[X + \overline{Z} + \overline{Y}] \quad\quad\quad \text{Distributive law}$$
$$= X + Y\overline{Z}(\overline{Z} + \overline{Y}) \quad\quad\quad\quad\quad \text{Factoring; AND and OR laws}$$
$$= X + Y\overline{Z}$$

EXAMPLE 2-14

Evaluate the following MIXED LOGIC expressions in minimum form and name the Boolean laws or equations to be used for each step.

a. $[X + \overline{X + Y} + Z](H) = (?)(L)$
b. $[1 \cdot (\overline{\overline{0} + 1 \cdot \overline{1}}) \cdot \overline{0}](L = (?)(H)$
c. $[\overline{A}(A \oplus \overline{B})](L) = (?)(H)$

SOLUTION **a.** $(?)(L) = (X + \overline{X + Y} + Z)(H) \quad\quad$ DeMorgan's law
$\quad\quad\quad\quad\quad = (X + \overline{X}\overline{Y} + Z)(H) \quad\quad\quad$ Absorptive law [Eq. (2-15)]
$\quad\quad\quad\quad\quad = (X + \overline{Y} + Z)(H) \quad\quad\quad\quad$ Eqs. (2-3) and DeMorgan's law
$\quad\quad\quad\quad\quad = (\overline{X}Y\overline{Z})(L)$

b. $(?)(H) = [1 \cdot (\overline{\overline{0} + 1 \cdot \overline{1}}) \cdot \overline{0}](L) \quad\quad$ NOT and AND laws
$\quad\quad\quad\quad = (\overline{1 + 0})(L) \quad\quad\quad\quad\quad\quad\quad$ OR laws
$\quad\quad\quad\quad = \overline{1}(L) \quad\quad\quad\quad\quad\quad\quad\quad\quad\quad$ Eqs. (2-3)
$\quad\quad\quad\quad = 1(H)$

c. $(?)(H) = [\overline{A}(A \oplus \overline{B})](L) \quad\quad$ XOR Absorptive law [Eq. (2-28)]
$\quad\quad\quad\quad = (\overline{A}\,\overline{B})(L) \quad\quad\quad\quad\quad$ Eqs. (2-3) and DeMorgan's law
$\quad\quad\quad\quad = (A + B)(H)$

EXAMPLE 2-15

Prove each of the following expressions in two different ways and indicate the Boolean laws or equations to be used for each step.

a. $(\overline{X} \cdot Z) \oplus (X + Z) = X$
b. $[\overline{a}c + (\overline{a} + b) \odot (\overline{a} + bc)](H) = (ab\overline{c})(L)$

SOLUTION **a.** Solution 1

$$X = (\overline{X} \cdot Z) \oplus (X + Z) \quad\quad \text{Eqs. (2-32) and DeMorgan's law}$$
$$= (X + \overline{Z}) \odot (X + Z) \quad\quad \text{Distributive law [Eq. (2-27)]}$$
$$= X + (\overline{Z} \odot Z) \quad\quad\quad\quad \text{EQV and OR laws}$$
$$= X$$

Solution 2

$$X = (\overline{\overline{X} \cdot Z}) \oplus (X + Z) \qquad \text{Eqs. (2-32) and DeMorgan's law}$$
$$= (X + \overline{Z}) \odot (X + Z) \qquad \text{Corollary II [Eq. (2-35)]}$$
$$= (X + \overline{Z}) \cdot (X + Z) \qquad \text{Distributive law [Eq. (2-13)]}$$
$$= X + \overline{Z} \cdot Z \qquad \text{AND and OR laws}$$
$$= X$$

b. Solution 1

$$(a \cdot b \cdot \overline{c})(L) = [\overline{a}c + (\overline{a} + b) \odot (\overline{a} + bc)](H) \qquad \text{Eqs. (2-32), DeMorgan's law}$$
$$= [\overline{a}c + (a \cdot \overline{b}) \oplus (\overline{a} + bc)](H) \qquad \text{Corollary I [Eq. (2-34)]}$$
$$= [\overline{a} \cdot c + (a\overline{b}) + (\overline{a} + bc)](H) \qquad \text{Factoring and OR laws}$$
$$= [\overline{a} + a\overline{b} + bc](H) \qquad \text{Absorptive law [Eq. (2-15)]}$$
$$= (\overline{a} + \overline{b} + c)(H) \qquad \text{Eqs. (2-3), DeMorgan's law}$$
$$= (ab\overline{c})(L)$$

Solution 2

$$(ab\overline{c})(L) = [\overline{a}c + (\overline{a} + b) \odot (\overline{a} + bc)](H) \qquad \text{Distributive law [Eq. (2-27)] and AND laws}$$
$$= [\overline{a}c + \overline{a} + b \odot (bc)](H) \qquad \text{Factoring law, OR laws, and Eqs. (2-32)}$$
$$= [\overline{a} + \overline{b} \oplus (bc)](H) \qquad \text{Corollary I and absorptive law [Eq. (2-15)]}$$
$$= (\overline{a} + \overline{b} + c)(H) \qquad \text{Eqs. (2-3), DeMorgan's law}$$
$$= (ab\overline{c})(L)$$

EXAMPLE 2-16

A digital device has three inputs A, B, and C and one output Y. The input signals are synchronous rectangular waveforms.

a. Construct a truth table indicating that the output is ACTIVE only if input A is ACTIVE or input B is ACTIVE but not both ACTIVE, and if input C is ACTIVE.

b. From the truth table of (a) write the Boolean expression representing the output Y and implement it by using the fewest number of gates possible. Assume that the inputs arrive from positive logic sources and that the output is ACTIVE HIGH.

SOLUTION

a. By taking logic 1 as the ACTIVE state there results the truth table shown in Fig. 2-65(a), where the Boolean interpretation is given to the right of the table.

b. The truth table is read as

$$Y = \overline{A}BC + A\overline{B}C$$
$$= (\overline{A}B + A\overline{B})C$$
$$= (A \oplus B)C$$

A	B	C	Y	
0	0	0	0	
0	0	1	0	
0	1	0	0	
0	1	1	1	$\overline{A} \cdot B \cdot C$
1	0	0	0	
1	0	1	1	$A \cdot \overline{B} \cdot C$
1	1	0	0	
1	1	1	0	

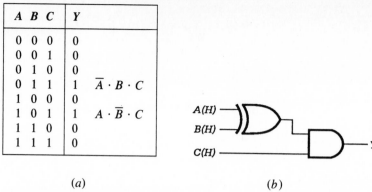

(a)   (b)

Fig. 2-65. *(a) Truth table and (b) gate-minimum circuit for Example 2-16.*

for which the factoring law [Eq. (2-12)] and Eq. (2-6) have been applied. Notice that the expression $(A \oplus B)C$ states: signal A or signal B but not both and input C in either case. The gate-minimum circuit for this function is given in Fig. 2-65(b).

EXAMPLE 2-17

Shown in Fig. 2-66(a) is a transistor circuit in which a TTL portion is used to drive a CMOS portion—TTL driving CMOS.

a. Construct a physical truth table for points X, Y, and Z, taking into account all possible combinations of LV and HV inputs.
b. If the inputs and output Z are all assumed to be ACTIVE HIGH, determine the logic functions at points X, Y, and Z and construct the logic diagram for this circuit, all in MIXED LOGIC notation.
c. Comment on the reliability of this circuit as a logic device if the CMOS noise margins are both 1.5 V and if V_{OL} and V_{OH} for the TTL output are 0.2 V and 3.5 V, respectively.

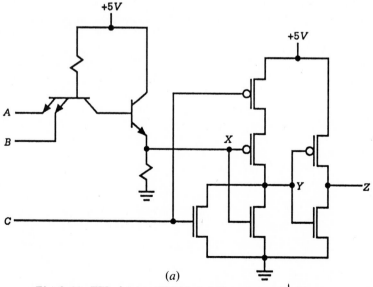

(a)

Fig. 2-66. *TTL driving CMOS. (a) Transistor circuit.*

2 / Fundamentals of Digital Design

A	B	C	X	Y	Z
LV	LV	LV	LV	HV	LV
LV	LV	HV	LV	LV	HV
LV	HV	LV	LV	HV	LV
LV	HV	HV	LV	LV	HV
HV	LV	LV	LV	HV	LV
HV	LV	HV	LV	LV	HV
HV	HV	LV	HV	LV	HV
HV	HV	HV	HV	LV	HV

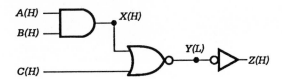

(b) (c)

Fig. 2-66 (contd.). *(b) Physical truth table for points* X, Y *and* Z *in Figure 2-66(a). (c) Logic diagram for the circuit in (a).*

SOLUTION

a. First, we recognize that point X is the output of an AND gate (see Sect. 2.4.5) in variables A and B, that point Y is the output of a NOR gate in variables X and C, and that point Z is the output of an INVERTER in variable Y. With this in mind, we construct the truth table shown in Fig. 2-66(b). Thus, by recognizing the gate configurations in such a circuit, the tedium of a voltage-by-voltage construction of the truth table can be avoided. We simply fill in the truth table as a composite of the physical truth tables for the AND and NOR gates and the INVERTER.

b. Again noting the gate composition of the circuit of Fig. 2-66(a), we write the following MIXED LOGIC expressions for points X, Y, and Z:

$$X(H) = (AB)(H)$$

$$Y(L) = (AB + C)(L)$$

$$Z(H) = (AB + C)(H)$$

The resulting logic circuit is given in Fig. 2-66(c), where we show the location of the three points (X, Y, and Z) in agreement with the circuit of Fig. 2-66(a).

c. Introducing the given data into the defining expressions for the noise margins [Eqs. (1-12)], we find that

$$NML = 1.5 \text{ V} = V_{IL(\text{max})} - 0 \quad \text{or} \quad V_{IL(\text{max})} = 1.5 \text{ V}$$

and

$$NMH = 1.5 \text{ V} = 5 - V_{IH(\text{min})} \quad \text{or} \quad V_{IH(\text{min})} = 3.5 \text{ V}$$

From these results we conclude that the LOW TTL output $V_{OL} = 0.2$ V is compatible with the input requirement $V_{IL(\text{max})} = 1.5$ V. However, the HIGH TTL output $V_{OH} = 3.5$ V is marginal with respect to the CMOS input requirement of $V_{IH(\text{min})} = 3.5$ V. This means that small negative fluctuations in the HV TTL output could result in failure of the CMOS to recognize the HV state. To compensate for this potential problem, a pull-up resistor is usually placed between point X and the supply connection. A resistor of a few kilo-ohms will usually pull V_{OH} up to approximately 5 V.

EXAMPLE 2-18

a. Construct the AND/NOR logic circuit for the function

$$F(L) = (\overline{A}\,\overline{B} + (AB)(L)$$
$$= (A \odot B)(L)$$

by omitting INVERTERs.

b. Construct the CMOS version for the logic circuit of (a) remembering that NMOS passes 0 (LV) well but not 1 (HV) and vice versa for PMOS (see Sect. 1.2.5). Also, construct the physical and logic truth tables for this circuit.

c. Compare the results of (b) with that of Fig. 2-39.

SOLUTION

a. The logic circuit for the function $F(L) = (\overline{A}\overline{B} + AB)(L)$, excluding INVERTERs, is shown in Fig. 2-67(a).

b. Because NMOS passes LV well while PMOS passes HV well, the CMOS combinational logic should be organized as shown in Fig. 2-67(b), where the functions F and $\overline{F}$ are defined by the shaded logic maps in Fig. 2-67(c). Notice that the voltage drop across the n section is made relative to ground potential as are all the CMOS circuits presented in this chapter.

The CMOS circuit for the EQV function is constructed in Fig. 2-67(d). MIXED LOGIC notation is used with the electrical circuit to help the reader identify its logic portions. The physical and logic truth tables for this circuit are given in Fig. 2-67(e).

c. The circuit of Fig. 2-39(a) is a simple MOS version of the EQV function with ACTIVE HIGH output. It requires that the PMOS gate pass LV, which it does not do well. So while this circuit is simpler than that of Fig. 2-67(d), it will distort the output signal somewhat. Note that the

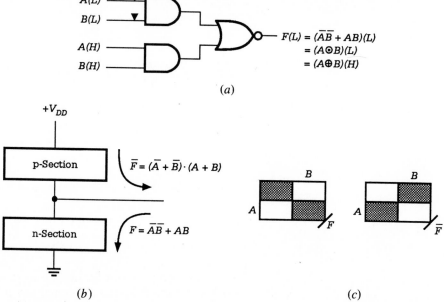

(a)

(b) (c)

Fig. 2-67. *CMOS design of the XOR function. (a) Logic circuit excluding INVERTERs. (b) CMOS organization. (c) K-maps for complementary function F and F̄.*

2 / Fundamentals of Digital Design

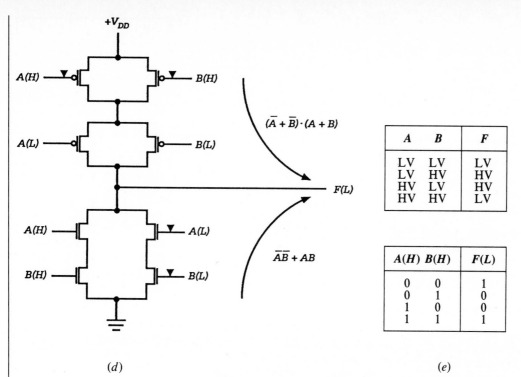

$$(\overline{A} + \overline{B}) \cdot (A + B)$$

$$\overline{A}\overline{B} + AB$$

A	B	F
LV	LV	LV
LV	HV	HV
HV	LV	HV
HV	HV	LV

A(H)	B(H)	F(L)
0	0	1
0	1	0
1	0	0
1	1	1

(d)

(e)

Fig. 2-67 (contd.). (d) CMOS circuit. (e) Physical and logic truth tables.

circuit of Fig. 2-67(d) requires that two CMOS INVERTERs be added if signals A and B arrive from ACTIVE HIGH sources. This is accomplished by connecting the output of the A INVERTER to both flagged A inputs and by connecting the output of the B INVERTER to both flagged B inputs.

EXAMPLE 2-19

This last example demonstrates the use of INVERTERs in CMOS implementation of logic function.

a. Construct the AND/NOR/INV logic for the function

$$F(H) = (\overline{A}B + C)(H)$$

Assume that inputs A, B, and C arrive from positive logic sources.
b. Construct the CMOS version of the logic circuit in (a).
c. Obtain the physical and positive logic truth tables for the circuit of part (b).

SOLUTION

a. The logic circuit for the function $F(H) = (\overline{A}B + C)(H)$ is given in Fig. 2-68(a).
b. Because NMOS passes LV well but not HV and vice versa for PMOS, the CMOS combinational logic should be organized as in Fig. 2-67(b). This is done, and the result is shown in Fig. 2-68(b). Note that two IN-VERTERs are required: one on the A input since A is complemented in the output expression, and the other on the output to convert F(L) to

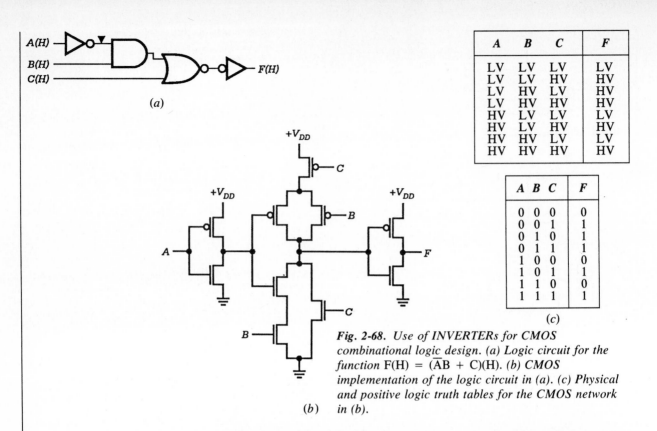

A	B	C	F
LV	LV	LV	LV
LV	LV	HV	HV
LV	HV	LV	HV
LV	HV	HV	HV
HV	LV	LV	LV
HV	LV	HV	HV
HV	HV	LV	LV
HV	HV	HV	HV

A	B	C	F
0	0	0	0
0	0	1	1
0	1	0	1
0	1	1	1
1	0	0	0
1	0	1	1
1	1	0	0
1	1	1	1

(c)

Fig. 2-68. *Use of INVERTERs for CMOS combinational logic design. (a) Logic circuit for the function F(H) = ($\overline{AB}$ + C)(H). (b) CMOS implementation of the logic circuit in (a). (c) Physical and positive logic truth tables for the CMOS network in (b).*

F(H). Notice also that the output of the *A* INVERTER is connected to both the PMOS and NMOS *A* inputs.

c. The physical and positive logic truth tables, which are derived from the CMOS circuit in part (b), are shown in Fig. 2-68(c).

ANNOTATED REFERENCES

Numerous texts offer an adequate coverage of the subject matter contained in this chapter. However, only the text by Fletcher appears to make extensive use of the MIXED LOGIC symbology. In fact, Fletcher's text may have been responsible for popularizing the MIXED LOGIC notation. Appendix C.2 of the book by Langdon reviews the use of the MIXED LOGIC notation and discusses two variations of it that are used by IBM and Digital Equipment Corporation (DEC).

Fletcher, W. I., *An Engineering Approach to Digital Design*, Prentice Hall, Englewood Cliffs, N.J., 1980.

Langdon, G. G., Jr., *Computer Design*, Computeach Press, San Jose, Calif., 1982.

Virtually every text on digital or logic design provides a suitable treatment of Boolean algebra. Most of these present the laws of Boolean algebra in theorem form with examples. Well-known examples are the texts of Nagle et al., Dietmeyer, Mano, and Kohavi. The book of Roth offers a somewhat different approach by using switching circuits to illustrate the laws. Texts by Roth, Dietmeyer, and McCluskey provide limited coverage of the algebra of XOR functions.

Nagle, H. T., Jr., B. D. Carroll, and J. D. Irwin, *An Introduction to Computer Logic*, Prentice Hall, Englewood Cliffs, N.J., 1975.

Dietmeyer, D. L., *Logic Design of Digital Systems*, 2nd ed., Allyn & Bacon, Boston, 1978.

Mano, M. M., *Digital Logic and Computer Design*, Prentice Hall, Englewood Cliffs, N.J., 1979.

Kohavi, Z., *Switching and Finite Automata Theory*, McGraw-Hill, New York, 1978.

Roth, C. H., *Fundamentals of Logic Design*, 2nd ed., West Publishing, St. Paul, Minn., 1979.

A refreshingly brief history of Boolean algebra is provided in Chapter 2 of Hill and Peterson, well-known authors in the field of digital design.

Hill, F. J., and G. R. Peterson, *Digital Logic and Microprocessors*, John Wiley, New York, 1984.

For CMOS logic, which is emphasized in this text, we recommend the book by Weste and Eshraghian. Chapter 1 and portions of Chapter 5 are pertinent to the needs of the present text.

Weste, N. H. E., and K. Eshraghian, *Principles of CMOS VLSI Design*, Addison-Wesley, Reading, Mass., 1985.

PROBLEMS

2.1 Define the following:

(a) MIXED LOGIC

(b) POLARIZED MNEMONIC

(c) LOGIC LEVEL CONVERSION

(d) ACTIVE and INACTIVE states

(e) Gate

2.2 Identify the gate appropriate to each of the following physical truth tables.

2.3 (a) Rearrange the position of the resistor R_C and the output V_o in Fig. 1-7(a) so as to produce a noninverting switch. Explain how this device operates.

(b) Suggest a way that the CMOS INVERTER in Fig. 2-5(a) could be converted to a noninverting switch. What would be the disadvantage (if any) of such a device?

(c) Convert the TTL INVERTER in Fig. 2-7(a) to a noninverting switch.

A B	Y	A B	Y	A B	Y	A B	Y
LV LV	HV	LV LV	LV	LV LV	LV	LV LV	HV
LV HV	HV	LV HV	HV	LV HV	LV	LV HV	LV
HV LV	HV	HV LV	HV	HV LV	LV	HV LV	LV
HV HV	LV	HV HV	HV	HV HV	HV	HV HV	LV
(a)		(b)		(c)		(d)	

2.4 Explain the differences that exist between the logic levels for the LOAD(H) and LOAD(L) waveforms in Fig. 2-2(b) and those for the $Z(H)$ and $Z(L)$ waveforms in Fig. 2-57(b). Do these differences represent a discrepancy in the definition of positive and negative logic?

2.5 A logic waveform $X_o(L)$ is derived from a logic device and is given here. Determine the logic waveforms for $X_o(H)$ and $\overline{X}_o(H)$ and the voltage waveform X for this logic device.

$$X_o(L) \quad {}^1_0 \underline{}\sqcap\underline{}\sqcap\sqcap\underline{}\sqcap\underline{}$$

2.6 A square wave of 1-kHZ frequency, ±5 V peak to peak, and 50% duty cycle (ACTIVE and INACTIVE portions are equal) is applied to the input X_i of the switch in Fig. 2-7(a). With reference to the input waveform, describe the output waveform X_o that will be measured.

2.7 Explain in what way the PMOS indicator bubbles in Fig. 2-11(a) are related to the ACTIVE LOW indicator bubbles on the circuit symbol of Fig. 2-11(b).

2.8 Construct the physical truth table for the RTL (resistor-transistor logic) gate given in Fig. P2-1. Indicate the name of the gate and the logic functions it performs, and give its two conjugate logic symbols.

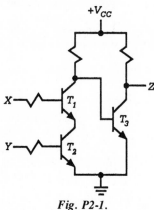

Fig. P2-1.

2.9 Construct the physical truth table for the gate shown in Fig. P2-2. Indicate the name of the gate and the logic functions it performs, and give its two conjugate logic symbols. To what logic family (RTL, TTL, PMOS, etc.) does this gate belong?

2.10 Obtain the physical and positive logic truth tables for the gate shown in Fig. P2-3. Indicate the name of the gate and the logic functions it performs, and give its two conjugate logic symbols. To what logic family does this gate belong?

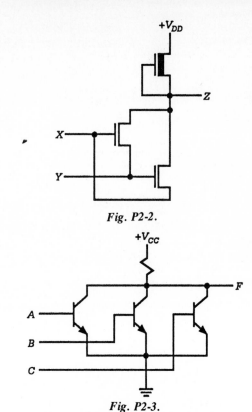

Fig. P2-2.

Fig. P2-3.

2.11 Determine the physical and positive logic truth tables for the gate shown in Fig. P2-4. Indicate the name of the gate and the logic functions it performs, and give its two conjugate logic symbols. To what logic family does this gate belong?

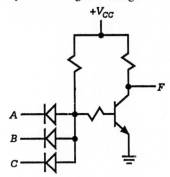

Fig. P2-4.

2.12 Construct the physical truth table for the gate shown in Fig. P2-5. Indicate the name of the gate and the logic functions it performs, and give its two conjugate logic symbols. To what logic family does this gate belong?

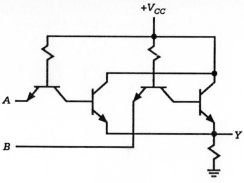

+V_CC

A

B

Y

Fig. P2-5.

2.13 Figure 2-54(a) features the switch analog circuit for a two-input NOR gate. Obtain the switch analog circuits for the following:

(a) A two-input OR gate

(b) A three-input NAND gate

(c) A three-input AND gate

(d) The function $F = \overline{AB + C}$

(e) The function $F = (A + B)(C + D)$

2.14 A room has two doors and a light that is controlled by three switches, A, B, and C. There is a switch beside each door and a third switch in another room. The light is turned on (LTON) any time an odd number of switches are closed (ACTIVE). Construct the truth table that indicates the light is on (LTON is ACTIVE). (Use Appendix 2.1 if necessary.)

2.15 A two-seat sports car has a seat-belt system that is designed so that the car cannot be started until all occupants have buckled their seat belts. Construct a truth table indicating the requirements for starting the car (START). Note that the car may be started without the passenger (PAS) present but not without a driver (DRV). Let DRVBKL and PASBKL represent the driver buckled and passenger buckled, respectively. (Use Appendix 2.1 if necessary.)

2.16 Archie (A), Betty (B), Cathy (C), and David (D) may attend a school dance. Archie likes to dance and will dance with either Betty or Cathy. However, Cathy will dance with Archie only if Betty is not present at the dance. David will dance only with Betty. Construct the truth table which defines the state of dancing for a couple. Let the ACTIVE state for A, B, C, and D represent the presence of Archie, Betty, Cathy, and David at the dance. (Use Appendix 2.1 if necessary.)

2.17 Obtain the logic expressions for the action indicated in each of the following:

(a) Bob will go fishing if the boat does not leak, if

it is not windy, and if the fish are biting, but may fish from the bank if the boat leaks.

(b) Sean will ski only if there is a snow base of at least 2 feet and if the temperature is not below 0°F, but will ski with that snow base in any temperature if Rusty accompanies him.

(c) Lissa has a learner's permit but can drive a car only if Sophia is not present and if Richard, Gloria, or Scott is present in the passenger seat.

(d) A robot is activated only if a majority of its three switches (X, Y, and Z) are turned on and is deactivated only if a majority of its three switches are turned off.

(e) A laboratory class consists of five students (A, B, C, D, and E) each from a different discipline. An experiment has been assigned that must be carried out with any one of the following combinations of students:

A and C but not with D.

A or B but not both.

D but only if E is present.

2.18 Use minimum AND/OR/INV logic gates to implement the following functions:

(a) $W(H) = (a \cdot \overline{b} + cd)(H)$

All inputs are ACTIVE HIGH.

(b) $X(L) = [(REQA \cdot REQB) + (\overline{GRNTC \cdot GRNTD})](L)$

All inputs arrive from negative logic sources.

(c) $Y(H) = (\overline{(AD)} + \overline{\overline{BC}})(H)$

Inputs A and C are ACTIVE HIGH.

Inputs B and D are ACTIVE LOW.

(d) $Z(L) = (\overline{Q_A} \overline{Q_C} + Q_B + Q_D)(L)$

Inputs are $Q_A(L)$, $Q_B(H)$, $Q_C(H)$, and $Q_D(L)$.

(e) Repeat parts (a), (b), (c), and (d) by using the ANSI/IEEE Standard for gate symbols given in Appendix 2.2.

2.19 Use minimum AND/OR/INV logic gates to implement the functions that follow exactly as presented. The inputs are A, B, and $\overline{C}$ from positive logic sources and D and E from negative logic sources.

(a) $F(H) = [(\overline{\overline{AD}}) \cdot (B + \overline{E})](H)$

(b) $G(L) = [(A\overline{B} + C)](\overline{D + E})(L)$

(c) $H(H) = [(B + \overline{D})(\overline{AC})(\overline{B} + E)](H)$

(d) Repeat parts (a), (b), and (c) by using the ANSI/IEEE Standard gate symbols in Appendix 2.2.

2.20 Use NAND/NOR logic as needed for a minimum implementation of the functions in parts (a), (b), (c), (d), and (e) in Problem 2.18. Note that INVERTERs must be implemented with either NAND or NOR gates.

2.21 Use NAND/NOR/INV logic as needed for a gate-minimum implementation of the functions in parts (a), (b), (c), and (d) of Problem 2.19.

2.22 Construct logic circuits for the expressions obtained in parts (a), (b), (c), (d), and (e) of Problem 2.17. Assume that all inputs and outputs are ACTIVE HIGH.

2.23 For each logic circuit given in Fig. P2-6, identify the numbered gates and read the circuit at the points indicated (X, Y, Z, etc.) in MIXED LOGIC notation. Note that conjugate logic symbols may be interchanged as needed.

2.24 Write the logic expressions for the logic circuits shown in Fig. P2-7; then determine the output waveforms, given the input waveforms shown. Neglect propagation delay of the gates. (Hint: Use of Appendix 2.1 and truth tables may be helpful in constructing the waveforms.)

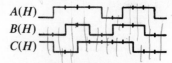

2.25 Read the logic circuits in Fig. P2-8 and obtain their MIXED LOGIC output expressions.

2.26 A staged CMOS network is given in Fig. P2-9. Find its logic function and logic diagram.

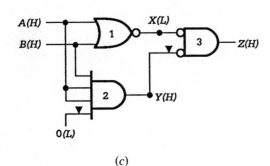

(a)

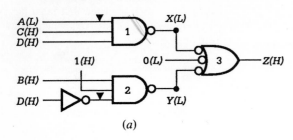

(b)

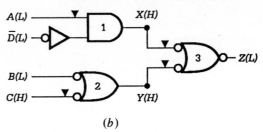

(c)

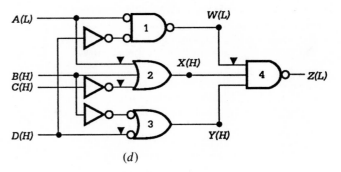

(d)

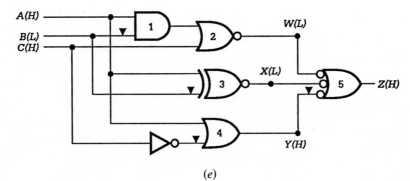

(e)

Fig. P2-6.

2 / Fundamentals of Digital Design

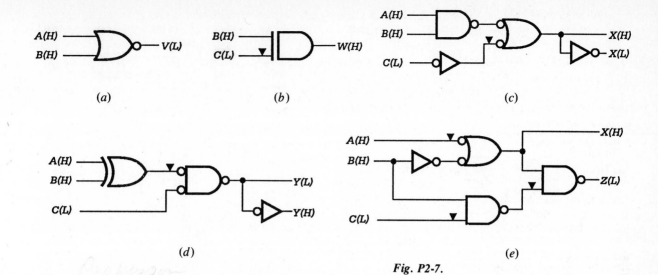

(a)

(b)

(c)

(d)

(e)

Fig. P2-7.

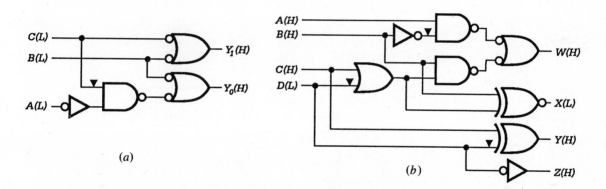

(a)

(b)

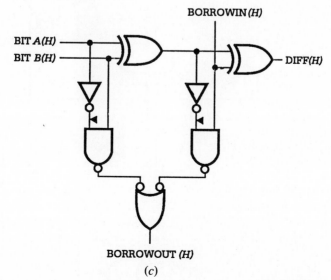

(c)

Fig. P2-8.

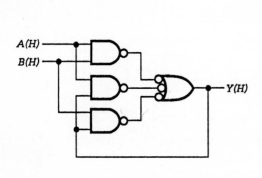

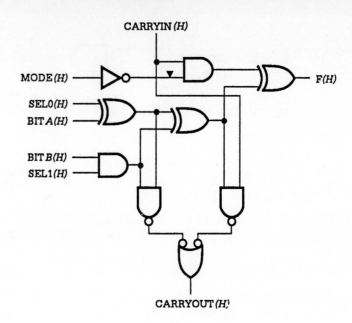

CARRYIN *(H)*

MODE *(H)*

SEL0 *(H)*

BIT *A(H)*

BIT *B(H)*

SEL1 *(H)*

F *(H)*

CARRYOUT *(H)*

(d) **Fig. P2-8 (contd.).** (e)

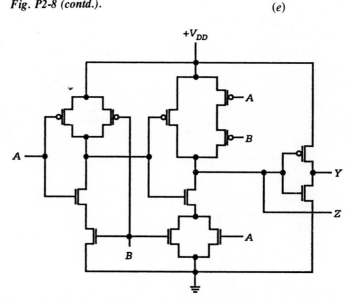

$+V_{DD}$

Fig. P2-9.

2.27 Given in Fig. P2-10 is the logic diagram for the function $F(L)$. Construct the CMOS circuit for this logic diagram exactly as it is given.

2.28 Reduce the following expressions to their simplest form and name the Boolean laws to be used in each step.

(a) $ab(c + \bar{c}) + a\bar{b}$

(b) $\overline{A + \overline{AB}} + A$

(c) $(\overline{X} + \overline{Y})(\overline{X} + Z)$

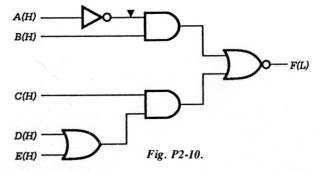

A *(H)*

B *(H)*

C *(H)*

D *(H)*

E *(H)*

F *(L)*

Fig. P2-10.

2 / Fundamentals of Digital Design

(d) $\overline{1\cdot(\overline{1} + 0\cdot1)\cdot\overline{0}}$

(e) $\overline{A + \overline{AC}} + B$

2.29 Reduce the following expressions to their simplest form:

(a) $\overline{A} + AB\overline{C} + \overline{A} + C$

(b) $\overline{a}\,\overline{b}c + ac + bc$

(c) $\overline{(X + \overline{YZ})(X\overline{Y} + \overline{XYZ})}$

(d) $(a + b)\,(\overline{a} + c)\,(b + c)$

(e) $\overline{A}B\overline{C} + ABC + AB\overline{C}$

2.30 Reduce the following expressions to their simplest form and name the Boolean laws to be used in each step.

(a) $(a \oplus b + \overline{b})\cdot(a + b)$

(b) $(XY) \oplus (\overline{X + Y})$

(c) $\overline{A} + A \odot B + \overline{A}B$

(d) $[(\overline{X} + Y) \odot (\overline{X} + Z)] + X$

(e) $\overline{B \oplus \overline{C} + (\overline{AB})(\overline{A} + C)}$

2.31 Use the laws of Boolean algebra, including the corollaries given by Eqs. (2-34) and (2-35), to prove whether the following equations are true or false.

(a) $X \oplus (\overline{X}Y) = X + (\overline{X} \odot Y)$

(b) $X \odot (\overline{X} + Y) = X \cdot \overline{Y}$

(c) $(A\overline{B} + C) \odot (\overline{A} + B + AC) = C$

(d) $[\overline{A}(A \oplus B)][A + (A \oplus \overline{B})] = \overline{A} \cdot B$

(e) $a \oplus b \oplus (ab) = a + b$

2.32 Use shaded logic maps to verify the validity of the following relations:

(a) $\overline{A} + AB = \overline{A}\overline{B}$

(b) $(X + \overline{Y})(\overline{Y} + Z) = \overline{Y} + XZ$

(c) $\overline{A} + A \odot B = \overline{A} + B$

(d) $(X\overline{Y}) \oplus (X \odot Y) = X + \overline{Y}$

(e) $\overline{a}bcd + ab\overline{c}d + c + \overline{d} = b + c + \overline{d}$

2.33 Use truth tables to verify the validity of parts (a), (b), (c), (d), and (e) of Problem 2.32. (Use Appendix 2.1 if necessary.)

2.34 Evaluate the following MIXED LOGIC expressions in their simplest form:

(a) $(A\overline{B} + C)(L) = (?)(H)$

(b) $[(RUN)(CNT) + \overline{(CNT)(\overline{RES})}](H) = (?)(L)$

(c) $[a \oplus (\overline{ab})](L) = (?)(H)$

(d) $[(\overline{X + Y}) \cdot XY](L) = (?)(H)$

(e) $[1 \cdot (\overline{\overline{1} + 0 \cdot 1}) + \overline{1}](H) = (?)(L)$

2.35 Implement the following function with NAND/EQV logic only (no INVERTERs):

$$F(H) = [(A \oplus S_o) \oplus (S_1B) \oplus (\overline{MC_{in}})](H)$$

Assume that the inputs are all ACTIVE HIGH except for A and M, which arrive ACTIVE LOW.

2.36 The logic circuits shown in Fig. P2-11 are redundant circuits since they contain logic gates in excess of those needed to yield the output function. Read each circuit of Fig. 2-11 in MIXED LOGIC notation and use the laws of Boolean algebra to determine the minimum logic required for each output function. (Hint: In some cases it may be advisable to change a gate to its conjugate form so as to eliminate the need to apply DeMorgan's laws. If this is done, take care to preserve the integrity of the function.)

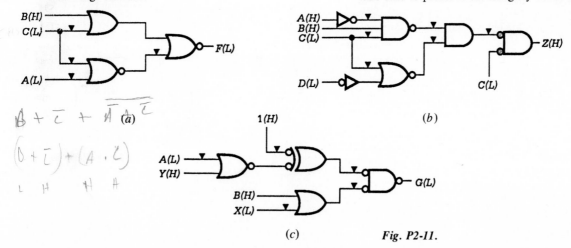

Fig. P2-11.

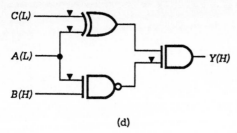

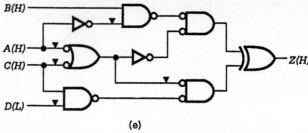

(d)   (e)

Fig. P2-11 (contd.).

2.37 **(a)** Use whatever gate forms are necessary to obtain a gate-minimum implementation of the following function without the use of INVERTERs. Take the inputs as $A(H)$, $B(H)$, $C(L)$, and $D(L)$.

$$F(H) = \{[A \oplus B] \cdot [(B\overline{C}) \odot D]\}(H)$$

(b) Use the laws of Boolean algebra and whatever gate forms are necessary to obtain a two-level (ANDing and ORing operations only) implementation of the function F(H) in part (a).

(c) Count the number of gates (exclusive of INVERTERs) and the number of gate inputs to obtain a gate/input tally for each of the logic circuits in parts (a) and (b). On the basis of these gate/input tallies, determine which of the logic circuits is simpler. Reevaluate the gate/input tallies by taking into account the presence of INVERTERs.

2.38 Use Boolean algebra and whatever gate forms are appropriate to obtain the simplest possible logic circuits for the following:

(a) Problem 2.14

(b) Problem 2.15

(c) Problem 2.16

Assume that all inputs and outputs are ACTIVE HIGH.

2.39 Use the laws of Boolean algebra to prove the following relations:

(a) $A X \overline{Y} + A \overline{X} Y + \overline{A} Y = (AX) \oplus Y$

(b) $\overline{A} \overline{X} \overline{Y} + \overline{A} X Y + A Y = (A + X) \odot Y$

2.40 Use shaded logic maps to verify the validity of the relations given in Problem 2.39.

2.41 A MOS bridge network is shown in Fig. P2-12.

(a) Find the logic function and logic circuit for this network exactly as it is presented.

(b) Reduce the logic expression of (a) to minimum form by using the laws of Boolean algebra and then construct an alternative NMOS network from this reduced expression. Which MOS network is the simpler?

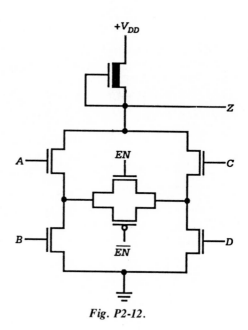

Fig. P2-12.

APPENDIX 2.1

Decimal-to-Binary Conversion Tables

Two Bit		Three Bit	
Decimal	**Binary**	**Decimal**	**Binary**
0	00	4	100
1	01	5	101
2	10	6	110
3	11	7	111

Four Bit		Five Bit		Six Bit	
Decimal	**Binary**	**Decimal**	**Binary**	**Decimal**	**Binary**
0	0000	16	10000	32	100000
1	0001	17	10001	33	100001
2	0010	18	10010	34	100010
3	0011	19	10011	35	100011
4	0100	20	10100	36	100100
5	0101	21	10101	37	100101
6	0110	22	10110	38	100110
7	0111	23	10111	39	100111
8	1000	24	11000	40	101000
9	1001	25	11001	41	101001
10	1010	26	11010	42	101010
11	1011	27	11011	43	101011
12	1100	28	11100	44	101100
13	1101	29	11101	45	101101
14	1110	30	11110	46	101110
15	1111	31	11111	47	101111

APPENDIX 2.2

The IEEE Standard Gate Symbols

The new standard, ANSI/IEEE Std. 91-1984, is extensive and will not be covered in its entirety in this text. The standard specifies certain rectangular symbols to represent logic gates, and we present these in this appendix. A recent review and explanation of the complete standard symbology is cited in the Annotated References of this chapter.

The rectangular gate symbols, as specified by the new standard, are not used in this text. We believe that the distinctive gate forms are better for teaching purposes. The standard does permit the use of the distinctive gate symbols but, of course, recommends the use of the rectangular forms. The new standard, it turns out, is used by the Defense Department, but has gained little acceptance in the academic field.

Shown in Fig. A2-1 is a comparison of the IEEE rectangular gate forms and the distinctive symbols used in this text. Note that the standard permits the triangle to be replaced by the traditional bubble as the ACTIVE LOW indicator symbol.

In Fig. A2-2 we illustrate the application of the IEEE standard gate symbols by implementing a function $F(H)$ having both ACTIVE HIGH and ACTIVE LOW inputs. In Fig. A2-2(a) is given the distinctive gate implementation of this function, and in Figs. A2-2(b) and (c) two versions which satisfy the standard. Notice that the latter representation involves some ambiguity with respect to the presence of logic incompatibilities.

LOGIC LEVEL CONVERSION

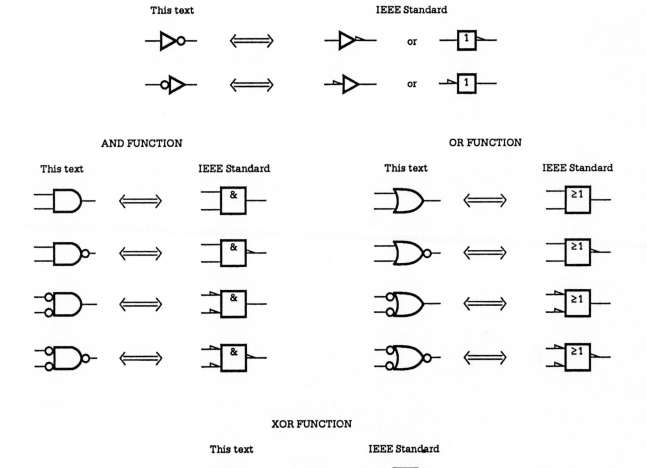

Fig. A2-1. *Comparison of the IEEE Standard gate symbols with the distinctive gate symbols used in this text.*

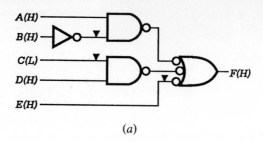

(a)

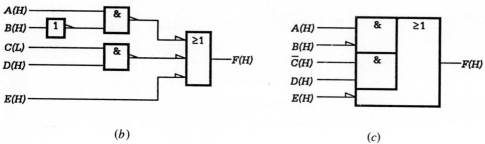

(b)

(c)

Fig. A2-2. A comparison of symbolic representations for the function F = A$\overline{\text{B}}$ + $\overline{\text{C}}$D + $\overline{\text{E}}$ *if* A, B, D *and* E *are ACTIVE HIGH inputs and* C *is an ACTIVE LOW input. (a) Distinctive gate representation. (b) and (c) Alternative IEEE Standard gate symbol representations.*

CHAPTER 3

Function Representation and Reduction

3.1 INTRODUCTION

This chapter deals mainly with the function reduction aspect. Here, we will consider logic function reduction and minimization by using three approaches: Boolean manipulation, logic tabulation, and graphics. This is done in preparation for Chapter 4, where we will "put it all together" and actually design combinational logic circuits and systems from the device criteria stage through the implementation stage.

So as to give the reader a perspective of what is to come, we provide in Fig. 3-1 a broad overview of the combinational logic design procedure. All four stages are usually considered important to any design problem. The importance of stage 1 should seem self-evident since no device or machine can be designed without specifications and a functional description. Stage 2 is a very important part of any design problem, especially one of some complexity. Its omission could make function reduction very difficult if not impossible—hence the reason for the question mark between stages 1 and 3. Stage 3 is necessary if the hardware requirements are to be kept to a minimum or reduced level. Bypassing stage 3 could result in costly design redundancies. Finally, stage 4 deals with the actual physical realization of the device. By the end of Chapter 4 the reader will be able to configure physically a given logic function in a variety of ways.

Shown in Fig. 3-2 are the three options available to the designer for logic function representation. Although this chapter deals with all three of these options, the main thrust is in logic function graphics which has the greatest power and versatility of the three. Like Chapter 2, the subject matter of this chapter is considered to be of vital importance to the remainder of this text and, hence, to an understanding of digital design.

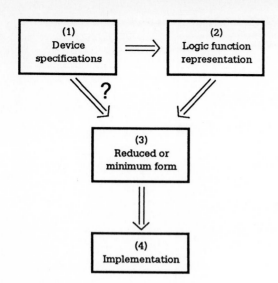

Fig. 3-1. *Normal step sequence for combinational logic design.*

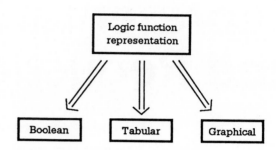

Fig. 3-2. *Options for logic function representation.*

3.2   SOP and POS FORMS

Earlier, in Sect. 2.4, the logic operations represented by $X \cdot Y$ and $X + Y$ were defined as the AND and OR operations of logic variables X and Y, respectively. By these definitions one could clearly differentiate these operations from the product and sum operations of conventional algebra. Now, however, it is convenient to refer to the AND operation as the Boolean PRODUCT and the OR operation as the Boolean SUM, terms that we will use periodically throughout this text. So, as long as the true meaning of these logic operations is kept in mind, the encroachment into conventional algebraic terminology should pose no particular problem for the reader.

3.2.1   The SOP Representation

Consider the function of three variables given by the Boolean expression

$$f(A,B,C) = A + \overline{B}C + \overline{A}BC \tag{3-1}$$

This function is written in *sum-of-products* (*SOP*) form, meaning ORing of ANDed terms. Although there are three *p-terms* (PRODUCT-terms) in this expression, only the term $\overline{A}BC$ is called a MINTERM. A MINTERM is defined as follows:

> *Any ANDed term containing all the variables of a function in complemented or uncomplemented form is a MINTERM.*

We will use the symbol

$$m_i = m_i(A,B,C, \ldots) \qquad (3\text{-}2)$$

to represent the ith MINTERM of a function. Notice that two of the three terms in Eq. (3-1) cannot be MINTERMs according to the definition.

To simplify the MINTERM representation, a shorthand notation has been established and is based on the following logic variable code:

> *Complemented variables: logic 0*
> *Uncomplemented variables: logic 1*

Once the logic 0's and 1's have been assigned to all variables in a given MINTERM, a *MINTERM code* is established where the subscript in m_i becomes the decimal equivalent of the binary code formed by the logic state assignments. As an example, the MINTERM in Eq. (3-1) is represented by

$$\overline{A}BC = m_3$$

$$0\,1\,1$$

since the binary 011 has a decimal value of 3. A complete MINTERM code table for four variables is given in Fig. 3-3. A similar MINTERM code table can be constructed for any number of variables.

SOP Term	Binary	Decimal	m_i	SOP Term	Binary	Decimal	m_i
$\overline{A}\,\overline{B}\,\overline{C}\,\overline{D}$	0 0 0 0	0	m_0	$A\overline{B}\,\overline{C}\,\overline{D}$	1 0 0 0	8	m_8
$\overline{A}\,\overline{B}\,\overline{C}D$	0 0 0 1	1	m_1	$A\overline{B}\,\overline{C}D$	1 0 0 1	9	m_9
$\overline{A}\,\overline{B}C\overline{D}$	0 0 1 0	2	m_2	$A\overline{B}C\overline{D}$	1 0 1 0	10	m_{10}
$\overline{A}\,\overline{B}CD$	0 0 1 1	3	m_3	$A\overline{B}CD$	1 0 1 1	11	m_{11}
$\overline{A}B\overline{C}\,\overline{D}$	0 1 0 0	4	m_4	$AB\overline{C}\,\overline{D}$	1 1 0 0	12	m_{12}
$\overline{A}B\overline{C}D$	0 1 0 1	5	m_5	$AB\overline{C}D$	1 1 0 1	13	m_{13}
$\overline{A}BC\overline{D}$	0 1 1 0	6	m_6	$ABC\overline{D}$	1 1 1 0	14	m_{14}
$\overline{A}BCD$	0 1 1 1	7	m_7	$ABCD$	1 1 1 1	15	m_{15}

Fig. 3-3. MINTERM code table for four variables.

A function composed completely of a logical sum of MINTERMs is said to be in CANONICAL SOP form. A typical example is given by the following equations, where use has been made of the MINTERM code shorthand notation:

$$Y(A,B,C) = (\overline{A}B\overline{C}) + (AB\overline{C}) + (\overline{A}BC) + (ABC)$$

$$010 \qquad 110 \qquad 011 \qquad 111$$

$$= m_2 + m_6 + m_3 + m_7$$

$$= \sum m(2, 3, 6, 7). \tag{3-3}$$

Notice that the sum sign $\sum$ is used to represent the ORing (Boolean SUM) of MINTERMs m_2, m_3, m_6, and m_7.

A reduced SOP function can be expanded to CANONICAL SOP form by applying the factoring law and OR laws given in Sect. 2-9. We demonstrate this by expanding the expression of Eq. (3-1) as follows:

$$f(A,B,C) = A + (\overline{B}C) + (\overline{A} \cdot B \cdot C)$$

$$= A(B + \overline{B}) \cdot (C + \overline{C}) + (A + \overline{A}) \cdot (\overline{B}C) + \overline{A}BC$$

$$= (ABC) + (AB\overline{C}) + (A\overline{B}C) + (A\overline{B}\overline{C}) + (A\overline{B}C)$$
$$\qquad + (\overline{A}\overline{B}C) + (\overline{A}BC)$$

$$= (ABC) + (AB\overline{C}) + (A\overline{B}C) + (A\overline{B}\overline{C}) + (\overline{A}\overline{B}C)$$
$$\qquad + (\overline{A}BC)$$

$$= m_7 + m_6 + m_5 + m_4 + m_1 + m_3$$

$$= \sum m(1, 3, 4, 5, 6, 7) \tag{3-4}$$

Note that the OR law, $X + \overline{X} = 1$, has been applied three separate times and that two identical MINTERMs, $A\overline{B}C$, were combined in agreement with the OR law, $X + X = X$.

The MINTERMs representing a given reduced function can be read directly from the truth table for that function. This, the tabular method, is easily demonstrated by constructing the truth table for Eq. (3-1) and by applying the MINTERM code of Fig. 3-3, as shown in Fig. 3-4. Here, the logic values for $f(A,B,C)$ are obtained by substituting all combinations of logic values for variables A, B, and C into Eq. (3-1). Of course, the same information is readily available from the CANONICAL SOP expression of Eq. (3-4), but a considerable effort went into obtaining this result.

Not surprisingly, the function expansion process just described can be reversed to yield a reduced or minimum expression. In most cases, however, this is not easily accomplished. In the equations that follow we show a step-by-step reduction of the CANONICAL SOP form given by Eq. (3-4). It will be instructive for the reader to follow along naming the Boolean algebraic law(s) used in each step.

A	B	C	f
0	0	0	0
0	0	1	$1 \to m_1$
0	1	0	0
0	1	1	$1 \to m_3$
1	0	0	$1 \to m_4$
1	0	1	$1 \to m_5$
1	1	0	$1 \to m_6$
1	1	1	$1 \to m_7$

Fig. 3-4. *Truth table for the logic function* f = A + ($\overline{B}$C) + ($\overline{A}$BC) *showing MINTERMs.*

$$f(A,B,C) = \sum m(1, 3, 4, 5, 6, 7)$$

$$= ABC + AB\overline{C} + A\overline{B}C + A\overline{B}\overline{C} + \overline{A}\overline{B}C + \overline{A}BC$$

$$= A \cdot (BC + B\overline{C} + \overline{B}C + \overline{B}\overline{C}) + \overline{A}\overline{B}C + \overline{A}BC$$

$$= A \cdot [B \cdot (C + \overline{C}) + \overline{B}(C + \overline{C})] + \overline{A} \cdot (\overline{B}C + BC)$$

$$= A \cdot (B + \overline{B})(C + \overline{C}) + \overline{A}C \cdot (\overline{B} + B)$$

$$= A + \overline{A}C$$

$$= A + C \tag{3-5}$$

Comparing Eq. (3-5) with the original expression, Eq. (3-1), leads to the conclusion that the latter expression is a reduced function but not a minimum form and that the result expressed by Eq. (3-5) is an *absolute minimum* function since it cannot be reduced further. Although arriving at the result given by Eq. (3-5) seems tedious, and it is, the reader should not despair. It will be shown later in this chapter that these results could have been arrived at far more easily by using the power of logic function graphics—that is, by using logic maps. It is in the subject area of function reduction and minimization that logic maps achieve their fullest potential for usefulness.

3.2.2 The POS Representation

An alternative means of representing a logic expression is to cast it in *product-of-sums* (*POS*) form, meaning the ANDing of ORed terms also called *s-terms* (SUM-terms). An example of POS representation is given by the function

$$f(A,B,C,D) = (A + \overline{B})(\overline{A} + \overline{B} + C + \overline{D})(B + \overline{C} + D) \tag{3-6}$$

where only the term $(\overline{A} + \overline{B} + C + \overline{D})$ is called a MAXTERM. A MAXTERM is defined as follows:

> *Any ORed term containing all the variables of a function in complemented or uncomplemented form is a MAXTERM.*

The symbol

$$M_i = M_i(A,B,C, \ldots) \tag{3-7}$$

will be used to represent the ith MAXTERM of a function.

MAXTERM representation can be simplified considerably by using the following logic variable code:

> *Complemented variables: logic 1*
> *Uncomplemented variables: logic 0*

The assignment of the logic 0's and logic 1's in this manner to all variables in each MAXTERM establishes the *MAXTERM code*, where the subscript in M_i becomes the decimal equivalent of the binary number formed by the logic state assignments. The MAXTERM code table for four variables is presented in Fig. 3-5. To illustrate its use, the MAXTERM in Eq. (3-6) would be represented by

POS Term	Binary	Decimal	M_i	POS Term	Binary	Decimal	M_i
$A + B + C + D$	0 0 0 0	0	M_0	$\overline{A} + B + C + D$	1 0 0 0	8	M_8
$A + B + C + \overline{D}$	0 0 0 1	1	M_1	$\overline{A} + B + C + \overline{D}$	1 0 0 1	9	M_9
$A + B + \overline{C} + D$	0 0 1 0	2	M_2	$\overline{A} + B + \overline{C} + D$	1 0 1 0	10	M_{10}
$A + B + \overline{C} + \overline{D}$	0 0 1 1	3	M_3	$\overline{A} + B + \overline{C} + \overline{D}$	1 0 1 1	11	M_{11}
$A + \overline{B} + C + D$	0 1 0 0	4	M_4	$\overline{A} + \overline{B} + C + D$	1 1 0 0	12	M_{12}
$A + \overline{B} + C + \overline{D}$	0 1 0 1	5	M_5	$\overline{A} + \overline{B} + C + \overline{D}$	1 1 0 1	13	M_{13}
$A + \overline{B} + \overline{C} + D$	0 1 1 0	6	M_6	$\overline{A} + \overline{B} + \overline{C} + D$	1 1 1 0	14	M_{14}
$A + \overline{B} + \overline{C} + \overline{D}$	0 1 1 1	7	M_7	$\overline{A} + \overline{B} + \overline{C} + \overline{D}$	1 1 1 1	15	M_{15}

Fig. 3-5. *MAXTERM code table for four variables.*

$$\overline{A} + \overline{B} + C + \overline{D} = M_{13}$$

$$\begin{matrix} 1 & 1 & 0 & 1 \end{matrix}$$

where the binary 1101 has a decimal value of 13.

A comparison of the MINTERM and MAXTERM code tables in Figs. 3-3 and 3-5 indicates that

$$M_i = \overline{m}_i$$

and

$$m_i = \overline{M}_i \tag{3-8}$$

revealing a complementation relationship between MINTERMs and MAX-TERMs. The validity of relations (3-8) is easily demonstrated by the following examples:

$$\overline{m}_5 = \overline{A\overline{B}C} = \overline{A} + B + \overline{C} = M_5$$

and

$$\overline{M}_{13} = \overline{\overline{A} + \overline{B} + C + \overline{D}} = AB\overline{C}D = m_{13}$$

where use has been made of DeMorgan's laws given by Eqs. (2-18) and (2-19).

The relationship between MINTERM and MAXTERM codes makes it possible to read truth tables (and later logic graphics) in either SOP or POS form, a feature which can be very useful in digital circuit design. However, before this can be done, we must complete our discussion of the POS representation.

A function whose terms are all MAXTERMs is said to be given in CA-NONICAL POS form as indicated next by using MAXTERM code.

$$f(A,B,C) = (A + B + \overline{C})(\overline{A} + B + \overline{C})(\overline{A} + B + C)(A + B + C)$$

$$= M_1 \cdot M_5 \cdot M_4 \cdot M_0$$

$$= \prod M(0, 1, 4, 5) \tag{3-9}$$

Observe that the product sign $\prod$ is used to represent the ANDing (Boolean PRODUCT) of MAXTERMs M_0, M_1, M_4, and M_5.

3 / Function Representation and Reduction

Expansion of a reduced POS function to CANONICAL POS form can be accomplished as indicated by the following example:

$$f(A,B,C) = (A + C)(\overline{B} + \overline{C})$$
$$= (A + B\overline{B} + C)(A\overline{A} + \overline{B} + \overline{C})$$
$$= (A + B + C)(A + \overline{B} + C)(A + \overline{B} + \overline{C})(\overline{A} + \overline{B} + \overline{C})$$
$$= \prod M(0,2,3,7) \tag{3-10}$$

As can be seen, the process of POS expansion is a tedious one (as was true for SOP expansion) usually involving multiple applications of the distributive, OR and AND laws in the form of $(X + Y)(X + \overline{Y}) = X$. We will show later that the use of logic maps offers a far easier method of identifying the MAXTERMs that are present in a given reduced form. Both MINTERMs and MAXTERMs can be read directly from the logic function graphics.

Reducing or minimizing a function from its CANONICAL POS form can also be accomplished by reversing the process just described. As an exercise in Boolean manipulation, the reader should prove the equivalence of Eqs. (3-3) and (3-9) expressed by

$$f(A,B,C) = \sum m(2,3,6,7) = \prod M(0,1,4,5) = B \tag{3-11}$$

Both CANONICAL forms reduce to the minimum B, and the missing coded terms of the SOP CANONICAL form become the coded terms of the POS CANONICAL form, and vice versa.

The results expressed by Eqs. (3-11) can be deduced from the truth table for Eqs. (3-3) and (3-9). Such a truth table is shown in Fig. 3-6, where use is made of both MINTERM and MAXTERM codes. Function f values equal to logic 1 are read as MINTERMs (using Fig. 3-3), and when ORed together give Eq. (3-3). Similarly, function f values equal to logic 0 are read as MAXTERMs (using Fig. 3-5) and when ANDed together yield Eq. (3-9). The function f is equal to B since the columns for f and B in Fig. 3-6 are the same.

By using DeMorgan's laws and Eqs. (3-8), one can easily show that the complement of Eqs. (3-11) is

$$\overline{f}(A,B,C) = \prod M(2,3,6,7) = \sum m(0,1,4,5) \tag{3-12}$$

A B C	f	
0 0 0	0	$M_0 = A + B + C$
0 0 1	0	$M_1 = A + B + \overline{C}$
0 1 0	1	$m_2 = \overline{A} \cdot B \cdot \overline{C}$
0 1 1	1	$m_3 = \overline{A} \cdot B \cdot C$
1 0 0	0	$M_4 = \overline{A} + B + C$
1 0 1	0	$M_5 = \overline{A} + B + \overline{C}$
1 1 0	1	$m_6 = A \cdot B \cdot \overline{C}$
1 1 1	1	$m_7 = A \cdot B \cdot C$

Fig. 3-6. Truth table showing the MINTERMs and MATERMs in Eqs. (3-11).

This follows from the result

$$\bar{f} = \overline{\sum m(2,3,6,7)}$$

$$= \overline{m_2 + m_3 + m_6 + m_7}$$

$$= \bar{m}_2 \cdot \bar{m}_3 \cdot \bar{m}_6 \cdot \bar{m}_7$$

$$= M_2 \cdot M_3 \cdot M_6 \cdot M_7$$

$$= \prod M(2,3,6,7)$$

and a similar set of equations for $\bar{f} = \overline{\prod M(0,1,4,5)}$. Equations (3-11) and (3-12), viewed as a set, illustrate the type of interrelationship that always exists between CANONICAL forms.

There is more information that develops from the interrelationship between CANONICAL forms. Applying the OR law, $X + \bar{X} = 1$ and the OR form of the commutative laws, we obtain from Eqs. (3-11) and (3-12) the result

$$f + \bar{f} = \sum m(2,3,6,7) + \sum m(0,1,4,5)$$

$$= \sum m(0,1,2,3,4,5,6,7)$$

$$= 1$$

which may be generalized for N variables by

$$\sum_{i=0}^{2^N-1} m_i = 1 \tag{3-13}$$

Similarly, by using the AND law, $X \cdot \bar{X} = 0$, and the AND form of the commutative laws, there results

$$f \cdot \bar{f} = [\prod M(0,1,4,5)][\prod M(2,3,6,7)]$$

$$= \prod M(0,1,2,3,4,5,6,7)$$

$$= 0$$

Or generally, for N variables

$$\prod_{i=0}^{2^N-1} M_i = 0 \tag{3-14}$$

Equations (3-13) and (3-14) are said to have a *dual* relationship following the definition of duality given in Sect. 2.9.2.

In summary, we may state the following:

> *Any function ORed with its complement is logic 1 definite, and any function ANDed with its complement is logic 0 definite—it does not matter that the function be in its CANONICAL form.*

Furthermore, the reader should remember that if *all* 2^N MINTERMs of an N variable function are ORed together, the result must be logic 1, or if *all* 2^N MAXTERMs of an N variable function are ANDed together, the result must be logic 0. Use will be made of the above information in the later sections on logic graphics.

3.3 TABULAR MINIMIZATION

A reduced or minimized form of a logic function can be read somewhat directly from the CANONICAL truth table for that function. Any truth table which gives the function logic levels for all possible logic levels of N variables of that function is said to be a *CANONICAL truth table* of the Nth order. Such truth tables contain 2^N rows of logic values (for the N variables) which are normally arranged in a binary sequence equivalent to decimal 0 through decimal $2^N - 1$. All truth tables presented so far have been of this type. We shall continue to refer to these tables simply as truth tables. Other special types of truth tables will be assigned special names.

To minimize a function in SOP form, use is made of the *SOP adjacency law* given by

$$X \cdot Y = Y \qquad (3\text{-}15)$$

where $X = A + \overline{A} = B + \overline{B} = \cdots = 1$. Similarly, to minimize a function in POS form, use is made of the dual of Eq. (3-15), the *POS adjacency law*,

$$X + Y = Y \qquad (3\text{-}16)$$

applied in the form $(A + Y)(\overline{A} + Y) = (B + Y)(\overline{B} + Y) = \cdots = X + Y = Y$, and so on, where $X = A \cdot \overline{A} = B \cdot \overline{B} = \cdots = 0$.

Properly employed, the adjacency laws lead to *collapsed truth tables* from which minimized functions can be read directly in either SOP or POS form. To illustrate, consider the truth table of Fig. 3-6 now recast stepwise into the collapsed forms of Fig. 3-7. Note that the truth table in Fig. 3-7(c) can be collapsed no further, thus permitting the minimum function to be read directly as B in either SOP or POS form. Here, use is made of the MINTERM and MAXTERM codes together with Eqs. (3-15) and (3-16), where the symbol X has the same meaning as in the adjacency laws.

As a second example consider the CANONICAL POS function

$$f = \prod M(0, 2, 3) \qquad (3\text{-}17)$$

The truth table and its collapsed form for this function are shown in Fig. 3-8. Again, use has been made of the MINTERM and MAXTERM codes and Eqs. (3-15) and (3-16), permitting the minimum function to be read directly from the collapsed truth table in either SOP or POS form. The equivalence of these two forms is established by the following Boolean manipulation:

$$f = (A + C)(A + \overline{B})$$
$$= A + A\overline{B} + AC + \overline{B}C$$
$$= A + \overline{B}C$$

A B C	f		
0 0 0	0	00X	POS
0 0 1	0		
0 1 0	1	01X	SOP
0 1 1	1		
1 0 0	0	10X	POS
1 0 1	0		
1 1 0	1	11X	SOP
1 1 1	1		

A B C	f		
0 0 X	0	XOX	POS
0 1 X	1		
1 0 X	0	X1X	SOP
1 1 X	1		

A B C	f		
X 0 X	0 → B	POS
X 1 X	1 → B	SOP

Fig. 3-7. *Stepwise collapse of the truth table of Fig. 3-6 resulting in the minimization of the function.*

A B C	f	
0 0 0	0	0X0 POS
0 0 1	1	
0 1 0	0	01X POS
0 1 1	0	X01 SOP
1 0 0	1	10X SOP
1 0 1	1	1XX SOP
1 1 0	1	11X SOP
1 1 1	1	

A B C	f		
0 X 0	0	$(A + C)(A + \overline{B})$	POS
0 1 X	0		
X 0 1	1	$A + \overline{B}C$	SOP
1 X X	1		

Fig. 3-8. *Minimization of the function* $f = \Pi\, M(0, 2, 3)$ *by the stepwise collapse of its truth table.*

Collapsed truth tables of the type given in Figs. 3-7 and 3-8 will later be used in Chapter 4 to simplify the design process. There, the symbol X is used to represent an *irrelevant input*—meaning one taking logic 0 or logic 1, it makes no difference.

3.3.1 Logic Adjacencies

Two terms are *logically adjacent* if they differ only in the complement orientation of a single variable. Thus, each X shown in Figs. 3-7 and 3-8 represents a pair (see arrows) of logically adjacent SOP terms or logically adjacent POS terms, as the case may be. Examples taken from Fig. 3-8 are given together with the Boolean algebraic laws involved for each.

$$X01 = A\overline{B}C + \overline{A}\,\overline{B}C$$
$$= (A + \overline{A})\overline{B}C = \overline{B}C \qquad \text{Factoring law and OR laws}$$

$$01X = (A + \overline{B} + C)(A + \overline{B} + \overline{C})$$

$$= A + \overline{B} + C\overline{C} = A + \overline{B} \qquad \text{Distributive law and AND laws}$$

$$1XX = ABC + A\overline{B}C + AB\overline{C} + A\overline{B}\overline{C}$$

$$= A(B + \overline{B})(C + \overline{C}) = A \qquad \text{Factoring law and OR laws}$$

The MINTERMs $A\overline{B}C$ and $\overline{A}\,\overline{B}C$, contained in the coded term X01, are logically adjacent, and when ORed together are read as $\overline{B}C$ by virtue of Eq. (3-15). Similarly, the MAXTERMs $(A + \overline{B} + C)$ and $(A + \overline{B} + \overline{C})$, contained in the coded term 01X, are logically adjacent, and when ANDed together are read as $A + \overline{B}$ according to Eq. (3-16). Because the relations given by Eqs. (3-15) and (3-16) form the basis for the logic adjacency concept, they are often referred to simply as the *logic adjacency laws*. These relations will again be invoked in a later discussion on function reduction by using logic graphics.

3.3.2   Why Minimization

Before moving to the next section on logic function graphics, let us pause briefly to reflect on a point or two which may or may not be obvious. The reader probably has sensed by now that function reduction or minimization is important. That this notion is a correct one can be easily demonstrated. Take, for example, the CANONICAL SOP function given by Eq. (3-4). To implement this function as written would require no fewer than six three-input AND gates to drive a six-input OR gate, or the equivalent using conjugate NAND gates. On the other hand, the minimized form of this function, given by Eq. (3-5), tells us that a single two-input OR gate would suffice. Thus, it should now be evident that one reduces or minimizes a given logic function so as to minimize or eliminate costly design redundancies, assuming the function in question is to be implemented. In addition to the implementation aspect, a reduced or minimized form often reveals information not apparent from the original expression. In the example given, we see that the input B has no influence whatsoever on the function f, a fact that could not easily have been deduced from the original CANONICAL form.

3.4   LOGIC FUNCTION GRAPHICS

Graphical representation of logic truth tables are called *Karnaugh maps* (*K-maps*) after M. Karnaugh who, in 1953, established the map method for combinational logic circuit synthesis. The shaded logic maps used in Figs. 2-48, 2-49, and 2-52 are actually a type of K-map. Karnaugh maps are very important in logic design for the following reasons:

1. K-maps offer a simple method of identifying the MINTERMs and MAXTERMs inherent in any given minimized or reduced function.
2. K-maps provide the designer with a relatively effortless means of minimization or reduction through pattern recognition.

These two advantages, particularly the latter, make K-maps extremely useful tools in logic circuit design. Not until we have covered Chapter 4 will the impact of this statement be fully realized.

While much has been said exalting the worth of Karnaugh maps (logic maps), they have yet to be defined adequately. Simply stated, a K-map is the graphical representation of a CANONICAL truth table for a given function whose variables form the logic axes of the map. The *order* of a standard K-map, like the order of a CANONICAL truth table, is equal to the number of variables in the function. For example, a second-order K-map is the graphical representation of a two-variable function, a third-order map for a three-variable function, and so on. Alternatively put, MINTERMs and MAXTERMs derived from an Nth-order K-map will each have N variables. As will be shown in the discussions that follow, K-maps are a natural but necessary extension of the shaded logic maps used previously.

3.4.1   First-Order K-maps

Karnaugh maps of a single variable are easily developed (as are all K-maps) from a MINTERM code table similar to that of Fig. 3-3. Shown in Fig. 3-9(a) is the MINTERM code table for one variable, and in Figs. 3-9(b), (c), and (d) are three formats for a first-order K-map. All three map formats shown have logic coordinates 0 and 1 on the single variable axis and all indicate the possible MINTERM positions. However, the format of Fig. 3-9(d) is preferred since the MINTERM positions are indicated by using MINTERM code in the lower right-hand corner of each cell, leaving room for truth table information (1's and 0's) representing a given function.

Consider the three examples given in Fig. 3-10 for functions $f_1 = A$, $f_2 = \overline{X}$, and $f_3 = A + \overline{A} = 1$. Notice that all the information contained in a given truth table is present in the corresponding K-map or shaded logic map, and that a shaded cell or the presence of a logic 1 in a cell identifies the MINTERM. Had logic 0's been present in f_3 instead of logic 1's, the function would be logic 0 since neither MINTERM would exist.

Use will be made of first-order K-maps in a later section on entered variable mapping and in all subsequent chapters of this text.

3.4.2   Second-Order K-maps

A second-order Karnaugh map is the graphical representation of a truth table for a function of two variables and is developed from the MINTERM code table shown in Fig. 3-11(a). The logic levels for A and B, which are assigned

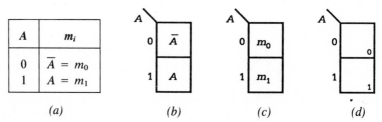

(a)   (b)   (c)   (d)

Fig. 3-9. (a) *MINTERM code table for one variable. (b), (c) and (d) First-order K-maps showing MINTERM positions.*

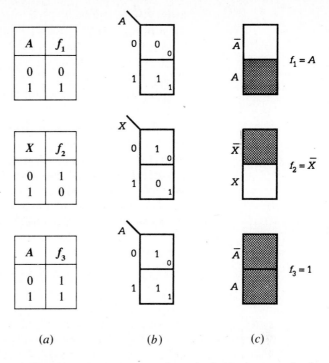

Fig. 3-10. *(a) Truth tables, (b) K-maps, and (c) shaded logic maps for functions* $f_1 = A$, $f_2 = \overline{X}$ *and* $f_3 = A + \overline{A} = 1$.

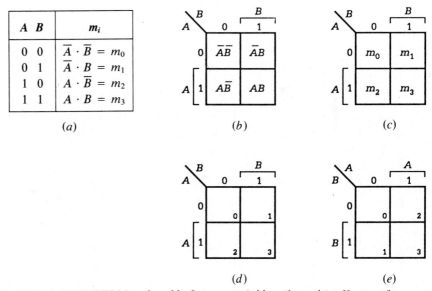

Fig. 3-11. *(a) MINTERM code table for two variables. (b) and (c) K-maps for two variables showing MINTERM positions. (d) and (e) Alternative map formats.*

in a binary sequence equivalent to decimals, 0, 1, 2, and 3, form the "logic coordinates" in MINTERM code for the four possible MINTERM positions shown in Figs. 3-11(b) and (c). Observe that there are 2^N cells in each K-map to accommodate each of the possible 2^N MINTERM positions for a second-order ($N = 2$) map. The K-map format of Fig. 3-11(d) is preferred since the possible MINTERM positions are indicated by MINTERM code decimal values in the lower right-hand corner of each cell leaving room for truth table information. An alternative map format shown in Fig. 3-11(e) differs from that of Fig. 3-11(d) only in the MINTERM coordinates.

The logic variables A and B have been placed on the sides of the K-maps in Fig. 3-11 as a reminder that the domain concept remains valid—recall that this was done for the shaded logic maps presented in Figs. 2-48, 2-49, and 2-52. For example, the A domain covers MINTERM $m_2 = A \cdot \overline{B}$ with logic coordinates 1 and 0 and MINTERM $m_3 = A \cdot B$ with logic coordinates 1 and 1; the B domain covers MINTERM $m_1 = \overline{A} \cdot B$ with logic coordinates 0 and 1 and m_3. Looking at it another way, MINTERM m_0 with logic coordinates 0 and 0 is the intersection of the $\overline{A}$ and $\overline{B}$ domains, and MINTERM m_1 is the intersection of the $\overline{A}$ and B domains, and so on. Karnaugh maps that use the domain format exclusively (no logic coordinates) are sometimes called *Vietch diagrams*.

To illustrate the use of second-order K-maps, consider the functions f_1 and f_2 defined in Fig. 3-12(a) by the truth tables and represented graphically in Figs. 3-12(b) and (c) by K-maps and shaded logic maps. By reading the 1's in the K-maps in MINTERM code, there result the CANONICAL SOP expressions given by

$$f_1(A,B) = \sum m(2,3) = A\overline{B} + AB$$
$$f_2(A,B) = \sum m(1,3) = \overline{A}B + AB$$

(3-18)

Similarly, by reading the 0's in the K-maps in MAXTERM code, we obtain the CANONICAL POS expressions

$$f_1(A,B) = \prod M(0,1) = (A + B)(A + \overline{B})$$
$$f_2(A,B) = \prod M(0,2) = (A + B)(\overline{A} + B)$$

(3-19)

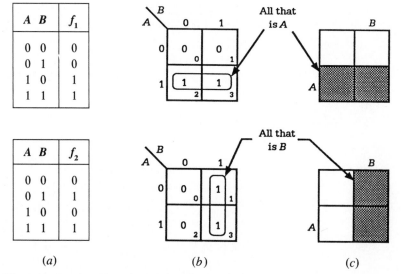

(a)   (b)   (c)

Fig. 3-12. *Truth tables (a), K-maps (b), and shaded logic maps (c) for functions* $f_1(A,B) = A$ *and* $f_2(A,B) = B$.

Furthermore, it is evident at a glance that MINTERMs m_2 and m_3 in f_1 are logically adjacent and that MINTERMs m_1 and m_3 in f_2 are logically adjacent. Similarly, the MAXTERMs M_0 and M_1 are logically adjacent in f_1 as are M_0 and M_2 logically adjacent in f_2.

That the presence of logic adjacencies in Eqs. (3-18) and (3-19) leads to function reduction is evident when one recalls that in Figs. 3-7 and 3-8 it was the logic adjacencies that permitted the formation of collapsed truth tables which ultimately yielded logic minima. Thus, by applying the adjacency law of Eq. (3-15) to the CANONICAL SOP expressions given by Eqs. (3-18), or by applying the adjacency law of Eq. (3-16) to the CANONICAL POS expressions given by Eqs. (3-19), we obtain the results

$$f_1 = A \quad \text{and} \quad f_2 = B$$

Of course, these reduced forms could have been obtained directly from the respective K-maps in Fig. 3-12(b) by reading "all that is A" and "all that is B," as indicated by the loops, or by reading the shaded logic maps of Fig. 3-12(c) in the same manner.

As another example, let us reverse the process just described by mapping the minimum function

$$Y = A + \overline{B} \tag{3-20}$$

to obtain Y in CANONICAL SOP form. This is accomplished simply by mapping "all that is A" and "all that is NOT B," as indicated by the loops in Fig. 3-13(a). Then by reading the 1's off the map in MINTERM code we obtain

$$Y = \sum m(0, 2, 3) = \overline{A}\,\overline{B} + A\overline{B} + AB \tag{3-21}$$

in CANONICAL SOP form. Observe that there are two sets of logic adjacencies, (m_0, m_2) and (m_2, m_3), which can be read as $(A + \overline{A}) \cdot \overline{B} = \overline{B}$, and $A \cdot (B + \overline{B}) = A$, respectively. A shaded logic map of the function $Y = A + \overline{B}$ is provided in Fig. 3-13(b) for comparison purposes.

There is more information that derives from the K-map of Fig. 3-13(a). Reading cell 1 (containing logic 0) in MAXTERM code gives

$$Y = \prod M(1) = A + \overline{B} \tag{3-22}$$

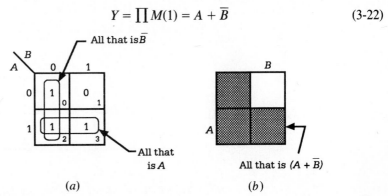

(a) (b)

Fig. 3-13. (a) K-map, and (b) shaded logic map for the function $Y = A + \overline{B}$.

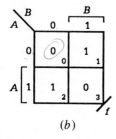

A B	f
0 0	0
0 1	1
1 0	1
1 1	0

(a)

(b)

Fig. 3-14. *Truth table (a) and K-map (b) for the function* f = A ⊕ B.

which is the original function. Evidently Eq. (3-20) is both a CANONICAL form and a minimum form. Furthermore, it is again apparent, comparing Eqs. (3-21) and (3-22), that the missing coded terms of one CANONICAL form become the coded terms of the other CANONICAL form, producing the equivalent relationship that one would expect.

To illustrate further the equivalence between CANONICAL forms, let us map the truth table in Fig. 3-14(a), which we recognize as the XOR function $f = A \oplus B$ represented by the truth table in Fig. 2-50(a) and graphically in Fig. 2-52. Shown in Fig. 3-14(b) is the K-map for the XOR function. Notice that there are *no* logic adjacencies, so reduction of this function beyond CANONICAL form is not possible. Reading the 1's in MINTERM code and the 0's in MAXTERM code results in the CANONICAL SOP and POS forms

$$f = \sum m(1,2) = \overline{A}B + A\overline{B} \qquad \text{SOP form}$$

$$f = \sum M(0,3) = (A + B)(\overline{A} + \overline{B}) \qquad \text{POS form}$$

respectively, which were given earlier by Eqs. (2-6) and (2-7) after applying DeMorgan's law [Eq. (2-16)]. The equivalence between these two CANON-ICAL forms is easily established by using the factoring law and the AND and OR laws given in Sect. 2.9.

3.4.3 Third-Order K-maps

Quite predictably, a third-order K-map is the graphical representation of a truth table for a function of three variables. However, there now arises the question of map format. Extrapolating from the development of first- and second-order K-maps, one would reasonably expect a three-variable graph-ical display to be three dimensional, that is, one logic axis for each variable. This could be done and has been. But such a graphical representation is difficult to construct and equally difficult to read. There is a much better way. In Fig. 3-15 are shown two common and equivalent K-map formats

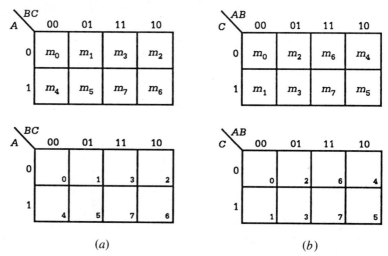

(a)   (b)

Fig. 3-15. *Two commonly used K-map formats for functions of three logic variables.*

3 / Function Representation and Reduction

for functions of three variables, both of which are based on the MINTERM code given in Fig. 3-3. Here, we continue our established practice of assigning the *most significant bit* (*MSB*) to the literal *A* and the *least significant bit* (*LSB*) to the last literal in the alphabetic sequence—*B* for two variables, *C* for three variables, and so on. So the format of Fig. 3-15(a) is assigned a MINTERM code sequence from left to right and top to bottom (normal reading direction), while the format of Fig. 3-15(b) is read top to bottom and left to right.

Some explanation of the two-literal logic axes used in Fig. 3-15 is necessary. The striking difference between the second- and third-order K-map formats is immediately apparent. The appearance of the two-literal axis for the third-order K-maps, *AB* or *BC* in Fig. 3-15, is the small cost that must be paid to avoid using a three-dimensional graphical display. In this case the logic coordinates are unfolded in the *unit distance code* order shown in Fig. 3-16(a) for coordinates *BC*. A unit distance code is one in which only one bit at a time is permitted to change between neighboring terms in the code. Now, it is observed that the 1's and 0's are logically adjacent to the 1's and 0's of neighboring pairs which has the effect of maintaining intact the *B* and *C* domains. On the other hand, unfolding the *BC* logic coordinates in the binary sequence shown in Fig. 3-16(b) would split up the *C* domain making very difficult the identification of the logic adjacencies, an essential aspect of function reduction by mapping. Note that in the case of Fig. 3-16(b) the 1's and 0's are no longer all adjacent to the 1's and 0's of neighboring pairs. For this reason the format in Fig. 3-16(b) is never used.

To help the reader gain some familiarity with three variable mapping and map reading, we present three examples in Fig. 3-17. For each of these examples the function is first represented in a truth table and then displayed in a K-map. A shaded logic map of the function is also provided but as a visual aid only, not as a substitute for the K-map. In fact, once the K-map methods are learned, shaded logic maps must no longer be used.

The three functions F_1, F_2, and F_3, represented in Fig. 3-17, are now read directly from their respective K-maps in CANONICAL SOP form by reading the 1's in MINTERM code and in CANONICAL POS form by reading the 0's in MAXTERM code, as shown:

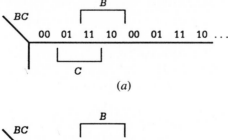

Fig. 3-16. *Two-variable logic coordinates.* (a) *Unfolded in unit distance code order.* (b) *Unfolded in standard binary code order showing split* C *domain.*

A	B	C	F_1
0	0	0	0
0	0	1	0
0	1	0	0
0	1	1	0
1	0	0	1
1	0	1	1
1	1	0	1
1	1	1	1

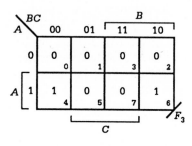

(a)

A	B	C	F_2
0	0	0	1
0	0	1	0
0	1	0	1
0	1	1	0
1	0	0	1
1	0	1	0
1	1	0	1
1	1	1	0

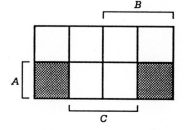

(b)

A	B	C	F_3
0	0	0	0
0	0	1	0
0	1	0	0
0	1	1	0
1	0	0	1
1	0	1	0
1	1	0	1
1	1	1	0

(c)

Fig. 3-17. *Truth tables, K-maps and shaded logic maps for three functions.*

$$F_1 = \sum m(4,5,6,7) = A\overline{B}\,\overline{C} + A\overline{B}C + AB\overline{C} + ABC$$
$$= \prod M(0,1,2,3) = (A + B + C)(A + B + \overline{C})(A + \overline{B} + C)$$
$$\cdot (A + \overline{B} + \overline{C})$$
$$F_2 = \sum m(0,2,4,6) = \overline{A}\,\overline{B}\,\overline{C} + \overline{A}B\overline{C} + A\overline{B}\,\overline{C} + AB\overline{C}$$
$$= \prod M(1,3,5,7) = (A + B + \overline{C})(A + \overline{B} + \overline{C})(\overline{A} + B + \overline{C})$$
$$\cdot (\overline{A} + \overline{B} + \overline{C})$$

(3-23)

$$F_3 = \sum m(4,6) = A\overline{B}\,\overline{C} + AB\overline{C}$$

$$= \prod M(0,1,2,3,5,7) = (A + B + C)(A + B + \overline{C})(A + \overline{B} + C)$$

$$\cdot (A + \overline{B} + \overline{C})(\overline{A} + B + \overline{C})(A + B + C) \Bigg)$$

Notice, as was pointed out for the case of second-order K-maps, that one CANONICAL form covers the remaining cells of the other form for a given function. This must be so since it is the 1's that are read to obtain the CANONICAL SOP expression while the 0's produce the CANONICAL POS form. Also, observe that $F_1 \cdot F_2 = F_3$ as is indicated by the K-maps or shaded logic maps and by the expression

$$F_1 \cdot F_2 = [\sum m(4, 5, 6, 7)] \cdot [\sum m(0, 2, 4, 6)] = \sum m(4, 6)$$

which is easily verified by using the AND laws: the two MINTERMs remaining are those common to both F_1 and F_2.

Now let us take the CANONICAL SOP expressions for the three functions given by Eqs. (3-23) and reduce them by using Boolean manipulation. The results are

$$F_1 = A\overline{B}\,\overline{C} + A\overline{B}C + AB\overline{C} + ABC$$

$$= A(\overline{B}\,\overline{C} + \overline{B}C + B\overline{C} + BC)$$

$$= A \cdot (B + \overline{B})(C + \overline{C})$$

$$= A \tag{3-23a}$$

and, similarly,

$$F_2 = \overline{A}\,\overline{B}\,\overline{C} + \overline{A}B\overline{C} + A\overline{B}\,\overline{C} + AB\overline{C}$$

$$= (\overline{A}\,\overline{B} + \overline{A}B + A\overline{B} + AB)\overline{C}$$

$$= [(A + \overline{A})(B + \overline{B})] \cdot \overline{C}$$

$$= \overline{C} \tag{3-23b}$$

and

$$F_3 = A\overline{B}\,\overline{C} + AB\overline{C}$$

$$= A(B + \overline{B})\overline{C}$$

$$= A\overline{C} \tag{3-23c}$$

But one observes that the results given by Eqs. (3-23a), (3-23b), and (3-23c) are exactly the same as those which can be read directly from the K-maps in Fig. 3-17 by grouping adjacent MINTERMs. Thus, we note that F_1 is read as "all that is A," which covers MINTERMs m_4, m_5, m_6, and m_7, all of which are logically adjacent. Similarly, F_2 maps out as "all that is NOT C" covering the logically adjacent MINTERMs m_0, m_2, m_4, and m_6, and finally F_3 is read as "all that is A AND NOT C" which covers logically adjacent MINTERMs m_4 and m_6.

The fact that MINTERM pairs (m_0, m_2) and (m_4, m_6) are easily seen to be logically adjacent results from the unit distance code order used for the BC logic coordinates shown in Fig. 3-16(a). This leads to an important conclusion:

> *Third-order K-maps whose two-variable axis is laid out in unit distance code must be continuous about the single variable axis.*

Such K-maps may be thought of as forming an imaginary cylinder about the single variable axis. The result for Fig. 3-15(a), as an example, would be that MINTERM pairs (m_0, m_2) and (m_4, m_6) are logically adjacent but MINTERM pairs (m_0, m_6) and (m_2, m_4) are not since they bear a diagonal relationship to one another. Three variable K-maps whose two-variable axes are unfolded in a binary sequence, as in Fig. 3-16(b), will not be continuous about the single variable axis and will not form an imaginary cylinder about that axis.

As final example, consider the following partially reduce function

$$Y = ABC + A\overline{C} + \overline{B}C + \overline{A}\,\overline{B} \qquad (3\text{-}24)$$

which is mapped in Fig. 3-18 by using the alternative format of Fig. 3-15(b). Applying the domain concept, MINTERM ABC is mapped as the intersection of the A, B, and C domains or MINTERM m_7, $A\overline{C}$ covers MINTERMs m_4 and m_6, $\overline{B}C$ is mapped as "all that is B NOT AND C" and covers MINTERMs m_1 and m_5, and $\overline{A}\,\overline{B}$ maps as "all that is NOT A AND NOT B" and covers MINTERMs m_0 and m_1. The two CANONICAL forms are now read directly off the K-map of Fig. 3-18 as

$$Y = \sum m(0, 1, 4, 5, 6, 7) \qquad \text{CANONICAL SOP form} \qquad (3\text{-}24a)$$

$$= \overline{A}\,\overline{B}\,\overline{C} + \overline{A}\,\overline{B}C + A\overline{B}\,\overline{C} + A\overline{B}C + AB\overline{C} + ABC$$

and

$$Y = \prod M(2, 3) \qquad \text{CANONICAL POS form}$$

$$= (A + \overline{B} + C)(A + \overline{B} + \overline{C}) \qquad (3\text{-}24b)$$

The CANONICAL SOP form given by Eq. (3-24a) can also be generated by direct expansion of Eq. (3-24) as indicated by the expression

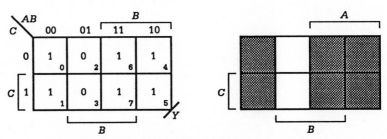

Fig. 3-18. *K-map and shaded logic map for Eq. (3-24).*

3 / Function Representation and Reduction

$$Y = ABC + A(B + \overline{B})\overline{C} + (A + \overline{A})\overline{B}C + \overline{A}\,\overline{B}(C + \overline{C})$$

but it is far easier to read the CANONICAL form from the map.

By now it should be clear that the horizontal and vertical lines (including the edges) of a K-map are all *domain boundaries*, each boundary separating some domain X from the domain of its complement $\overline{X}$. Two MINTERMs separated by a single boundary (line or edge) are logically adjacent, thus permitting the boundary variable X to be dropped from the term expressing that adjacency. If four MINTERMs are logically adjacent, two boundaries are crossed, and hence two boundary variables can be dropped, and so on, all due to the OR law $X + \overline{X} = 1$ for SOP mapping or due to the AND law $X \cdot \overline{X} = 0$ for POS mapping. This being understood, the reader should be able to read the function in Fig. 3-18 as "all that is A OR NOT B," that is, $Y = A + \overline{B}$. Much more will be said regarding logic function minimization later following a discussion of fourth-order K-maps.

3.4.4 Fourth-Order K-maps

At this point it is expected that the reader is familiar with the K-map format for first-, second-, and third-order maps. Following the same development, we present in Fig. 3-19 two commonly used formats for fourth-order K-maps where use of the MINTERM code table is implied and where A is taken to be the MSB and D the LSB. Here, both two-variable axes have logic coordinates which are unfolded in unit distance code order so that all juxtaposed MINTERMs and MAXTERMS (that is, those separated by any single map line or edge) are logically adjacent. This leads to the following important conclusion:

> *Fourth-order K-maps whose two-variable axes are both laid out in unit distance code order must be continuous about both sets of axes (horizontal and vertical).*

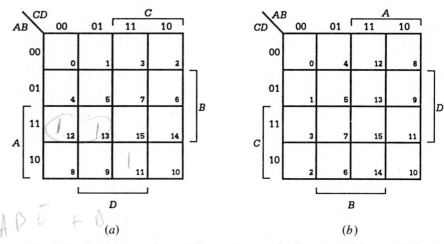

Fig. 3-19. *Two commonly used K-map formats for functions of four variables.*

Therefore, just as a third-order K-map forms an imaginary cylinder about its single variable axis, a fourth-order K-map with unit distance coded axes will form an imaginary toroid (doughnut-shaped figure), the result of trying to form two cylinders about perpendicular axes. Thus, cells (0, 8) and (8, 10) and (1, 9) in Fig. 3-19 are examples of logically adjacent pairs while cells (0, 1, 4, 5) and (3, 7, 11, 15) and (0, 2, 8, 10) are examples of logically adjacent groups of four.

An inspection of the two K-map formats of Fig. 3-19 reveals that domains for variables A, B, C, and D or their complements each cover eight adjacent MINTERM positions (cells). For example, domain A covers cells 8 through 15 while domain $\overline{C}$ covers cells 0, 1, 4, 5, 8, 9, 12, and 13. Also revealed is the fact that the intersection of any two domains covers four adjacent MINTERM positions, the intersection of any three domains covers two adjacent MINTERM positions, and the intersection of any four domains covers but one MINTERM position.

To illustrate the domain/adjacency relationship inherent in the K-maps of Fig. 3-19, consider the following possible SOP terms, where use of the MINTERM code is implied:

$A \cdot \overline{B} \cdot C$	Covers the intersection of domains A, $\overline{B}$, and C, or adjacent MINTERMs m_{10} and m_{11}.
$\overline{B} \cdot \overline{D}$	Covers the intersection of the $\overline{B}$ and $\overline{D}$ domains or adjacent MINTERMs m_0, m_2, m_8, and m_{10}.
$A \cdot B \cdot \overline{C} \cdot D$	Covers MINTERM m_{13}.
$\overline{A}$	Covers the $\overline{A}$ domain or adjacent MINTERMS m_0 through m_7.

The domain/adjacency relationship is equally valid for possible POS terms in MAXTERM code, as is demonstrated by the following examples:

$\overline{A} + B$	Covers the intersection of domains $\overline{A}$ and B or adjacent MAXTERMS M_8 through M_{11}.
$\overline{B} + C + \overline{D}$	Covers the intersection of domains $\overline{B}$, C, and $\overline{D}$ or adjacent MAXTERMS M_5 and M_{13}.
B	Covers the B domain or adjacent MAXTERMs M_0 through M_3 and M_8 through M_{11}.
$\overline{A} + B + C + \overline{D}$	Covers MAXTERM M_9.

Let us now apply what we have just learned by extracting both the SOP and POS CANONICAL forms from the following minimum function:

$$F(A,B,C,D) = \overline{A}\,\overline{B}\,\overline{D} + \overline{B}C + CD + AB\overline{C}\overline{D} \qquad (3\text{-}25)$$

Each of the terms in this expression is shown looped in the K-map of Fig. 3-20(a) and by the appropriate shaded areas in Fig. 3-20(b). Notice that the term $\overline{A}\,\overline{B}\,\overline{D}$ represents the two adjacent MINTERMs m_0 and m_2 and that $\overline{B}C$ and CD cover adjacent MINTERMs $m(2, 3, 10, 11)$ and $m(3, 7, 11, 15)$, respectively, and $AB\overline{C}\overline{D}$ is the MINTERM m_{12}. The CANONICAL SOP and POS expressions are read directly from the K-map and are

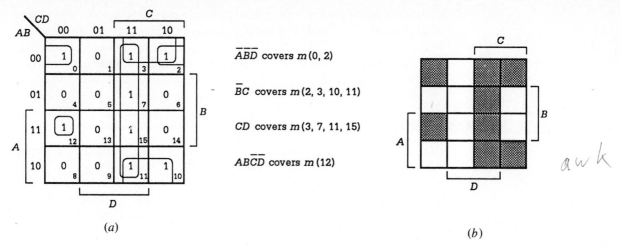

ABD covers $m(0, 2)$

$\overline{B}C$ covers $m(2, 3, 10, 11)$

CD covers $m(3, 7, 11, 15)$

$AB\overline{C}\overline{D}$ covers $m(12)$

Fig. 3-20. (a) K-map and (b) shaded logic map for the function of Eq. (3-25).

$$F = \sum m(0, 2, 3, 7, 10, 11, 12, 15) \quad \text{CANONICAL SOP}$$
$$= \overline{A}\,\overline{B}\,\overline{C}\,\overline{D} + \overline{A}\,\overline{B}C\overline{D} + \overline{A}\,\overline{B}CD + \overline{A}BCD + A\overline{B}C\overline{D}$$
$$+ A\overline{B}CD + AB\overline{C}\,\overline{D} + ABCD$$

and

$$F = \prod M(1, 4, 5, 6, 8, 9, 13, 14) \quad \text{CANONICAL POS}$$
$$= (A + B + C + \overline{D})(A + \overline{B} + C + D)(A + \overline{B} + C + \overline{D})$$
$$\cdot (A + \overline{B} + \overline{C} + D)(\overline{A} + B + C + D)(\overline{A} + B + C + \overline{D})$$
$$\cdot (\overline{A} + \overline{B} + C + \overline{D})(\overline{A} + \overline{B} + \overline{C} + D)$$

Of course, one can generate these CANONICAL forms from an expansion of the minimum expression given by Eq. (3-25). To do this, it is necessary to recognize which boundaries (if any) have been crossed. For example, $\overline{A}\overline{B}\overline{D}$ crosses the $C/\overline{C}$ boundary and can, therefore, be written as $\overline{A}\overline{B}\overline{D}(C + \overline{C})$, generating MINTERMs m_0 and m_2. Also, CD crosses both the $A/\overline{A}$ and $B/\overline{B}$ boundaries and, accordingly, is written as $(A + \overline{A})(B + \overline{B})CD$, which generates MINTERMs m_3, m_7, m_{11}, and m_{15}, and similarly for $\overline{B}C$.

But the important point is that MINTERM or MAXTERM formulation is by far more easily determined from the K-map than by function expansion. Furthermore, it must be remembered that once one CANONICAL form has been determined, the other follows immediately by the process of elimination. That is, the 1's generate the CANONICAL SOP terms and the 0's produce the CANONICAL POS terms so that the code numbers (cells) not covered by one CANONICAL form become the code numbers of the other. Algebraically, this means that

$$Y_{\text{SOP}} = Y_{\text{POS}} \tag{3-26}$$

for the CANONICAL forms of any function regardless of the complexity of that function. As will be demonstrated in the next section, the algebraic equivalence between the SOP and POS forms extends to reduced forms as well, the one exception being the case of incompletely specified functions discussed in Sect. 3-6.

3.5 K-MAP FUNCTION REDUCTION

Use of a Karnaugh map offers a simple and reliable method of reducing (often minimizing) logic expressions. In fact this is the most important aspect of a K-map. In the discussion that follows we will show how reduced functions can be read in either of two ways, depending on whether it is the 1's that are read or the 0's, and review the simple rules and procedures which must be followed in function reduction by mapping.

As was indicated in the previous section, each line or edge of a K-map forms the boundary between two domains. As a result, MINTERMs or MAXTERMs which are separated by a line or edge are logically adjacent and can be combined to form a reduced logic function. Examples of adjacency groupings and function reduction are shown in Fig. 3-21. Single adjacency pairs called *diads* are given in Figs. 3-21(a), (b), and (c), and a grouping of four adjacencies, called a *quad*, is presented in Fig. 3-21(d). For

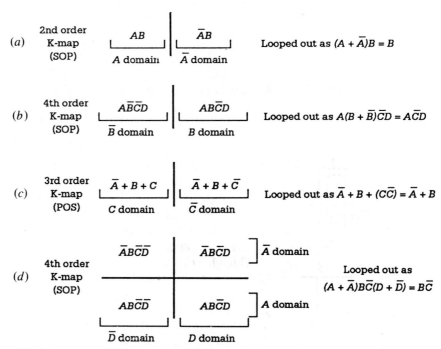

Fig. 3-21. *Examples of adjacency groupings and function reduction. (a) and (b) Reduction of MINTERM adjacency pairs (diads). (c) Reduction of a MAXTERM adjacency pair (diad). (d) Reduction of a group of four adjacent MINTERMs (quad).*

each example the loop-out (reduced) function is indicated to the right. Implied is the application of the factoring law and the OR law ($X + \overline{X} = 1$) to the SOP terms, and the application of the distributive law and the AND law ($X \cdot \overline{X} = 0$) to the POS terms. So for every boundary that is crossed in looping out MINTERM or MAXTERM adjacencies, a reduction of one variable will result. In this way 2^n adjacencies ($n = 0, 1, 2, 3, \ldots$) can be mapped (looped) out to produce an ($N - n$) variable term of an N variable function.

3.5.1 Function Reduction and the Loop-out Protocol

There is no single loop-out procedure which we consider to be a "must" procedure for logic function reduction tasks. However, to ensure reduction of a given function to its simplest form avoiding possible costly redundancies, we recommend that the largest 2^n group of logically adjacent MINTERMs or MAXTERMs be looped out in the order of increasing n ($n = 0, 1, 2, 3, \ldots$) according to the following protocol:

Loop-out Protocol

1. *Monads ($n = 0$). Single* MINTERMs or MAXTERMs which have no logic adjacencies should be looped out first.
2. *Diads ($n = 1$).* Groups of *two* logically adjacent MINTERMs or MAXTERMs which cannot be grouped in any other way to form larger 2^n groups should be looped out following the monads. A reduction of one variable for each diad will result.
3. *Quads ($n = 2$).* Any group of *four* logically adjacent MINTERMs or MAXTERMs which cannot be grouped in any other way to form a larger 2^n group should be looped out following the diads. A reduction of two variables for each quad will result.
4. *Octads ($n = 3$).* Any group of *eight* logically adjacent MINTERMs or MAXTERMs which cannot be further combined to form a hexadecad (sixteen adjacencies) should be looped out next. A reduction of three variables per octad will result.

As an example of the type of pitfalls which can result from failure to follow the loop-out protocol, consider the function represented in the map of Fig. 3-22. Instinctively, one might be tempted to first loop out the quad $m(5, 7, 13, 15)$ because it is so conspicuous. However, to do so immediately creates a redundancy since all MINTERMs of the quad are covered by the four diads required to cover MINTERMs m_3, m_4, m_9, and m_{14}—the quad is not needed and, therefore, is redundant.

Though not explicitly stated, it should be clear that groups of MINTERMs or MAXTERMs other than 2^n (e.g., three, five, six, seven) are forbidden since such groups are not continuously adjacent. Examples of improper groupings are illustrated in Fig. 3-23. The diagonal loops shown will later be used to identify XOR patterns which occur often in the design of certain devices but which do not represent logically adjacent groups.

Just as CANONICAL forms can be read from a K-map in two ways (SOP and POS), so also can a function be read from a K-map in either minimum SOP form or minimum POS form. Let us illustrate by minimizing the function

$$f = \sum m(0, 2, 3, 4, 6) \tag{3-27}$$

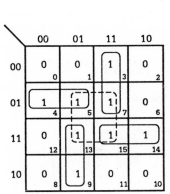

Fig. 3-22. *Minimum cover following the loopout protocol which avoids the redundant quad shown by the dashed loop.*

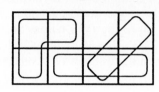

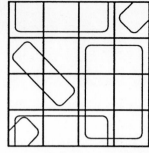

Fig. 3-23. Examples of improper groupings of MINTERMs or MAXTERMS.

in both SOP and POS form. To accomplish this, we first map the function by using MINTERM code as shown in Fig. 3-24. Then noting that the 1's are mapped out as a diad and a quad, while the 0's are looped out as two diads, we read the two minimum forms directly as follows:

Reading 1's $\qquad f = \overline{A}B + \overline{C} \qquad$ Minimum SOP cover

Reading 0's $\qquad f = (\overline{A} + \overline{C})(B + \overline{C}) \qquad$ Minimum POS cover

Two separate maps were used in Fig. 3-24 to illustrate clearly minimum SOP and POS loop cover. Had we wished to do so, however, a single K-map could have been used to obtain both minimum SOP and POS cover. If the reader wishes to use only a single map for both types of cover, it is advisable to use a two-color system or one in which continuous and dashed loops are used.

As a second example, consider the partially reduced function

$$Y = (A + \overline{B} + \overline{D})(B + \overline{C})(A + \overline{C} + D)(\overline{A} + \overline{B} + C + \overline{D}) \qquad (3\text{-}28)$$

To map this function, one simply maps 0's in MAXTERM code, as shown in Fig. 3-25(a). Remember that K-maps are formatted in MINTERM code so that the A domain in MAXTERM code is the $\overline{A}$ domain on the K-map and the $\overline{C}$ domain in MAXTERM code is the C domain on the K-map, and

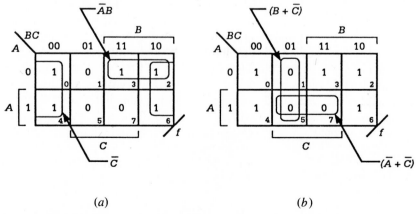

(a) $\qquad\qquad\qquad\qquad\qquad$ (b)

Fig. 3-24. K-map for Eq. (3-27). (a) SOP cover. (b) POS cover.

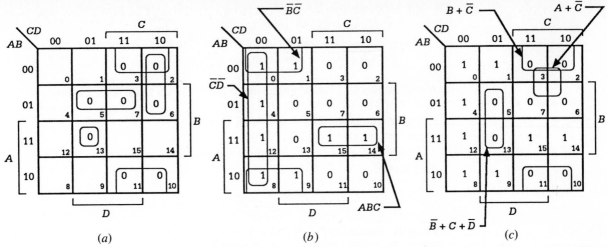

Fig. 3-25. K-maps for Equation (3-28). (a) POS cover for equation as given. (b) Minimum SOP cover. (c) Minimum POS cover.

so on. For example, the POS term $(A + \overline{B} + \overline{D})$ is the diad $M(5, 7)$, and the term $(B + \overline{C})$ is the quad $M(2, 3, 10, 11)$, all cells of which take 0's.

The minimum SOP and minimum POS cover are now read from Figs. 3-25(b) and (c), respectively, yielding

$$Y = ABC + \overline{C}\overline{D} + \overline{B}\overline{C} \qquad \text{Minimum SOP cover}$$

$$Y = (\overline{B} + C + \overline{D})(A + \overline{C})(B + \overline{C}) \qquad \text{Minimum POS cover}$$

Notice that in mapping out the 0's we begin with M_{13} since it has only one adjacency (M_5) and proceed to map out the remaining two MAXTERMs as quads. Had we not done so, we might have been tempted to add the redundant POS diad $M(5, 7)$ which appeared in Eq. (3-28). Similar reasoning applies to m_{15} and the looping of 1's, where $m(12, 14)$ would have been the redundant SOP diad.

3.5.2 Prime Implicants

Groups of 2^n adjacent MINTERMs or MAXTERMs which are sufficiently large that they cannot be combined with other 2^n adjacent groups in any way to produce terms of fewer variables are called PRIME IMPLICANTs (PIs). The loop-out protocol described in the previous section offers a procedure for achieving minimum cover by systematically extracting PIs in the order of increasing n ($n = 0, 1, 2, 3, \ldots$).

But the task of achieving minimum cover following the loop-out protocol (or any procedure for that matter) is not quite as straightforward as we have led the reader to believe. Difficulties can arise when optional and redundant groupings of adjacent MINTERMs or MAXTERMs are present. To deal with these problems, it will be helpful to identify the following three subsets of PIs:

ESSENTIAL PRIME IMPLICANTs (EPIs). Single-way PIs, which *must* be used to achieve minimum cover.

OPTIONAL PRIME IMPLICANTs (OPIs). Optional-way PIs, which are used for alternative minimum cover.

REDUNDANT PRIME IMPLICANTs (RPIs). Superfluous PIs, which cannot be used if minimum cover is to result.

Any grouping of 2^n adjacencies is an IMPLICANT, including a single MINTERM or MAXTERM, but it may not be a PI. For example, a solitary quad EPI contains eight REDUNDANT IMPLICANTs (RIs), four monads and four diads, none of which are PIs.

To illustrate a simple mapping problem with optional coverage, consider the function

$$f(A,B,C,D) = \sum m(0, 2, 4, 5, 7, 10, 11, 14, 15) \qquad (3\text{-}29)$$

which is mapped in Fig. 3-26 by using the format of Fig. 3-19(b). Noting first the MINTERM adjacencies which form diads (no SOP monads exist) and then those which form the single quad, the 1's are mapped out to yield the two optional minimum SOP expressions

$$f = \overline{A}\,\overline{B}\overline{D} + \overline{A}B\overline{C} + BCD + AC$$

and

$$f = \overline{A}\,\overline{C}\overline{D} + \overline{A}BD + \overline{B}C\overline{D} + AC \qquad (3\text{-}30)$$

which correspond to the respective maps in Figs. 3-26(a) and (b). Observe that the quad AC is the only EPI common to both expressions and that there are six OPIs, three for each K-map in Fig. 3-26. Of course, once one alternative minimum cover has been selected, none of the remaining OPIs can be used for minimum cover in that expression. Referring to Eqs. (3-30), we see that the OPI diads given in one expression become RPI diads of the other.

The minimum POS representation for Eq. (3-29) does not have optional cover as is evident from an inspection of Fig. 3-26. Following the loop-out protocol, we read the 0's as the monad M_6, the diad $M(1, 3)$ since M_3 has

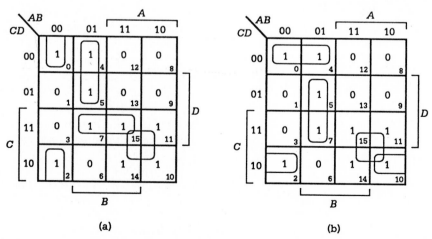

(a)   (b)

Fig. 3-26. *Optional minimum cover for Eq. (3-29).*

3 / Function Representation and Reduction

but one adjacency, and the quad $M(8, 9, 12, 13)$ to yield the minimum POS cover

$$f = (A + \overline{B} + \overline{C} + D)(A + B + \overline{D})(\overline{A} + C) \qquad (3\text{-}31)$$

which obviously contains only EPIs. Had we not followed the loop-out protocol, we might have been tempted to include the RPI diad $M(1, 9)$ read as $(B + C + \overline{D})$. Observe that the minimum POS cover given by Eq. (3-31) is simpler than the minimum SOP cover represented by Eqs. (3-30). That is, the POS expression involves the ANDing of three ORed terms, whereas the SOP forms require the ORing of four ANDed terms.

3.6   INCOMPLETELY SPECIFIED FUNCTIONS (DON'T CARES)

In the design of logic circuits *nonessential* MINTERMs or MAXTERMs may be introduced so as to simplify the circuit. Such nonessential MINTERMs or MAXTERMs are called *don't cares* (take them or leave them) which we represent by the symbol

$$\phi = \text{MIN/MAX} = \text{don't care}$$

Thus, the don't care can be taken as logic 0 or logic 1, take your choice. Other commonly used symbols for don't cares include X, $(-)$, and d. We will reserve the symbol X for use as an irrelevant input and use ϕ to represent a nonessential MINTERM or MAXTERM in the output function. The symbol ϕ may be thought of as a logic 0 with a logic 1 superimposed on it.

Don't cares can arise under the following two conditions:

1. When certain combinations of input logic variables never occur, the output function for such combinations are nonessential and are assigned don't cares.
2. When all combinations of input logic variables occur but certain combinations of these variables are nonessential, the output function for such combinations are assigned don't cares.

As an example of the second condition, any complete set of 8421 binary-coded decimal (BCD) numbers represents decimals 0 through 9 with the result that there are six combinations of 1's and 0's, equivalent to decimals 10 through 15, which are never used. That is, we "don't care" about them and, accordingly, will assign the symbol ϕ to their MINTERMs (or MAXTERMs) in a fourth-order K-map. A reader who is unfamiliar with the BCD number system, should refer to Fig. 3-3 where the binary numbers equivalent to decimal 0 through 9 become the 8421 BCD numbers, the remaining six being unused or irrelevant.

In looping out a K-map with ϕ's, one should make use of the don't cares if, and only if, their use results in greater simplification of the function. This is important since to use don't cares unnecessarily would be to create RPIs in the output expression. This leads to design redundancies which, on a large scale, can be costly.

To illustrate the use of don't cares, consider the three-variable function

$$f(A,B,C) = \underbrace{\sum m(0, 1, 5, 7)}_{\substack{\text{Essential} \\ \text{MINTERMs}}} + \underbrace{\phi(2, 4)}_{\substack{\text{Nonessential} \\ \text{MINTERMs} \\ \text{(don't cares)}}} \qquad (3\text{-}32)$$

written in CANONICAL SOP form showing essential MINTERMs and non-essential MINTERMs (don't cares). Without regard to how this function came about, let us map it in MINTERM code and then reduce it to minimum form following the loop-out protocol. The K-maps representing minimum SOP cover and minimum POS cover are presented in Fig. 3-27. For the SOP cover in Fig. 3-27(a), the diad $m(5, 7)$ is looped out first since m_7 has only one adjacency (m_5). This leaves MINTERMs m_0 and m_1 to be looped out as a quad by using the don't care in cell 4 (ϕ_4). Notice that it is not necessary to use ϕ_2 and that diad $m(0, 1)$ is made redundant due to the presence of ϕ_4. Similarly, for the POS cover given in Fig. 3-27(b), the 0's are mapped out in MAXTERM code as diads $M(2, 3)$ and $M(4, 6)$ thereby utilizing both don't cares. The resulting SOP and POS expressions representing minimum cover are

and
$$f = AC + \overline{B} \qquad \text{SOP}$$
$$f = (A + \overline{B})(\overline{A} + C) \qquad \text{POS} \qquad (3\text{-}33)$$

Observe that algebraically $f_{\text{POS}} \neq f_{\text{SOP}}$, a direct result of the existence and shared use of don't cares in obtaining SOP and POS cover. Still, the two expressions of Eqs. (3-33) are logically equivalent since they will yield the same output logic level (0 or 1) for any given set of input conditions exclusive of that corresponding to don't care $\phi_4 = 100$.

Had we chosen not to use the two don't cares in Eq. (3-32), the extent of function reduction would have been significantly less as in evident from the result

$$f_{\text{SOP}} = AC + \overline{A}\,\overline{B}$$
$$f_{\text{POS}} = (A + \overline{B} + \overline{C})(\overline{A} + \overline{B} + C) \qquad (3\text{-}34)$$

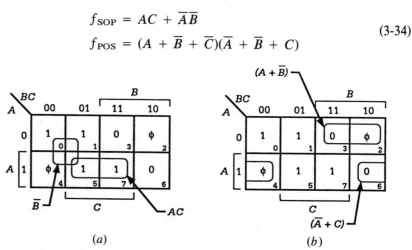

(a)   (b)

Fig. 3-27. *K-maps for Eq. (3-32). (a) Minimum SOP cover. (b) Minimum POS cover.*

In this case the diad $m(0, 1)$ is used in place of the quad $m(0, 1, 4, 5)$ in the SOP expression, and the monads M_3 and M_4 are looped out instead of the diads in the POS expression. Thus, the results expressed by Eqs. (3-33) are preferred over those of Eqs. (3-34) because Eqs. (3-33) are simpler.

Consider, as a second example, the following four-variable function in CANONICAL POS form showing the ESSENTIAL and nonessential MAXTERMs:

$$Y(A, B, C, D) = \underbrace{\prod M(0, 1, 4, 6, 8, 14, 15)}_{\substack{\text{Essential} \\ \text{MAXTERMs}}} \cdot \underbrace{\phi(2, 3, 9)}_{\substack{\text{Nonessential} \\ \text{MAXTERMs} \\ \text{(don't cares)}}} \quad (3\text{-}35)$$

In Fig. 3-28 we map the 0's and ϕ's in MAXTERM code and subsequently reduce the function to its simplest POS and SOP forms following the loop-out protocol described earlier. The resulting POS and SOP expressions, representing minimum cover for Eq. (3-35), are

$$
\begin{aligned}
Y_{\text{POS}} &= (\overline{A} + \overline{B} + \overline{C})(B + C)(A + D) \\
Y_{\text{SOP}} &= AB\overline{C} + \overline{A}BD + \overline{B}C
\end{aligned} \quad (3\text{-}36)
$$

Again we notice that while these expressions are logically equivalent, they are also algebraically unequal due to the shared use of don't cares in the minimization process.

Let us review the procedure used in obtaining the results given by Eqs. (3-36). In Fig. 3-28(a), showing POS cover, the diad $M(14, 15)$ is looped out first as an EPI since M_{15} has but one adjacency (M_{14}). Next, MAXTERM M_8 is combined with MAXTERMs M_0, M_1, and ϕ_9, to form the quad EPI $(B + C)$, thereby eliminating the need to use the diad $M(0, 8)$. The remaining

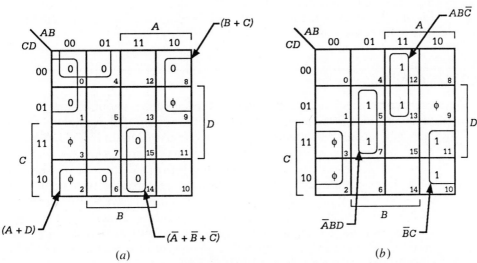

Fig. 3-28. *K-maps for the function of Eq. (3-35). (a) Minimum POS cover. (b) Minimum SOP cover.*

3.6 / Incompletely Specified Functions (Don't Cares)

131

essential MAXTERMs, M_4 and M_6, are looped out as a quad by using ϕ_2 together with M_0 as shown. Notice that the don't care ϕ_3 is left unused, that there are no OPIs present, and that the quad $M(0, 1, 2, 3)$ is an RPI.

The minimum SOP cover given in Eqs. (3-36) is found in similar fashion. Referring to the K-map in Fig. 3-28(b), we see that the MINTERM m_{12} has only one adjacency, m_{13}, with which it is looped out to give the diad EPI $A B \overline{C}$. This forces use of diad grouping $m(5, 7)$ and avoids the use of the RPI diad $m(5, 13)$. The remaining two MINTERMs, m_{10} and m_{11}, are looped out with ϕ_2 and ϕ_3 to form the quad $\overline{B} C$. Again notice that one don't care (ϕ_9) remains unused.

The Gate/Input Tally. A quantity called the *gate/input tally* is used as a measure of the complexity of a logic expression or logic circuit. If there are no fan-in limitations, the gate/input tally (exclusive of INVERTERs) can be obtained directly from an SOP expression. The gate tally is obtained by counting the number of Boolean PRODUCT-terms (*p*-terms) and by adding one to this count for their Boolean SUM (hence, *p*-terms + 1). The input tally is determined by counting all *p*-term variables and by adding to that count the number of *p*-terms to be ORed. Similarly, if no fan-in limitations exist, the gate/input tally can be obtained directly from a POS expression. In this case the gate tally is determined by counting the number of Boolean SUM-terms (*s*-terms) and by adding one to that count for their BOOLEAN PRODUCT (hence, *s*-terms + 1). The input tally is obtained by counting all *s*-term variables and by adding to that count the number of *s*-terms to be ANDed. If the ACTIVATION LEVELs for the inputs are known, then IN-VERTERs can be included in the gate/input tally for either an SOP or POS expression. To do this, one counts the number of unique complemented inputs arriving ACTIVE HIGH (or unique uncomplemented inputs arriving ACTIVE LOW) and adds this count to both the gate and input tally for the expression.

As an example of the use of the gate/input tally, consider the expressions in Eqs. (3-33). The gate/input tally for the SOP expression is 2/4 while that for the POS expression is 3/6, both tallies being exclusive of INVERTERs. The SOP expression is seen to be the simpler of the two.

3.7 MULTIPLE-OUTPUT MINIMIZATION

Frequently, logic design problems require minimization of multiple-output functions all of which are functions of the same input variables. This is generally regarded as a tedious task to accomplish without the aid of a computer, and for this reason computer programs have been written to obtain the optimum cover for multioutput functions. In this section we will describe a simple search method which will illustrate one process used for multiple output minimization. Our treatment will be limited to systems of three outputs or fewer, each output being a function of four variables or less.

Consider the four-input/three-output system illustrated by the block diagram in Fig. 3-29. Let us suppose that the object is to minimize each of the three functions in such a way as to retain as many shared terms between them as possible, thus optimizing the combinational logic of this system.

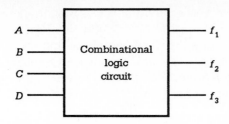

Fig. 3-29. Block diagram for a four-input, three-output combinational logic system.

The procedure we recommend is given in four steps:

Step 1. Map each function to obtain the CANONICAL SOP or POS forms. If the CANONICAL forms are already known, skip to step 2.

Step 2. AND the CANONICAL SOP forms or OR the CANONICAL POS forms in some systematic way (for example, $f_1 \cdot f_2$, $f_2 \cdot f_3$, $f_3 \cdot f_4, \ldots$, or $f_1 + f_2$, $f_2 + f_3$, $f_3 + f_4, \ldots$) and map each ANDed or ORed expression separately, looping out all shared PIs (common terms).

The MINTERM ANDing rules are given by

$$m_i \cdot m_i = m_i$$
$$m_i \cdot m_j = 0 \qquad (i \neq j) \tag{3-37}$$

and the MAXTERM ORing rules are

$$M_i + M_i = M_i$$
$$M_i + M_j = 1 \qquad (i \neq j) \tag{3-38}$$

where Eqs. (3-37) and (3-38) are duals of each other.

Step 3. Make a table of the results of step 2 giving all shared PIs in literal form.

Step 4. Map the original functions, if they have not been mapped previously, then immediately loop out (in the appropriate maps) the shared PIs given in step 3. Finally, loop out the remaining EPIs following the loop-out protocol with one possible exception. If the adjacencies of the shared PIs are part of a larger 2^n grouping of adjacencies, use the larger grouping but only if it leads to a simpler form.

A shortened version of the foregoing optimization procedure may also work effectively, especially if the functions are relatively simple. To do this one loops out minimum cover for each function while, at the same time, comparing them for common terms to be used later in implementing the system. This process may involve the sacrifice of absolute minimum cover for any given function so as to yield the common terms necessary for an optimized system. An example would be the use of two diads instead of a quad because one of the diads is common to two other output functions. Decisions of this type are best made on K-maps and are, of course, discretionary and applicable to either version (long or short) of the optimization procedure.

To illustrate the simplicity of the four-step procedure given, let us find a set of functions that will optimally cover the three output expressions given below, each a function of three variables A, B, and C.

$$f_1 = \sum m(0, 3, 4, 5, 6)$$

$$f_2 = \sum m(1, 2, 4, 6, 7) \tag{3-39}$$

$$f_3 = \sum m(1, 3, 4, 5, 6)$$

Since the expressions given by Eqs. (3-39) are already in CANONICAL SOP form, we may skip to step 2 and continue through step 4: the ANDed output functions, each with their respective K-map and minimum cover, are shown in Fig. 3-30. The minimum cover for each ANDed K-map indicates the common terms which must be included in the optimized expressions. Next, we construct a table of shared PIs for each of the ANDed forms and transfer these shared terms onto the K-maps for the original functions, as in Fig. 3-31. Notice that the diad $A\overline{C}$ is common to all three ANDed functions, a conclusion we would have arrived at had we chosen to include the ANDed function $f_1 \cdot f_2 \cdot f_3 = m(4, 6)$ in Fig. 3-31.

By looping out the shared PIs first in Fig. 3-31 then all remaining EPIs next, there results the optimal expressions

$$f_1 = \overline{A}BC + A\overline{C} + A\overline{B} + \overline{B}C$$

$$f_2 = \overline{A}\overline{B}C + A\overline{C} + AB + B\overline{C} \tag{3-40}$$

$$f_3 = \overline{A}BC + \overline{A}\overline{B}C + A\overline{C} + A\overline{B}$$

Here it is observed that the diad $m(1, 3)$ in the f_3 map is avoided so that the expression for f_3 can be completely generated from the terms in f_1 and f_2, the optimal solution to this problem. The optimum gate/input tally is 10/28 for this system of three functions.

When the original output functions contain don't cares, then, for optimal multifunction cover, one proceeds as done earlier but with the following additional ANDing and ORing rules:

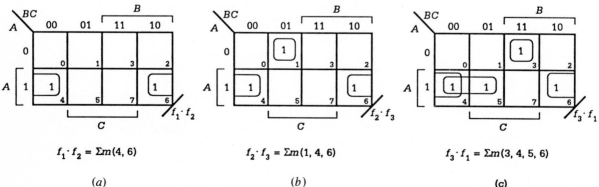

$$f_1 \cdot f_2 = \Sigma m(4, 6)$$

(a)

$$f_2 \cdot f_3 = \Sigma m(1, 4, 6)$$

(b)

$$f_3 \cdot f_1 = \Sigma m(3, 4, 5, 6)$$

(c)

Fig. 3-30. *ANDed functions, their K-maps and minimum cover for Eq. (3-39).*

ANDed Functions	Shared Prime Implicants
$f_1 \cdot f_2$	$A\overline{C}$
$f_2 \cdot f_3$	$\overline{A}\overline{B}C, \ A\overline{C}$
$f_3 \cdot f_1$	$\overline{A}\overline{B}C, \ A\overline{C}, \ A\overline{B}$

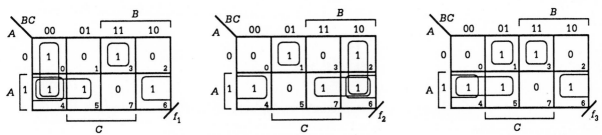

Fig. 3-31. *Table of shared PIs and the K-maps for the original functions showing optimal cover.*

$$m_i \cdot \phi_i = m_i$$

$$\phi_i \cdot \phi_i = \phi_i \qquad (3\text{-}41)$$

$$m_i \cdot \phi_j = \phi_i \cdot \phi_j = 0 \qquad (i \neq j)$$

and their dual

$$M_i + \phi_i = M_i$$

$$\phi_i + \phi_i = \phi_i \qquad (3\text{-}42)$$

$$M_i + \phi_j = \phi_i + \phi_j = 1 \qquad (i \neq j)$$

The use of the ANDing rules of Eqs. (3-41) may be demonstrated by finding the optimal cover for the functions

$$f_1 = \sum m(0, 3, 7, 8, 9, 10, 14) + \phi(1, 5, 11, 12, 15)$$

$$f_2 = \sum m(2, 4, 7, 8, 9, 11, 13, 15) + \phi(3, 5, 12) \qquad (3\text{-}43)$$

Beginning with steps 2 and 3, we find the ANDed function and its minimum cover as shown in Fig. 3-32. Observe that neither don't care (ϕ_5 or ϕ_{12}) is used in looping out the minimum cover for the ANDed function.

Now we proceed to step 4 where we map the original functions as shown in Fig. 3-33. In this case, the optimal cover is achieved by avoiding the shared diad $m(8, 9)$ in favor of the quads $m(0, 1, 8, 9)$ and $m(8, 9, 12, 13)$ for f_1 and f_2, respectively. Thus, only the shared PI, quad CD, is preserved in the optimization process. Looping out the EPIs shown in each map yields the optimal SOP results

$$f_1 = CD + A\overline{D} + \overline{B}\overline{C}$$

$$f_2 = CD + \overline{A}\overline{B}C + B\overline{C} + A\overline{C} \qquad (3\text{-}44)$$

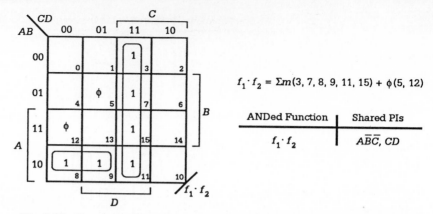

$$f_1 \cdot f_2 = \Sigma m(3, 7, 8, 9, 11, 15) + \phi(5, 12)$$

ANDed Function	Shared PIs
$f_1 \cdot f_2$	$A\overline{B}\overline{C}, CD$

Fig. 3-32. *K-map, ANDed functions, and table of shared PIs for Eqs. (3-43).*

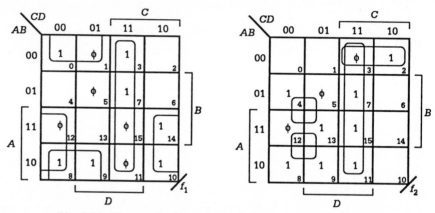

Fig. 3-33. *K-maps for Eqs. (3-43) showing optimal SOP cover.*

where each expression represents, in this case, minimum cover. Notice that minimum cover for f_1 can also be achieved by use of the quads $m(1, 3, 5, 7)$ and $m(10, 11, 14, 15)$, but their use would preclude the use of the shared PI, CD, and optimum cover would not result. The optimum gate/input tally for this system of two functions is 8/20.

The simple search method just outlined becomes quite tedious when applied to multiple-output systems more complicated than those just described. For example, a four-input/four-output SOP optimization problem would require ten ANDed fourth-order K-maps, including one for each of six ANDed pairs ($f_1 \cdot f_2$, $f_2 \cdot f_3$, $f_3 \cdot f_4$, $f_4 \cdot f_1$, $f_2 \cdot f_4$, $f_3 \cdot f_1$), and a map for each of four ANDed triplets ($f_1 \cdot f_2 \cdot f_3$, $f_2 \cdot f_3 \cdot f_4$, $f_3 \cdot f_4 \cdot f_1$, $f_4 \cdot f_1 \cdot f_2$). An eleventh map could also be constructed for $f_1 \cdot f_2 \cdot f_3 \cdot f_4$, but such "end" maps are not usually necessary since sufficient information to complete the optimization most likely already exists. Two points should be made at this time. First, the table of ANDed functions versus shared PIs becomes a very important part of such a search process due to the obvious complexity of the problem—it becomes a matter of good "bookkeeping." Second, complex systems are best optimized either by the shortened version described earlier, or better perhaps, by computer methods. Quite obviously,

3 / Function Representation and Reduction

if a guaranteed optimum solution is required for a complex system, a computer optimization program must be used.

Optimum cover, as used here, means the least number of gates required for implementation. Since this includes INVERTERs, a knowledge of the input ACTIVATION LEVELs is necessary. For alternative designs requiring an equal number of gates, optimum coverage could be determined by practical considerations such as fan-in limitations.

3.8   ENTERED VARIABLE MAP REDUCTION

Conspicuously absent in the foregoing discussion on mapping and function reduction was the treatment of function reduction in K-maps of lesser order than the number of variables of the function. An example of this would be the function reduction of five or more variables in a fourth-order K-map. In this section we will address these problems by presenting the subject of *entered variable (EV) mapping*, which is a "logical" and very useful extension of the conventional methods (using 1's and 0's) developed earlier.

Properly used, an EV K-map can significantly facilitate the function reduction process. But function reduction is not the only use to which EV K-maps can be put advantageously. Frequently, the specifications of a logic design problem lends itself quite naturally to EV map representation from which useful information can be obtained directly. Many examples of this will be provided in subsequent chapters. In fact, EV mapping will be seen as the most common form of graphical representation used in this text.

In addition, there is another way in which EV mapping finds significant usage. Up to now we have looped out of K-maps minimum cover in either pure SOP or pure POS form. Occasionally, it is highly advantageous to map out EV functions so as to produce hybrid (mixed) forms of representation. Such EV functions may represent subsystems, the mixed form of which is required to be preserved in the overall expression. To "purify" the Boolean form as an intermediate step only for the purpose of using conventional mapping methods would be, in this case, an unnecessary, time-consuming, and even error-prone process. In Chapter 4 we will have several opportunities to demonstrate these and other useful features of EV mapping.

3.8.1   Origin of Map Entered Variables

Simply stated, *map entered variables (MEVs)* originate when a conventional N^{th}-order K-map is compressed to one of lesser order. It will be recalled that conventional K-maps are constructed by using the logic variables (literals or mnemonics) as logic axes with 1's and 0's entered in the appropriate cells. But if a function of N variables is compressed into a K-map of order $n < N$, then terms of $(N - n)$ variables must be entered into the appropriate cells of the nth-order K-map. In effect, each cell of the nth-order map now becomes a submap of the $(N - n)^{th}$ order, hence maps within maps.

To illustrate the map compression process and the MEVs that result, consider the three-variable function

$$f(A,B,C) = \sum m(2, 5, 6, 7) \tag{3-45}$$

which we have placed on a truth table and mapped into a third-order K-map, as shown in Fig. 3-34. The truth table is divided into four two-MIN-TERM sections, each represented by a first-order submap in variable C placed at the right of each section. The submaps are read as 0, $\overline{C}$, C, and 1, respectively, for code pairs (0, 1), (2, 3), (4, 5), and (6, 7). That the four submaps collectively form the third-order K-map of Fig. 3-34(b) is easily seen by comparing the submap cells with those having the same MINTERM code decimals as in the third-order map.

The three-variable function of Eq. (3-45) is now compressed into a second-order truth table and K-map simply by entering the four submap functions of Fig. 3-34(a) into the appropriate table positions and map cells provided in Fig. 3-35. Truth tables of the type shown in Fig. 3-35(a) are called, not surprisingly, *EV truth tables*. The cell numbers for the second-order map of Fig. 3-35(b) are given at the extreme left of the EV truth table of Fig. 3-34(a) and are the decimal equivalents of the AB logic coordinates. Note that the single literal entries here represent diads, not monads.

The reduced SOP and POS functions can now be read directly from the EV K-map of Fig. 3-35(b) by following the loop-out protocol given earlier. However, to do so one must understand that in this map $1 = C + \overline{C}$ and $0 = C \cdot \overline{C}$, since C is the only map entered variable. In Fig. 3-36 we present the K-maps showing minimum SOP cover and minimum POS cover which have been looped out in agreement with the logic adjacency laws [Eqs. (3-15) and (3-16) given in Sect. 3.3]. In the SOP map the logic 1 (meaning $C + \overline{C}$) is covered since it is used to loop out both the C and $\overline{C}$ terms in MINTERM code. Similarly, the logic 0 (meaning $C \cdot \overline{C}$) is covered since it

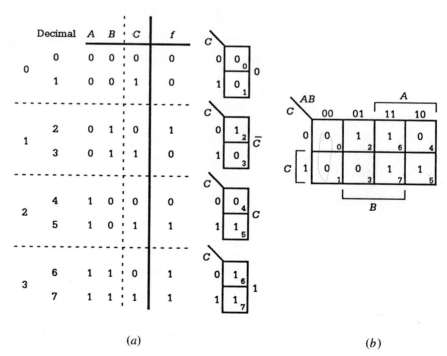

(a)   (b)

Fig. 3-34. *(a) Truth table and submaps, (b) standard K-map for Eq. (3-45).*

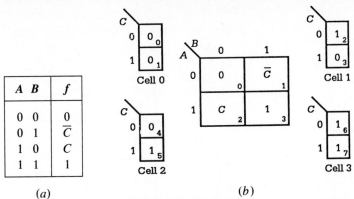

A	B	f
0	0	0
0	1	$\overline{C}$
1	0	C
1	1	1

(a)

(b)

Fig. 3-35. *Entered variable representation of the function given by Eq. (3-45). (a) Second-order EV truth table. (b) Second-order EV K-map showing the submaps for each cell.*

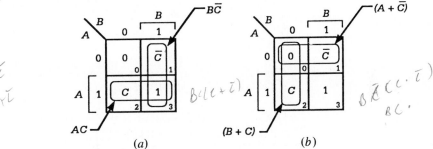

(a)

(b)

Fig. 3-36. *Entered variable K-map minimization of the function given by Eq. (3-45). (a) Minimum SOP cover. (b) Minimum POS cover.*

is used to loop out both the C and $\overline{C}$ POS terms in MAXTERM code. The resulting minimum SOP and POS expressions are given by

$$f = AC + B\overline{C} \qquad \text{SOP}$$
$$= (A + \overline{C})(B + C) \qquad \text{POS} \tag{3-46}$$

As an exercise the reader should verify Eqs. (3-46) by looping out the minimum SOP and POS cover from the conventional third-order K-map of Fig. 3-34(b).

To illustrate map compression further, let us take the four-variable ($N = 4$) function

$$Y(A, B, C, D) = \prod M(0, 1, 6, 8, 9, 11, 14, 15) \tag{3-47}$$

and compress it into a third-order ($n = 3$) K-map. Recalling that submaps of order ($N - n$) are entered into the cells of the nth order map, we can (and should) bypass the truth table stage and proceed directly to the K-maps. Shown in Fig. 3-37 are the standard fourth-order map and third-order EV map which derive from the function given by Eq. (3-47). The EV map of Fig. 3-37(b) is constructed by entering first-order ($N - n = 1$) submap functions into each cell as indicated. For example, the submap for MAX-

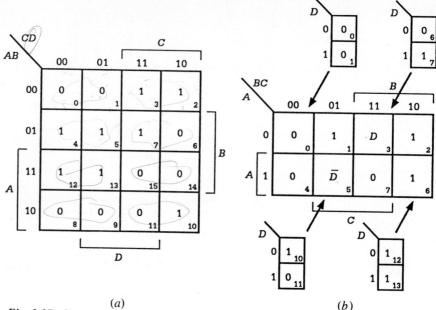

(a)   (b)

Fig. 3-37. *K-map representation of the function given by Eq. (3-47). (a) Standard K-map. (b) Third-order EV K-map showing sample submaps.*

TERMs M_0 and M_1 requires that a logic 0 (meaning $D \cdot \overline{D}$) be entered in cell 0, the submap for MINTERMs m_2 and m_3 requires that a logic 1 (meaning $D + \overline{D}$) be placed in cell 1, the submap for M_6 and m_7 requires that the EV (D) be entered in cell 3, the submap for m_{10} and M_{11} requires that the EV ($\overline{D}$) be entered in cell 5, and so on. Collectively, the eight first-order submaps, represented by the third-order EV map, form the fourth-order map, as is evident by comparing the submap cells with those of the same coded cell numbers in the fourth-order map given by Fig.3-37(a).

Let us now loop out the minimum cover for the function represented by the third-order EV map of Fig. 3-37(b). In Figs. 3-38(a) and (b) are given the K-maps representing minimum SOP and POS cover, respectively. The results, looped out as shown, are

$$
\begin{aligned}
Y &= \overline{B}C\overline{D} + \overline{A}CD + B\overline{C} & \text{SOP} \\
&= (\overline{B} + \overline{C} + D)(\overline{A} + \overline{C} + \overline{D})(B + C) & \text{POS}
\end{aligned}
\qquad (3\text{-}48)
$$

where the terms of both forms are written in the order required by the loop-out protocol. Observe that the logic 1 in cell 1 of Fig. 3-38(a) is covered by the two SOP diads $\overline{A}CD$ and $\overline{B}C\overline{D}$ and that the logic 0 in cell 7 of Fig. 3-38(b) is covered by the diads $\overline{B} + \overline{C} + D$ and $\overline{A} + \overline{C} + \overline{D}$. It is easily shown that minimum cover for the fourth-order map of Fig. 3-37(a) yields the same results given by Eqs. (3-48).

So as to make a rather important point, we carry the function reduction process one step further. We compress the four variable function of Eq. (3-47) into a second-order K-map by entering into its four cells the reduced functions (in variables C and D) from submaps of the second order as indicated in Fig. 3-39. Here, it is seen that the SOP and POS maps are identical

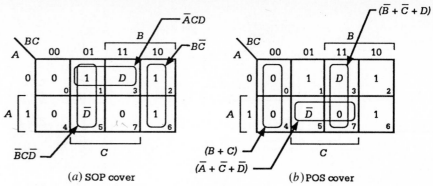

(a) SOP cover   (b) POS cover

Fig. 3-38. *Third-order EV K-maps for the four-variable function represented by Eq. (3-47) showing minimum SOP and POS cover.*

but are looped out differently in agreement with the MINTERM and MAX-TERM codes. Note that the EV subfunctions in Fig. 3-39 could have been obtained from first-order submaps derived from the third-order EV K-maps in Fig. 3-38.

The reduced SOP and POS cover for the K-maps of Fig. 3-39 is

$$Y = \overline{B}C\overline{D} + \overline{A}\,\overline{B}C + \overline{A}BD + B\overline{C} \qquad\qquad \text{SOP} \qquad (3\text{-}49)$$
$$= (\overline{B} + \overline{C} + D)(\overline{A} + \overline{B} + \overline{C})(\overline{A} + B + \overline{D})(B + C) \quad \text{POS}$$

Comparing Eqs. (3-49) with those of Eqs. (3-48) clearly indicates that Eqs. (3-49) are not minimum forms, but they are greatly reduced relative to the CANONICAL expression of Eq. (3-47). It is true that absolute minimum SOP and POS cover could have been read from the second-order EV K-maps of Fig. 3-39 had we recognized from the submaps that the diad $\overline{A}CD$ covers the terms $\overline{A}BD$ and $\overline{A}\,\overline{B}C$ for the SOP map and that the diad $\overline{A}$ +

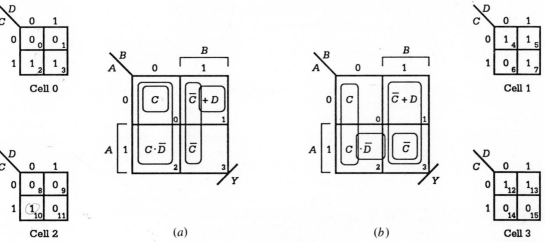

(a)   (b)

Fig. 3-39. *Second-order EV K-maps and second-order submaps for the four-variable function given by Eq. (3-47). (a) Reduced SOP cover. (b) Reduced POS cover.*

3.8 / Entered Variable Map Reduction   **141**

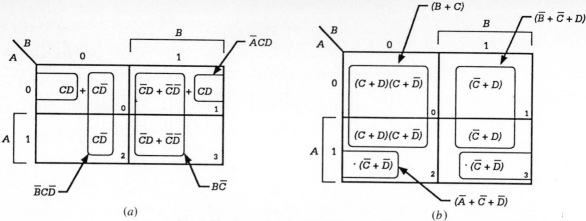

Fig. 3-40. *Second-order EV K-maps containing CANONICAL subfunctions of Eq. (3-47). (a) Minimum SOP cover. (b) Minimum POS cover.*

$\overline{C} + \overline{D}$ covers the terms $\overline{A} + B + \overline{D}$ and $\overline{A} + \overline{B} + \overline{C}$ for the POS map. Here, we have chosen to utilize the simplicity of the reduced entries rather than reading the CANONICAL data in the submaps, but at the sacrifice of absolute minimum cover. Apparently, by looping out reduced EV functions in a compressed map, there results a tendency to depart increasingly from absolute minimum form with increasing compression, that is, with increasing submap order $(N - n)$. As we shall see next, the EV mapping methods can, in fact, be used to obtain an absolute minimum form, provided we make the effort to alter the form of the EV terms.

Instead of entering reduced EV functions into the second-order map cells, as was done in Fig. 3-39, we can enter MINTERMs (or MAXTERMs) into the appropriate cells. Reading the submaps in Fig. 3-39, one obtains the second-order EV maps shown in Fig. 3-40 and the minimum functions of Eqs. (3-47). Notice that once the loop-out process has been completed, the subfunctions must be reduced to obtain the minimum terms $B\overline{C}$ and $B + C$ in the SOP and POS maps, respectively, a procedure that must be followed if an absolute minimum form is to be obtained. With a little practice, the reader can use the submaps of Fig. 3-39 to accomplish the same results as in Fig. 3-40 but without the need to reenter CANONICAL data—the submaps already represent CANONICAL data.

The point to be made here is that it is not always desirable to require a guaranteed absolute minimum function. A greatly reduced function may suffice for the particular design problem and may be obtained with minimum effort. However, if it is necessary to obtain a guaranteed absolute minimum function from an EV K-map, then it may be necessary that all entries be in CANONICAL MINTERM or MAXTERM form. In this case the use of submaps can advantageously reduce the work required.

3.9 PRACTICAL EV MAPPING CONSIDERATIONS

At this time we need to clarify a few points which ostensibly emerge from the previous section. First, we ask, "Why would one want to compress the four-variable function of Eq. (3-47) by two orders into a second-order map,

3 / Function Representation and Reduction

as in Fig. 3-39 or Fig. 3-40?'' The answer is that one probably would never do this. In fact, given a four-variable function in CANONICAL or reduced form, it is common practice to loop out minimum cover from a fourth-order map, or perhaps occasionally a third-order map, but not one of second order. However, there does arise the occasion when the design specifications of a problem lead quite naturally to a multiorder compression. For example, a four-variable function might naturally be represented in a second-order EV truth table as a result of the design specifications. In this case transferring the information from a second-order truth table to a second-order K-map for function reduction is the simplest and most reasonable approach to take. We will provide many examples of the use of multiorder compression methods in the remaining chapters.

A second point of clarification needs to be made with regard to the subject of EV mapping and the requirement for a guaranteed absolute minimum. The power and utility of the EV map method lies in the simplicity with which reduced forms can be obtained. To use the EV method to achieve absolute minimum form for complex functions when there are other more efficient means available (e.g., computer methods) would be improper. Likewise, it would be improper to use computer methods to obtain function minima which are more easily obtained with pencil and paper. Therefore, the following statement can be made:

> *Unless there is a compelling reason to do otherwise, all subfunctions entered into EV K-maps should be reduced or minimized forms with the goal of achieving reduced cover, not necessarily absolute minimum cover.*

Map compressions of one order often lead to absolute minimum cover even when reduced subfunction entries are used. But for most EV K-maps absolute minimum cover is *not* guaranteed, even if the loop-out protocol is strictly followed.

Also, there is the question of how the loop-out protocol is affected by the presence of entered variables, if at all. Actually, the loop-out protocol, as given in Sect. 3.5, remains in effect but *extended* in the following way:

> *In looping out EV K-maps, first loop out all EVs following the loop-out protocol outlined in Sect. 3-5; then loop out the 1's for SOP representation or the 0's for POS representation as a "clean-up" operation, also following the loop-out protocol.*

Thus, the loop-out protocol may be thought of as being applied twice for EV K-maps, first to EVs and then to the 1's or 0's. The reason for doing this is simply to avoid possible redundant cover which is the reason for using the loop-out protocol in the first place. As will be shown in the following discussion, it happens occasionally that a logic 1 (or logic 0) will be covered by using it to loop out EV adjacencies. Only when the 1's or 0's are not completely covered will one want to loop them out as a clean-up operation. To do otherwise would create redundant cover, a condition to be avoided.

A useful observation emerges from the map compression process described earlier. As we have already seen, each cell of the compressed nth-order map represents a submap of order $(N - n)$ for an $N > n$ variable function. Furthermore, it is also observed that each submap covers 2^{N-n} possible MINTERMs (or MAXTERMs). This leads us to conclude that any compressed n^{th} order K-map, representing a function of $N > n$ variables, has a MAP KEY defined by

$$\text{MAP KEY} = 2^{N-n} \qquad N > n \qquad (3\text{-}50)$$

The MAP KEY has the special property that when it is multiplied by a cell code number of the compressed n^{th}-order map, there results the code number of the first MINTERM or MAXTERM possible for that cell. For future reference we will call the statement of Eq. (3-50) the *MAP KEY relation* and express it in the following way:

$$\left(\begin{array}{c}\text{Compressed map} \\ \text{Cell number}\end{array}\right) (\text{MAP KEY}) = \left(\begin{array}{c}\text{Code number of the first} \\ \text{MINTERM or MAXTERM} \\ \text{possible for that} \\ \text{compressed map cell}\end{array}\right) \quad (3\text{-}51)$$

cell in compressed map covers map key cells of an compressed map

As an example of the use of Eq. (3-51), let us refer back to Fig. 3-39. Here, it is easily seen that the MAP KEY for this second-order EV map is $2^{4-2} = 4$. Therefore, according to the MAP KEY relation, cell 1 of the second-order EV map must cover code numbers 4, 5, 6, and 7 just as cell 3 must cover code numbers 12, 13, 14, and 15, and so on. With a little practice the MAP KEY relation will prove to be of considerable value in making reduced submap entries into EV maps directly from the CANONICAL forms or the reverse.

We will further illustrate the EV mapping method and the use of the MAP KEY relation by considering the following four variable function containing don't cares:

$$T(A,B,C,D) = \sum m(3, 4, 6, 7, 11, 14) + \phi(0, 2, 15) \qquad (3\text{-}52)$$

Let us compress this function into a third-order EV map and then find its minimized SOP and POS forms. To do this we must first enter the reduced function from submaps of order $4 - 3 = 1$ into each cell of the third-order EV map which has a MAP KEY of $2^{4-3} = 2$. Thus we are, in effect, counting by 2's (the MAP KEY). Next we construct on paper or in our minds eight first-order submaps containing the six essential MINTERMs and the three nonessential MINTERMs (don't cares) represented by Eq. (3-52). The resulting SOP and POS EV K-maps are shown in Fig. 3-41 together with five sample submaps to illustrate their SOP and POS interpretation. *Don't care submap entries are treated as entered variables* and are assigned logic values ($\phi = 0$ or $\phi = 1$) as needed. For example, the subfunction in cell 0 is interpreted as $\phi D = D$ ($\phi = 1$) for minimum SOP cover but is $\phi D = 0$ ($\phi = 0$) for minimum POS cover. Similarly, the subfunction in cell 1 is interpreted as $\phi + D = 1$ ($\phi = 1$) for minimum SOP cover, but is taken as $\phi + D = D$ ($\phi = 0$) for minimum POS cover.

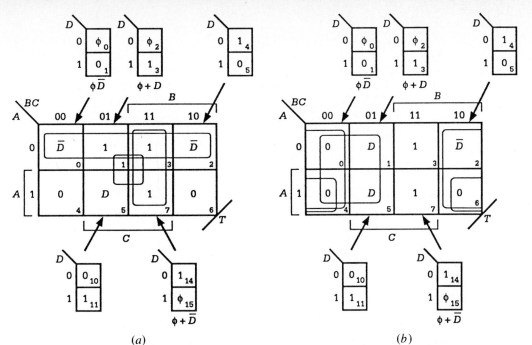

(a)

(b)

Fig. 3-41. *Third-order EV K-maps and sample submaps for the function given by Eq. (3-52). (a) Minimum SOP cover. (b) Minimum POS cover.*

Following the loop-out protocol and remembering that $1 = D + \overline{D}$ and $0 = D \cdot \overline{D}$ for the maps of Fig. 3-41, there results the minimum SOP and POS cover

$$T_{\text{SOP}} = \overline{A}\,\overline{D} + CD + BC$$

$$T_{\text{POS}} = (B + D)(C + \overline{D})(\overline{A} + C)$$

(3-53)

It is important to understand that algebraically $T_{\text{SOP}} \neq T_{\text{POS}}$ but that logically both expression yield the same results for any given set of input logic levels consistent with Eq. (3-52). This algebraic inequality results directly from the fact that the don't cares are interpreted differently in generating the minimum SOP and minimum POS cover. Consequently, it is understandable that the SOP and POS K-maps of Fig. 3-41 are not congruent but that they are logically equivalent.

The EV K-map methods described here can also be used for partially reduced functions. Take, for example, the partially reduced four-variable function

$$S = A\overline{B}C + \overline{A}C\overline{D} + \overline{B}CD + ABC\overline{D} + \overline{A}BD$$

(3-54)

This function can be compressed into a third-order EV map simply by selecting an EV, say, D as before, and entering the reduced terms (in D) into the appropriate cells. The results are given in Fig. 3-42, where the 1's and 0's are interpreted as $D + \overline{D}$ and $D \cdot \overline{D}$, respectively. Observe that the term $A\overline{B}C$ places a logic 1 in cell 5 and that the term $\overline{A}C\overline{D}$ is mapped as $\overline{D}$ for "all that is $\overline{A}$ and C," and so on. The logic 1 in cell 1 results from the intersection of the two diads, $\overline{A}C\overline{D}$ and $\overline{A}BD$.

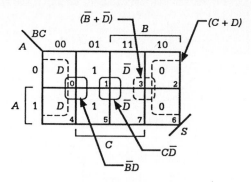

Fig. 3-42. *Third-order EV K-map for the partially reduced function of Eq. (3-54) showing minimum SOP and POS cover.*

Looping out the minimum SOP and POS cover we obtain

$$S = \bar{B}D + C\bar{D}$$
$$= (C + D)(\bar{B} + \bar{D}) \tag{3-55}$$

and note that the 1's are covered by two SOP quads, as are the 0's covered by two POS quads (see dashed loops). These minimized functions could have been looped out of conventional fourth-order map. As to which method is easier, is a matter of personal preference or, perhaps, familiarity with the EV map methods. However, it must be born in mind that the EV map method with reduced subfunction entries does not, in general, yield a guaranteed absolute minimum function whereas conventional mapping (when practical to use) will, provided that the loop-out protocol is obeyed.

The CANONICAL SOP and POS forms are easily obtained from Fig. 3-42 by reversing the submapping process and by applying the MAP KEY relation. For example, noting that the MAP KEY for this map is 2, we see that the D in cell 0 can be accounted for only if MINTERM m_1 exists but not m_0. Similarly, the $\bar{D}$ term in cell 3 results only if MINTERM m_6 exists but not m_7. The 1's represent the existance of both MINTERMs covered by a cell, and the 0's represent the existance of both MAXTERMs (or the absence of both MINTERMs) covered by a cell. In this way, taking the cells of the EV map in sequence one at a time, we can obtain directly the CANONICAL forms:

$$S = \sum m(1, 2, 3, 6, 9, 10, 11, 14)$$
$$= \prod M(0, 4, 5, 7, 8, 12, 13, 15) \tag{3-56}$$

These CANONICAL forms could have been obtained from the partially reduced function of Eq. (3-54) but only by Boolean expansion, a more tedious task. Remember that once one CANONICAL form is obtained the other follows immediately—the code numbers absent in one CANONICAL form become the code numbers of the other.

3.10 FUNCTION REDUCTION OF FIVE OR MORE VARIABLES

Perhaps the most powerful application of the EV mapping method is the reduction of functions having five to eight variables. Beyond eight variables the EV method could become too tedious to be of value, given the computer

3 / Function Representation and Reduction

methods available. We will give a few examples of function reduction through six variables, and then indicate how one would proceed in cases where seven or eight variables are present.

Consider the following five-variable function

$$f(A,B,C,D,E) = \sum m(0, 1, 2, 3, 8, 9, 10, 11, 14, 20, 21, 22, 25) \quad (3\text{-}57)$$

To compress this function into a fourth-order EV map requires that submaps of order $5 - 4 = 1$ be entered into each cell of the EV map which has a MAP KEY of $2^{5-4} = 2$. The results are shown in Fig. 3-43 together with four sample submaps.

By following the loop-out protocol, by using the MAP KEY relation, and by noting that the 1's and 0's represent $E + \overline{E}$ and $E \cdot \overline{E}$, respectively, we find

$$f = A\overline{B}C\overline{E} + \overline{A}BD\overline{E} + B\overline{C}\overline{D}E + \overline{A}\overline{B}C\overline{D} + \overline{A}\,\overline{C}$$
$$= (\overline{A} + \overline{B} + E)(\overline{A} + \overline{D} + \overline{E})(\overline{B} + \overline{C} + \overline{E})(\overline{A} + B + C) \quad (3\text{-}58)$$
$$\cdot (A + B + \overline{C})(A + \overline{C} + D)$$

which, in this case, represent minimum SOP and POS cover, the former being simpler than the latter. Since minimum POS cover is not shown in Fig. 3-43, it will be instructive for the reader to verify the POS expression. Notice that the EV terms are looped out first followed by the 1's or 0's, all in agreement with the extended loop-out protocol given in Sect. 3-9.

The five-variable function of Eq. (3-57) can be minimized by conventional means. This requires the use of two fourth-order maps, one for $\overline{A}$ and one for A, A being called the *map heading variable*. In effect, this is a three-dimensional mapping problem, where one map is imagined to be placed on top of the other in what is called the stack format. By using this method and the loop-out protocol, a guaranteed absolute minimum function can be achieved, but only at the expense of considerable work. This is demonstrated by Example 3.14(b) in Sect. 3.13.

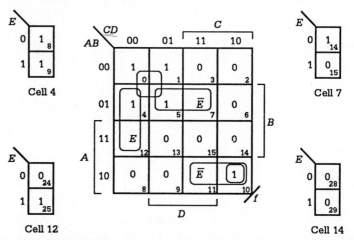

Fig. 3-43. *Fourth-order EV K-map, sample submaps and minimum SOP cover for the function given by Eq. (3-57).*

As another example, let us compress the six-variable function

$$Y(A,B,C,D,E,F) = \sum m(0,2,4,6,8,10,12,14,16,20,23,32,34,36, \quad (3\text{-}59)$$
$$38,40,42,44,45,46,49,51,57,59,60,61,62,63)$$

into a fourth-order EV map, a second-order compression. In this case, we must enter submaps of order $6 - 4 = 2$ into each cell of the compressed EV map which has a MAP KEY of $2^{6-4} = 4$. The results are shown in Fig. 3-44, along with four sample submaps. Note that minimum subfunction forms have been read from the submaps and have been placed in the appropriate cells of the fourth-order EV map. The entries in cells 4 and 5 are, of course, given in CANONICAL form since no adjacencies exist for their SOP representation. Cell 5 could have been represented as $E \odot F$ or $\overline{E} \oplus F$, but such forms would not be helpful in the search for minimum cover.

Again following the loop-out protocol and noting that in this map $1 = \overline{E}\overline{F} + \overline{E}F + E\overline{F} + EF = (E + \overline{E}) = (F + \overline{F})$, the absolute minimum SOP cover indicated in Fig. 3-44 is read as

$$Y = \overline{A}\overline{B}\overline{C}DEF + \overline{A}\,\overline{C}EF + AB\overline{D}F + ACD\overline{E} \quad (3\text{-}60)$$
$$+ \ \overline{B}\overline{F} + ABCD$$

The EF subfunction in cell 5 is looped out as a monad (it has no logic adjacencies) to yield the first term, the only MINTERM in Eq. (3-60). The second term results from looping out the subfunction $\overline{E}\overline{F}$ in cells 4 and 5 with the subfunctions $\overline{F}$ in cells 0 and 1 to form a quad. Had we not noticed

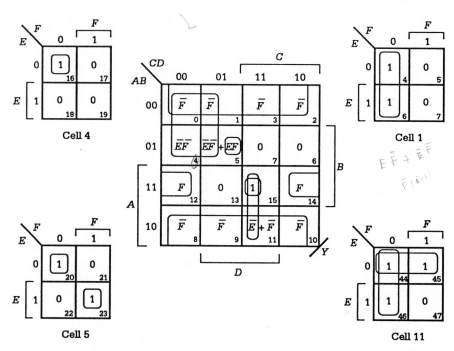

Fig. 3-44. *Fourth-order EV K-map, four sample submaps, and minimum SOP cover for the six-variable function of Eq. (3-59).*

3 / Function Representation and Reduction

that the $\overline{E}\overline{F}$ terms in cells 4 and 5 were contained in the $\overline{F}$ terms of cells 0 and 1, we would have looped out the term $\overline{E}\overline{F}$ as the diad $\overline{A}B\overline{C}\,\overline{E}\overline{F}$, an increase of one input. The term $\overline{B}\overline{F}$ results from looping out the subfunction $\overline{F}$ as a hexadecad from cells 0, 1, 2, 3, 8, 9, 10, and 11. The logic 1 quad in cell 15 is looped out last as a "clean-up" term to cover the remaining logic adjacencies contained in that cell. Thus, the procedure is one of first applying the loop-out protocol to the EVs and then applying it to the 1's as a clean-up stage. Such a procedure helps to eliminate possible redundant coverage.

The absolute minimum result given by Eq. (3-60) could have been obtained by looping out SOP cover by conventional means. The conventional approach would require four fourth-order K-maps, one each for $\overline{A}\overline{B}$, $\overline{A}B$, $A\overline{B}$ and AB, assuming that the variables A and B are chosen as the map heading variables. But this task is a formidable one since these maps must be looped out as a three-dimensional problem. Our point is that the minimum form of Eq. (3-60) is easily obtained from reduced EV subfunctions providing one takes the time to learn the EV mapping method. Another example of six-variable EV map reduction is provided by Example 3-16 in Sect. 3.13.

Representing the function of Eq. (3-59) in reduced or minimum POS form requires some special consideration. Presented in Fig. 3-45 is the POS map which is congruent with the SOP map except for the EQV subfunction of cell 5 which is now given in POS form. Following the loop-out protocol, first for the EVs and proceeding to the logic 0's for the clean-up terms, and recognizing that for this POS map $0 = (E + F)(E + \overline{F})(\overline{E} + F)(\overline{E} + \overline{F}) = E\overline{E} = F\overline{F}$, one possible reduced POS result is given by the following expression:

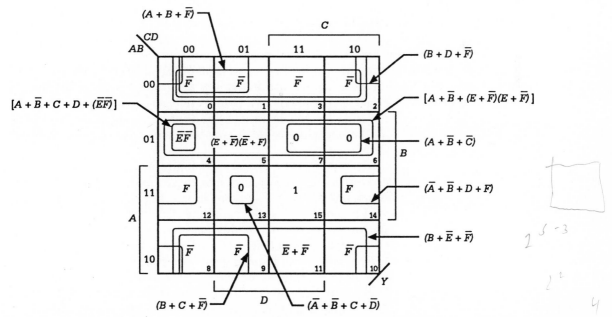

Fig. 3-45. *Second-order compression of the function given by Eq. (3-59) showing reduced POS cover with hybrid subfunctions.*

$$Y = (A + \bar{B} + C + D + \bar{E} \cdot \bar{F})(\bar{A} + \bar{B} + D + F)$$
$$\cdot [A + \bar{B} + (E + \bar{F})(\bar{E} + F)](B + D + \bar{F})$$
$$\cdot (B + C + \bar{F})(A + B + \bar{F})(B + \bar{E} + \bar{F})$$
$$\cdot (\bar{A} + \bar{B} + C + \bar{D})(A + \bar{B} + \bar{C}) \tag{3-61}$$
$$= (A + \bar{B} + C + D + \bar{E})(A + \bar{B} + C + D + \bar{F})$$
$$\cdot (\bar{A} + \bar{B} + D + F)(A + \bar{B} + E + \bar{F})$$
$$\cdot (A + \bar{B} + \bar{E} + F)(B + D + \bar{F})(B + C + \bar{F})$$
$$\cdot (A + B + \bar{F})(B + \bar{E} + \bar{F})$$
$$\cdot (\bar{A} + \bar{B} + C + \bar{D})(A + \bar{B} + \bar{C})$$

The first expression for Y in Eqs. (3-61) is a *hybrid* form since it contains both POS and SOP terms. It is obtained by reading the cell groupings as indicated in Fig. 3-45. The hybrid term $(A + \bar{B} + C + D + \bar{E}\bar{F})$ is the result of looping out cell 4 as an "island" and represents two POS diads, one in $\bar{E}$ and the other in $\bar{F}$, according to the distributive law [Eq. (2-13)]. The second hybrid term $[A + \bar{B} + (E + \bar{F})(\bar{E} + F)]$ is looped out of cells 4, 5, 6, and 7, and represents two POS quads, one in $(E + \bar{F})$ and the other in $(\bar{E} + F)$, after applying Eq. (2-13). That these four cells contain the term $(E + \bar{F})(\bar{E} + F)$ is evident when one compares the submaps for cells 4 and 5 and acknowledges that cells 6 and 7 contain the four MAXTERMs needed to cover any adjacencies.

Also noteworthy is the fact that the subfunction $(\bar{E} + \bar{F})$ is contained in cells 0, 1, 2, 3, 8, 9, 10, and 11, leading to the octad $(B + \bar{E} + \bar{F})$. This is obvious when one compares the submaps for cells 1 and 11. Further, the octad terms, $(A + B + \bar{F})$ and $(B + D + \bar{F})$, result from the need to cover $\bar{F}$ in cells 3 and 10, respectively. Finally, the octad $(A + \bar{B} + \bar{C})$ is a cleanup term covering all remaining adjacencies between cells 6 and 7. The remainder of the terms follow straightforwardly.

The hybrid terms in the first expression of Eqs. (3-61) can be expanded to POS form by using the distributive law as was indicated. When this is done the second expression of Eqs. (3-61) results and is seen to be a reduced POS function of 11 PRIME IMPLICANTs. Further study of the K-map in Fig. 3-45 will be necessary to determine whether or not it is an absolute minimum form.

A little reflection on the adjacency groupings in Fig. 3-45 reveals that other reduced forms are possible, including those more closely approaching the absolute minimum cover. For example, an inspection of the submaps for cells 4, 5, 12, and 13 indicates that all four contain the POS subfunction $(\bar{E} + F)$ and can be looped out as the quad $(\bar{B} + C + \bar{E} + F)$. Furthermore, cells 0 through 7 all contain an $(E + \bar{F})$ term, leading to the octad $(A + E + \bar{F})$, that, with the foregoing quad, completes the cover of cell 5. Since an $(\bar{E} + \bar{F})$ POS term yet remains in cell 4 (refer to its submap), one need only loop out this term in cells 0, 2, 4, and 6 to achieve complete cover for cell 4. Combining these results with those given previously leads us to the absolute minimum POS cover given by

$$Y = (\overline{A} + \overline{B} + D + F)(\overline{B} + C + \overline{E} + F)(A + E + \overline{F})$$
$$\cdot (A + D + \overline{F})(B + C + \overline{F})(B + D + \overline{F}) \tag{3-62}$$
$$\cdot (B + \overline{E} + \overline{F})(\overline{A} + \overline{B} + C + \overline{D})(A + \overline{B} + \overline{C})$$

Now we have an expression of nine ESSENTIAL PRIME IMPLICANTs (EPIs) which cannot be reduced further. Thus, many reduced expressions can be derived from the K-map of Fig. 3-45, but only one is an absolute minimum form.

The third-order compression of a seven-variable function into a fourth-order K-map requires that submap functions of order $7 - 4 = 3$ be entered into each cell of the compressed map having a MAP KEY of $2^{7-4} = 8$. Also, by Eq. (3-51), it follows that $8 \times$ (cell number of the 4th-order map) gives the code number of the first MINTERM or MAXTERM possible in any given cell. To illustrate, consider the subfunction

$$Y(A,B,C,D,E,F,G) = \sum m(89, 90, 91, 93, 95) \tag{3-63}$$

which could represent a portion of a much larger function. The third-order submap for Eq. (3-63) is shown in Fig. 3-46, where the cells are numbered appropriately from 88 to 95. Looping out the minimum SOP and POS terms for this subfunction gives the results

$$\begin{array}{lll} Y_{11} = \overline{E}F + G & \text{SOP} & \\ = (\overline{E} + G)(F + G) & \text{POS} & \end{array} \tag{3-64}$$

which can now be entered into cell 11 (since $8 \times 11 = 88$) of the compressed fourth-order SOP or POS map with axes AB and CD. The process just described would continue until all the minimized subfunctions from the third-order submaps are entered into the appropriate cells of the compressed K-map which covers MINTERM (or MAXTERM) code numbers from 0 to ($8 \times 16) - 1 = 127$. At this point a reduced function could be looped out following the loop-out protocol, beginning first with the EVs and proceeding to the 1's or 0's as clean-up terms for the SOP or POS maps. The results almost certainly would not be absolute minimum expressions.

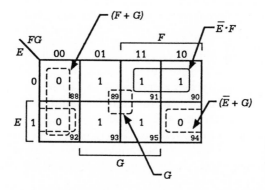

Fig. 3-46. Third-order submap for the subfunction given by Eq. (3-63).

The fourth-order compression of an eight-variable function into a fourth-order EV map follows in a similar fashion. In this case, reduced submap functions of order $8 - 4 = 4$ must be entered into the compressed map having a MAP KEY at $2^{8-4} = 16$. Thus, each cell of the compressed EV map would represent the minimum (or reduced) subfunction obtained from a fourth-order submap.

We leave it as an exercise for the reader to show that the CANONICAL POS subfunction

$$Y(A,B,C,D,E,F,G,H) = \prod M(49, 50, 51, 54, 57, 58, 59, 60, 62) \quad (3\text{-}65)$$

yields the minimum SOP and POS subfunctions given by

$$\begin{aligned} Y &= \overline{F}\,\overline{G}\,\overline{H} + \overline{E}F\overline{G} + FH \\ &= (\overline{E} + \overline{F} + H)(F + \overline{H})(\overline{G} + H) \end{aligned} \quad (3\text{-}66)$$

which must be entered into cell 3 of the appropriate compressed fourth-order map (with axes AB, CD) covering MINTERM (or MAXTERM) code numbers from 0 to 255. Also, the reader should verify that this subfunction must be represented in a submap having cells numbered from 48 to 63.

Completion of an eight-variable reduction problem would require that the minimized subfunctions from 16 fourth-order submaps be entered into the appropriate cells of the compressed EV map. Then by following the loop-out protocol, a greatly reduced function could be read quite easily from the map. This is really not a difficult task when one considers the alternatives short of a computer solution. For example, to minimize such a function by conventional mapping methods would require that 16 fourth-order maps be read simultaneously, a nearly impossible three-dimensional problem. A seven-variable minimization would require that 8 fourth-order maps be read simultaneously, also a task too difficult to be practical. Entering CANONICAL EV data for seven or eight variables into the 16 cells of a compressed map is still an unreasonable approach to the function reduction problem. Thus, considering its simplicity, the EV map reduction method with reduced or minimized EV subfunctions can be a powerful tool provided that it is properly understood and properly used. We will reflect further on the use aspect of the EV mapping method in the next section.

3.11 PERSPECTIVE ON MAP COMPRESSION AND MINIMIZATION

The EV mapping method just discussed is of great value in logic design provided that we are satisfied to accept reduced expressions that are not necessarily in absolute minimum form. An absolute minimum result is guaranteed only if an exhaustive MINTERM (or MAXTERM) adjacency search method is used. But for complex multiinput/multioutput systems, such exhaustive search methods require the use of computers. Even then, an optimum solution may not be possible. The problem is that the exhaustive search method may not converge in a reasonable period of time to give an optimum solution for a complex system—a computer has limited memory capability and speed.

There are tabular methods which will achieve an optimum solution for relatively simple systems and which can be programmed for a digital computer. One, the Quine-McCluskey (Q-M) tabular reduction method, uses an exhaustive search procedure to determine all possible combinations of logically adjacent cell entries, and then uses this information in a systematic way to obtain an optimum solution. The Q-M method can also be used to optimize multiple-output systems. Section 3.7 describes a similar but less formal approach to the multiple-output optimization problem, an approach that does not guarantee an optimum solution. Used without the aid of a computer, the Q-M method is tedious and, in most cases, unwarranted given the advantages that the EV method offers. The EV (pencil-and-paper) method, described in the previous sections, is useful in obtaining minimized or near minimized solutions to single-output or multioutput systems of up to eight input variables. With some skill the EV method can even be used to obtain greatly reduced functions of up to 12 variables but with little chance of achieving optimum results.

If computer optimized solutions are required for complex multiinput/multioutput systems, it is not necessary to write software programs for the Quine-McCluskey method because a variety of excellent computer optimization programs are available that will suffice, and will do so in much less computer time than is possible by using the Q-M algorithm. Some of these programs will accept entered variables and will yield optimum two-level solutions in either SOP and POS form for single- or multioutput functions or many variables. Thus, beyond its historical significance as the first systematic, exhaustive search, minimization algorithm, the Q-M method is of little practical importance.

Computer optimization of logic functions has limitations also. It is an established fact that the generalized optimal solution for an n-variable function is impossible. The only reason why a 12-variable function, for example, can be minimized is because the computer in which the minimized algorithm is applied has sufficient memory and speed to accomplish the task. The minimization of a 20-variable function might not be possible in the same computer considering the fact that 2^{20} MINTERMS would have to be exhaustively searched for logic adjacencies. The Annotated References at the end of this chapter contain the references for the Q-M method and useful computer optimization programs.

The EV mapping method is particularly useful for up to three or four orders of map compression. However, with increasing compression order beyond third order, the gap usually widens between the reduced forms obtained and the absolute minimum result—a sacrifice that one must be prepared to make if the simple EV method is to be used. For this reason, we set as a practical upper limit four orders of compression and strongly recommend the use minimized or reduced subfunction K-map entries. It is possible to extend this limit, but to do so would necessitate the use of submaps within the fourth-order submaps, a task which is seemingly manageable but somewhat tedious nevertheless.

Proper SOP and POS interpretation of subfunctions containing don't cares is essential if the EV method is to be correctly applied. This was demonstrated in Fig. 3-41 for the case of a first-order map compression. To

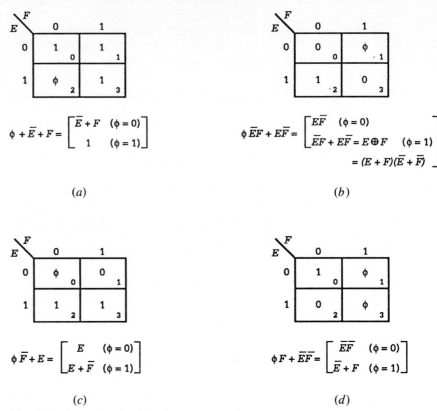

$$\phi + \overline{E} + F = \begin{bmatrix} \overline{E} + F & (\phi = 0) \\ 1 & (\phi = 1) \end{bmatrix}$$

(a)

$$\phi\,\overline{E}F + E\overline{F} = \begin{bmatrix} E\overline{F} & (\phi = 0) \\ \overline{E}F + E\overline{F} = E \oplus F & (\phi = 1) \\ = (E + F)(\overline{E} + \overline{F}) \end{bmatrix}$$

(b)

$$\phi\,\overline{F} + E = \begin{bmatrix} E & (\phi = 0) \\ E + \overline{F} & (\phi = 1) \end{bmatrix}$$

(c)

$$\phi F + \overline{E}\overline{F} = \begin{bmatrix} \overline{E}\overline{F} & (\phi = 0) \\ \overline{E} + F & (\phi = 1) \end{bmatrix}$$

(d)

Fig. 3-47. *Second-order submaps containing don't cares and their interpretations for second-order map compressions.*

gain a better perspective of this type of EV problem, consider the four second-order submaps in Fig. 3-47 which are typical of those that might be encountered in second-order map compressions, say, from sixth to fourth order. Included with each submap are the subfunction and their minimum forms for values $\phi = 0$ and $\phi = 1$. The accent here is on simplicity and the option to interpret the $(\phi \cdot X)$- and $(\phi + X)$-type terms as needed to produce minimum SOP or POS cover. Once minimized, such subfunctions will not reveal their CANONICAL origin—all don't care information is lost. It is important to remember that a don't care (ϕ) entry is treated just like an entered variable, one that becomes a logic 0 or logic 1 as needed.

To maintain a reasonable level of simplicity, an important feature of the EV method, submaps (of any order) containing don't cares should usually be looped out in minimum or reduced form. Such subfunctions will, of course, be exclusive of all $(\phi \cdot X)$- and $(\phi + X)$-type terms, even though don't cares may have been used in their formation. As a result of using these reduced subfunctions, the chance is lessened that the final reduced expressions will be an absolute minimum form. An absolute minimum result by the EV method can be guaranteed only if all subfunctions are represented in CANONICAL form and if they are looped out to yield minimum cover.

As a final thought on the use of submaps, we offer the following suggestion:

> *If there is any doubt as to whether or not one subfunction is contained in another for adjacency grouping purposes, refer to the submaps of those subfunctions. A brief comparison of the adjacent submaps will usually yield the answer immediately.*

Remember that the use of CANONICAL subfunctions offers the surest means of obtaining an absolute minimum cover, if an absolute minimum is required. Since conventional submaps (those containing 1's, 0's, and ϕ's) represent CANONICAL data, they can be used for this purpose. Example 3-15 in Sect. 3.13 illustrates the use of conventional submaps in achieving absolute minimum results. Although these submaps are of second order, it is easy to see how fourth-order conventional submaps could be used in the same manner. In fact, a five-variable subfunction could be represented in a fourth-order submap with a reasonable expectation of achieving minimum cover. Thus, a nine-variable function, compressed into a fourth-order K-map, could be minimized in this way.

3.12 MULTILEVEL LOGIC MINIMIZATION

The function reduction and minimization methods discussed to this point apply mainly to two-level (two-stage) logic networks. Two-level logic implies the ORing of ANDed terms or the ANDing of ORed terms of a function without regard to fan-in limitations (limiting number of gate inputs) or the presence of INVERTERs required to provide the physical realization of the complemented (or uncomplemented) variables. The term level also refers to the propagation delay associated with a gate. Thus, a two-level implementation of a function implies two gate path delays, one through the input stage and one through the output stage of a logic circuit. In this section we will consider minimization methods that yield multiple-level logic circuits of minimized gate count.

3.12.1 Minimization by Factoring

INVERTERs provide the physical realization of the complemented (or uncomplemented) variables depending on the input sources and the type of gates that are used. It is possible to eliminate the need for these INVERTERs and, at the same time, reduce the fan-in requirements of the gates by using a technique called *factoring*. We illustrate this technique with the following simple example.

Consider the minimum SOP function

$$F = AB\overline{C} + AD + \overline{B}D + \overline{C}D$$

which requires a gate/input tally (excluding INVERTERs) of 5/13. This includes four ANDing operations with a maximum of three inputs per gate, and one ORing operation requiring four inputs. Also, if all inputs are assumed to arrive ACTIVE HIGH, then two INVERTERs are required, thereby making a total gate/input tally of 7/15. Now, let us assume that all gates have a fan-in limitation of three inputs and that it is desirable to eliminate the need to use LOGIC LEVEL CONVERTERs (INVERTERs) to generate the complements of the B and C variables. To do this we need to factor the foregoing function as follows:

$$F = AB\overline{C} + AD + \overline{B}D + \overline{C}D$$
$$= AB(\overline{B} + \overline{C}) + AD + (\overline{B} + \overline{C})D$$
$$= AB(\overline{BC}) + AD + (\overline{BC})D$$

where the $AB\overline{C}$ term is factored as $AB\overline{C} = AB(\overline{B} + \overline{C}) = AB(\overline{BC})$. The result is a NAND gate/input tally of 5/12 requiring no INVERTERs. It is left as an exercise for the reader to verify this result.

The technique of function minimization by factoring can be extended to multioutput systems of the type considered in Sect. 3.7. This is accomplished by matching up factored terms within the same function and between different functions of an optimized system. We illustrate the technique with the following multioutput system.

Consider the set of three optimized functions which are factored as indicated:

$$f_1 = \overline{A}\,\overline{B} + \overline{A}\,\overline{C} + AB = \overline{A}(\overline{B} + \overline{C}) + AB = \overline{A}(\overline{BC}) + AB$$
$$f_2 = \overline{A}BC + B\overline{C} + AB = B(\overline{A} + \overline{C}) + AB = B(\overline{AC}) + AB$$
$$f_3 = \overline{B}C + \overline{A}BC + B\overline{C} = C(\overline{B} + \overline{C}) + B(\overline{A} + \overline{C}) = C(\overline{BC}) + B(\overline{AC})$$

Here, two terms are factored as $\overline{A}BC + B\overline{C} = B(\overline{A}C + \overline{C}) = B(\overline{A} + \overline{C})$ $= B(\overline{AC})$ and $\overline{B}C = C(\overline{B} + \overline{C}) = C(\overline{BC})$. Then, assuming that all inputs arrive ACTIVE HIGH, we see that the initial expressions require a gate/input tally of 12/25 which includes three INVERTERs. Also, the initial expressions require a fan-in of three inputs for two of the ANDing operations and three inputs for all three of the ORing operations. In contrast, the factored expressions require a gate/input tally of 10/19, including one IN-VERTER, and have fan-in requirements that are now reduced to two inputs for all gates.

3.12.2 Minimization by Using XOR Patterns

Combining entered variable subfunctions offers a special and very useful form of function reduction and minimization which we regard as an important extension of the standard EV mapping methods presented in Sects. 3.8, 3.9, and 3.10. If true minimum cover (gatewise) is to be obtained from a K-map, XOR-type patterns must be considered along with those that yield two-level logic (AND and OR functions). In this section we will consider various types of XOR patterns, XOR minimization, and the multilevel logic that results.

The XOR Patterns. There are four types of XOR patterns that can occur in EV K-maps. They are

1. Diagonal patterns.
2. Adjacent patterns.

3. Offset patterns.
4. Associative patterns.

In the discussion that follows, each of these EV patterns will be considered, and typical examples will be given to illustrate their character.

We are already familiar with one type of diagonal pattern—that given in the conventional, second-order K-map of Fig. 3-14. Here, the 1's and 0's represent MINTERMs and MAXTERMs, respectively, resulting in SOP and hybrid POS expressions of the type given by Eqs. (2-6) and (2-7). The complement of the K-map in Fig. 3-14 is the K-map for the EQUIVALENCE (EQV) function which results in an SOP expression of the type given by Eq. (2-8). The reader should become reacquainted with Eqs. (2-6), (2-8), and (2-32) because they will be helpful in reading the XOR cover in second- and third-order EV K-maps, as we shall demonstrate.

Presented in Fig. 3-48 are simple examples of diagonal and adjacent EV patterns represented in terms of the multivariable subfunctions X and Y. Here, it can be seen that the diagonal XOR patterns are formed from *like* terms in diagonal cells while the adjacent XOR patterns are formed by combining a term in one cell with its complement in an adjacent cell. With this information in mind, the reader should now learn how to read these patterns directly in XOR form. For example, in Fig. 3-48(a) the diagonal pattern is read in MINTERM code as $A \oplus B = \overline{A}B + A\overline{B}$ for "all that is X," hence $X(A \oplus B)$. Likewise, the adjacent pattern is read as $A \odot Y = \overline{A}Y + AY$ for "all that is B," or $B(A \odot Y)$. In Fig. 3-48(b) one reads the diagonal pattern in MAXTERM code as $A \odot B\,[= (\overline{A} + \overline{B})(A + B)]$ for "all that is Y," and the adjacent pattern is read as $A \odot X\,[= (A + \overline{X})(\overline{A} + X)]$ for "all that is B," with a combined POS result of $(Y + A \oplus B)(B + A \odot X)$. Remember that Eqs. (2-32) are always applicable, permitting $A \odot Y = A \oplus \overline{Y} = \overline{A \oplus Y}$ or $A \oplus B = \overline{A} \odot B = \overline{A \odot B}$, and the like.

In Fig. 3-49 are shown examples of diagonal and adjacent patterns in a third-order EV map. In this case, we disregard the empty cells 2 and 5 so as to center our attention on the patterns which are looped out in both MINTERM and MAXTERM code. The diagonal pattern is read in MINTERM code as $A \oplus C\,(= \overline{A}C + A\overline{C})$ for "all that is B and Y," that is,

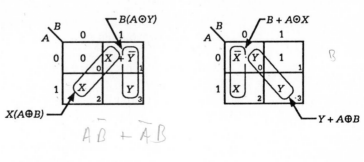

$$F = X(A \oplus B) + B(A \odot Y) \quad \text{SOP} \qquad\qquad G = (Y + A \oplus B)(B + A \odot X) \quad \text{POS}$$

(*a*)   (*b*)

Fig. 3-48. *Second-order EV K-maps showing diagonal and adjacent XOR patterns and the pseudo SOP and POS expressions obtained from them.*

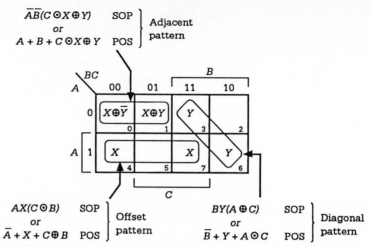

Fig. 3-49. *Third-order EV K-map showing typical diagonal, adjacent, and offset type XOR patterns and the pseudo SOP and POS expressions derived from them.*

$BY(A \oplus C)$, and in MAXTERM code is read as $A \odot C\, [= (A + \overline{C})(\overline{A} + C)]$ for "all that is $\overline{B}$ or Y," meaning $\overline{B} + Y + A \odot C$. The adjacent pattern between $X \oplus Y$ and its complement is read in MINTERM code as $C \odot (X \oplus Y)\, [= \overline{C}(X \oplus Y) + C(X \oplus Y)]$ for "all that is $\overline{A}$ and $\overline{B}$," or $\overline{A}\,\overline{B}(C \odot X \oplus Y)$, but in MAXTERM code is read as $C \odot X \oplus Y\, [= (C + \overline{Z})(\overline{C} + Z)$, where $Z = X \oplus Y]$ for "all that is $A + B$," hence $A + B + C \odot X \oplus Y$.

The Boolean expressions derived from the XOR patterns in Figs. 3-48 and 3-49 have been labeled SOP or POS. Actually, these are really *pseudo*-SOP and -POS forms. Just as the XOR and OR operations are called EX-CLUSIVE OR and INCLUSIVE OR, respectively, so also can the AND and EQV operations be called EXCLUSIVE AND and INCLUSIVE AND, respectively. Thus, XOR is to OR as AND is to EQV, and the use of the terms pseudo-SOP and pseudo-POS is justified. We will occasionally refer to expressions containing the XOR or EQV functions simply as SOP or POS but with this understanding. More accurately, such expressions should be labeled as hybrid forms since they are not explicitly represented in either SOP or POS form.

Offset patterns are formed by combining like terms which exist in non-diagonal cells removed from one another by only one cell, a change of two axis code bits. Figure 3-49 gives one example in terms of the multivariable subfunction X. This pattern is read in MINTERM code as $C \odot B\, (= \overline{B}\,\overline{C} + BC)$ for "all that is A AND X," or $AX(C \odot B)$. In MAXTERM code this same pattern would be read as $C \oplus B\, [= (\overline{B} + \overline{C})(B + C)]$ for "all that is $\overline{A}$ OR X," meaning $\overline{A} + X + C \oplus B$. Offset patterns can exist only in third- and fourth-order K-maps.

Associative-type XOR patterns offer one of the best examples of the multilevel logic advantage over two-level logic, and considerable savings in hardware can be realized by the use of such patterns. Typical associative patterns are presented in Figs. 3-50 and 3-51. They are formed between XORed (or EQVed) terms and like terms in adjacent cells, similar to the familiar loop-out patterns for ORed (or ANDed) subfunctions. In fact, the XOR and EQV operator symbols may be treated as though they were the

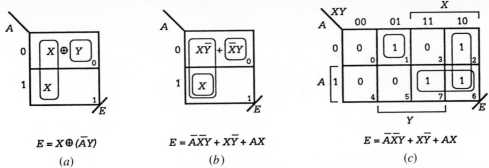

Fig. 3-50. *Comparison of multilevel and two-level cover for a function E. (a) Simple associative pattern in a first-order K-map looped out in MINTERM code. (b) Two-level EV cover and (c) conventional cover for the function E.*

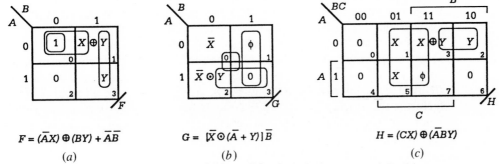

Fig. 3-51. *More complex associative patterns. (a) Pattern with residual cover looped out in MINTERM code. (b) Pattern with residual cover looped out in MAXTERM code. (c) Pattern looped out in MINTERM code.*

OR and AND operator symbols, respectively, for the purpose of looping out pseudo-SOP and -POS cover, but the XOR or EQV operators must be preserved. Take, for example, the simple associative pattern shown in the first-order map of Fig. 3-50(a), a second-order compression. It is read as $X \oplus (\overline{A}Y)$, similar to the SOP form $X + \overline{A}Y$ that would have resulted had the XOR operator been replaced by the OR operator. The two-level minimum forms for this function are represented in Figs. 3-50(b) and (c). Notice that the associative pattern of Fig. 3-50(a) is not easily recognized from the conventional cover in Fig. 3-50(c), a characteristic of most XOR-type patterns but especially the associative pattern.

The three associative patterns given in Fig. 3-51 are worthy of a few additional comments. First, let it be made clear that a given associative pattern can involve any 2^n adjacent cells containing the XORed (or EQVed) variable(s). For example, in Figs. 3-51(b) and (c) are associative patterns which require the use of both two- and four-cell associations to form the XOR logic minima as indicated by the loops. Second, notice that in Figs. 3-51(a) and (b) the logic 1 and logic 0 contain residual terms which must be looped out as clean-up operations. Again, we point to the striking similarity between conventional EV mapping and the associative XOR patterns.

That the associative patterns represent a significant savings in the number of gates required to implement the functions they represent is easily demonstrated. The multilevel (XOR) representation of the function E in Fig. 3-50(a) would require a gate/input tally of 2/4, meaning two gates (exclusive of any INVERTERs) having a total of four inputs. To implement the function E in the minimum two-level logic representation of Figs. 3-50(b) or (c) would require a gate/input tally of 4/10, a significant increase over the multilevel representation. Similarly, the gate/input tally for function G in Fig. 3-51(b) is 3/6, but would be 4/11 in minimum POS form or 4/12 in minimum SOP form. The two-level gate/input tallies were obtained from Fig. 3-51(b) by appropriately expanding the XOR term $[\overline{X} \odot Y = (X + Y)(\overline{X} + \overline{Y}) = \overline{X}Y + X\overline{Y}]$ and by looping out the resulting two-level minima $G_{\text{POS}} = (\overline{A} + X + Y)(\overline{X} + \overline{Y})(A + \overline{X})\overline{B}$ and $G_{\text{SOP}} = A\overline{B}X\overline{Y} + \overline{B}XY + \overline{A}\,\overline{X}$, respectively.

The four types of XOR patterns may be combined in a variety of ways. One example of compound XOR patterns is shown in Fig. 3-52, where use is again made of an associative pattern, but now in combination with diagonal, adjacent, and offset patterns. Here, the last three patterns are nested with respect to the associative pattern. The combined patterns of cells 1, 2, 5, and 6 can be easily understood by the reader if they are viewed as being composed of two parts. Thus, taking cells 1 and 5, the adjacent pattern is $\overline{B}C(A \oplus X)$, while for cells 2 and 6, we read $B\overline{C}(A \oplus X)$. Combining and factoring then yields $(A \oplus X)(B \oplus C)$ as indicated in Fig. 3-52(a). The gate/input tally for minimum multilevel cover is 6/13, compared to a two-level SOP minimum tally of 7/31. We leave it for the reader to prove the validity of these gate/input tallies.

Although we have not specifically mentioned it previously, the multilevel minimum may represent an even greater savings in the hardware needed for implementation when account is taken of the INVERTER requirements. Generally speaking, the number of INVERTERs required for a given function increases in proportion to the number of inputs indicated by the gate/input tally. For example, in the case of Fig. 3-52, assuming all inputs arrive as A, B, C, X, and Y from positive logic sources, two INVERTERs would be required for the multilevel representation, whereas five INVERTERs

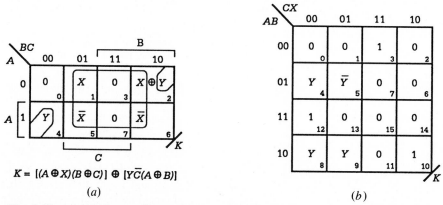

$K = [(A \oplus X)(B \oplus C)] \oplus [Y\overline{C}(A \oplus B)]$

(a)

(b)

Fig. 3-52. (a) EV K-map illustrating compound XOR type patterns for a function K. (b) Representation of the function K in a map of lesser compression showing loss of pattern recognition.

3 / Function Representation and Reduction

would be required for the two-level form. The proof of this is left for the reader.

The four associative XOR patterns presented in Figs. 3-51 and 3-52 could have been illustrated by using conventional mapping methods—that is, with 1's and 0's. However, this would have been a most difficult and tedious task. To demonstrate, the reader is invited to prove any of the associative patterns of Fig. 3-51 by conventional means. In fact, representing the functions of Figs. 3-51 and 3-52 by maps of lesser or greater compression than second order would certainly complicate the task of recognizing XOR-type patterns. This is demonstrated by Fig. 3-52(b) where it can be seen that a compression of only one order greatly obscures the patterns needed to loop out a multilevel minimum. Apparently, there exists an optimum compression for the XOR pattern recognition of a given function, and in most cases this is second order.

Why Multilevel Logic Minimization? XOR minimum functions, particularly those of the type shown in Figs. 3-48 through 3-52(a), have some practical applications which are limited mainly to multioutput arithmetic circuits. This is so, at least in part, because the algorithms for most arithmetic manipulations of binary numbers involve addition operations, and the XOR operation is fundamental to the addition process as is later indicated in Sect. 4.4. Also, there exists the possibility of improving the optimization of a given multioutput arithmetic circuit by using XOR logic. This follows from the discussion given in Sect. 3-7. But there are other factors of which account must be taken.

An XOR gate is usually thought of as a two-unit time delay device. If a single-level gate (in, say, the TTL logic family) is viewed as having a propagation time delay of one unit, then the XOR gate, implemented according to Fig. 2-30 or Fig. 2-32 in TTL logic, is a device having two units of path delay. For better or worse, modern terminology has tagged circuits of this type as *two-level* circuits, meaning those having two units of time delay— whether the time units are equal or not. An important exception to this is the use of the MOS family XOR (or EQV) IC gates which have time delays only slightly larger than one unit. Thus, such MOS XOR or EQV gates will have propagation time delays only slightly larger than, say, MOS family NAND or NOR gates. Furthermore, a MOS XOR–type gate will occupy about the same IC chip real estate (on the average) as a MOS NAND or NOR gate. Examples of MOS XOR and EQV gates are shown in Figs. 2-36 and 2-39.

What this discussion is leading up to is the fact that XOR-minimized designs usually represent multilevel circuits with higher time delays than conventional designs with single-level gates (NAND, NOR, AND, and OR). For such XOR designs one must be prepared to weigh the advantages of any reduced real estate requirements against the disadvantages of reduced circuit speed. However, these disadvantages are apparently becoming much less important with the advancement of new IC technology. This means that circuit design considerations need no longer be limited to two-level logic but, in fact, should now be extended to include XOR logic when significant savings in the gate/input tally can be realized. In Chapter 4 that follows, we will have the opportunity to explore the various approaches to combinational logic design, including multilevel logic minimization, and make some interesting comparisons.

3.13 WORKED EXERCISES

In tabulating, mapping, and minimizing Boolean expressions, one can encounter a variety of subtle difficulties. Consider that there are SOP and POS representations in CANONICAL or reduced form, that the CANONICAL forms may contain don't cares, and that these expressions may be minimized in two-level or multilevel form by using K-maps of various orders of compression and format. Add to these the subtleties of the loop-out process itself and suddenly the subject matter no longer seems simple, and it is not. In this section we cover many of these subtle difficulties by using worked exercises, but at the same time we provide a chapter overview and study guide.

EXAMPLE 3-1

Place the following three-variable CANONICAL POS function in a truth table and represent it in a third-order K-map:

$$f(A,B,C) = (A + \overline{B} + C)(\overline{A} + \overline{B} + \overline{C})(A + B + C)$$
$$\cdot (\overline{A} + B + C)(\overline{A} + B + \overline{C}) \tag{3-67}$$

SOLUTION

The truth table in Fig. 3-53 is constructed by placing a logic 0 in the f column for each MAXTERM represented by the function in Eq. (3-67) and by using the MAXTERM code table given in Fig. 3-5. The absence of a MAXTERM is a MINTERM, which, accordingly, is assigned logic 1. The K-map is a graphical representation of the CANONICAL truth table and is constructed directly from the truth table as shown in Fig. 3-53(b).

A B C	f
0 0 0	0
0 0 1	1
0 1 0	0
0 1 1	1
1 0 0	0
1 0 1	0
1 1 0	1
1 1 1	0

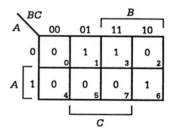

(a) (b)

Fig. 3-53. *Conventional truth table and K-map for the three variable function of Eq. (3-67).*

EXAMPLE 3-2

Place the following four-variable CANONICAL SOP function in a truth table and represent it in a fourth-order K-map.

$$f(A,B,C,D) = \sum m(0, 1, 3, 5, 6, 8, 9, 10, 13) \tag{3-68}$$

SOLUTION

The procedure is similar to that followed in Example 3-1 except that, in this case, a logic 1 is placed in the f column and K-map cell for each MINTERM

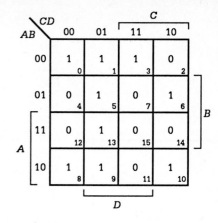

A	B	C	D	f	A	B	C	D	f
0	0	0	0	1	1	0	0	0	1
0	0	0	1	1	1	0	0	1	1
0	0	1	0	0	1	0	1	0	1
0	0	1	1	1	1	0	1	1	0
0	1	0	0	0	1	1	0	0	0
0	1	0	1	1	1	1	0	1	1
0	1	1	0	1	1	1	1	0	0
0	1	1	1	0	1	1	1	1	0

(a)   (b)

Fig. 3-54. (a) Truth table and (b) K-map for the four variable function given by Eq. (3-68).

present in Eq. (3-68), as required by the MINTERM code table of Fig. 3-3. The absence of a MINTERM is a MAXTERM and is assigned logic 0. The results are shown in Fig. 3-54.

EXAMPLE 3-3

Convert the reduced SOP function given in this example to CANONICAL SOP and POS form by using a fourth-order K-map. Represent the CANONICAL expressions by using both literal and coded notation.

$$f(A,B,C,D) = ABCD + A\overline{D} + \overline{B}\,\overline{C}\,\overline{D} + \overline{A}\,\overline{B}C$$
$$+ \overline{A}B\overline{C}D + BC\overline{D} + \overline{A}\,\overline{B}\,\overline{D} \qquad (3\text{-}69)$$

SOLUTION The MINTERMs covered by each term in Eq. (3-69) are represented by logic 1's in the fourth-order K-map of Fig. 3-55. For example, all that is $A\overline{D}$, covers MINTERMs m_8, m_{10}, m_{12}, and m_{14} while all that is $\overline{A}\,\overline{B}C$, covers MINTERMs m_2 and m_3, and so on. The CANONICAL expressions are read directly from the K-map and are given as follows:

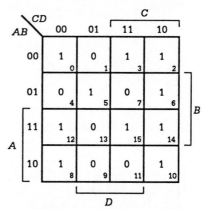

Fig. 3-55. K-map for Eq. (3-69).   163

$$f = \sum m(0, 2, 3, 5, 6, 8, 10, 12, 14, 15)$$
$$= \overline{A}\,\overline{B}\,\overline{C}\,\overline{D} + \overline{A}\,\overline{B}C\overline{D} + \overline{A}\,\overline{B}CD + \overline{A}B\overline{C}\,\overline{D} + \overline{A}BC\overline{D} + A\overline{B}\,\overline{C}\,\overline{D}$$
$$+ A\overline{B}C\overline{D} + AB\overline{C}\,\overline{D} + ABC\overline{D} + ABCD$$
$$f = \prod M(1, 4, 7, 9, 11, 13)$$
$$= (A + B + C + \overline{D})(A + \overline{B} + C + D)(A + \overline{B} + \overline{C} + \overline{D})$$
$$\cdot (\overline{A} + B + C + \overline{D}) \cdot (\overline{A} + B + \overline{C} + \overline{D})(\overline{A} + \overline{B} + C + \overline{D})$$

Notice that the missing coded SOP terms form the coded POS expression, and vice versa.

EXAMPLE 3-4

By using the tabular minimization procedure given in Sect. 3.3, minimize the following CANONICAL function in both SOP and POS form.

$$f = \overline{A}BC + A\overline{B}C + AB\overline{C} + \overline{A}\,\overline{B}\,\overline{C} + \overline{A}\,\overline{B}C \qquad (3\text{-}70)$$

SOLUTION

In agreement with the MINTERM code table given in Fig. 3-3, a truth table is constructed for Eq. (3-70), as shown in Fig. 3-56(a). From this table MIN-TERM and MAXTERM adjacencies are identified by placing an X in the appropriate position, thus forming the first-level IMPLICANTs, that is, groups of $2^1 = 2$ adjacencies. For example, the MINTERMs $\overline{A}\,\overline{B}\,\overline{C}$ and $\overline{A}\,\overline{B}C$ in Eq. (3-70) are logically adjacent and are represented by $\overline{A}\,\overline{B}X$, meaning $\overline{A}\,\overline{B}\,\overline{C} + \overline{A}\,\overline{B}C = \overline{A}\,\overline{B}X$, where $X = C + \overline{C} = 1$. In binary MINTERM code this result is represented by 00X. Similarly, the MAXTERMs $(A + \overline{B} + C)$ and $(\overline{A} + \overline{B} + C)$ are logically adjacent, meaning that $(A + \overline{B} + C)(\overline{A} + \overline{B} + C) = X + \overline{B} + C$, where $X = A\overline{A} = 0$. In binary MAXTERM code this result is expressed as X10.

Continuing this process for the second-level IMPLICANTs (i.e., $2^2 = 4$ adjacencies) yields the final results shown in Fig. 3-56(b), which cannot be further reduced. Notice that the SOP and POS results are algebraically equal.

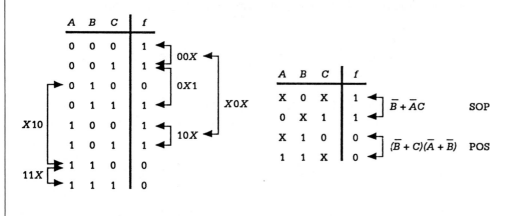

(a)   (b)

Fig. 3-56. Tabular minimization of Eq. (3-70) by the stepwise collapse of its truth tables.

3 / Function Representation and Reduction

EXAMPLE 3-5

Minimize the function given by Eq. (3-67) in both SOP and POS form by using third-order K-maps. Identify all OPIs present in both forms.

SOLUTION

K-maps for both SOP and POS cover are shown in Fig. 3-57. The EPIs are identified by solid line loops, while the OPIs are indicated by dashed line loops.

The results, read directly off the the K-maps, are

$$f = AB\overline{C} + \overline{A}C \qquad \text{SOP}$$

$$= (A + C)(\overline{A} + \overline{C}) \begin{cases} (B + C) \\ \\ (\overline{A} + B) \end{cases} \qquad \text{POS}$$

where the two POS OPIs have been placed in braces. Remember that once one OPI is selected, the other becomes an RPI for that expression. We leave it to the reader to prove that the SOP and POS forms are algebraically equal.

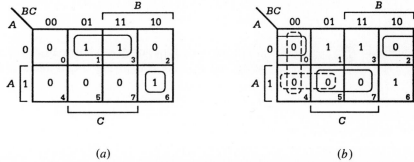

(a)   (b)

Fig. 3-57. *K-maps for Eq. (3-67). (a) Minimum SOP cover. (b) Minimum POS cover with OPTIONAL PRIME IMPLICANTS indicated by dashed loops.*

EXAMPLE 3-6

Minimize the function of Eq. (3-68) in both SOP and POS form by using fourth-order K-maps having the format given in Fig. 3-19(b). Identify all OPIs for the two minimum expressions.

SOLUTION

The K-maps for both SOP and POS cover are shown in Fig. 3-58. Reading the maps, we obtain the results

$$f = \overline{A}BC\overline{D} + A\overline{B}\overline{D} + \overline{A}BD + \overline{B}\overline{C} + \overline{C}D \qquad \text{SOP}$$

$$= (A + B + \overline{C} + D)(\overline{B} + \overline{C} + \overline{D})(\overline{B} + C + D)$$

$$\begin{cases} (\overline{A} + \overline{B} + D) \\ \\ (\overline{A} + \overline{B} + \overline{C}) \end{cases} (\overline{A} + \overline{C} + \overline{D}) \qquad \text{POS}$$

where the POS terms $(\overline{A} + \overline{B} + D)$ and $(\overline{A} + \overline{B} + \overline{C})$ are OPIs for the minimum POS cover. No OPIs exist for the minimum SOP cover. Note that the terms in the minimum cover expressions are written in the order of 2^n adjacencies, where $n = 0, 1, 2, 3, \ldots$, in accordance with the loop-out

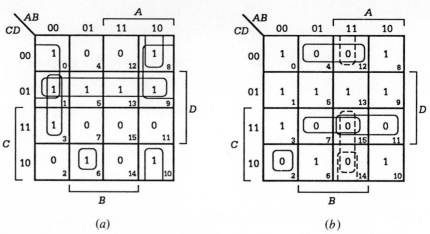

Fig. 3-58. *K-maps for Eq. (3-68). (a) Minimum SOP cover. (b) Minimum POS cover showing OPIs as dashed loops.*

protocol. Must the SOP and POS results given here be algebraically equal? Explain.

EXAMPLE 3-7

Represent the function given in the truth table of Fig. 3-59 in CANONICAL SOP and POS form by using coded and literal notation.

A	B	C	f
0	0	0	0
0	0	1	ϕ
0	1	0	ϕ
0	1	1	1
1	0	0	1
1	0	1	1
1	1	0	0
1	1	1	0

Fig. 3-59. *Truth table for Example 3-7.*

SOLUTION

Reading the truth table in MINTERM code and MAXTERM code, we obtain

$$f = \sum m(3, 4, 5) + \phi(1, 2)$$
$$= \overline{A}BC + A\overline{B}\,\overline{C} + A\overline{B}C + \phi(\overline{A}\,\overline{B}C + \overline{A}B\overline{C})$$

SOP

$$f = \prod M(0, 6, 7) \cdot \phi(1, 2)$$
$$= (A + B + C)(\overline{A} + \overline{B} + C)(\overline{A} + \overline{B} + \overline{C})$$
$$\cdot \phi[(A + B + \overline{C})(A + \overline{B} + C)]$$

POS

EXAMPLE 3-8

Minimize the function represented in the truth table of Fig. 3-59 in both SOP and POS form by using third-order K-maps having the format given in Fig. 3-15(b). Identify any OPIs that are present.

SOLUTION

The K-maps for SOP and POS cover are shown in Fig. 3-60. The EPIs are represented by solid line loops while the OPIs are identified by dashed line loops.

3 / Function Representation and Reduction

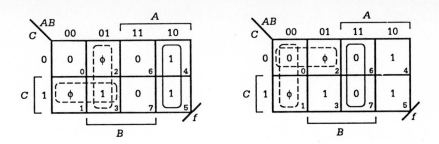

(a) SOP cover   (b) POS cover

Fig. 3-60. *K-maps for the function represented in Figure 3-59 showing minimum SOP and POS cover with OPIs indicated by dashed loops.*

Reading the maps, we obtain

$$f = A\bar{B} + \left\{\begin{array}{c}\overline{AB} \\ \overline{AC}\end{array}\right\} \qquad \text{SOP}$$

$$f = \left\{\begin{array}{c}(A + B) \\ (A + C)\end{array}\right\} \cdot (\bar{A} + \bar{B}) \qquad \text{POS}$$

where the OPIs have been placed in braces. As an exercise the reader should prove that these SOP and POS expressions are not algebraically equal but they are logically equivalent.

EXAMPLE 3-9   The output functions for the four-input/two-output logic system shown in Fig. 3-61 are given by the CANONICAL expressions

$$f_1(A,B,C,D) = \prod M(1,2,3,4,5,9,10) \cdot \phi(6,11,13)$$
$$= \sum m(0,7,8,12,14,15) + \phi(6,11,13) \tag{3-71}$$

$$f_2(A,B,C,D) = \prod M(2,5,9,10,11,15) \cdot \phi(3,4,13,14)$$
$$= \sum m(0,1,6,7,8,12) + \phi(3,4,13,14) \tag{3-72}$$

Find the optimum cover (SOP or POS) for this system and implement it in NAND/INV or NOR/INV logic.

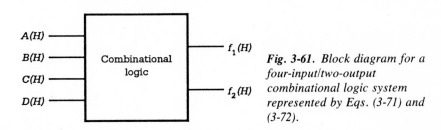

Fig. 3-61. *Block diagram for a four-input/two-output combinational logic system represented by Eqs. (3-71) and (3-72).*

SOLUTION   Following the procedure given in Sect. 3.7, we find the optimal POS cover and the optimal SOP cover, and then compare the two to determine which (if either) is the more minimum.

1. Optimal POS Cover

ORing of the CANONICAL POS forms of Eqs. (3-71) and (3-72) gives, by inspection, the results

$$f_1 + f_2 = \prod M(2, 3, 4, 5, 9, 10, 11) \cdot \phi(13)$$

where use has been made of the ORing rules given by Eqs. (3-38) and (3-42). The K-map for the ORed function, the resultant table of shared POS PIs, and the K-maps for the individual functions f_1 and f_2 are given in Fig. 3-62.

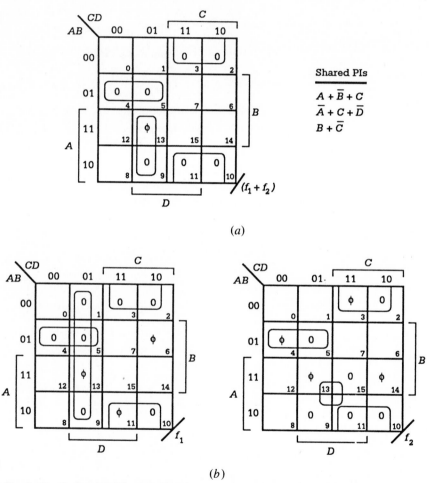

(a)

(b)

Fig. 3-62. *Optimal POS cover for the two-output system of Figure 3-61. (a) K-map and shared PIs for the ORed function. (b) K-maps for the individual functions* f_1 *and* f_2 *showing optimal POS cover for the two-output system.*

The final step involves looping out the individual functions by following the loop-out protocol in such a manner as to incorporate as many shared PIs as necessary to achieve optimum cover for the two outputs. Reading the K-maps for f_1 and f_2, we find the optimal POS cover to be

$$f_1 = (A + \overline{B} + C)(B + \overline{C})(C + \overline{D})$$

$$f_2 = (A + \overline{B} + C)(B + \overline{C})(\overline{A} + \overline{D})$$

Notice that the shared PI diad $(\overline{A} + C + \overline{D})$ is covered by the quad $(C + \overline{D})$ in the expression for f_1 and that the optimum coverage for both f_1 and f_2 is, in this case, that of the individual minimum forms. As we shall see next, optimum cover and minimum cover will not be the same for the SOP results.

2. Optimal SOP Cover

ANDing the CANONICAL SOP forms of Eqs. (3-71) and (3-72) by using the ANDing rules given by Eqs. (3-37) and (3-41), gives us the results

$$f_1 \cdot f_2 = \sum m(0, 6, 7, 8, 12, 14) + \phi(13)$$

The K-map for the ANDed functions, the resulting table of shared SOP PIs, and the K-maps for the individual functions are presented in Fig. 3-63.

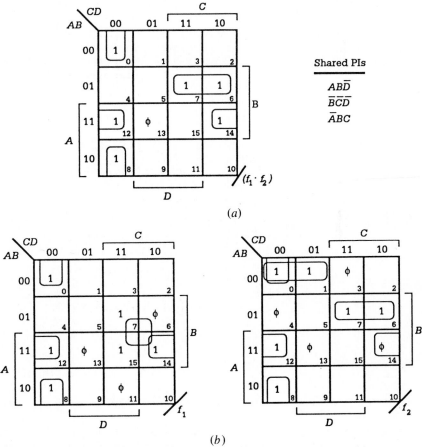

Fig. 3-63. *Optimal SOP cover for the four-input/two-output system represented by Figure 3-61. (a) K-map and table of shared PIs for the ANDed function. (b) K-maps for the individual functions showing optimal SOP cover for the two-output system.*

Following the loop-out protocol but including the shared PIs so as to produce optimum SOP cover, we read the K-maps for f_1 and f_2 to give

$$f_1 = AB\overline{D} + \overline{B}\,\overline{C}\,\overline{D} + BC$$

$$f_2 = AB\overline{D} + \overline{B}\,\overline{C}\,\overline{D} + \overline{A}BC + \overline{A}\,\overline{B}\,\overline{C}$$

Notice that neither expression by itself is an absolute minimum function, but when taken together, the two functions represent optimum SOP cover for the system.

By inspection of the optimal POS and SOP results, it is easily seen that six NOR gates and four INVERTERs are required for the POS cover, while seven NAND gates and four INVERTERs are needed for SOP cover. Thus, the POS design is simpler and yields the circuit presented in Fig. 3-64.

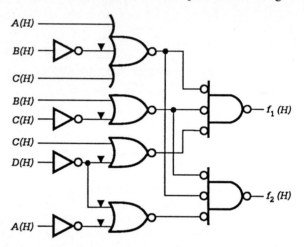

Fig. 3-64. *Optimum NOR logic circuit for the output functions represented by Eqs. (3-71) and (3-72).*

EXAMPLE 3-10

A four-variable function is compressed by one order and is represented by the EV truth table in Fig. 3-65. Give this function in CANONICAL SOP and POS form by using only coded notation.

A	B	C	f
0	0	0	0
0	0	1	1
0	1	0	S
0	1	1	1
1	0	0	0
1	0	1	$\overline{S}$
1	1	0	1
1	1	1	S

Fig. 3-65. *EV truth table for Example 3-10.*

SOLUTION   The truth table of Fig. 3-65 is read just as one would read a K-map with MAP KEY = 2^{4-3} = 2. In MINTERM code, term 1 is read as 0010 + 0011

$= m_2 + m_3$, term 2 is read as $0101 = m_5$ (visualize a first-order K-map in S to get the least significant bit (LSB)), and term 5 is read as $1010 = m_{10}$, and so on. In MAXTERM code terms 0, 2, and 5 are read as $(M_0 \cdot M_1)$, M_4, and M_{11}, respectively. Proceeding in this manner the results are

$$f(A,B,C,S) = \sum m(2, 3, 5, 6, 7, 10, 12, 13, 15)$$
$$= \prod M(0, 1, 4, 8, 9, 11, 14)$$

EXAMPLE 3-11

Minimize the function of Example 3-10 in both SOP and POS form by using a third-order K-map.

SOLUTION

Shown in Fig. 3-66 are the compressed K-maps representing SOP and POS cover for the function of Fig. 3-65.

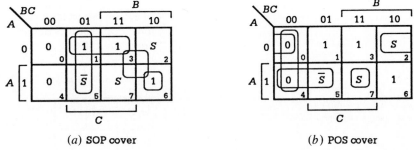

(a) SOP cover   (b) POS cover

Fig. 3-66. *Minimum SOP and POS cover for the function represented in Figure 3-65.*

Following a procedure similar to that used for Figs. 3-44 and 3-45, we apply the loop-out protocol twice, first, to the entered variables and, then, to the 1's and 0's as "clean-up" operations. From Fig. 3-66 the results are read directly as

$$f(A,B,C,S) = \overline{B}C\overline{S} + BS + AB\overline{C} + \overline{A}C \qquad \text{SOP}$$
$$= (\overline{A} + \overline{B} + \overline{C} + S)(\overline{A} + B + \overline{S})(A + C + S)(B + C) \qquad \text{POS}$$

which can be shown to be algebraically equal.

EXAMPLE 3-12

A four-variable function containing don't cares is shown in the compressed third-order SOP K-map of Fig. 3-67.

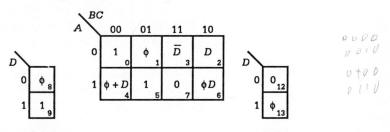

Cell 4   Cell 6

Fig. 3-67. *Compressed SOP K-map for Example 3-12 showing sample first-order submaps.*

3.13 / Worked Exercises

171

a. Represent the function in CANONICAL SOP and POS form by using coded notation.

b. Find the minimum SOP and POS cover for the function by following the loop-out protocol as applied in Example 3-11.

SOLUTION   **a.** Noting that the MAP KEY is $2^{4-3} = 2$, we may write directly the CANONICAL forms by making use of first-order submaps for D and by applying the MINTERM and MAXTERM codes. Thus, cell 0 represents $(m_0 + m_1)$, cell 2 represents m_5 or M_4, cell 3 represents m_6 or M_7, cell 4 represents $(\phi m_8 + m_9)$ or ϕM_8, and so on. Proceeding in this manner, we find the results

$$f(A,B,C,D) = \sum m(0, 1, 5, 6, 9, 10, 11) + \phi(2, 3, 8, 13)$$
$$= \prod M(4, 7, 12, 14, 15) \cdot \phi(2, 3, 8, 13)$$

b. Since don't care terms are present in the EV map of Fig. 3-67, the POS interpretation may not be the same as the SOP interpretation. For this reason both maps should be constructed as shown in Fig. 3-68.

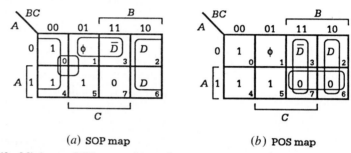

(a) SOP map   (b) POS map

Fig. 3-68. *Minimum SOP cover (a) and minimum POS cover (b) for the function given by Figure 3-67.*

Reading these maps, we obtain

$$f = \overline{A}C\overline{D} + \overline{C}D + \overline{B} \qquad\qquad \text{SOP}$$
$$f = (\overline{B} + C + D)(\overline{B} + \overline{C} + \overline{D})(\overline{A} + \overline{B}) \qquad \text{POS}$$

which are found to be algebraically unequal but logically equivalent. Notice that the 1's in the SOP map are looped out as the octad $\overline{B}$ by using the don't care term in cell 4 of Fig. 3-67 as $\phi + D = 1$ ($\phi = 1$), and that the 0 in the POS map is looped out as the quad $\overline{A} + \overline{B}$ by using the don't care term in cell 6 of Fig. 3-67 as $\phi D = 0$ ($\phi = 0$).

EXAMPLE 3-13   A four-variable function $f(A,B,C,D)$ is compressed into the truth table given in Fig. 3-69. This table represents a second-order compression in variables C and D.

a. Represent the function f in a second-order K-map, and express f in CANONICAL SOP and POS form by using coded notation.

b. By proper interpretations of the don't care subfunctions, loop out the minimum SOP and POS cover from the second-order K-map.

A B	f
0 0	$C \cdot D$
0 1	$C \cdot (\phi + \overline{D})$
1 0	$(\phi + C + D)$
1 1	0

Fig. 3-69. *Compressed truth table for a function of four variables.*

SOLUTION

a. The simplest means of obtaining the CANONICAL forms from Fig. 3-69 is to use a second-order K-map. Shown in Fig. 3-70 is the second-order compressed K-map for this function together with its submaps for a MAP KEY of $2^{4-2} = 4$.

By reading the submaps directly, we obtain the CANONICAL forms

$$f = \prod M(0, 1, 2, 4, 5, 12, 13, 14, 15) \cdot \phi(7, 8)$$

$$= \sum m(3, 6, 9, 10, 11) + \phi(7, 8)$$

Notice that the subfunctions in cells 1 and 2 can also be written in equivalent SOP form as $(\phi C + C\overline{D})$ and $(\phi \overline{C}\overline{D} + C + D)$, respectively.

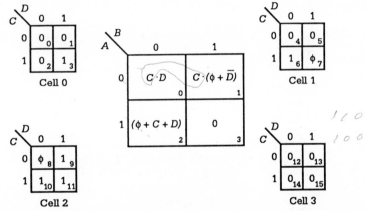

Fig. 3-70. *Second-order compressed K-map and its submaps for the four variable function given in the EV truth table of Figure 3-69.*

b. Given in Fig. 3-71 are the compressed second-order K-maps showing both POS and SOP cover. Notice that minimum POS and SOP cover results from these K-maps by taking $\phi_7 = 1$ to give $C \cdot (\phi_7 + \overline{D}) = C$ in cell 1, and by taking $\phi_8 = 1$ to give $(\phi_8 + C + D) = 1$ in cell 2. Thus, in this case, the SOP and POS K-maps, required for minimum cover, are congruent.

By following the loop-out protocol, first for the EV terms then for the 0's (POS cover) and 1's (SOP) cover, we obtain the minimum POS and SOP forms as follows:

$$f(A,B,C,D) = (A + B + D)(A + C)(\overline{A} + \overline{B}) \quad \text{POS}$$

$$f(A,B,C,D) = \overline{A}BC + \overline{B}CD + A\overline{B} \quad \text{SOP}$$

3.13 / Worked Exercises

173

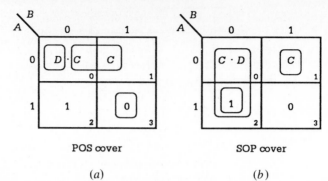

POS cover

SOP cover

(a)

(b)

Fig. 3-71. *Minimum POS and SOP cover for the function represented in Figure 3-69.*

That these results are absolute minimum expressions may be shown by looping out minimum cover with the aid of the submaps provided in Fig. 3-70, or by looping out the CANONICAL data (given earlier) in fourth-order K-maps. Note that the p-terms $\overline{B}CD$ and $\overline{A}CD$ are OPIs for the SOP expression.

EXAMPLE 3-14

A five-variable function is given by the CANONICAL form

$$f(A,B,C,D,E) = \sum m(3, 9, 10, 12, 13, 16, 17, 24, 25, 26, 27, 29, 31) \quad (3\text{-}73)$$

a. Use a fourth-order EV K-map to minimize this function in both SOP and POS form.
b. Minimize this function with conventional mapping methods by using two fourth-order K-maps, one for $\overline{A}$ and one for A.

SOLUTION

a. A compression of one order requires that the MAP KEY be 2. Therefore, each cell of the fourth-order EV map represents a first-order submap covering two possible MINTERM or MAXTERM positions. Keeping this in mind, we construct the fourth-order K-maps for SOP and POS cover, as shown in Fig. 3-72.

To avoid possible redundancy, the loop-out protocol is applied first to the EVs and then to the 1's and 0's for SOP and POS cover, respectively. The results, read directly off the maps in Fig. 3-72, are

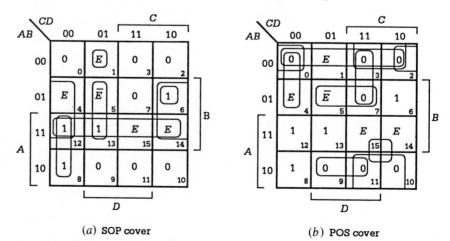

(a) SOP cover

(b) POS cover

Fig. 3-72. *Minimum SOP and POS cover for the five-variable function given by Eq. (3-73).*

$$f = \overline{A}\,\overline{B}\,\overline{C}DE + B\overline{C}D\overline{E} + BDE + ABE + \overline{A}BC\overline{D} + A\overline{C}\overline{D}$$

$$= (A + C + D + E)(A + \overline{B} + \overline{D} + \overline{E})(\overline{A} + \overline{C} + E)$$

$$\cdot\, (A + B + E)(\overline{A} + B + \overline{D})(A + \overline{C} + \overline{D})$$

$$\cdot\, (A + B + D)(B + \overline{C})$$

where for the POS result it is seen that $(B + \overline{D} + E)$ is an OPI for the $(A + B + E)$ term.

    b. To use the conventional mapping method, two fourth-order maps must be constructed, one for the map heading variable and the other for its complement. Although any of the five variables can be selected as the map heading variable, it is usually much easier to use the most significant bit for that purpose. Hence, in this case, we would construct one map for $A = 0$ ($\overline{A}$ map) and the other for $A = 1$ (A map). This permits all possible MINTERMs (or MAXTERMs), 0 through 15, to be placed in the $\overline{A}$ map with the remaining, 16 through 31, assigned to the A map.

    There are two possible formats which can be used to carry out this conventional minimization. The first is the *stack format*, shown in Fig. 3-73(a),

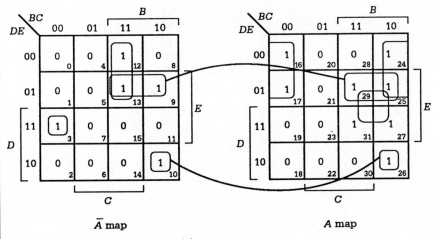

(a) Stack format

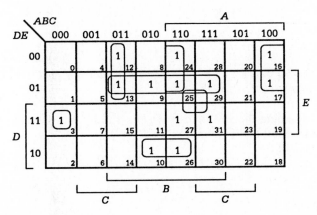

(b) Planar format

Fig. 3-73. *Conventional K-map formats for minimizing the function given by Eq. (3-73).*

where the maps are viewed as being placed one above the other to produce a two-layer, three-dimensional array of cells. The other is the *planar format* presented in Fig. 3-73(b). This latter format features one three-variable axis unfolded in such manner as to permit the $\overline{A}$ and A maps to be placed in juxtaposition to one another with adjacency cells located horizontally across the $\overline{A}/A$ boundary. Notice that the C domain is split into two parts, the cost that one must pay to use the planar format. The SOP cover, leading to the same results as in part (a), is given by both map formats of Fig. 3-73.

EXAMPLE 3-15

Compress the function of Eq. (3-68) in Example 3-2 into a second-order K-map by using reduced (not CANONICAL) EV subfunctions, and obtain an absolute minimum SOP and POS cover. Compare the results with those obtained in Example 3-6.

SOLUTION

The procedure to be followed here differs significantly from that followed in Example 3-13 since, in this case, the second-order EV maps are used with the aid of second-order submaps to obtain the absolute minimum expressions. Without the use of uncompressed submaps, or CANONICAL subfunctions, an absolute minimum cover cannot be guaranteed as was indicated in Sect. 3-9. The result in Example 3-13 turned out to be an absolute minimum expression only because the reduced subfunctions were such as to permit it by direct reading.

Required for this problem is a second-order compression with a MAP KEY of $2^{4-2} = 4$. In Fig. 3-74 is shown the second-order EV maps for SOP and POS cover together with the second-order submaps.

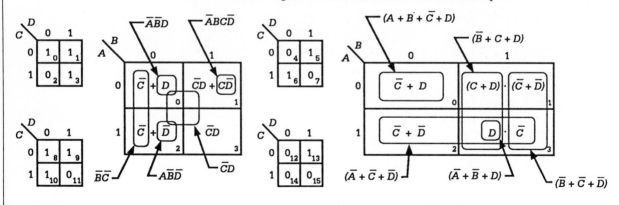

(a) SOP cover   (b) POS cover

Fig. 3-74. Second-order EV K-maps and submaps for Eq. (3-68) showing minimum SOP and POS cover.

Reading the EV maps with the aid of the submaps, we obtain the minimum SOP and POS cover

$$f = \overline{A}BC\overline{D} + \overline{A}BD + A\overline{B}\overline{D} + \overline{B}\overline{C} + \overline{C}D$$
$$= (A + B + \overline{C} + D)(\overline{B} + \overline{C} + \overline{D})(\overline{B} + C + D)$$
$$\cdot (\overline{A} + \overline{B} + D)(\overline{A} + \overline{C} + \overline{D})$$

as in Example 3-6. Notice how easy it is to read a reduced EV subfunction when accompanied by a submap. Thus, the SOP term $\overline{C}D$ is readily observed to be present in each of the four submaps of Fig. 3-74. Similarly, $\overline{C}D$, read as a POS term in Fig. 3-74, is easily seen to contain both the $(C + D)$ and $(\overline{C} + \overline{D})$ terms by a cursory inspection of the submaps.

EXAMPLE 3-16

A six-variable function is represented by the CANONICAL POS expression

$$f(A,B,C,D,E,F) = \prod M(1, 3, 4, 6, 7, 9, 10, 11, 12, 13, 14,$$
$$15, 20, 22, 24, 32, 33, 34, 35, 39, \qquad (3\text{-}74)$$
$$41, 43, 44, 45, 46, 47, 48, 49, 56, 57)$$

a. Use a fourth-order EV map with reduced subfunctions to minimize this function in both POS and SOP form and comment on which is the simpler.
b. Demonstrate how one would arrive at these results by using conventional mapping methods.

SOLUTION **a.** Following the procedure outlined in Sect. 3-10 and noting that the MAP KEY is 4, we construct in MAXTERM code the fourth-order EV maps shown in Fig. 3-75. Although the SOP and POS maps are congruent, separate maps are used here for the sake of clarity.

The minimum cover indicated in the K-maps is easily obtained, for the most part, by direct inspection of each map. However, the use of second-order submaps is helpful to deduce some of the less obvious adjacencies. Reading the K-maps, we find the minimum results to be

$$f_{\text{POS}} = (\overline{B} + \overline{C} + D + E + F)(\overline{A} + \overline{B} + D + E)(A + C + \overline{D} + F)$$
$$\cdot (A + B + \overline{C} + \overline{E})(B + \overline{D} + \overline{E} + \overline{F})(B + D + \overline{F})$$
$$\cdot (\overline{A} + B + C + D)(B + \overline{C} + \overline{D})$$

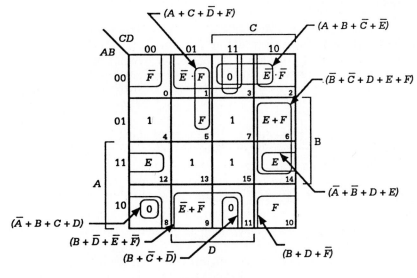

(a) **POS cover**

Fig. 3-75. Fourth-order EV K-maps showing minimum POS and SOP cover for the six-variable function represented by Eq. (3-74).

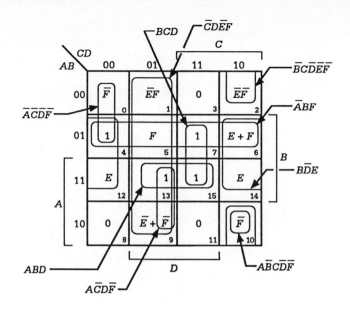

(b) SOP cover

Fig. 3-75 (contd.).

and

$$f_{\text{SOP}} = A\overline{B}C\overline{D}\,\overline{F} + \overline{B}\,\overline{C}\,\overline{D}\,\overline{E}\,F + \overline{C}D\overline{E}F + \overline{A}\,\overline{C}D\overline{F} + A\overline{C}D\overline{F}$$

$$+ B\overline{D}E + \overline{A}BF + ABD + BCD$$

Both expressions are written in accordance with the loop-out protocol, applied first to the EV subfunctions and then to the 1's and 0's as clean-up operations. Note that the POS form is the simpler of the two since it requires the ANDing of eight ORed terms, whereas the SOP expression requires the ORing of nine ANDed terms.

b. Conventional minimization of Eq. (3-74) requires the use of four fourth-order K-maps, each representing two variables called the map heading variables. The map heading variables are usually taken to be the two most significant bits so that the cells of the fourth-order K-maps can be numbered in the conventional manner from 0 through 63 while, at the same time, retaining the familiar adjacency patterns for each map.

In Fig. 3-76 are presented the two conventional formats which can be used to minimize the six-variable function given by Eq. (3-74). To use the stack format, the maps are viewed as being placed one above (or below) the other in the map heading order $\overline{A}\,\overline{B}$, $\overline{A}B$, $A\overline{B}$, and AB, to produce a four-layer three-dimensional array of cells—somewhat like space chess. The planar format requires the use of two three-variable axes, each unfolded in such a manner as to permit the maps of the stack format to be placed in juxtaposition to one another with adjacent cells located, in this case, vertically across the $\overline{A}/A$ boundary and horizontally across the $\overline{D}/D$ boundary. So, while the A, B, D, and E domains remain intact, the domains for C and F are split into two parts, the penalty one must pay to use this format.

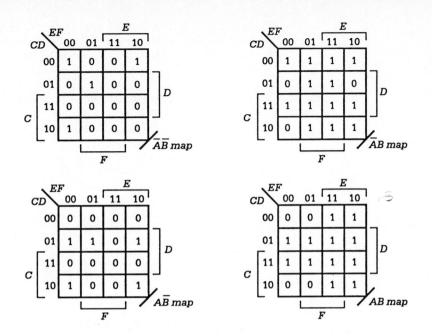

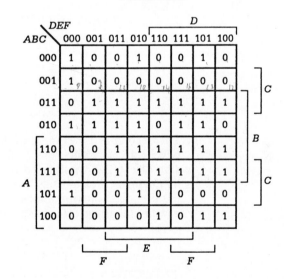

(b) Planar format

Fig. 3-76. *Stack and planar formats for the conventional minimization of the six-variable function of Eq. (3-74).*

EXAMPLE 3-17

A three-variable function is represented by the CANONICAL expression

$$Y(A,B,C) = \sum m(1, 3, 4, 5, 6) \qquad (3\text{-}75)$$

a. Minimize the function of Eq. (3-75) in SOP and POS form by using a second-order K-map with A (the MSB) as the entered variable.

b. Repeat (a) by using a first-order K-map with A and B as the entered variables.

SOLUTION   **a.** The simplest approach to the solution of this problem is to construct a CANONICAL truth table and read from it the map entries in variable A. This is done in Fig. 3-77, where we show four first-order submaps in A taken from the truth table, one each for $BC = 00, 01, 10$, and 11, as indicated by the arrows to the right of the truth table. Notice that the minimum SOP expression has OPIs, whereas the simpler POS form does not.

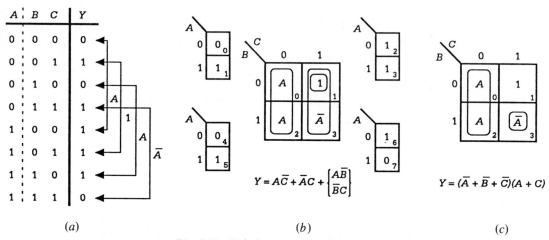

A	B	C	Y
0	0	0	0
0	0	1	1
0	1	0	0
0	1	1	1
1	0	0	1
1	0	1	1
1	1	0	1
1	1	1	0

$$Y = A\bar{C} + \bar{A}C + \begin{bmatrix} A\bar{B} \\ \overline{BC} \end{bmatrix}$$

$$Y = (\bar{A} + \bar{B} + \bar{C})(A + C)$$

(a)   *(b)*   *(c)*

Fig. 3-77. *Tabular representation and second-order EV K-maps in variable A for the function given by Eq. (3-75). (a) CANONICAL truth table. (b) Minimum SOP cover showing the four submaps and two OPIs. (c) Minimum POS cover.*

b. The second-order compression of Eq. (3-75) can be obtained from the CANONICAL truth table of Fig. 3-77(a) or from the first-order compression of Fig. 3-77(b). The results are given in Fig. 3-78 together with the minimum SOP and POS cover. Use of the CANONICAL truth table requires the construction of two second-order submaps, one for $C = 0$ and the other for $C = 1$, as shown.

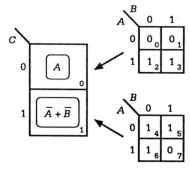

$$Y = A\bar{C} + \bar{A}C + \bar{B}C$$
$$= (\bar{A} + \bar{B} + \bar{C})(A + C)$$

Fig. 3-78. *Second-order compression in variables A and B showing submaps and minimum SOP and POS cover.*

EXAMPLE 3-18

Given in Fig. 3-79 is a five-variable function represented in a third-order K-map.

a. Keeping in mind the various XOR-type patterns discussed in Sect. 3.12.2, find the *gate-minimum* cover for this function.
b. Determine the gate/input tally for the minimum cover of part (a) if implementation is limited to gates having three inputs, a fan-in of three.
c. Compare the results of part (b) with the gate/input tally for minimum two-level logic cover obtained from a third-order EV K-map.

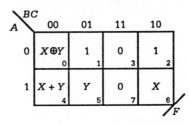

Fig. 3-79. *Second-order compression of a five-variable function showing an XOR subfunction entry.*

SOLUTION

a. The gate-minimum cover for the function of Fig. 3-79 is shown in Fig. 3-80(a) and is read as

$$F = (\overline{C}X) \oplus (\overline{B}Y) + \begin{Bmatrix} A\overline{B}Y \\ A\overline{C}X \end{Bmatrix} + \overline{A}(B \oplus C)$$

which includes two OPIs. Inclusion of one of the OPIs is necessary to remove the residual term in the $X + Y$ entry of cell 4. As an exercise the reader should show by use of submaps that inclusion of both X and Y (of cell 4) in the associative pattern does not cover all the terms contained in the cell. Thus, when looping out associative XOR patterns (or any EV patterns for that matter), it is important to check for any residual terms that might exist. If present, they must be looped out as a clean-up stage. Notice that the 1's in Fig. 3-80(a) are looped out as an offset pattern to give the $\overline{A}(B \oplus C)$ term.

b. By count, the gate/input tally for the results of part (a) is 7/16, including two XOR gates and five three-input NAND gates (four for ANDing and one for ORing).

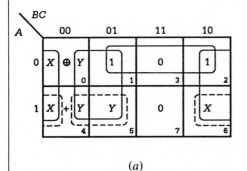

(a)

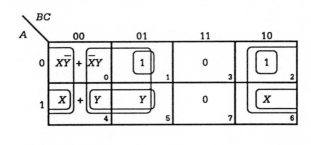

(b)

Fig. 3-80. *(a) Gate minimum cover and (b) two-level minimum cover for the function given by Figure 3-79.*

c. The minimum cover for two-level logic is shown in Fig. 3-80(b) and is read as

$$F = \overline{C}X\overline{Y} + \overline{B}XY + A\overline{C}X + A\overline{B}Y + \overline{A}\,\overline{B}C + \overline{A}B\overline{C}$$

Given a fan-in of three limitation, the gate/input tally for two-level implementation would be 9/26, significantly larger than the multilevel result. If account must be taken of the INVERTERs required for any given set of input conditions, the gate/input tally differential would certainly rise.

EXAMPLE 3-19

The operators $(+)$ and $\oplus$ connecting any two terms of an expression can be interchanged if, and only if, the two terms meet the requirement of corollary I [Eqs. (2-34) given in Sect. 2.9.6]. Similarly, the operators $(\cdot)$ and $\odot$ connecting any two terms of an expression can be interchanged providing the two terms meet the requirements of corollary II [Eqs. (2-35)].

a. Interchange the operators of the CANONICAL expressions

$$f(A,B,C,D) = \sum m(0, 2, 3, 4, 6)$$
$$= \prod M(1, 5, 7) \tag{3-76}$$

b. Map Eqs. (3-76) in second-order K-maps and loop out the most minimum cover which will permit interchange of the operators connecting the product and sum terms.

SOLUTION
a. Since any CANONICAL expression must necessarily satisfy either corollary I or corollary II (refer to Sect. 2.9.6), there results

$$f(A,B,C,D) = \sum m(0, 2, 3, 4, 6)$$
$$= \sum \text{XOR}(0, 2, 3, 4, 6)$$
$$= (\overline{A}\,\overline{B}\,\overline{C}) \oplus (\overline{A}B\overline{C}) \oplus (\overline{A}BC) \oplus (A\overline{B}\,\overline{C}) \oplus (AB\overline{C})$$

and

$$f(A,B,C,D) = \prod M(1, 5, 7)$$
$$= \prod \text{EQV}(1, 5, 7)$$
$$= (A + B + \overline{C}) \odot (\overline{A} + B + \overline{C}) \odot (\overline{A} + \overline{B} + \overline{C})$$

Observe that the corollaries are satisfied in each case since $(\overline{A}\,\overline{B}\,\overline{C})(\overline{A}B\overline{C}) = 0$ and $(\overline{A}B\overline{C})(\overline{A}BC) = 0$, and so on, and since $(A + B + \overline{C}) + (\overline{A} + B + \overline{C}) = 1$ and $(\overline{A} + B + \overline{C}) + (\overline{A} + \overline{B} + \overline{C}) = 1$.
b. The second-order K-maps for Eqs. (3-76) are given in Fig. 3-81. The minimum permissible cover to satisfy corollaries I and II is read from each map to be

$$f_{\text{SOP}} = \overline{C} + \overline{A}BC$$
$$= \overline{C} \oplus \overline{A}BC$$

since $\overline{C}(\overline{A}BC) = 0$

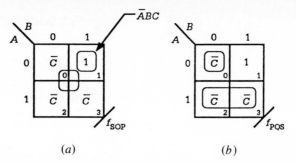

Fig. 3-81. *K-maps for Eqs. (3-76) showing the minimum SOP cover (a) and minimum POS cover (b) which will satisfy Corollaries I and II given by Eqs. (2-34) and (2-35), respectively.*

and
$$f_{POS} = (\overline{A} + \overline{C})(A + B + \overline{C})$$
$$= (\overline{A} + \overline{C}) \odot (A + B + \overline{C})$$

since $(\overline{A} + \overline{C}) + (A + B + \overline{C}) = 1$. Notice that in each case the most minimum cover will not satisfy the corollaries. That is, $f_{SOP} = \overline{C} + \overline{A}B$ does not satisfy corollary I, nor does $f_{POS} = (A + \overline{C})(B + \overline{C})$ satisfy corollary II. So in this case, operator interchange is not permissible for minimum cover.

ANNOTATED REFERENCES

Most of the references cited in the Annotated References of Chapter 2 provide adequate coverage of logic function representation and manipulation and will not be cited here.

Only two texts are known to describe and make significant use of the entered variable map method of function reduction. These are the texts of Fletcher and Comer.

Fletcher, W. I., *An Engineering Approach to Digital Design*, Prentice Hall, Englewood Cliffs, N.J., 1980.

Comer, D. J., *Digital Logic and State Machine Design*, Holt, Rinehart and Winston, New York, 1984.

There are no known references on XOR entered variable map patterns.

Conventional (1's and 0's) mapping and function reduction methods are covered adequately by most of the references given in the Annotated References of Chapter 2.

The Quine-McCluskey tabular minimization method and the entered variable map method are reviewed in Chapter 3 of the text by Unger. Also, in Chapter 7 on software tools, Unger provides a short review of logic minimization methods generally. Some useful references are given at the end of Chapters 3 and 7.

Unger, S. H., *The Essence of Logic Circuits*, Prentice Hall, Englewood Cliffs, N.J., 1989.

An excellent software tool for minimizing Boolean functions, called BOOZER (for BOOlean ZEro-one Reduction), is available. The basic BOOZER algorithm is described in Chapter 3 of Fletcher. The algorithm has been extended to include map entered variables and can handle up to 16 functions of 16 variables when used on modern mini- and mainframe computers. For information contact M. L. Manwaring, Dept. of Electrical and Computer Engineering, Washington State University, Pullman, Wash. 99164-2752.

A popular software tool for minimizing Boolean functions, called Espresso, is available through the University of California, Berkeley, 1986 VLSI tools distribution. It supports advanced algorithms for minimization of two-level, multiple-output Boolean functions but does not accept MEVs. The algorithms are described in an article by Rudell.

Rudell, R., "Espresso-MV: Algorithms for Multiple-Valued Logic Minimization," Proc. Int. Circ. Conf., May 1985.

PROBLEMS

3.1 Expand each of the following expressions into CANONICAL (literal) form by using the appropriate Boolean laws:

(a) $e(a,b) = a + \bar{b}$

(b) $f(x,y) = x + \bar{x}y$

(c) $g(A,B,C) = A\bar{B}C + \bar{A}BC + AB + BC + \bar{A}B\bar{C}$

(d) $h(X,Y,Z) = (X + Z)(\bar{X} + \bar{Y} + Z)(Y + \bar{Z}) \cdot (\bar{X} + Y + \bar{Z})$

3.2 Expand each of the following expressions into CANONICAL code form by using the appropriate Boolean laws:

(a) $E(A,B,C,D) = [\bar{A} + (\bar{B}C)](B + D) \cdot (\bar{A} + C + D) \cdot (A + \bar{B} + C + \bar{D})(B + D)$

(b) $F(w,x,y,z) = wxy\bar{z} + \bar{w}\bar{x}z + x\bar{y}z + w\bar{x}yz + xyz + w\bar{x}\bar{y}\bar{z} + \bar{w}x\bar{y}z$

(c) $G(a,b,c,d,e) = (a + \bar{b} + c + \bar{d}) \cdot (b + \bar{c} + \bar{d}) \cdot (\bar{a} + b + d + e) \cdot (\bar{b} + d + \bar{e})$

(d) $H(V,W,X,Y,Z) = VW\bar{X}\bar{Y}Z + \bar{X}Z + WX\bar{Y}Z + VW\bar{X}Y\bar{Z} + VWXY + \overline{VWX}Y\bar{Z} + \overline{VWXYZ}$

3.3 Place each of the following functions in a CANONICAL truth table and in a conventional (1's and 0's) K-map:

(a) $Q(A,B,C) = (A + B + \bar{C})(A + \bar{B} + \bar{C}) \cdot (A + B + C)(\bar{A} + B + \bar{C}) \cdot (\bar{A} + \bar{B} + \bar{C})$

(b) $R(u,v,w,x) = \sum m(0, 2, 3, 7, 8, 9, 10, 11, 13)$

(c) $S(a,b,c,d) = (a + \bar{b}) [\bar{a} + (b\cdot\bar{c})] \cdot (\bar{b} + c) (\bar{a} + b + c)$

(d) $T(W,X,Y,Z) = YZ + \overline{WX}Y + WX\bar{Y}Z + \bar{X}YZ + \bar{W}YZ + WXY\bar{Z} + X\bar{Y}Z$

3.4 Place each of the following functions in a CANONICAL truth table and in a conventional (1's and 0's) K-map:

(a) $W(A,B,C) = \bar{A}B\bar{C} + ABC + A\bar{B}C + AB\bar{C} + \bar{A}BC$

(b) $X(A,B,C,D) = \prod M(0, 5, 8, 9, 11, 12, 15)$

(c) $Y(A,B,C) = AB + \bar{B}C + A(B \oplus C) + \bar{A}\bar{C}$

(d) $Z(A,B,C,D) = (B + C + \bar{D}) \cdot (\bar{A} + \bar{C} + \bar{D}) \cdot (\bar{B} + \bar{C} + D) (A + B) \cdot (A + \bar{B} + D)$

3.5 Minimize the following functions in both SOP and POS form by using the method of tabular minimization:

(a) $f(w,x,y) = \sum m(0, 1, 3, 5, 7)$

(b) $g(a,b,c) = \prod M(2, 3, 4, 6)$

(c) $F(W,X,Y,Z) = \sum m(0, 2, 4, 5, 6, 8, 10, 11, 13, 14)$

(d) $G(A,B,C,D) = \prod M(1, 2, 3, 5, 7, 9, 11, 12, 14)$

3.6 Minimize the following functions in both SOP and POS form by using third-order K-maps. For each, determine which form is the simpler.

(a) $f(a,b,c) = \sum m(1, 2, 4, 5, 6)$

(b) $g(x,y,z) = \prod M(0, 1, 2, 6, 7)$

(c) $h(A,B,C) = A\overline{B}C + \overline{A}\,\overline{B}\,\overline{C} + AB\overline{C} + ABC + \overline{A}B\overline{C}$

(d) $i(X,Y,Z) = (X + \overline{Y} + \overline{Z})(\overline{X} + Y + \overline{Z}) \cdot (\overline{X} + \overline{Y} + Z)(X + Y + Z)$

3.7 Minimize the following functions in both SOP and POS form by using fourth-order K-maps. List the EPIs and OPIs in each case.

(a) $W(A,B,C,D) = \prod M(1, 2, 3, 4, 5, 7, 10, 11, 12, 14, 15)$

(b) $X(A,B,C,D) = \sum m(0, 4, 5, 7, 8, 9, 13, 15)$

(c) $Y(A,B,C,D) = \prod M(0, 2, 4, 5, 8, 9, 10, 12, 15)$

(d) $Z(A,B,C,D) = \sum m(2, 3, 8, 9, 10, 11, 12, 13, 14, 15)$

3.8 Minimize the functions Q, R, S, and T of Problem 3 in both SOP and POS form by using conventional (1's and 0's) K-maps and the loop-out protocol.

3.9 Minimize the functions W, X, Y, and Z of Problem 4 in both SOP and POS form by using conventional (1's and 0's) K-maps and the loop-out protocol.

3.10 Reduce the following functions to their *simplest* SOP and POS forms by using conventional (1's and 0's) K-maps with and without the use of the don't cares. Indicate with the gate/input tally which of the forms (SOP or POS) is the simpler in each case.

(a) $e(X,Y,Z) = \sum m(0, 1, 2, 7) + \phi(3, 5)$

(b) $f(X,Y,Z) = \prod M(3, 4, 6) \cdot \phi(0, 2)$

(c) $g(A,B,C,D) = \sum m(4, 5, 7, 12, 14, 15) + \phi(3, 8, 10)$

(d) $h(A,B,C,D) = \prod M(0, 1, 2, 5, 7, 9) \cdot \phi(4, 6, 10)$

3.11 Reduce the following functions to their *simplest* SOP and POS forms by using conventional (1's and 0's) K-maps. Indicate with the gate/input tally which of the forms (SOP or POS) is the simpler in each case.

(a) $P(a,b,c,d) = \sum m(6, 11, 12, 13, 14) + \phi(0, 1, 2, 3, 4, 5)$

(b) $Q(A,B,C,D) = \prod M(0, 3, 6, 11, 13, 15) \cdot \phi(5, 8, 10, 14)$

(c) $R(W,X,Y,Z) = \sum m(0, 4, 6, 8, 9, 10, 11, 14, 15) + \phi(1, 5)$

(d) $T(w,x,y,z) = \prod M(1, 2, 3, 9, 10, 14) \cdot \phi(11, 13)$

3.12 Given the following set of three functions, find $f_1 \cdot f_2$, $f_2 \cdot f_3$, $f_3 \cdot f_1$, $f_1 + f_2$, $f_2 + f_3$, and $f_3 + f_1$. Present your answers in CANONICAL and minimum form.

$$f_1(a,b,c,d) = \sum m(0, 1, 5, 7, 8, 10, 14, 15)$$

$$f_2(a,b,c,d) = \sum m(0, 2, 4, 5, 6, 7, 8, 10, 12)$$

$$f_3(a,b,c,d) = \sum m(0, 1, 2, 3, 4, 6, 8, 9, 10, 11)$$

3.13 Find the *optimum cover* (either SOP or POS) for the following four-input/two-output logic system. Base your choice on the gate/input tally including IN-VERTERs. Assume the inputs and outputs are all ACTIVE HIGH.

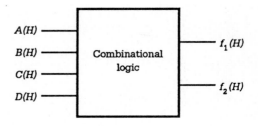

$$f_1 = \sum m(0, 2, 4, 5, 9, 10, 11, 13, 15)$$

$$f_2 = \sum m(2, 5, 10, 11, 12, 13, 14, 15)$$

3.14 Find the optimum SOP cover for the following four-input/two-output logic system. Give the gate/input tally for this optimum cover.

$$f_1(A,B,C,D) = \sum m(1, 7, 8, 10, 14, 15)$$
$$+ \phi(2, 5, 6)$$

$$f_2(A,B,C,D) = \sum m(5, 7, 8, 11, 14, 15)$$
$$+ \phi(2, 3, 10)$$

3.15 Find the optimum cover (either SOP or POS) for the system of three functions given in Problem 12. Base your choice on the gate/input tally excluding IN-VERTERs.

3.16 Find the optimum cover (either SOP or POS) for the following four-input/two-output logic system and implement in either NAND/INV or NOR/INV logic. Base your choice on the gate/input tally including INVERTERs. Assume ACTIVE HIGH inputs.

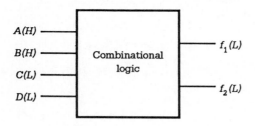

$$f_1 = \prod M(0, 3, 4, 11, 12, 13, 15) \cdot \phi(2, 5, 6)$$
$$f_2 = \prod M(0, 1, 9, 12, 13) \cdot \phi(2, 3, 4, 10)$$

3.17 Compress the functions f, g, h, and i in Problem 6 into second-order EV K-maps and extract minimum SOP and POS cover for each.

3.18 Compress the functions W, X, Y, and Z in Problem 7 into third-order EV K-maps and extract minimum SOP and POS cover for each.

3.19 Compress the following five-variable functions into a fourth-order EV K-maps and extract minimum SOP and POS cover for each:

(a) $q(A,B,C,D,E) = \prod M(0, 1, 2, 5, 14, 16, 17, 18,$
$$19, 21, 26, 27, 30)$$

(b) $r(A,B,C,D,E) = A\overline{B}CE + \overline{A}CDE + \overline{B}C\overline{D}E$
$$+ \ \overline{A}B C\overline{E} + \overline{A}BDE$$
$$+ \ \overline{A}B\overline{C}\overline{D}E + BC\overline{D}\overline{E}$$
$$+ \ AB\overline{D}E + ABCD$$
$$+ \ A\overline{B}\overline{C}\overline{D}\overline{E}$$

(c) $s(A,B,C,D,E) = \sum m(0, 2, 4, 5, 7, 10, 13, 15,$
$$21, 23, 24, 25, 28, 29, 30)$$

(d) $t(A,B,C,D,E) = (A + B + D + \overline{E})$
$$\cdot (\overline{B} + \overline{C} + \overline{D} + E)$$
$$\cdot (\overline{A} + \overline{B} + E)$$
$$\cdot (\overline{A} + C + D + \overline{E})$$
$$\cdot (\overline{B} + C + D)$$
$$\cdot (B + C + \overline{D} + E)$$
$$\cdot (\overline{A} + B + \overline{C} + D + \overline{E})$$
$$\cdot (\overline{B} + \overline{C} + D + E)$$
$$\cdot (A, + \overline{B} + C)\ (\overline{B} + C + \overline{D})$$
$$\cdot (B + \overline{C} + D + E)$$

3.20 Minimize in both SOP and POS the functions of Problem 19 by using conventional (1's and 0's) K-maps.

3.21 Compress the functions W, X, Y, and Z of Problem 4 into second-order EV K-maps and extract minimum SOP and POS cover for each.

3.22 Compress the functions e, f, g, and h of Problem 10 into second-order EV K-maps and extract minimum SOP and POS cover for each.

3.23 Compress the functions P, Q, R, and T of Problem 11 into third-order EV K-maps and extract minimum SOP and POS cover for each.

3.24 Compress the functions E, F, G, and H in Problem 2 into third-order EV K-maps and extract minimum SOP and POS cover for each.

3.25 Compress the functions e, f, g, and h in Problem 1 into first-order EV K-maps and extract minimum SOP and POS cover for each.

3.26 Write the CANONICAL SOP and POS forms for the K-map that follows by using MINTERM and MAXTERM codes; then loop out minimum SOP and POS cover.

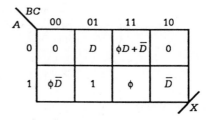

3.27 Write the CANONICAL SOP and POS forms for the K-map that follows by using MINTERM and MAXTERM codes; then extract minimum SOP and POS cover.

A \ BC	00	01	11	10
0	1	$\overline{X} + Y$	$X \cdot Y$	Z
1	1	$\overline{X} \cdot Y$	0	Z

F

3.28 Write the CANONICAL SOP and POS forms for the K-map that follows by using MINTERM and MAXTERM codes; then loop out minimum SOP and POS cover.

186

A \ BC	00	01	11	10
0	$\phi + D$	0	$\bar{D} \cdot (D + \phi)$	D
1	1	ϕ	0	0

Z

3.29 Compress the following fourth-order K-map into a second-order EV K-map and loop out minimum SOP and POS cover.

AB \ CD	00	01	11	10
00	1	0	0	1
01	1	1	0	0
11	0	1	1	0
10	1	0	0	1

F

3.30 Compress the following functions into fourth-order K-maps and loop out minimum SOP and POS forms. Indicate which form is the simpler in each case.

(a) $e(A,B,C,D,E) = \sum m(2, 10, 11, 14, 15, 21, 26, 27, 28, 30, 31) + \phi(1, 6, 16, 17)$

(b) $f(A,B,C,D,E,F) = \sum m(4, 6, 10, 11, 14, 15, 20, 22, 26, 27, 30, 31, 36, 38, 39, 52, 54, 56, 57, 60, 61)$

(c) $g(A,B,C,D,E,F) = \prod M(0, 1, 5, 7, 9, 15, 16, 18, 21, 24, 29, 31, 35, 37, 39, 40, 45, 49, 50, 56, 58, 60, 61, 63)$

(d) $h(A,B,C,D,E) = \prod M(0, 4, 7, 8, 12, 16, 20, 24, 27, 28) \cdot \phi(1, 3, 5, 11, 14, 21, 25, 31)$

3.31 Use a fourth-order K-map to find the minimum SOP and POS cover for the following subfunctions and give their cell location in the fourth-order map.

(a) $P(A,B,C,D,E,F,G) = \sum m(33, 34, 36, 38) + \phi(32, 39)$

(b) $Q(a,b,c,d,e,f,g,h) = \sum m(114, 116, 118, 122, 124, 126)$

(c) $R(A,B,C,D,E,F,G) = \prod M(105, 107, 108, 109, 110)$

(d) $S(a,b,c,d,e,f,g,h) = \prod M(176, 181, 182, 183, 184, 189, 191) \cdot \phi(177, 185, 190)$

3.32 Reduce the following function to its simplest form (SOP or POS) in a fourth-order EV K-map.

$Y(A,B,C,D,E,F,G) = \sum m(4, 5, 10, 12, 13, 14, 15, 20, 21, 26, 28, 29, 30, 31, 33, 35, 37, 39, 40, 42, 44, 45, 46, 47, 49, 51, 53, 55, 56, 58, 60, 61, 62, 63, 64, 65, 68, 69, 70, 71, 74, 75, 76, 77, 78, 79, 80, 81, 84, 85, 90, 91, 94, 95, 96, 97, 100, 101, 102, 103, 106, 107, 108, 109, 110, 111, 112, 113, 115, 116, 117, 119, 121, 122, 123, 125, 126, 127)$

3.33 Compress the function below into a first-order K-map of axis A and loop out a gate-minimum cover by using XOR-type patterns. Use the gate/input tally to compare this result with those for the two-level minimum results (SOP and POS).

$$E = \overline{A}\,\overline{X}\,\overline{Y} + \overline{A}XY + AY$$

3.34 Use XOR type patterns to loop out a gate-minimum cover from each of the second-order EV K-maps that follow. Indicate the gate/input tally for each.

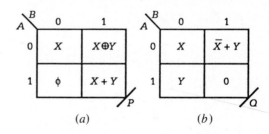

(a)   (b)

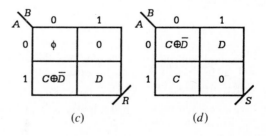

(c)   (d)

Problems

187

3.35 Compress the following functions into second-order K-maps and loop out a gate-minimum cover by using XOR-type patterns. Use the gate/input tally to compare the result for each with those obtained for the two-level minimum SOP and POS results.

(a) $W(A,B,C,D) = \sum m(3, 6, 9, 12)$

(b) $X(A,B,C,D) = \prod M(2, 3, 4, 5, 7, 8, 9, 11, 14, 15)$

(c) $Y(A,B,C,D) = \sum m(1, 2, 4, 7, 11, 13, 14)$

(d) $Z(A,B,C,D) = \prod M(0, 3, 4, 6, 10, 11, 13)$

3.36 Compress the following function into a second-order EV K-map of axes A and B, and loop out a gate-minimum cover by using XOR-type patterns. Use the gate/input tally to compare this result with those for the two-level minimum results (SOP and POS).

$$F(A,B,C,D) = \prod M(0, 1, 2, 3, 8, 11, 12, 13)$$
$$\cdot \phi(4, 5, 6, 7)$$

3.37 Use XOR-type patterns to loop out a gate-minimum cover from each of the third-order EV K-maps shown. Indicate the gate/input tally for each.

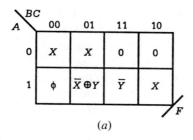

(a)

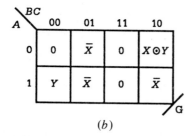

(b)

3.38 Find the minimum two-level SOP and POS cover and the CANONICAL forms for each of the functions F and G in Problem 37.

3.39 From the EV truth table that follows, construct the EV K-map and loop out a gate-minimum cover by using XOR-type patterns. Determine the gate/input tally for this result and compare it with that obtained for the two-level result.

A	B	C	J
0	0	0	0
0	0	1	$\overline{X} \oplus Y$
0	1	0	$\overline{Y}$
0	1	1	0
1	0	0	X
1	0	1	Y
1	1	0	0
1	1	1	Y

3.40 From the EV truth table that follows, construct the EV K-map and loop out a gate-minimum cover by using XOR-type patterns. (Hint: Look for compound associative XOR patterns.) Determine the gate/input tally for this result and compare it with that obtained for the two-level result.

A	B	C	K
0	0	0	$Z \oplus \overline{X}$
0	0	1	0
0	1	0	0
0	1	1	$\overline{X}$
1	0	0	X
1	0	1	0
1	1	0	$Y \oplus Z$
1	1	1	$X \oplus Y$

3.41 A computer program has been written which will yield a minimum solution to a combinational logic function, but only in SOP form. It accepts the data in either conventional (1's or 0's) form or in two-level EV form—it does not recognize the XOR and EQV operators.

Given the function G in the K-map of Fig. 3-51(b), show the minimum two-level result that would be obtained from the computer and the final step necessary to obtain the desired minimum POS result. Give the minimum POS result and its gate/input tally. Plan to enter the data into the program from the second-order K-map. Also, briefly outline the procedure used to obtain the POS result.

CHAPTER 4
Combinational Logic Design

4.1 INTRODUCTION

The short discussion on the synthesis of MIXED LOGIC circuits in Sect. 2.7 was actually a short discussion of very simple *combinational logic design*. There, the object was to implement certain logic expressions, for example, those of the XOR function, using discrete gates. Indeed, it is the purpose of combinational logic design to build larger more sophisticated logic circuits by using the most adaptable and versatile building blocks available. The choice of discrete gates as the building blocks is not always a good one owing to the complex nature of the circuits that must be designed and to the fact that there are integrated circuit (IC) packages available which are much more adaptable. This chapter deals with the design of basic combinational logic circuits from both a gate perspective and a modular perspective. It is our plan to develop certain manageable packages, or building blocks as we call them, and demonstrate their use in constructing larger combinational logic systems.

Before proceeding to a detailed discussion of combinational logic design, it will be worthwhile to characterize combinational logic circuits in more general terms and classify them according to established practice.

Shown in Fig. 4-1 are the block diagrams with complete and abridged I/O notation for a generalized combinational logic circuit having *n* inputs and *m* outputs. Use will be made of the abridged notation throughout the remainder of this text whenever it is necessary to simplify the circuit drawings.

Not shown in Fig. 4-1 are the propagation delay times associated with each output signal. Except for occasional passing mention, these time delays

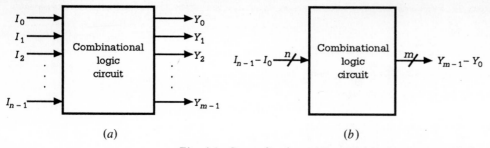

Fig. 4-1. *Generalized combinational logic circuit symbols. (a) Complete input/output notation. (b) Abridged I/O notation.*

will be neglected until the end of this chapter, where full consideration will be given to them.

Also conspicuously absent in Fig. 4-1 are any feedback loops connecting outputs to inputs. This means that the outputs do not depend on any previous output conditions and, consequently, depend only on the present state of the inputs. Logic circuits of this type, that is, those without feedback, are appropriately called *combinational logic circuits*. These are the subject of this chapter.

4.1.1 The Building Blocks

Now that we have given a general characterization of combinational logic circuits, let us develop some understanding of how they are used and of what kinds of building blocks they are constructed. Those that are familiar with the design and operation of digital systems know that the digital designer must be able to create combinational circuits which will perform a large variety of tasks. Typical examples of these tasks are

    Data manipulation (arithmetically or logically)
    Data code conversion
    Data selection from various sources
    Data routing from source to various destinations
    Data error detection
    Data busing from one part of a digital system to another part

To implement circuits that will perform tasks of the type listed, the logic designer can draw upon an impressive and growing list of combinational logic devices which are commercially available in the form of IC packages called *chips*. Shown in Fig. 4-2 is a listing of some of the more common combinational logic chips. It is the object of this chapter to develop the logic equivalent of these devices and demonstrate how they can be used as manageable and versatile building blocks for large-scale circuit design.

4.1.2 Classification of IC Chips

IC chips for devices of the type listed in Fig. 4.2 can be classified as small-scale integrated (SSI) circuits, medium-scale integrated (MSI) circuits, large-scale integrated (LSI) circuits, very-large-scale integrated (VLSI) circuits,

Arithmetic-Type Circuits	Multipurpose Circuits

ADDERs
SUBTRACTORs
MULTIPLIERs
DIVIDERs
COMPARATORs
PARITY GENERATORS and
   Error detectors
	DECODERs and ENCODERs
	CODE CONVERTERS
	MULTIPLEXERs and
	   DEMULTIPLEXERS
	Asynchronous SHIFTERS
	Arithmetic and logic units (ALUs)
	Read-only memories (ROMs)
	Programmable logic arrays
	   (PLAs and PALs)

Fig. 4-2. *Partial list of available combinational logic IC devices.*

and wafer-scale integrated (WSI) circuits. It has become customary to assign one of the preceding acronyms to a given IC circuit on the basis of number of equivalent *fundamental gates* (meaning AND, OR, INVERTER or NAND, NOR, INVERTER) that are required to implement it. By one convention, we may associate the acronyms with the following gate count:

SSI circuits—up to 12 gates

MSI circuits—13 to 99 gates

LSI circuits—100 to 1,000 gates

VLSI circuits—greater than 1,000 gates

and WSI chips might have equivalent gates numbering in the millions. This classification system, although attractive in appearance, is of questionable value, since it reveals nothing of the true complexity of the IC relative to the digital system within which it operates.

A more useful classification system should take into account the true complexity of the IC. As an example, an LSI chip might be a moderately complex microprocessor or some other type of *application-specific integrated circuit* (*ASIC*), or a VLSI chip might function as a WSI device. We are not advocating a new set of acronyms but merely suggesting that there are better ways to classify ICs other than to use discrete gate count (or, even worse, transistor count) as a measure of their operational complexity. In any case, the reader should exercise caution when evaluating the complexity of a chip based on some count system.

4.2 BINARY WORDS: NOMENCLATURE AND PHYSICAL REALIZATION

Except for that part of Chapter 2 dealing with the electrical equivalent of logic functions (i.e., physical logic gates), we have worked mainly with logic 1's and 0's and with single independent logic functions. This was necessary in order to cover the important background material of Chapters 2 and 3 in the most efficient and effective manner possible. Now we must move on to the design of building blocks of the type listed in Fig. 4-2. To do this will require that we be able to manipulate functionally arrays of binary digits using electrical circuits, a seemingly formidable task. However, with a workable nomenclature and a perspective of how binary digit arrays are physically realized, the task should be reduced to a simple exercise in communication.

The logic level (1 or 0) assigned to a logic variable of a digital circuit is called a *binary digit*, abbreviated as *bit*. A linear array of juxtaposed bits which represents a number or which conveys an item of information is called a binary WORD, or simply WORD. A WORD of 4 bits is sometimes called a *nibble* or *nybble*, and, more commonly, a WORD of 8 bits is termed a *byte*. A 16-bit WORD is, therefore, one of 2 bytes, and a 32-bit WORD is one of 4 bytes, and so on. In general, the total number of bits in a WORD is equal to some power of 2.

In *natural* binary code, bits are positioned in a binary WORD array according to their positional weight in polynomial notation. For example, the binary number 1010 would be represented by the polynomial

$$N_2 = (1 \times 2^3) + (0 \times 2^2) + (1 \times 2^1) + (0 \times 2^0)$$

$$= (1 \times 8) + (0 \times 4) + (1 \times 2) + (0 \times 1) = 8 + 0 + 2 + 0 = 10_{10}$$

Or if a radix point is indicated as in 10111.110, the polynomial representation would be

$$N_2 = (1 \times 2^4) + (0 \times 2^3) + (1 \times 2^2) + (1 \times 2^1) + (1 \times 2^0)$$

$$+ (1 \times 2^{-1}) + (1 \times 2^{-2}) + (0 \times 2^{-3})$$

$$= 16 + 0 + 4 + 2 + 1 + \tfrac{1}{2} + \tfrac{1}{4} = 23.75_{10}$$

The bit of highest positional weight (usually the extreme left bit) is called the *most significant bit* (MSB) while the bit of lowest positional weight (usually the bit on the extreme right) is called the *least significant bit* (LSB). The following is a byte of information:

$$
\begin{array}{cccccccc}
128 & 64 & 32 & 16 & 8 & 4 & 2 & 1 \\
1 & 0 & 1 & 1 & 0 & 1 & 0 & 0 \\
\uparrow & & & & & & & \uparrow \\
\text{MSB} & & & & & & & \text{LSB}
\end{array}
$$

This WORD is shown with the bits in descending positional weight from left to right and has a decimal value of $128 + 32 + 16 + 4 = 170_{10}$. Unless otherwise indicated, all WORDS will be written by using this bit format, whether in literal form (e.g., A B C . . .) or in logic level form (e.g., 0 1 1 0 1 0 . . .).

To denote positional weight in literal form, we will use subscript notation such as $(B_3 \times 2^3) + (B_2 \times 2^2) + (B_1 \times 2^1) + (B_0 \times 2^0) + (B_{-1} \times 2^{-1}) + (B_{-2} \times 2^{-2}) + (B_{-3} \times 2^{-3}) + \cdots$. Here, it is understood that B_3 is the MSB and that the positional weight decreases with decreasing subscript number. Thus, this binary WORD is represented as $B_3 \, B_2 \, B_1 \, B_0 \, B_{-1} \, B_{-2} \, B_{-3} \ldots$.

A linear array of *n* bits can be arranged to form 2^n different binary WORDS. When these WORDS are taken to represent numerals, letters, symbols, or a different sequence of these WORDS, a *code* is said to exist. Two examples of a three-bit binary code are 000, 001, 010, 011, 100, 101,

110, 111 and 000, 001, 011, 010, 110, 111, 101, 100—both corresponding to decimals 0, 1, 2, 3, 4, 5, 6, 7. These codes are commonly referred to as natural binary code and Gray code, respectively. There are a variety of binary codes several of which are given in Sect. 4.7.1, written in 4-bit binary. During our discussions in this chapter, we will have cause to become familiar with a few of the more frequently used codes. More important, we will learn how to design circuits that will perform electronic tasks equivalent to converting these codes from one to another.

The Word ACTIVE Revisited. Before proceeding it will be instructive to review the use of the word ACTIVE as it is used in this text. Recall that the bits of a WORD are the logic levels assumed by an array of elements (e.g., leads) of a physical circuit. An element is called ACTIVE when it assumes the logic 1 value and is called INACTIVE (or NOT ACTIVE) when it assumes the logic 0 value. Therefore, an element that is ACTIVE HIGH is one for which $HV = 1(H)$. Conversely, an element that is ACTIVE LOW is one for which $LV = 1(L)$. Thus, an ACTIVE HIGH element is also INACTIVE LOW, and vice versa. For a review of this subject matter the reader is referred to Sect. 2.2.

4.2.2 Physical Realization of Binary WORDs

We already know that each output of an operating logic circuit is either at high voltage (e.g., $HV = 5.1$ V) or low voltage (e.g., $LV = 0.02$ V), both taken with respect to some reference potential, usually ground (zero volts). But voltage information by itself is not sufficient for proper logic representation. To represent correctly a given element in MIXED LOGIC notation, we need to know (1) its polarization level, that is, whether the voltage of that element is from a positive or negative logic source, and (2) either its voltage level (HV or LV) or its assumed logic value (1 or 0). We illustrate this in Fig. 4-3 for a circuit N with and without INVERTERs on the output elements, assuming that the input conditions are the same for each.

Notice that the placement of ACTIVE LOW indicator bubbles in Fig. 4-

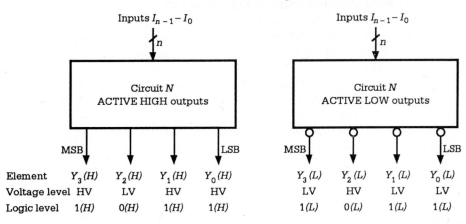

Element	$Y_3(H)$	$Y_2(H)$	$Y_1(H)$	$Y_0(H)$		$Y_3(L)$	$Y_2(L)$	$Y_1(L)$	$Y_0(L)$
Voltage level	HV	LV	HV	HV		LV	HV	LV	LV
Logic level	1(H)	0(H)	1(H)	1(H)		1(L)	0(L)	1(L)	1(L)

(a)   (b)

Fig. 4-3. *Circuit N with output WORD 1011. (a) ACTIVE HIGH outputs. (b) ACTIVE LOW outputs.*

3(b) requires the inversion of the voltage levels and polarization indicator symbols but not the logic values—a feature of the MIXED LOGIC notation. Thus, for either circuit of Fig. 4-3 the WORD may be written as 1011, reading from MSB to LSB, but with the stipulation that each element in the array is either ACTIVE HIGH in one case or ACTIVE LOW in the other. This is important to know in design work and it applies to inputs as well as outputs. What this means is that once a circuit is designed, any combination of AC-TIVE LOW and ACTIVE HIGH inputs or outputs can be established simply by placing INVERTERs or ACTIVE LOW indicator bubbles at those places where the polarization indicator symbol is reversed. Hence, there is no need to use MIXED LOGIC truth tables and MIXED LOGIC K-maps in designing logic circuits. Section 4.6 offers some good examples of the application of this principle.

4.3 DESIGN PROCEDURE

The design of a combinational logic circuit generally begins with the description of and specifications for the device and ends with a suitable logic implementation. To assist the reader in developing good design practices, we recommend the following six-part sequence:

Part I. Understand the Device. Describe the function of the device; then clearly indicate its input/output (I/O) specifications, and construct its block diagram.

Part II. State the Algorithms. Indicate all algorithms and/or binary manipulations necessary for the design. Include a general operations format if necessary.

Part III. Construct a Truth Table. From Part II construct a truth table that details the I/O relationships. The truth table is usually presented in positive logic form.

Part IV. Obtain the Output Function. Map and minimize (or reduce), or otherwise read directly, the information in the truth table of Part III to obtain the logic expressions for the outputs.

Part V. Construct the Logic Diagram. Use either a gate or modular level approach (or both) to implement the logic expressions obtained in Part IV. Implement from output to input and take into account any required MIXED LOGIC I/O conditions.

Part VI. Check the Results. Where it is reasonable to do so, check the final logic circuit at the functional or logic level.

It is our plan to follow this six-part sequence where appropriate without specifically mentioning each step.

4.4 ARITHMETIC CIRCUITS

In this section we will design digital circuits with electrical capabilities that can be interpreted as performing the basic arithmetic operations of binary numbers. These basic operations include

Addition

Subtraction

Multiplication

Division

Now, the Boolean equations are uniquely defined so as to perform specific arithmetic operations, and the 1's and 0's, which we have previously used only as logic levels, must take on a numerical significance. Thus, we will use the 1's and 0's to perform arithmetic operations. However, we must always keep in mind that any arithmetic circuit we design is only the electrical equivalent of the arithmetic operation it represents. It is the interpretation of the electrical circuits behavior that bridges the gap between pure logic and the physical world. Frequently, special algorithms must be used to make possible a workable physical realization for a given arithmetic operation. As we shall see, a good example of this is the division operation.

The treatment of arithmetic circuits in this section is intended only as an introduction to this subject matter. Actual IC circuits performing the electrical equivalent of the basic arithmetic operations may be much more sophisticated and may even be based on different operating principles.

4.4.1 The HALF ADDER

The HALF ADDER is the simplest of all the arithmetic circuits. In fact, it may be regarded as the smallest building block for the modular design of such circuits. Electronically, the HALF ADDER must perform an operation which can be interpreted as the addition of two input bits, bit A plus bit B, to generate two output bits, SUM (S) and CARRY (C). The operation format and block diagram symbol for the HALF ADDER are presented in Fig. 4-4. The inputs and outputs are all assumed to be ACTIVE HIGH.

The design of the HALF ADDER is easily accomplished by following the procedure outlined in Sect. 4.3. If input bits A and B take on all possible combinations of logic values, then the general format for the addition operation, given in Fig. 4-4(a), generates the truth table shown in Fig. 4-5(a) as follows:

0	0	1	1	Augend bit A
+0	+1	+0	+1	Addend bit B
0 0	0 1	0 1	1 0	SUM A plus B

MSB ⌐ ⌐ LSB

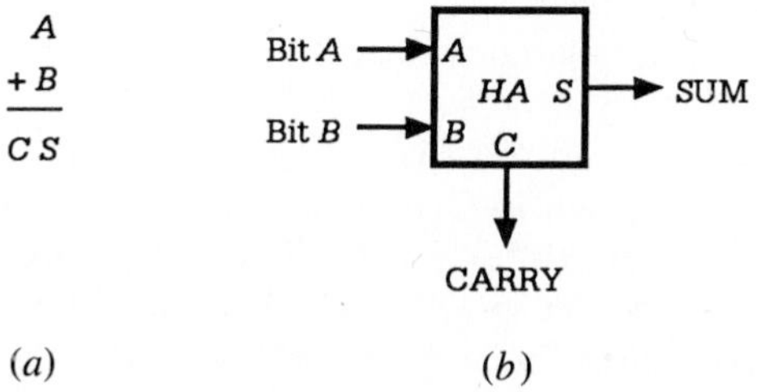

Fig. 4-4. HALF ADDER. (a) Operation format. (b) Block diagram symbol.

(a)   (b)

A	B	C	S
0	0	0	0
0	1	0	1
1	0	0	1
1	1	1	0

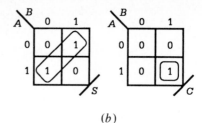

(a) (b)

Fig. 4-5. (a) Truth table for a HALF ADDER. (b) K-maps for SUM and CARRYout outputs of a HALF ADDER.

In each of the four addition operations the MSB is the CARRY bit, while the LSB is the SUM bit. Remember that for a positive logic system the physical interpretation of logic 0 and logic 1 is LV and HV, respectively.

The second-order K-maps for S and C are constructed in Fig. 4-5(b). An XOR pattern (diagonal loop) is recognized in the K-map for output S and the Boolean expressions are read directly as

$$S = A \oplus B$$

and

$$C = A \cdot B$$

(4-1)

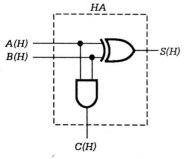

Fig. 4-6. Implementation of a HALF ADDER by using Eqs. (4-1).

The logic circuit for the HALF ADDER can be implemented using Eqs. (4-1). When this is done we obtain the circuit shown in Fig. 4-6. Of course, had we desired to do so, we could have replaced the XOR gate with any circuit derived from Eqs. (2-6) or (2-7), as, for example, the circuit of Fig. 2-31. However, XOR gates are readily available and are designed to be used in circuits of this type, among their many other uses.

As a final step the circuit of Fig. 4-6 is functionally tested to see if it conforms to the requirements of a HALF ADDER. Indeed, Eqs. (4-1) are read directly from the circuit.

4.4.2 The FULL ADDER

The HALF ADDER we just designed has severe limitations in modular design applications, since it cannot accept a CARRY bit from the previous stage called CARRYin. Hence, the HALF ADDER can be used only in addition operations without CARRYin requirements.

The limitations of the HALF ADDER are easily overcome by using the FULL ADDER (*FA*) presented in Fig. 4-7. The FULL ADDER can accept and add the CARRYin (C_{in}) bit from the CARRYout (C_{out}) of the previous stage. With this feature the FULL ADDER becomes a useful device as a modular building block in arithmetic circuits and is used extensively for that purpose.

Let us design the FULL ADDER by following the procedure given in Sect. 4.3. Shown in Figs. 4-7(a) and (b) are the operation format and the block diagram symbol for the FULL ADDER. It is required that the FULL ADDER function as the electrical equivalent for the binary sum of three input bits (bit A and plus bit B plus CARRYin) to produce two output bits

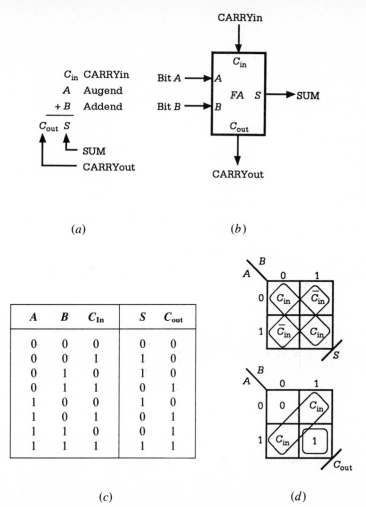

$$C_{in} \quad \text{CARRYin}$$
$$A \quad \text{Augend}$$
$$\underline{+B} \quad \text{Addend}$$
$$C_{out} \quad S$$

SUM

CARRYout

(a)

(b)

A	B	C_{In}	S	C_{out}
0	0	0	0	0
0	0	1	1	0
0	1	0	1	0
0	1	1	0	1
1	0	0	1	0
1	0	1	0	1
1	1	0	0	1
1	1	1	1	1

(c)

(d)

Fig. 4-7. *FULL ADDER. (a) Operation format. (b) Block diagram symbol. (c) Truth Table. (d) Karnaugh maps for SUM and CARRYout.*

(SUM and CARRYout). The truth table given in Fig. 4-7(c) is the logic interpretation for the function of the FULL ADDER. It is generated by the operation format [Fig. 4-7(a)] when the three inputs take on all possible combinations of logic values. We will assume that all inputs arrive from positive logic sources and that the outputs are likewise ACTIVE HIGH.

By using a MAP KEY of 2, we can compress the truth table of Fig. 4-7(c) into a second-order K-map yielding the results shown in Fig. 4-7(d). Noting the diagonal XOR type patterns indicated be the diagonal loops in the EV K-maps (see Fig. 3-48), we can easily read the SUM and CARRYout functions as follows:

$$S = C_{in}(\overline{A \oplus B}) + \overline{C}_{in}(A \oplus B)$$

$$= A \oplus B \oplus C_{in} \tag{4-2}$$

$$C_{out} = C_{in}(A \oplus B) + (AB)$$

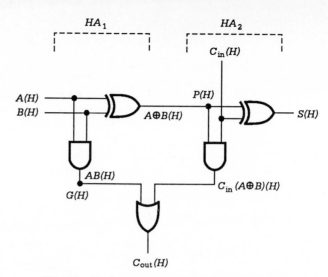

Fig. 4-8. *Implementation of the FULL ADDER.*

A gate-level design exclusive of XOR gates can be obtained simply by looping out the EVs in Fig. 4-7(d) as islands. We leave this task as an exercise for the reader.

Proceeding with our gate-level approach, we use the expressions of Eqs. (4-2) to implement the FULL ADDER. The results, given in Fig. 4-8, easily show that a FULL ADDER is composed of two HALF ADDERs, a conclusion that would not be obvious from an inspection of a two-level gate design.

Included in Fig. 4-8 for the convenience of the reader are the intermediate functions written in MIXED LOGIC notation. Completion of the reading process yields directly the SUM and CARRYout expressions of Eqs. (4-2). Notice that the AND/OR logic of Fig. 4-8 can be replaced by the appropriate conjugate NAND gate forms without changing any of the logic functions, but that the intermediate polarization levels must change from (H) to (L). Also included in Fig. 4-8 are the P and G nodes which will be discussed later in Sect. 4.4.4.

4.4.3 The *n*-Bit PARALLEL ADDER

As we indicated in Sect. 4.4.2, the FULL ADDER is a useful device for modular construction of arithmetic circuits. Now it is desirable that we make use of this feature to design a PARALLEL ADDER (PA) of *n*-bits.

An *n*-bit PARALLEL ADDER is defined as a $(2n + 1)$-input/$(n + 1)$-output combinational logic device which can add two *n*-bit numbers. The general operation format for such an ADDER is presented in Fig. 4-9(a). It represents the familiar addition algorithms used in conventional arithmetic. Notice that the subscripts are consistent with the powers of 2 to the left of the radix point in polynomial notation. Hence, the bits of each WORD are written in ascending order of positional weight from right to left.

Figure 4-9(b) shows the *ripple-carry* effect which characterizes the addition algorithm for this PARALLEL ADDER. ADDERs of this type are

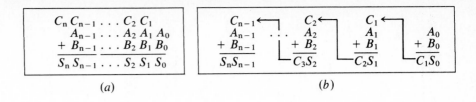

$$\begin{array}{r} C_n\,C_{n-1}\ldots C_2\,C_1 \\ A_{n-1}\ldots A_2\,A_1\,A_0 \\ +\;B_{n-1}\ldots B_2\,B_1\,B_0 \\ \hline S_n\,S_{n-1}\ldots S_2\,S_1\,S_0 \end{array}$$

(a)

(b)

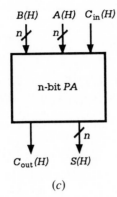

B(H)   A(H) C_{in}(H)

n-bit PA

C_{out}(H)   S(H)

(c)

Fig. 4-9. *An* n-*bit PARALLEL ADDER. (a) Operation format. (b) Carry-ripple effect. (c) Block diagram.*

sometimes called RIPPLE-CARRY ADDERs. Observe that the CARRYin bit for each stage is the CARRYout bit from the previous stage and that the CARRYout bit for the $(n-1)$th stage is the final SUM bit, that is, $C_{out} = C_n = S_n$. The block diagram representing an n-bit PA is given in Fig. 4-9(c).

The modular design for the n-bit PARALLEL ADDER follows directly from Fig. 4-9. All that is required is an array of n FULL ADDERs designated $FA_0, FA_1, \ldots, FA_{n-1}$, one for each bit, connected such that the C_{in} input of each stage is taken from the C_{out} output of the preceding stage. This is presented in Fig. 4-10, where it is assumed that all inputs arrive from positive logic sources. Notice that the condition of no CARRYin to the first stage is

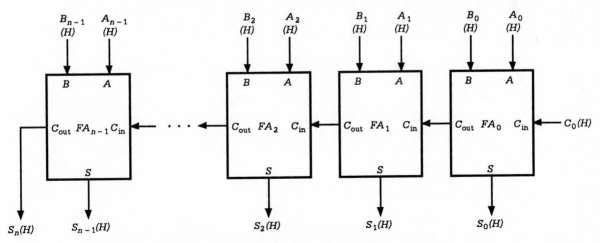

Fig. 4-10. *An* n-*bit PARALLEL ADDER implemented with FULL ADDERs.*

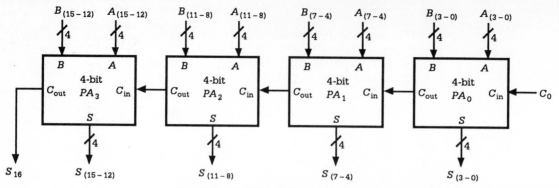

Fig. 4-11. *A 16-bit PARALLEL ADDER composed of four 4-bit PARALLEL ADDERs with inputs and outputs all ACTIVE HIGH.*

satisfied by grounding the C_{in} input terminal of FA_0 or by using a HALF ADDER for that stage.

An *n*-bit PARALLEL ADDER is more likely to be designed by using n/m ADDER modules rather than individual FULL ADDERs. One example, represented in Fig. 4-11, uses four 4-bit PARALLEL ADDERs in a ripple-carry configuration to produce a 16-bit ADDER. Here, we use the abridged I/O notation (introduced in Sect. 4.1) so as to simplify the circuit drawing. Other examples include a 16-bit ADDER composed of two 8-bit PARALLEL ADDERs, or a 32-bit ADDER made up of four 8-bit ADDERs, and so on.

It must be understood that a PARALLEL ADDER cannot correctly perform the electrical equivalent of a given binary addition operation unless *all* A and B input data are presented simultaneously to the ADDER terminals. For example, if the input signals are pulse trains, but all pulse trains are not synchronized with each other, the ADDER may not function properly. To overcome this problem, storage registers (i.e., PIPO registers as in Fig. 5-98) are used to present, on command, all the *A*- and *B*-bit information to the ADDER at the same time. Furthermore, storage registers can also be used on the output of ADDERs so that the data bits resulting from the addition operation can be presented simultaneously to the next stage. Storage registers belong to the general class of registers which are discussed in Sect. 5.10.1.

Even if the *A*- and *B*-bit information is supplied to the ADDER terminals at the same time, there still remains the problem of the CARRY bit that must ripple through the entire ADDER before the sum process is complete. Thus, by specifying a maximum time period between successive addition operations, we are forced to place a limit on the number of stages that a PARALLEL ADDER can have—a limitation that results from the ripple-carry effect.

4.4.4 The LOOK-AHEAD-CARRY ADDER

The PARALLEL (or RIPPLE-CARRY) ADDER just described is satisfactory for most applications up to about 16 bits at moderate to high speeds. Where larger numbers of bits are to be added together at very high speeds, fast adder configurations must be devised. One that comes to mind immediately is a gate-level design obtained directly from the truth table and K-maps for an *n*-bit ADDER. Such an ADDER would have no ripple-carry feature and would be cascadable, but would also be tedious to design.

A clever—actually intuitive—design has been devised to make use of the modular approach but at the same time reduce the propagation time of the ripple-carry effect. This design has become known as the LOOK-AHEAD-CARRY (LAC) ADDER. In effect, the LAC ADDER electronically "anticipates" the need for a carry and then generates and propagates it more directly than does the PARALLEL ADDER, as we will now demonstrate.

One frequently encountered LAC ADDER design begins with a generalization of Eqs. (4-2). For the *n*th stage of a PARALLEL ADDER we may write

$$S_n = A_n \oplus B_n \oplus C_n$$
$$= \text{SUM of } n\text{th stage} \tag{4-3}$$
$$C_{n+1} = (A_n \oplus B_n)C_n + A_nB_n$$
$$= \text{CARRYout of } n\text{th stage}$$

We now show that there are only four possible combinations of A_n, B_n, and C_n that can produce a CARRYout (C_{n+1}) of the *n*th stage. Following the format of Fig. 4-9, they are

$$
\begin{array}{cccccl}
1 & 1 & 0 & 1 & & C_n \\
0 & 1 & 1 & 1 & & A_n \\
+1 & +0 & +1 & +1 & & +B_n \\
\hline
10 & 10 & 10 & 11 & & C_{n+1}S_n
\end{array}
$$

$$\underbrace{}\qquad\underbrace{}$$
$$(A_n \oplus B_n)C_n = 1 \qquad A_nB_n = 1$$

Therefore, $C_{n+1} = 1$ is assured if $(A_n \oplus B_n) = 1$ and $C_n = 1$, or if $A_nB_n = 1$.

Next, it is desirable to expand Eqs. (4-3) for each of the *n* stages, beginning with the first (or 0th) stage. To accomplish this task, it is convenient to define the two quantities

$$P_n \equiv A_n \oplus B_n = \text{CARRY PROPAGATE}$$

and $\tag{4-4}$

$$G_n \equiv A_nB_n = \text{CARRY GENERATE}$$

which are seen to be the intermediate functions of the FULL ADDER shown in Fig. 4-8. By introducing these equations into Eqs. (4-3), we have

$$
\text{1}^{\text{st}}\text{ stage}
\begin{cases}
S_0 = P_0 \oplus C_0 \\
C_1 = P_0C_0 + G_0
\end{cases}
$$

$$\tag{4-5}$$

$$
\text{2}^{\text{nd}}\text{ stage}
\begin{cases}
S_1 = P_1 \oplus C_1 \\
C_2 = P_1C_1 + G_1 \\
 = P_1P_0C_0 + P_1G_0 + G_1
\end{cases}
$$

$$\text{3rd stage} \begin{cases} S_2 = P_2 \oplus C_2 \\ C_3 = P_2 C_2 + G_2 \\ \quad = P_2 P_1 P_0 C_0 + P_2 P_1 G_0 + P_2 G_1 + G_2 \\ \quad \vdots \end{cases}$$

$$\text{(4-5)}$$

$$n^{\text{th}} \text{ stage} \begin{cases} S_n = P_n \oplus C_n \\ C_{n+1} = P_n C_n + G_n \\ \quad = P_n P_{n-1} P_{n-2} \cdots P_0 C_0 + P_n P_{n-1} \cdots P_1 G_0 + \cdots + G_n \end{cases}$$

To implement the LAC ADDER, we will use FULL ADDERs of the type shown in Fig. 4-8 which have leads from the P and G nodes. Whereas the S_n terms are produced within the ADDERs, the C_{n+1} terms must be formed by external combinational logic. Although it may not be obvious, the first carry expressions in Eqs. (4-5) ($C_1 = P_0 C_0 + G_0$, $C_2 = P_1 C_1 + G_1$, etc.) represent the ripple-carry effect in a PARALLEL ADDER. However, it is the last carry expression for each stage of Eqs. (4-5) that must be implemented so as to produce the LOOK-AHEAD-CARRY configuration we seek. By using these last carry expressions for a 3-bit LAC ADDER, we obtain the modular level circuit given in Fig. 4-12. The combinational logic circuit below the dashed line is called the LAC GENERATE circuit.

One interesting facet of the FULL ADDER is that it can function in an LAC ADDER circuit without the use of C_{out}. Actually, only that portion of the FULL ADDER that supplies the P_n, G_n, and S_n functions is necessary. Commercial LAC ADDERs take advantage of this fact.

The modular approach to LAC ADDER design can be taken one step further by using the LAC feature between m-bit LAC ADDER modules to

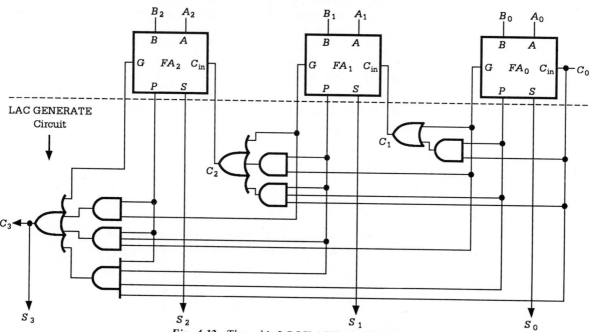

Fig. 4-12. *Three-bit LOOK-AHEAD-CARRY ADDER implemented with FULL ADDERs and showing the LAC GENERATE circuit below the dashed line.*

4 / Combinational Logic Design

produce a larger ADDER, similar to that shown in Fig. 4-11 for a modular PARALLEL ADDER. To do this one simply expands Eqs. (4-3) in the following way:

$$1^{\text{st}} \; m^{\text{th}} \text{ stage} \quad \left\{ \begin{array}{l} S_{m0} = P_{m0} \oplus C_{m0} \\ C_{m1} = P_{m0}C_{m0} + G_{m0} \end{array} \right.$$

$$2^{\text{nd}} \; m^{\text{th}} \text{ stage} \quad \left\{ \begin{array}{l} S_{m1} = P_{m1} \oplus C_{m1} \\ C_{m2} = P_{m1}C_{m1} + G_{m1} \\ \quad\; = P_{m1}P_{m0}C_{m0} + P_{m1}G_{m1} + G_{m1} \end{array} \right. \qquad (4\text{-}6)$$

$$3^{\text{rd}} \; m^{\text{th}} \text{ stage} \quad \left\{ \begin{array}{l} S_{m2} = P_{m2} \oplus C_{m2} \\ C_{m3} = P_{m2}C_{m2} + C_{m2} \\ \quad\; = P_{m2}P_{m1}P_{m0}C_{m0} + P_{m2}P_{m1}G_{m0} \\ \qquad + P_{m2}G_{m1} + G_{m2} \\ \qquad \vdots \end{array} \right.$$

$$r^{\text{th}} \; m^{\text{th}} \text{ stage} \quad \left\{ \begin{array}{l} S_{mr} = P_{mr} \oplus C_{mr} \\ C_{m,r+1} = P_{mr}C_{mr} + G_{mr} \\ \qquad\quad = S_{n+1} \end{array} \right.$$

In this case there are r stages each of m bits denoted by subscripts r and m, respectively. It follows that this LAC ADDER will add two binary numbers, each up to $n = mr$ bits in length.

As an example of the application of Eqs. (4-6), let us design a 12-bit LAC ADDER made up of 4-bit LAC modules. Such an ADDER is presented in Fig. 4-13, where a block diagram has been used to represent the LAC GENERATE circuit. The outputs P_m and G_m represent the CARRY PROPAGATE and CARRY GENERATE, respectively, from the mth (4th) stage of each of the three modules. Notice that the final carry bit (C_3) is the most significant sum bit S_{12}.

The reader should not be left with the impression that the LAC ADDER is the panacea for all adder design ills. Indeed, the use of the LAC feature involves a trade-off. For large numbers of bits, an LAC ADDER design can

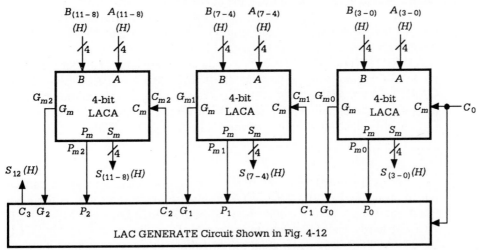

Fig. 4-13. A 12-bit LAC ADDER composed of three 4-bit LAC ADDER stages.

result in substantially reduced ripple delay times compared to a comparable PARALLEL ADDER design, but only at the expense of additional logic hardware. Whether the gain in reduced ripple delay times justifies the increased board or chip real estate requirements is a matter that must be weighed by the designer. Generally speaking, PARALLEL ADDER configurations satisfy most of the designer's needs. This is especially true for IC PARALLEL ADDER packages composed of "fast" transistors.

4.4.5 The CARRY-SAVE ADDER

Addition of binary numbers is not restricted to two numbers as the reader might have inferred from the past discussion. Actually many binary numbers can be added together simultaneously. One way of accomplishing this is to use a type of "fast" ADDER called a CARRY-SAVE ADDER. This ADDER is conceptually simple, but, as might be expected, it requires additional hardware.

The algorithm for a CARRY-SAVE ADDER is stated as follows:

> *Sum the bits exclusive of carries, then add to this result the carries shifted one place to the left.*

A numerical example illustrating the application of the CARRY-SAVE (CS) algorithm is given below. In this example four 3-bit numbers are added together.

$$
\begin{array}{rl}
1\ 1\ 1 & A \\
0\ 1\ 0 & B \\
1\ 0\ 1 & C \\
\underline{+0\ 1\ 1} & D \\
0\ 1\ 1 & \text{Sum without carries} \\
\underline{+1\ 1\ 1} & \text{Carries shifted left one place} \\
1\ 0\ 0\ 0\ 1 & \text{Final sum}
\end{array}
$$

Shown in Fig. 4-14 is the modular design of a four 3-bit number CS ADDER by using a 3×3 matrix of FULL ADDERs with a tenth ADDER added to provide the two most significant sum bits. The reader may find it helpful to trace the numerical example through the circuit.

CARRY-SAVE ADDERs can be designed to add larger numbers of larger bit WORDS simply by expanding the matrix of FULL ADDERs. Thus, CARRY-SAVE ADDERs offer a means of speeding up the addition process but only at the expense of increased hardware. Problem 4.47 at the end of this chapter illustrates the use of the CARRY-SAVE ADDER.

4.4.6 The FULL SUBTRACTOR

The electrical equivalent of the binary subtraction operation can be carried out by using FULL SUBTRACTOR hardware much the same as the electrical equivalent of the addition operation is accomplished with FULL ADDER hardware. Although this is rarely done, the design of the FULL SUBTRACTOR provides a good introduction to the more useful subject of

4 / Combinational Logic Design

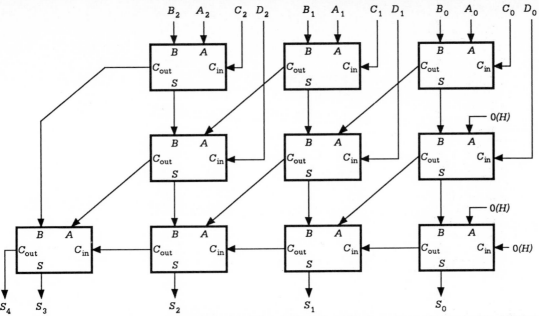

Fig. 4-14. *Modular design of a 3-bit 4-number CARRY-SAVE ADDER by using FULL ADDERS.*

ADDER/SUBTRACTORs (devices which can serve in either capacity) and to the subject of DIVIDERs.

The design of a FULL SUBTRACTOR follows in a manner similar to the design of a FULL ADDER with, of course, some important differences. The inputs now become the minuend bit A, the subtrahend bit B, and the previous-BORROW bit B_{in} (BORROWin), and the outputs become the difference bit D and the borrow-from-next-stage bit B_{out} (BORROWout). Shown in Fig. 4-15 are the essential features of the FULL SUBTRACTOR, namely, the operation format (a), the block diagram symbol (b), the I/O truth table (c), the output bit K-maps and Boolean expressions (d), and the resulting logic circuit (e). The truth table is generated by the operation format when the three inputs are assigned all possible combinations of logic values and it is understood that D is equal to $(A - B - B_{in})$. Notice that the familiar XOR patterns in the K-maps yield an expression for D identical to that for the S bit of a FULL ADDER, and an expression for B_{out} which is similar but not equal to that for C_{out}.

The logic circuit for the FULL SUBTRACTOR, shown in Fig. 4-15(e), is constructed directly from the Boolean expression for D and B_{out}. Observe that it is composed essentially of two HALF SUBTRACTORs (HS_1 and HS_2) and that two INVERTERs are appropriately positioned to create the logic incompatibilities necessary to yield the B_{out} expression. The fact that the circuits for the FULL ADDER (Fig. 4-8) and FULL SUBTRACTOR are identical except for the absence or presence of the two INVERTERs, raises the possibility of a FULL ADDER/SUBTRACTOR circuit. Indeed, if the two INVERTERs of Fig. 4-15(e) are replaced by CONTROLLED INVERTERs (see Fig. 2-42), such a device would result. When the control leads for the CONTROLLED INVERTERs are held at $0(H)$, that is, at ground potential, the circuit functions as a FULL ADDER. Conversely,

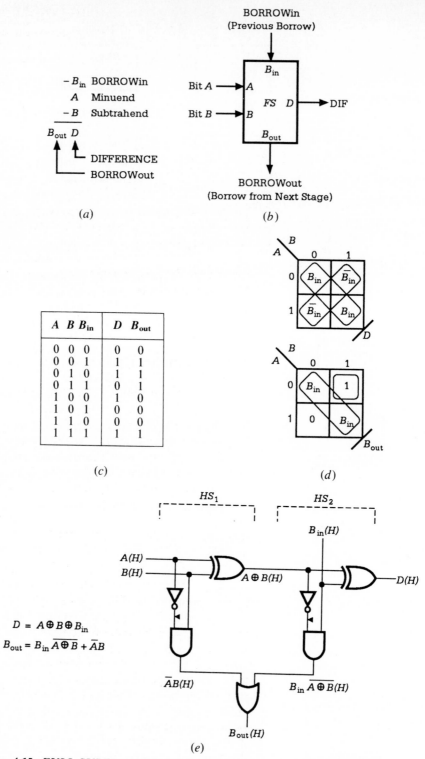

$D = A \oplus B \oplus B_{in}$

$B_{out} = B_{in} \overline{A \oplus B} + \overline{A}B$

Fig. 4-15. *FULL SUBTRACTOR. (a) Operation format. (b) Block diagram symbol. (c) I/O truth table. (d) EV K-maps for DIFFERENCE and BORROWout. (e) Boolean expressions and logic circuit for a FULL SUBTRACTOR.*

206

when the control leads are held at $1(H)$, that is, at HV, then the XOR gates become INVERTERs (LOGIC LEVEL CONVERTERs) and the circuit functions as a FULL SUBTRACTOR. In the following section we will explore the ADDER/SUBTRACTOR concept more fully.

4.4.7 ADDERs/SUBTRACTORs—Sign-Complement Arithmetic

Up to this point we have established that FULL ADDERs and FULL SUBTRACTORs function as the electrical analogs for their respective binary addition and subtraction operations. This includes various modular designs using these arithmetic units. However, the simplicity of the binary number system offers an important alternative which is simple and manageable—an approach that permits the use of one set of hardware to accomplish both addition- and subtraction-type operations. As we shall demonstrate, the binary subtraction operation is best carried out by *adding* together *signed binary numbers* using arithmetic operations collectively termed *sign-complement arithmetic*. To use sign-complement arithmetic for this purpose, three requirements must be met:

1. The MSB for the augend and addend must be designated as the *sign bit* which is either logic 0 for a positive number or logic 1 for a negative number.
2. The arithmetic operation must yield the correct sum which can be read as either a positive or negative number independent of the most significant carry bit.
3. The algorithm used for the operation must be one suitable for logic implementation.

The use of either 1's complement or 2's complement arithmetic easily satisfies the three requirements just listed. So, we will establish useful generating algorithms for these two sign-complement notations and show how they are used in signed-complement arithmetic to produce a difference operation. The algorithms which generate 1's and 2's complement numbers from binary, and which are suitable for logic implementation, are now given.

1's complement generating algorithm: Complement each bit of a binary number, N_1. Let this be known as the $\overline{N}_2$ algorithm.

2's complement generating algorithm: Complement each bit of a binary number N_2 and add 1 to the LSB. Let this be known as the $(\overline{N}_2$ plus 1) algorithm.

A further discussion of complement generation algorithms is given in Sect. 4.7.1.

The application of these complement generating algorithms to any positive binary number yields a negative binary number represented in signed-complement notation. As an example, let us find the binary equivalent of -12_{10} (negative decimal 12) by using 8-bit binary, but keeping in mind that the MSB is the sign bit. If the $\overline{N}_2$ and $(\overline{N}_2$ plus 1) algorithms are applied to $N_2 = 00001100_2$, there results the following negative binary forms:

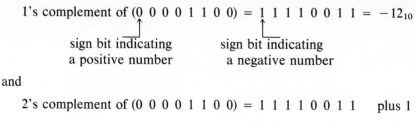

1's complement of $(0\ 0\ 0\ 0\ 1\ 1\ 0\ 0) = 1\ 1\ 1\ 1\ 0\ 0\ 1\ 1 = -12_{10}$

sign bit indicating
a positive number

sign bit indicating
a negative number

and

2's complement of $(0\ 0\ 0\ 0\ 1\ 1\ 0\ 0) = 1\ 1\ 1\ 1\ 0\ 0\ 1\ 1$ plus 1

$$= 1\ 1\ 1\ 1\ 0\ 1\ 0\ 0 = -12_{10}$$

sign bit indicating
a negative number

Notice that in any positive or negative number the sign bit is an integral part of that number and is included in its numerical value or in any operation involving that number.

One interesting facet of sign-complement generation is that the process is reversible. That is, when the appropriate generating algorithm is applied to a negative number, its positive representation results. The reader may wish to test this on the two examples given.

If an independent sign bit is assigned to a binary number, there results the *sign-magnitude* representation. For example, in 8-bit binary the sign-magnitude representation of -12_{10} would be

$$1 \quad 0\ 0\ 0\ 1\ 1\ 0\ 0$$

sign bit   magnitude

where the MSB is the sign bit and the remaining seven bits represent the magnitude. In sign-magnitude arithmetic only the magnitude is manipulated, analogous to decimal arithmetic.

All that remains at this point is to establish the procedures for signed-complement arithmetic and then proceed to the design of an ADDER/SUBTRACTOR. These procedures are simply stated:

> *1's complement arithmetic.* With all negative numbers in 1's complement notation, add the two numbers as in standard binary addition. If there is a carry overflow of the MSB (sign bit), it must be carried around and added to the LSB of the resultant number. This is called the carry-end-around-and-add step. Any overflow from this step is discarded.
>
> *2's complement arithmetic.* With all negative numbers represented in 2's complement notation, add the two numbers as in standard binary addition. If there is a carry overflow of the MSB (sign bit), it is discarded.

Although both 1's and 2's complement arithmetic are used in computer systems, 2's complement is the more common of the two for arithmetic applications. The reason for this is twofold. First, the carry-end-around-and-add requirement of 1's complement arithmetic requires additional circuit propagation delay time to produce the electrical equivalent of a final sum. Second, the reader should note that the 1's complement notation has two arithmetic zeros, namely, one with logic 0's only and one with logic 1's only. Two's complement arithmetic overcomes both of these disadvantages since any carry overflow is discarded and since there is but a single zero—all logic 0's. As we shall see, the $(\overline{N}_2 + 1)$ algorithm requirement for 2's complement possesses no significant problem for the implementing hardware.

The design of an ADDER/SUBTRACTOR now follows directly by using the ADDER hardware developed in Sect. 4.4.3 together with the signed-complement generating algorithms and signed-complement arithmetic procedures given. To illustrate, we will design two versions of a simple PARALLEL ADDER/SUBTRACTOR.

In Fig. 4-16 are shown the combined general operation formats, and the modular logic circuits for 1's and 2's complement versions of a 2-bit PARALLEL ADDER/SUBTRACTOR, where the B input is chosen to represent either the addend or subtrahend. The two versions differ only with respect

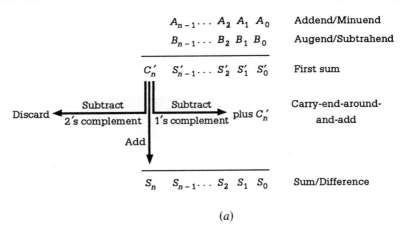

(a)

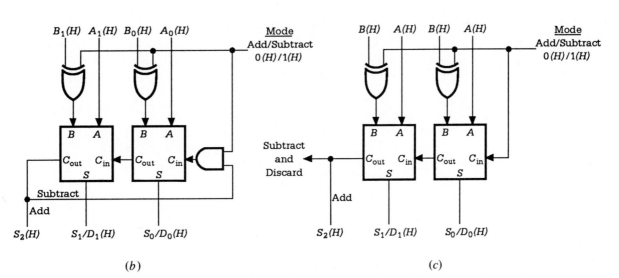

(b)   (c)

Fig. 4-16. PARALLEL ADDER/SUBTRACTORs. (a) General operations format for binary addition and 1's and 2's complement arithmetic. (b) Two-bit 1's complement PARALLEL ADDER/SUBTRACTOR. (c) Two-bit 2's complement PARALLEL ADDER/SUBTRACTOR.

to the disposition of the final carryout as required by the 1's and 2's complement arithmetic procedures. Observe that the arithmetic mode (ADD or SUBTRACT) of either circuit is determined via XOR gates, again used as CONTROLLED INVERTERs, and that the most significant carry bit is the final sum bit in the ADD mode. Also, it must always be remembered that if the final sum bits represent a negative number they are in either 1's or 2's complement notation.

All the modular design configurations discussed regarding PARALLEL ADDERS also apply to PARALLEL ADDER/SUBTRACTORs. However, one cautionary note could be added at this time. Again, it is necessary that the input data be presented simultaneously to the ADDER/SUBTRACTOR. This could be done by placing storage registers (see Fig. 5-98) immediately outside of the inputs to the FULL ADDERs or PARALLEL ADDER modules, as the case may be. Since there are ripple-carry effects involved in any PARALLEL ADDER–type configuration, storage registers might also be necessary at the output stage to route simultaneously the sum data elsewhere in a digital system. Much more will be said about the subject of registers and their applications following the treatment of flip-flops in Chapter 5.

4.4.8 MULTIPLIERs

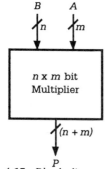

Fig. 4-17. *Block diagram symbol for a* n × m *bit MULTIPLIER.*

An $n \times m$ bit MULTIPLIER is an $(n + m)$-input/$(n + m)$-output device that will perform electronically in a manner that can be interpreted as the binary multiplication of an n-bit WORD with an m-bit WORD. Its block diagram symbol is presented in Fig. 4-17. The design detail for such a device depends on the algorithm used to execute the multiplication operation and on the type of implementation used (gate or modular level). The algorithm we will use in this section is the familiar one used in decimal arithmetic.

To illustrate the function of the MULTIPLIER, let us design a 2×2 bit MULTIPLIER at the gate level. Consider the operation format in Fig. 4-18(a) which represents the algorithm we wish to use. Here, the multiplicand A of two bits is multiplied by the multiplier B, also of two bits, to produce the partial product bits (P_{ij}) and carry bits (C). The P_{ij} and C bits then add (just as in decimal arithmetic) to produce the final product represented by the four P_k bits. The double subscript notation P_{ij} is used to identify the

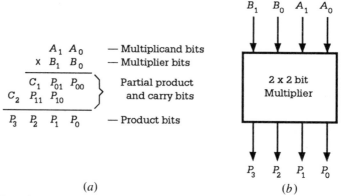

Fig. 4-18. *A 2 × 2 bit MULTIPLIER. (a) Operation format. (b) Block diagram symbol.*

4 / Combinational Logic Design

multiplicand and multiplier bits involved ($P_{01} = B_0A_1$, $P_{11} = B_1A_1$, and so on). The block diagram symbol for this four-input/four-output MULTIPLIER is given in Fig. 4-18(b).

When the inputs $B_1B_0A_1A_0$ are assigned all possible combinations of logic values, the operation format generates the I/O truth table for $A \times B = P$ displayed in Fig. 4-19(a). Since it is usually easier to convert back and forth from binary to decimal when generating such tables, we have included decimal values adjacent to the input and output entries. For example, $2 \times 3 = 6$ is equivalent to inputs 1011 (10×11), yielding outputs 0110, and so forth.

To illustrate the usefulness of the EV mapping method, the A variables are compressed into the second-order EV K-maps shown in Fig. 4-19(b). These represent second-order map compressions with a MAP KEY of $2^{4-2} = 4$. Hence, each cell of each map is a second-order submap in variables A_1 and A_0. Tick marks are placed on the vertical line of the truth table in Fig. 4-19(a) to indicate the proper subfunction groupings. Notice the simplicity of the K-maps and the reoccurring term B_0A_0. Also, observe that the K-map for P_1 produces an associative XOR pattern, a pattern that could not

$B_1B_0A_1A_0$	$P_3P_2P_1P_0$	Product (P) (Decimal)	$B_1B_0A_1A_0$	$P_3P_2P_1P_0$	Product (P) (Decimal)
0 0 0 0	0 0 0 0	0	1 0 0 0	0 0 0 0	0
0 0 0 1	0 0 0 0	0	1 0 0 1	0 0 1 0	2
0 0 1 0	0 0 0 0	0	1 0 1 0	0 1 0 0	4
0 0 1 1	0 0 0 0	0	1 0 1 1	0 1 1 0	6
0 1 0 0	0 0 0 0	0	1 1 0 0	0 0 0 0	0
0 1 0 1	0 0 0 1	1	1 1 0 1	0 0 1 1	3
0 1 1 0	0 0 1 0	2	1 1 1 0	0 1 1 0	6
0 1 1 1	0 0 1 1	3	1 1 1 1	1 0 0 1	9

(a)

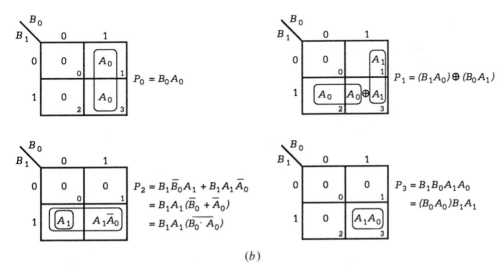

(b)

Fig. 4-19. (a) Truth table for a 2×2 bit MULTIPLIER. (b) Second-order EV K-maps for a 2×2 bit MULTIPLIER.

easily have been extracted from a map of higher order. The reader is referred to Sect. 3.12.2 for a review of XOR patterns in EV K-maps.

There is no single correct configuration for a given gate-level design unless it would be that for absolute minimum cover. Therefore, we will make use of the XOR function for P_1 in our gate-level design. The result is the relatively simple circuit given in Fig. 4-20 which consists of only seven gates including an INVERTER, assuming that all inputs and outputs are ACTIVE HIGH. That the circuit outputs yield the Boolean expressions in Fig. 4-18 is easily established by reading the functions input to output. Notice that three of the P outputs make use of the B_0A_0 term and that as a result we have optimized a multiple-output circuit by inspection.

For the purpose of comparison, we will now design the same 2×2 bit MULTIPLIER using a modular-level approach. Let us assume that we have access to FULL ADDERS (sometimes called *arithmetic units*) and a variety of discrete gates. Our design begins with the operation format of Fig. 4-18(a) which represents the algorithm we wish to apply. Recalling that the AND operation is the Boolean product and referring to Eqs. (4-2) for a FULL ADDER, we conclude that the partial product bits (P_{ij}) carry bits (C), and product bits (P_k) are given by

$$P_{00} = B_0A_0 \qquad\qquad C_1 = P_{01}P_{10}$$
$$P_{01} = B_0A_1 \qquad\qquad C_2 = C_1P_{11}$$
$$P_{10} = B_1A_0 \qquad\qquad\quad\; = P_{01}P_{10}P_{11}$$
$$P_{11} = B_1A_1$$

$$(4\text{-}7)$$

$$P_0 = P_{00}$$
$$P_1 = P_{01} \oplus P_{10}$$
$$P_2 = C_1 \oplus P_{11}$$
$$P_3 = C_2$$

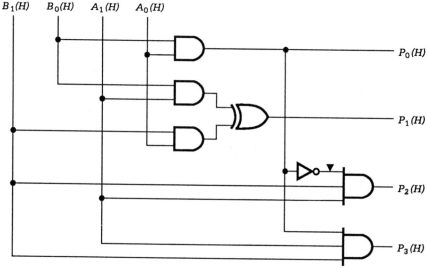

Fig. 4-20. *Gate-level circuit for a 2 × 2 bit MULTIPLIER.*

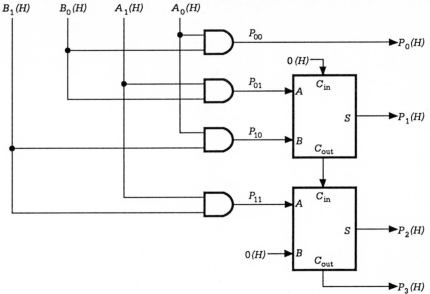

Fig. 4-21. *Modular design of a 2 × 2 bit MULTIPLIER by using two FULL ADDERs.*

A brief inspection of Fig. 4-18(a) reveals that these equations meet the requirements of the operation format for a 2 × 2 bit MULTIPLIER.

Implementation now follows directly from Eqs. (4-7) beginning at the output stage (P terms) and working toward the input. The result is shown in Fig. 4-21, where $C_1 = P_{01}P_{10}$ is C_{out} for FA_1, and $C_2 = C_1P_{11}$ is C_{out} for FA_2, according to Eqs. (4-2). Notice that in this particular design HALF ADDERs could have been used in place of the FULL ADDERs, since only two inputs are used on any given arithmetic unit.

Comparing the modular design of Fig. 4-21 with the gate-level design of Fig. 4-20, we see some similarities and some striking differences. The input line pattern is identical, and the outputs P_0 and P_1 are configured the same if one thinks of the FULL ADDER as performing the function of the XOR gate. The manner in which outputs P_2 and P_3 are produced exposes the major differences between the two types of design.

Another interesting comparison between the modular and gate-level designs is the equivalent gate count. If we take each FULL ADDER as being comprised of five discrete gates, then an equivalent gate count for the modular design would come to 14. Compared to a gate count of seven for the gate-level approach, it is clear that the modular design, although simpler to build, is far from a minimum. This is typical of most all modular design configurations.

Were we to extend the modular approach of Fig. 4-21 to larger binary WORDs, it would soon become obvious that the hardware requirements would increase rapidly. In fact, for an $n \times n$ bit MULTIPLIER the AND gate count would increase as n^2, while the FULL ADDER count would increase approximately as $5n - 10$ for MULTIPLIERs larger that 2 × 2 but smaller thatn 10 × 10. For example, a 4 × 4 bit MULTIPLIER would require 16 AND gates and 9 FULL ADDERs, and a 8 × 8 bit MULTIPLIER would require 64 AND gates and 30 FULL ADDERs. Thus, while these

MULTIPLIERS are conceptually simple and fast, they have space requirements which make them impractical for many applications, especially where space rather than speed is the primary consideration. Other algorithms can be used to help solve the space limitation, but only at the expense of operation time—a space/time trade-off. This trade-off is often desirable since with "fast" ICs the penalty in operation time does not seriously limit their application. But the savings in real estate (chip space) can be considerable. Such algorithms are usually implemented by synchronous sequential machines.

There are approaches to the modular design of a binary MULTIPLIER other than that illustrated in Fig. 4-21. The simplest of these is to generate the partial products, P_{ij}, as in Fig. 4-21, and then use a CARRY SAVE ADDER to obtain the product bits. This involves the use of more FULL ADDERs than would be required by the method of Fig. 4-21, but the simplicity and speed of this approach more than offsets the additional hardware required. As an example, a 4 × 4 bit binary MULTIPLIER would require 22 FULL ADDERs (three rows of seven FAs plus a final carry FA) in a CARRY SAVE configuration similar to that of Fig. 4-14. The design of a cascadable 4 × 4 bit NBCD MULTIPLIER, required in Problem 4.47 at the end of this chapter, illustrates this and will not be considered further here. More information in this subject area can be obtained from sources cited in the Annotated References of this chapter.

4.4.9 DIVIDERs

An $n \div m$-bit DIVIDER is an $(n + m)$-bit input/variable-bit output device that will perform the electrical analog of the binary operation symbolized by $A \div B = Q$ with remainder R. Here, A and B are the dividend and divisor WORDs, respectively, and Q is the quotient. The general block diagram symbol for the DIVIDER is given in Fig. 4-22.

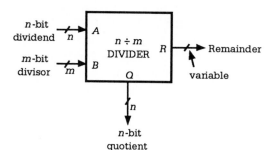

Fig. 4-22. Block diagram symbol for an n ÷ m bit DIVIDER.

The details of the electronic circuitry in a DIVIDER depends on the algorithm used to produce the division operation and on the type of circuit implementation used, gate level or modular level. For our treatment in this section we will design two simple DIVIDERs by using both gate-level and modular designs together with a common comparison algorithm for division. The algorithm for binary division which is closest to that used for decimal arithmetic can be stated in the following way:

1. Successively subtract the subtrahend B from the minuend A by starting from the MSB end of A and shifting 1 bit toward the LSB after each subtraction stage:

   a. When the most significant (MS) BORROW for the present stage is 0, the minuend for the next stage (remainder from the present stage) is the difference of the present stage.

   b. When the MS BORROW is 1, the minuend for the next stage (remainder) is the minuend of the present stage.

2. Complement the MS BORROW bit for each stage and let it become the quotient bit for that stage.

We demonstrate this division algorithm below by dividing 10001 by 11 (i.e., $17 \div 3$):

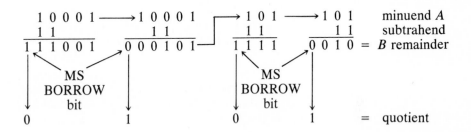

What results from this is that 10001 divided by 11 is a quotient of 101 (decimal 5) with a remainder of 10 (decimal 2). The reader may recognize this algorithm as the one used in decimal arithmetic.

Let us now design a simple DIVIDER at the gate level. Since these circuits are somewhat complicated, we will limit this design to a $2 \div 2$ bit DIVIDER which will adequately serve to illustrate the design procedure. Consider the I/O truth table in Fig. 4-23(a) which is generated by the division algorithm just described or, more simply, by converting back and forth between decimal and binary. Notice that don't cares (ϕ's) are placed in output

Value A B	Inputs $A_1 A_0 B_1 B_0$	Outputs $Q_1 Q_0 R_1 R_0$	Value Q R	Value A B	Inputs $A_1 A_0 B_1 B_0$	Outputs $Q_1 Q_0 R_1 R_0$	Value Q R
0 0	0 0 0 0	0 0 0 0	0 0	2 0	1 0 0 0	ϕ ϕ ϕ ϕ	—
0 1	0 0 0 1	0 0 0 0	0 0	2 1	1 0 0 1	1 0 0 0	2 0
0 2	0 0 1 0	0 0 0 0	0 0	2 2	1 0 1 0	0 1 0 0	1 0
0 3	0 0 1 1	0 0 0 0	0 0	2 3	1 0 1 1	0 0 1 0	0 2
1 0	0 1 0 0	ϕ ϕ ϕ ϕ	—	3 0	1 1 0 0	ϕ ϕ ϕ ϕ	—
1 1	0 1 0 1	0 1 0 0	1 0	3 1	1 1 0 1	1 1 0 0	3 0
1 2	0 1 1 0	0 0 0 1	0 1	3 2	1 1 1 0	0 1 0 1	1 1
1 3	0 1 1 1	0 0 0 1	0 1	3 3	1 1 1 1	0 1 0 0	1 0

(a)

Fig. 4-23. *Design of a $2 \div 2$ bit DIVIDER. (a) I/O truth table showing decimal equivalents for the dividend (A), divisor (B), quotient (Q), and remainder (R).*

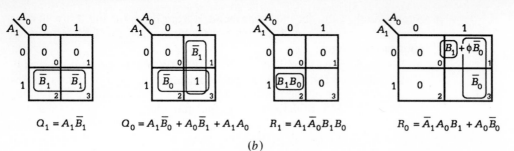

$$Q_1 = A_1\overline{B}_1 \qquad Q_0 = A_1\overline{B}_0 + A_0\overline{B}_1 + A_1A_0 \qquad R_1 = A_1\overline{A}_0B_1B_0 \qquad R_0 = \overline{A}_1A_0B_1 + A_0\overline{B}_0$$

(b)

Fig. 4-23 (contd.). (b) EV K-maps for a 2 ÷ 2 bit DIVIDER.

states for which the outputs cannot be represented as, for example, 2 ÷ 0, and so on.

The K-maps shown in Fig. 4-23(b) are constructed from the truth table by using a MAP KEY of 4, hence second-order compressions in variables B_1 and B_0. Again, tick marks are located in the truth table of Fig. 4-23(a) to indicate the subfunction groupings. The NAND/AND implementation of this DIVIDER is shown in Fig. 4-24. The functions for the quotients, Q_1 and Q_0, and the functions for the remainders, R_1 and R_0, are easily read from the circuit diagram using MIXED LOGIC notation.

The modular level design of a SUBTRACTOR follows directly from the binary division algorithm that was stated earlier. The first requirement is that the subtrahend be successively subtracted from the minuend, and shifted from the MSB toward the LSB by one bit after each subtraction stage. This is easily accomplished by staging FULL SUBTRACTORs so that the subtrahend bits are presented to each successive stage shifted by one bit toward the LSB.

The second requirement is that the remainder R, which is the new minuend for the next stage, be properly gated. Taking B_{out} to mean the MS BORROW for any given stage, the division algorithm requires that $R = D$

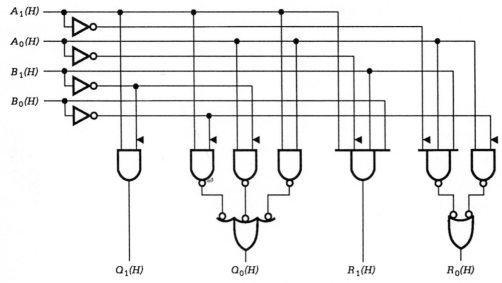

Fig. 4-24. NAND/AND implementation of the 2 ÷ 2 bit DIVIDER of Fig. 4-23.

4 / Combinational Logic Design

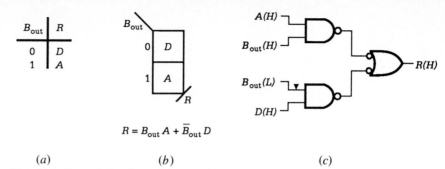

$$R = B_{out} A + \overline{B}_{out} D$$

B_{out}	R
0	D
1	A

(a)   (b)   (c)

Fig. 4-25. *Remainder for any stage of a DIVIDER. (a) EV truth table. (b) EV K-map. (c) Gating circuit.*

for $B_{out} = 0$, and $R = A$ for $B_{out} = 1$, where D and A are the difference and minuend, respectively, for the present stage. In Fig. 4-25 are presented the truth table, K-map, and gating circuit for each remainder R, that is, for each successively new minuend.

The third and last requirement for the modular design of a DIVIDER is that we take the complemented MS BORROW ($\overline{B}_{out}$) of each stage as the quotient bit for that stage. Since $\overline{B}_{out}$ appears in the R gating circuit, we need only tap off that lead for each Q bit.

By attaching the gating circuit of Fig. 4-25(c) to each FULL SUB-TRACTOR we can construct a DIVIDER of any size, limited only by the number of divider units (FULL SUBTRACTORs each with a gating circuit) we wish to use. We illustrate the design method by constructing the simple $3 \div 2$ bit DIVIDER shown in Fig. 4-26 which has two quotient (Q) bits and

Fig. 4-26. *Modular design of a $3 \div 2$ bit DIVIDER showing a sample division of 111 divided by 10.*

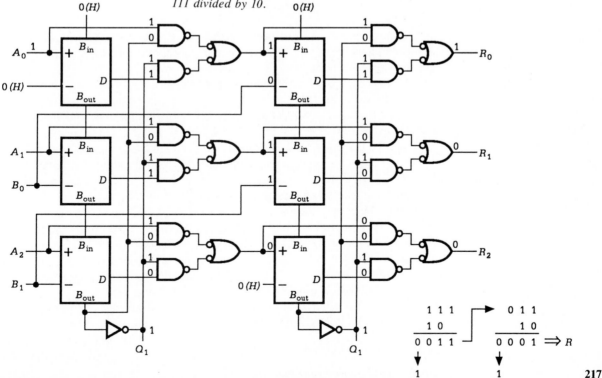

three remainder (R) bits. A larger divider can be produced by adding rows and columns of divider units as needed. To demonstrate its operation, 1's and 0's, representing 111 divided by 10, are appropriately placed throughout the circuit. The equivalent binary operation is shown to the right of Fig. 4-26 for comparison. Notice that with only two quotient bits, this circuit is incapable of performing any division operation requiring three quotient bits, as for example $111 \div 01$.

DIVIDER circuits of the type given in Fig. 4-24 and in Fig. 4-26 can be classified as "fast" DIVIDERs. This is because they are asynchronous in nature. However, it is characteristic of such circuits that with increasing WORD size the hardware requirements increase rapidly making them impractical for many applications where space requirements are important. There are other methods of achieving the division operation with less hardware but at the expense of operation time, as expected. These methods are sequential in nature and fall outside the treatment of this text.

4.5 COMPARATORS AND PARITY GENERATORS

There are special logic circuits which perform the electrical equivalent of bit-pattern comparison operations. When the particular operation is one of comparing the magnitude of two n-bit numbers, the device is called a MAGNITUDE COMPARATOR or simply COMPARATOR. When the operation involves generating or detecting an even or odd number of bits of the same logic value, the device is called a PARITY GENERATOR or PARITY DETECTOR. We will explore the details of these bit-pattern comparison operations in this section.

4.5.1 COMPARATORS

The comparison operation performed by a COMPARATOR may be one determining which of two numbers is the larger, or it may be just an equality comparison, or a combination of these two operations. We will design a COMPARATOR of the combined type which is classified as a $2n$-input/3-output device. This device is represented by the block diagram symbol shown in Fig. 4-27, where the inputs are n-bit numbers A and B and the outputs are variables represented by $(A > B)$, $(A < B)$, and $(A = B)$. Therefore, this COMPARATOR is designed to perform the electrical analog of an operation that can determine which of two binary numbers is the larger or if they are equal.

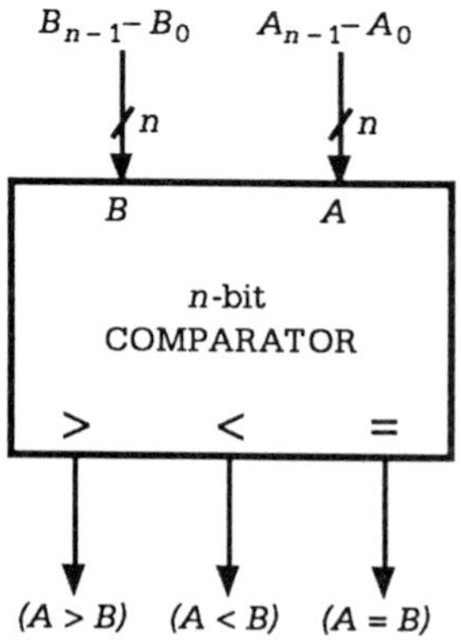

Let us further stipulate that our design be at the gate level but with cascading capability so that any size COMPARATOR can be produced simply by adding or subtracting COMPARATOR units. To do this requires that we establish for each of the three outputs a predictable pattern of logic terms which depends only on the number of cascadable units (hence input bits) used. This task is most easily accomplished by designing progressively larger COMPARATORs beginning with the simplest one possible.

Consider first the truth table, K-maps and output logic expressions for the 1-bit COMPARATOR given in Fig. 4-28. Notice that the output $(A = B)$ is the EQUIVALENCE (EQV) function. Also, from an inspection of the K-maps or truth table it can be seen that the outputs $(A > B)$, $(A < B)$, and

Fig. 4-27. *Block diagram symbol for an* n-*bit COMPARATOR.*

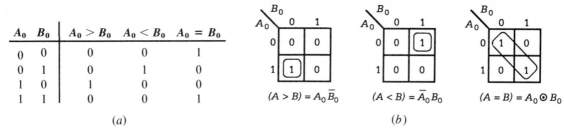

A_0	B_0	$A_0 > B_0$	$A_0 < B_0$	$A_0 = B_0$
0	0	0	0	1
0	1	0	1	0
1	0	1	0	0
1	1	0	0	1

(a)

$(A > B) = A_0\overline{B}_0$ $(A < B) = \overline{A}_0 B_0$ $(A = B) = A_0 \odot B_0$

(b)

Fig. 4-28. *A 1-bit COMPARATOR. (a) Truth table. (b) K-maps and output expressions.*

$(A = B)$ are related by the expression $X = \overline{Y + Z}$, where X is any one of the three outputs and Y and Z are the remaining two.

Next, let us follow the same procedure for a 2-bit COMPARATOR. Making use of the decimal equivalents to the left of the input columns, we construct the I/O truth table presented in Fig. 4-29(a). The truth table is then compressed into the third-order EV K-maps shown in Fig. 4-29(b) by using a MAP KEY of 2. These compressed K-maps reveal functional patterns which could not be easily detected by using conventional mapping methods. Notice, for example, that $A\overline{B}-$, $\overline{A}B-$, and $(A \odot B)$–type terms continue to appear and that, once again, the outputs are related by $X = \overline{Y + Z}$. This latter fact can also be verified by a brief inspection of the truth table.

Value A	B	A_1	A_0	B_1	B_0	$A > B$	$A < B$	$A = B$	Value A	B	A_1	A_0	B_1	B_0	$A > B$	$A < B$	$A = B$
0	0	0	0	0	0	0	0	1	2	0	1	0	0	0	1	0	0
0	1	0	0	0	1	0	1	0	2	1	1	0	0	1	1	0	0
0	2	0	0	1	0	0	1	0	2	2	1	0	1	0	0	0	1
0	3	0	0	1	1	0	1	0	2	3	1	0	1	1	0	1	0
1	0	0	1	0	0	1	0	0	3	0	1	1	0	0	1	0	0
1	1	0	1	0	1	0	0	1	3	1	1	1	0	1	1	0	0
1	2	0	1	1	0	0	1	0	3	2	1	1	1	0	1	0	0
1	3	0	1	1	1	0	1	0	3	3	1	1	1	1	0	0	1

(a)

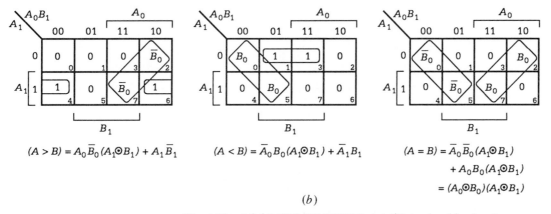

$(A > B) = A_0\overline{B}_0(A_1 \odot B_1) + A_1\overline{B}_1$

$(A < B) = \overline{A}_0 B_0(A_1 \odot B_1) + \overline{A}_1 B_1$

$(A = B) = \overline{A}_0\overline{B}_0(A_1 \odot B_1)$
$+ A_0 B_0(A_1 \odot B_1)$
$= (A_0 \odot B_0)(A_1 \odot B_1)$

(b)

Fig. 4-29. *A 2-bit COMPARATOR. (a) I/O truth table showing a magnitude comparison of 2-bit WORDs. (b) EV K-maps and output logic expressions for a 2-bit MAGNITUDE COMPARATOR.*

If one were to continue this procedure for COMPARATORs of larger WORDs, it would become clear in comparing the results that a predictable pattern of logic terms emerges for each output variable. Indeed, for an n-bit COMPARATOR the output expressions are

$$(A > B) = A_0\overline{B}_0(A_1 \odot B_1)(A_2 \odot B_2)(A_3 \odot B_3) \cdots (A_{n-1} \odot B_{n-1})$$
$$+ A_1\overline{B}_1(A_2 \odot B_2)(A_3 \odot B_3) \cdots (A_{n-1} \odot B_{n-1})$$
$$+ \cdots + A_{n-2}\overline{B}_{n-2}(A_{n-1} \odot B_{n-1}) + A_{n-1}\overline{B}_{n-1}$$
$$(A < B) = \overline{A}_0B_0(A_1 \odot B_1)(A_2 \odot B_2)(A_3 \odot B_3) \cdots (A_{n-1} \odot B_{n-1}) \quad \text{(4-8)}$$
$$+ \overline{A}_1B_1(A_2 \odot B_2)(A_3 \odot B_3) \cdots (A_{n-1} \odot B_{n-1})$$
$$+ \cdots + \overline{A}_{n-2}B_{n-2}(A_{n-1} \odot B_{n-1}) + \overline{A}_{n-1}B_{n-1}$$
$$(A = B) = (A_0 \odot B_0)(A_1 \odot B_1)(A_2 \odot B_2) \cdots (A_{n-1} \odot B_{n-1})$$

As in the previous two examples, the outputs for an n-bit COMPARATOR are composed exclusively of $(A\overline{B})$–, $(\overline{A}B)$–, and $(A \otimes B)$–type terms and are related by the expression $X = \overline{Y + Z}$, where X is any one of the outputs with Y and Z the remaining two.

Implementation of Eqs. (4-8) is best accomplished by reading each output expression from end to beginning and by recalling that $A \odot B(H) = A \oplus B(L) = (A\overline{B} + \overline{A}B)(L)$. To illustrate, let us design a 3-bit COMPARATOR by assuming that all inputs arrive from positive logic sources. The result is shown in Fig. 4-30, where subfunctions have been provided to assist the reader in following the operation of the circuit, and where X, Y, and Z now specifically represent $(A > B)$, $(A < B)$, and $(A = B)$, respectively. Observe that the basic COMPARATOR unit (enclosed by the dashed line in Fig. 4-30) is an XOR-type circuit similar to that of Fig. 2-30, but with an additional input for cascading purposes.

The 3-bit COMPARATOR of Fig. 4-30 can be expanded to accommodate binary WORDs of larger size by adding additional COMPARATOR units in

$$P = \overline{A}_2B_2(L) \qquad\qquad S = \overline{A}_1B_1(A_2 \odot B_2)(L)$$
$$Q = A_2\overline{B}_2(L) \qquad\qquad T = A_1\overline{B}_1(A_2 \odot B_2)(L)$$
$$R = (A_2 \oplus B_2)(L) \qquad U = [(A_1 \oplus B_1)(A_2 \odot B_2) + (A_2 \oplus B_2)](L)$$
$$V = \overline{A}_0B_0(A_1 \odot B_1)(A_2 \odot B_2)(L) \quad W = A_0\overline{B}_0(A_1 \odot B_1)(A_2 \odot B_2)(L)$$

$$(A > B)(H) = [W + T + Q](H)$$
$$= [A_0\overline{B}_0(A_1 \odot B_1)(A_2 \odot B_2) + A_1\overline{B}_1(A_2 \odot B_2) + A_2\overline{B}_2](H)$$
$$(A < B)(H) = [V + S + P](H)$$
$$= [\overline{A}_0B_0(A_1 \odot B_1)(A_2 \odot B_2) + \overline{A}_1B_1(A_2 \odot B_2) + \overline{A}_2B_2](H)$$
$$(A = B)(H) = [\overline{U} \cdot \overline{V} \cdot \overline{W}](H)$$
$$= [(A_0 \odot B_0)(A_1 \odot B_1)(A_2 \odot B_2)](H)$$

(a)

Fig. 4-30. A 3-bit COMPARATOR. (a) Subfunctions and output expressions.

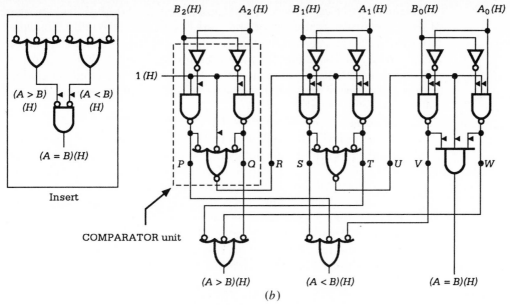

$(A > B)$ $(A < B)$
(H) (H)

$(A = B)(H)$

Insert

COMPARATOR unit

$(A > B)(H)$ $(A < B)(H)$ $(A = B)(H)$

(b)

Fig. 4-30 (contd.). *(b) NAND/AND logic circuit. Insert is an alternative for* (A = B)(H).

cascade fashion at either end. The $(A = B)(H)$ lead may be taken off of the last AND gate, as in Fig. 4-30, or it may be taken off of a NOR gate performing the operation $(A = B)(H) = \overline{(A > B) \cdot (A < B)}(H)$, which is consistent with $X = \overline{Y + Z} = \overline{Y} \cdot \overline{Z}$ (see the insert of Fig. 4-30).

Commercial IC COMPARATORs are designed with a combination of two types of inputs, namely, WORD inputs (as in Fig. 4-27) and cascading inputs $[(A > B), (A < B),$ and $(A = B)]$. The main difference between a commercial COMPARATOR and one of the type shown in Fig. 4-30 is the manner in which they are cascaded. Commercial units are cascaded by feeding the three outputs of the stage handling the LSBs into the cascading inputs of the stage handling the MSBs, and so on. For our design, cascading is achieved by adding COMPARATOR units to either end of the circuit given in Fig. 4-30. When this is done, each additional stage adds an additional input to the NAND gates of the X and Y outputs, that is, one input per stage.

4.5.2 PARITY GENERATORs and Error Detection

A parity bit can be generated for any WORD of n-bits and is based on either an even or odd number of WORD bits of the same logic value, say, logic 1. Thus, at some point in time a parity bit represents a specific binary WORD in the sense that it reads the even or odd count of bits of the same logic value. During data transport or storage within a digital system something could happen to cause an error—an undesirable single bit change—in that WORD and this error could go undetected. However, if the parity bit of the original WORD is compared with that of the bused (transported) or stored WORD, any single or odd number of bit changes (errors) can be detected. For example, if the WORD is the byte 10111010 then an even parity generate bit of logic 1 (based on an even total number of 1's including the parity bit itself) would be added as a ninth bit. The result would be 101110101 or

110111010, depending on whether the parity bit is placed as the LSB or MSB, respectively. If any one or odd number of the eight WORD bits were to be changed, the parity bit would be changed to logic 0. Comparison of the two parity generate bits, one representing the original WORD and the other representing the bused or stored WORD would reveal immediately the presence of the error. It should be obvious that an even number of bit changes will go undetected by this method.

Let us design an even/odd PARITY GENERATOR based on logic 1. This describes a device that will perform the electrical analog of generating, on command, an even or odd number of 1's including the parity generate bit. Presented in Fig. 4-31 are the truth table, K-maps, and logic circuit for a 4-bit even/odd PARITY GENERATOR based on logic 1. The K-maps, which are second-order compressions (MAP KEY = 4), easily reveal the XOR and EQV functions inherent in parity circuits. Notice that the choice of an even or odd parity generate bit is determined by the action of a controlled source and that the parity bit columns (not the headings) in Fig. 4-31(a) are interchanged for an even/odd PARITY GENERATOR based on logic 0. Furthermore, an even PARITY GENERATOR is an odd PARITY DETECTOR, and vice versa. This is true because a parity detect bit takes on a logic value as a result of the like-bit count but does not contribute to that count as does the parity generate bit. So the choice of name for the device, PARITY GENERATOR or DETECTOR, depends on whether the parity bit detects or generates a like-bit count.

Larger PARITY GENERATOR circuits can be constructed by cascading in various ways parity modules of the type shown in Fig. 4-31(c). For example, if the outputs of two parallel parity modules are the inputs to an XOR gate, an 8-bit PARITY GENERATOR results. Or, if one to four of the inputs to a parity module are each the output of a similar module, a 7-, 10-, 13-, or 16-bit PARITY GENERATOR would result. So by using various cascading configurations, most any size PARITY GENERATOR (or DETECTOR) can be built.

Error detection in a digital system usually amounts to a parity checking procedure where the parity bit of a reference WORD is compared with the parity bit of that same WORD following data busing and/or storage. One scheme to accomplish this is shown in Fig. 4-32. Here, a parity bit is generated from an 8-bit data bus, merged with the data bus to produce a 9-bit bus with parity, then later compared with a parity bit generated from the same 8-bit data lines following transport and/or storage. Comparison of the two parity bits is accomplished by using an EQV gate as an even PARITY DETECTOR. Hence, if no odd number of errors occurs during busing and/or storage, a $0(L) = 1(H)$ output is produced validating the data. An even number of bit errors would go undetected using this scheme.

Further explanation of Fig. 4-32 is warranted. A data bus *tap* is usually nothing more than simple lead taps like those used on numerous occasions in previous sections. A *merge* refers to the association of discrete data lines and a *split* is the opposite operation. Merged groups of lines are denoted by commas. Last, the PARITY GENERATOR symbols used in Fig. 4-32 may represent cascaded parity modules of the type shown in Fig. 4-31.

To the extent that an error detection circuit compares the parity bits of two WORDs, it may be regarded as a degenerate form of COMPARATOR.

A B C D	P_{even}	P_{odd}	A B C D	P_{even}	P_{odd}
0 0 0 0	0	1	1 0 0 0	1	0
0 0 0 1	1	0	1 0 0 1	0	1
0 0 1 0	1	0	1 0 1 0	0	1
0 0 1 1	0	1	1 0 1 1	1	0
0 1 0 0	1	0	1 1 0 0	0	1
0 1 0 1	0	1	1 1 0 1	1	0
0 1 1 0	0	1	1 1 1 0	1	0
0 1 1 1	1	0	1 1 1 1	0	1

(a)

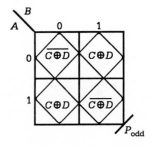

$$P_{\text{even}} = (\overline{C \oplus D})(A \oplus B) + (C \oplus D)(\overline{A \oplus B})$$
$$= (A \oplus B) \oplus (C \oplus D)$$

$$P_{\text{odd}} = (\overline{C \oplus D})(\overline{A \oplus B}) + (C \oplus D)(A \oplus B)$$
$$= (A \oplus B) \odot (C \oplus D)$$

(b)

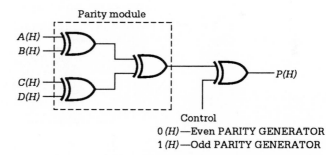

Parity module

$A(H)$
$B(H)$
$C(H)$
$D(H)$

$P(H)$

Control
0 (H)—Even PARITY GENERATOR
1 (H)—Odd PARITY GENERATOR

(c)

Fig. 4-31. *A 4-bit even/odd PARITY GENERATOR. (a) Truth table. (b) K-maps and output logic expressions. (c) Logic circuit.*

The COMPARATOR treats the bit patterns as magnitudes, whereas the error detection circuit looks only at the like-bit count of the patterns. It is because of this similarity that we discuss these two types of circuits together as a single class of devices.

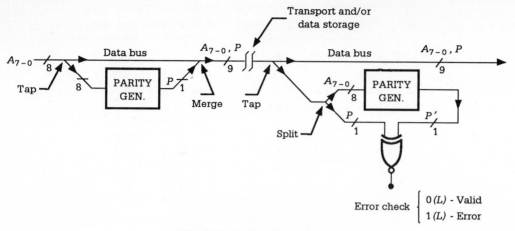

Fig. 4-32. *A possible error detection scheme for an 8-bit data bus showing tap, merge, and split operations.*

4.6 DECODERS AND ENCODERS

A DECODER is an n-input/2^n-output combinational logic device which has the function of activating one of the 2^n outputs for every unique input pattern (WORD) of n bits. Each output is identified by the MINTERM CODE, m_i, of the input WORD pattern A it represents. For this reason the DECODER can be called a MINTERM CODE GENERATOR or MINTERM RECOGNIZER. Its block diagram symbol is presented in Fig. 4-33. DECODERs normally come with an ENABLE LOW [$EN(L)$] input so that they can be turned ON or OFF for stacking purposes.

Commercial IC DECODERs usually have ACTIVE LOW outputs, as shown in Fig. 4-33. This means that the outputs of such DECODERs represent ACTIVE LOW MINTERMs or ACTIVE HIGH MAXTERMs according to the first of the relations

$$m_i(L) = M_i(H)$$
$$M_i(L) = m_i(H)$$

(4-9)

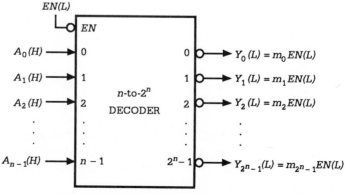

Fig. 4-33. *Block diagram symbol for an* n-*line to* 2^n-*line DECODER with ACTIVE LOW outputs and ENABLE control,* EN(L).

which are the MIXED LOGIC forms of Eqs. (3-8). Thus, DECODERs with ACTIVE LOW outputs are more accurately described as ACTIVE LOW MINTERM CODE GENERATORs or ACTIVE HIGH MAXTERM CODE GENERATORs.

DECODERs are important combinational logic devices that find wide application in digital systems. Also they are conceptually simple. To illustrate their simplicity, let us design a 3- to 8-line DECODER with outputs and ENABLE control all ACTIVE LOW, and with inputs from positive logic sources—an ACTIVE LOW ENABLE $[EN(L)]$ simply means that the device will function only when $EN(L)$ is connected to LV [i.e., $EN(L) = 1(L) = 0(H)$], but will be disabled when $EN(L)$ is connected to HV [i.e., $EN(L) = 0(L) = 1(H)$]. Remember that $0(H)$ and $1(H)$ represent low and high voltage, respectively, and that $0(H)$ usually means ground potential (zero).

Displayed in Fig. 4-34(a) is the truth table for this DECODER constructed in positive logic MINTERM code form, where ENABLE is included as an input. Also shown are the outputs $Y(7 - 0)$ meaning Y_7 through Y_0. The function of the DECODER is easily deduced from the truth table—that is, any given combination of logic values for the three inputs activates a single output line $Y_i = m_i EN$, but only if EN is ACTIVE $[EN(L) = 1(L)]$. The ACTIVE LOW and ACTIVE HIGH requirements are not included in the homologic truth table but are introduced during the implementation stage. The symbol X represents an *irrelevant input*, either logic 1 or logic 0—it doesn't matter. Its use, equivalent to the don't care (ϕ) for MINTERM or MAXTERM outputs, results in a collapsed truth table similar to those shown in Figs. 3-7 and 3-8.

Each output column (Y_i) of Fig. 4-34(a) represents a third-order K-map containing one MINTERM ANDed with ENABLE. However, it is not necessary to construct eight EV K-maps to obtain the Boolean expressions, since this information can be read directly from the truth table as indicated in Fig. 4-34(a). When the requirements of ACTIVE LOW outputs and EN-ABLE are introduced, there results the NAND circuit of Fig. 4-34(b). The block diagram for this DECODER is shown in Fig. 4-34(c). Notice that the

$EN\ A_2\ A_1\ A_0$	$Y_7\ Y_6\ Y_5\ Y_4\ Y_3\ Y_2\ Y_1\ Y_0$	
0 X X X	0 0 0 0 0 0 0 0	
1 0 0 0	0 0 0 0 0 0 0 1	$Y_0 = \overline{A_2}\,\overline{A_1}\,\overline{A_0}EN$
1 0 0 1	0 0 0 0 0 0 1 0	$Y_1 = \overline{A_2}\,\overline{A_1}A_0 EN$
1 0 1 0	0 0 0 0 0 1 0 0	$Y_2 = \overline{A_2}A_1\overline{A_0}EN$
1 0 1 1	0 0 0 0 1 0 0 0	$Y_3 = \overline{A_2}A_1 A_0 EN$
1 1 0 0	0 0 0 1 0 0 0 0	$Y_4 = A_2\overline{A_1}\,\overline{A_0}EN$
1 1 0 1	0 0 1 0 0 0 0 0	$Y_5 = A_2\overline{A_1}A_0 EN$
1 1 1 0	0 1 0 0 0 0 0 0	$Y_6 = A_2 A_1\overline{A_0}EN$
1 1 1 1	1 0 0 0 0 0 0 0	$Y_7 = A_2 A_1 A_0 EN$

(a)

Fig. 4-34. A 3-line-to-8-line DECODER with ENABLE. (a) Collapsed truth table.

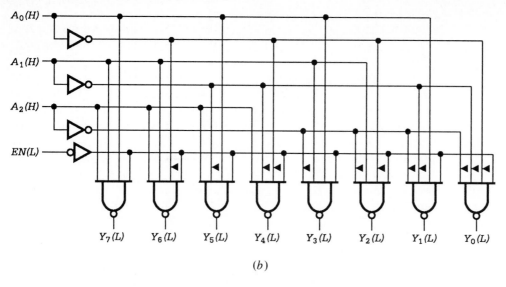

(b)

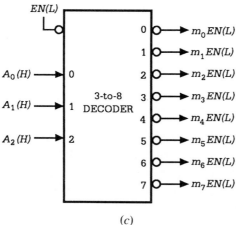

(c)

Fig. 4-34. (contd.) (b)
Implementation of the output
expressions given in part (a).
(c) Block diagram symbol for a
3-to-8-line DECODER.

NAND gates provide the AND operation required by the truth table and the ACTIVE LOW outputs.

DECODERs can be *stacked* to produce much larger DECODERs. This requires that the 2^n outputs of one DECODER drive the ENABLE controls of 2^n other DECODERs, all having n inputs. The result is a $2n$ line to 2^{2n} line DECODER. As an example, four 2 line to 4 line DECODERs are stacked in Fig. 4-35 to produce a 4-to-16 line DECODER. Similarly, 8 3-to-8 line DECODERs can be stacked to produce a 6-to-64 line DECODER, and 16 4-to-16 line DECODERs can be stacked to produce an 8-to-256 line DE-CODER, and so on. This does not include special DECODER packages which are used as CODE CONVERTERS, a subject that will be discussed later in this chapter.

4.6.1 Combinational Logic Design with DECODERs

DECODERs can be used effectively to implement any function in CANON-ICAL SOP or CANONICAL POS form. All that is needed is the external logic required to OR MINTERMs for SOP representation or to AND MAX-

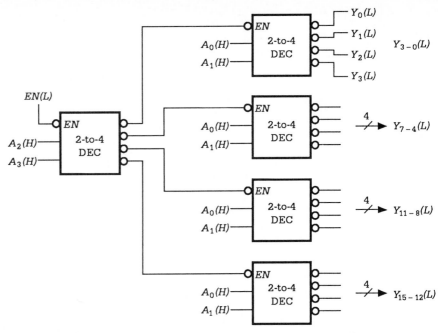

Fig. 4-35. *Stacking of 2-to-4-line DECODERs to produce a 4-to-16-line DECODER.*

TERMs for POS representation. Consider the following functions in CANONICAL form:

$$F(A,B,C) = \sum m(1, 3, 4, 7)(H) \qquad \text{SOP}$$
$$G(A,B,C) = \prod M(1, 3, 4, 7)(H) \qquad \text{POS} \tag{4-10}$$

Implementation of Eqs. (4-10) is accomplished by using the 3-to-8 line DECODER of Fig. 4-34. Since both CANONICAL expressions are ACTIVE HIGH and the DECODER outputs are ACTIVE LOW, NAND logic is required for the POS form. The results are given in Fig. 4-36. When any combination of voltages equivalent to 001, 011, 100, or 111 appears at the input terminals, the F_{SOP} output becomes HV (logic 1) and the G_{POS} output becomes LV (logic 0); for all other patterns, the F_{SOP} and G_{POS} outputs become the equivalent of logic 0 and logic 1, respectively.

Some observations can be made regarding Fig. 4-36. First, observe the simplicity with which CANONICAL SOP and POS functions can be implemented. All that is necessary is to connect the output for each MINTERM to the appropriate gate. Second, notice that the DECODER was enabled by connecting ENABLE (L) to LV (ground), commonly done for TTL ICs. Third, note that if the F_{SOP} and G_{POS} outputs are INVERTED, there results the relationships

$$F_{\text{SOP}}(L) = \overline{F}_{\text{SOP}}(H) \quad \text{and} \quad G_{\text{POS}}(L) = \overline{G}_{\text{POS}}(H) \tag{4-11}$$

which are the general MIXED LOGIC forms for equations of the type given by Eqs. (3-11) and (3-12). Thus, CANONICAL SOP representation CONVERTs to CANONICAL POS representation (of the same code terms), and vice versa.

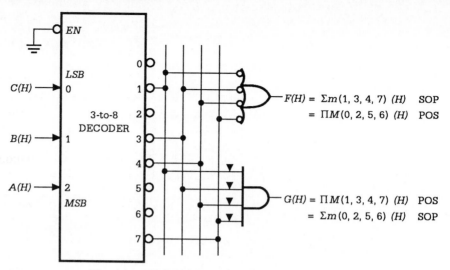

Fig. 4-36. DECODER implementation of Eqs. (4-10).

Now, there arises the question of MIXED LOGIC inputs. How does one use a DECODER chip to implement a CANONICAL SOP or POS expression when a portion of the inputs arrive from negative logic sources, that is, are ACTIVE LOW? One solution might be to reconfigure the circuit to meet the needs of the input conditions. If we had access to the DECODER's internal logic circuitry, as in Fig. 4-34, we could easily rewire it so that it could output functions like Eqs. (4-10) for any given set of MIXED LOGIC inputs. However, DECODER IC chips are sealed packages, as are all IC chips, and alteration of their internal circuitry would be difficult if not impossible. There is a far better solution to this problem, one that is best described by the following example.

Referring to Fig. 4-36, let us assume that inputs B and C arrive at the DECODER from negative logic sources with input A arriving from a positive logic source. Thus, the inputs become $A(H)$, $B(L)$, and $C(L)$. Functionally, we want

$$F_{SOP}[A(H),B(L),C(L)] = F_{SOP}[A,\bar{B},\bar{C}](H)$$

and
$$G_{POS}[A(H),B(L),C(L)] = G_{POS}[A,\bar{B},\bar{C}](H)$$

which means that we must invert each of the B and C bits:

$$m_1 = 0\ 0\ 1 \ \Rightarrow\ 0\ 1\ 0 = m_2$$
$$m_3 = 0\ 1\ 1 \ \Rightarrow\ 0\ 0\ 0 = m_0$$
$$m_4 = 1\ 0\ 0 \ \Rightarrow\ 1\ 1\ 1 = m_7$$
$$m_7 = 1\ 1\ 1 \ \Rightarrow\ 1\ 0\ 0 = m_4,$$

where the double-edged arrow is read as "becomes." Therefore, it follows that Eqs. (4-10) must be restated as

$$F_{SOP}[A(H), B(L), C(L)] = \sum m(0, 2, 4, 7)$$

and

$$G_{POS}[A(H), B(L), C(L)] = \prod M(0, 2, 4, 7)$$

(4-12)

to accommodate the MIXED LOGIC input conditions.

The DECODER implementation of Eqs. (4-12) can now be carried out in a manner similar to that shown in Fig. 4-36 except that, in this case, the external logic is connected to leads 0, 2, 4, and 7 instead of leads 1, 3, 4, and 7.

There is a third, and even simpler, solution to the problem of MIXED LOGIC inuts, provided one is willing to use additional hardware. This solution involves the use of an INVERTER on each ACTIVE LOW input to the DECODER. Applied to the implementation of Eqs. (4-10), the external logic would then be connected to output DECODER leads 1, 3, 4, and 7, exactly as in Fig. 4-36.

Finally, one more point is worth mentioning at this time. In using DE-CODERs for combinational logic design purposes, special care must be taken to properly connect the inputs to the DECODER. Since, we have become accustomed to writing a binary WORD as *A B C D* . . . , we must connect line *A* to the MSB lead specified for the DECODER, with all others connected in order down to and including the LSB. The manufacturers of DE-CODERs traditionally use *A* as the LSB and the last literal in alphabetical order as the MSB. Thus, one must connect the leads up in reverse literal order.

Other applications of DECODERs include their use as CODE CON-VERTERs and DEMULTIPLEXERs. These and other uses of DECODERs will be discussed at some length later in this chapter.

4.6.2 PRIORITY ENCODERs

By definition, an ENCODER performs an operation which is the opposite of that for a DECODER. That is, an ENCODER must generate a different output bit pattern for each input element (line) that becomes ACTIVE. Since this requirement can be enforced only for binary WORDs equivalent to decimal 2^n (0, 1, 2, 4, 16, . . .), ENCODERs are useful only in the form of PRIORITY ENCODERs which permit only the highest priority data to be encoded.

PRIORITY ENCODERs are combinational logic devices designed to permit any number of input elements to be ACTIVE simultaneously but with a preassigned priority which permits the device to select a particular AC-TIVE input element. In fact, it is common practice to assign to the EN-CODER an entire set of priorities from highest priority to lowest. For a PRIORITY ENCODER of *m* outputs, the number of priority assignments, as well as the number of inputs, cannot exceed 2^m. The block diagram symbol for a typical PRIORITY ENCODER is shown in Fig. 4-37.

Usually, the inputs and outputs for PRIORITY ENCODERs are all AC-TIVE LOW (as indicated in Fig. 4-37) for reasons that will be made clear later. The input *EI*, and outputs *EO* and *GS*, representing ENABLEIN, ENABLEOUT and GROUP SIGNAL, respectively, provide the means for

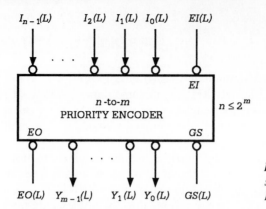

$I_{n-1}(L)$ $I_2(L)$ $I_1(L)$ $I_0(L)$ $EI(L)$

EI

n-to-m
PRIORITY ENCODER

$n \le 2^m$

EO   GS

$EO(L)$ $Y_{m-1}(L)$ $Y_1(L)$ $Y_0(L)$ $GS(L)$

Fig. 4-37. *Block diagram symbol for an* n-to-m *PRIORITY ENCODER.*

cascading (stacking) ENCODERs. These features are included in the design of a simple PRIORITY ENCODER that follows.

In case the reader is wondering why we could not simply use a DE-CODER in reverse to produce an ENCODER, we would point out that the DECODER and ENCODER perform entirely different operations with no direct relationship between them. Furthermore, devices composed of transistors cannot be operated in reverse; the operation of the simple BJT IN-VERTER switch in Fig. 1-7 indicates why this is so.

We illustrate the design of a PRIORITY ENCODER (PE) by the following simple example. Let the address inputs I_2, I_1, and I_0 have the priority assignments and encoded data given in Fig. 4-38(a). Therefore, the I_2 input has been assigned the highest priority and is encoded to 11. This means that regardless of the logic values of I_1 and I_0, the outputs Y_1 and Y_0 will both become ACTIVE (i.e., both assume logic 1 values) when I_2 becomes AC-TIVE. Similarly, I_1 is assigned the next highest priority (middle) and is encoded to 10, meaning that regardless of the logic value of I_0 the outputs, Y_1 and Y_0, will assume logic values of 1 and 0, respectively, when I_1 becomes ACTIVE (logic 1) when I_2 is INACTIVE. The input I_0 is assigned the lowest priority and is encoded to 01. Finally, any well-designed PRIORITY EN-CODER will identify the INACTIVE state (no ACTIVE address input). We do this by assigning the only remaining output logic values (00) to this condition.

The collapsed truth table representative of these priority assignments is given in Fig. 4-38(b), where the symbol X signifies both logic 1 and logic 0, that is, an irrelevant input. Observe that when EI is NOT ACTIVE all I inputs are irrelevant and the outputs are all logic 0 (INACTIVE). To simplify the design, we have chosen EI as the entered variable, which compresses the K-maps by one order. These maps, together with the resulting logic circuit, are shown in Figs. 4-38(c) and (d). Notice that the logic circuit represents a multiple-output optimization which was deduced by inspection of the shared PIs.

We stipulated at the beginning of this example that all inputs and outputs must be ACTIVE LOW, consistent with commercial PEs. This is accomplished by placing LOGIC LEVEL CONVERTERs (INVERTERs) on the inputs and ACTIVE LOW indicator bubbles on the output gates. If necessary, we can also be selective about our choice of which inputs or outputs would be ACTIVE LOW.

Priority Schedual	EI I_2 I_1 I_0	GS Y_1 Y_0 EO
I_2 (highest)—encoded to 11	0 X X X	0 0 0 0
I_1 (middle)—encoded to 10	1 1 X X	1 1 1 0
I_0 (lowest)—encoded to 01	1 0 1 X	1 1 0 0
INACTIVE state—assigned to 00	1 0 0 1	1 0 1 0
	1 0 0 0	0 0 0 1

(a) (b)

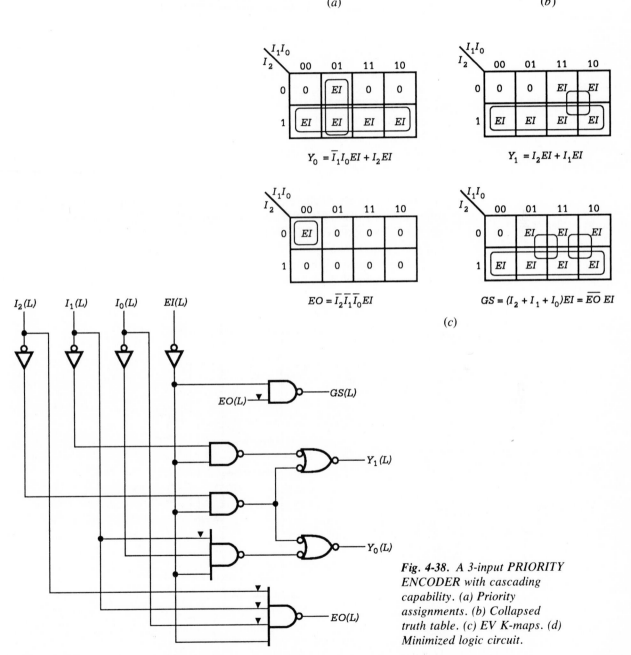

$$Y_0 = \overline{I_1}I_0 EI + I_2 EI$$

$$Y_1 = I_2 EI + I_1 EI$$

$$EO = \overline{I_2}\,\overline{I_1}\,\overline{I_0} EI$$

$$GS = (I_2 + I_1 + I_0)EI = \overline{EO}\, EI$$

(c)

Fig. 4-38. *A 3-input PRIORITY ENCODER with cascading capability. (a) Priority assignments. (b) Collapsed truth table. (c) EV K-maps. (d) Minimized logic circuit.*

(d)

The outputs *EO* and *GS* in Fig. 4-38(b) require some special attention since their logic values have been specifically chosen to make cascading of PEs possible. When the address inputs for the *n*th stage are *all* logic 0 (IN-ACTIVE), it is the function of EO_n to drive EI_{n-1} (of the preceding stage) ACTIVE. Therefore, *EO* can be ACTIVE only for INACTIVE address and ACTIVE *EI*. It is the primary function of the *GS* output to recognize any legitimate encoding condition for that stage (meaning *EI* ACTIVE with an ACTIVE address input) and to provide an additional output bit (MSB) for each stage added in cascade fashion.

Shown in Fig. 4-39 are the collapsed truth table and modular design for two stacked PEs of the type given in Fig. 4-38. Here, the EO output of the higher positionally weighted stage is connected to the EI input of the lower positionally weighted stage. Note that the *Y* outputs of the individual stages are ORed ACTIVE LOW by using AND gates so as to produce the required outputs for the stacked units. Notice also that, in this case, one output state (100) cannot occur.

The outputs of PRIORITY ENCODERs are designed to be ACTIVE LOW so that they can be connected directly to conjugate NAND and NOR gates to produce ACTIVE HIGH signals. Similarly, the inputs to PEs are designed to be ACTIVE LOW so that they can be connected directly to other devices like DECODERs which have ACTIVE LOW outputs. The ENABLEIN input is ACTIVE LOW permitting the ACTIVE condition $1(L)$ to be accomplished by $0(H)$ which is ground potential in most cases. For cascading purposes it is obviously an advantage to design both *EO* and *EI* with the same polarization level—ACTIVE LOW in this case.

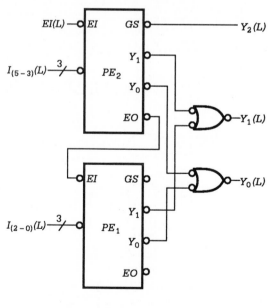

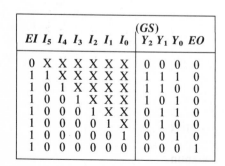

EI	I_5	I_4	I_3	I_2	I_1	I_0	(GS) Y_2	Y_1	Y_0	EO
0	X	X	X	X	X	X	0	0	0	0
1	1	X	X	X	X	X	1	1	1	0
1	0	1	X	X	X	X	1	1	0	0
1	0	0	1	X	X	X	1	0	1	0
1	0	0	0	1	X	X	0	1	1	0
1	0	0	0	0	1	X	0	1	0	0
1	0	0	0	0	0	1	0	0	1	0
1	0	0	0	0	0	0	0	0	0	1

(a)   (b)

Fig. 4-39. *Two stacked PRIORITY ENCODERs of the type shown in Figure 4-37: (a) Collapsed truth table. (b) Modular level logic circuit.*

PRIORITY ENCODERs of the type shown in Fig. 4-37 are commercially available as IC chips. Typically, they are produced in the 8-to-3 line size which can be stacked to produce 16-to-4 line and 32-to-5 line PEs, and so on. Their applications include code conversion, code generation, and *N*-bit encoding in digital systems having a hierarchy of subsystems which must be prioritized.

PRIORITY ENCODERs which do not have *EO* and *GS* outputs and which cannot be stacked are also available commercially as IC chips. Their design follows closely the design of PEs of the type shown in Fig. 4-38. They are commonly produced in the 9-to-4 line size and are used for decimal to BCD (binary-coded decimal) priority encoding, keyboard encoding, and range selection. Problems 4-47 and 4-49 at the end of this chapter illustrate the design of both types of PRIORITY ENCODERs.

4.7   CODES AND CODE CONVERTERS

A digital device is usually designed to operate with a specific binary code. But in putting together a complicated digital system, it may be necessary for the designer to use a variety of devices and subsystems which operate with different codes. Of concern in this section is the problem that arises when a device or subsystem, operating with one type of code, must communicate with another device or subsystem based on a different code. To permit the passage of intelligible information between these two devices or subsystems, a device known as a CODE CONVERTER must be installed which can convert one code to the other. For example, suppose an arithmetic unit of some sort (say, an ADDER), designed to operate with natural binary, must communicate its results to another device which is designed to operate in natural binary-coded decimal (NBCD) code. What is needed is a binary-to-NBCD CODE CONVERTER positioned between the devices.

The reader should be forewarned that there is not uniform agreement on nomenclature in this area—a situation not uncommon in the fast-growing field of digital electronics. Unhappily, some leading IC chip manufacturers refer to CODE CONVERTERs as DECODERs, and perhaps as a result of this, the subject of code conversion is covered along with DECODERs in some texts and reference books. In this text we will do neither of these. We view code conversion as a distinct operation and, accordingly, treat it separately.

Most binary codes of significant value in digital design fall into one or more of the following ten categories:

Weighted binary codes	Unit distance codes
Unweighted binary codes	Reflective codes
Biased codes	Unit distance reflective codes
Binary coded decimal codes	Alphanumeric codes
Self-complementing codes	Error detecting codes

Before providing a few examples of typical code conversion problems, it will be worthwhile to consider the subject of digital machine coding and discuss the properties of some of the better known codes.

Digital machines are designed to operate with just two stable states, HV and LV. So, it is natural that machine codes be based on the binary system of true or false, or logic 1 or logic 0, and so on. However, human beings are accustomed to the decimal system for calculation and to the alphabet for word formation. For day-to-day operations the binary system would be very awkward for us to use. One could imagine the difficulty we would experience in writing or reading a report or letter in some binary code. Consequently, a number of codes have been devised which make communication with digital machines, like computers, easier and more efficient for humans. Such codes as the binary-coded decimal (BCD) codes and the alphanumeric codes are used for that purpose. The BCD codes have the added advantage that the hardware required for BCD to decimal conversion is significantly less than that required for converting other binary codes to decimal.

Still, there are other codes designed to facilitate certain computer operations such as subtraction, error detection, and correction. Many of these codes are also of the BCD type which can be easily converted to decimal form by using the familiar seven-segment display. Shown in Fig. 4-40 are a few examples of BCD codes. Notice that codes (1) through (4) require just 10 of the 16 possible states for a 4-bit WORD, while codes (5) and (6) use only 10 of the 32 possible states for a 5-bit WORD. Obviously, a minimum of 4 bits is required to represent decimal 0 through decimal 9, but there is no upper limit to the number of bits in each code WORD other than that imposed by practical considerations such as implementation.

The seven BCD codes given in Fig. 4-40 can be classified as either *weighted* or *unweighted* codes. All but one of these codes are weighted as indicated in the figure. A weighted code can be converted directly to its decimal equivalent by using positional code weights in an arithmetic series. An unweighted code cannot be converted by any such mathematical weighting procedure. A good example of a weighted code is the natural BCD (NBCD) code which has code weights of 8, 4, 2, and 1. It is converted to its decimal equivalent by the polynomial

	Weighted Codes						Unweighted
Dec.	(1) (NBCD)	(2) (XS3)	(3) 2421	(4) 84-2-1	(5) 86421	(6) 51111	(7) Creeping Code
0	0 0 0 0	0 0 1 1	0 0 0 0	0 0 0 0	0 0 0 0 0	0 0 0 0 0	0 0 0 0 0
1	0 0 0 1	0 1 0 0	0 0 0 1	0 1 1 1	0 0 0 0 1	0 0 0 0 1	1 0 0 0 0
2	0 0 1 0	0 1 0 1	0 0 1 0	0 1 1 0	0 0 0 1 0	0 0 0 1 1	1 1 0 0 0
3	0 0 1 1	0 1 1 0	0 0 1 1	0 1 0 1	0 0 0 1 1	0 0 1 1 1	1 1 1 0 0
4	0 1 0 0	0 1 1 1	0 1 0 0	0 1 0 0	0 0 1 0 0	0 1 1 1 1	1 1 1 1 0
5	0 1 0 1	1 0 0 0	1 0 1 1	1 0 1 1	0 0 1 0 1	1 0 0 0 0	1 1 1 1 1
6	0 1 1 0	1 0 0 1	1 1 0 0	1 0 1 0	0 1 0 0 0	1 1 0 0 0	0 1 1 1 1
7	0 1 1 1	1 0 1 0	1 1 0 1	1 0 0 1	0 1 0 0 1	1 1 1 0 0	0 0 1 1 1
8	1 0 0 0	1 0 1 1	1 1 1 0	1 0 0 0	1 0 0 0 0	1 1 1 1 0	0 0 0 1 1
9	1 0 0 1	1 1 0 0	1 1 1 1	1 1 1 1	1 0 0 0 1	1 1 1 1 1	0 0 0 0 1

Fig. 4-40. Weighted and unweighted BCD codes.

4 / Combinational Logic Design

$$N_{10} = (B_3 \times 2^3) + (B_2 \times 2^2) + (B_1 \times 2^1) + (B_0 \times 2^0)$$

$$= (B_3 \times 8) + (B_2 \times 4) + (B_1 \times 2) + (B_0 \times 1)$$

for any $B_3B_2B_1B_0$ code state. Thus, decimal 6 would be represented as $(0 \times 8) + (1 \times 4) + (1 \times 2) + (0 \times 1)$ or 0110 in NBCD code. Like natural binary code (weighted $\ldots$, 16, 8, 4, 2, 1, $\frac{1}{2}$, $\frac{1}{4}$, $\ldots$), NBCD code is called "natural" because its code weights are derived from integer powers of 2 (2^n). The NBCD code is currently the most widely used of the BCD codes.

It may be recalled that decimal numbers greater than nine can be represented by *any* BCD code if each digit is given in that BCD code and if the results are combined. As an example, let us represent decimal 36.9 in NBCD code:

$$
\begin{array}{cccc}
3 & 6 & \cdot & 9 \\
|----| & |-~--| & & |----| \\
36.9_{10} = \quad 0011 & 0110 & \cdot & 1001
\end{array}
$$

$$= [00110110.1001]_{\text{NBCD}}$$

Here, the code weights are 80, 40, 20, 10; 8, 4, 2, 1; and 0.8, 0.4, 0.2, 0.1 for the tens digit, units digit, and tenths digit, respectively, representing three decades. To reverse the conversion process (i.e., NBCD to decimal), groups of 4 bits are separated out, mentally at least, beginning from the right or from the radix point, if present. If there are *not* 4 bits in the leftmost group, 0's are implied. Negative BCD numbers can be represented either in sign-magnitude notation or 1's 2's complement notation.

Code (2) in Fig. 4-40, called Excess 3 (or XS3 NBCD or simply XS3), is an example of a *biased-weighted* (or biased-natural) code. It is formed from NBCD by adding the binary equivalent of decimal 3, that is, 0011, to each of the ten states. This code, said to have a bias of 3, has some interesting and useful properties which will soon be discussed. Codes of any bias can be generated in the same way.

Not all weighted BCD codes are natural, in the sense that their code weights cannot be derived from positive integer powers of two. Codes (3) through (6) in Fig. 4-40 are of this type. Code weights such as -1, -2, 5, and 6 cannot be generated by any positive integer power of 2, but they can still serve as code weights. To demonstrate, consider how decimal 5 is represented by codes (4) and (5):

$$\text{Decimal equivalent} = 5$$

$$84-2-1 \text{ code representation} = (1 \times 8) + (0 \times 4)$$
$$+ [1 \times (-2)] + [1 \times (-1)]$$
$$= 1\ 0\ 1\ 1$$

$$86421 \text{ code representation} = (0 \times 8) + (1 \times 4) + (0 \times 2) + (1 \times 1)$$
$$= 0\ 1\ 0\ 1$$

Notice that there may be more than one combination of weighted bits that produce a given state. When this occurs, the procedure is usually to use the fewest number of 1's. For example, decimal 7 can be represented by 00111 in 86421 BCD code, but 01001 is preferred. An exception to this rule is the 2421 code discussed next.

The BCD codes of Fig. 4-40 have some interesting properties which are worth mentioning at this time. We have already discussed the nature of the code weights for NBCD code. Codes (2), (3), and (4) are examples of BCD codes that have the unusual property of being *self-complementing*. This means that the *1's complement of the code number is the code for the 9's complement of the corresponding decimal number*. In other words, the 1's complement of any state N (in decimal) is the same as the $(9 - N)$ state in that same self-complementing code. As an example, the 1's complement of state 3 in XS3 (0110) is state 6 (1001) in that same code. This is demonstrated in Fig. 4-41 where arrows have been drawn to help the reader follow the example. For reference purposes the 9's complement NBCD code is also included.

We have just mentioned that the 9's complement of a number is found by subtracting that number from 9. The 10's complement of a number can be formed in a similar manner, that is, by subtracting the number from decimal 10. Alternatively, the 10's complement of a number can be obtained by adding 1 to the 9's complement. This suggests a relationship between 1's and the 9's complement codes and the 2's and 10's complement codes when one recalls the $\overline{N}_2$ and ($\overline{N}_2$ plus 1) algorithms discussed in Sect. 4.4.7. In fact, the 2's and 10's complement of a number are called the *radix complements* defined mathematically by

$$\text{Radix complement of } N \equiv r^n - N \qquad (4\text{-}13)$$

Similarly, the 1's complement for binary numbers and the 9's complement for decimal numbers are called the *diminished (or radix − 1) complements* defined mathematically by

$$\text{Radix } - 1 \text{ complement of } N \equiv r^n - N - 1 \qquad (4\text{-}14)$$

In these equations, N is the number for which the complement is to be found, r is the radix (2 for binary, 10 for decimal), and n is the number of bits (or digits) in N.

Decimal	XS3 Code	1's Complement of XS3	9's Complement Decimal	9's Complement NBCD
0	0 0 1 1	1 1 0 0	9	1 0 0 1
1	0 1 0 0	1 0 1 1	8	1 0 0 0
2	0 1 0 1	1 0 1 0	7	0 1 1 1
3	0 1 1 0 →	1 0 0 1	6	0 1 1 0
4	0 1 1 1	1 0 0 0	5	0 1 0 1
5	1 0 0 0	0 1 1 1	4	0 1 0 0
6	1 0 0 1	0 1 1 0	3	0 0 1 1
7	1 0 1 0	0 1 0 1	2	0 0 1 0
8	1 0 1 1	0 1 0 0	1	0 0 0 1
9	1 1 0 0	0 0 1 1	0	0 0 0 0

Fig. 4-41. The self-complementing property of the XS3 BCD code and the 9's complement NBCD code.

It should be understood that for 10's and 9's complement, $r = 10$ and N is in decimal, but they may be represented in binary by conversion from decimal. In this sense they are binary codes just as are the 2's and 1's complements.

As an example, let us find the 2's and 9's complement of $23_{10} = 10111_2$. By applying Eq. (4-13), the 2's complement is $2^5 - 10111$ or $10000 - 10111 = 01001$, a result which is more easily obtained using the $(\overline{N}_2 + 1)$ algorithm. Now, by applying Eq. (4-14), the 9's complement of 23 is found to be $10^2 - 23 - 1$ or 76, which, of course, is the difference between 99 and 23. Conversion to natural binary or NBCD (as in Fig. 4-41) now follows from the decimal value and is $76_{10} = [01110110]_{\text{NBCD}}$ for this example.

BCD subtraction can be carried out using either 9's or 10's complement arithmetic—procedures similar to but slightly more complicated than the 1's and 2's complement arithmetic discussed in Sect. 4.4.7. These methods for BCD subtraction require the use of additional hardware to generate the 9's or 10's complement in binary. However, this additional complexity can be avoided by using a self-complementing code, thereby yielding the 9's complement when each bit is inverted, as described earlier. The XS3 BCD code is the most widely used code for this purpose.

There is another class of weighted or semiweighted codes with the special property that their states contain either an even number or an odd number of logic 1's. Four examples of such codes are presented in the table of Fig. 4-42. This unique feature makes these codes attractive as error-detecting (parity-checking) codes. Notice that both the 2-out-of-5 code (semiweighted 74210) and the biquinary code (weighted 50 43210) must have two 1's in each of their ten states and are, therefore, even parity codes. In contrast, the one-hot code (weighted 9876543210) is an odd parity code, since it is permitted to have only a single logic 1 in any given state. Code (d) is no more than an NBCD code with an attached odd parity-generating bit, P.

The advantage of using an error-detecting code is that single-bit errors (those most likely to occur) are easily detected by a PARITY DETECTOR placed at the receiving end of a data bus. If an error is detected, further

Decimal Value	(a) Even Parity 2-out-of-5 (7 4 2 1 0)	(b) Even Parity Biquinary 5 0 4 3 2 1 0	(c) Odd Parity One-Hot 9 8 7 6 5 4 3 2 1 0	(d) Odd Parity NBCD P 8 4 2 1
0	1 1 0 0 0	0 1 0 0 0 0 1	0 0 0 0 0 0 0 0 0 1	1 0 0 0 0
1	0 0 0 1 1	0 1 0 0 0 1 0	0 0 0 0 0 0 0 0 1 0	0 0 0 0 1
2	0 0 1 0 1	0 1 0 0 1 0 0	0 0 0 0 0 0 0 1 0 0	0 0 0 1 0
3	0 0 1 1 0	0 1 0 1 0 0 0	0 0 0 0 0 0 1 0 0 0	1 0 0 1 1
4	0 1 0 0 1	0 1 1 0 0 0 0	0 0 0 0 0 1 0 0 0 0	0 0 1 0 0
5	0 1 0 1 0	1 0 0 0 0 0 1	0 0 0 0 1 0 0 0 0 0	1 0 1 0 1
6	0 1 1 0 0	1 0 0 0 0 1 0	0 0 0 1 0 0 0 0 0 0	1 0 1 1 0
7	1 0 0 0 1	1 0 0 0 1 0 0	0 0 1 0 0 0 0 0 0 0	0 0 1 1 1
8	1 0 0 1 0	1 0 0 1 0 0 0	0 1 0 0 0 0 0 0 0 0	0 1 0 0 0
9	1 0 1 0 0	1 0 1 0 0 0 0	1 0 0 0 0 0 0 0 0 0	1 1 0 0 1

Fig. 4-42. Error detection codes. (a) 2-out-of-5 code (even parity). (b) Bi-quinary code (even parity). (c) One-hot code (odd parity). (d) P8421 code (odd parity).

Dec.	(1) BCD Code	(2) 4-Bit Gray Code	(3) XS3 Gray BCD Code
0	0 0 0 0	0 0 0 0	0 0 1 0
1	0 0 0 1	0 0 0 1	0 1 1 0
2	0 0 1 1	0 0 1 1	0 1 1 1
3	0 0 1 0	0 0 1 0	0 1 0 1
4	0 1 1 0	0 1 1 0	0 1 0 0
5	1 1 1 0	0 1 1 1	1 1 0 0
6	1 1 1 1	0 1 0 1	1 1 0 1
7	1 1 0 1	0 1 0 0	1 1 1 1
8	1 1 0 0	1 1 0 0	1 1 1 0
9	0 1 0 0	1 1 0 1	1 0 1 0
10	——	1 1 1 1	——
11	——	1 1 1 0	——
12	——	1 0 1 0	——
13	——	1 0 1 1	——
14	——	1 0 0 1	——
15	——	1 0 0 0	——

Fig. 4-43. *Unit distance codes. (1) A BCD code (nonreflective). (2) Four-bit Gray code (reflective). (3) XS3 Gray BCD code (reflective).*

processing can be delayed until the error is corrected. Error-detecting schemes of the type shown in Fig. 4-32 are necessary only when nonerror-detecting codes are used.

The last class of codes we will consider are the *unit distance* codes, so called because only *one* bit is permitted to change between any two of their adjacent states—in natural binary, adjacent states can differ by one or more bits. Three examples of unit distance codes are given in the table of Fig. 4-43: (1) a BCD code, (2) a *reflective* unit distance code usually called a Gray code, and (3) an XS3 Gray code obtained from the inner 10 states of code (2) as indicated by the bracket in Fig. 4-43. The reflective character of the Gray codes is easily revealed by the fact that all bits except the MSB are reflected across an imaginary mirror plane (dotted line) located midway in the 16 (or 10) states represented.

We are already familiar with 2-bit Gray code because it was used for the logic coordinates in third- and fourth-order K-maps. The unit distance property maximizes logic adjacency in neighboring states thus permitting a third-order K-map to be continuous about its single variable axis, and a fourth-order K-map to be continuous about both of its axes, as discussed in Sect. 3.4. Also, the unit distance and reflective properties of the Gray code make it uniquely suitable as a position indicator code for rotating disks and shafts. Encoding errors produced by rotational irregularities can be minimized in this way. In any case, it is usually necessary to convert Gray code to natural binary or NBCD, since most devices are designed to operate with one of these natural binary codes.

4.7.2 CODE CONVERTERS

The numbers of different binary codes possible are limited only by our imagination to create them. In the last section we explored some of the better known codes to gain a clearer understanding of the properties that make codes useful. We learned that the use of codes is necessary to satisfy two important needs: a need for intermachine and human-machine communication, and a need to reduce logic hardware complexity. Not surprisingly some codes work better than others in satisfying one, but usually not both, of these needs. Therefore, the design of a digital system may require the use of more than one binary code together with the appropriate code conversion hardware. Devices designed for the purpose of converting one code to another are called CODE CONVERTERs.

The design of a CODE CONVERTER is usually straightforward and uncomplicated. The design process begins with a conversion algorithm (if needed) and an I/O truth table for the input and output codes. If a modular-level approach is desired, then implementation usually follows directly from the algorithm and truth table. If a gate-level design is required, use of K-maps is almost always necessary. To construct the truth table and K-maps, we recommend the following procedure:

1. If the conversion involves *any* BCD code input, only ten states can be used. The six unused input states are called *false data* inputs. For these six states, the outputs must be represented either by don't cares (ϕ's) or by logic 0's (or logic 1's) corresponding to unused output states. That is, if it is required that the false data be rejected, the K-map entries must correspond to at least one unused output state; if not, enter ϕ's and use them to loop out a minimum of reduced cover. Thus, false data rejection means that the outputs must never correspond to a used-output state when any one of the six unused input states arrives at the input terminals. If false data are not to be rejected, then each unused output state becomes a don't care state, and don't care states can have logic values that correspond to a used output state when any one of the six unused states arrives at the input terminals.

2. If the input code is any other than natural binary-coded decimal (NBCD), construct the truth table such that the input code is arranged in the order of increasing MINTERM code number, taking care to match each output state with its assigned input state. This procedure greatly facilitates the construction of EV K-maps.

Now that we have established a workable procedure, let us illustrate its use by designing five representative CODE CONVERTERs. Included in the examples will be those having BCD and nonnatural code inputs, and those with binary, NBCD, decimal, and seven-segment outputs, all with specifically chosen input/output combinations.

Gray-to-Binary Conversion. The Gray-to-natural binary conversion table for 4-bit codes is given in Fig. 4-44(a), where the Gray code has been rearranged in the order of increasing MINTERM code number. An inspection of the two codes reveals a simple algorithm for converting from natural binary to Gray code: if a logic 0 is imagined to be placed in the 32 weight position of each state of the natural binary code, then successive XOR operations between adjacent bits, from MSB to LSB, generate the Gray code. For example, state 9 is converted as

$$9_{10} = 0\ 1\ 0\ 0\ 1 \quad \text{in natural binary code}$$
$$9_{10} = 1\ 1\ 0\ 1 \quad \text{in Gray code}$$

Gray A B C D	Binary A' B' C' D'	Decimal Value	Gray A B C D	Binary A' B' C' D'	Decimal Value
0 0 0 0	0 0 0 0	0	1 0 0 0	1 1 1 1	15
0 0 0 1	0 0 0 1	1	1 0 0 1	1 1 1 0	14
0 0 1 0	0 0 1 1	3	1 0 1 0	1 1 0 0	12
0 0 1 1	0 0 1 0	2	1 0 1 1	1 1 0 1	13
0 1 0 0	0 1 1 1	7	1 1 0 0	1 0 0 0	8
0 1 0 1	0 1 1 0	6	1 1 0 1	1 0 0 1	9
0 1 1 0	0 1 0 0	4	1 1 1 0	1 0 1 1	11
0 1 1 1	0 1 0 1	5	1 1 1 1	1 0 1 0	10

(*a*)

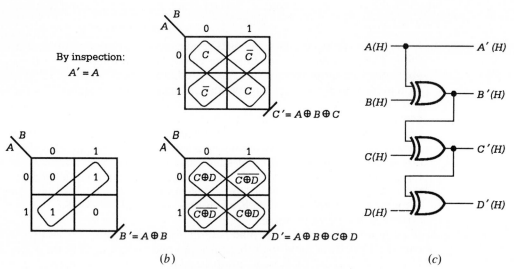

(*b*)   (*c*)

Fig. 4-44. *Gray-to-binary code conversion. (a) I/O truth table. (b) EV K-maps. (c) Logic circuit using XOR gates.*

By using this algorithm, we can obtain the required I/O relations without the need for K-maps. Thus,

$$A = A'$$
$$B = A' \oplus B'$$
$$C = B' \oplus C'$$
$$D = C' \oplus D'$$

But intercode relationships are usually more complex than the example given above, and require the use of K-maps. To illustrate the procedure to be followed, we will use EV K-maps to design a gate level Gray-to-binary CON-VERTER.

Presented in Fig. 4-44(b) are the EV K-maps which represent second-order compressions (MAP KEY = 4) of the truth table given in Fig. 4-44(a). The XOR-type patterns, inherent in this code conversion, are easily rec-ognized as indicated by the diagonal loops. The final result is the relatively simple logic circuit shown in Fig. 4-44(c). It should seem evident that these results could not have been deduced as easily from either the truth table or from standard noncompressed K-maps.

XS3-to-NBCD Conversion. The XS3 code is a biased NBCD code with the six input states 0, 1, 2, 13, 14, 15 left unused. Let us design this CODE CONVERTER by using a gate-level approach *without* false data rejection. We will assume that all input data arrive from positive logic sources but that the outputs are to be issued ACTIVE LOW.

Since the false data will *not* be rejected, we may enter φ's in the truth table for those outputs which correspond to the six unused input states. In Figs. 4-45(a), (b), and (c) are shown the truth table, EV K-maps, and logic circuit for this CONVERTER. Notice that the flexible use of φ's [see submap in Fig. 4-45(b)] together with the XOR type patterns discussed in Sect. 3.12.2 can greatly simplify the combinational logic required. Also, observe that the CONVERTER has been implemented with NAND/EQV logic to satisfy the requirement of ACTIVE LOW outputs.

The conversion of NBCD code to any biased NBCD code, or vice versa, can be achieved simply by using a 4-bit ADDER/SUBTRACTOR. For example, the NBCD-to-XS3 conversion can be carried out by adding 0011 to each state of the NBCD code. The reverse conversion process would be accomplished by subtracting 0011 from the XS3 states (or by adding 1101

Dec.	A B C D	A' B' C' D'	Dec.	A B C D	A' B' C' D'
	0 0 0 0	↑	5	1 0 0 0	0 1 0 1
	0 0 0 1	φ	6	1 0 0 1	0 1 1 0
	0 0 1 0	↓	7	1 0 1 0	0 1 1 1
0	0 0 1 1	0 0 0 0	8	1 0 1 1	1 0 0 0
1	0 1 0 0	0 0 0 1	9	1 1 0 0	1 0 0 1
2	0 1 0 1	0 0 1 0		1 1 0 1	↑
3	0 1 1 0	0 0 1 1		1 1 1 0	φ
4	0 1 1 1	0 1 0 0		1 1 1 1	↓

(*a*)

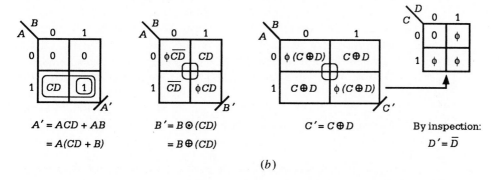

$A' = ACD + AB$
$\quad = A(CD + B)$

$B' = B \odot (CD)$
$\quad = B \oplus (CD)$

$C' = C \oplus D$

By inspection:
$D' = \bar{D}$

(*b*)

Fig. 4-45. *XS3-to-NBCD CODE CONVERTER without false data rejection. (a) Truth table. (b) EV K-maps and output expressions.*

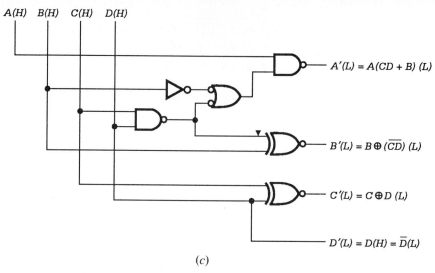

A(H) B(H) C(H) D(H)

$A'(L) = A(CD + B) \ (L)$

$B'(L) = B \oplus (\overline{CD}) \ (L)$

$C'(L) = C \oplus D \ (L)$

$D'(L) = D(H) = \overline{D}(L)$

(c)

Fig. 4-45 (contd.). (c) Logic circuit for an XS3-to-NBCD CODE CONVERTER with outputs ACTIVE LOW.

in 2's complement to the XS3 states) to produce the NBCD code. Use of ADDERs/SUBTRACTORs in this way constitutes a form of modular design.

XS3 Gray-to-Decimal Conversion. Here is an example of a code conversion scheme which leads quite naturally to a DECODER-type implementation. The term "decimal," as used in code conversion nomenclature, refers to any one of ten output lines that becomes ACTIVE when the electrical analog of the corresponding NBCD state appears at the input terminals. Consequently, this CODE CONVERTER is actually a 4-to-10 line DECODER.

Since only 10 of the 16 input states are used, we must decide whether the CODE CONVERTER is to be designed with or without false data rejection. This provides us with the opportunity to demonstrate a somewhat different use for the K-map representation.

The truth table for XS3 Gray-to-decimal conversion is shown in Fig. 4-46(a), where only the ten valid input states are given. Presented in Figs. (b) and (c) are two K-map representations of a DECODER, one with and the other without false data rejection (FDR). A K-map used in this way is called a *reference matrix*, since it contains multiple-output entries. Thus, Figs. 4-46(b) and (c) are fourth-order reference matrices each representing ten separate K-maps, one entry per K-map. A reference matrix for the DECODER in Fig. 4-34 would be of a third order and would represent eight K-maps.

The I/O relations needed for implementation can be read directly from the reference matrices in Fig. 4-46. For a design with false data rejection each of the ten outputs (Y_9 through Y_0) represented in Fig. 4-46(b) is the MINTERM given by its cell coordinates. The result is a 4-to-10 line DE-CODER similar to that of Fig. 4-34(b). A design without FDR, as in Fig. 4-46(c), makes use of the ϕ's in looping out the individual output variables [see Fig. 4-46(b)]. This has the effect of reducing the number of inputs required for all but two of the Y outputs. The I/O relations for this design are given in Fig. 4-46(d). Observe that only outputs Y_1 and Y_8 are MINTERMs.

Dec.	XS3 Gray				Output
0	0	0	1	0	Y_0
1	0	1	1	0	Y_1
2	0	1	1	1	Y_2
3	0	1	0	1	Y_3
4	0	1	0	0	Y_4
5	1	1	0	0	Y_5
6	1	1	0	1	Y_6
7	1	1	1	1	Y_7
8	1	1	1	0	Y_8
9	1	0	1	0	Y_9

(a)

Fig. 4-46. Design of an XS3-to-decimal CONVERTER. (a) I/O truth table. (b), (c) Reference matrices with and without false data rejection (FDR). (d) and (e) I/O relations and logic circuit for an XS3 gray-to-decimal CONVERTER design without (FDR).

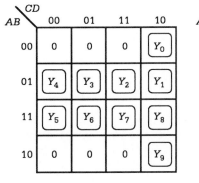

Reference matrix with FDR.

(b)

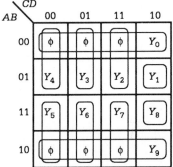

Reference matrix without FDR.

(c)

$$Y_0 = \overline{A}\,\overline{B} \qquad Y_2 = \overline{A}CD \qquad Y_4 = \overline{A}\,\overline{C}\,\overline{D} \qquad Y_6 = A\overline{C}D \qquad Y_8 = ABC\overline{D}$$

$$Y_1 = \overline{A}BC\overline{D} \qquad Y_3 = \overline{A}C\overline{D} \qquad Y_5 = A\overline{C}\,\overline{D} \qquad Y_7 = ACD \qquad Y_9 = A\overline{B}$$

(d)

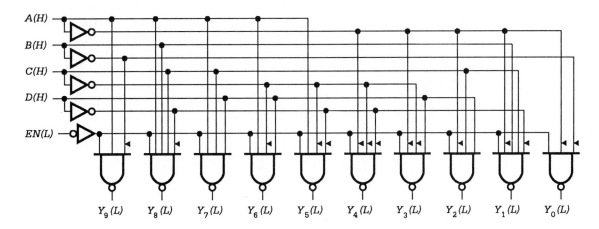

(e)

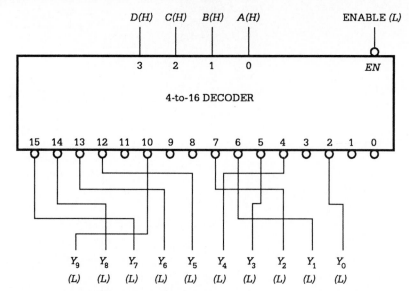

Fig. 4-47. *DECODER implementation of an XS3 Gray-to-decimal CODE CONVERTER with false data rejection.*

When the I/O relations of Fig. 4-46(d) are implemented, the result is the logic circuit of Fig. 4-46(e), where an ENABLE (L) control has been added and the outputs have been made to be ACTIVE LOW.

The design of an XS3 Gray-to-decimal CONVERTER with false data rejection is best accomplished by using a 4-to-16 line DECODER since all outputs are MINTERMs. The results are shown in Fig. 4-47, where the connections are determined from the reference matrix of Fig. 4-46(b). The reader should understand that a DECODER can be used for *any* BCD-to-decimal conversion with FDR but cannot be used for designs without FDR unless additional logic is used external to the DECODER.

NBCD-to-Binary Conversion. Natural BCD and natural binary are two of the most widely used codes in digital design, so it is fitting that we consider the conversion of one to the other. The simplest approach is to establish a workable algorithm and then use this algorithm to produce a modular design. A gate-level design, which requires the use of truth tables and K-maps, would be possible but tedious, considering that it takes a 4-bit NBCD WORD to represent each digit. Thus, two 4-bit NBCD WORDs are required to represent any 7-bit binary number, three NBCD WORDs for 8-bit binary, and so on.

Conversion from NBCD to binary is easier than the reverse, and one of the simplest algorithms for the NBCD-to-binary conversion is stated in three steps as follows:

1. Convert the value of each bit in the NBCD number to its binary equivalent.
2. Add the binary numbers as in binary arithmetic.
3. The result is the binary equivalent of the NBCD number.

This algorithm will hereafter be called the *addition algorithm*.

To illustrate the use of the addition algorithm, let us find the binary equivalent of the two-digit NBCD number 00110101, where $0011 = 3_{10}$ is the most significant digit (MSD) and $0101 = 5_{10}$ is the least significant digit (LSD). The procedure is as follows:

	NBCD								
	MSD weights (tens digit)				LSD weights (units digit)				
	80	40	20	10	8	4	2	1	Binary	Decimal
$35_{10} =$	0	0	1	1	0	1	0	1		
									0 0 0 0 0 0 1	1
									0 0 0 0 1 0 0	4
									0 0 0 1 0 1 0	10
									0 0 1 0 1 0 0	20
									0 1 0 0 0 1 1	35

Already it should be obvious that the modular design will make use of ADDERs. But to see how the ADDERs are connected, it will be necessary to construct a conversion table giving the binary representation for each NBCD bit. Such a table is given Fig. 4-48, where, for the purpose of illustration, the NBCD number has been limited to two digits—the units digit and the tens digit.

Based on the addition algorithm established earlier, we can write the I/O relations directly from the conversion table by summing the 1's in each column. The XOR terms that result are shown in Fig. 4-48. The symbol C, it may be recalled, represents the CARRYin for a given stage but is the CARRYout of the previous state. Thus, all outputs except Y_1 will generate CARRY bits. A CARRY-generating scheme is shown in Fig. 4-48.

This CONVERTER can be implemented by using eight FULL ADDERs connected according to the I/O relations given in Fig. 4-48. When this is done, there results the configuration shown in Fig. 4-49. This CONVERTER could also have been implemented with two 4-bit PARALLEL ADDERs. In either case, cascading of the ADDER units will produce CONVERTERs having NBCD inputs exceeding two decades.

Another algorithm which is useful for NBCD-to-binary conversion could be called the shift-right/subtract 3 [or $(\div 2)/(-3)$] algorithm. It is based on the principle that shifting right by n bits is equivalent to dividing by 2^n, a subject which will be discussed further in Sect. 4.9.2. For reference purposes we will state the shift-right/substract 3 algorithm as follows:

1. Place the NBCD number in imaginary storage cells (hence a storage register). For example, a two-decade NBCD number will originally occupy eight imaginary storage cells.
2. Shift the NBCD number to the right one bit at a time *keeping account of those bits which are shifted out of the least significant digit storage cells.*

	Binary Outputs						
	Y_{64}	Y_{32}	Y_{16}	Y_8	Y_4	Y_2	Y_1
LSD $\{\,I_1$	0	0	0	0	0	0	1
I_2	0	0	0	0	0	1	0
I_4	0	0	0	0	1	0	0
I_8	0	0	0	1	0	0	0
MSD $\{\,I_{10}$	0	0	0	1	0	1	0
I_{20}	0	0	1	0	1	0	0
I_{40}	0	1	0	1	0	0	0
I_{80}	1	0	1	0	0	0	0

Two-Digit NBCD Inputs

$$Y_1 = I_1$$

$$Y_2 = I_2 \oplus \overset{\frown{C_4}}{I_{10}}$$

$$Y_4 = I_4 \oplus I_{20} \oplus \overset{\overbrace{C_8}}{C_4}$$

$$Y_8 = I_8 \oplus I_{10} \oplus I_{40} \oplus C_8$$

$$Y_{16} = I_{20} \oplus I_{80} \oplus C'_{16} \oplus C_{16}$$

$$Y_{32} = I_{40} \oplus C'_{32} \oplus C_{32}$$

$$Y_{64} = I_{80} \oplus C_{64}$$

Fig. 4-48. Conversion table and I/O relations for a 2-digit NBCD-to-binary code converter showing a carry gererating scheme.

3. Subtract 0011 from the LSD positions (or add 1101 to the LSD positions and discard the carry) when, and only when, the new LSD number within the storage cells is greater than 7 (0111). After subtracting 3, shift right immediately, even if the new LSD number is greater than 7.

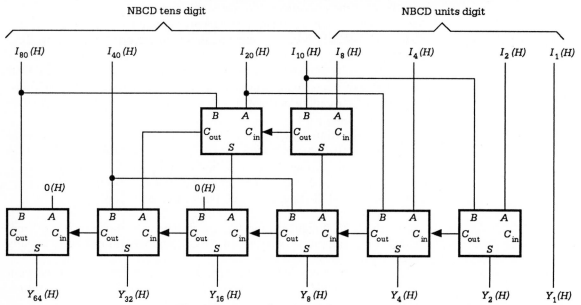

Fig. 4-49. A two-decade NBCD-to-binary converter implemented with FULL ADDERs.

246

4. Repeat steps 2 and 3 until the new LSD number can no longer be greater than decimal 7. The answer, now in binary, includes all bits inside and outside of the imaginary storage cells.

To provide a better understanding of the $(\div 2)/(-3)$ algorithm, we will convert the NBCD number 00110101 to natural binary:

	MSD	LSD		
$35_{10} =$	$\underline{0\ 0\ 1\ 1}$	$\underline{0\ 1\ 0\ 1}$		
	$0\ 0\ 1$	$1\ 0\ 1\ 0$	1	Shift right
		$\underline{-1\ 1}$		Subtract 3
	$0\ 0\ 1$	$0\ 1\ 1\ 1$	1	New LSD
	$0\ 0$	$1\ 0\ 1\ 1$	$1\ 1$	Shift right
		$\underline{-1\ 1}$		Subtract 3
	$0\ 0$	$1\ 0\ 0\ 0$	$1\ 1$	New LSD
	0	$0\ 1\ 0\ 0$	$0\ 1\ 1$	Shift right

Answer in binary $=\ 0\ 1\ 0\ 0\ 0\ 1\ 1_2$

Unlike the addition algorithm, the $(\div 2)/(-3)$ algorithm has a reciprocal which is applicable to the reverse code conversion process, binary to NBCD. This algorithm could be called the shift-left/add 3 or $(\times 2)/(+3)$ algorithm, since shifting left by n bits is equivalent to multiplying the number by 2^n, as discussed in Sect. 4.9.2. We now state this algorithm for reference purposes:

1. Place the natural binary number *outside and to the right* of the empty imaginary storage cells which will eventually hold the NBCD number. Shift left one bit at a time.
2. Add 3 (0011) to the digit being formed when it becomes greater than 4 (0100). After adding 3 shift left immediately.
3. Repeat steps 2 and 3. When all external bits have been shifted into the imaginary storage cells, the process ceases and the answer is the number in the storage cells.

The use of the $(\times 2)/(+3)$ algorithm is illustrated by finding the NBCD code number for 0101110_2:

MSD	LSD		
		$0\ 1\ 0\ 1\ 1\ 1\ 0 = 46_{10}$	
	0	$1\ 0\ 1\ 1\ 1\ 0$	Shift left
	$0\ 1$	$0\ 1\ 1\ 1\ 0$	Shift left
	$0\ 1\ 0$	$1\ 1\ 1\ 0$	Shift left
	$0\ 1\ 0\ 1$	$1\ 1\ 0$	Shift left
	$+1\ 1$		Add 3
	$\overline{1\ 0\ 0\ 0}$	$1\ 1\ 0$	New LSD
1	$0\ 0\ 0\ 1$	$1\ 0$	Shift left
$1\ 0$	$0\ 0\ 1\ 1$	0	Shift left
$0\ 1\ 0\ 1$	$0\ 1\ 1\ 0$		Shift left

Answer $=\ 0\ 1\ 0\ 0\ 0\ 1\ 1\ 0$ in NBCD code

The two shift/add (or subtract) algorithms just described are frequently used to convert NBCD to binary, or vice versa. The reason is that these algorithms can be used easily to modularize conversion units for *n*-digit conversion. Problems 4.39 and 4.40 at the end of this chapter illustrate this feature. Problem 4.47 requires the use of a two-digit binary-to-NBCD conversion unit to design a one-digit NBCD multiplier.

NBCD-to-Seven-Segment Conversion. It would seem that no discussion of code conversion is complete without including NBCD-to-seven-segment CONVERTERs. We are all aware that light-emitting diodes (LEDs) and liquid crystal displays (LCDs) are used extensively to produce the familiar Arabic numerals. The use of seven-segment display devices offers one means of accomplishing this.

Presented in Fig. 4-50(a) is the seven-segment display format which has been labeled *a* through *g*, in agreement with popular convention. An example of how the display is used is also provided in the figure.

All the information needed for implementation of the NBCD-to-seven-segment CONVERTER can be obtained directly from the truth table of Fig. 4-50(b), where the outputs *a* through *g* correspond to those in the display format. For example, decimal 4 would require activating segments *b*, *c*, *f*, and *g*, and so 1's are placed at the intersection of each of these outputs and the input state 0100. Since each logic entry in the truth table represents a MINTERM, the outputs can be written in CANONICAL SOP form giving the results shown in Fig. 4-50(b). This suggests a DECODER design. Of course, each output could be mapped, and minimum cover could be used for a gate-level design. But little would be gained in doing so, since an off-the-shelf DECODER could do the job much more easily. Shown in Fig. 4-50(c) is the implementation of the CANONICAL SOP expressions of Fig. 4-50(b) by using a 4-to-16 line DECODER with external AND logic for ACTIVE LOW outputs. Notice that this design automatically includes false data rejection.

There are 4-to-10 line DECODER IC chips which are made expressly for NBCD-to-seven-segment conversion. They are usually designed with an optional *zero-blanking* feature which permits any seven-segment display (connected to it) to be blanked out if all inputs are zero. These ICs are designed with a *zero-blanking input* (*ZBI*) and a *zero-blanking output* (*ZBO*) so that when the decade stages are connected, ZBO to ZBI, the zero-blanking effect can *ripple* in the direction of radix point terminal. Thus, the MSD will be blanked out if it is zero permitting, in turn, the next MSD to be blanked out if it is zero, and so on, but never in the reverse direction. Applied to decades less than zero, zero blanking would ripple from LSD to MSD—again in the direction of the radix point terminal.

The zero-blanking feature can be added to CONVERTER circuits of the type shown in Fig. 4-50(c). All that is required is that the inputs, *A*, *B*, *C*, *D*, and ZBI, be ORed LOW to the EN(*L*) terminal. This is demonstrated in Fig. 4-51 for two intermediate decades. To simplify the drawing, the OR stages are represented in modular form and the abridged notation is used throughout. For a better understanding of the zero-blanking feature, the reader should trace zero and nonzero inputs through this circuit.

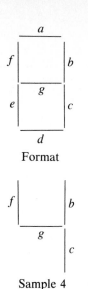

Format

Sample 4

(a)

	NBCD											
Dec.	8	4	2	1	a	b	c	d	e	f	g	
0	0	0	0	0	1 1 1 1 1 1 0							$a = \sum m(0, 2, 3, 5, 7, 8, 9)$
1	0	0	0	1	0 1 1 0 0 0 0							$b = \sum m(0, 1, 2, 3, 4, 7, 8, 9)$
2	0	0	1	0	1 1 0 1 1 0 1							$c = \sum m(0, 1, 3, 4, 5, 6, 7, 8, 9)$
3	0	0	1	1	1 1 1 1 0 0 1							$d = \sum m(0, 2, 3, 5, 6, 8)$
4	0	1	0	0	0 1 1 0 0 1 1							$e = \sum m(0, 2, 6, 8)$
5	0	1	0	1	1 0 1 1 0 1 1							$f = \sum m(0, 4, 5, 6, 8, 9)$
6	0	1	1	0	0 0 1 1 1 1 1							$g = \sum m(2, 3, 4, 5, 6, 8, 9)$
7	0	1	1	1	1 1 1 0 0 0 0							
8	1	0	0	0	1 1 1 1 1 1 1							
9	1	0	0	1	1 1 1 0 0 1 1							

(b)

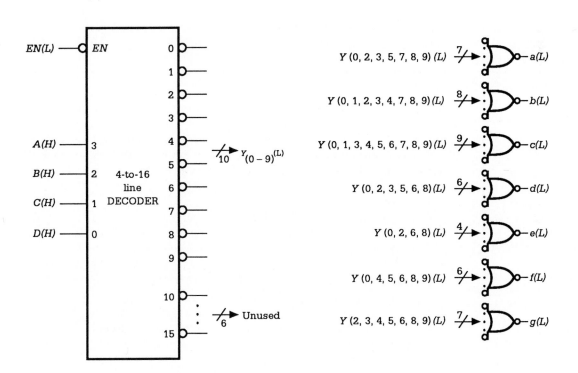

(c)

Fig. 4-50. *Design of an NBCD-to-seven-segment CONVERTER. (a) 7-segment display format and sample 4. (b) I/O truth table and CANONICAL SOP output relations. (c) Logic circuit with FDR and ACTIVE LOW outputs.*

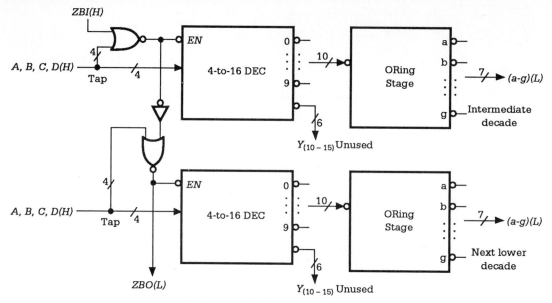

Fig. 4-51. *Two intermediate decade stages of an NBCD-to-7-segment CONVERTER showing the zero blanking circuitry connected to standard 4-to-16 line DECODERs.*

4.8   MULTIPLEXERS AND DEMULTIPLEXERS

There is a type of digital device that performs the function of selecting one of many data input lines for transmission (busing) to some destination. This device is called a MULTIPLEXER (MUX for short) or DATA SELECTOR. It requires n DATA SELECT lines to control 2^n DATA INPUT lines. Hence, a MUX is a $(2^n + n)$-input/1-output device identified by the block diagram given in Fig. 4-52(a). Shown in Fig. 4-52(b) is the mechanical switch equiv-

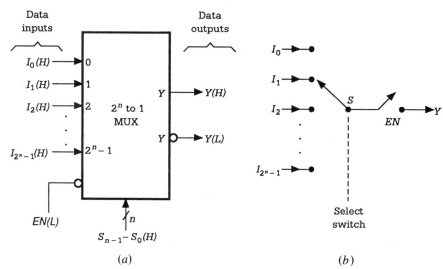

Fig. 4-52. *A 2^n-to-1 MULTIPLEXER (MUX) or DATA SELECTOR. (a) Block diagram symbol. (b) Mechanical switch analog.*

alent for this MUX. Notice that the function of ENABLE is to provide DISABLE capability to the device. Commercial MUX ICs usually come with ACTIVE LOW ENABLE.

The MUX is one of the most useful devices available to the designer. It is used extensively to select and steer to the output one specific input function from any number of other input functions. But of particular value is the fact that the required connections to accomplish this are easily deduced directly from an EV truth table or EV K-map without the need for reduced or minimum cover. MUXs, together with DEMULTIPLEXERs (DMUXs), are also used extensively in time-multiplexing applications as discussed in Sect. 4.8.3. The DMUX performs the opposite function of a MUX and is described in Sect. 4.8.2.

4.8.1 MULTIPLEXER Design

The easiest and most comprehensible way to design a MUX is to use an entered variable (EV) truth table and EV K-map. We illustrate this in Fig. 4-53 for the design of a 4-to-1 line MUX with ACTIVE LOW ENABLE [EN(L)]. The choice of the DATA SELECT code to issue I_0, I_1, I_2, and I_3 is arbitrary but is usually standard binary, as given in the EV truth table of Fig. 4-53(a). Use of the irrelevant input condition (symbolized by X) creates a collapsed truth table similar to that of Fig. 4-38(b), except now the data inputs are presented as EVs in the output column. A conventional truth table for this 4-to-1 MUX would require seven input columns and $2^7 = 128$ rows of logic values. Construction of such a truth table would not only be tedious, but the result would not convey the required information nearly as well as the collapsed EV truth table of Fig. 4-53(a).

The K-map of Fig. 4-53(b) is the graphics representation of the EV truth table in Fig. 4-53(a). It represents a compression of five orders. The utility of the entered variable representation can again be appreciated when it is realized that this EV K-map replaces $2^3 = 8$ conventional fourth-order K-maps—a seven-variable problem. Of course, no one of sound mind would use seven fourth-order K-maps to solve this problem when one K-map would suffice.

Also included in Fig. 4-53(b) is the Boolean expression for output Y

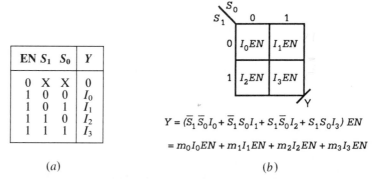

EN S_1 S_0	Y
0 X X	0
1 0 0	I_0
1 0 1	I_1
1 1 0	I_2
1 1 1	I_3

(a)

$$Y = (\overline{S}_1\overline{S}_0 I_0 + \overline{S}_1 S_0 I_1 + S_1\overline{S}_0 I_2 + S_1 S_0 I_3)\, EN$$

$$= m_0 I_0 EN + m_1 I_1 EN + m_2 I_2 EN + m_3 I_3 EN$$

(b)

Fig. 4-53. *A 4-to-1 MULTIPLEXER with ENABLE LOW control. (a) Collapsed EV truth table. (b) EV K-map and output logic expressions.*

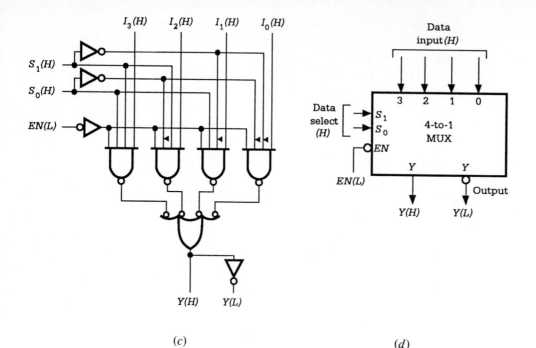

(c)

(d)

Fig. 4-53 (contd.). (c) Logic circuit. (d) Block diagram symbol.

obtained either by reading the truth table or by looping out the four terms of the K-map as islands. When this expression is implemented in NAND logic, there results the 4-to-1 line MUX circuit presented Fig. 4-53(c). Its block diagram symbol is shown in Fig. 4-53(d).

MULTIPLEXERs can be stacked to produce 2^n-to-1 MUX stacks or "trees." Suppose, for example, it is desirable to produce a 16-to-1 MUX from four 4-to-1 MUXs. This is achieved by connecting the outputs from the four MUXs to the inputs of a fifth (leading) 4-to-1 MUX. The result is shown in Fig. 4-54, where the abridged notation has been used to simplify the circuit drawing. The circuit above the dashed line is called a quad four-input (or four-wide) MUX which is another form of MUX—that is, one with four different sets of inputs controlled by two common DATA SELECT input lines. Both types of MUX, multiple output and single output, are commercially available as IC chips in various sizes. Notice that the MUX of Fig. 4-54 can also be DISABLED by using the EN(L) line to the leading MUX.

There are many variations of the stacked MUX configuration possible limited only by the availability of different MUX sizes. For example, two 16-to-1 line MUXs can be stacked in the manner of Fig. 4-54 to produce a 32-to-1 line MUX, but a 2-to-1 line MUX (or one larger operating as a 2-to-1 line MUX) must be used to complete the stack configuration.

4.8.2 DEMULTIPLEXERs

A device that has one data input line, n data select lines, and 2^n output lines, and which can route data from the single input source to one of several destinations is called a DEMULTIPLEXER (DMUX) or DATA DISTRIB-

4 / Combinational Logic Design

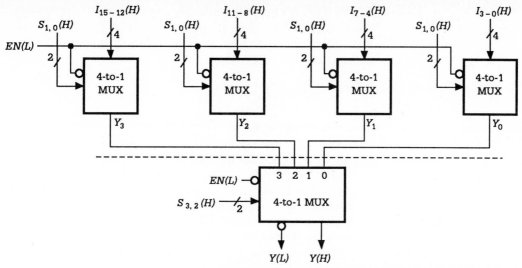

Fig. 4-54. *A 16-to-1 MUX produced by stacking four 4-to-1 MUXs.*

UTOR—the opposite function of a MUX. DEMULTIPLEXERs are *not* commercially available because DECODERs can be adapted easily to this use. In fact, manufacturers describe this device as a DECODER/DEMULTIPLEXER, meaning that it is given one or the other of these two names depending on how it is used. Use of a MUX in reverse to produce a DMUX is not possible for the same reason that DECODERs cannot be used in reverse to produce ENCODERs.

All that is necessary to operate a DECODER as a DEMULTIPLEXER is to use one of the inputs (usually the ENABLE input) as the DATA INPUT line with the remaining inputs used as DATA SELECT lines. This arrangement is illustrated in Fig. 4-55(a), where a 2-to-4 line DECODER (the smallest commercially available) is used as a 1-to-4 line DMUX with input $EN(L)$. If it is desirable to have the data input $EN(H)$ instead of $EN(L)$, an INVERTER can be placed in the input line. The mechanical equivalent of a DMUX is shown in Fig. 4-55(b).

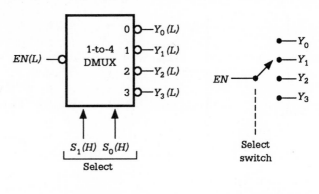

(a)   *(b)*

Fig. 4-55. *(a) A 2-to-4 DECODER used as a 1-to-4 DEMULTIPLEXER with data input* En(L). *(b) The mechanical analog of (a).*

Larger DEMULTIPLEXERs can be produced by stacking DECODERs. For example, the stacked 2-to-4 line DECODERs of Fig. 4-35 can be used as a 1-to-16 line DMUX with $EN(L)$ as the DATA INPUT line, and A_3, A_2, A_1, and A_0 as the DATA SELECT lines.

4.8.3 Use of MUXs and DMUXs in Combinational Logic Design

Transport (busing) of data from one location to another in a digital system is an important consideration in digital design. This is especially true if circuit real estate (chip space) is severely limited. One way to alleviate an over-crowding condition is to reduce the number of data transport lines by using a busing scheme called *time multiplexing*. Time multiplexing is the process by which many input signals can be bused on a single transmission line (data bus) by time-shared data selection from a transmitting MUX to a receiving DMUX. This scheme is illustrated in Fig. 4-56, where a 4-to-1 MUX/DMUX system is used. Here, the history of two pulse trains is shown with one pulse train initially separated in time from the other by $\Delta t = t_2 - t_1$. Thus, if the DATA SELECT inputs are set to $S_1 = 0$ and $S_0 = 1$ at time t_1, the signal on line I_1 is subsequently selected, transmitted, received, and distributed to output Y_1 before the pulse on line I_3 is selected. Then, if the DATA SELECT inputs are changed to $S_1 = 1$ and $S_0 = 1$ at time t_2, the pulse train on line I_3 is subsequently selected, transmitted, received, and distributed to output Y_3. In this way, many signals (one for each input line) can be distributed by a receiving DMUX to their final destination, all at different times. In the foregoing example, the propagation time delays of the MUX and DMUX are represented as $t' - t$ and $t'' - t'$, respectively, assuming that the data bus delay time is negligible, which it may not be.

Combinational logic design with MUXs is not limited to time-multiplexing applications. Like DECODERs, MULTIPLEXERs can be used to generate any CANONICAL SOP or POS expression, often at a savings of both space and time. Unlike DECODERs, MUXs are also readily adapted to multi-function generation which makes them one of the most versatile devices available to the designer. In the simple example that follows, both types of applications will be demonstrated.

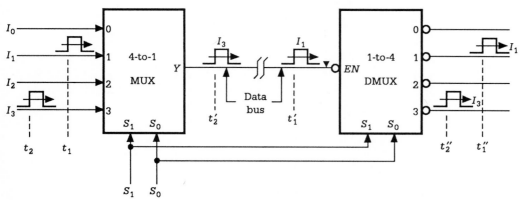

Fig. 4-56. *A MUX/DMUX time-multiplexing system showing the history of two-pulse trains $t_2 - t_1$ apart in time. Propagation delay times are not shown.*

To understand how a MUX can be used to generate a CANONICAL form, it will be helpful to write the output expression for a MUX. A generalization of the output expression given in Fig. 4-53(b), exclusive of EN, would read

$$Y_{SOP}(H) = [m_0 I_0 + m_1 I_1 + m_2 I_2 + \cdots + m_{2^n-1} I_{2^n-1}](H)$$

$$= \sum m_i I_i(H) \tag{4-15}$$

$$= Y_{POS}(L)$$

where m_i represents the DATA SELECT MINTERMs

$$m_0 = S_0 S_1 S_2 \cdots S_{2^n-1}$$

$$m_1 = S_0 S_1 S_2 \cdots S_{2^n-2} S_{2^n-1}$$

$$\vdots$$

etc.

and I_i represents the corresponding inputs with coded subscsripts.

An examination of Eqs. (4-15) leads to the following steps we recommend be taken for the MUX implementation of a CANONICAL expression:

1. If the number of DATA SELECT inputs (n) is equal to the number of variables (N) in the CANONICAL expression, identify the code numbers of the existing MINTERMs (or MAXTERMs) and connect their associated I_i inputs to HV [1(H)] with all others connected to LV [0(H)]. Take $Y_{SOP}(H)$ off the $Y(H)$ output of the MUX and $Y_{POS}(H)$ off the $Y(L)$ output.

2. If the n DATA SELECT inputs number less than the N variables of the CANONICAL expression, then construct an nth order K-map having ($N - n$) entered variables selected to be those of least positional weight. Compare the cell entries (subfunctions) with the corresponding cells of the MUX K-map [like that of Fig. 4-53(b)] to determine the external logic required for each of the I_i inputs to the MUX. The cell numbers correspond to the MUX input line numbers.

3. If it is assumed that the internal logic circuitry of the MUX is not accessible to the designer, then there remains the problem of dealing with MIXED LOGIC inputs. In this case, complement the bit value of any DATA SELECT input variable of the CANONICAL expression that arrives from a negative logic source (hence is ACTIVE LOW), or use an INVERTER on the DATA SELECT line. ACTIVE LOW DATA inputs are handled in the usual manner by external logic to the MUX. When EV K-maps are used, the following procedure is recommended:

   a. Complete the nth order EV K-map as in step 2.
   b. Either complement any axis bit (of the K-map) corresponding to an ACTIVE LOW DATA SELECT input or use an INVERTER on that input line.
   c. Reassign the cell numbers of the EV K-map to agree with the new axis or axes.
   d. Take the new cell numbers to be the subscripts of the MUX inputs I_i, and take the entries of these cells to be the combination logic required for these I_i inputs.

The four-step procedure may sound complicated, but it is really quite straightforward. To illustrate, let us implement the function

$$Y_{SOP}(A,B,C,D) = \sum m(3, 4, 5, 6, 7, 9, 10, 12, 14, 15) \qquad (4\text{-}16)$$

by using an 8-to-1 line MUX and by assuming that all inputs and outputs are ACTIVE HIGH. Since there are only three DATA SELECT inputs ($n = 3$), we must construct a third-order K-map in variable D. By using a MAP KEY of $2^{N-n} = 2^{4-3} = 2$, there results the K-map given in Fig. 4-57(a). When the cell entries of this map are compared with those of the MUX EV

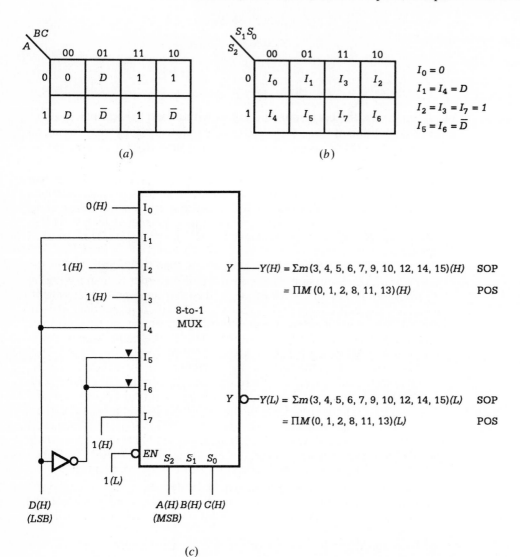

(a)   (b)

(c)

Fig. 4-57. *Use of an 8-to-1 MUX to implement Eq. (4-16). (a) EV K-map. (b) EV K-map for MUX showing input connections relative to the K-map in Figure (a). (c) MUX logic circuit based on Figures 4-57(a) and (b) showing CANONICAL SOP and POS forms available at the terminals.*

K-map in Fig. 4-57(b), we see that I_0 must be connected to LV $[0(H)]$, I_1 must be connected to D, I_2 must be connected to HV $[1\ (H)]$, and so on. By completing these connections, we obtain the logic circuit shown in Fig. 4-57(c) together with the CANONICAL SOP and POS expressions available at the output terminals. It is well to remember that the DATA SELECT inputs must be properly matched with the function variables—that is, the MSB A must be matched with the MSB S_2 input of the MUX, B with S_1 and C with S_0. The entered variable D is required to be the LSB.

When the K-map cells of Fig. 4-57(a) are read in both MINTERM and MAXTERM code, we obtain the reduced expressions

$$Y_{\text{SOP}} = \overline{A}\,\overline{B}CD + \overline{A}B\overline{C} + \overline{A}BC + A\overline{B}\,\overline{C}D + A\overline{B}C\overline{D} + AB\overline{C}D + ABC$$

$$Y_{\text{POS}} = (A + B + C)(A + B + \overline{C} + D)(\overline{A} + B + C + D)$$
$$\cdot\ (\overline{A} + B + \overline{C} + \overline{D})(\overline{A} + \overline{B} + C + \overline{D})$$

which are also available at the output terminals of Fig. 4-57(c). As an exercise the reader may wish to find the minimum SOP and POS forms that are also available at the terminals of the MUX.

Next, suppose we wish to implement Eq. (4-16) by using a 4-to-1 line MUX instead of a 8-to-1 line MUX as before. Because there are only two DATA SELECT inputs for this MUX, the procedure given earlier directs us to assign $S_1 = A$ and $S_0 = B$, and to construct a second-order K-map with entered variables C and D. Such an EV K-map is presented in Fig. 4-58(a). By comparing the cell entries of this map with the corresponding cells of the MUX K-map in Fig. 4-58(b), we conclude that the external combinational logic must be $I_0 = CD$, $I_1 = 1$, $I_2 = C \oplus D$, and $I_3 = C + \overline{D}$ as indicated. The result is the logic circuit shown in Fig. 4-58(c), where we have included both CANONICAL and reduced hybrid expressions available at the output terminals. Obviously, a reduction in MUX size has increased considerably the complexity of the external logic required to generate the function given by Eq. (4-16).

A few additional remarks regarding Fig. 4-58 are worthwhile at this time. The circuit in Fig. 4-58(c) can be used as a FUNCTION GENERATOR. To do this requires that the DATA SELECT inputs [now called $A(H)$ and $B(H)$] be used in the SELECT mode to generate any one of the four functions of Fig. 4-58(a). For example, setting $S_1 = 1$ and $S_0 = 0$ generates $Y = C \oplus D$, or setting $S_1 = 1$ and $S_0 = 1$ generates $Y = C + \overline{D}$, and so on, as shown

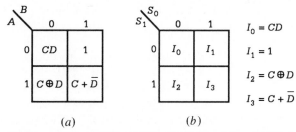

(a)　　　　　　　　　　(b)

Fig. 4-58. *Implementation of Eq. (4-16) with a 4-to-1 line MUX. (a) Second-order EV K-map. (b) EV K-map for MUX showing input connections relative to Figure (a).*

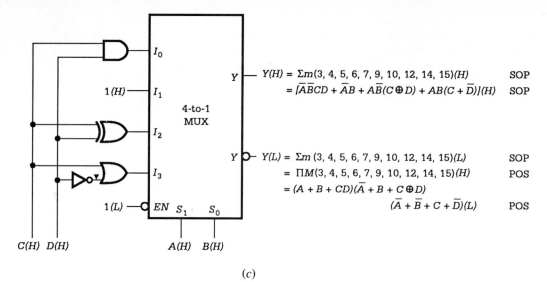

$$Y(H) = \Sigma m(3, 4, 5, 6, 7, 9, 10, 12, 14, 15)(H) \qquad \text{SOP}$$
$$= [\overline{A}\overline{B}CD + \overline{A}B + A\overline{B}(C \oplus D) + AB(C + \overline{D})](H) \qquad \text{SOP}$$

$$Y(L) = \Sigma m(3, 4, 5, 6, 7, 9, 10, 12, 14, 15)(L) \qquad \text{SOP}$$
$$= \Pi M(3, 4, 5, 6, 7, 9, 10, 12, 14, 15)(H) \qquad \text{POS}$$
$$= (A + B + CD)(\overline{A} + B + C \oplus D)$$
$$(\overline{A} + \overline{B} + C + \overline{D})(L) \qquad \text{POS}$$

(c)

Fig. 4-58 (contd.). (c) MUX logic circuit based on Figures 4-58(a) and (b) showing some CANONICAL and reduced expressions available at the outputs.

by the output expressions in Fig. 4-58(c). In fact, any 2^n-to-1 MULTI-PLEXER can be used to generate 2^n different functions of any complexity. Before concluding the discussions in this chapter, we will have other opportunities to use this feature of the MUX.

As a final variation of the foregoing example, let us implement Eqs. (4-16) for the case of MIXED LOGIC inputs by using both an 8-to-1 and a 4-to-1 line MUX. Assume that inputs A and C arrive at the MUX from positive logic sources while inputs B and D arrive from negative logic sources. According to the procedure set forth earlier, we must first construct the EV K-maps as in any positive logic problem. This has already been done in Figs. 4-57(a) and 4-58(a). Next, we complement the B axis bits in both maps (since B is an ACTIVE LOW input), and then renumber the cells to agree with the new axes. The results are shown in Fig. 4-59. Notice that in this particular case the BC axis in Fig. 4-59(a) and the B axis in Fig. 4-59(b) are now numbered in the reverse direction, that is, from right to left.

Implementation of the results given in Fig. 4-59 follows directly by comparing the cells of these K-maps with those cells of the same number in the MUX K-maps given in Figs. 4-57(b) and 4-58(b). Thus, for the 8-to-1 line MUX the external combinational logic is $I_0 = 1$, $I_1 = 1$, $I_2 = 0$, $I_3 = D$, $I_4 = \overline{D}$, and so on, and for the 4-to-1 line MUX $I_0 = 1$, $I_1 = CD$, $I_2 = C + \overline{D}$, and $I_3 = C \oplus D$, as shown in Fig. 4-59(c).

An alternative means of dealing with MIXED LOGIC DATA SELECT inputs amounts simply to adding a LOGIC LEVEL CONVERTER (INVERTER) to each ACTIVE LOW DATA SELECT input line and then proceeding as in the case of ACTIVE HIGH inputs. As to which approach is best for a given design problem is often left as a matter of choice for the designer. However, if it is the intent to minimize the external logic required to implement a given function with a MUX, the MIXED LOGIC method is the obvious choice—no additional INVERTERs are needed.

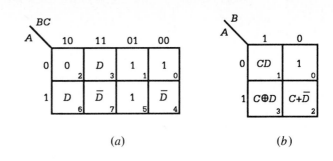

(a) (b)

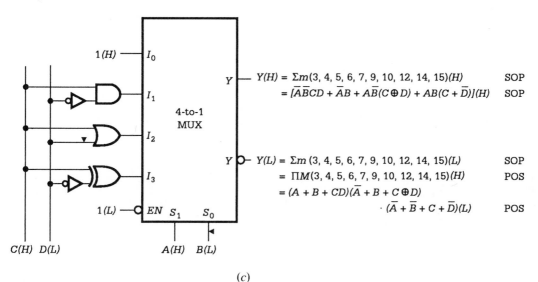

$Y(H) = \Sigma m(3, 4, 5, 6, 7, 9, 10, 12, 14, 15)(H)$ SOP

$= [\overline{A}\,\overline{B}CD + \overline{A}B + A\overline{B}(C \oplus D) + AB(C + \overline{D})](H)$ SOP

$Y(L) = \Sigma m (3, 4, 5, 6, 7, 9, 10, 12, 14, 15)(L)$ SOP

$= \Pi M(3, 4, 5, 6, 7, 9, 10, 12, 14, 15)(H)$ POS

$= (A + B + CD)(\overline{A} + B + C \oplus D)$

$\cdot (\overline{A} + \overline{B} + C + \overline{D})(L)$ POS

(c)

Fig. 4-59. *EV K-maps for Eq. (4-16) when inputs* A *and* C *are ACTIVE HIGH, and inputs* B *and* D *are ACTIVE LOW. (a) First-order compression. (b) Second-order compression. (c) MUX implementation of Figure 4-59(b).*

There are many useful applications for the MULTIPLEXER in combinational logic design. We have touched on only two. During the course of our discussions in the remainder of this chapter and in the chapters that follow, we will have the opportunity to explore other uses of this most versatile device.

4.9 SHIFTERS

Shifting of WORD bits to the right or left can be accomplished by using combinational logic exclusive of any master clock triggering mechanism. Shifting devices which operate in this manner are called COMBINATIONAL SHIFTERs, or simply SHIFTERs, and are the subjects for discussion in this section. Shifting of WORD bits can also be accomplished synchronously by devices which must be triggered by clock pulses. These devices, called SHIFT REGISTERs, will be covered in depth later in Chapter 5. Both SHIFTERs and SHIFT REGISTERs are used for bit handling, transport, editing, data modification, and arithmetic manipulation, among other applications.

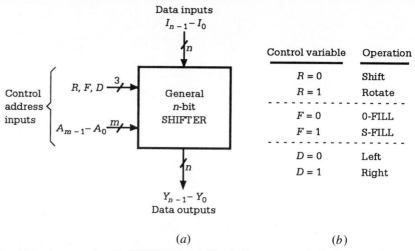

Data inputs
$I_{n-1} - I_0$

Control address inputs { R, F, D $\xrightarrow{3}$ $A_{m-1} - A_0$ $\xrightarrow{m}$

General n-bit SHIFTER

$Y_{n-1} - Y_0$
Data outputs

Control variable	Operation
$R = 0$	Shift
$R = 1$	Rotate
$F = 0$	0-FILL
$F = 1$	S-FILL
$D = 0$	Left
$D = 1$	Right

(a)   (b)

Fig. 4-60. *A general* n-*bit SHIFTER. (a) Block diagram symbol. (b) Possible interpretation of the control inputs* R, F *and* D.

A general n-bit SHIFTER can be perceived to be an $(n + m + 3)$-input/n-output device represented by the block diagram symbol given in Fig. 4-60(a), where all inputs and outputs are assumed to be ACTIVE HIGH. A SHIFTER of this type accepts n data input bit values ($I_{n-1} - I_0$) and either passes these values straight through to the data outputs ($Y_{n-1} - Y_0$), or shifts or rotates them by one or more bit positions to the right or left on command of the $m + 3$ control/address inputs. The control/address inputs consist of a ROTATE control (R), a FILL control (F), a DIRECTION control (D), and m ADDRESS lines. The control/address input WORD is given the positional weights $RFDA_{m-1} \ldots A_0$.

It is the function of the m ADDRESS inputs to determine the number of bit positions to be shifted or rotated—usually binary encoded to 0, 1, 2, 3, . . . positions, where 0 is interpreted as a straight-through (or transparent) operation. The D control sets the shift direction to either shift left (sl) or shift right (sr), while the F control determines whether the FILL (vacated) position in the output is to receive a logic 0 (0-FILL), or the value of the input sign bit (S-FILL) as required for sign-complement right shifting discussed in Sec. 4.9.2. The R control, positioned as the most significant control bit, sets the SHIFTER to either the shift mode or the rotate mode. Figure 4-60(b) gives one possible interpretation of the control inputs R, F, and D.

Further explanation is needed of Fig. 4-60(b) and the notation which is used. In SHIFTER terminology, for logic left shifting, the MSB position is called the SPILL, and the LSB position is called the FILL. Conversely, for right shifting, the MSB and LSB positions become the FILL and SPILL, respectively. In the rotate mode, R is ACTIVE ($R = 1$), the F setting is immaterial, and D determines the direction of rotation—rotation right (rr) or rotation left (rl). Omission of both R and F control data would imply that $R = 0$ and $F = 0$ and that the operation is a 0-FILL shift. Shifts and rotations of multiple-bit positions are denoted by psr or psl, and prr or prl, where p is the number of bit positions shifted or rotated.

There are three basic types of shifting operations commonly used. They are *SPILL-off/zero-FILL* shifting, *rotate or end-around* shifting, and *arithmetic* shifting. SHIFTERs can be designed with the capability of performing any of these operations. The general type of SHIFTER shown in Fig. 4-60 is one such example.

In this section we will design simple examples of arithmetic and non-arithmetic SHIFTERs, beginning first with the latter type. We will also demonstrate both gate- and modular-level approaches to SHIFTER design.

4.9.1 Nonarithmetic SHIFTERs

Simple nonarithmetic SHIFTERs lack the capability to generate and preserve a dedicated sign bit as required for shifting sign-magnitude and sign-complement binary numbers. Consequently, the main function of nonarithmetic SHIFTERs is to carry out SPILL-off/0-FILL and rotation operations. Such SHIFTERs are commonly used to transfer data bit by bit from an ALU (arithmetic and logic unit) on to a data bus for transport to some destination, an operation not usually designed into an ALU. Other uses of nonarithmetic SHIFTERs include operations that modify existing WORDs by SPILL and FILL changes, operations that rotate portions of a binary WORD to more accessible positions, and the rotation operations required for serial arithmetic and cyclic code generation.

As the term implies, SPILL-off (or sometimes end-off) means that the SPILL bit values are lost—there is no place for them to go. Furthermore, each FILL bit position usually receives a logic 0 (hence 0-FILL) unless the operation is one of rotation, in which case the FILL position receives the SPILL value.

Figure 4-61 provides two examples of simple shift operations in which the SPILL bit values are lost (SPILL-off) and the FILL bit positions receive logic 0. A 4-bit WORD 1101 is shifted left by one bit position in Fig. 4-61(a) and is shifted right by two bit positions in Fig. 4-61(b). In SHIFTER notation, the shifts for these two figures are, respectively, sl or $Y_n \leftarrow I_{n-1}$ and 2sr or $Y_{n-2} \leftarrow I_n$, where the arrow ($\leftarrow$) is read as "receives the contents of." Thus,

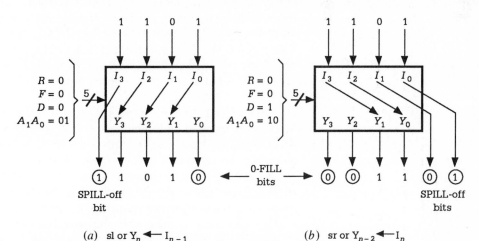

(a) sl or $Y_n \leftarrow I_{n-1}$ (b) sr or $Y_{n-2} \leftarrow I_n$

Fig. 4-61. *Examples of non-arithmetic SPILL-off/0-FILL shifting operations for a 4-bit SHIFTER. (a) Shift left by one bit. (b) Shift right by two bits.*

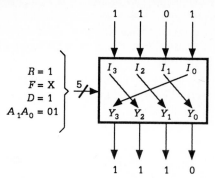

Fig. 4-62. A 1-bit right rotation of a 4-bit WORD.

$Y_{n-2} \leftarrow I_n$ is read as Y_{n-2} receives the contents of I_n, meaning that the input is shifted two bit positions to the right in the output.

The control settings for the two SPILL-off/0-FILL shift operations in Fig. 4-61 are given to the left of the block symbols. They are consistent with the interpretations given in Fig. 4-60(b). Notice that the SPILL bits are I_3 for the sl operation, and I_1, I_0 for 2sr. The 0-FILL requirements shown are satisfied by the internal logic circuitry of the SHIFTER.

Shown in Fig. 4-62 is a 4-bit SHIFTER used to rotate the WORD 1101 to the right by one bit (hence rr). In this case the FILL position Y_3 receives the SPILL value I_0. The control settings for this operation, shown to the left of the figure, are also consistent with the interpretations given in Fig. 4-60(b). Rotation operations of up to $n - 1$ bit positions can be carried out with a SHIFTER of this type. An irrelevant input X is assigned to the FILL parameter since the operation is one of rotation only.

Now, let us design a SHIFTER that can perform shift and rotation operations of the type shown in Figs. 4-61 and 4-62. To maintain a reasonable degree of circuit simplicity, we will limit our design to a 4-bit SHIFTER having the capability of left shifting and rotation by one or two bits but without ENABLE control.

Presented in Figs. 4-63(a) and (b) are the block diagram symbol and collapsed EV truth table for a 4-bit SHIFTER capable of shifting left or rotating left up to two bit positions. R is the MSB, and omission of the control variables D and F is meant to imply left shifting and 0-FILL, all in agreement with Fig. 4-60(b). Notice that the number of bit positions shifted (or rotated) are binary encoded as 0, 1, and 2 and that don't cares are assigned to those outputs which correspond to unused input states, namely, states 3 and 7.

The reader can again appreciate the utility of the EV map method by considering that the four outputs, each a function of seven variables, have been easily mapped and minimized in Fig. 4-63(c) by using only four third-order K-maps. How many fourth-order K-maps would be required to accomplish this same task by conventional means?

For the sake of illustration, we will implement the SHIFTER of Fig. 4-63 at both the gate level and the modular level. Implementation at the gate level will be accomplished by using the minimum SOP cover given in Fig. 4-63(c). The result is the circuit shown in Fig. 4-64, where we have chosen to use NAND logic and have assumed that all inputs and outputs are AC-TIVE HIGH. For the modular approach, DECODERs could be used by reading MINTERMs (not minimum SOP cover) off of the EV K-maps of

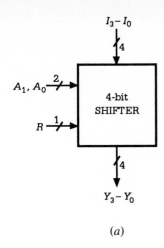

Decimal	R A_1 A_0	Y_3 Y_2 Y_1 Y_0	Operation
0, 4	X 0 0	I_3 I_2 I_1 I_0	Transparent
1	0 0 1	I_2 I_1 I_0 0	sl
2	0 1 0	I_1 I_0 0 0	2sl
3	0 1 1	ϕ ϕ ϕ ϕ	Don't care
5	1 0 1	I_2 I_1 I_0 I_3	rl
6	1 1 0	I_1 I_0 I_3 I_2	2rl
7	1 1 1	ϕ ϕ ϕ ϕ	Don't care

(b)

(a)

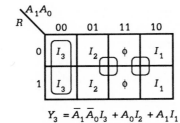

$$Y_3 = \bar{A}_1\bar{A}_0 I_3 + A_0 I_2 + A_1 I_1$$

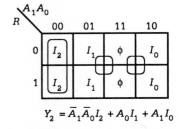

$$Y_2 = \bar{A}_1\bar{A}_0 I_2 + A_0 I_1 + A_1 I_0$$

Fig. 4-63. A 4-bit non-arithmetic SHIFTER capable of left shifting or rotation up to two bit positions. (a) Block diagram symbol. (b) Collapsed EV I/O truth table showing the operation for each input state. (c) EV K-maps and minimum SOP cover.

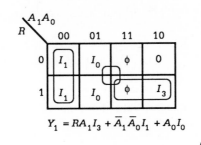

$$Y_1 = RA_1 I_3 + \bar{A}_1\bar{A}_0 I_1 + A_0 I_0$$

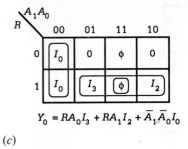

$$Y_0 = RA_0 I_3 + RA_1 I_2 + \bar{A}_1\bar{A}_0 I_0$$

(c)

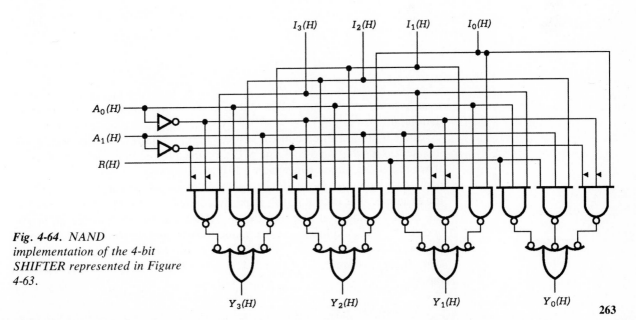

Fig. 4-64. NAND implementation of the 4-bit SHIFTER represented in Figure 4-63.

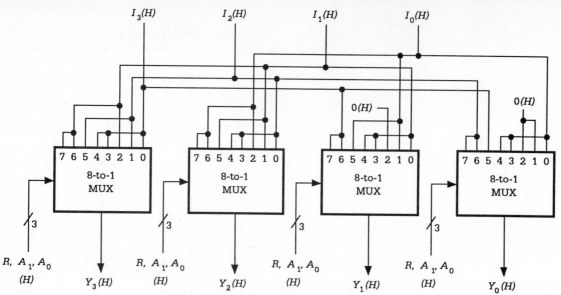

Fig. 4-65. *MUX implementation of the 4-bit SHIFTER represented in Figure 4-63.*

Fig. 4-63(c). In this case a stacked 7-to-128 line DECODER is required, but only 21 of the 128 output lines would be used, an inefficient use of the modular device.

The SHIFTER of Fig. 4-63 can also be implemented at the modular level by using MULTIPLEXERs. That the MULTIPLEXER is an obvious choice as a module for this purpose is easily established by comparing Fig. 4-64 with Fig. 4-53(c). If we take C_i to represent the ANDed term $(RA_1A_0)_i$ associated with each data input in the K-maps of Fig. 4-63(c), the output expressions can be read from the maps as

$$Y_3 = (C_0 + C_4)I_3 + (C_1 + C_5)I_2 + (C_2 + C_6)I_1$$
$$Y_2 = (C_0 + C_4)I_2 + (C_1 + C_5)I_1 + (C_2 + C_6)I_0$$
$$Y_1 = C_6I_3 + (C_0 + C_4)I_1 + (C_1 + C_5)I_0$$
$$Y_0 = C_5I_3 + C_6I_2 + (C_0 + C_4)I_0$$

(4-17)

where $C_0 = \overline{R}\,\overline{A}_1\overline{A}_0$, $C_1 = \overline{R}\,\overline{A}_1A_0$, $C_2 = \overline{R}A_1\overline{A}_0$, and so on. Here, we have used hybrid forms to emphasize the C terms which are common to a given data input.

Since there are three control/address inputs required for this SHIFTER, it is necessary that an 8-to-1 line MUX be used for each of the four outputs. If we recall the manner in which a MUX functions (see Sec. 4-8), the connections which must be made are easily determined. Equations (4-17) show, for example, that data input I_3 must be connected to input lines 0 and 4 of the Y_3 MUX, to input line 6 of the Y_1 MUX, and to line 5 of the Y_0 MUX, and so on. By completing these connections we obtain the results shown in Fig. 4-65, exclusive of an ENABLE control. Notice that the don't care control/address states, 011 and 111, are connected to any convenient data input.

A brief inspection of Figs. 4-64 and 4-65 reveals that both circuits perform the operations exactly as required by the I/O function table of Fig. 4-63(b). When the control/address WORD is 000 or 100 [0 meaning $0(H)$ and 1 meaning $1(H)$], the SHIFTER is transparent and the inputs pass directly to the outputs (that is, $Y_3 \leftarrow I_3$, $Y_2 \leftarrow I_2$, $Y_1 \leftarrow I_1$, and $Y_0 \leftarrow I_0$). However, if the control/address inputs are, say, 010, then $Y_3 \leftarrow I_2$, $Y_2 \leftarrow I_1$, $Y_1 \leftarrow I_0$, and $Y_0 = 0$, indicating a sl/0-FILL shift. And similarly, if the control/address inputs are 101, then an rr operation is produced requiring that $Y_3 \leftarrow I_2$, $Y_2 \leftarrow I_1$, $Y_0 \leftarrow I_0$, and $Y_0 \leftarrow I_3$.

4.9.2 Arithmetic SHIFTERs

Arithmetic shifting must include the capability of generating and preserving a sign bit, the feature which distinguishes arithmetic shifting from nonarithmetic shifting. Furthermore, since rotation is nonarithmetic, in the usual sense, we can limit our consideration to purely shifting operations.

We know that a binary WORD (a number in this case) is made up of bits having positional weights . . . 16, 8, 4, 2, 1, 0.5, 0.25, 0.125, . . . , that is, . . . 2^4, 2^3, 2^2, 2^1, 2^0, 2^{-1}, 2^{-2}, 2^{-3}, ... Therefore, each bit position to the left of a given reference bit has a positional weight larger than the preceding one by a factor of 2^1. Also, each bit position to the right of that reference bit has a positional weight smaller than the preceding one by a factor of 2^{-1}. So, a shift to the left by one bit multiplies the original binary number by 2^1, whereas a shift to the right by one bit divides the original number by 2^1. Similarly, a shift of two bits would multiply or divide the original number by 2^2 for left and right shifting, respectively. Providing no 1's are SPILLed or FILLed, we may characterize arithmetic shifting of a binary number N by p-bit positions as follows:

$$\text{Shift left by } p\text{-bit positions (psl)} \Rightarrow N \times 2^p$$

$$\text{Shift right by } p\text{-bit positions (psr)} \Rightarrow N \div 2^p$$

where the double-edged arrow is read as "gives." For example, shifting the binary number 00001100 by 2sr gives 00000011, which in the arithmetic sense is equivalent to the decimal result $12 \div 2^2 = 3$. Or shifting 00001100 by 3sl gives 01100000, which is equivalent to $12 \times 2^3 = 96$ in decimal. Multiplication of binary numbers in a calculator is usually accomplished by the "add and shift left" algorithm just as we multiply decimal numbers. Likewise, division is usually carried out by the "subtract and shift right" algorithm.

The requirements for sign-complement and sign-magnitude shifting are somewhat different than for logic shifting due to the existence of a sign bit located at the MSB position. Depending on how the SHIFTER is designed, the sign bit may be an integral part of the shifted WORD or it may be a separate, dedicated bit unaffected by shifting manipulations. In the latter case the bit to the right of the sign bit (hence the next MSB) becomes the SPILL and FILL bit for left and right shifting, respectively.

A separate, dedicated sign bit is usually the preferred format for sign-magnitude and sign-complement manipulations. Consider, as examples, left shifting the 8-bit sign-magnitude and 2's complement representations of -3.25_{10}:

Sign-Magnitude Left Shifting

$$-3.25_{10} = \overset{\underset{\text{Sign}}{\text{bit}}}{\underset{\downarrow}{1}} \underbrace{0\ 0\ 1\ 1\ .\ 0\ 1\ 0}_{\text{Magnitude}}{}_2 \quad \overset{\text{sl}}{\underset{(\times 2)}{\longrightarrow}} \quad \overset{\underset{\text{Sign}}{\text{bit}}}{\underset{\downarrow}{1}} \underbrace{0\ 1\ 1\ 0\ .\ 1\ 0\ 0}_{\text{Magnitude}}{}_2 = -6.50_{10}$$

SPILL
bit

0-FILL

$$\overset{2\text{sl}}{\underset{(\times 4)}{\longrightarrow}} \quad 1\ 1\ 1\ 0\ 1\ .\ 0\ 0\ 0_2 = -13.00_{10}$$

Sign Magnitude
bit

2's Complement Left Shifting

$$-3.25_{10} = \overset{\underset{\text{Sign bit}}{\downarrow}}{1}\ 1\ 1\ 0\ 0\ .\ 1\ 1\ 0_2 \quad \overset{\text{sl}}{\underset{(\times 2)}{\longrightarrow}} \quad \overset{\underset{\text{Sign bit}}{\downarrow}}{1}\ 1\ 0\ 0\ 1\ .\ 1\ 0\ 0_2 = -6.50_{10}$$

SPILL
bit

0-FILL

$$\overset{2\text{sl}}{\underset{(\times 4)}{\longrightarrow}} \quad 1\ 0\ 0\ 1\ 1\ .\ 0\ 0\ 0_2 = -13.00_{10}$$

Sign bit

As used in these examples, the double-edged arrow is read "becomes." Notice that a logic 0 is spilled off (lost) in each of the sign-magnitude shifts, whereas a logic 1 is spilled off in each of the 2's complement shifts. Also, observe that a 3-bit left shift would, in either case, result in an *overflow* error since a false magnitude would be produced. Had we chosen to make the sign bit part of the shifted WORD, a 3-bit left shift would have produced both a false sign and false magnitude as the overflow error. Of course, overflow errors can be avoided simply by adding more WORD bit outputs to the SPILL end.

Right shifting of a signed binary number requires that the sign bit be retained. Here, the FILL bit (next MSB) receives the value of the sign bit (S-FILL) for sign-complement representation but receives a logic 0 (0-FILL) for sign-magnitude representation. To illustrate, let us shift right the 8-bit sign-magnitude and 2's complement representations of -3.25_{10}:

Sign-Magnitude Right Shifting

$$-3.25_{10} = \overset{\underset{\text{Sign}}{\text{bit}}}{\underset{\downarrow}{1}} \underbrace{0\ 0\ 1\ 1\ .\ 0\ 1\ 0}_{\text{Magnitude}}{}_2 \quad \overset{\text{sr}}{\underset{(\div 2)}{\longrightarrow}} \quad \overset{\underset{\text{Sign}}{\text{bit}}}{\underset{\downarrow}{1}} \underbrace{0\ 0\ 0\ 1\ .\ 1\ 0\ 1}_{\text{Magnitude}}{}_2 = -1.625_{10}$$

SPILL
bit

0-FILL

$$\overset{2\text{sr}}{\underset{(\div 4)}{\longrightarrow}} \quad 1\ 0\ 0\ 0\ 0\ .\ 1\ 1\ 0_2 = -0.75_{10}$$

Sign Magnitude
bit

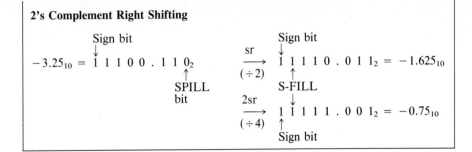

2's Complement Right Shifting

$$-3.25_{10} = 1\ 1\ 1\ 0\ 0\ .\ 1\ 1\ 0_2 \quad \xrightarrow[(\div 2)]{sr} \quad 1\ 1\ 1\ 1\ 0\ .\ 0\ 1\ 1_2 = -1.625_{10}$$

SPILL bit — S-FILL

$$\xrightarrow[(\div 4)]{2sr} \quad 1\ 1\ 1\ 1\ 1\ .\ 0\ 0\ 1_2 = -0.75_{10}$$

Sign bit

Notice that a 2-bit right shift results in an underflow or *round-off* error, since a logic 1 is spilled off, producing a magnitude which is too small by an amount equal to 2^{-4} or 0.0625. Further shifts to the right increases the round-off error ultimately reaching, as it must, zero magnitude. Round-off errors can be minimized by adding more WORD bit outputs to the SPILL end.

Before proceeding to an actual design, we offer a cautionary note regarding arithmetic SHIFTERs. If the SHIFTER design requires that the sign bit be an integral part of the binary WORD subject to all manipulations, then the sign bit becomes the SPILL bit for left shifting and the FILL bit for right shifting. This poses no serious problem, providing the designer understands the obvious problems, mainly those associated with shifting sign-magnitude numbers. Of course, these complications are easily avoided by designing the SHIFTER with a separate dedicated sign bit, a practice commonly followed.

We now illustrate the design procedure by designing a 5-bit, sign-magnitude, arithmetic SHIFTER with a dedicated sign bit, S. To make it interesting, we will require that it be capable of left or right shifts of up to three bit positions and that it have a transparent state and an ENABLE control.

Shown in Fig. 4-66(a) is the collapsed EV truth table for this SHIFTER. The values for the control/address inputs DA_1A_0 are consistent with those given in Fig. 4-60(b), where $F = 0$ and $R = 0$ are implied by their absence. Drawing on the experience we have gained in the previous example, we can read the output expressions directly from the truth table with the result given

$EN\ D\ A_1\ A_0$	$Y_4\ Y_3\ Y_2\ Y_1\ Y_0$	
0 X X X	0 0 0 0 0	
1 X 0 0	$S\ I_3\ I_2\ I_1\ I_0$	$Y_4 = S \cdot EN$
1 0 0 1	$S\ I_2\ I_1\ I_0\ 0$	$Y_3 = [(C_0 + C_4)I_3 + C_1I_2 + C_2I_1 + C_3I_0] \cdot EN$
1 0 1 0	$S\ I_1\ I_0\ 0\ 0$	$Y_2 = [C_5I_3 + (C_0 + C_4)I_2 + C_1I_1 + C_2I_0] \cdot EN$
1 0 1 1	$S\ I_0\ 0\ 0\ 0$	$Y_1 = [C_6I_3 + C_5I_2 + (C_0 + C_4)I_1 + C_1I_0] \cdot EN$
1 1 0 1	$S\ 0\ I_3\ I_2\ I_1$	$Y_0 = [C_7I_3 + C_6I_2 + C_5I_1 + (C_0 + C_4)I_0] \cdot EN$
1 1 1 0	$S\ 0\ 0\ I_3\ I_2$	
1 1 1 1	$S\ 0\ 0\ 0\ I_3$	

Fig. 4-66. *Design of a 5-bit sign-magnitude arithmetic SHIFTER with an ENABLE control and separate sign bit. (a) Collapsed EV truth table. (b) Output logic expressions read directly from the truth table, where the C's represent the ANDed control/address terms associated with each data input.*

in Fig. 4-66(b). As in Eqs. (4-17), we have used C_i to represent the ANDed term $(DA_1A_0)_i$ associated with each data input. Thus, $C_0 = \overline{D}\,\overline{A}_1\overline{A}_0$, $C_1 = \overline{D}\,\overline{A}_1A_0$, and so on. It should be apparent that the irrelevant input X used in the transparent state $(DA_1A_0 = X00)$ leads to the $(C_0 + C_4)$ terms appearing in the expressions of Fig. 4-66(b).

Implementation of the I/O expressions of Fig. 4-66(b) follows in similar fashion to that of Eqs. (4-17). Again, for the modular approach, one would use four 8-to-1 line MUXs as in Fig. 4-65 but connected according to the requirements of Fig. 4-66(b). The data inputs are now S, I_3, I_2, I_1, and I_0, and the Y_4 output requires that we AND the sign bit $S(H)$ to $EN(H)$. Completion of this design is left as an exercise for the reader.

More complicated SHIFTERs can be designed with only a little more effort but greatly increased hardware requirements. For example, to add a fourth control/address variable to the SHIFTER of Fig. 4-66 would require the use of 16-to-1 line MUXs. Therefore, at some point the designer must weigh the simplicity of the modular approach against the increased hardware requirements. Where space is a critical factor, it may be necessary to have a manufacturer produce an IC SHIFTER to meet the specific requirements of the design.

4.10 ARITHMETIC AND LOGIC UNITS

As the name implies, the *arithmetic and logic unit (ALU)* is a universal combinational logic device capable of performing both arithmetic and logic operations. It is this versatility that makes the ALU an attractive building block in the *central processing unit (CPU)* of a computer or microprocessor. The main objective of this section is to develop the techniques required to design and cascade devices of this type.

The number and complexity of the operations that a given ALU is capable of performing is a matter of the designer's choice and may vary widely from ALU to ALU. However, for two operands (say, A and B), each a WORD of several bits, the choice of operations is usually drawn from the following lists:

Arithmetic Operations	Logic Operations
Negation	INACTIVE
Increment	NOT
Decrement	AND
Addition	OR
Subtraction	XOR

Other possible operations include zero, unity, transparency (inputs for an operand pass straight through to the outputs), sign-complement representation, magnitude comparison, parity generation, multiplication, division, powers, shifting, and roots. Multiplication, division, and the related operations such as arithmetic shifting are complex and are usually found only in the more sophisticated LSI ALU chips. Also, the AND, OR, and XOR operations will normally be applied to various combinations of comple-

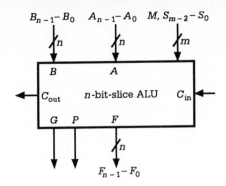

Fig. 4-67. *Block diagram symbol for a general 1-bit-slice ALU with LAC and cascading capabilities.*

mented and uncomplemented operands making possible a wide assortment of such operations.

Presented in Fig. 4-67 is the block diagram symbol for a general n-bit slice ALU. This ALU accepts two n-bit input operands, $B_{n-1} - B_0$ and $A_{n-1} - A_0$, and a CARRYin bit, C_{in}, and operates with them in some predetermined way to output an n-bit function WORD, $F_{n-1} - F_0$ and a CARRYout bit, C_{out} (if applicable). Here, the term *n-bit slice* indicates a logic partition of identical n-bit modules or stages which can be cascaded in parallel. Thus, a FULL ADDER in an n-bit PARALLEL ADDER could be called a 1-bit slice for that ADDER. Also, we assume the use of sign-complement arithmetic so as to avoid the need for both CARRY and BORROW parameters.

The choice of operation between the two operand WORDs is determined by m mode/select inputs, namely, a mode input M, and $m - 1$ function-select inputs, $S_{m-2} - S_0$, shown in Fig. 4-67. The mode input M determines whether an arithmetic or logic function is to be generated, and the function-select inputs, $S_{n-2} - S_0$, determine the particular operation. Just as the CARRYout bit is required for cascading standard arithmetic units, the CARRY PROPAGATE and CARRY GENERATE bits, P and G, are required for cascading LOOK-AHEAD CARRY (LAC) units. Commercial chips are available which have all these features.

4.10.1 Standard ALU Design and Cascading Methods

Consider the EV function table in Fig. 4-68 representing a simple one-bit slice ALU capable of performing four arithmetic functions and four logic functions on command of the three mode/select inputs M, S_1, and S_0. Presented in the two columns on the right side is a brief description of the operation and the CARRYout function for each mode/select input state. The C_{out} functions for the arithmetic operations are taken from Eqs. (4-2), where Eqs. (2-32) have been applied. Don't cares are placed in the C_{out} column for the four logic operations since the CARRYout function has no relevance in a logic operation. Notice further that the two possible logic states for C_{in} lead to different interpretations for each of the first two arithmetic operations. Thus, $A \oplus C_{in}$ means the transfer of A for $C_{in} = 0$, but represents the decrement of A if the LSB $C_{in} = 1$ ($B = 0$ is implied). Or, $\overline{A} \oplus C_{in}$ represents the increment of A if the LSB $C_{in} = 0$ but represents the transfer of A if

	M S_1 S_0	F	Operation*	C_{out}
Arithmetic operations	0 0 0	$A \oplus C_{in}$	A minus 1 if LSB $C_{in} = 1$	$A \cdot C_{in}$
	0 0 1	$\overline{A} \oplus C_{in}$	A plus 1 if LSB $C_{in} = 0$	$A + C_{in}$
	0 1 0	$A \oplus B \oplus C_{in}$	A plus B plus C_{in}	$C_{in}(A \oplus B) + AB$
	0 1 1	$\overline{A} \oplus B \oplus C_{in}$	B minus A if LSB $C_{in} = 1$	$C_{in}(\overline{A \oplus B}) + \overline{A}B$
Logic operations	1 0 0	A	Transfer A	ϕ
	1 0 1	$\overline{A}$	Complement and transfer A	ϕ
	1 1 0	$A \oplus B$	XOR	ϕ
	1 1 1	$A \odot B$	EQV	ϕ

* Subtraction operations assume 2's complement arithmetic

Fig. 4-68. *I/O function truth table for a simple one-bit slice ALU showing output functions, F and C_out, for four arithmetic operations (M = 0) and four logic operations (M = 1).*
$C_{in} = 1$ ($B = 1$ is implied). The arithmetic operations assume 2's complement arithmetic.

Since the ALU represented in Fig. 4-68 has three mode/select inputs controlling eight operations, it is convenient to map the output function F in a third-order K-map, as shown in Fig. 4-69(a). Here the XOR cover indicated by the loops results in the reduced SOP expression given to the right of the figure, where use has been made of the identity $\overline{A \oplus B \oplus C_{in}} = \overline{A} \oplus B \oplus C_{in}$. Since the XOR operation is commutative, we may present F in the more convenient form

$$F = \overline{M}\,\overline{S}_1(A \oplus S_0 \oplus C_{in}) + \overline{M}S_1(A \oplus S_0 \oplus B \oplus C_{in}) \qquad (4\text{-}18)$$
$$+ M\overline{S}_1(A \oplus S_0) + MS_1(A \oplus S_0 \oplus B)$$

Recall that the adjacent patterns in Fig. 4-69(a) form one of four types of XOR patterns as discussed in Sect. 3.12.2.

The third-order K-map of Fig. 4-69(a) can be further compressed into a second-order K-map, a fourth-order compression with a MAP KEY of 2^4 = 16. This has been done in Fig. 4-69(b), where minimum XOR cover is produced by looping out associative patterns of the type discussed in Sect. 3.12.2. The pseudo-SOP and -POS expressions that result are given by

$$F_{SOP} = (A \oplus S_0) \oplus (S_1B) \oplus (\overline{M}C_{in})$$
$$= (A \oplus S_0) \odot (S_1B) \odot (\overline{M}C_{in})$$
$$F_{POS} = (A \oplus S_0) \oplus (\overline{S}_1 + \overline{B}) \oplus (M + \overline{C}_{in}) \qquad (4\text{-}19)$$
$$= (A \oplus S_0) \odot (\overline{S}_1 + \overline{B}) \odot (M \oplus \overline{C}_{in}),$$

where the pseudo-POS forms are either read directly off the map of Fig. 4-69(b) or are obtained by repeated application of Eqs. (2-32). In either case, F_{SOP} is seen to be algebraically equal to F_{POS}. The reader should appreciate the fact that Eqs. (4-18) and (4-19) could not easily have been obtained by conventional (1's and 0's) mapping methods.

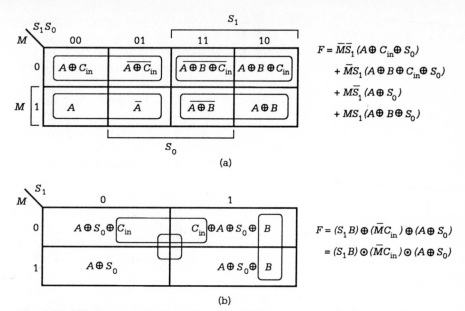

$$F = \overline{M}\,\overline{S}_1\,(A \oplus C_{in} \oplus S_0)$$
$$+ \overline{M}S_1\,(A \oplus B \oplus C_{in} \oplus S_0)$$
$$+ M\overline{S}_1\,(A \oplus S_0)$$
$$+ MS_1\,(A \oplus B \oplus S_0)$$

(a)

$$F = (S_1\,B) \oplus (\overline{M}\,\overline{C}_{in}) \oplus (A \oplus S_0)$$
$$= (S_1\,B) \odot (\overline{M}\,\overline{C}_{in}) \odot (A \oplus S_0)$$

(b)

Fig. 4-69. *EV K-maps showing adjacency XOR patterns for the function F in Figure 4-68. (a) Third-order map. (b) Second-order map and minimum XOR cover.*

If it is the object to implement only the function F, a simple noncascadable FUNCTION GENERATOR would result. Either Fig. 4-69(a) or Eq. (4-18) leads quite naturally to the MUX implementation presented in Fig. 4-70(a). However, Fig. 4-69(b), or the SOP form of Eqs. (4-19), is best suited to give the minimum gate-level circuit shown in Fig. 4-70(b). In the latter case, the POS form could be implemented simply by replacing the AND gates with NAND gates. As we shall soon see, the circuit of Fig. 4-70(b) forms the basis for the ALU we wish to design.

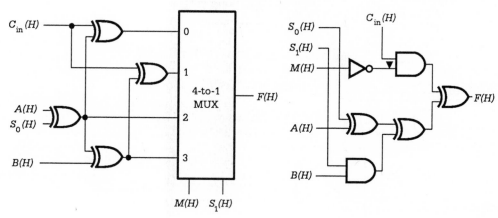

Fig. 4-70. *Design of a simple FUNCTION GENERATOR. (a) MUX implementation of Eq. (4-18). (b) Minimum gate-level design using the SOP form of Eq. (4-19).*

To design an ALU that can perform the operations of a FULL ADDER and can be cascaded, it is necessary that we include a CARRYout function. Drawing on our experience we have gained in this section and noting the inherent XOR patterns between adjacent MS_1S_0 states, we can construct a second-order K-map for C_{out} directly from the function table of Fig. 4-68. The result is given in Fig. 4-71 which has a minimum cover given by

$$C_{out} = C_{in}(A \oplus S_0) \oplus (S_1B) + (S_1B)(A \oplus S_0) + \overline{S}_1S_0A \qquad (4\text{-}20)$$

Notice how the don't cares are used to produce the minimum XOR form and, in particular, notice how the associative patterns are used to loop out the $C_{in}(A \oplus S_0)$ term and B.

We can now assemble a gate-level ALU circuit from Eqs. (4-19) and (4-20). Because of the terms common to both F and C_{out}, the circuit of Fig. 4-70(b) need be only slightly modified to accommodate the CARRYout bit, C_{out}. The result is the minimum gate-level circuit and its block diagram shown in Fig. 4-72. This circuit represents a two-output multilevel optimization with a gate/input tally of 9/20 exclusive of INVERTERs.

An n-bit PARALLEL ALU can be produced by cascading the 1-bit slice ALU of Fig. 4-72 in a manner similar to that used in Fig. 4-10 to produce

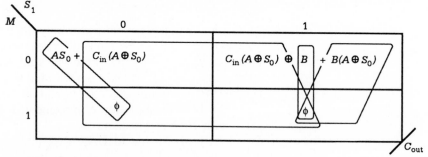

Fig. 4-71. *Second-order EV K-map and minimum SOP XOR cover for the CARRYout functions given in Figure 4-68.*

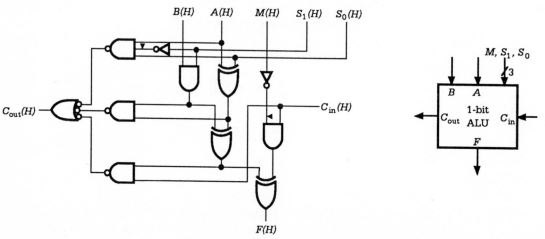

Fig. 4-72. *Minimum gate-level design and block diagram symbol for the 1-bit-slice ALU represented in the I/O table of Figure 4-68.*

4 / Combinational Logic Design

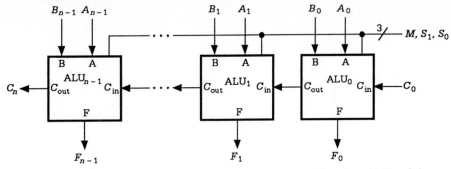

Fig. 4-73. An n-*bit slice ALU produced by cascading* n *1-bit-slice ALUs of the type shown in Figure 4-72.*

an n-bit PARALLEL ADDER from n FULL ADDERs. This is done in Fig. 4-73, but with the added requirement of connecting the three mode/select inputs to all stages as shown. It is also possible to construct an n-bit PARALLEL ALU by using 2^m m-bit-slice ALUs, similar to the PARALLEL ADDER shown in Fig. 4-11.

The n-bit PARALLEL ALU, like the PARALLEL ADDER, suffers a size limitation due to the ripple-carry effect as discussed in Sect. 4.4.3. For this reason the design of our simple ALU will now be extended to include the LOOK-AHEAD-CARRY (LAC) feature. (See Sect. 4.4.4 for a discussion of the LAC ADDER.) In Fig. 4-74 we reconstruct the I/O function table of Fig. 4-68 by replacing the C_{out} functions with those for CARRY PROPAGATE (P) and CARRY GENERATE (G) derived from Eqs. (4-4). Don't cares are again used to represent irrelevant mode/select input states.

The design of our ALU with LAC capability follows from the table in Fig. 4-74. Presented in Fig. 4-75 are the second-order EV K-maps and minimum XOR cover for the P and G parameters. Here, as in Fig. 4-71, associative XOR patterns are produced by using the don't cares in adjacent cells. By combining the logic expressions for P and G with Eqs. (4-19) and noting the similarity of terms, we obtain the ALU circuit and its block diagram symbol given in Fig. 4-76. This is a cascadable 1-bit-slice ALU with the LAC feature. Observe that no additional circuitry is needed to generate the CARRY PROPAGATE function from the circuit of Fig. 4-70(b), a con-

$M\ S_1\ S_0$	F	P	G
0 0 0	$A \oplus C_{in}$	A	0
0 0 1	$\overline{A} \oplus C_{in}$	$\overline{A}$	A
0 1 0	$A \oplus B \oplus C_{in}$	$A \oplus B$	$A \cdot B$
0 1 1	$\overline{A} \oplus B \oplus C_{in}$	$\overline{A} \oplus B$	$\overline{A} \cdot B$
1 0 0	A	ϕ	ϕ
1 0 1	$\overline{A}$	ϕ	ϕ
1 1 0	$A \oplus B$	ϕ	ϕ
1 1 1	$\overline{A \oplus B}$	ϕ	ϕ

Fig. 4-74. *I/O function truth table for a simple ALU with LAC capability.*

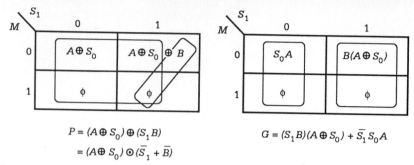

$$P = (A \oplus S_0) \oplus (S_1 B)$$
$$= (A \oplus S_0) \odot (\overline{S}_1 + \overline{B})$$

$$G = (S_1 B)(A \oplus S_0) + \overline{S}_1 S_0 A$$

Fig. 4-75. *EV K-maps and minimum XOR cover for the CARRY PROPAGATE (P) and CARRY GENERATE (G) parameters given in Figure 4-74.*

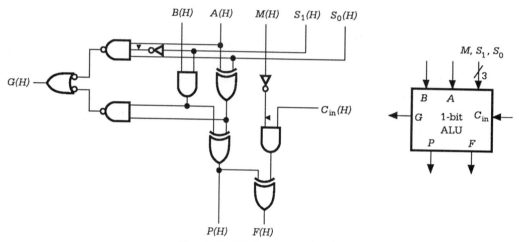

Fig. 4-76. *Minimum gate-level circuit and block diagram symbol for a 1-bit-slice ALU having the function and LAC capabilities given in Figure 4-74.*

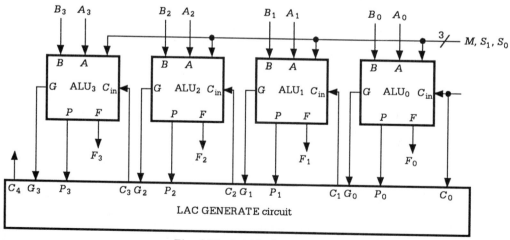

Fig. 4-77. *A 4-bit-slice ALU produced by cascading four 1-bit-slice ALUs of the type shown in Figure 4-76.*

clusion that could have been deduced directly from Fig. 4-74 by comparing the P and F columns.

The ALU module in Fig. 4-76 can be cascaded by using an LAC GENERATE circuit of the type shown in Fig. 4-12. We do this in Fig. 4-77 for a 4-bit LAC ALU capable of performing the functions given in Fig. 4-74, or any other similar set of operations controlled be three mode/select inputs. Cascadable units of this type can themselves be cascaded to produce even larger units. This is accomplished by connecting the CARRYout of one stage to the CARRYin of another and connecting the mode/select inputs to all stages.

4.10.2 The MUX Approach

There are occasions when the designer needs an ALU with unique capabilities but does not need a minimum circuit and cannot afford the time or costs required for the production of a commercial IC chip. A circuit board design using off-the-shelf components is the answer, and the use of MUXs provides one of the simplest and most effective means of achieving this. In fact, the MUX approach can provide design versatility not achievable by any other means.

As a simple example, suppose we wish to design a 1-bit slice ALU with the capability of generating one of 32 arithmetic functions, or one of 16 logic functions. Let the arithmetic operations include addition or subtraction of CANONICAL forms relative to bit A. Also, let us assume that the combined 48 operations are to be generated from a single 8-to-1 line MUX. Shown in Fig. 4-78(a) is the logic circuit and block diagram symbol for this ALU. The four gates together with the mixed-rail outputs of the MUX provide the CANONICAL SOP and POS forms required. The 2-to-1 MUX and 1-to-2 DMUX/DECODER act to steer the mixed-rail outputs of the 8-to-1 function MUX to either the B input of the FULL ADDER or to the OR gate directly. Note that the four AND gates could be replaced by a 2-to-4 DECODER.

The 48 arithmetic and logic operations that are possible with the ALU in Fig. 4-78(a) are given in the table of Fig. 4-78(b). Here, 2's complement arithmetic is assumed for all subtraction operations. Thus, for such operations, the subtrahend bit is complemented and the LSB C_{in} is set at logic 1 as required by the $\overline{N}_2 + 1$ algorithm (see Sect. 4.4.7). In contrast, all arithmetic operations for which the LSB $C_{in} = 0$ must be addition operations. As an example, consider the arithmetic operations produced by the control WORD $MS_3S_2S_1S_0 = 01011$. If the LSB $C_{in} = 1$, the operation is interpreted as A_J minus (A_JB_J); otherwise, it is interpreted as A_J plus $(\overline{A}_J + \overline{B}_J)$. Or, if the control WORD is $MS_3S_2S_1S_0 = 00000$, then the operation is the decrement of A_J if the LSB $C_{in} = 1$, or is the transfer of A_J if the LSB $C_{in} = 0$. The 1-bit slice ALU of Fig. 4-78 can be cascaded as in Fig. 4-73.

Modifications can be made to the ALU of Fig. 4-78 that make it even more versatile. As it stands, this ALU is asymmetric with respect to its arithmetic operations since only the function of the B input to the FULL ADDER can be altered. Another system of MUXs, similar to that shown in Fig. 4-78(a) with the same or different functions, can be used for input A to the FULL ADDER, thereby increasing the scope of the arithmetic and logic

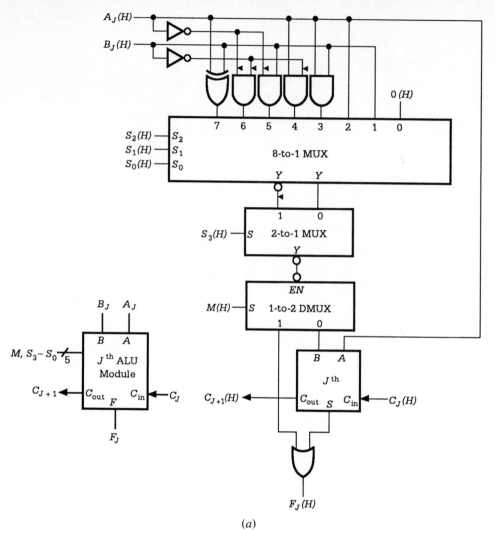

(a)

Fig. 4-78. *A 1-bit-slice ALU capable of performing 48 arithmetic and logic operations. (a) Block circuit symbol and logic circuit for the Jth module.*

operations significantly. In fact, by using a DECODER to activate one of several MUX systems, it is possible to create an ALU of tremendous capability. Problem 4.57 at the end of this chapter illustrates a very simple DECODER/MUX scheme for an ALU.

ALUs are not limited to binary addition and subtraction operations nor are they limited to simple AND-, OR-, and XOR-type logic functions. NBCD addition and subtraction, and binary or NBCD multiplication are examples of other arithmetic operations that can be included in an ALU. And complex logic functions can replace the simple gates shown in Fig. 4-78(a).

The list of possible modifications continues on and on, limited mainly by the designer's imagination. But one fact stands out—the circuit board approach offers considerable flexibility to the designer. Alterations can be made easily and quickly, and the individual units can be tested separately from the ALU to ensure proper operation before they are installed. Fur-

	Function			
Control Inputs $S_3\ S_2\ S_1\ S_0$	**Arithmetic*** $(M = 0)$		**Logic** $(M = 1)$
	LSB $C_{in} = 0$	LSB $C_{in} = 1$	
0 0 0 0	Transfer A_J	A_J minus 1	0
0 0 0 1	A_J plus B_J	A_J plus B_J plus 1	B_J
0 0 1 0	A plus $A_J = 2A_J$	A_J plus A_J plus 1	A_J
0 0 1 1	A_J plus (A_JB_J)	A_J plus (A_JB_J) plus 1	A_JB_J
0 1 0 0	A_J plus $(A_J\overline{B}_J)$	A_J plus $(A_J\overline{B}_J)$ plus 1	$A_J\overline{B}_J$
0 1 0 1	A_J plus $(\overline{A}_JB_J)$	A_J plus $(\overline{A}_JB_J)$ plus 1	$\overline{A}_JB_J$
0 1 1 0	A_J plus $(\overline{A}_J\overline{B}_J)$	A_J plus $(\overline{A}_J\overline{B}_J)$ plus 1	$\overline{A}_J\overline{B}_J$
0 1 1 1	A_J plus $(A_J \oplus B_J)$	A_J plus $(A_J \oplus B_J)$ plus 1	$A_J \oplus B_J$
1 0 0 0	A_J plus 1	Transfer A_J	1
1 0 0 1	A_J plus $\overline{B}_J$	A_J minus B_J	$\overline{B}_J$
1 0 1 0	A_J plus $\overline{A}_J$	A_J minus $A_J = 0$	$\overline{A}_J$
1 0 1 1	A_J plus $(\overline{A}_J + \overline{B}_J)$	A_J minus (A_JB_J)	$\overline{A}_J + \overline{B}_J$
1 1 0 0	A_J plus $(\overline{A}_J + B_J)$	A_J minus $(A_J\overline{B}_J)$	$\overline{A}_J + B_J$
1 1 0 1	A_J plus $(A_J + \overline{B}_J)$	A_J minus $(\overline{A}_JB_J)$	$A_J + \overline{B}_J$
1 1 1 0	A_J plus $(A_J + B_J)$	A_J minus $(\overline{A}_J\overline{B}_J)$	$A_J + B_J$
1 1 1 1	A_J plus $(A_J \odot B_J)$	A_J minus $(A_J \oplus B_J)$	$A_J \odot B_J$

* Subtraction operations assume 2's complement arithmetic

(b)

Fig. 4-78 (contd.). (b) Possible arithmetic and logic operations.

thermore, the arithmetic and logic sections can be implemented in a variety of ways by combining gate- and modular-level circuits as desired. The design of a programmable ALU in Appendix 4.2 offers a good example. This ALU is capable of performing shifter and comparator operations in addition to a variety of arithmetic and logic operations.

If an ALU is to be designed with unique capabilities, it will need to be tested, and a circuit board approach using off-the-shelf components like MUXs and DECODERs is the simplest way to begin. Having done this, the designer must then take into account other factors such as physical size and speed. If it is intended that the ALU be used in a compact environment with high-speed requirements, a greatly reduced or minimum gate-level design may be necessary. In this case, the EV K-map methods described in this section can be very useful. On the other hand, if the ALU is to be used as part of a one-time-only circuit, the initial off-the-shelf version may suffice. Finally, if the ALU is to be commercially manufactured as part of a much larger system, say, a microprocessor or computer, greater programming capability may be called for. The programmable ALU, illustrated in Appendix 4.2 is one example. Also, the use array logic for ALU design in an IC circuit environment should not be overlooked by the designer.

4.11 ARRAY LOGIC—ROMs AND PLAs

Read-only memories (ROMs) and programmable logic arrays (PLAs) each belongs to an important class of general-purpose IC array logic devices which find extensive use in combinational logic design. As we shall soon discover, ROMs and PLAs share much in common but differ mainly in the way they are programmed. In this section we will consider their logic structure, a brief description of the various types and sizes of ROMs and PLAs that are commercially available, and some simple examples of their application in combinational logic design.

4.11.1 Logic Structure and Types of ROMs

A ROM is an n-input/m-output device composed of a DECODER (ANDing) stage and a "memory" array (OR matrix) stage as illustrated by the block diagram of Fig. 4-79. Bit combinations of the n input variables are called *addresses,* and in a ROM there are 2^n possible addresses each representing a coded MINTERM on the output side of the DECODER stage. Bit combinations of the m outputs are called WORDs. Since there are 2^n possible addresses, there are 2^n possible WORDs that can be stored in the ROM, each WORD being m bits in length. Any unique output WORD of m bits programmed into the ROM can be selected by the appropriate input address and is nonvolatile—it is stored permanently in the ROM.

The *dimensions* and *size* of an n-input/m-output ROM are usually given by

$$\underbrace{2^n \times m}_{\text{dimensions}} = \underbrace{(2^n)(m) \text{ bits}}_{\text{size}}$$

meaning 2^n WORDs by m bits for a total of $(2^n)(m)$ bits. The size of the ROM is often rounded to the nearest integer power of 2 in kbits (1,000 bits) of ROM. For example, an 8-input/4-output ROM is represented as a

$$2^8 \times 4 = 256 \times 4 \text{ ROM}$$

$$= 1{,}024 \text{ bit ROM}$$

$$= 1\text{K bit ROM}$$

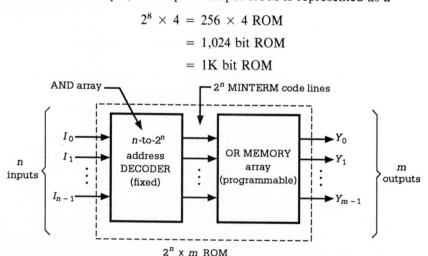

Fig. 4-79. *Block diagram showing the fixed address DECODER stage and programmable OR MEMORY array stage for a 2^n WORD $\times$ m bit ROM.*

4 / Combinational Logic Design

Or a 12-input/8-output ROM is sized as a

$$2^{12} \times 8 = 4{,}096 \times 8 \text{ bit ROM}$$
$$= 32{,}768 \text{ bit ROM}$$
$$= 32\text{K bit ROM}.$$

Commercial ROMs are available in a variety of dimensions and sizes commonly ranging from $2^4 \times 1 = 16$ bit ROMs to $2^{16} \times 8 = 512$ kbit ROMs with the strong majority having 4 and 8 outputs.

ROMs are constructed of interconnecting arrays of switching devices, such as transistors or diodes, which perform the AND and OR operations required by the DECODER and memory stages, respectively. They may differ in a variety of ways but the main differences center about the manner in which they are programmed, and on whether or not they can be erased and reprogrammed and how this is accomplished. Generally speaking, members of the ROM family may be divided into three main categories:

1. Read-only memories (ROMs)—mask programmable OR plane only
2. Programmable read-only memories (PROMs)—user programmable ROMs
3. Erasable programmable read-only memories (EPROMs)—user erasable PROMs

Mask programmable ROMs are programmed during the fabrication process by selectively including or omitting the switching elements (transistors or diodes) which form the memory array stage of the ROM. Because the masking process is expensive, the use of mask programmable ROMs is economically justifiable only if large numbers (perhaps thousands) are required with the same I/O requirements.

When one or a few ROM-type devices are needed with I/O requirements which will never need to be altered, PROMs can be very useful. Most PROMs are fabricated with metal or semiconductor *fusible links* on all transistors (or diodes) in the memory stage thereby permitting user programming of the ROM—a write-once capability. We illustrate this by using the logic equivalent circuit for an unprogrammed $2^n \times m$ PROM shown in Fig. 4-80. Here, the symbols m_i represent the 2^n possible MINTERMs which can be generated by the DECODER stage of the ROM. Each OR gate of the OR (memory) array has 2^n inputs which are connected initially to the outputs of the DECODER stage as shown.

In effect, the PROM is programmed by disconnecting the appropriate inputs to the OR array. However, in an actual PROM, this must be accomplished by "blowing" or "burning" the fusible links connected to the appropriate DECODER MINTERM code lines. A special device called a PROM programmer selectively blows the fusible links by supplying the necessary overvoltage pulses across the appropriate terminals of the PROM. The appropriate terminals for a given link can be thought of as coordinates. For example, the encircled fusible link in Fig. 4-80 has coordinates $m_1(H)$ and $Y_1(H)$, where the MINTERM code line m_1 is selected, just as in any DECODER, by inputs equivalent to decimal 1. The result of programming is a set of m outputs each representing a specific CANONICAL SOP expres-

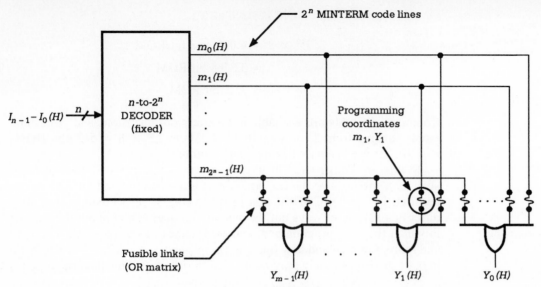

Fig. 4-80. *Logic equivalent circuit for an unprogrammed $2^n \times$ m PROM showing the fusible links in the OR matrix required for user programming.*

sion ACTIVE HIGH, or a CANONICAL POS expression ACTIVE LOW. Once programmed, the PROM cannot be reprogrammed since the blowing of the fusible links is a destructive process and cannot be reversed.

The logic equivalent circuit of Fig. 4-80 can also be used to represent one of the various types of EPROMs. In this case the fusible links would represent special nonvolatile charge-storage mechanisms which can enable or disable the MOSFETs that are used as the switching devices in the memory array of the ROM. Thus, data bits are represented by the presence or absence of stored charge. Removal of the stored charge erases the data. The programming cycle of an EPROM involves removal of stored charge on all MOSFETs and the creation of stored charge on those MOSFETs targeted to meet specifications. EPROMs can be subjected to a large number of program cycles. Those EPROMs which are electrically erasable are called EE-PROMs while others which are erased by ultraviolet light are labeled UVE-PROMs. In any case, EPROMs offer the designer or researcher the opportunity to alter the storage content of the ROM. This feature is of particular value in the construction of prototype circuits preparatory for IC fabrication where alteration is nearly inevitable.

Regardless of what kind of ROM is to be used, it is necessary for the user to construct a CANONICAL I/O program table or have access to CANONICAL SOP or POS code data. This information is necessary before the ROM can be programmed. Remember that the outputs of a ROM represent specific CANONICAL SOP (or POS) expressions. By reading the 1's and 0's from the program table (or the MINTERM or MAXTERM code numbers from CANONICAL expressions or K-maps), the programmer can determine which fusible links are to be burned and which are to be left intact. Shown in Fig. 4-81 is a typical I/O program table for a $2^n \times m$ ROM. Notice that the input variables are given in the order of increasing decimal value just

	n Input Variables			m Output Variables		
	I_{n-1} $\cdots$	I_1	I_0	Y_{m-1} $\cdots$	Y_1	Y_0
m_0	0 $\ldots$	0	0	1 $\ldots$	0	1
m_1	0 $\ldots$	0	1	0 $\ldots$	1	1
m_2	0 $\ldots$	1	0	1 $\ldots$	0	1
m_3	0 $\ldots$	1	1	1 $\ldots$	0	0
$\vdots$	$\vdots$			$\vdots$		
m_{2^n-2}	1 $\ldots$	1	0	0 $\ldots$	0	1
m_{2^n-1}	1 $\ldots$	1	1	1 $\ldots$	1	0

Fig. 4-81. *Typical I/O program table for a $2^n \times$ m ROM.*

as is done in any CANONICAL truth table and that I_{n-1} and Y_{m-1} are the MSBs for the input and output WORDs, respectively.

Various abridged symbolisms have been used to represent the logic equivalent circuit of a programmed ROM. One such abridged symbolism is used in Fig. 4-82 to represent the PROM programmed to the specifications of Fig. 4-81. Here, a dot (node) at the intersection of two lines represents a connected fusible link (in Fig. 4-80) and the storage of a 1(H), while a circle represents a disconnected (blown) fusible link or the storage of a 0(H). Thus, the pattern of 1's and 0's in the I/O program table translates as a pattern of dots and circles, respectively, in the abridged ROM circuit symbolism. Accordingly, the circles which lie along any vertical line in Fig. 4-81 represent "blown" fusible links associated with the appropriate OR gate in the logic equivalent circuit of Fig. 4-80.

Whether or not the intact fusible links of Fig. 4-80 represent the storage of a 1(H) in a commercial PROM (or EPROM) depends on the array technology that has been used in its manufacture. For example, if an intact fusible link represents an enabled transistor (TTL or MOSFET) or a disabled Schottky diode, it most likely stores a logic 0(H) instead of a 1(H). In such

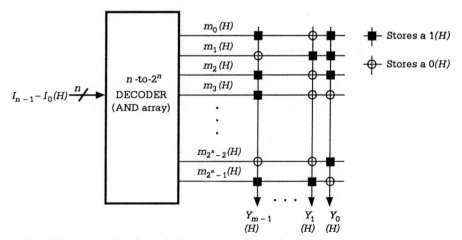

Fig. 4-82. *An abridged symbolic representation of a $2^n \times$ m PROM programmed to the specifications of Figure 4-81.*

cases the OR array is referred to as a NOR plane, but performing the OR memory functions. What this means is that the programmer must know what array technology has been used to program the PROM (or EPROM) correctly from the program table. In fact, there are several different types of arrays which have been used for both the ANDing and ORing stages of a PROM, and each type may differ from the other as to the switching device used or the manner in which it is connected into the array, or both.

4.11.2   Logic Structure and Types of PLAs

Like the ROM, the PLA is an n-input/m-output device composed of an input decoding (ANDing) stage and memory (ORing) stage. Unlike the ROM, however, both stages of the PLA are programmable as indicated by the block diagram given in Fig. 4-83. The AND matrix (array) produces the Boolean product terms, while the OR matrix ORs the appropriate product terms together to produce the desired SOP expressions.

The dimensions of a PLA are specified by using three numbers:

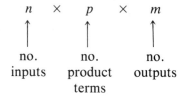

$$ n \quad \times \quad p \quad \times \quad m $$

no. inputs   no. product terms   no. outputs

The number p gives the maximum number of product terms (p-terms) permitted by the PLA. The magnitude of p is set by the manufacturer based on expected user needs and is usually much less than 2^n, the DECODER output for a ROM. For example, a PLA specified by dimensions $16 \times 48 \times 8$ would have 16 possible inputs and could generate 8 different outputs (representing 8 different SOP expressions) composed of up to 48 unique ORed p-terms. A p-term may or may not be a MINTERM. In contrast, a 16-input ROM could generate the specified number of outputs with up to $2^{16} = 65,536$ ORed MINTERMs. The comparison here is 48 AND "gates"

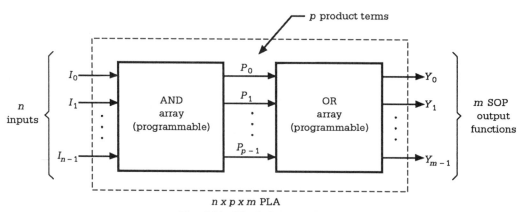

$n \times p \times m$ PLA

Fig. 4-83. Block diagram for an n-input/m-output PLA showing the programmable AND and OR arrays and p product-term lines.

4 / Combinational Logic Design

and 48 OR "gate" inputs for the unprogrammed PLA versus 65,536 AND "gates" and 65,536 OR "gate" inputs for the unprogrammed ROM, a vast difference in hardware. Typically, commercial PLAs will have inputs ranging in number from 8 to 20, with 20 to 80 addressable p-terms and 8 to 20 possible outputs. However, PLA IC chips of most any $n \times p \times m$ size can be manufactured to user specifications.

PLAs, like ROMs, are constructed of interconnecting arrays of switching elements (diodes or transistors) which perform the AND and OR operations. Members of the PLA family fall generally into four main classes:

1. Programmable logic arrays (PLAs)—mask programmable AND and OR planes
2. Field programmable logic arrays (FPLAs)—user programmable PLAs
3. Programmable array logic (PALs)—mask programmable AND plane only
4. Field programmable array logic (FPALs)—user programmable PALs

PLAs and PALs are programmed during fabrication in a manner similar to ROMs, while FPLAs and FPALs are write-once programmed by the user.

Shown in Fig. 4-84 is the logic equivalent circuit for an unprogrammed FPLA of arbitrary dimensions, $n \times p \times m$. Comparing it with the logic equivalent circuit for a ROM (Fig. 4-80) reveals to us their main differences, namely, the FPLA has write-once capabilities (fusible links) in both the AND and OR arrays and has p-term lines numbering much less than 2^n. The logic equivalent circuit for an FPAL of arbitrary dimensions would be similar to

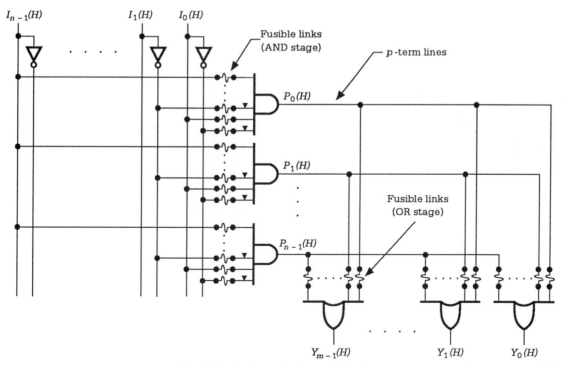

Fig. 4-84. *Logic equivalent circuit for an unprogrammed* $n \times p \times m$ *FPLA showing fusible links in both the OR and AND stages.*

that of Fig. 4-84 but with mask programmed (hence fixed) inputs to the OR "gates," that is, AND programmable only.

A device called an FPLA programmer provides the overvoltage pulses required to "blow" the appropriate fusible links in either stage of an FPLA. Programming an FPAL is simpler and can also be accomplished by using an FPLA programmer. In either case, the blowing of a fusible link to an AND gate is equivalent to disconnecting the AND gate from its external source and then reconnecting that "gate" input lead to $1(H)$. As with the PROM, once the FPLA or FPAL has been programmed, the process cannot be reversed.

Because of the limited number of p-terms in an FPLA, reduced or minimum SOP (or POS) data is nearly always necessary for programming. The data is best presented in the form of a *p-term table* of the type shown in either Fig. 4-85(a) or 4-85(b). A 1 or 0 in an input column of Fig. 4-85(b) represents the connection of the input variable or its complement, respectively, to a p-term line. A dash represents no connection and is interpreted logically as the absence of an input. The 1's and 0's in the output columns of Fig. 4-85(b) represent the presence and absence of p-term storage, respectively. In addition to providing a visual perspective of the I/O patterns that are to be programmed into the FPLA, the use of a p-term table permits the designer to determine easily whether or not the capabilities of the FPLA have been exceeded, a matter of great importance. Although it appears we have done so in Fig. 4-85, it is not necessary that the input, output, or p-term capability of a PLA always be fully utilized.

p-terms	Outputs
$P_0(H) = \bar{I}_1 \cdot I_2 \cdots I_{n-1}(H)$ $P_1(H) = I_0 \cdot \bar{I}_1 \cdots (H)$ $P_2(H) = I_1 \cdot I_2 \cdots (H)$ . . . $P_{p-1}(H) = I_0 \cdot \bar{I}_1 \cdot \bar{I}_2 \cdots \bar{I}_{n-1}(H)$	$Y_0(H) = [P_0 + P_1 + \cdots + P_{p-1}](H)$ $Y_1(H) = [P_0 + \cdots](H)$ $Y_2(H) = [P_1 + P_2 + \cdots](H)$ . . . $Y_{m-1}(H) = [P_0 + P_2 + \cdots](H)$

(a)

p-terms	Inputs I_{n-1}	$\cdots$	I_2	I_1	I_0	Outputs Y_{m-1}	$\cdots$	Y_2	Y_1	Y_0
$P_0 = \bar{I}_1 \cdot I_2 \cdots I_{n-1}$	1	$\cdots$	1	0	$-$	1	$\cdots$	0	1	1
$P_1 = I_0 \cdot \bar{I}_1 \cdots$	$-$	$\cdots$	$-$	0	1	0	$\cdots$	1	0	1
$P_2 = I_1 \cdot I_2 \cdots$	$-$	$\cdots$	1	1	$-$	1	$\cdots$	1	0	0
. . .			.					.		
$P_{p-1} = I_0 \cdot \bar{I}_1 \cdot \bar{I}_2 \cdots \bar{I}_{n-1}$	0	$\cdots$	0	0	1	0	$\cdots$	0	0	1

(b)

Fig. 4-85. *Typical* p-*term tables suitable for programming an* n $\times$ p $\times$ m *FPLA. (a) Expression form with polarization levels. (b) Short form without polarization levels.*

It may be necessary to implement one or more reduced POS expressions with a PLA-type device. This can happen if the POS cover is simpler and more adaptable to PLA design than the SOP cover, or if the design leads naturally to the POS form. In either case the problem is to implement POS (sum) terms with logic that is SOP oriented. To do this, first recall that in MIXED LOGIC notation the relationship between POS and SOP forms is given by

$$Y_{\text{POS}}(H) = \overline{Y}_{\text{POS}}(L) = \overline{Y}_{\text{SOP}}(L) \qquad (4\text{-}21)$$

Next, it is necessary to construct an *s-term* (*s* for sum) *table* similar to the *p*-term table of Fig. 4-85(b). Now, however, terms such as $I_0 + \overline{I}_2 + \cdots + I_{s-1}$ are entered into the *s*-term column on the left, the input columns are completed in MAXTERM code, and the outputs are taken ACTIVE LOW. Thus, in following this procedure, each POS function $Y_{\text{POS}}(H)$ is converted to $Y_{\text{SOP}}(L)$ by the SOP-based PLA device. If it is necessary that these functions be presented ACTIVE HIGH to the next stage, each output $Y_{\text{SOP}}(L)$ must be complemented according to Eqs. (4-21). To do this requires either that INVERTERs be placed on the output lines or that use be made of a PLA device with ACTIVE LOW outputs.

Reduced or minimum cover can be obtained from K-maps provided that the numbers of inputs are not too large as to render this method intractable. The number of outputs does not become a factor unless a multioutput optimization is required to meet the *p*-term number limitation or unless, of course, the number of outputs required exceeds the FPLA output capability. For large numbers of inputs, say, 9 or more, computer-aided reduction or optimization may be required.

An abridged symbolism is useful in representing a programmed FPLA just as it was useful in representing a programmed PROM. In Fig. 4-86 we

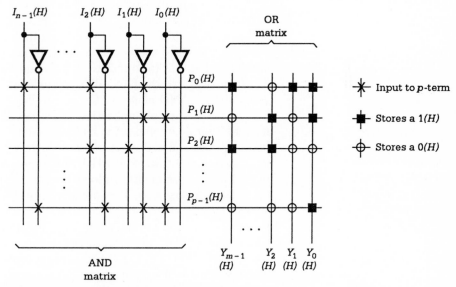

Fig. 4-86. *An abridged symbolic representation of a* n × p × m *FPLA programmed according to the p-term table in Figure 4-85.*

illustrate the use of the abridged symbolism for a general $n \times p \times m$ FPLA programmed in compliance with the reduced SOP data presented in Fig. 4-85. Here, an X along a given horizontal line in the AND matrix represents a connection and hence the presence of an input (complemented or uncomplemented) in a particular p-term; the nodes and circles in the OR matrix have the same meaning as in Fig. 4-82. Thus, the absence of an X in the AND matrix represents a "blown" fusible link in the sense discussed earlier. Also, note that I_{n-1} and Y_{m-1} may or may not represent the MSBs for the input and output WORDs, respectively.

4.11.3 Application of ROMs and PLAs in Combinational Logic Design

Members of the ROM and PLA logic array families are important universal implementing devices which can serve as powerful design tools. So versatile are they that a designer could indiscriminately choose any one array type from either of these two families and make it work. However, when design constraints such as size, speed, and cost are present the selection process is not nearly so simple. We will briefly review some of the important selection criteria involved but only after we have presented a few simple examples of their applications including modular configurations. This order of treatment is followed so that the reader can develop a better understanding of these devices and, in so doing, have a better appreciation for the considerations that must be made in the selection process.

The MAGNITUDE COMPARATOR discussed in Sect. 4.5.1 offers a good example of a device that lends itself nicely to ROM or PROM implementation. We can see from an inspection of Fig. 4-29 that there are no don't care states and hence little opportunity for minimization other than that possible via the EQV operations. So the choice of a FPLA over a PROM would be of no particular benefit. To maintain a reasonable degree of simplicity, we will implement the 2-bit COMPARATOR of Fig. 4-29 with a $2^4 \times 4$ ROM, thus utilizing all 16 of the stored MINTERMs but leaving one of the four outputs unused. The CANONICAL I/O truth table in Fig. 4-29(a) is the program table for the PROM and the results are shown in Fig. 4-87, where the abridged symbology is used. Here, we see that the patterns of 1's and 0's in Fig. 4-29(a) translate as dots and circles, respectively, in Fig. 4-87. Remember that the DECODER "box" represents the nonprogrammable AND matrix (AND plane) of the IC PROM.

The most efficient application of an FPLA would be one for which significant Boolean minimization makes possible the use of most of the FPLA's limited p-term capability. A good example would be the FPLA (or FPAL) implementation of the 4-bit SHIFTER represented in Fig. 4-63. This is a 7-input/4-output combinational logic device capable of left shifting or rotation up to two bit positions. It is obvious from Fig. 4-63(c) that the don't care states permit considerable minimization of the individual outputs but that further multioutput optimization is not possible—there are no shared EPIs. The minimum SOP cover presented in Fig. 4-63(c) is used to construct the p-term table shown in Fig. 4-88, where it is clear that there are a total of 12 p-terms in the four output expressions.

Presented in Fig. 4-89 is the abridged symbolic representation of an $8 \times 16 \times 8$ FPLA (or $8 \times 12 \times 4$ FPAL) which has been programmed by using

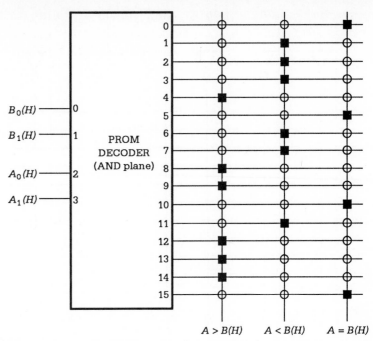

Fig. 4-87. *An abridged symbolic representation of a $2^4 \times 4$ PROM programmed to function as the 2-bit COMPARATOR of Figure 4-29(a).*

p-term	Inputs R A_1 A_0 I_3 I_2 I_1 I_0	Outputs Y_3 Y_2 Y_1 Y_0	p-term	Inputs R A_1 A_0 I_3 I_2 I_1 I_0	Outputs Y_3 Y_2 Y_1 Y_0
$\overline{A_1}\overline{A_0}I_3$	$-$ 0 0 1 $-$ $-$ $-$	1 0 0 0	RA_1I_3	1 1 $-$ 1 $-$ $-$ $-$	0 0 1 0
A_0I_2	$-$ $-$ 1 $-$ 1 $-$ $-$	1 0 0 0	$\overline{A_1}\overline{A_0}I_1$	$-$ 0 0 $-$ $-$ 1 $-$	0 0 1 0
A_1I_1	$-$ 1 $-$ $-$ $-$ 1 $-$	1 0 0 0	A_0I_0	$-$ $-$ 1 $-$ $-$ $-$ 1	0 0 1 0
$\overline{A_1}\overline{A_0}I_2$	$-$ 0 0 $-$ 1 $-$ $-$	0 1 0 0	RA_0I_3	1 $-$ 1 1 $-$ $-$ $-$	0 0 0 1
A_0I_1	$-$ $-$ 1 $-$ $-$ 1 $-$	0 1 0 0	RA_1I_2	1 1 $-$ $-$ 1 $-$ $-$	0 0 0 1
A_1I_0	$-$ 1 $-$ $-$ $-$ $-$ 1	0 1 0 0	$\overline{A_1}\overline{A_0}I_0$	$-$ 0 0 $-$ $-$ $-$ 1	0 0 0 1

Fig. 4-88. *The p-term table for the 4-bit SHIFTER represented in Fig. 4-63.*

the *p*-term table of Fig. 4-88, thus meeting the requirements of the SHIFTER given in Fig. 4-63(c). Notice that in the case of the FPLA, 12 of the 16 *p*-term lines are used while four of the eight outputs are left unused. Obviously more efficient use would be made of the FPAL than the FPLA.

The SHIFTER of Fig. 4-63 could be implemented by using a $2^8 \times 4$ PROM instead of an FPLA or FPAL. However, that would be a very inefficient application of the PROM since only 21 of the $2^8 = 256$ MINTERMs are actually used leaving 235 unused MINTERMs to be stored in the PROM's exhaustive *n*-to-2^n DECODER.

The PROM or FPLA implementation of functions containing XOR terms can be accomplished only if the functions are first converted to CANONICAL or reduced two-level form. For example, the I/O function table of Fig. 4-68 is unsuitable for array logic programming as it stands. However, when the terms for F and C_{out} are converted to two-level SOP form and a mul-

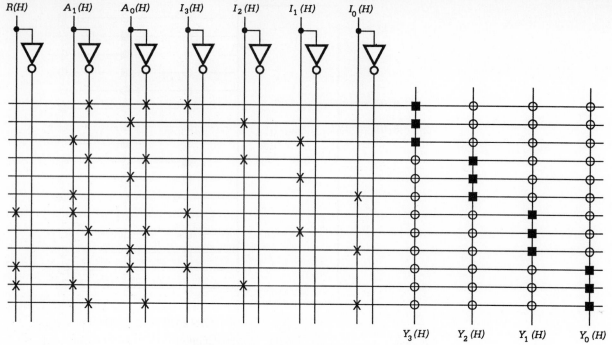

Fig. 4-89. *An abridged symbolic representation of an 8 × 16 × 8 FPLA (or an 8 × 12 × 4 FPAL) programmed to function as the non-arithmetic SHIFTER of Figure 4-63.*

tioutput optimization is carried out, there results 18 p-terms for F and 5 p-terms for C_{out}. Thus, an 8 × 32 × 4 FPLA would easily suffice for the implementation of this ALU. A 2^8 × 2 ROM could also be used but not nearly as efficiently as the FPLA due to the large number of unused MIN-TERMs that must be stored in the ROM's DECODER.

4.11.4 MIXED LOGIC Inputs and Outputs

Although we have not specifically addressed the problem of MIXED LOGIC inputs to array logic devices, the solution is easily deduced from information we have already. For members of the ROM family, CANONICAL data, as in Fig. 4-81, can be altered so as to accommodate the MIXED LOGIC input and output conditions. Or, alternatively, INVERTERs can be placed on the lines of ACTIVE LOW inputs or on the lines of ACTIVE LOW outputs. If the choice is to alter the CANONICAL data, then each 1 or 0 appearing in an ACTIVE LOW input column or ACTIVE LOW output column of the program table must be complemented. Once the program table has been altered in this way to accommodate all MIXED LOGIC I/O conditions, INVERTERs cannot be added for that purpose. In any case, after the AC-TIVE LOW I/O requirements have been satisfied, programming of the ROM proceeds as usual.

Members of the PLA family can also be programmed for MIXED LOGIC inputs simply by complementing each 1 and 0 appearing in an ACTIVE LOW input column of the p-term table in Fig. 4-85(b). The alteration of the AND plane data is equivalent to applying $I(L) = \bar{I}(H)$ to all ACTIVE LOW input variables in the p-term column of Fig. 4-85(a). An ACTIVE LOW output, however, requires the presence of an INVERTER on that output line. The 1's and 0's in an ACTIVE LOW output column of the *p*-term table must *not* be complemented since they do not represent CANONICAL data.

4.11.5 The Multiple Programmable Logic Device Approach to Combinational Logic Design

Occasions arise when the I/O requirements of a design exceed the capabilities of the available array logic device. When this happens the designer may have no alternative but to combine array logic devices in some suitable fashion so as to meet the design requirements.

A useful scheme which will be considered later requires that the outputs of different *programmable logic devices (PLDs)* be ORed together. To accomplish this, the array logic devices (those considered so far) would require the use of OR gates. While this is a workable procedure, it would suffer the disadvantage of limiting the number of ORing operations to the OR gate fan-in capability. A far better scheme makes use of *tristate* and *wired-OR* technology which permits nearly an unlimited number of outputs of different PLDs to be ORed together. This matter is of considerable importance and will be given special attention at this time.

A switching device that operates in either a "transfer" mode or a "disconnect" mode is called a TRISTATE DRIVER (or BUFFER) as discussed in Sect. 1.2.6. A TRISTATE DRIVER with ACTIVE LOW ENABLE is illustrated in Fig. 4-90(a). In the transparent mode $[EN(L) = 1(L)]$ the device can output a digital signal ACTIVE HIGH or ACTIVE LOW depending on the input ACTIVATION level. In the disconnect mode $[EN(L) = 0(L)]$ the TRISTATE DRIVER has a very high impedance, as sensed from the output terminal, thus electrically isolating itself from any device to which it might be wire connected. The two "ON" output states of the transparent mode (ACTIVE HIGH or ACTIVE LOW) and the "OFF" state of the disconnect mode comprise the three states which give the device its name. TRISTATE DRIVERs are also available with ACTIVE LOW outputs or ACTIVE HIGH ENABLEs.

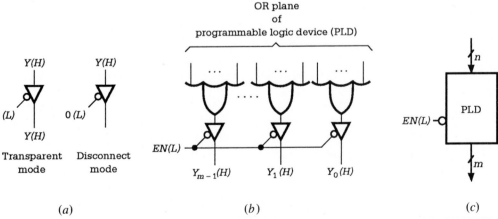

(a)   *(b)*   *(c)*

Fig. 4-90. *TRISTATE DRIVERs with ACTIVE LOW ENABLE. (a) Modes of operation. (b) Use on the OR plane outputs of a programmable logic device (PLD). (c) Block diagram symbol for an n-input/m-output PLD with tristate ACTIVE LOW ENABLE.*

Most commercial PLDs are manufactured with TRISTATE DRIVERs on the outputs as shown in Fig. 4-90(b). The block diagram symbol for such PLDs is given in Fig. 4-90(c). Because of the "ON/OFF" feature of the TRISTATE DRIVER, the outputs of one PLD can be wired (ORed) together with any other PLD and operated in a multiplexed fashion by using a DECODER. In this case the inputs to the DECODER act as the DATA (PLD) SELECT inputs of the multiplexed system. Shown in Fig. 4-91 is one such multiplexed scheme designed to accommodate variables numbering more than the number of PLD inputs that are permitted. As indicated in the figure, not all the available inputs to a given PLD need be used. Not shown but implied is the fact that the PLDs need not be of the same array logic family. Also, each PLD output may (or may not) be connected (wire-ORed) with the corresponding outputs of other PLDs to form *tristate bus* lines. If there are m outputs for each PLD, there can be no more than m tristate bus outputs as is seen from an inspection of Fig. 4-91.

Although Fig. 4-91 satisfies the need to augment the input capability of PLDs, it does not address the problem of output capability. When the number of output functions of a design exceed the output capability of the PLDs in use, a parallel arrangement of the type shown in Fig. 4-92 can be used. This scheme applies to any stage of the multiplexed configuration of Fig. 4-91. It indicates that k PLDs, each of m outputs, yield a maximum of $(k \times m)$ possible outputs per stage. However, it is important for the reader to understand that *these outputs must not be wire-ORed together,* since they are from PLDs which are activated by the same DECODER output; that is, the PLDs are not multiplexed. Interconnecting these outputs could result in

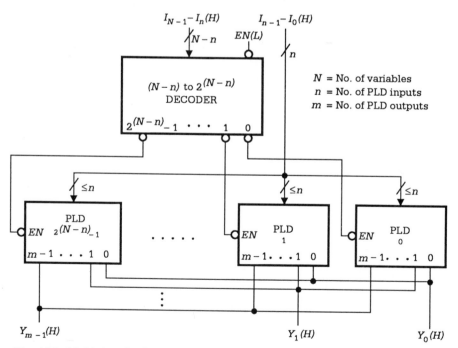

Fig. 4-91. *Multiplexed scheme to generate* m *functions of* N *variables by using* $2^{(N-n)}$ *PLDs each having* n *inputs and* m *outputs.*

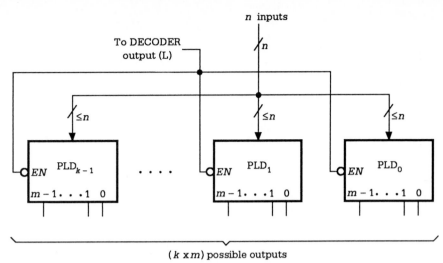

Fig. 4-92. *Scheme for increasing the output function capability from* m *to* k $\times$ m *for any stage of the PLD configuration shown in Figure 4-91.*

permanent damage to the devices involved. Note that, as before, the PLDs need not be of the same array logic family and so need not have the same number of inputs or outputs. The number of inputs to each PLD must be equal to or less than the number n as used in the figure.

Let us illustrate the use of the modular scheme given in Fig. 4-91 with a simple example. Suppose it is necessary to generate three output functions of ten variables by using $2^8 \times 4$ PROMs. Since the number of variables exceeds the number of PLD inputs by two, a 2-to-4 DECODER is required to activate four PROMs one at a time. Presented in Fig. 4-93(a) is one format of a program table that satisfies the requirements of this example. Since four PROMs are required $[2^{(N-n)} = 4]$, the table must be partitioned into four parts, each part being a PROM program. Once programmed, the four PROMs are configured with the 2-to-4 DECODER as shown in Fig. 4-93(b). The

Inputs $I_9\ I_8\ I_7\ \cdots\ I_0$	Outputs $Y_2\ Y_1\ Y_0$	Inputs $I_9\ I_8\ I_7\ \cdots\ I_0$	Outputs $Y_2\ Y_1\ Y_0$
0 0 0 $\cdots$ 0 . . . 0 0 1 $\cdots$ 1	Output data PROM$_0$	1 0 0 $\cdots$ 0 . . . 1 0 1 $\cdots$ 1	Output data PROM$_2$
0 1 0 $\cdots$ 0 . . . 0 1 1 $\cdots$ 1	Output data PROM$_1$	1 1 0 $\cdots$ 0 . . . 1 1 1 $\cdots$ 1	Output data PROM$_3$

(a)

Fig. 4-93. *Modular scheme to generate three output functions of 10 variables using $2^8 \times 4$ PROMs. (a) Partitioned program table.*

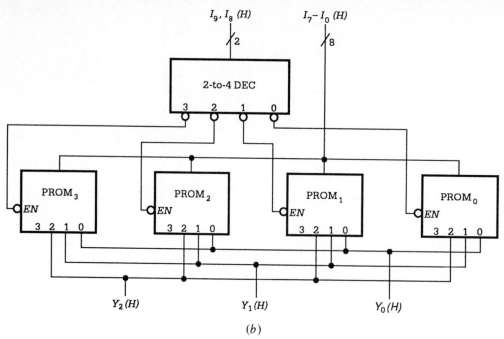

$$I_9, I_8 \ (H) \qquad I_7 - I_0 \ (H)$$

(b)

Fig. 4-93 (contd.). (b) Multiplexed scheme required by the program table of Figure 4-93(a) showing three tristate bus outputs from four $2^8 \times 4$ PROMs.

outputs are wired-ORed together as indicated so as to produce the multiplexed system.

An Example. Application of the scheme given in Fig. 4-91 to PLAs not only increases the input capability of the PLA system but also increases the *p*-term capability in proportion to the number of PLAs that are wired ORed (multiplexed) together. For the purpose of illustration let us suppose it is necessary to implement the simple ALU represented in Fig. 4-68 by using $5 \times 12 \times 4$ FPLAs, assuming these devices exist. Since there are six variables and only five inputs are available, two FPLAs of this size must be used. In the practical sense, one would not implement the ALU in this manner but would instead use a single FPLA of larger dimensions. Following a two-level multioutput optimization of the output functions F and C_{out}, we obtain the *p*-term table of Fig. 4-94(a) which has been partitioned to show how the two FPLAs should be programmed. An examination of the table reveals that there are a total of 24 *p*-terms (relative to input S_0), 11 with $S_0 = 1$, 11 with $S_0 = 0$, and 1 ($S_1 BC_{in}$) which can be expanded as $S_1 BC_{in} S_0 + S_1 BC_{in} \overline{S}_0$ making a total of 24 *p*-terms. So, to use the two $5 \times 12 \times 4$ FPLAs, the 12 *p*-terms for $S_0 = 1$ are assigned to FPLA$_1$ while those for $S_0 = 0$ are assigned to FPLA$_0$. After programming the FPLAs in agreement with the *p*-term table of Fig. 4-94(a), we obtain the configuration shown in Fig. 4-94(b), where the INVERTER serves as a 1-to-2 DECODER.

It should be noted that in the case of the ALU of Fig. 4-68 a two-level, multioutput optimization is necessary to achieve the results of Fig. 4-94 when $5 \times 12 \times 4$ FPLAs are used. Reading the expanded SOP terms from the

P-term	Inputs M S_1 S_0 A B C_{in}						Outputs F C_{out} FPLA			p-term	Inputs M S_1 S_0 A B C_{in}						Outputs F C_{out} FPLA		
$\overline{M}S_1S_0\overline{A}BC_{in}$	0	1	1	0	1	1	1	0	1	$S_1\overline{S}_0\overline{A}BC_{in}$	−	1	0	0	1	0	1	0	0
$\overline{M}\overline{S}_1S_0ABC_{in}$	0	1	0	1	1	1	1	0	0	$S_1S_0AB\overline{C}_{in}$	−	1	1	1	1	0	1	0	1
$\overline{M}S_1S_0AC_{in}$	0	0	1	1	−	1	1	0	1	$\overline{S}_0AB\overline{C}_{in}$	−	−	0	1	0	0	1	0	0
$\overline{M}S_0ABC_{in}$	0	−	1	1	0	1	1	0	1	$\overline{S}_1\overline{S}_0A\overline{C}_{in}$	−	0	0	1	−	0	1	0	0
$\overline{M}\overline{S}_1\overline{S}_0\overline{A}C_{in}$	0	0	0	0	−	1	1	0	0	$S_0\overline{A}\overline{B}\overline{C}_{in}$	−	−	1	0	0	0	1	0	1
$\overline{M}S_0\overline{A}\overline{B}C_{in}$	0	−	0	0	0	1	1	0	0	$\overline{S}_1S_0\overline{A}\,\overline{C}_{in}$	−	0	1	0	−	0	1	0	1
$MS_1\overline{S}_0\overline{A}B$	1	1	0	0	1	−	1	0	0	$S_1\overline{S}_0AB$	−	1	0	1	1	−	0	1	0
MS_1S_0AB	1	1	1	1	1	−	1	0	1	$S_1S_0\overline{A}B$	−	1	1	0	1	−	0	1	1
$M\overline{S}_0A\overline{B}$	1	−	0	1	0	−	1	0	0	S_1BC_{in}	−	1	−	−	1	1	0	1	0, 1
$M\overline{S}_1\overline{S}_0A$	1	0	0	1	−	−	1	0	0	$\overline{S}_0AC_{in}$	−	−	0	1	−	1	0	1	0
$MS_0\overline{A}\overline{B}$	1	−	1	0	0	−	1	0	1	$S_0\overline{A}C_{in}$	−	−	1	0	−	1	0	1	1
$M\overline{S}_1S_0\overline{A}$	1	0	1	0	−	−	1	0	1										

(a)

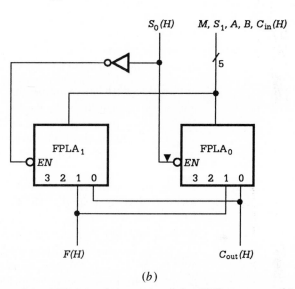

(b)

Fig. 4-94. *A modular FPLA implementatin of the ALU represented in Fig. 4-68. (b) Multiplexed scheme for the* p-*term table of Figure 4-94(a) by using two 5 × 12 × 4 FPLAs.*

table of Fig. 4-68, one obtains a total of 26 *p*-terms, 20 of which correspond to the arithmetic operations ($M = 0$) and the remaining six to the logic operations ($M = 1$). Obviously, without optimizing, two 5 × 12 × 4 FPLAs cannot be configured in any way to satisfy the *p*-term requirements of this ALU—three would be required. On the other hand, a single 6 × 30 × 4 FPLA (if available) would suffice for the generation of F and C_{out}, regardless of whether or not optimized cover is used.

The parallel configuration shown in Fig. 4-92 is useful when it is necessary to increase the output capability of an PLD system. We will illustrate its

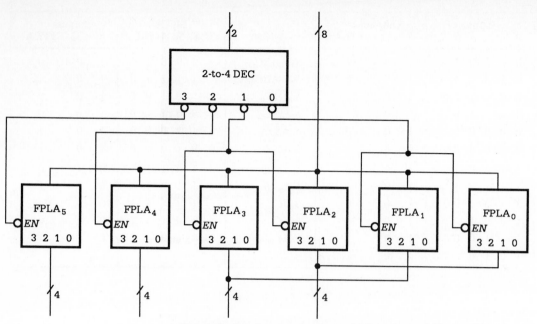

Fig. 4-95. *A scheme to generate 16 functions of 10 variables using 8 × 20 × 4 FPLAs. The p-term requirements are: 8 functions for which 20 < p ≤ 40, and 8 functions for which p is 20 or fewer.*

application with an example that requires augmentation of both the input and output capabilities. Consider how one would implement 16 functions of ten variables by using $8 \times 20 \times 4$ FPLAs under the following p-term constraints: Eight of the functions each require that $20 < p \leq 40$, while the remaining 8 require that $p \leq 20$. Implementation of these 16 functions requires six $8 \times 20 \times 4$ FPLAs. One possible arrangement is given in Fig. 4-95. Notice that the tristate ENABLEs for $FPLA_0$ and $FPLA_1$ are connected together and to a single DECODER output as are those for $FPLA_2$ and $FPLA_3$, in accordance with the scheme of Fig. 4-92. By wire-ORing the outputs together as indicated, we see that the multiplexed configuration is complete, thereby permitting 8 of the output functions to each have up to 40 p-terms. How many partitions of the p-term table would be necessary to program this PLA system properly?

The number of different modular schemes that can be devised continues on and on, limited only by the designer's creativity. It is hoped that the treatment provided in this section will assist the reader in developing imaginative but sound modular design practices. One can expect that the modular approach to design will become more and more important as PLD selectivity improves and costs diminish.

4.12 PROPAGATION TIME DELAY AND HAZARDS

Except for an introductory treatment of propagation time delay given in Sect. 1.3.1, and a few passing remarks made in Sect. 3.12 regarding propagation time delay in multilevel logic circuits, little else has been said concerning this subject. In fact, in this chapter we have treated combinational logic circuits as though they were composed of "ideal" gates in the sense of having no propagation path delays. Now it is necessary that we take one step into the real world and consider that each gate has associated with it a propagation time delay and that, as a result of this delay, undesirable effects may occur.

Under certain conditions unwanted "spikes" can occur in the output signals of combinational logic circuits. These spikes have become known generally as *glitches,* a term that is traceable to the German *glitsche* meaning a "slip" (hence, the slang, error or mishap). Glitches are undesirable because their presence may initiate unwanted processes in a later stage. In some circuits glitches can be avoided through good design practices while in other circuits they are unavoidable and must be dealt with accordingly. In this section we will consider how some glitches occur, demonstrate how they may be avoided, and discuss the more common cases where they cannot be avoided. Since the subject of glitches is also of considerable importance to sequential machine design, we will return to a discussion of these undesirable effects in Chapters 5 and 6.

STATIC HAZARDs. HAZARDs that are produced as a result of asymmetric path delays through INVERTERs (or other gates) are called STATIC HAZARDs. The term "static" is used to indicate that the HAZARD appears in an otherwise steady-state output signal. In Chapter 6 we will need to discuss another important type of HAZARD, called an ESSENTIAL HAZARD, that is unique to asynchronous sequential machines. Both types of HAZARDs require logic operations and an asymmetric path delay, but they differ as to the manner in which they respond to an asymmetric path delay. In Chapter 6 we discuss these differences in considerable detail. For our purposes in this section we will be concerned only with STATIC-type HAZARDs.

Shown in Fig. 4-96(a) is the STATIC HAZARD associated with an AND gate. This HAZARD (a positive glitch in this case) is produced by a $0 \rightarrow 1$

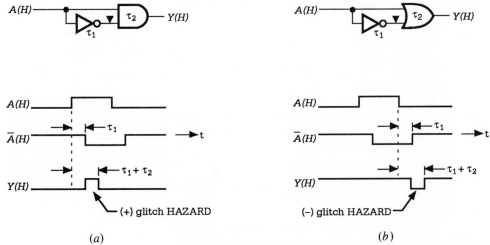

(a)   (b)

Fig. 4-96. *Distributed path delay notation and timing diagrams showing HAZARDs associated with ANDing and ORing operations. (a) Positive glitch HAZARD produced by a $0 \rightarrow 1$ input transition to an AND gate. (b) Negative glitch HAZARD produced by a $1 \rightarrow 0$ input transition to an OR gate.*

transition of the input, A. Here, the INVERTER produces the asymmetric path delay required for HAZARD production. The circuit of Fig. 4-96(a) is also presented in *distributed path delay notation*, meaning that each gate is assigned a certain path delay. Notice that the AND operation is incapable of producing a HAZARD on a $1 \rightarrow 0$ transition of the input.

If we replace the AND gate in Fig. 4-96(a) with an OR gate, we obtain a negative glitch HAZARD on a $1 \rightarrow 0$ transition of the input, A, as shown by the timing diagram in Fig. 4-96(b). The HAZARD associated with an OR gate cannot be produced by a $0 \rightarrow 1$ transition of the input.

The AND gate featured in Fig. 4-96(a) can be replaced in turn by any other gate which performs the AND operation (NAND, NOR, or OR in Fig. 2-25). When this is done the reader can easily verify that, regardless of whether the inputs or outputs are ACTIVE HIGH or ACTIVE LOW, the STATIC HAZARD can be produced only by a $0 \rightarrow 1$ input transition. Similarly, when the OR gate of Fig. 4-96(b) is replaced in turn by any other gate which performs the OR operation (NOR, NAND, or AND in Fig. 2-25), it is observed that HAZARD production occurs only on a $1 \rightarrow 0$ input transition. This leads to the following important conclusion:

- HAZARDs associated with an ANDing operation can be produced *only* on a $0 \rightarrow 1$ input transition. Such HAZARDs may be called STATIC POS HAZARDs.

- HAZARDs associated with an ORing operation can be produced *only* on a $1 \rightarrow 0$ input transition. These HAZARDs can be termed STATIC SOP HAZARDs.

Hence, HAZARD production is associated *not* with a specific gate but rather with the operation that the gate performs. Of course, asymmetric path delays must always be present before a HAZARDous transition is possible.

To illustrate the implications of the conclusions just reached, consider the two-level SOP circuit presented in Fig. 4-97(a). The presence of the INVERTER provides the asymmetric path delay required for HAZARD production. The timing diagram of Fig. 4-97(b) shows the development of the HAZARD on a $1 \rightarrow 0$ transition of the input A. This HAZARD is called an SOP HAZARD because it derives from the ORing (SUM) of ANDed (PRODUCT) terms. Note that the circuit of Fig. 4-97(a) is incapable of POS HAZARD production.

In the K-map of Fig. 4-97(c) is shown the HAZARDous transition from STATE 111 to STATE 011, caused by a $1 \rightarrow 0$ transition of input A. Notice that the HAZARDous transition (see arrow) develops between adjacent states of the two diads representing minimum cover. This suggests that the SOP HAZARD can be eliminated by adding the REDUNDANT diad, BC, to the output expression. Figure 4-98 demonstrates that this is indeed the case. Now, the HAZARD produced in Fig. 4-97 is no longer possible because the redundant input (BC) to the OR stage is maintained HIGH during the $111 \rightarrow 011$ transition. Figure 4-98 also demonstrates that the presence of HAZARD cover hardware does not affect the output response to input data change other than to remove the HAZARD.

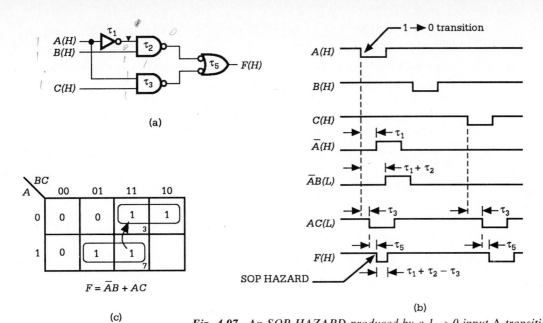

Fig. 4-97. *An SOP HAZARD produced by a 1 → 0 input A transition. (a) Two-level circuit indicating path delays. (b) Timing diagrams showing development of the HAZARD. (c) K-map showing the HAZARDous transition from state 111 to state 011.*

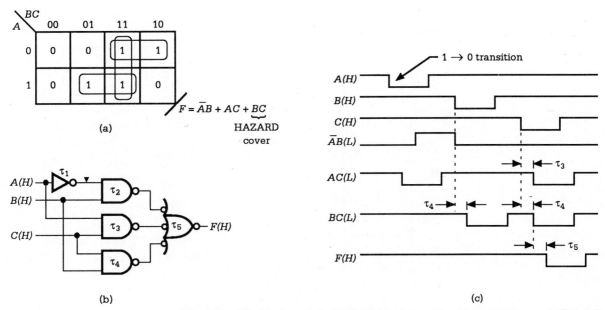

Fig. 4-98. *Elimination of the HAZARDous transition shown in Figure 4-97. (a) K-map showing HAZARD cover (RPI diad BC). (b) Implementation with HAZARD cover. (c) Resulting timing diagram showing HAZARDless output, F(H).*

The procedure for eliminating STATIC HAZARDs suggested by Fig. 4-98 can be generalized as follows:

> *When it is ascertained from a minimum (or reduced) K-map cover that one or more HAZARDous transitions exist, loop out the largest grouping of adjacent cells that will cover each HAZARD and include all such redundant cover in the expression to be implemented.*

An example of the use of this procedure is given in Fig. 4-99. Here, the $0 \rightarrow 1$ (POS) HAZARDous transition in variable A, indicated by the arrow in Fig. 4-99(a), occurs during the state change $010 \rightarrow 110$ and is covered (hence eliminated) by the diad $(\overline{B} + C)$ shown mapped in Fig. 4-99(b).

Similar results are obtained for the SOP HAZARDs shown in Fig. 4-100. The four possible $1 \rightarrow 0$ HAZARDs, indicated by the arrows in Fig. 4-100(a), are eliminated by looping out the RPI diads, $\overline{A}\,\overline{B}\overline{D}$ and $\overline{A}BC$, together with

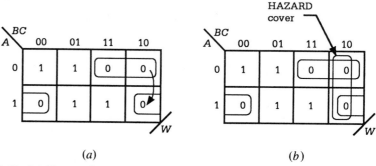

(a)   (b)

Fig. 4-99. *(a) K-map showing a HAZARDous $0 \rightarrow 1$ (POS) transition of the variable* A. *(b) K-map showing HAZARD cover produced by the RPI diad* $(\overline{B} + C)$.

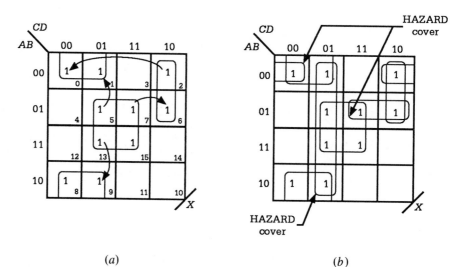

(a)   (b)

Fig. 4-100. *K-maps showing (a) HAZARDous $1 \rightarrow 0$ (SOP) transitions of variables* B, C, *and* D, *and (b) their cover for the function* X.

the RPI quad $\overline{C}D$, all shown mapped in Fig. 4-100(b). Thus, the two HA-ZARDous $1 \to 0$ transitions in variable B occur for state changes $0101 \to 0001$ and $1101 \to 1001$ (cell 5 to cell 1 and cell 13 to cell 9) and are both eliminated by the HAZARD cover $\overline{C}D$. The HAZARDous SOP transition in variable C is produced by a $0010 \to 0000$ state change and is eliminated by the cover $\overline{A}\,\overline{B}\,\overline{D}$. Similar reasoning follows for the HAZARDous transition in variable D from state 0111 to state 0110 for which the cover $\overline{A}BC$ eliminates the HAZARD.

Entered variable K-maps can also be used to reveal and eliminate possible HAZARDous transitions, but their use for such purposes is *not* recommended. Beyond a first-order compression, identification of the HAZARDous transitions in K-maps can be tricky at best and often very difficult, especially when these transitions involve EV inputs. Consider the K-maps in Figs. 4-101 and 4-102 as examples. The function Y in Fig. 4-101(a) has two HAZARDous $0 \to 1$ (POS) transitions, one in variable C, as shown by the arrow, and the other in variable OSC which is not easily identified. Likewise, the function Z in Fig. 4-102(a) has HAZARDous $1 \to 0$ (SOP) transitions in variables D and E, neither of which is easily identified by the K-map method. Determination of the HAZARD cover for the functions Y_{POS} and Z_{SOP} is equally difficult by using the map method. So clearly, simpler

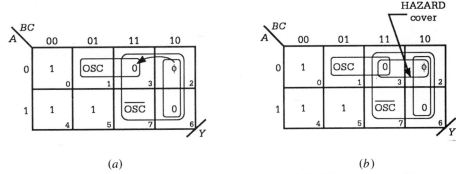

(a) (b)

Fig. 4-101. *K-maps showing (a) HAZARDous $0 \to 1$ (POS) transitions of inputs OSC and C, and (b) their cover by the single RPI diad (A + $\overline{B}$).*

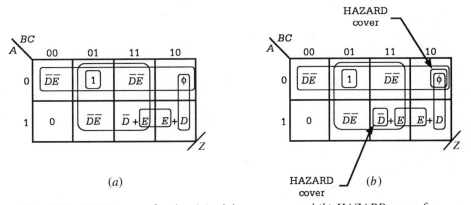

(a) (b)

Fig. 4-102. *EV K-maps showing (a) minimum cover and (b) HAZARD cover for the function Z, and demonstrating that HAZARDous transitions cannot easily be identified by using the K-map method.*

means must be devised for identifying and eliminating HAZARDs in functions which are represented in compressed K-maps, that is, means other than expansion of these functions into conventional maps. We recommend the following procedure which is based on the use of the Boolean expressions representing minimum or reduced cover obtained from K-maps.

Referring back to Figs. 4-97, 4-98, and 4-99, we see a pattern emerge which may or may not be obvious to the reader. The Boolean expression representing any logic circuit having a HAZARDous transition is composed of at least two terms one of which contains an uncomplemented variable and another with the same variable complemented. For example, Fig. 4-97(c) yields a minimum cover given by

$$F = (\overline{A}B) + (AC)$$

Here, the variable A is contained in one term and is complemented in the other. We call the variable A a *coupled variable,* and the terms $\overline{A}B$ and AC are called *coupled terms*. Thus, a coupled term is one that contains a *single* coupled variable. The SOP HAZARD is produced by a $1 \rightarrow 0$ transition of the variable A as indicated by the arrow in the expression for F. The $A \rightarrow \overline{A}$ transition is read in MINTERM code as $1 \rightarrow 0$.

The HAZARD cover required to remove a STATIC SOP HAZARD is determined by ANDing the *residues* of the coupled terms, that is, by ANDing what remains of the two terms after the coupled variable is removed from each. Once obtained, this HAZARD cover must be added to the original expression to be implemented so as to remove the HAZARD. To illustrate, consider the expression for the function F given. Since the coupled variable is A, the residues of the terms $\overline{A}B$ and AC are B and C, respectively. By ANDing the residues and adding the result to the original expression, we obtain

$$F = (\overline{A}B) + (AC) + (BC) \qquad (4\text{-}23)$$

HAZARD
cover

which is the same result obtained in Fig. 4-98.

A similar but dual procedure is required to remove a POS HAZARD. Now, the HAZARD cover is determined by ORing the residues of the coupled terms, that is, by ORing what remains of the two (coupled) terms after the coupled variable is removed from each. For example, consider the minimum cover from Fig. 4-99(a) given by

$$W = (A + \overline{B})(\overline{A} + C)$$

Again the coupled variable is A but this time the HAZARDous transition is produced by a $0 \rightarrow 1$ change in A as read in MAXTERM code. By ORing the residues of the $(A + \overline{B})$ and $(\overline{A} + C)$ terms and by adding the result to the original expression, we obtain the results

4 / Combinational Logic Design

$$W = (A + \overline{B})(\overline{A} + C)\underbrace{(\overline{B} + C)}_{\substack{\text{HAZARD} \\ \text{cover}}}$$

as indicated graphically in Fig. 4-99(b).

The minimum SOP cover as read from Fig. 4-100(a) together with the resulting HAZARD cover give

$$X = (\overline{A}\,C\overline{D}) + (BD) + (\overline{B}\,\overline{C}) + \underbrace{(\overline{A}\,B\overline{D}) + (\overline{A}BC) + (\overline{C}D)}_{\text{HAZARD cover}}$$

Here, the HAZARDous $1 \rightarrow 0$ transitions for variables B, C, and D are indicated by arrows together with the ANDed residues for each.

This same procedure can be applied to functions represented in EV K-maps. All that is necessary is to extract minimum or reduced cover from the map, identify the coupled terms (those containing *single* coupled variables) and obtain the HAZARD cover by ANDing the residues for SOP functions or by ORing the residues for POS cover. By applying this procedure to functions Y and Z of Figs. 4-101 and 4-102, we obtain the results

$$Y = (A + \overline{C} + OSC)(\overline{B} + \overline{OSC})(\overline{B} + C) \cdot \underbrace{(A + \overline{B})}_{\substack{\text{HAZARD} \\ \text{cover}}}$$

and

$$Z = (B\overline{C}D) + (C\overline{D}\overline{E}) + (\overline{A}\,\overline{D}\overline{E}) + (ABE) + (\overline{A}\,\overline{B}C) + \underbrace{(ABC\overline{D}) + (\overline{A}\,B\overline{C})}_{\text{HAZARD cover}}$$

Notice that HAZARD cover is taken as the most minimum cover permitted by the K-map. Thus, for Y both HAZARDous transitions are covered by the RPI quad $A + \overline{B}$ as indicated by the looping of cells 2 and 3 in Fig. 4-101(b). For the function Z, $\overline{A}B\overline{C}$ is the most minimum HAZARD cover for the $D \rightarrow \overline{D}$ transition as shown in Fig. 4-102(b).

STATIC HAZARDs can be eliminated by HAZARD cover only if the internal logic of the device can be altered. MSI and LSI devices such as DECODERs, MUXs, and ROMs present unique cases where alteration of the cover is not possible—all have exhaustive DECODER stages that are not user-accessible. This points to yet another advantage in the use of PLAs.

By the proper programming of a PLA, we can eliminate all HAZARDs assuming, of course, that the HAZARD cover can be found with a reasonable effort.

It is important for the reader to understand that the presence of a STATIC HAZARD in a combinational logic circuit may not pose a problem. This is so for two possible reasons. First, a known HAZARD in a combinational logic circuit may never produce a glitch simply because that particular transition never occurs. This can be the case in sequential circuits where only certain state-to-state transitions are allowed. Second, even if a HAZARDous transition does occur, the glitch may not develop sufficiently to cause errors in the next stage. We will return to the subject of HAZARDs in Sect. 5.8.3, but a detailed treatment of this subject is left to Chapter 6.

FUNCTION HAZARDs. In the expression for Z just given, it is observed that pairs of terms such as $B\overline{C}D$ and $C\overline{D}\overline{E}$, or $\overline{A}\overline{D}\overline{E}$ and ABE each contain two coupled variables. These pairs of terms cannot produce STATIC HAZARDs of the type we have just described. Also, their ANDed residues are always logic 0—as are the ORed residues logic 1 for such pairs of terms in POS expressions. But these pairs of terms can produce another type of HAZARD called a FUNCTION HAZARD, which is also static in the sense that it occurs in an otherwise steady-state output signal. FUNCTION HAZARDs result when an attempt is made to change two or more coupled variables in a pair of terms at the same time. Potential HAZARDs of this type are very common. However, they can be easily avoided if care is taken not to permit the coupled variables to change too close together in time. Section 5.8.2 deals with a type of FUNCTION HAZARD that can be produced in sequential machines.

ANNOTATED REFERENCES

For combinational logic design of arithmetic circuits and code conversion, there appears to be no better single source of information than the book of Kostopoulos. Beginning with Chapter 7, the author covers small-scale and medium-scale designs of adders, subtractors, multipliers, and dividers. Powers and roots are also considered. Chapter 12 is devoted to the binary-coded decimal system while Chapter 14 deals mainly with code conversion.

Kostopoulos, G. K., *Digital Engineering*, John Wiley, New York, 1975.

The subjects of adders, subtractors, comparators, multiplexers, decoders, encoders, wired logic, and propagation delay are adequately covered in Chapter 4 of the book by Fletcher. Chapter 8 of Fletcher's book provides a good treatment of array logic devices and their application.

Fletcher, W. I., *An Engineering Approach to Digital Design*, Prentice Hall, Englewood Cliffs, N.J., 1980.

A fair treatment of combinational logic design is given in Chapter 3 of the book by Comer. The author uses entered variable mapping methods and the IEEE standard symbology.

Comer, D. J., *Digital Logic and State Machine Design*, 2nd ed., Holt, Rinehart and Winston, New York, 1990.

The design of arithmetic and logic units (ALUs) is covered in texts by Mano (Chapter 9) and by Johnson and Karim (Chapter 5). The Mead-Conway (M-C) ALU, designed by VLSI implementation, is described by Mead and Conway in their book. The M-C ALU provides the basis for the ALU described in Appendix 4.2 of this text.

Mano, M. M., *Computer Engineering Hardware Design*, Prentice Hall, Englewood Cliffs, N.J., 1988.

Johnson, E. L., and M. A. Karim, *Digital Design*, Prindle, Weber and Schmidt, Boston, Mass., 1987.

Mead, C., and L. Conway, *Introduction to VLSI Systems*, Addison-Wesley, Reading, Mass., 1980.

An adequate coverage of combinational shifters seems to be somewhat scarce. This subject is covered in the book by Ercegovac and Lang (Chapter 4) and to a lesser extent in the book by Shiva (Chapter 6).

Ercegovac, M. D., and T. Lang, *Digital Systems and Hardware/Firmware Algorithms*, John Wiley, New York, 1985.

Shiva, S. G., *Introduction to Logic Design*, Scott, Forsman, Glenview, Ill., 1985.

It is worth mentioning that the book by Ercegovac and Lang covers most of the subjects associated with combinational logic design. However, this book uses a high-level, algorithmic specification notation that may not be familiar to the reader.

The subject of combinational hazards is adequately presented in books by McCluskey and Dietmeyer. The book by Unger, which is lemma oriented, provides an extensive treatment of this subject.

McCluskey, E. J., *Logic Design Principles with Emphasis on Testable SemiCustom Circuits*, Prentice Hall, Englewood Cliffs, N.J., 1986.

Dietmeyer, D. L., *Logic Design of Digital Systems*, 2nd ed., Allyn & Bacon, Boston, 1978.

Unger, S. H., *Asynchronous Sequential Switching Circuits*, Wiley-Interscience, New York, 1969 (reissued by Krieger and Malabar, Florida, 1983).

References for software tools used in the minimization of Boolean functions are given in the Annotated References at the end of Chapter 3.

A number of useful data manuals and handbooks are available for semicustom MSI and LSI logic design. These include

Texas Instruments, *The TTL Data Book*, in several volumes, Texas Instruments, Inc., Dallas, Tex., 1986.

Texas Instruments, *HIGH-Speed CMOS Logic*, Texas Instruments, Inc., Dallas, Tex., 1988.

National Semiconductor, *CMOS Databook*, National Semiconductor Corp., Santa Clara, Calif., 1984.

LSI Logic Corporation, *CMOS Macrocell Manual*, LSI Logic Corporation, Milpitas, Calif., 1985.

PROBLEMS

4.1 Define the following terms:
(a) Combinational logic (f) Radix
(b) MIXED LOGIC (g) ACTIVE HIGH
(c) VLSI (h) Chip
(d) Natural binary (i) Byte
(e) Most significant bit (j) Positional weight
(MSB)

4.2 Convert the following decimal numbers to binary by using positional weights (see Sect. 4.2.1):
(a) 5 (b) 14 (c) 39 (d) 107.25 (e) 0.6875

4.3 Convert the following binary numbers to decimal by using positional weights:
(a) 0110 (d) 110110.11
(b) 1011 (e) 0.0101
(c) 11101

4.4 Expand the binary numbers of Problem 4.3 in polynomial form.

4.5 Use two NAND gates and one OR gate (nothing else) to design a circuit that will indicate the increment of a 2-bit binary number by 3. Let the inputs be ACTIVE LOW and outputs be ACTIVE HIGH.

4.6 Find the minimum AND/INV logic circuit that will indicate the square of a 2-bit binary number, AB. Assume that the inputs and outputs are all ACTIVE HIGH.

4.7 Design a gate-minimum circuit that will indicate the product of a positive 2-bit binary number by 3. Assume that the inputs arrive ACTIVE LOW and that the outputs are ACTIVE HIGH.

4.8 Use a FULL ADDER (nothing else) to implement a circuit that will indicate the binary equivalent of $(x^2 + x + 1)$, where $x = AB$ is a 2-bit binary number. Assume that the inputs and outputs are all AC-TIVE HIGH.

4.9 Design a three-input logic circuit that will cause an output F to go ACTIVE under the following conditions:

All inputs are logic 1
An odd number of inputs are logic 1
None of the inputs are logic 1

Assume that the inputs are $A(H)$, $B(H)$, and $C(H)$ and that the output is issued ACTIVE LOW. Use only the following hardware.

One FULL ADDER

Two NOR gates

4.10 Prove that the logic circuit in Fig. P4-1 is that of a FULL ADDER. [Hint: Read the circuit, convert the results to CANONICAL form, and then compare with the truth table of Fig. 4.7(c).]

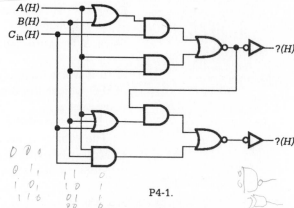

P4-1.

4.11 Use the logic circuit in Fig. P4-1 (exactly as it is given) to construct a staged CMOS implementation of the FULL ADDER. Refer to Examples 2-18 and 2-19 in Sect. 2.10 if assistance is needed.

4.12 Add the following binary numbers and give each answer in both binary and decimal (see Sect. 4.4):
(a)   10
    + 11
(b)   101
    + 11
(c)   10111
    + 01110

(d)   101101.11
    + 011010.10
(e)   0.1100
    + 1.1101

4.13 Perform the binary subtractions below by using 2's complement arithmetic and give the result in 2's complement notation (where appropriate) and decimal. (See Sect. 4.4.7.)

(a) $11 - 01$ (b) $1100 - 0101$ (c) $111011 - 11001$
(d) $10000 - 10101$ (e) $00100 - 10001$

4.14 Repeat Problem 4.13 in 1's complement arithmetic.

4.15 Use OR/XOR logic (nothing else) to design a gate-minimum circuit for a 4-bit binary-to-2's-complement (or vice versa) converter. Let $B_3B_2B_1B_0$ and $T_3T_2T_1T_0$ represent the binary and 2's complement WORDs, respectively. (Hint: Construct second-order EV K-maps from the truth table and use XOR patterns.)

4.16 Without working too hard, design the binary-to-2's complement converter of Problem 4.15 by using four HALF ADDERs and four INVERTERs (nothing else).

4.17 Find the product of the following binary numbers by using the "multiply, shift, and add" algorithm used in decimal arithmetic. Give each answer in both binary and decimal. Check each result with decimal arithmetic. [Hint: In multiplying binary numbers with a radix point, carry out the multiplication as though the radix point were absent then place it in the appropriate position in the binary number by counting bits beginning with the LSB (analogous to decimal arithmetic).]

(a) $\begin{array}{r} 11 \\ \times\ 10 \\ \hline \end{array}$ (b) $\begin{array}{r} 0101 \\ \times\ 11 \\ \hline \end{array}$ (c) $\begin{array}{r} 1111 \\ \times\ 111 \\ \hline \end{array}$

(d) $\begin{array}{r} 0101.11 \\ \times\ \ \ \ 10 \\ \hline \end{array}$ (e) $\begin{array}{r} 1001.101 \\ \times\ \ \ 100.1 \\ \hline \end{array}$

4.18 Find the quotient for each of the following binary numbers by using the "divide, subtract, and borrow" algorithm as used in decimal long division. Give the results in both binary and decimal. Check the results with decimal arithmetic.

(a) $110 \div 10$ (b) $1100 \div 100$
(c) $111111 \div 1001$ (d) $10010 \div 1000$
(e) $11001.1 \div 011.11$

(Note: In part (e) carry out the quotient to the 2^{-2} bit and indicate the remainder.)

4.19 Analyze the following circuits by filling in the logic levels (1's and 0's) at all stages as in the example of Fig. 4-26. Check each result by carrying out the indicated arithmetic. Assume that all inputs and outputs are ACTIVE HIGH.

(a) Fig. 4-10: $A = 1011$, $B = 0111$, $C_o = 0$
(b) Fig. 4-12: $A = 011$, $B = 111$, $C_o = 1$

(c) Fig. 4-14: $A = 110$, $B = 101$, $C = 011$, $D = 100$
(d) Fig. 4-16(c): $A = 01$, $B = 11$, $A/S = 1$
(e) Fig. 4-21: $A = 11$, $B = 10$

4.20 Design a 4-bit comparator by using two 4-bit subtracters and one NOR gate (nothing else). [Hint: In a subtracter, a final borrow of 1 indicates (minuend) < (subtrahend), but a final borrow of 0 indicates (minuend) $\geq$ (subtrahend).]

4.21 Design a gate-minimum combinational logic circuit that will indicate the binary equivalent of $N > 9$, $N < 9$, or $N = 9$, where N is a 4-bit binary number.

4.22 Use XOR/EQV logic (nothing else) as needed to design the following 8-bit PARITY circuits based on logic 1. Assume that all inputs arrive from positive logic sources.

(a) An odd PARITY DETECTOR
(b) An even PARITY DETECTOR
(c) An odd PARITY GENERATOR
(d) As in parts (a), (b), and (c) but based on logic 0

4.23 Design a circuit that will detect any odd number of errors in a 4-bit data bus from source to receiver. Assume that a $1(H)$ from an error check output indicates a single error.

4.24 Implement the following system of three functions with a single 4-to-16 DECODER. Assume that the DECODER outputs are all ACTIVE LOW, that the inputs arrive ACTIVE HIGH, and that all external gates have input capabilities fixed at 6.

$$F_1(H) = [AB\overline{C}D + B(\overline{A}C + D)$$
$$+ \ \overline{B}CD + \overline{A}CD](H)$$

$$F_2(L) = [BC\overline{D} + (A \oplus B) + AB\overline{C}D$$
$$+ \ \overline{A}\,\overline{B}\,\overline{C}D](L)$$

$$F_3(H) = [(A + B + \overline{D})(\overline{A} + \overline{C} + D)(C + \overline{D})$$
$$\cdot(B + \overline{C} + D)(\overline{A} + \overline{B})](H)$$

4.25 A function is given by the CANONICAL form

$$F(A,B,C) = \sum m(0, 1, 6, 7)$$

(a) Implement this function by using only a 3-to-8 DECODER and one NAND gate (nothing else). Assume that the inputs are $A(H)$, $B(L)$, and $C(H)$ and that the output is issued ACTIVE LOW. The DECODER has ACTIVE LOW outputs.

(b) Use two 2-to-4 DECODERs, a NAND gate, and one INVERTER to accomplish the same result as in part (a).

(c) List other CANONICAL and minimum forms for the function above that can be read from the circuits in parts (a) and (b).

4.26 Stack 3-to-8 line DECODERs to produce the following:

(a) A 4-to-16 line DECODER

(b) A 5-to-32 line DECODER

Use whatever additional hardware is needed to stack the DECODERs.

4.27 Design with minimum NOR/INV logic a three-input PRIORITY ENCODER that will operate according to the following schedule:

Input C—highest priority, encoded as 10

Input A—middle priority, encoded as 01

Input B—lowest priority, encoded as 00

Assign 11 to the INACTIVE state. Assume that the inputs arrive ACTIVE HIGH and that the outputs are ACTIVE LOW.

4.28 Shown in Fig. P4-2 is the block diagram and truth table for a priority interrupt. A priority interrupt is a special PRIORITY ENCODER that is used in microprocessors and computers. Its function is to issue an interrupt request signal when prioritized input signals arrive and are encoded. Design the priority interrupt by using NAND/INV logic.

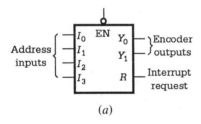

(a)

EN	I_3 I_2 I_1 I_0	Y_1 Y_2 R
0	X X X X	0 0 0
1	1 X X X	1 1 1
1	0 1 X X	1 0 1
1	0 0 1 X	0 1 1
1	0 0 0 1	0 0 1
1	0 0 0 0	X X 0

(b)

Fig. P4-2.

4.29 Design a six-input PRIORITY ENCODER that can be bit-sliced and that will operate according to the schedule shown in the accompanying collapsed truth table. Use only NAND/NOR/AND logic (no INVERTERs) to obtain a gate-minimum result. Assume that all inputs and outputs are ACTIVE LOW.

EI I_5 I_4 I_3 I_2 I_1 I_0	GS Y_3 Y_2 Y_1 EO
0 X X X X X X	0 0 0 0 0
1 1 X X X X X	1 1 1 0 0
1 0 1 X X X X	1 1 0 1 0
1 0 0 1 X X X	1 1 0 0 0
1 0 0 0 1 X X	1 0 1 1 0
1 0 0 0 0 1 X	1 0 1 0 0
1 0 0 0 0 0 1	1 0 0 1 0
1 0 0 0 0 0 0	0 1 1 1 1

X = irrelevant input.

4.30 Convert the decimal numbers in Problem 4.2 to NBCD by using positional weights. (See Sect. 4.7.1.)

4.31 Convert the binary numbers in Problem 4.3 to NBCD. (See Sects. 4.2.1 and 4.7.1.)

4.32 Convert the following NBCD numbers to binary by using positional weights.

(a) 00010011

(b) 01010111

(c) 000101000110

(d) 01001000.00100101

(e) 0000.100001110101

4.33 Prove that a self-complementing unit-distance code is not possible.

4.34 Use gate-minimum logic to design the decimal-to-NBCD converter shown in Fig. P4-3.

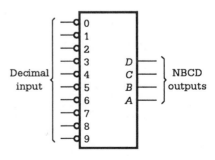

Fig. P4-3.

4.35 Without working too hard, use four FULL AD-DERs (nothing else) to design a circuit that will convert XS3 to NBCD. Assume that all inputs and outputs are ACTIVE HIGH.

4.36 (a) Design a gate-minimum logic circuit that will generate 4-bit 9's complement binary from NBCD code. Assume that the inputs and outputs are all ACTIVE HIGH.

(b) Design the converter of part (a) by using a single 4-to-16–line DECODER together with the necessary external hardware. Assume that the inputs are $A(H)$, $B(H)$, $C(L)$, and $D(L)$.

4.37 Design a gate-minimum NBCD-to-10's complement converter. Assume that the inputs and outputs are ACTIVE HIGH and that all unused input states are indicated by INACTIVE outputs. (Hint: Check for XOR patterns in second-order EV K-maps.)

4.38 Design a gate-minimum NBCD to 84-2-1 code converter. Assume that the inputs and outputs are ACTIVE HIGH and that false data is not rejected. (Hint: A minimum gate/input tally of 5/10 (excluding INVERTERS) results by using XOR patterns in second-order EV K-maps.)

4.39 (a) Design an 8-bit binary-to-NBCD converter based on the shift-left/add 3 algorithm given in Sect. 4.7.2. To accomplish this, first design from a truth table the minimum NAND/INV logic required to convert the binary numbers equivalent of decimal 0 through 19 to NBCD. Note that the shift-left/add 3 algorithm is inherent in the truth

table and that the LSB in binary is the same as the LSB in NBCD.

Next, cascade these modules as in Fig. P4-4 so as to carry out the shift-left/add 3 algorithm for an 8-bit binary WORD. The modules are offset, as shown, to satisfy the shift left requirement of the modules. All inputs and outputs are assumed to be ACTIVE HIGH.

(b) Use the module logic expressions obtained in part (a) to analyze the converter of Fig. P4-4 for a binary input equivalent to 159_{10}.

4.40 (a) Design an 8-bit NBCD-to-binary converter based on the shift-right/subtract 3 algorithm given in Sect. 4.7.2. To do this, first design from a truth table the minimum NAND/INV logic required to convert NBCD to the binary equivalent of decimal 0 through 19. Note that the shift-right/subtract 3 algorithm is inherent in the truth table and that the LSB in NBCD is the same as the LSB in binary.

Next, cascade four of these modules as shown in Fig. P4-5 so as to carry out the shift-right/subtract 3 algorithm for an 8-bit binary WORD. Assume that all inputs and outputs are ACTIVE HIGH.

(b) Use the logic expressions obtained for the module in part (a) to analyze the converter of Fig. P4-5 for an NBCD input equivalent to 93_{10}.

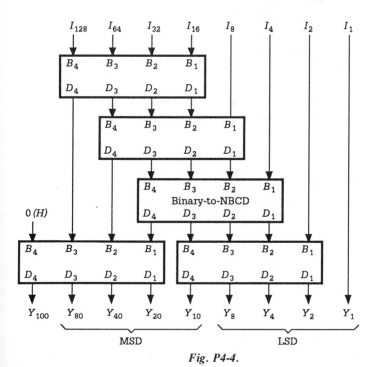

Fig. P4-4.

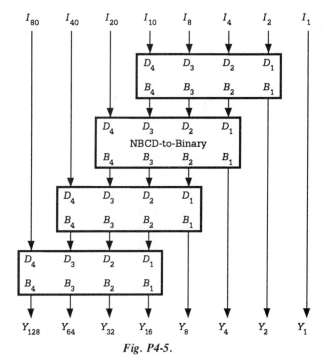

Fig. P4-5.

4.41 Use the following algorithm for NBCD arithmetic to carry out the operations represented here in decimal:

Add two NBCD numbers by using binary addition. If their sum is 1001 (9_{10}) or less, the sum is valid and no correction is made. If their sum is greater than 9, add 0110 (6_{10}) to the sum to give the correct NBCD result, and send the carry to the next most significant digit (MSD) stage.

Remember that the sum of two NBCD numbers cannot exceed 10010 (18_{10}) and that a carry cannot be generated in NBCD subtraction.

(a)	3	(b)	19	(c)	87
	+ 4		+ 9		+ 48

(d)	59	(e)	247
	− 27		− 151

4.42 Use the following algorithm for XS3 arithmetic to carry out the operations represented in Problem 4.41:

Add XS3 numbers by using binary addition. If there is no 1 carry from the four-bit sum, correct that sum by subtracting 0011 (3_{10}). If a 1 carry is generated from the 4-bit sum, correct that sum by adding 0011.

Remember that XS3 states less than 3 or greater than 12 are not used and that the sum of two XS3 numbers cannot exceed 24_{10}. Also, note that a carry cannot be generated in XS3 subtraction.

4.43 Use two 4-bit binary ADDERs and the necessary AND/OR correction logic to design a single-digit NBCD ADDER (shown in Fig. P4-6) that can be bit-sliced to produce an n-bit NBCD ADDER. Use the algorithm stated in Problem 4.41 to accomplish this. Remember that the sum of two NBCD numbers can never exceed 10010 (18_{10}). (Hint: First find the minimum logic for a correction parameter X that will indicate when the sum is greater than 9 but less than 19. Then use one of the two 4-bit ADDERs to add 0110 to the sum conditionally on X.)

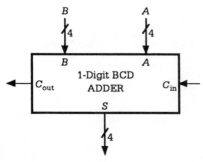

Fig. P4-6.

4.44 Alter the design of Problem 4.43 so as to create a one-digit NBCD ADDER/SUBTRACTOR that can be bit-sliced to form an n-digit NBCD ADDER/SUBTRACTOR. (Hint: Add an ENABLE to the correction logic of Problem 4.43 and use 2's complement arithmetic to perform NBCD subtraction operations. Remember that a 1 carry cannot be generated in NBCD subtraction.)

4.45 Use two 4-bit binary ADDERs and the necessary correction and external logic to design a single-digit XS3 ADDER (as in Fig. P4-6) that can be cascaded to form an n-digit XS3 ADDER. To do this, use the algorithm stated in Problem 4.42. Remember that XS3 states less than 3 or greater than 12 are not used and that the sum of two XS3 numbers cannot exceed decimal 24. (Hint: First, find the minimum logic for the correction parameter X that will indicate when the sum is greater than 12 but less than 25, the range over which a 1 carry is generated.)

4.46 Alter the design of Problem 4.45 so as to create a one-digit XS3 ADDER/SUBTRACTOR that can be bit-sliced to form an n-digit XS3 ADDER/SUBTRACTOR. Note that a 1 carry cannot be generated in XS3 subtraction.

4.47 (a) Use the results of Problem 4-39 to design a one-digit NBCD multiplier that can be bit-sliced. To do this, first design a 4×4 binary multiplier; then use the binary-to-NBCD converter to convert the result to 2-digit NBCD form. The multiplier should be designed according to the scheme illustrated in Fig. P4-7. Follow a multiplication format similar to that given in Fig. 4-18(a) except now with eight product bits instead of four. Assume that the inputs and outputs are all ACTIVE HIGH. (Hint: The partial products can be generated as in Fig. 4-21.)

(b) Analyze the results of part (a) by using the NBCD equivalent of 15×15.

4.48 Use the NBCD multiplier symbol given in Fig. P4-7 and the results of Problem 4.43 to construct a 2×2 digit NBCD multiplier. To do this, form a 2×2 array of one-digit multipliers and connect them properly to a 4-digit NBCD ADDER. Indicate the digit orders of magnitude (10^0, 10^1, 10^2, and 10^3) at all stages of the multiplier.

4.49 Analyze the network shown in Fig. P4-8 by obtaining the output logic expressions for Y_3, Y_2, Y_1, and Y_0.

4.50 Implement a 16-to-1 line MUX by using four 4-to-1 line MUXs with ACTIVE LOW ENABLEs, a 2-to-4 line DECODER, and an OR gate (nothing else).

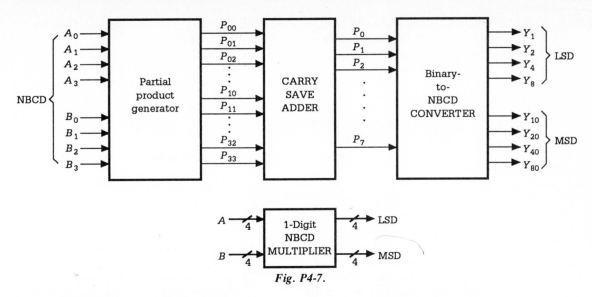

Fig. P4-7.

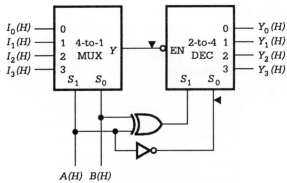

Fig. P4-8.

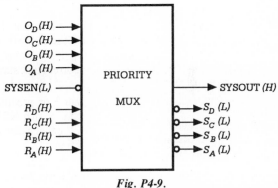

Fig. P4-9.

4.51 Design the priority MULTIPLEXER (PM) in Fig. P4-9 so that it will connect, on request, the appropriate line O_A, O_B, O_C, or O_D from any one of four devices to a system output line (SYSOUT) subject to the following priority schedule:

> Device A—First (highest) priority
>
> Device B—Second priority
>
> Device C—Third priority
>
> Device D—Fourth (lowest) priority

The priority MULTIPLEXER receives device request signals on lines R_A, R_B, R_C, and R_D and sends back to the devices signals S_A, S_B, S_C, or S_D, indicating that a given device request was accepted. The PM must have an ACTIVE LOW system ENABLE [SYSEN(L)]. (Hint: Use a PRIORITY ENCODER, a 4-to-1 line MUX, and a 2-to-4 DECODER all with ACTIVE LOW ENABLEs.)

4.52 Design a Gray-to-NBCD code converter by using four 4-to-1 line MUXs and a gate-minimum external logic. The inputs are $A(H)$, $B(L)$, $C(H)$, and $D(H)$, and the outputs are all ACTIVE HIGH.

4.53 Consider the four-variable function Z and the available hardware given.

C D	Z	Available Hardware
00	$\overline{A}$	One 2-to-1 line MUX
01	$\overline{A} \cdot \overline{B}$	One NAND gate
10	$\overline{A} + B$	One AND gate
11	$A \odot B$	

Use the available hardware (nothing else) to implement the function Z. Assume that the inputs arrive as $A(L)$, $B(H)$, $C(H)$, and $D(H)$ and that the output is ACTIVE HIGH. (Hint: First find the absolute minimum expression for Z.)

4.54 Carry out the following binary shifts and indicate the decimal value before and after the shift. Where appropriate indicate the round-off error (see Sect. 4.9.2).

(a) 0110   1sl

(b) 101100   2sr

(c) 00011.011   3sl

(d) 0110.110   2sr

```
     ↑
     └─spill bit ─────┐
                      ↓
```

(e) $-21.4375 =$ 01010.1001   2sr
                (2's complement)

4.55 In Fig. P4-10 is shown a network containing several combinational logic devices.

(a) Complete the following truth table:

A	B	C_{out} S_3 S_2 S_1 S_0
0	0	
0	1	
1	0	
1	1	

(b) Use the results of part (a) to construct a DE-CODER implementation with a single OR gate that will accomplish the same task as the network. Assume that the DECODER has AC-TIVE HIGH outputs.

4.56 Find the gate-minimum external logic for a 4-bit shifter that will cause it to operate according to the following table. Show the resulting logic circuit connected to the shifter. (Hint: Make use of XOR-type patterns in second-order K-maps.)

Shifter Input	Action
Even 1 parity	Shift right 1 position
Odd 1 parity	Shift left 2 positions
1111 (exception)	Transfer

4.57 Shown here is the I/O function truth table for a special purpose 1-bit ALU which has four arithmetic operations ($M = 0$) and four logic operations ($M = 1$). The arithmetic operations are expressed in 2's complement notation.

(a) Design this ALU by using two 4-to-1 MUXs, a FULL ADDER, and a gate-minimum external logic. Assume that all inputs and outputs are ACTIVE HIGH.

(b) Design this ALU by using a gate-minimum logic without MUXs. Note that this design includes the use of compound XOR type patterns which lead to gate/input tallies of 9/18 and 7/16 for F and C_{out}, respectively.

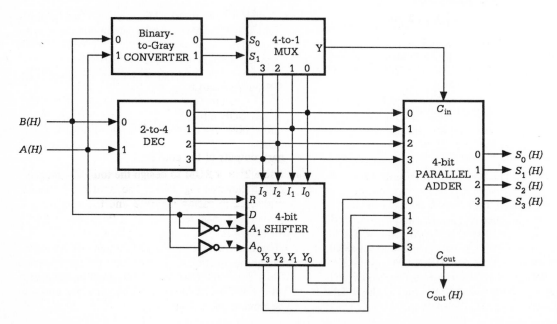

Fig. P4-10.

M	S_1	S_0	F	C_{out}	Operation
0	0	0	$A \oplus B \oplus C_{in}$	$C_{in}(A \oplus B) + AB$	A plus B
0	0	1	$A \oplus \overline{B} \oplus C_{in}$	$C_{in}(A \odot B) + A\overline{B}$	A minus B*
0	1	0	$(A + B) \oplus C_{in}$	$C_{in}(A + B)$	A plus ($\overline{A}B$)
0	1	1	$(\overline{A}\,\overline{B}) \oplus C_{in}$	$C_{in}\overline{B} + A$	A plus ($A + \overline{B}$)
1	0	0	$A \oplus B$	ϕ	A XOR B
1	0	1	$A \odot B$	ϕ	A EQV B
1	1	0	$A + B$	ϕ	A OR B
1	1	1	$\overline{A} \cdot \overline{B}$	ϕ	$\overline{A}$ AND $\overline{B}$

* Assumes 2's complement arithmetic.

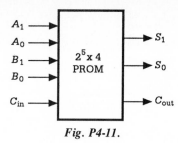

Fig. P4-11.

4.58 Analyze the PALU of Fig. A4-2 (Appendix 4.2) and complete the following truth table for the conditions $D = 1011$, $E = 1001$, $R = 0110$, and $C_{in} = 1$. In the analysis, follow the example in Fig. 4-26 by placing the correct logic levels at each stage.

A	B	F	C_{out}
0	0		
0	1		
1	0		
1	1		

4.59 A $2^4 \times 4$ ROM is to be used to implement the following system of three functions:

$$F_1(A,B,C) = \prod M(0, 2, 4)$$

$$F_2(A,B,C,D) = \prod M(3 - 12)$$

$$F_3(A,B,C) = \prod M(2, 3, 5, 7)$$

(a) Construct the PROGRAM TABLE for this ROM if the inputs are $A(H)$, $B(H)$, $C(L)$, and $D(H)$ and the outputs are $F_1(H)$, $F_2(L)$, and $F_3(H)$. You are not permitted to use INVERTERs.

(b) Use the program table in part (a) to construct an abridged symbolic representation of the ROM (see Fig. 4-82).

4.60 A $2^5 \times 4$ PROM is to be used to implement the 2-bit (fast) ADDER in Fig. P4-11 which can be bit-sliced. Construct the program table for this ROM. Assume that all inputs and outputs are ACTIVE HIGH. (See Appendix 2.1 for decimal-to-binary conversion.)

4.61 The following three functions are to be implemented by using a $4 \times 8 \times 4$ FPLA.

$$F_1 = A\overline{B} + \overline{A}B\overline{C} + \overline{B}C + AC$$

$$F_2 = A \oplus (\overline{B}C)$$

$$F_3 = \prod M(1, 3, 6)$$

(a) Construct the p-term table for the three functions. Assume that the inputs arrive as $A(H)$, $B(H)$, and $C(L)$ and that the outputs are issued as $F_1(H)$, $F_2(L)$, and $F_3(H)$.

(b) Use the abridged symbolism in Fig. 4-86 and the P-term table in part (a) to represent the programmed FPLA circuit.

4.62 Design an XS3-to-Gray code converter by using a $4 \times 8 \times 4$ FPLA. Assume that all inputs and outputs are ACTIVE HIGH and that false data are not rejected.

4.63 Design an NBCD-to-XS3 code converter by using a $4 \times 12 \times 4$ PAL. Assume that all inputs and outputs are ACTIVE HIGH and that all false data are encoded as 0000.

4.64 Use a multiplexed scheme of $2^6 \times 4$ ROMs to design a circuit that will convert 8-bit one-hot code (see Fig. 4-42) to 4-bit binary. Assume that all false data are rejected and indicated by binary 1000. In place of a ROM program table give the coded CANONICAL SOP forms for each binary output. Next, show how the outputs of the multiplexed ROM scheme must be wire-ORed to produce the four binary outputs of the converter.

4.65 Use a $2^4 \times 4$ ROM to design the four-bit sine generator shown in Fig. P4-12. The input x is the angle expressed in radian measure and the output Z is $\sin(\pi x)$. Let x be represented by the binary WORD $x_3x_2x_1x_0$ with weights $\frac{1}{2}, \frac{1}{4}, \frac{1}{8}, \frac{1}{16}$, respectively, and let Z be the binary WORD $Z_3Z_2Z_1Z_0$ also with weights $\frac{1}{2}, \frac{1}{4}, \frac{1}{8}, \frac{1}{16}$, respectively (see Sect. 4.7.1 for a discussion of weighted binary codes). Note that the input angle (radian measure) and the output may be given approximate representation. For example, 30°

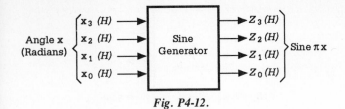

Angle x (Radians)
$\left\{ \begin{array}{l} x_3\ (H) \\ x_2\ (H) \\ x_1\ (H) \\ x_0\ (H) \end{array} \right.$ → Sine Generator → $\left. \begin{array}{l} Z_3\ (H) \\ Z_2\ (H) \\ Z_1\ (H) \\ Z_0\ (H) \end{array} \right\}$ Sine πx

Fig. P4-12.

must be represented as 3/16 (0011) since 1/6 cannot be represented accurately with four bits. Or, similarly, Sin 45° = 0.7071 must be given by 1011 or 0.6875. Greater accuracy results only by using a larger number of bits as, for example, 8 or 12 bits.

4.66 Design the 4-bit sine generator of Problem 4.65 by using the following:

(a) A 4-to-16 DECODER and the necessary external NAND logic.

(b) A 4 × 20 × 4 FPLA.

4.67 A function Y is represented in the K-map of Fig. P4-13.

(a) Loop out minimum SOP and POS cover and then use arrows in separate K-maps to indicate the HAZARDous SOP and POS transitions that are present in the function.

(b) List the terms for the SOP and POS HAZARD cover and show this cover on the K-maps.

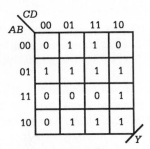

Fig. P4-13.

4.68 The following absolute minimum function has one or more STATIC HAZARDs:

$$F = \overline{A}B\overline{C} + BD + AC\overline{D} + \overline{B}\,\overline{C}D$$

(a) Map and loop out this minimum function in a fourth-order K-map. Indicate each HAZARDous transition on the map by using an arrow. Show the HAZARD cover on the K-map and in the expression for F.

(b) Repeat part (a) for POS HAZARDs in the absolute minimum POS representation of the function F.

(c) Use the gate/input tally to compare parts (a) and (b). Which result is simpler?

4.69 (a) Loop out minimum POS cover from the K-map in Fig. 4-97(c) and identify the coupled variable and associated coupled terms for the POS HAZARD that exists in this function. In the K-map, indicate with an arrow the HAZARDous transition and show the HAZARD cover.

(b) Construct a timing diagram that will illustrate the presence and removal of the POS HAZARD identified in part (a). To do this, construct the logic circuit (say with NOR gates) and assign a unit path delay τ_p to each gate and INVERTER. Assume that all inputs and the output are ACTIVE HIGH.

4.70 (a) Loop out minimum POS cover from Fig. 4-100 and analyze the result for POS HAZARDs. Specifically, identify each coupled variable and associated coupled terms that will cause a POS HAZARD, and indicate each HAZARDous transition with an arrow in the K-map. Show the HAZARD cover for each HAZARDous transition in the K-map and in the expression for F.

(b) Compare the gate/input tally obtained from the results of parts (a) with that obtained for the SOP results from Fig. 4-100(b). Which form (SOP or POS) is the simpler after HAZARD cover has been added to each?

4.71 Shown in Fig. P4-14 is the logic circuit for the function $F = \overline{A}\,\overline{B}C + AB\overline{D}$.

(a) Demonstrate with a timing diagram that a FUNCTION HAZARD is formed if $D(H) = C(H) = 1(H)$ and if $A(H)$ and $B(H)$ are identical pulses of ACTIVE duration much larger than any gate path delay. To do this, assign a unit path delay τ_p to each gate and INVERTER and construct the waveforms for $\overline{A}\,\overline{B}C(L)$, $AB\overline{D}(L)$, and $F(H)$.

(b) Discuss how this HAZARD can be avoided.

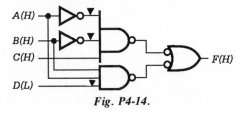

Fig. P4-14.

Design of a Bitwise Logic Function Generator

By using a 4-to-1 MUX, we design a bitwise logic function generator that will generate any one of 16 possible functions of data bits, A and B, as shown in Fig. A4-1(a). We assume that all inputs to the MUX are from positive logic sources.

The design of this bitwise logic function generator calls for a MUX application only slightly different from that in Fig. 4-53. In this case the roles of the data select inputs and data inputs are reversed as indicated in Figs. A4-1(b) and (c). Thus, the four inputs I_3–I_0 are now the control address inputs, while the two inputs S_1 and S_0 become the data inputs. With the data inputs maintained $A(H)$ and $B(H)$, any one of the 16 functions given in Fig. A4-1(a) can be generated by the appropriate combination of control address inputs I_3, I_2, I_1, and I_0 according to the MUX output expression

$$Y = I_3AB + I_2A\overline{B} + I_1\overline{A}B + I_0\overline{A}\,\overline{B} \qquad \text{(A4-1)}$$

The control address WORD $I_3I_2I_1I_0$, written in the order of decreasing positional weight, is used to indicate which of the 16 functions is generated. For example, $Y_2 = \overline{A}B$ is generated if $I = 0010_2 = 2_{10}$, meaning that I_1 is 1 while I_3, I_2, and I_0 are all 0. Or $Y_6 = A\overline{B} + \overline{A}B = A \oplus B$ is generated if $I = 0110_2 = 6_{10}$, which means that only I_2 and I_1 are 1. Thus, to indicate which function is to be generated, one need only specify the decimal equivalent of the I WORD. Equivalently, any one of the 16 functions can be generated by combining the positional weights indicated in Fig. A4-1(d). For example, $(8 + 1)$ gives $Y_9 = A \odot B$, or $(4 + 2 + 1)$ yields $Y_7 = \overline{A} + \overline{B}$, and so on.

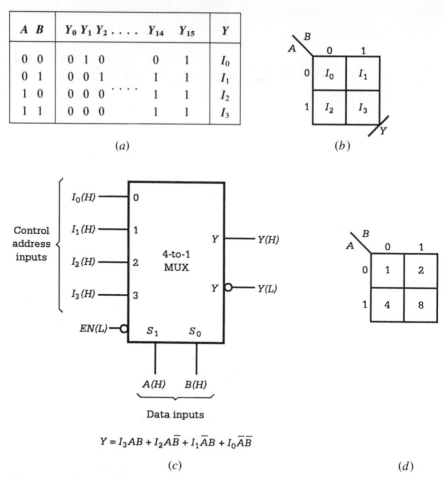

A B	Y_0 Y_1 Y_2	Y_{14}	Y_{15}	Y
0 0	0 1 0	0	1	I_0
0 1	0 0 1	1	1	I_1
1 0	0 0 0	1	1	I_2
1 1	0 0 0	1	1	I_3

(a)

(b)

(c)

(d)

$$Y = I_3 AB + I_2 A\overline{B} + I_1 \overline{A}B + I_0 \overline{A}\overline{B}$$

Fig. A4-1. *Design of a bit-wise logic function generator by using a 4-to-1 MUX. (a) Truth table showing the 16 logic functions. (b) EV K-map of the truth table. (c) MUX implementation and output expression. (d) K-map showing positional weighting of control address inputs.*

APPENDIX 4.2

Design of a Programmable Arithmetic and Logic Unit

We now design a programmable arithmetic and logic unit (PALU) which can be used to perform the following operations on two n-bit WORDs:

1. Arithmetic operations
   a. Add with carry or increment
   b. Subtract with borrow or decrement
   c. Multiply or divide steps
   d. Ones and 2's complements
2. COMPARATOR operations
   a. Equal
3. Bitwise logic operations
   a. AND, OR, XOR, and EQV
   b. Logic-level conversion
   c. Transmit or complement data bit
4. Shifter operations
   a. Shift or rotate left or right one bit

It is required that each 1-bit slice of the PALU have a resulting function output F and that the PALU be cascaded with respect to the CARRYin and CARRYout parameters, C_{in} and C_{out}. Also, it is required that C_{out} be disabled for all nonarithmetic operations.

The 1-bit slice PALU will consist of three 4-to-1 MUXs, each used the same way as the MUX in Fig. A4-1(c), and an AND-OR-INVERT (A-O-I) gate, which is a commercially available IC device described shortly. For reference purposes we name the three MUXs as follows:

Disable carry MUX (D-MUX)
Extend carry MUX (E-MUX)
Result MUX (R-MUX)

Each of these MUXs will have four control address inputs, hence a total of 12 control address inputs altogether. As we shall see, it is the ability to program these 12 control inputs that gives the PALU its power and versatility. For comparison purposes, the ALU represented in Fig. 4-78 has only five control inputs.

Shown in Fig. A4-2(a) is the circuit diagram for a 1-bit slice PALU configured to perform any of the operations listed above. Its circuit symbol is given in Fig. A4-2(b). Notice that along with the C_{in} input the outputs of the D-MUX and E-MUX drive the A-O-I gate represented in Fig. A4-2(c), but that only the E-MUX output combines with C_{in} in the R-MUX to produce the resultant function, F. The C_{out} output of the A-O-I gate, on the other hand, is a function of both the D-MUX and E-MUX outputs as well as C_{in}. A single INVERTER is used to CONVERT $C_{in}(L)$ to $C_{in}(H)$ for use in the A-O-I gate and R-MUX as explained next.

To help understand how the PALU is able to perform such a variety of tasks, it will be useful to write the Boolean expressions for the outputs from the three MUXs and from the A-I-O gate. Referring to Fig. A4-1 and Eq. (A4-1), we write the results

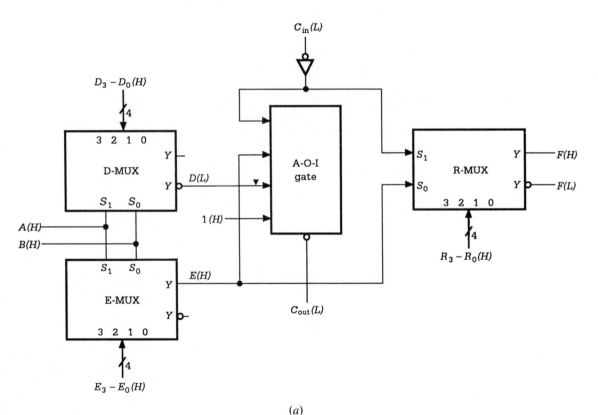

(a)

Fig. A4-2. Implementation of a 1-bit-slice PALU. (a) Circuit diagram.

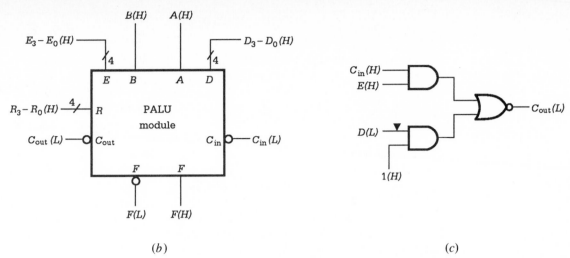

(b) (c)

Fig. A4-2 (contd.). *(b) Block diagram symbol. (c) The A-O-I gate.*

$$D = D_3AB + D_2A\overline{B} + D_1\overline{A}B + D_0\overline{A}\,\overline{B}$$

$$E = E_3AB + E_2A\overline{B} + E_1\overline{A}B + E_0\overline{A}\,\overline{B} \qquad \text{(A4-2)}$$

$$F = R = R_3C_{in}E + R_2C_{in}\overline{E} + R_1\overline{C}_{in}E + R_0\overline{C}_{in}\overline{E}$$

and

$$C_{out} = C_{in}E + \overline{D} \qquad \text{(A4-3)}$$

The form of Eq. (A4-3) is required for the adding and subtracting operations—compare with C_{out} in Eqs. (4-2). The $\overline{D}$ in Eq. (A4-3) can be used to DISABLE C_{out} for nonarithmetic operations if C_{out} is issued ACTIVE LOW from the A-O-I gate. So, by setting $D = 0$ for each nonarithmetic operation, $\overline{D} = 1$ and $C_{out}(L) = 1(L) = 0(H)$, which can be interpreted as a CARRYout DISABLE for that operation. In effect, the output of the D-MUX performs the same mode control operation as does M in the ALU represented in Fig. 4-68. However, the CARRYout DISABLE of the PALU is a false data rejection feature not shared with the ALU of Fig. 4-68. Recall that in Fig. 4-68 don't cares are assigned to C_{out} outputs for all nonarithmetic operations.

The use of the 1-bit slice PALU is illustrated by ten examples in Fig. A4-3. Here, each MUX control address is encoded in the order of decreasing positional weight but represented in decimal equivalent, as was done in Appendix 4-1. Thus, $D = \overline{A} + \overline{B}$ is represented as $D = 0111 = 7$, or $E = A\overline{B}$ is represented by $E = 0100 = 4$, and $F = E \oplus C_{in}$ by $R = 6$, and so on.

Some explanation of the results given in Fig. A4-3 seems necessary. For operation 2 (A minus B), C_{in} is the CARRYin parameter for 2's complement arithmetic. For this operation $D = \overline{A} + B$, $E = A \odot B$, $R = E \oplus C_{in}$, and $F = A \odot B \oplus C_{in} = A \oplus \overline{B} \oplus C_{in}$. Operations 1, 3, and 5 are explained in the discussion of Fig. 4-68. Operation 10 shifts the A bit one bit place to the left. Rotate left is possible in a cascaded PALU if the C_{out} for the MSB PALU module is connected to the C_{in} input for the LSB module.

	Operation*	F	C_{out}	D	E	R
1	A plus B plus C_{in}	$A \oplus B \oplus C_{in}$	$C_{in}(A \oplus B) + AB$	7	6	6
2	A minus B	$A \oplus \overline{B} \oplus C_{in}$	$C_{in}(A \odot B) + A\overline{B}$	11	9	6
3	$\begin{Bmatrix} B \text{ plus } 1 \text{ if} \\ \text{LSB } C_{in} = 0 \end{Bmatrix}$	$\overline{B} \oplus C_{in}$	$C_{in} + B$	5	5	6
4	$A = B$	$(A \odot B)\overline{C}_{in}$	$C_{in} + A \oplus B$	9	9	2
5	$\begin{Bmatrix} A \text{ minus } 1 \text{ if} \\ \text{LSB } C_{in} = 1 \end{Bmatrix}$	$A \oplus C_{in}$	$C_{in} \cdot A$	15	12	6
6	$A \oplus B$	$A \oplus B$	1	0	6	10
7	$A \cdot B$	$A \cdot B$	1	0	8	10
8	$\overline{A} + B$	$\overline{A} + B$	1	0	11	10
9	Complement of A	$\overline{A}$	1	0	3	10
10	Shift A left	C_{in}	A	3	0	12

* Subtraction operations assume 2's complement arithmetic (LSB $C_{in} = 1$)

Fig. A4-3. *Sample functions generated by the 1-bit PALU represented in Fig. A4-2.*

Operation 4 in Fig. A4-3 is the COMPARATOR operation, $A = B$, which appears on the LSB output (F) of a cascaded PALU unit. The results shown for this operation derive from the circuit of Fig. 4-30. The CARRYout of one stage of the COMPARATOR is the CARRYin for the following stage, which yields

$$F = (A = B) = (A\overline{B}\,\overline{C}_{in}) \cdot C_{in} \cdot (\overline{A}B\overline{C}_{in})$$
$$= (\overline{A} + B + C_{in}) \cdot \overline{C}_{in} \cdot (A + \overline{B} + C_{in})$$
$$= (A \odot B)\overline{C}_{in}$$

and

$$C_{out} = A\overline{B}\,\overline{C}_{in} + C_{in} + \overline{A}B\overline{C}_{in}$$
$$= C_{in} + A \oplus B$$
$$= C_{in}(A \odot B) + A \oplus B$$
$$= C_{in}E + \overline{D}$$

From these results we see that $D = E = A \odot B$ and $F = E \cdot \overline{C}_{in}$.

A 1-bit slice PALU, like a 1-bit slice PARALLEL ADDER, is of little use until it is cascaded. Shown in Fig. A4-4 is the Jth module for a cascaded PALU of arbitrary size. The modules of a cascaded PALU are operated off common D, E, and R buses as shown. Notice the similarity between Fig. A4-4 and 4-73.

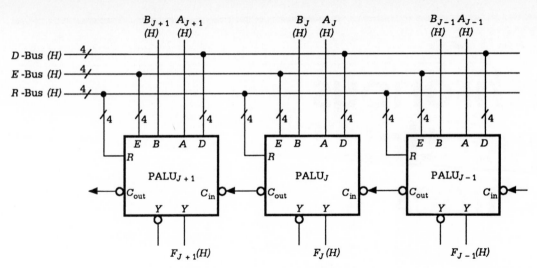

Fig. A4-4. *A cascaded segment showing the* J^{th} *module of a PALU of arbitrary size.*

CHAPTER 5

Synchronous Sequential Machines

5.1 INTRODUCTION

Up to this point we have considered only combinational logic machines, those whose outputs depend solely on the present state of the inputs. AD-DERs, DECODERs, MUXs, ROMs, ALUs, and the many other combinational logic machines are remarkable and very necessary machines in their own right. But they all suffer the same limitation. They cannot perform operations that depend on past as well as present input information. A ROM cannot remember the result of a previous set of input instructions, and an ADDER cannot count without changing the input data after each addition operation. In short, combinational logic devices lack true memory and, so, lack the ability to perform sequential operations. Yet their presence in a sequential machine may be indispensible.

We deal with sequential devices all the time. In fact our experience with such devices is so commonplace we take them for granted. For example, many of us have had the experience of being delayed by a modern four-way traffic control light system which is vehicle actuated with pedestrian overrides and the like. Once at the light we must wait for a certain sequence of events to take place before we are allowed to proceed. The controller for such a traffic light system is a fairly complex digital sequential machine.

Then there is the familiar elevator system for a multistory building. We may push the button to go down only to find that upward-bound stops have priority over our command. But once in the elevator and downward bound we are likely to find the elevator stopping at floors preceding ours in sequence, again demonstrating a sequential priority. Added to these features are the usual safety and emergency overrides, and the fact that the elevator motor itself is probably controlled by a stepping motor controller. Obviously, modern elevator systems are controlled by rather sophisticated sequential machines.

The list of sequential machines that touch our daily lives is vast. The car we drive, the home we live in, or our place of employment all use sequential machines of one type or another. Automobiles use digital sequential machines to control starting, braking, fuel injection, cruise control, and safety features. Most homes have automatic washing machines, microwave ovens, audio and video devices of various types, and, of course, computers. Some homes have complex security and energy control systems. It is common to find audio playback equipment that will permit search, scan, programming, queuing, and auto repeat or stop of the recordings. All these remarkable but now commonplace gifts of modern technology are made possible by the use of digital sequential machines—the subject of this and the remaining chapter of this text.

The machines we have just mentioned are called *sequential machines* because they possess true memory capability and can issue time-dependent sequences of logic signals controlled by present and past input information. These sequential machines may also be *synchronous* because the data path is controlled by a system CLOCK. In synchronous sequential machines input data are introduced into the machine and are processed sequentially according to some algorithm, and outputs are generated—all regulated by a system CLOCK. Sequential machines whose data path is CLOCK independent (i.e., self-timed) are called *asynchronous sequential machines*, the subject of Chapter 6.

This chapter is about synchronous sequential machines—their design, analysis, and operation. Moreover, it is in this chapter that we establish a firm foundation for the design of synchronous controllers, which are the machines that control traffic light systems, elevators, computers, and audio playback equipment, to name but a few. At chapter's end we will introduce the reader to system-level design.

5.2 A SEQUENCE OF LOGIC STATES

Let us suppose we have built a synchronous sequential machine by some means and that this machine is a multiple-input/three-output device represented by the block symbol of Fig. 5-1(a). Suppose also we were to fix the inputs, activate the machine, and detect (say, with an oscilloscope) the voltage waveforms from its output terminals as shown in Fig. 5-1(a). This sequential machine is synchronous in its operation, since all signals change synchronously with respect to one another.

Now, assuming that the voltage signals represent positive logic sources, we construct the logic waveforms as in Fig. 5-1(b). A group of logic waveforms such as these is commonly known as a *timing diagram*. Notice that the logic waveforms of a timing diagram are always represented as having zero rise and fall times, whereas the physical voltage waveforms must have finite rise and fall times. Finite rise and fall times are a consequence of the fact that changes in the physical state of anything cannot occur instantaneously. Logic-level transitions, on the other hand, are nonphysical and so

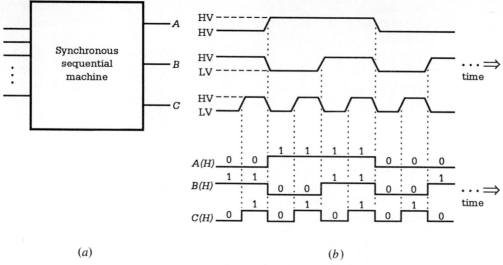

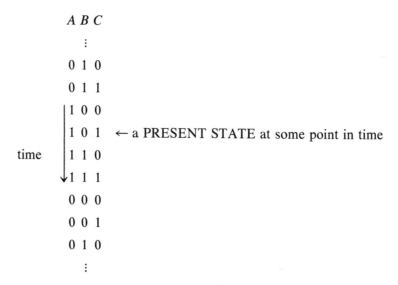

(a)   *(b)*

Fig. 5-1. *A repetitive sequence of logic events from a synchronous sequential machine. (a) Block diagram and output voltage waveforms. (b) Timing diagram representing the positive logic interpretation of the voltage waveforms.*

occur abruptly at the ACTIVE and INACTIVE transition points of the physical waveform as indicated by the vertical dotted lines in Fig. 5-1.

From the timing diagram of Fig. 5-1(b) we can determine the sequence of *states* through which this machine must transit. By reading the logic levels of the timing diagram in the direction of increasing time, we observe the state sequence

$$
\begin{array}{c}
A\ B\ C \\
\vdots \\
0\ 1\ 0 \\
0\ 1\ 1 \\
1\ 0\ 0 \\
1\ 0\ 1 \quad \leftarrow \text{a PRESENT STATE at some point in time} \\
1\ 1\ 0 \\
1\ 1\ 1 \\
0\ 0\ 0 \\
0\ 0\ 1 \\
0\ 1\ 0 \\
\vdots
\end{array}
$$

time

which is a binary count assumed to be repetitive as long as the sequential machine is operating properly.

A *logic state* is defined as follows:

> *The unique set of logic values that characterize the logic status of a machine at some point in time is called a logic state.*

Any variable whose logic values contribute to the logic status of a machine is called a *state variable*. The sequential machine illustrated in Fig. 5-1 has three state variables whose logic values define the eight unique states of the sequence. We have chosen the STATE 101 in the sequence (see arrow) as the PRESENT STATE of this machine at some point in time. Knowing that 101 is the PRESENT STATE, then we also know that 100 was the PREVIOUS STATE and that 110 must be the NEXT STATE. Thus, each state in the sequence of Fig. 5-1(c) is a PRESENT STATE, at one time or another, with its own associated PREVIOUS and NEXT states. In contrast, any of the combinational logic machines discussed in Chapter 4 have but *one* state corresponding to a unique set of inputs—there are *no* associated PREVIOUS and NEXT states for a combinational (nonsequential) logic machine.

A sequential machine, such as the one represented in Fig. 5-1, has a finite number of states into which it can reside during its operation. Such sequential machines are commonly known as *finite state machines* (FSMs). An FSM of N state variables can have no more than 2^N states and no less than two. That is,

$$2 \leq \text{number of states} \leq 2^N$$

Usually, the number of state variables to be used is determined by the number of states required by the FSM. For example, two states require one state variable, three or four states require two state variables, five to eight states require three state variables, and so on. More state variables can be used than are needed to satisfy the 2^N state requirement, but this is done only rarely to overcome certain design limitations.

The simple synchronous FSM illustrated in Fig. 5-1 turns out to be a binary counter. It is one of a countless number of FSMs that could be designed given no restriction on the number of state variables the machine can have. In this chapter we will design and analyze sufficient numbers and complexities of FSMs so as to demonstrate the methodology to be used.

5.3 MODELS FOR SEQUENTIAL MACHINES

The use of models in designing sequential machines is important because they permit the design process to be organized and standardized. The use of models also provides a means of communicating design information from one person to another. References can be made to specific parts of a design by using standard model nomenclature. Without having specifically stated so, we used two forms of a single model to design combinational logic in Chapter 4. These forms, commonly known as sum-of-products (SOP) and product-of-sums (POS), are used universally by logic designers to communicate design information.

In this section we will heuristically develop the most general model beginning with the most elemental forms. We will show that all other models are degenerate forms of the general model.

5.3.1 The Basic Model

In Sect. 5.2 we described an FSM (a binary counter in this case) that issues a sequence of eight unique states before repeating the sequence. Each state in that sequence becomes the PRESENT STATE at some time and has associated with it a NEXT STATE which is predictable given the PRESENT STATE. Now, the question is: What logic elements are required to do what this machine does? To answer this question, let us examine the thinking processes we use to carry out sequences of events each day. Whether the sequence is the daily routine of going to work or just giving our telephone number to someone, we must be able to remember our present position in the sequence and know what the next step must be. It is no different for a sequential machine. There must be a MEMORY section, as in Fig. 5-2(a),

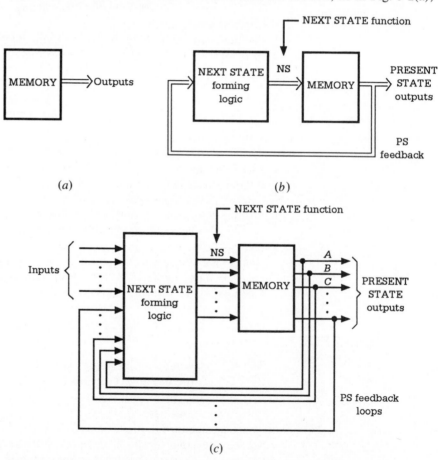

Fig. 5-2. Development of the basic model for a sequential machine. (a) The MEMORY section only. (b) The MEMORY and NEXT STATE forming logic sections showing feedback of the PRESENT STATE information. (c) The basic model.

which gives the PRESENT STATE. And there must be a NEXT STATE logic section, as in Fig. 5-2(b), which determines the NEXT STATE given the PRESENT STATE information from the MEMORY section. Then, to be able to vary the sequential behavior of the FSM, we need external inputs. Putting all this together leads to the basic model for a sequential machine given in Fig. 5-2(c).

5.3.2 The General Model

The basic model of Fig. 5-2(c) represents a minimum configuration if the FSM is to issue a sequence of logic states which depends on the settings of the external inputs as well as the PRESENT STATE (feedback) information. The sequence may be as simple as a single logic variable that changes between logic 0 and logic 1, or it may be as complex as several juxtaposed logic variables that change through a sequence of logic states in accordance with some algorithm. In any case, an FSM that conforms to the basic model is limited to PRESENT STATE outputs only. If outputs other than the PRESENT STATE outputs are to be issued, output forming logic must be added. Shown in Fig. 5-3 is the basic model with an output-forming logic section. This model, commonly known as the Moore model, is characterized by a PRESENT STATE (PS), NEXT STATE (NS), INPUT STATE (IP), and OUTPUT STATE (OP), which have the functional relationships

$$PS = f(NS)$$
$$NS = g(IP, PS) \tag{5-1}$$
$$OP = h(PS)$$

Any FSM that conforms to the functional requirements of Eqs. (5-1) is called a Moore machine. The name Moore is attached in honor of E. F. Moore, a pioneer in sequential circuit design.

The OUTPUT STATE of an FSM may depend on the INPUT STATE as well as the PRESENT STATE. The model shown in Fig. 5-4 permits this. It is the most general model and is called the Mealy model after G. H. Mealy, another pioneer in the field of sequential machines. Here, the PRESENT STATE, NEXT STATE, and INPUT and OUTPUT states are functionally related by

$$PS = f(NS)$$
$$NS = g(IP, PS) \tag{5-2}$$
$$OP = h(IP, PS)$$

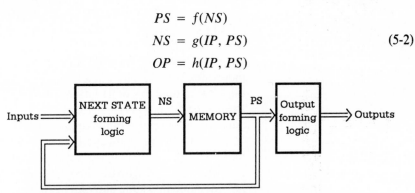

Fig. 5-3. Moore's model for a sequential machine.

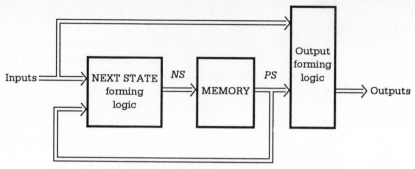

Fig. 5-4. *Mealy's model for a sequential machine.*

Any machine that conforms to these functional relationships, and hence to the model of Fig. 5-4, is called a Mealy machine. Thus, the output logic for a Mealy machine depends on the external inputs, whereas for the Moore machine it does not—the only difference between the two types of sequential machines. Hereafter, reference made to a Mealy machine or a Moore machine will imply this difference.

5.4 THE STATE DIAGRAM

The basic model of Fig. 5-2(c) and the Moore and Mealy models given in Figs. 5-3 and 5-4 permit external inputs to influence the sequential behavior of an FSM. A simple example would be to add an external input to the binary counter represented in Fig. 5-1 so that it could count down as well as up. A more complex example is the use of several inputs to cause an FSM to enter one of several possible sequences (or routines) each with subroutines and outputs, all controlled by external inputs whose values change at various times during the operation of the FSM. Obviously, some means must be found by which both simple and complex sequential FSM behavior can be represented in a precise and meaningful way. The fully documented *state diagram* discussed in this section provides this means.

5.4.1 Features of the State Diagram

The important features of a state diagram are illustrated in Fig. 5-5. Included are the PRESENT STATE (e), the PREVIOUS STATE (d), the NEXT STATEs (f) and (g), and the various input branching conditions. Branching from STATE (e) to STATE (f) can occur only under input conditions $\overline{X}Y$, whereas branching from STATE (e) to STATE (g) is conditional on input Y independent of X. The branching condition $X\overline{Y}$ is the input requirement for the FSM to hold (branch to itself) in STATE (e)—it is its own NEXT STATE. Branching from STATE (d) to STATE (e) is unconditional since it does not depend on a specific input requirement.

The identifying state symbols d, e, f, and g are useful in discussing the various states, but they are not essential. The state code assignment (e.g., . . . 010 . . . for PRESENT STATE (e)) is more important. Each state of a fully documented state diagram must be given a state code assignment before the design process can continue. The logic variables that represent the state

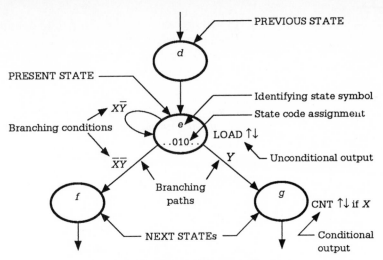

Fig. 5-5. *Features of the state diagram.*

code bits are called the *state variables* for the FSM. The names commonly given to state variables are $A, B, C, \ldots$ meaning $Q_A, Q_B, Q_C, \ldots$ We will learn that a wise choice of state code assignments can mean the difference between a good design and a poor design, or between one that will succeed and one that will fail.

Output notation is also straightforward. There are two kinds of outputs that can be represented in a state diagram. Referring to Fig. 5-5, the output

$$\text{LOAD} \uparrow \downarrow$$

is an *unconditional output* issued any time the FSM is in STATE ⓔ. The up/down arrows ($\uparrow \downarrow$) signify that LOAD becomes ACTIVE (up arrow, $\uparrow$) when the FSM enters STATE ⓔ and becomes INACTIVE (down arrow, $\downarrow$) when the FSM leaves STATE ⓔ. The second type of output, indicated by

$$\text{CNT} \uparrow \downarrow \text{ if } X$$

is an output that is generated any time the FSM is in STATE ⓖ but only if X is ACTIVE—hence, CNT is a *conditional or Mealy output*.

There are two important requirements that any fully documented state diagram must meet. The first is that *all* possible branching conditions from a given state must be accounted for. This "accountability" requirement is met if the *sum rule*

$$\sum \text{(outgoing branching conditions)} = 1 \qquad (5\text{-}3)$$

is satisfied. Equation (5-3) requires that all possible outgoing branching conditions from each state OR (Boolean SUM) to logic 1. This rule must be satisfied when it is the intent that no particular branching condition restrictions be imposed on the FSM design.

While satisfying the SUM ($\sum = 1$) rule is a necessary condition for state branching accountability in a fully documented state diagram, it is not sufficient to ensure that the branching conditions are nonoverlapping, that is, *mutually exclusive* or unique. This is the second requirement of any fully documented state diagram, which can be stated as follows:

> *Each possible branching condition from a given state must be associated with no more than one branching path.*

If the mutually exclusive requirement is not obeyed for a given state, then branching from that state could be ambiguous.

Testing for branching accountability and mutual exclusivity at STATE ⓔ in Fig. 5-5, indicates that Eq. (5-3) is satisfied since

$$X\overline{Y} + \overline{X}\,\overline{Y} + Y = 1$$

and that each branching condition is mutually exclusive of the others. Thus, the second requirement is met in STATE ⓔ since there is no possible set of input conditions that could force the FSM to transit along two different paths. We will return to the subjects of branching condition requirements in Sect. 6.3 where they will be considered on a more formal basis.

5.4.2 Flowcharts Versus the State Diagram

The state diagram is the single most important means of describing the sequential behavior of an FSM. Its use is essential to the design of most sequential machines. However, for the design of complex FSMs, it may be difficult to directly construct the state diagram from a verbal description of the FSM's operations. In such cases a "thinking aid" can be useful in conceptualizing and developing the detailed sequential behavior of an FSM. The flowchart is the thinking-aid tool we recommend for this purpose. Once the flowchart is constructed, it is a simple matter to transform it to the state diagram, the essential design tool.

Shown in Fig. 5-6 are the two symbols used in flowcharting an algorithm. A third symbol, the terminal symbol used to denote an initial or start-up condition, will not be used in this text. The action block is used to indicate one or more outputs or to indicate a delay of one time unit (no output). The decision symbol is used to indicate a binary answer such as yes or no, or true or false, to a question asked. Thus, with use of the latter two symbols, a complex algorithm can be broken down into decisions and actions which can then be converted into a state diagram.

Action block

Decision symbol

Fig. 5-6. Flow chart symbols.

Figure 5-7 demonstrates that a state begins with an action block and ends with the next action block, including any intervening decision paths. The decision symbol in Fig. 5-7(a) asks if X is ACTIVE. If the answer is yes (meaning X), then the process proceeds and the FSM transits from STATE ⓐ to STATE ⓑ. If the answer is no (meaning $\overline{X}$), then the process does not proceed forward and the decision on X must be made again after an additional unit of time has elapsed. In effect, the decision "no" requires the FSM to transit out of and immediately back into STATE ⓐ. This information is vividly depicted by the state diagram segment in Fig. 5-7(b). Remember that

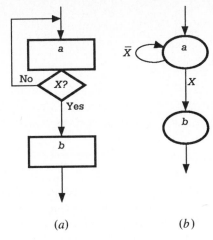

Fig. 5-7. Sequence of steps showing that a state begins with an action block and ends with the next action block. (a) Flow chart. (b) State diagram.

(a)   (b)

all decision symbols immediately following an action block are associated with that block.

The branching conditions from a given state are determined by tracing the decision paths from the reference action block to all other next nearest action blocks. This is demonstrated in Fig. 5-8. Decision symbols in Fig. 5-8(a) first ask if *X* is ACTIVE then if *Y* is ACTIVE. The resulting decisions determine the branching conditions for the state diagram shown in Fig. 5-8(b). Because each decision symbol provides both yes and no paths, the sum rule and mutually exclusive branching requirements are automatically satisfied, a fact easily verified from the state diagram.

Unconditional and conditional outputs are unambiguously stated in the action block where the algorithm requires that they be issued. Examples of both types of outputs are given in Fig. 5-8. Action block *b* contains the

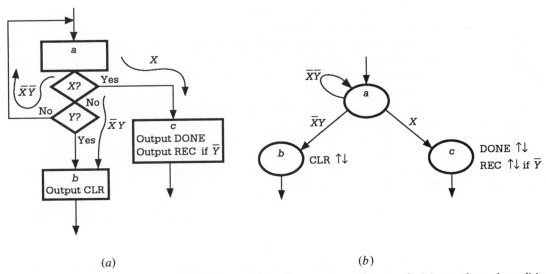

(a)   (b)

Fig. 5-8. (a) Flow chart segment showing decision paths and conditional and unconditional outputs. (b) State diagram for the flow chart of (a).

unconditional output CLR (for CLEAR) while action block *c* contains unconditional output DONE and output REC (for RECYCLE) conditional on input $\overline{Y}$. All such outputs are transferred directly to the state diagram, as can be seen from Fig. 5-8(b).

Again we emphasize that the flowchart is to be regarded only as a thinking aid. Most of the FSMs designed in this chapter are simple enough that flow charting is unnecessary. Still, a reader who finds that flowcharts are helpful in constructing fully documented state diagrams should use them.

5.5 THE BASIC MEMORY CELL

The concept of MEMORY was introduced in Sect. 5.3. There we showed that no sequential machine could operate without MEMORY. But what constitutes the MEMORY of an FSM? To answer this question, we will first need to know what kind of device remembers a logic 1 or a logic 0 indefinitely or until directed to change to the other value. This section is devoted to examining this specific issue.

5.5.1 A Heuristic Development of the BASIC CELL

Imagine there exists a loop of wire for which a time Δt is required for an electrical signal to traverse the loop. Let this wire loop be modeled by Fig. 5-9(a), where the traverse time (representing the real wire) is isolated as a lumped path delay element, and the connecting wire is treated as being ideal. Now suppose there is introduced by some means a positive voltage pulse of duration equal to or greater than the path delay Δt of the loop. Defining the state parameters

$$Q_t = \text{PRESENT STATE}$$

$$Q_{t+1} = \text{NEXT STATE}$$

and assuming a positive logic interpretation of the pulse, it follows that the NEXT STATE and PRESENT STATE are equal to logic 1 ($Q_{t+1} = Q_t = 1$) for as long as the signal exists. Of course, in this simple wire loop it could not be otherwise.

But the wire loop of Fig. 5-9(a) is a lossy system, meaning that power is dissipated and is not restored. So the voltage level of Q_t will in time fall to zero. To prevent this from happening, two INVERTERs can be added

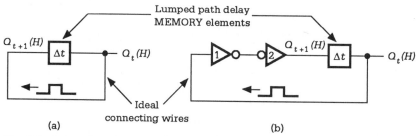

(a)   (b)

Fig. 5-9. *A wire loop used as a primitive form of MEMORY. (a) Wire loop with a lumped path delay,* Δt. *(b) Wire loop with a lumped path delay, and two INVERTERs used to restore the signal.*

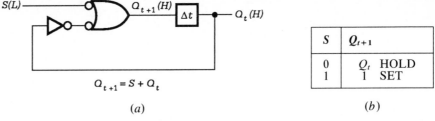

$$Q_{t+1} = S + Q_t$$

(a)

S	Q_{t+1}	
0	Q_t	HOLD
1	1	SET

(b)

Fig. 5-10. *SETting the MEMORY. (a) Circuit of Figure 5-9(b) with INVERTER 2 replaced by a NAND gate. (b) The Operation table derived from (a).*

to restore the signal as shown in Fig. 5-9(b). Now, $Q_{t+1} = Q_t = 1$ logically for as long as the INVERTERs are operating. Remember that INVERTERs are restoring switches since they are capable of amplification.

We have just demonstrated by Fig. 5-9 that a wire loop can function as a primitive form of MEMORY. The problem is that the loop has no input, and thus no user access, and so there is no means by which a logic signal can be introduced, or altered once introduced. This is remedied in Fig. 5-10(a) by replacing INVERTER 2 of Fig. 5-9(b) with an OR operation. Here, we choose a NAND gate for this purpose. As a result, the MEMORY can now be SET and maintained by setting $S(L) = 1(L)$. Reading for the NEXT STATE we find

$$Q_{t+1} = S + Q_t$$

Then, if we introduce the logic values for S one at a time into this expression, we obtain the *operation table* for this circuit given in Fig. 5-10(b).

The logic circuit of Fig. 5-10(a) is not yet suitable as a memory element. It cannot be RESET, meaning that the output cannot be brought to logic 0 once the circuit has been SET. This deficiency can be corrected by replacing the remaining INVERTER, shown in Fig. 5-10(a), with an AND operation performed by another NAND gate. The result is the NAND-centered (all NAND) BASIC CELL presented in Fig. 5-11(a), which is seen to conform to the basic model of Fig. 5-2. A $1(L) = 0(H)$ introduced at the $R(L)$ input simultaneously with a $0(L)$ at the $S(L)$ input RESETs the BASIC CELL, and a $1(L)$ on the $S(L)$ input SETs it. Further examination of this deceptively simple little machine is necessary.

Reading the BASIC CELL of Fig. 5-11(a) gives us the SOP results

$$Q_{t+1} = S + \overline{R}Q_t \tag{5-4}$$

for the NEXT STATE parameter in terms of the PRESENT STATE parameter, Q_t. Then by introducing all possible logic value combinations for S and R into Eq. (5-4), we generate the operation table for the BASIC CELL given in Fig. 5-11(b). Notice that this BASIC CELL can be SET for either the $(S, R) = (1, 0)$ or $(1, 1)$ input combination. It is for this reason we state the following:

The NAND-centered BASIC CELL of Fig. 5-11(a) is SET-dominant.

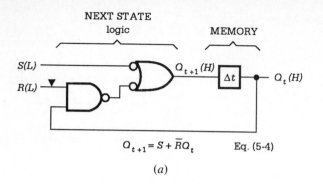

$$Q_{t+1} = S + \bar{R}Q_t \qquad \text{Eq. (5-4)}$$

(a)

S	R	Q_{t+1}	
0	0	Q_t	HOLD
0	1	0	RESET
1	0	1	SET
1	1	1	SET

(b)

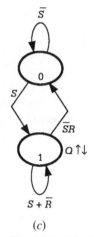

(c)

$Q_t \rightarrow Q_{t+1}$	S	R
$0 \rightarrow 0$	0	ϕ
$0 \rightarrow 1$	1	ϕ
$1 \rightarrow 0$	0	1
$1 \rightarrow 1$	$\begin{cases} 1 & \phi \\ \phi & 0 \end{cases}$	

(d)

Fig. 5-11. *The NAND-centered BASIC CELL. (a) Logic circuit showing combined SET and RESET capability. (b) Operation table derived from Eq. (5-4). (c) State diagram derived from the operation table in (b). (d) State transition table derived from the state diagram in (c).*

Note also that the term HOLD for the $(S, R) = (0, 0)$ input condition means "no change"; that is, $Q_{t+1} = Q_t$.

The state diagram for the BASIC CELL is given in Fig. 5-11(c). It is obtained directly from the operation table of Fig. 5-11(b) by looking for the S, R branching conditions required for SET, RESET, and HOLD. For example, the SET operation $(0 \rightarrow 1)$ is permitted for inputs $S\bar{R} + SR = S$, while for RESET $(1 \rightarrow 0)$ $\bar{S}R$ is required. A $(0 \rightarrow 0)$ transition, on the other hand, occurs for $\overline{SR} + \bar{S}R = \bar{S}$, since it is both a HOLD operation and a RESET; hence, HOLD RESET. Similarly, a $1 \rightarrow 1$ transition is permitted by $\overline{SR} + S\bar{R} + SR = S + \bar{R}$, since it is both a HOLD operation and a SET; hence, HOLD SET. Notice that once the SET and RESET branching conditions are known (S and $\bar{S}R$, respectively), the appropriate HOLD conditions follow immediately from the $\sum = 1$ rule given by Eq. (5-3).

There is another table in Fig. 5-11(d), called the *state transition table*, that is extremely important for design purposes. It is most easily obtained from the state diagram by reading the branching conditions for the $0 \rightarrow 0$, $0 \rightarrow 1$, $1 \rightarrow 0$, and $1 \rightarrow 1$ transitions in Q. Thus, the $0 \rightarrow 0$ transition is caused by $\bar{S}$ (i.e., $S = 0$, $R = \phi$), the $0 \rightarrow 1$ transition is produced by S (i.e., $S = 1$, $R = \phi$), the $1 \rightarrow 1$ transition results from $S + \bar{R}$ (meaning S

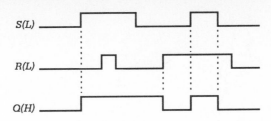

Fig. 5-12. Timing diagram illustrating the operation of the SET-dominant BASIC CELL.

$= 1, R = \phi$ or $S = \phi, R = 0$), and the $1 \to 0$ transition is forced by $\overline{S}R$ ($S = 0, R = 1$). The don't care (ϕ) represents an unspecified input condition.

The operation and state transition tables in Fig. 5-11 provide the same CELL operation information. The difference in their appearance lies in the fact that the operation table is input oriented (inputs explicitly given) while the state transition table is output oriented (outputs explicitly given). The latter table will prove to be the more useful in design work.

The operation of the SET-dominant BASIC CELL is best summarized by the MIXED LOGIC timing diagram shown in Fig. 5-12. This diagram is most easily constructed from the operation table by noting the branching conditions for SET, RESET, and HOLD. The path delays through the NAND gates are not included.

The NAND-centered BASIC CELL in Fig. 5-11 is not the only BASIC CELL form that can be heuristically derived from the wire loop circuit of Fig. 5-9(b). By replacing the INVERTERs with conjugate NOR gates one at a time to SET and RESET the MEMORY, there results the NOR-centered BASIC CELL in Fig. 5-13(a). From this circuit we read

$$Q_{t+1} = \overline{R} \cdot (S + Q_t) \tag{5-5}$$

for the NEXT STATE parameter in terms of the PRESENT STATE parameter. Now, by introducing all possible S, R input combinations into Eq. (5-5), we obtain the operation table shown in Fig. 5-13(b). Then, by using this table, we construct the state diagram given in Fig. 5-13(c), which, in turn, is used to produce the state transition table presented in Fig. 5-13(d).

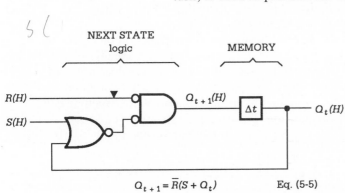

$Q_{t+1} = \overline{R}(S + Q_t)$ Eq. (5-5)

S	R	Q_{t+1}	
0	0	Q_t	HOLD
0	1	0	RESET
1	0	1	SET
1	1	0	RESET

(a)   (b)

Fig. 5-13. The NOR-centered BASIC CELL. (a) Logic circuit with lumped path delay element. (b) Operation table derived from Eq. (5-5).

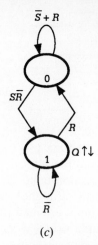

$Q_t \rightarrow Q_{t+1}$	S	R
$0 \rightarrow 0$	$\left\{\begin{matrix}0 \\ \phi\end{matrix}\right.$	$\left.\begin{matrix}\phi \\ 1\end{matrix}\right\}$
$0 \rightarrow 1$	1	0
$1 \rightarrow 0$	ϕ	1
$1 \rightarrow 1$	ϕ	0

(c)   (d)

Fig. 5-13 (contd.). *(c) State diagram derived from the operation table in (b).*
(d) State transition table derived from the state diagram in (c).

An inspection of the operation table, the state diagram, or the state transition table in Fig. 5-13 indicates that this BASIC CELL is RESET by one of two possible sets of input branching conditions, $\overline{S}R + SR = R$, whereas a single branching condition, $S\overline{R}$, SETs it. For this reason we make the following statement:

> *The NOR-centered BASIC CELL of Fig. 5-13(a) is RESET-dominant.*

The reader should compare the results given in Fig. 5-13 with those for the SET-dominant BASIC CELL in Fig. 5-11.

The timing diagram for the RESET-dominant BASIC CELL is shown in Fig. 5-14. For comparison purposes it is constructed by using the same waveforms for $S(H)$ and $R(H)$ as were used for $S(L)$ and $R(L)$ in Fig. 5-12. Notice that the output responses for the two types of BASIC CELLs are identical except for the $S, R = 1, 1$ input condition where their SET- or RESET-dominant character becomes evident.

When used for design purposes, the state transition tables for the SET- and RESET-dominant BASIC CELLs require that a specific BASIC CELL (NAND or NOR centered) be used. However, these tables are neither easy to remember nor convenient to use. To overcome this apparent difficulty we can combine them into one by using common S, R branching conditions for each $Q_t \rightarrow Q_{t+1}$ transition. For example, $(S, R) = (0, \phi)$ is common for the $0 \rightarrow 0$ transition; $(S, R) = (1, 0)$ is common for the $0 \rightarrow 1$ transition;

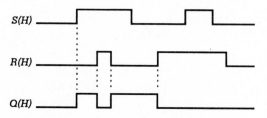

Fig. 5-14. *Timing diagram illustrating the operation of the RESET-dominant BASIC CELL.*

$Q_t \to Q_{t+1}$	S	R
$0 \to 0$	0	ϕ
$0 \to 1$	1	ϕ
$1 \to 0$	0	1
$1 \to 1$	$\begin{Bmatrix}\phi & 0\\ 1 & \phi\end{Bmatrix}$	

$Q_t \to Q_{t+1}$	S	R
$0 \to 0$	$\begin{Bmatrix}0 & \phi\\ \phi & 1\end{Bmatrix}$	
$0 \to 1$	1	0
$1 \to 0$	ϕ	1
$1 \to 1$	ϕ	0

(a)　　　　　　(b)

$Q_t \to Q_{t+1}$	S	R
$0 \to 0$	0	ϕ
$0 \to 1$	1	0
$1 \to 0$	0	1
$1 \to 1$	ϕ	0

(c)

Fig. 5-15. *State transition tables for the BASIC CELL. (a) SET-dominant format. (b) RESET-dominant format. (c) Generic (combined) format in which the* (S, R) = (1, 1) *condition is not represented.*

and so on. The result is the *generic* (combined) form of the state transition table given in Fig. 5-15. This table applies to either the NAND and NOR centered BASIC CELL.

It is important that the reader notice that the $(S, R) = (1, 1)$ condition is not represented in the generic form of the state transition table given in Fig. 5-15(c). Therefore, when this table is used for design purposes, the FSM resulting from this design will be incapable of logically producing the $(S, R) = (1, 1)$ condition. In most cases the simplicity and versatility of the generic form of the state transition table make it the form of choice for design purposes. We will use only the generic form in this text.

5.5.2 Mixed-Rail Outputs

There is another property of the two types of BASIC CELL we have yet to discuss. The outputs of two NAND (or NOR) gates are logically identical under certain conditions. This is demonstrated in Fig. 5-16 by using the NAND BASIC CELL. Here, we have neglected the lumped path delay element, Δt, and have oriented the two NAND gates so as to emphasize the "cross-coupling" appearance of this BASIC CELL. Reading the two outputs from the NAND gates, we obtain $(S + \overline{R}Q_t)(H)$ and $\overline{R} \cdot (S + \overline{R}Q_t)(L)$, as indicated in Fig. 5-16(a). Then by introducing all possible combinations of the S, R input conditions into these output expressions, we produce the operation table in Fig. 5-16(b). An inspection of this table reveals that the two outputs are logically identical except for the $(S, R) = (1, 1)$ input condition where they differ. This leads to the following important conclusion:

> *The BASIC CELL exhibits mixed-rail outputs under the condition that* (S, R) = (1, 1) *never occurs.*

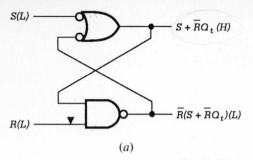

$S + \overline{R}Q_t \ (H)$

$\overline{R}(S + \overline{R}Q_t)(L)$

S R	$S + \overline{R}Q_t$	$\overline{R} \cdot (S + \overline{R}Q_t)$	
0 0	Q_t	Q_t	HOLD
0 1	0	0	RESET
1 0	1	1	SET
1 1	1	0	AMBIGUOUS

(a)   *(b)*

Fig. 5-16. *(a) The NAND-centered BASIC CELL with mixed-rail outputs. (b) Operation table for the NAND-centered BASIC CELL showing the requirements for mixed-rail outputs.*

Mixed-rail outputs are defined as any two outputs from a logic device that are identical logically but of opposite ACTIVATION levels when no account is taken of path delay.

The mixed-rail output response to input change for a NAND-centered BASIC CELL is shown in Fig. 5-17(a). This demonstrates that the output response is not instantaneous as the table in Fig. 5-16(b) might lead one to believe. Rather, following a gate path delay of τ_1, the OR operation responds to a SET condition and then the AND operation responds, with a gate path delay τ_2 separating the two responses. Conversely, following a gate path delay τ_2, the AND operation responds first to the RESET condition then the OR operation with a gate path delay τ_1 separating them. Notice that the mixed-rail output response for the NOR BASIC CELL in Fig. 5-17(b) is the dual of that for the NAND BASIC CELL in Fig. 5-17(a). The timing characteristics for the BASIC CELL will become important in later discussions of two-phase CLOCKing (Sect. 5.6.2) and of HAZARD production in synchronous FSMs (Sect. 5.8.3).

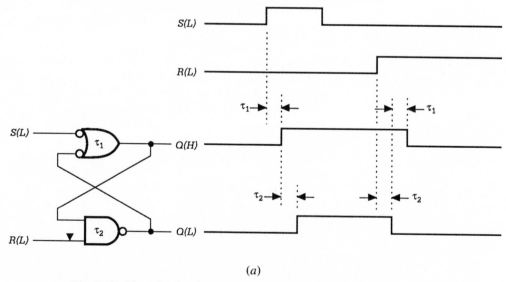

(a)

Fig. 5-17. *The mixed-rail output response of a BASIC CELL to an input change. (a) NAND-centered BASIC CELL.*

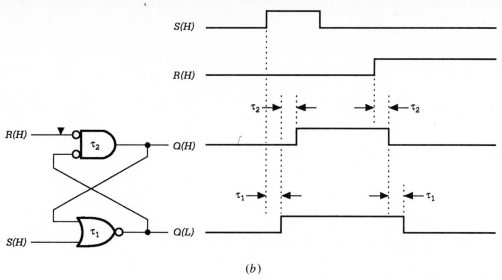

(b)

Fig. 5-17 (contd.). (b) NOR-centered BASIC CELL.

To summarize, we may represent the NAND and NOR (i.e., SET- and RESET-dominant) BASIC CELLs in MIXED LOGIC form as in Fig. 5-18 but with the understanding that the $(S, R) = (1, 1)$ input condition never be permitted. If $(S, R) = (1, 1)$ happens to occur, the NAND-centered BASIC CELL will SET with $Q(H) = 1(H)$ and $Q(L) = 0(L)$ showing loss of the mixed-rail output property. Similarly, if $(S, R) = (1, 1)$ occurs, the NOR-centered BASIC CELL will RESET giving $Q(H) = 0(H)$ with $Q(L) = 1(L)$, again showing loss of the mixed-rail output property. The circuit symbols, also shown in Fig. 5-18, will be used frequently to represent the BASIC CELLs.

One final but important point is noteworthy. Since the generic form of the state transition table of Fig. 5-15(c) does not contain the $(S, R) = (1, 1)$ input condition, its use in FSM design ensures the mixed-rail output character for the design regardless of which type of BASIC CELL is used. In

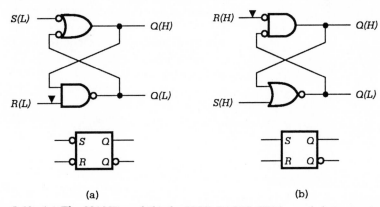

(a)   (b)

Fig. 5-18. (a) The NAND and (b) the NOR BASIC CELLs and their circuit symbols showing mixed-rail outputs that are valid only if the condition $(S, R) = (1, 1)$ never occurs.

the following section on FLIP-FLOP design, use of the generic form of the state transition table for the BASIC CELL will prove to have a distinct advantage since the resulting FLIP-FLOPs will inherit the mixed-rail output property.

5.6 DESIGN AND ANALYSIS OF FLIP-FLOPS

The BASIC CELL, to which we devoted the last section, is not by itself an adequate MEMORY element for a synchronous sequential machine. It lacks versatility, and, more important, its operation cannot be synchronized with other parts of a logic circuit or system. It is an asynchronous FSM without a timing control input. In this section we will use the BASIC CELL as the building block for the design of more versatile, CLOCK-regulated, MEMORY elements called FLIP-FLOPs—devices that can perform properly as the MEMORY elements for synchronous FSMs.

Before continuing it should be helpful to define FLIP-FLOP:

> *A FLIP-FLOP is a one-bit MEMORY element that exhibits sequential behavior regulated exclusively by a CLOCK input.*

A FLIP-FLOP samples input data by means of a CLOCK signal, stores the immediate results, and then outputs a signal at some later time determined by the CLOCK-triggering mechanism. CLOCK-triggering mechanisms constitute one important means by which FLIP-FLOPs (FFs) are classified.

5.6.1 CLOCK-Triggering Mechanisms

The CLOCK (CK), as the name implies, is usually thought of as a regular, rectangular logic waveform having some specific duty cycle as in Fig. 5-19(a). However, regularity need not be a requirement. The CLOCK pulses may be quite irregular as in Fig. 5-19(b). A regular CLOCK waveform has a period $T_{CK} = 1/f_{CK}$ associated with it, whereas an irregular CLOCK waveform does not. But whatever the case may be, it is the CLOCK that regulates the state-to-state transitions in a synchronous FSM through the action of FLIP-FLOPs—the MEMORY elements.

It is the function of CLOCK to *sample* the NEXT STATE signals of a synchronous FSM and effect a state transition based on those input conditions. The term *triggering* is used to describe this action. Because CLOCK

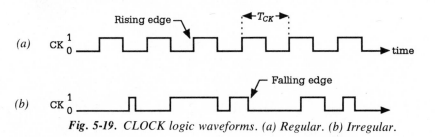

Fig. 5-19. CLOCK logic waveforms. (a) Regular. (b) Irregular.

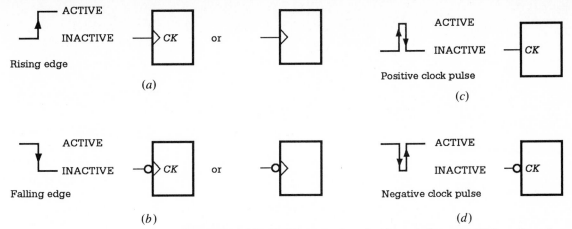

Fig. 5-20. *FLIP-FLOP triggering circuit symbology. (a) Rising edge-triggering. (b) Falling edge-triggering. (c) Positive pulse-triggering. (d) Negative pulse-triggering.*

is an input that acts somewhat like an ENABLE, the FLIP-FLOP is triggered on either the rising edge or falling edge of the CLOCK waveform. FLIP-FLOPs whose triggering mechanisms "lock out" any logic noise that is generated in the NEXT STATE function shortly after initiation of the state change are called *edge-triggered FLIP-FLOPs* (*ET-FFs*). Thus, in such FLIP-FLOPs CLOCK performs two important operations: it samples the NEXT STATE function(s) thereby initiating a state change, and it locks out any logic noise that may be produced in the NEXT STATE forming logic of the FSM during the initiation of that state change. The triggering edge indicator symbols used for FLIP-FLOPs are shown in Figs. 5-20(a) and (b), where the plain arrow symbol is used for *rising edge-triggered FLIP-FLOPs* (*RET FFs*) and the arrow with the ACTIVE LOW indicator bubble is used for *falling edge-triggered FLIP-FLOPs* (*FET FFs*).

Not all FLIP-FLOPs are edge triggered in the sense just described. There are FLIP-FLOPs that trigger on the rising or falling edge of the CLOCK waveform but that lack the triggering mechanism required to lock out the NEXT STATE function(s) following initiation of the state change. These FLIP-FLOPs, for one reason or another, require that the NEXT STATE function(s) be sampled by a very narrow CLOCK pulse to ensure proper operation. We call FLIP-FLOPs of this type *pulse-triggered* (*PT*) FLIP-FLOPs for want of a better name. Shown in Figs. 5-20(c) and (d) are the circuit symbols used for *positive pulse-triggered* (*PPT*) FLIP-FLOPs and *negative pulse-triggered* (*NPT*) FLIP-FLOPs, respectively. Here we notice the absence of the arrow symbol used to indicate edge triggering.

Another class of FLIP-FLOPs requires the use of both the rising edge and the successive falling edge of the CLOCK waveform to execute a state change. The name given to this class of memory element is *master/slave* (*MS*) FLIP-FLOP, a name that implies a two-step triggering mechanism.

In this section we will design pulse-triggered FLIP-FLOPs first, then move on to master/slave FLIP-FLOPs, and conclude with the design of the more complex edge-triggered FLIP-FLOPs.

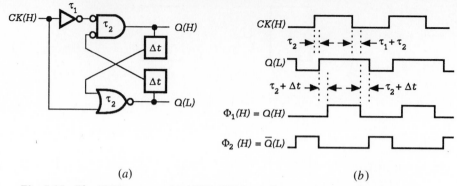

(a)　　　　　　　　　　　　　　　(b)

Fig. 5-21. *The NOR-centered BASIC CELL used to produce a two phase nonoverlapping CLOCK ACTIVE HIGH. (a) Logic circuit. (b) Timing diagram.*

5.6.2 Two-Phase CLOCKing

In VLSI (very-large-scale integrated circuit) technology there is a need to operate synchronous FSMs with what is called a two-phase (2Φ) nonoverlapping CLOCK. The 2Φ CLOCK can be obtained in a number of ways. One method uses a logic device already familiar to us—the BASIC CELL. Shown in Fig. 5-21(a) is a NOR-centered BASIC CELL configured with an INVERTER and additional path delays Δt in the cross-coupled leads. The purpose of the INVERTER is to preserve the mixed-rail output feature by preventing the $(S, R) = (1, 1)$ condition from ever occurring. The additional Δt path delays (if needed) increase the margin of safety for the nonoverlapping waveforms.

The resulting 2Φ CLOCK waveforms from the circuit of Fig. 5-21(a) are given in Fig. 5-21(b). This is called a two-phase nonoverlapping CLOCK ACTIVE HIGH, or 2Φ $CK(H)$, where we have taken $Q(H)$ as $\Phi_1(H)$ and $\overline{Q}(L)$ as $\Phi_2(H)$. Here, nonoverlapping means that the two waveforms, Φ_1 and Φ_2, will never be ACTIVE at the same time. The interval, $\tau_2 + \Delta t$, is a measure of the extent to which the two CLOCK phases are nonoverlapping. Obviously, both waveforms cannot be at 50% duty cycle at the same time or they would cease to be nonoverlapping. Notice that by taking $Q(L) = \Phi_1(L)$ and $\overline{Q}(H) = \Phi_2(L)$, a 2Φ $CK(L)$ is produced for nonoverlapping INACTIVE regions.

The additional path delays, Δt, shown in Fig. 5-21(a) may be necessary to prevent overlap of the waveforms. In a VLSI system such delays can be produced by normal inertial effects in leads, but these effects are usually unpredictable. If more control is needed over the Δt values, use can be made of a series of INVERTERs or a suitable delay circuit consisting of an RC (resistor/capacitor) component coupled with a noninverting driver (buffer).

5.6.3 Pulse-Triggered FLIP-FLOPs

In this section we will illustrate the design process by designing the so-called pulse-triggered FLIP-FLOPs. To do this, use will be made of the BASIC CELL that we heuristically developed in Sect. 5.5.1. The character of the PT FFs and the reasons why they are classified as pulse-triggered FLIP-FLOPs will be made clear in the treatment that follows.

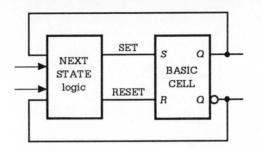

Fig. 5-22. *The basic model used to design pulse-triggered FLIP-FLOPs.*

The model to be used in the design of the PT FFs is shown in Fig. 5-22. It is seen to be the basic model of Fig. 5-2(c) but with a BASIC CELL as the MEMORY. Thus, it is a degenerate form of the general model given in Fig. 5-4.

The D FLIP-FLOP. The D FLIP-FLOP is characterized by the operation table in Fig. 5-23(a), which indicates that it can either SET or RESET. This table applies to any D-FF regardless of its triggering mechanism. The operation table may be regarded as the defining instrument for the FLIP-FLOP exclusive of any triggering information.

The operation of the PT D-FF is explicitly given by its state diagram presented in Fig. 5-23(b). This state diagram is constructed from the operation table by first assuming (in this case) a two-state FSM that can change state only when CLOCK goes ACTIVE. Next, the SET and RESET branching conditions are determined from the operation table by reading SET (0 $\rightarrow$ 1) as D with CLOCK or DCK, and by reading RESET (1 $\rightarrow$ 0) as $\overline{D}$ with CK or $\overline{D}CK$. The HOLD branching conditions, $\overline{D} + \overline{CK}$ for 0 $\rightarrow$ 0 and $D + \overline{CK}$ for 1 $\rightarrow$ 1, follow from the $\sum = 1$ rule [Eq. (5-3)] since $DCK + \overline{DCK} = 1$ and $\overline{DCK} + \overline{\overline{DCK}} = 1$, respectively. Thus, by knowing one

D	Q_{t+1}	
0	0	RESET
1	1	SET

(a)

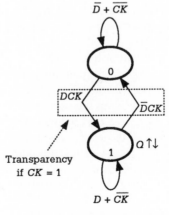

(b)

$Q_t \rightarrow Q_{t+1}$	D
0 $\rightarrow$ 0	0
0 $\rightarrow$ 1	1
1 $\rightarrow$ 0	0
1 $\rightarrow$ 1	1

(c)

Fig. 5-23. *Characterization of the D FLIP-FLOP. (a) Operation table. (b) State diagram for the pulse-triggered D-FF, derived from the operation table, showing the transparent property for* CK = 1. *(c) State transition table derived from the state diagram.*

5.6 / Design and Analysis of Flip-Flops

outward branching condition, we can find the second one by taking the complement of the first. Remember that Q is the state variable with logic values 0 (INACTIVE) and 1 (ACTIVE). Therefore, SET means Q goes or remains ACTIVE ($0 \rightarrow 1$ or $1 \rightarrow 1$) and RESET means Q goes or remains INACTIVE ($1 \rightarrow 0$ or $0 \rightarrow 0$).

The state diagram of Fig. 5-23(b) identifies an RET D-FF (FET if we complement CK and $\overline{CK}$ over the diagram). But on the state diagram we have drawn attention to the fact that when CLOCK is ACTIVE ($CK = 1$), Q tracks D, that is, the FLIP-FLOP is *transparent*. Thus, when CLOCK is ACTIVE, logic noise can be transferred from input to output. To prevent this from happening, narrow CLOCK pulses must be used to minimize CLOCK's ACTIVE duration. Accordingly, this FLIP-FLOP is called a positive pulse-triggered D-FF even though a state transition is initiated on the rising edge of the CLOCK waveform. The PT D-FF is also called a TRANSPARENT D LATCH or GATED D LATCH.

The state transition table for the D-FF is given in Fig. 5-23(c). It is derived directly from the state diagram by reading the HOLD, SET, and RESET branching requirements exclusive of CLOCK. For example, the $0 \rightarrow 0$ HOLD RESET branching condition is $\overline{D}$ or logic 0, the $0 \rightarrow 1$ SET condition is D hence logic 1, and so on. Like the operation table of Fig. 5-23(a), this state transition table is used to characterize any D-FF regardless of its triggering mechanism. Extensive use will later be made of it in the design of FSMs whose MEMORY elements are D-FFs.

Although the operation table, state diagram, and state transition table offer much of the same information regarding the operation of the FLIP-FLOP, only the state diagram provides the details of its sequential behavior. We already know that the state diagram for the PT D-FF predicts the transparent mode when CLOCK is HIGH, yet this information is not available from an inspection of either the operation table or the state transition table. Also, the state diagram defines the PT D-FF as a two-state FSM, information not obtainable from the tables. In short, the operation and state transition tables place no restriction on the CLOCKing mechanism or the number of states to be used in the design of the FLIP-FLOP—it is the function of the state diagram to do this.

Now that we have characterized the PT D-FF, let us continue with the design process. We have already identified the MEMORY element for this FSM as the BASIC CELL, which is characterized by the state transition table in Fig. 5-24(a). By using this table and the state diagram for the PPT D-FF, we plot the NEXT STATE logic K-maps for SET and RESET, as shown in Fig. 5-24(b). The procedure is simply to AND the 1, 0, or ϕ appearing in the S or R column of the S/R state transition table with the appropriate branching condition for SET ($0 \rightarrow 1$), RESET ($1 \rightarrow 0$), or HOLD ($0 \rightarrow 0$ or $1 \rightarrow 1$) obtained from the D-FF state diagram. By ANDing with the input requirements of Fig. 5-24(a), we identify the MEMORY element as the BASIC CELL. For example, the $0 \rightarrow 0$ HOLD RESET transition requires a $\overline{D} + \overline{CK}$ branching condition. Therefore, a $0 \cdot (\overline{D} + \overline{CK})$ is placed in cell 0 of the SET K-map while a $\phi \cdot (\overline{D} + \overline{CK})$ is placed in cell 0 of the RESET K-map. The $0 \rightarrow 1$ SET transition requires a DCK branching condition, so a $1 \cdot (DCK)$ must be ORed into cell 0 of the SET K-map while a

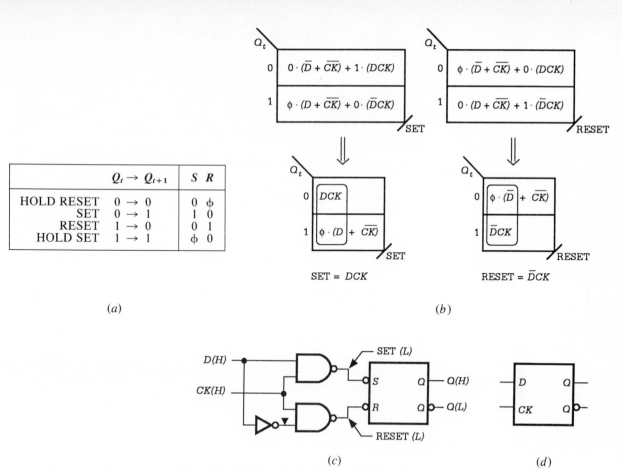

	$Q_t \rightarrow Q_{t+1}$	S	R
HOLD RESET	$0 \rightarrow 0$	0	ϕ
SET	$0 \rightarrow 1$	1	0
RESET	$1 \rightarrow 0$	0	1
HOLD SET	$1 \rightarrow 1$	ϕ	0

(a)

(b)

(c)

(d)

Fig. 5-24. *Design of the pulse-triggered D-FF. (a) State transition table for the MEMORY (BASIC CELL). (b) NEXT STATE logic K-maps and minimum cover. (c) Logic circuit based on the NAND BASIC CELL. (d) Circuit symbol.*

$0 \cdot (DCK)$ must be ORed into cell 0 of the RESET K-map. Similarly, a $1 \rightarrow 1$ HOLD SET transition requires that a $\phi \cdot (D + \overline{CK})$ be placed into cell 1 of the SET K-map while a $0 \cdot (D + \overline{CK})$ must go into cell 1 of the RESET K-map. This process continues until all the outgoing branching conditions have been accounted for in the K-maps. The final step in the plotting of the K-maps is, of course, to reduce the entries to their simplest form as indicated in Fig. 5-24(b). *Remember: The coordinates of a K-map cell represent the PRESENT STATE while the entry in that cell represents the NEXT STATE instruction for a given state variable bit.*

Once the mapping procedure is understood the reader may wish to use a "shortcut" approach. Observe that for every essential K-map entry [i.e., any $1 \cdot (\ )$ entry] in a given cell of Fig. 5-24(b), there exists a ϕ-term entry in the same cell of the other K-map that is the complement of the essential entry. Thus, a DCK in cell 0 of SET K-map calls for a $\phi \cdot (\overline{DCK}) = \phi \cdot (\overline{D} + \overline{CK})$ entry in cell 0 of the RESET K-map, and so on. By using this shortcut method, one can save time in the SET/RESET mapping process. However, the reader is cautioned not to lose sight of the reasons behind it.

Having plotted the K-maps, we now have an easy task to complete the design of the PPT D-FF. The minimum cover is found as shown in Fig. 5-24(b), which yields the logic circuit of Fig. 5-24(c) and its circuit symbol given in Fig. 5-24(d). Here, we have assumed that both the *D* and *CK* inputs arrive ACTIVE HIGH and that a NAND BASIC CELL is to serve as the MEMORY element. Had we wanted to implement this FF with a NOR-centered BASIC CELL, we would use the circuit symbol of Fig. 5-17(b) with two AND gates. This is equivalent to complementing on the "inside" of the logic circuit in Fig. 5-24(c) so as to remove the ACTIVE LOW indicator bubbles.

The operation of the PPT D-FF is best illustrated by the timing diagram in Fig. 5-25, where both the FLIP-FLOP and transparent modes are represented. Notice that when CLOCK is HIGH, *Q* follows *D*, meaning all logic transitions appearing on the *D* input line will be transferred to the output during the ACTIVE period of the CLOCK waveform. For the sake of simplicity, all propagation delays have been neglected.

The *CK* pulses required for the PPT D-FF can be produced by any one of several suitable *pulse-narrowing* circuits. One simple but effective circuit is shown in Figs. 4-96(a), a circuit which generates a POS HAZARD. The single INVERTER should be replaced by three INVERTERs so as to produce pulses of reliable width but still narrow enough to minimize the ACTIVE duration of the CLOCK signal. Of course, if it could be guaranteed that no unauthorized data transitions would occur in the *D* input signal during the time the CLOCK waveform is ACTIVE, then there would be no need for pulse-narrowing circuits.

The T FLIP-FLOP. The design of the "toggle" or T FLIP-FLOP follows in a manner similar to that of the D-FF. We begin by characterizing this FLIP-FLOP as in Fig. 5-26. The operation table in Fig. 5-26(a) indicates that a T-FF can either HOLD or TOGGLE. TOGGLE means that the FF alternates between SET and RESET on each successive triggering edge of the CLOCK waveform.

The state diagram for the PT T-FF is presented in Fig. 5-26(b). It is derived from the operation table by assuming a two-state FSM whose state transitions are initiated on the rising edge of the CLOCK pulse. The branching conditions for the $0 \rightarrow 1$ and $1 \rightarrow 0$ transitions are determined by reading

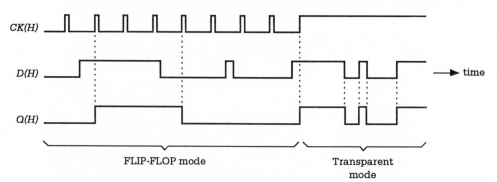

Fig. 5-25. *Timing diagram for the pulse-triggered D-FF showing FLIP-FLOP and transparent modes.*

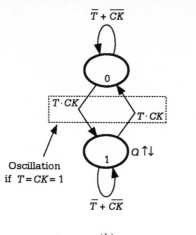

T	Q_{t+1}	
0	Q_t	HOLD
1	$\overline{Q}_t$	TOGGLE

(a)

$Q_t \rightarrow Q_{t+1}$	T
$0 \rightarrow 0$	0
$0 \rightarrow 1$	1
$1 \rightarrow 0$	1
$1 \rightarrow 1$	0

(b)

(c)

Fig. 5-26. *Characterization of the T FLIP-FLOP. (a) Operation table (b) State diagram for the pulse-triggered T-FF, derived from the operation table, showing an oscillation for the* (T, CK) = (1, 1) *condition. (c) State transition table derived from the state diagram.*

TOGGLE as TCK for SET and TCK for RESET. The HOLD condition for each state follows either from the SUM rule or from the operation table.

The state transition table, given in Fig. 5-26(c), is easily derived from either the operation table or from the state diagram. The state transition table, like the operation table, provides the same information about the operation of any T-FF independent of the triggering mechanism. Only the state diagram explicitly characterizes the PT T-FF triggering mechanisms.

Notice that the state diagram of Fig. 5-26(b) clearly predicts that this FLIP-FLOP will break into oscillation for the $T = CK = 1$ condition. That is, when CK samples T HIGH on the rising edge, the $0 \rightarrow 1$ transition takes place, but if T and CK remain HIGH, a $1 \rightarrow 0$ transition is required. This creates an endless cycle (oscillation) for as long as T and CK are both AC-TIVE. When one of the inputs goes INACTIVE, the FF will stabilize into one of the two states. To prevent the oscillation (if that is possible), this T-FF must be operated with positive CLOCK pulses of very short duration; hence the reason for calling it a PPT T-FF.

The design of the PPT T-FF continues by mapping the state diagram. As in the design of the PT D-FF, the SET and RESET K-map cell entries are determined by ANDing each branching condition in the state diagram with the appropriate S or R input requirement of the BASIC CELL given in the state transition table of Fig. 5-27(a). Thus, for the $0 \rightarrow 1$ SET transition a 1 $\cdot (TCK)$ is placed in cell 0 of the SET K-map, but for the $1 \rightarrow 1$ HOLD SET condition, a $\phi \cdot (\overline{T} + \overline{CK})$ must be placed in cell 1 of the SET K-map, and so on. In short, the mapping process involves matching a transition branching condition in the state diagram with the input requirement of the MEMORY element for that transition, and then placing the ANDed result in the appropriate cell of the K-map. Once this is understood, the reader is encouraged to simplify the mapping process by recognizing that each don't care entry in a cell is the complement of the essential entry in the same cell of the other K-map.

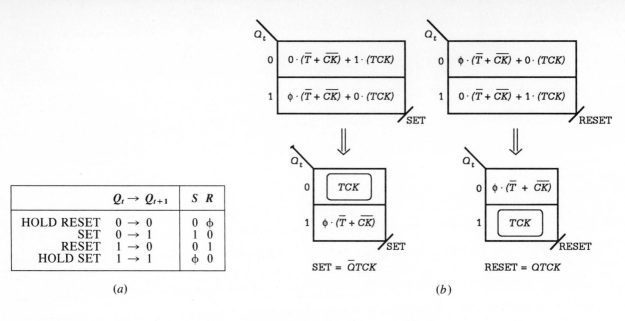

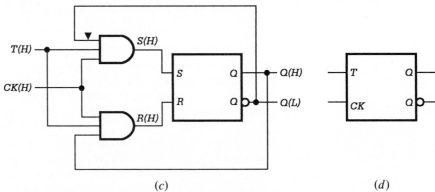

Fig. 5-27. *Design of the pulse-triggered T-FF. (a) State transition table for the MEMORY (BASIC CELL). (b) NEXT STATE logic K-maps and minimum cover. (c) Logic circuit based on the NOR BASIC CELL. (d) Circuit symbol.*

The design of the PPT T-FF is completed by extracting minimum cover from the K-maps and then implementing the results as shown in Figs. 5-27(b) and (c). In this case we have chosen to use a NOR-centered BASIC CELL to illustrate flexibility of the design process. Recall that use of the generic form of the state transition table for the BASIC CELL permits this since the $(S, R) = (1, 1)$ condition does not occur in the table. Whatever form of the BASIC CELL is used, the circuit symbol for the PPT T-FF remains the same as that in Fig. 5-27(d).

The timing diagram in Fig. 5-28 summarizes the operation of the PPT T-FF. Shown are the TOGGLE mode and the oscillatory behavior predicted from the state diagram. To prevent oscillation while in the TOGGLE mode, the ACTIVE durations of the CLOCK waveform must not exceed the worst case path delay for the FF, which is that for the AND gate and BASIC CELL combined. Use of a pulse-narrowing circuit (for CLOCK) of the type

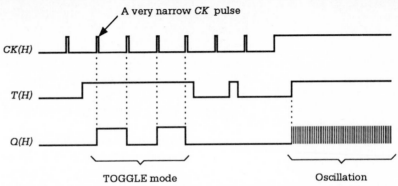

A very narrow CK pulse

CK(H)

T(H)

Q(H)

TOGGLE mode   Oscillation

Fig. 5-28. *Timing diagram for the PT T-FF showing the TOGGLE mode and oscillatory behavior under the* T = CK = 1 *condition.*

described in the previous example is obviously necessary if proper operation of the PT T-FF is to be made possible.

The JK FLIP-FLOP. The operation table that characterizes any JK FLIP-FLOP is given in Fig. 5-29(a). This table indicates that the JK-FF possesses HOLD, RESET, SET, and TOGGLE operations. A comparison of this table with those in Figs. 5-23(a) and 5-26(a) suggests that the JK-FF can perform the combined operations of the D and T FLIP-FLOPs, and this is true. In fact the JK-FF is sometimes called the *universal* FLIP-FLOP since only minor external logic is needed to convert it to either a D-FF or a T-FF. This subject will be explored further in Sect. 5.6.6.

The state diagram for the PPT JK-FF is shown in Fig. 5-29(b). Its construction is derived from the operation table by assuming a two-state machine and by assuming that CLOCK samples (triggers) on the rising edge of its waveform. The branching conditions are determined by reading the oper-

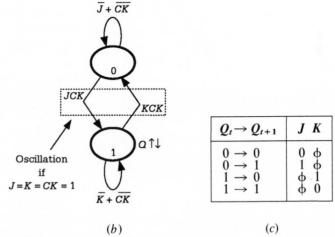

J K	Q_{t+1}	
0 0	Q_t	HOLD
0 1	0	RESET
1 0	1	SET
1 1	$\overline{Q}_t$	TOGGLE

(a)

$\overline{J} + \overline{CK}$

0

JCK   KCK

Oscillation
if
$J = K = CK = 1$ 1 $a \uparrow\downarrow$

$\overline{K} + \overline{CK}$

(b)

$Q_t \rightarrow Q_{t+1}$	J K
$0 \rightarrow 0$	0 ϕ
$0 \rightarrow 1$	1 ϕ
$1 \rightarrow 0$	ϕ 1
$1 \rightarrow 1$	ϕ 0

(c)

Fig. 5-29. *Characterization of the JK-FF. (a) Operation table. (b) State diagram for the PT JK-FF derived from the operation table showing oscillation under the* J = K = CK = 1 *condition. (c) State transition table derived from the state diagram.*

ation table for SET, RESET, and HOLD. For example, the $0 \rightarrow 1$ transition is produced by both the SET and TOGGLE operations, meaning that $J\overline{K} + JK = J$ with CK, or JCK, is the required branching condition for this transition. Similarly, the $0 \rightarrow 0$ HOLD RESET transition requires $\overline{J}\overline{K} + \overline{J}K + \overline{CK} = \overline{J} + \overline{CK}$ as the branching condition, a result that could have been deduced from the $\sum = 1$ rule knowing the $0 \rightarrow 1$ branching condition. The remaining two branching conditions for $1 \rightarrow 0$ and $1 \rightarrow 1$ transitions are determined in a like manner.

The state transition table given in Fig. 5-29(c) also characterizes any JK-FF since no triggering mechanism is specified. It is easily derived from the state diagram by noting the input branching requirements for the transitions exclusive of CLOCK. Thus, the $0 \rightarrow 0$ transition requires $\overline{J}$ (K is irrelevant), so a logic 0 is placed in the J column and a don't care is placed in the K column. Or the $1 \rightarrow 0$ transition occurs under input condition K, which places a logic 1 in the K column and a ϕ in the J column (J is irrelevant). Remembering this and the other state transition tables and knowing how to use them in sequential FSM design will be important to the reader in the sections to come.

The state diagram of Fig. 5-29(b) contains other useful information not available in either the operation table or the state transition table. Under the condition $J = K = CK = 1$, it is predicted that this JK-FF will break into an oscillation and will continue to oscillate until one of the inputs goes INACTIVE, the final state being unpredictable. As in the case of the PPT T-FF, the oscillatory behavior of the PPT JK-FF can be prevented only by using very narrow CLOCK pulses from a pulse-narrowing circuit of the type described previously for the PPT T-FF or by requiring that inputs J and K never be ACTIVE at the same time. This is the reason for classifying this FF as pulse triggered. Figure 6-55(b) shows one possible application of NPT JK-FFs in PULSE mode circuits.

The procedure for completing the design of the PPT JK-FF follows from the two previous design examples. The K-maps for SET and RESET are plotted by ANDing each outgoing branching condition with the associated S, R input requirements taken from the state transition table in Fig. 5-27(a). The results are presented in Fig. 5-30(a). Once plotted, minimum cover is extracted from the K-maps and then used to implement the NEXT STATE logic for the BASIC CELL (NAND in this case) as shown in Figs. 5-30(a) and (b). Notice the presence of the Q-output feedback loops which account for the TOGGLE character and oscillatory behavior of this FLIP-FLOP similar to that of the previous T-FF.

A fully detailed circuit is used to implement the PPT JK-FF in Fig. 5-30(b) so as to demonstrate the use of the asynchronous overrides, PRESET (PR) and CLEAR (CL), which are commonly found on FLIP-FLOPs of all types. Because a NAND-centered design is used, the overrides are presented ACTIVE LOW. For normal operation of the FLIP-FLOP both PR(L) and CL(L) must be INACTIVE [$0(L) = 1(H)$]. Then, if PR(L) goes ACTIVE [$1(L)$] with CL(L) INACTIVE [$0(L)$], the FLIP-FLOP is SET. That is, the $1(L)$ to NAND gate 3 forces its output to $1(H)$; the $1(L) = 0(H)$ presented to NAND gate 2 introduces a $0(L) = 1(H)$ to NAND gate 4 which causes its output, $Q(L)$, to be $1(L)$, thereby completing the SET condition of the FLIP-FLOP. Similarly, if CL(L) becomes ACTIVE [$1(L) = 0(H)$] while

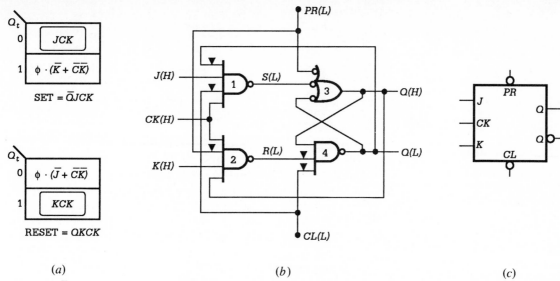

Fig. 5-30. *Design of a PT JK-FF. (a) NEXT STATE logic K-maps for SET and RESET. (b) Logic circuit with PRESET and CLEAR overrides. (c) Circuit symbol.*

PR(L) is INACTIVE, the FLIP-FLOP is RESET. Simultaneous ACTIVATION of both overrides SETs $Q(H)$ and RESETs $Q(L)$—a loss of the mixed-rail feature, as expected. The circuit symbol representing the PPT JK-FF with ACTIVE LOW asynchronous overrides is shown in Fig. 5-30(c).

If we had used a NOR-centered BASIC CELL instead of NAND based, NOR gates would be needed for the NEXT STATE forming logic. In this case the PR and CL asynchronous overrides would be ACTIVE HIGH and would be held INACTIVE [0(H)] for normal operation of the FLIP-FLOP. The SET and RESET of this FLIP-FLOP takes place if PR(H) and CL(H) are ACTIVATED separately. Without INVERTERs the inputs to this FLIP-FLOP would have to arrive ACTIVE LOW.

The operation of the PPT JK-FF is illustrated by the timing diagram in Fig. 5-31. Here, it is observed that ACTIVE J only causes a SET with CLOCK, ACTIVE K only causes a RESET with CLOCK, and when both J and K are ACTIVE the FLIP-FLOP TOGGLES with each pulse of CLOCK. Then when J, K, and CLOCK are all ACTIVE, the FLIP-FLOP breaks into oscillation as predicted from the state diagram. The asynchronous overrides, PR and CL, interrupt the "normal" operation as shown.

5.6.4 Master/Slave FLIP-FLOPs

It should seem obvious to the reader that the pulse-triggered FLIP-FLOPs have some severe limitations and problems relative to their use as MEMORY devices. For example, the PT D-FF becomes transparent when CLOCK is ACTIVE, allowing input data (logic noise and all) to be transferred to the output of the FLIP-FLOP. Worse yet, the PT JK-FF breaks into oscillation when J, K, and CLOCK are all ACTIVE. In this section we will consider a better alternative to the pulse-triggered FLIP-FLOPs.

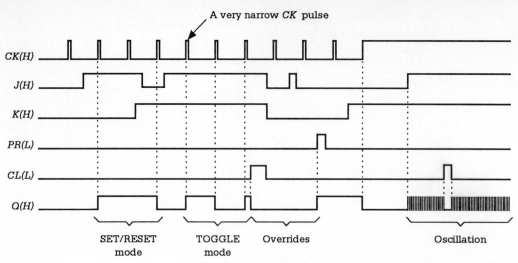

A very narrow CK pulse

SET/RESET mode   TOGGLE mode   Overrides   Oscillation

Fig. 5-31. *Timing diagram for the PT JK-FF showing the main features.*

The Master/Slave D FLIP-FLOP. One solution to the PT-FF problem is the use of a master/slave FLIP-FLOP configuration which involves the coupling of two PT FFs. Shown in Fig. 5-32(a) is the logic circuit for an MS D-FF, and in Fig. 5-32(b) its circuit symbol. It is seen to be composed of two PT D-FFs coupled together with an INVERTER such that the master D-FF is triggered on the rising edge of the CLOCK pulse while the slave D-FF is triggered on the falling edge of the same CLOCK pulse. So at no time is the

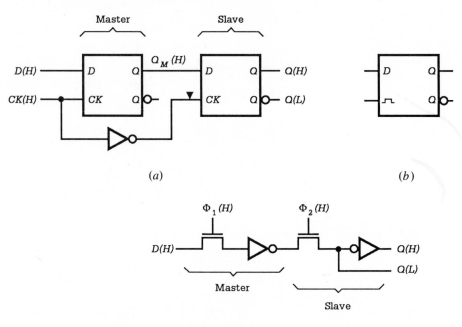

Fig. 5-32. *The master/slave D-FF. (a) Logic circuit composed of coupled PT D-FFs. (b) Circuit symbol. (c) Electronic realization.*

entire device transparent to the input D. Data are CLOCKed to the input of the slave D-FF on the ACTIVE transition of CLOCK but are not CLOCKed to the output of the FLIP-FLOP until the INACTIVE transition of CLOCK arrives, a two-stage process. However, during the ACTIVE period of CLOCK ($CK = 1$), the master D-FF is transparent to D. So, the only requirement is that the D input be stable at its proper logic level just before, during and just after the INACTIVE transition of the CLOCK waveform to ensure proper operation of the FLIP-FLOP. From this information we conclude that the MS D-FF is only a partial solution to the PT-FF problem. The transparency mode found in the PT D-FF has been eliminated but now there exists a half-CK-cycle delay in the output response to an input change, a feature that was not present in the PT D-FF.

Figure 5-33 summarizes the operation of the MS D-FF. Here, the half-cycle delay in output response to an input change is clearly observed. Notice that while the master "cell" is transparent to D for ACTIVE CLOCK, the slave "cell" is not. This, of course, is a consequence of the fact that the master and slave sections are never CLOCKed at the same time.

In Sect. 5.6.3 we took special care to emphasize the fact that the operation table and state transition table characterize a FLIP-FLOP independent of any triggering mechanism. Not surprisingly, then, the tables of Figs. 5-23(a) and (c) also characterize the MS D-FF. It is important for the reader to remember this fact, as the state transition table will become an important instrument for design.

Because of the simplicity of their physical configuration, MS D-FFs have become very important in VLSI design. Shown in Fig. 5-32(c) is one electronic realization of an MS D-FF. As can be seen, it is composed of just two NMOS pass transistor switches and two INVERTERs alternatively placed. To avoid the FLIP-FLOP transparency problem, the transistor switches are operated by a 2Φ nonoverlapping CLOCK of the type produced by the NOR-centered BASIC CELL in Fig. 5-21. Charge is stored long enough in the capacitances of the pass transistor and INVERTER switches to prevent loss of the data information between the ACTIVE periods of the two phases. A $Q(L)$ output can be generated by tapping off the input to the second INVERTER as shown in Fig. 5-32(c). This has the effect of producing a $Q(L)$ signal that always leads $Q(H)$ in time. Such is not the case for the mixed-rail outputs from a BASIC CELL (see Fig. 5-17).

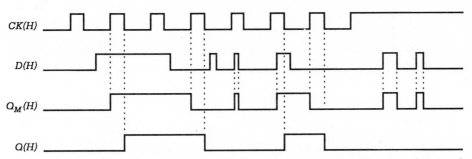

Fig. 5-33. *Timing diagram for the MS D-FF showing that it solves the transparency problem of the PT D-FF.*

The Master/Slave JK FLIP-FLOP. The master/slave configuration, as a solution to the PT-FF problem, is not as straightforward for the JK-FF as it was for the D-FF. We will now discover that the common MS JK-FF solves the oscillation problem encountered with the PT JK-FF, but it creates another problem almost as serious.

Usually, the MS JK-FFs are configured by two GATED BASIC CELLs (master and slave) as shown in Fig. 5-34—the GATED BASIC CELL is one where the S and R inputs are ENABLED by the CLOCK through the action of two ANDing operations. An inspection of the circuit in Fig. 5-34(a) reveals that oscillation is not possible for the $J = K = CK = 1$ condition, as was the case for the PT JK-FF. This is so because the INVERTER ensures that the slave CELL will never be CLOCKed at the same time that the master CELL is CLOCKed. The MS JK-FF operates by loading the master CLOCKed SR CELL on the rising edge of the CLOCK waveform and then updating the output via the slave CELL on the falling edge, a process no different than that of the MS D-FF. The MS JK-FF is sometimes called a pulse-triggered device and is given the circuit symbol shown in Fig. 5-34(b). The reason it is classified this way will be made evident following a more detailed description of its operation.

The normal operation of the MS JK-FF is illustrated by the timing diagram in Fig. 5-35. Here, the half-cycle delay in output response to an input change is evident. However, a closer inspection of the circuit in Fig. 5-34 reveals a serious problem called *error catching* (also called 1's or 0's catching), illustrated in Fig. 5-35. While CLOCK is ACTIVE ($CK = 1$), the master CELL can be irreversibly SET or RESET by logic noise (glitches) in the J and K input signals allowing erroneous data to be CLOCKed to the output by the slave CELL on the falling edge of CLOCK. For example, if Q and CLOCK are ACTIVE and K undergoes an unauthorized positive (ACTIVE) glitch, the master CELL will be irreversibly RESET causing the output to be RESET by the slave CELL on the next falling edge of the CLOCK waveform. Similarly, if Q is INACTIVE when CLOCK is ACTIVE, an

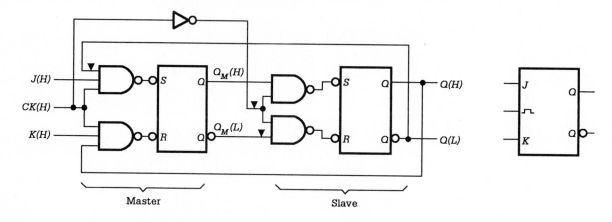

(a)   (b)

Fig. 5-34. *(a) The MS JK-FF configured with GATED BASIC CELLs. (b) Circuit symbol.*

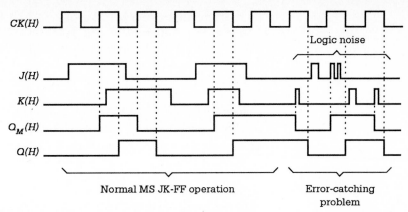

Fig. 5-35. *Timing diagram for the MS JK-FF of Figure 5-34 showing normal FLIP-FLOP operation and the error-catching problem.*

unauthorized positive glitch in *J* will irreversibly set the master CELL, causing, in turn, the FLIP-FLOP to SET (via the slave CELL) on the next falling edge of CLOCK.

The error-catching problem of the MS JK-FF just described is not shared by the MS D-FF. It will be recalled that the only requirement in the use of the MS D-FF is that the *D* input signal be stable at its proper logic level just before, during and just after the INACTIVE transition of CLOCK, thereby permitting the *D* signal to assume any logic level prior to that time. This is not the case for the MS JK-FF whose *J* and *K* input signals must be logically noise free throughout the ACTIVE period of CLOCK to avoid an irreversible error SET or RESET of the master CELL and the eventual error SET or RESET of the FLIP-FLOP on the next falling edge of CLOCK. In addition, the *J* and *K* input signals must also be stable at their proper logic levels just before, during and after both the rising and falling edges of CLOCK. Obviously, the requirements for proper operation of the MS JK-FF are more stringent than are those for the MS D-FF. We will return to this subject later in Sect. 5.6.7.

If FSMs are to be designed by using the MS JK-FFs as MEMORY elements, care must be taken to ensure that unauthorized transitions do not occur on *J* or *K* input lines during the ACTIVE period of CLOCK. One approach to accomplish this is to trigger the MS JK-FF by using very narrow pulses. But to do this would defeat the purpose in using the MS JK-FF. A better approach is to design the MS JK-FF around the MS D-FF rather than around the GATED BASIC CELL. This device will have the less stringent triggering requirements of the MS D-FF while still providing the *JK* operating features. Section 5.6.6 will explore this subject in greater detail.

5.6.5 Edge-Triggered FLIP-FLOPs

The complete solution to the pulse-triggered FLIP-FLOP problem is the edge-triggered FLIP-FLOP. Within these devices triggering occurs on either the rising edge or falling edge of the CLOCK waveform but thereafter "locks out" any changes that occur on the input data lines until the next triggering edge of CLOCK. The output is updated immediately following the triggering edge of CLOCK plus an intervening path delay. Therefore, the only requirement for proper operation of the ET FLIP-FLOP is that the input signals be stable at their proper logic levels just before, during, and just after the sampling interval (triggering edge) of the CLOCK waveform. In this section we will design a rising edge-triggered D-FF from first principles and then discuss the operation of a falling edge-triggered JK-FF.

The RET D FLIP-FLOP. The operation table and state transition table given in Figs. 5-23(a) and (c) characterize the RET D-FF, but they do not provide detailed information concerning its sequential behavior. This latter information is best obtained by considering its state diagram and positive logic timing diagram presented in Fig. 5-36. In STATE $\textcircled{a}$ the FLIP-FLOP is in a RESET condition (Q is INACTIVE) and CLOCK is INACTIVE ($\overline{CK}$). On the rising edge of CLOCK, the FLIP-FLOP will sample D. If CLOCK samples D LOW, hence $\overline{D}CK$, the FLIP-FLOP will transit to STATE $\textcircled{d}$ where it retains the RESET condition. If, instead, CLOCK samples D HIGH, hence DCK, then the FLIP-FLOP will make the transition $\textcircled{a} \rightarrow \textcircled{b}$ and issue the output Q—a SET condition. The FLIP-FLOP will remain in STATE $\textcircled{b}$ for as long as CLOCK is ACTIVE, hence the HOLD condition, CK, on STATE $\textcircled{b}$ in the state diagram. To sample D again, CLOCK must first go LOW ($\overline{CK}$) which accounts for the transition $\textcircled{b} \rightarrow \textcircled{c}$. In STATE $\textcircled{c}$ CLOCK is LOW but Q is still ACTIVE. On the rising edge of CLOCK, the

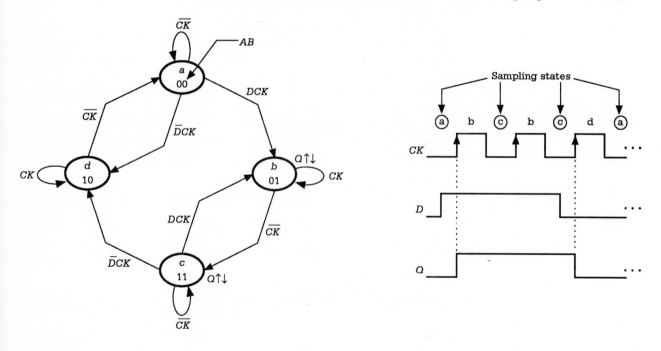

(a)　　　　　　　　　　　　　　　　　　　　(b)

Fig. 5-36. *Sequential description of an RET D-FF. (a) State diagram. (b) Positive logic timing diagram for (a).*

FLIP-FLOP will sample *D* again. If CLOCK samples *D* HIGH (*DCK*), then the FLIP-FLOP will transit back to ⓑ where *Q* remains ACTIVE. But if CLOCK samples *D* LOW ($\overline{D}CK$), the FLIP-FLOP must transit ⓒ→ⓓ, a RESET condition (*Q* is INACTIVE) which is maintained for as long as CLOCK is HIGH. To further sample *D*, CLOCK must first go INACTIVE forcing the FLIP-FLOP to transit from ⓓ to ⓐ thus completing the sequence of steps represented by the state diagram.

The operation of the RET D-FF is unique in the sense that once a transition occurs from a sampling state (STATE ⓐ or STATE ⓒ), variation in the logic level of *D* can have no effect on the output *Q* until the next rising edge of the CLOCK waveform. This is a consequence of the four state operation. The PT D-FF, which is defined by the two-state state diagram in Fig. 5-23(b), does not share this "lockout" feature. It should be remembered that for the PT D-FF, *Q* tracks *D* when *CK* is HIGH—that is, it becomes transparent.

Design of the RET D-FF follows the same procedure used for the PT FLIP-FLOPs in Sect. 5.6.3 except now there are two state variables (*A* and *B*) instead of one. We will use the basic model of Fig. 5-2(c) with two BASIC CELLs for the memory as indicated in Fig. 5-37. The procedure requires that we characterize the BASIC CELL by its state transition table in Fig. 5-38(a) and use it together with the state diagram of Fig. 5-36(a) to plot the NEXT STATE K-maps given in Fig. 5-38(b). Recall that the plotting procedure requires ANDing the *S*, *R* information in the state transition table for the BASIC CELL with the appropriate branching conditions for SET, RESET, and HOLD taken from the state diagram. A shortcut method for plotting these K-maps exists by realizing that each φ-term in a given cell is the complement of the essential entry in the same cell of the other K-map. Thus, the 0 cell entry for S_A is $\overline{D}CK$ so the 0 cell entry for R_A is $\phi \cdot (D + \overline{CK})$, and so on. The output K-map, also presented in Fig. 5-38(b), is plotted directly from the state diagram.

The design of the RET D-FF is completed by extracting minimum cover

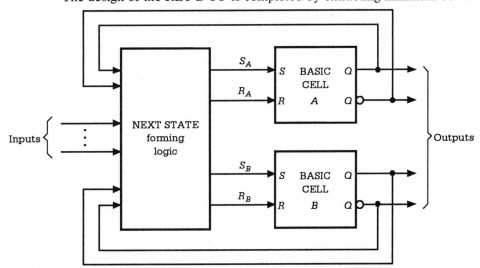

Fig. 5-37. *The basic model of Figure 5-2(c) with two BASIC CELLs as the MEMORY stage.*

FIGURE 5-38(a)

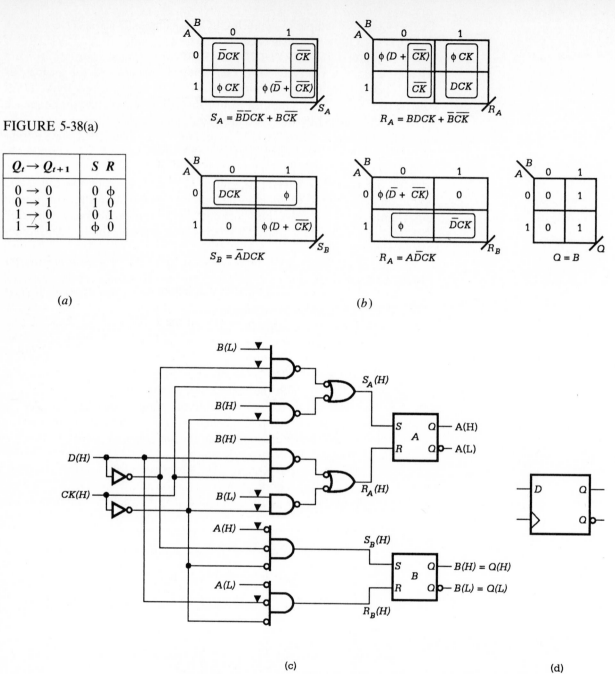

$Q_t \rightarrow Q_{t+1}$ table:

$Q_t \rightarrow Q_{t+1}$	S	R
$0 \rightarrow 0$	0	ϕ
$0 \rightarrow 1$	1	0
$1 \rightarrow 0$	0	1
$1 \rightarrow 1$	ϕ	0

$S_A = \overline{B}DCK + B\overline{CK}$

$R_A = BDCK + \overline{B}\,\overline{CK}$

$S_B = \overline{A}DCK$

$R_A = A\overline{D}CK$

$Q = B$

(a)

(b)

(c)

(d)

Fig. 5-38. *Design of the RET D-FF by using the basic model of Figure 5-37. (a) Characterization of the MEMORY element. (b) NEXT STATE and output K-maps. (c) Implementation by using NOR BASIC CELLs and mixed NEXT STATE logic. (d) Circuit symbol for an RET D-FF.*

from the NEXT STATE K-maps in Fig. 5-38(b) and implementing these results as shown in Fig. 5-38(c). Observe that we have chosen two NOR BASIC CELLs for the MEMORY and NAND/NOR gates for the NEXT STATE forming logic. Notice also that the output K-map is read as $Q = B$,

5 / Synchronous Sequential Machines

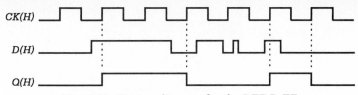

Fig. 5-39. *Timing diagram for the RET D-FF.*

which requires that the mixed-rail outputs are taken from BASIC CELL *B*. The circuit symbol for the RET D-FF is given in Fig. 5-38(d).

The operation of the RET D-FF is summarized by the MIXED LOGIC timing diagram presented in Fig. 5-39. The reader can gain a good understanding of the differences that exist between the three types of D-FF triggering mechanisms (PT, MS, and ET) considered so far by comparing the timing diagrams of Figs. 5-25, 5-33, and 5-39. In doing so the relative simplicity of the ET FF operation can be appreciated.

The FET JK FLIP-FLOP. The sequential behavior of the falling edge-triggered JK-FF is clearly indicated by the state diagram and timing diagram shown in Figs. 5-40(a) and (b). Assume that the FLIP-FLOP is RESET into STATE ⓐ at some point in time. On the falling edge of the CLOCK waveform, the FF samples the inputs. If CLOCK samples *J* LOW, hence $\overline{JCK}$, the FLIP-FLOP transits ⓐ → ⓐ, retaining the RESET condition. But if CLOCK samples *J* HIGH ($J\overline{CK}$), the FLIP-FLOP SETs into STATE ⓑ and HOLDs there for as long as CLOCK is LOW (INACTIVE). On the

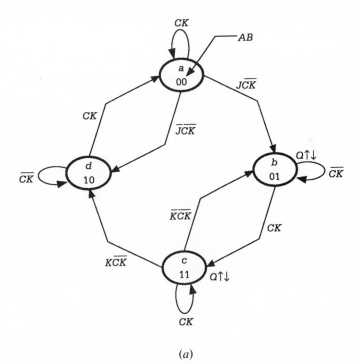

(a)

Fig. 5-40. *Sequential behavior and symbology for the FET JK-FF. (a) State diagram.*

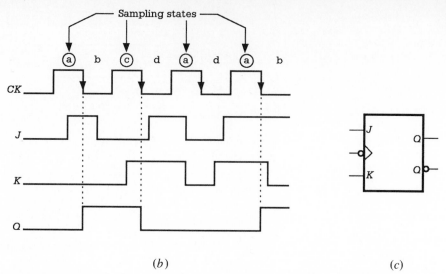

(b)   (c)

Fig. 5-40 (contd.). *(b) Positive logic timing diagram. (c) Circuit symbol for the FET JK-FF.*

rising edge of CLOCK the FLIP-FLOP will transit $\textcircled{b} \rightarrow \textcircled{c}$ and retain its SET condition for the ACTIVE CLOCK duration. In STATE $\textcircled{c}$ the FLIP-FLOP will again sample the inputs on the falling edge of CLOCK. If CLOCK samples K LOW, the FLIP-FLOP will SET back into STATE $\textcircled{b}$ with a branching condition $\overline{K}\overline{CK}$. Or if CLOCK samples K HIGH (hence, $K\overline{CK}$), the FLIP-FLOP will RESET into STATE $\textcircled{d}$ and remain there for the INACTIVE CK interval. Then, on the rising edge of CLOCK the FLIP-FLOP will transit $\textcircled{d} \rightarrow \textcircled{a}$, thereby retaining the RESET condition and completing the sequence. The circuit symbol for the FET JK-FF is given in Fig. 5-40(c).

Although the sequential behavior of the FET JK-FF is described by both the state diagram and the timing diagram in Fig. 5-40, the state diagram representation provides the more lucid description. For example, the TOGGLE mode is clearly indicated for J and K ACTIVE by the sequence . . . $\textcircled{a} \rightarrow \textcircled{b} \rightarrow \textcircled{c} \rightarrow \textcircled{d} \rightarrow \textcircled{a}$ . . . , each transition being initiated on the falling edge of the CLOCK waveform. Should the reader have difficulty deducing the branching conditions for the $\textcircled{a} \rightarrow \textcircled{d}$ and $\textcircled{c} \rightarrow \textcircled{b}$ transitions, the $\Sigma = 1$ rule can be applied. Thus, the branching condition for the $\textcircled{c} \rightarrow \textcircled{b}$ transition is found from $(\overline{K\overline{CK}} + CK) = \overline{K}\,\overline{CK}$, since for STATE $\textcircled{c}$ the three outgoing branching conditions must yield $(K\overline{CK} + CK) + \overline{K}\,\overline{CK} = 1$ according to the sum rule. Completion of this design is left as an exercise for the reader.

Some additional comments are in order regarding the operation of the FET JK-FF. First, observe from Fig. 5-40 that the FLIP-FLOP effectively "locks out" logic level variations in the J and K input signals that may occur between successive sampling intervals of CLOCK. Therefore, the only requirement is that the J and K inputs be stable at their proper logic levels during the sampling interval, that is, just before, during and just after the triggering edge of the CLOCK waveform. Also, notice that an RET JK-FF is defined by the state diagram of Fig. 5-40(a) if CK and $\overline{CK}$ are complemented throughout. Finally, remember that the operation and state transition

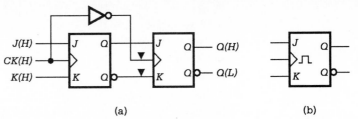

Fig. 5-41. *The data lockout FLIP-FLOP. (a) Logic circuit as configured from RET JK-FFs. (b) Circuit symbol.*

tables given in Figs. 5-29(a) and (c) characterize a JK-FF of any type (PT or MS or ET) because there is nothing in these tables that specifies a particular triggering mechanism.

There is another type of edge-triggered FLIP-FLOP that combines the characteristics of the MS-FF with those of the ET-FF. It is called the data-lockout (DL) FLIP-FLOP. The DL FLIP-FLOP has the half-*CK*-cycle delay in output response to input change of the MS FLIP-FLOP but with the data "lockout" feature of the edge-triggered FLIP-FLOP. The DL JK-FF can be configured from ET JK-FFs, as shown in Fig. 5-41(a) and has the circuit symbol given in Fig. 5-41(b). A DL D-FF version is similarly configured by using ET D-FFs. Data-lockout FFs can also be designed as single asynchronous FSMs but, because of their somewhat more complex sequential nature, such designs are postponed until Sect. 6.7.2.

5.6.6. FLIP-FLOPs Designed from Other FLIP-FLOPs

This section, unpretentious as it sounds, actually brings us to an important juncture in the development of a design methodology for synchronous FSMs. In the previous sections on FLIP-FLOP design we used the BASIC CELL as a MEMORY element, but the BASIC CELL has no CLOCKing mechanism associated with it. The CLOCKing mechanism, or triggering mechanism as we have called it, was introduced by the external logic required to satisfy the sequential behavior represented by the state diagram of the FLIP-FLOP. Now, we will design simple FSMs (FLIP-FLOPs) by using other FLIP-FLOPs which have a triggering mechanism associated with them.

In this section we will make use of the following very important fact:

> *The triggering mechanism of a device being designed inherits the triggering mechanism of the FLIP-FLOPs used as the MEMORY elements for the design.*

This means that a PT T-FF can be designed from a PT D-FF or an ET JK-FF can be designed by using an ET D-FF, and so on. The process of designing FLIP-FLOPs from other FLIP-FLOPs is sometimes called FLIP-FLOP *conversion*. Designing FLIP-FLOPs of different triggering mechanism than the MEMORY element FLIP-FLOP is far more difficult to accomplish and unnecessary, the exceptions being the design of the MS FFs or the DL FFs.

There is another related but important point that must be remembered.

To design an FSM by using a particular type of FLIP-FLOP (D, T, JK, etc.), it is necessary that the FLIP-FLOP be characterized and that this information be introduced into the design process. Therefore, we emphasize the following:

> *It is the state transition table that characterizes the FLIP-FLOP MEMORY element, and it is this information that must be introduced into the design process by ANDing the input requirements of the state transition table with the appropriate branching conditions taken from the state diagram of the FSM to be designed.*

The reader may recall that this is precisely what was done in the design of the PT- and ET-FFs where use was made of a MEMORY element (the BASIC CELL) with no built-in CLOCKing mechanism. What is different now is that the choice of FLIP-FLOP MEMORY element determines the triggering mechanism for the FSM (FLIP-FLOP in this case) to be designed.

Having digested this important information, we are ready to begin the design of FLIP-FLOPs by using other FLIP-FLOPs as the MEMORY. We will carry out three such designs, leaving a variety of other possible designs to the imagination of the reader.

T-FFs Designed from JK-FFs. Type T-FFs are usually not commercially available, and so they must be designed (converted) from other types of FLIP-FLOPs. The state diagram for the T-FF FSM is given in Fig. 5-42(a). The input CK has been excluded from this state diagram because it is understood to be an inherent part of the MEMORY element FLIP-FLOP, which, in this case, is chosen to be a JK-FF.

Shown in Fig. 5-42(b) is the state transition table for the JK-FF. It is this information that must be introduced into the design process so as to characterize the MEMORY element. This is done by ANDing the information in the J, K columns with the appropriate branching conditions taken from state diagram and then placing the results in the NEXT STATE K-map as shown in Fig. 5-42(c). Thus, the $0 \rightarrow 0$ (HOLD RESET) transition places a $0 \cdot \overline{T}$ in cell 0 of the J K-map and a $\phi \cdot \overline{T}$ in cell 0 of the K K-map. The $0 \rightarrow 1$ (SET) transition ORs a $1 \cdot T$ into cell 0 of the J K-map but ORs a $\phi \cdot T$ into cell 0 of the K K-map. Or the $1 \rightarrow 1$ (HOLD SET) transition places a $\phi \cdot \overline{T}$ in cell 1 of the J K-map while placing a $0 \cdot \overline{T}$ in cell 1 of the K K-map, and so on.

Once this JK mapping procedure is understood, the reader may wish to use a shortcut method by noting that for every essential entry in the J, K K-maps there exists a ϕ entry in the same cell of the other K-map. Notice that $\phi \cdot X$ terms will never be entered into the J, K K-maps as was the case for the S, R K-maps used in the design of FLIP-FLOPs from BASIC CELLs. To this extent J, K mapping is simpler than S, R mapping. Use of this information can lead to significant savings of time and effort during the J, K mapping process. Again, care should be taken not to forget the underlying reasons for the shortcut approach. The reader will have ample opportunity to practice the J, K mapping method in later sections of this chapter.

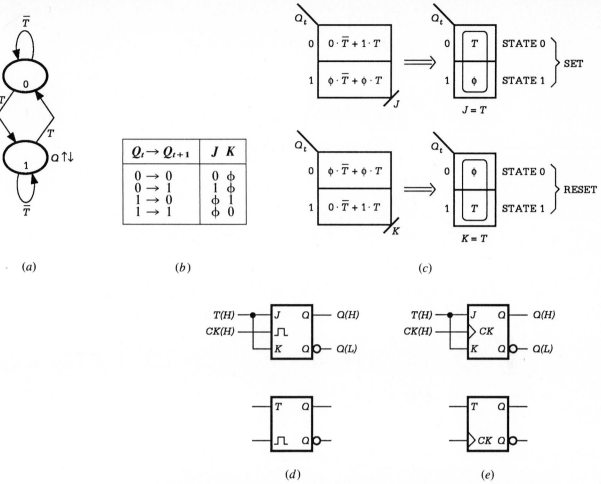

(a) (b) (c)

(d) (e)

Fig. 5-42. *Design of T-FFs from JK-FFs. (a) State diagram for the T-FF. (b) Characterization of the JK-FF MEMORY element. (c) NEXT STATE logic K-maps and minimum cover. (d) Logic circuit and circuit symbol for an MS T-FF. (e) Logic circuit and circuit symbol for a RET T-FF.*

The minimum cover for the J and K K-maps, given in Fig. 5-42(c), is implemented in Figs. 5-42(d) and (e) for an MS T-FF and an RET T-FF, respectively. In fact, any other triggering mechanism could have been chosen, for example, pulse triggering or data lockout. Observe that the external logic consists only of connecting the J and K leads together to produce the T-FF.

JK-FFs Designed from D-FFs. Here, we have the opportunity to design a FLIP-FLOP having two data inputs (*J* and *K*) from one having just one (*D*). The result of this design example will prove to be useful in several ways as we shall soon discover.

Shown in Fig. 5-43(a) is the state diagram for the FSM we want to design—a JK-FF. This state diagram is actually derived from the state diagram for a PT JK-FF exclusive of the *CK* input. But since this FSM will inherit

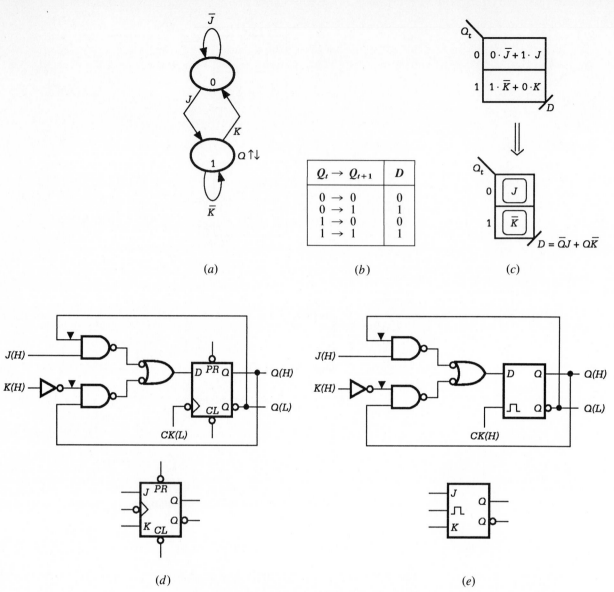

$$Q_t \to Q_{t+1} \quad D$$

$Q_t \to Q_{t+1}$	D
$0 \to 0$	0
$0 \to 1$	1
$1 \to 0$	0
$1 \to 1$	1

(a)

(b)

(c)

(d)

(e)

Fig. 5-43. *Design of JK-FFs from D-FFs. (a) State diagram for the JK-FF. (b) Characterization of the D-FF MEMORY element. (c) NEXT STATE logic K-map and minimum cover. (d) Logic circuit and circuit symbol for an FET JK-FF. (e) Logic circuit and circuit symbol for an MS JK-FF.*

the triggering character of the MEMORY element (a D-FF of our choice), it provides all the information necessary to determine the NEXT STATE logic for the JK-FF.

The NEXT STATE logic K-map for this FSM is found by ANDing the information in the D column of the state transition table in Fig. 5-43(b) with the associated branching conditions read from the state diagram given in Fig. 5-43(a). The result is the K-map given in Fig. 5-43(c), which has a minimum cover given by

$$D = \overline{Q}J + Q\overline{K} \qquad\qquad (5\text{-}6)$$

This expression is important because it forms the basis for converting D K-maps to J, K K-maps, and vice versa. We will have cause to use Eq. (5-6) in Sect. 5.7.

Implementation of the minimum cover expressed by Eq. (5-6) is shown in Figs. 5-43(d) and (e) for a FET JK-FF and an MS JK-FF. Notice that the external NEXT STATE logic is identical for both but that the MEMORY elements differ in their triggering character and the presence of PR and CL overrides.

The MS JK-FF shown in Fig. 5-43(e) is of interest because it does not have the error-catching problem that the MS JK-FF of Fig. 5-34(a) has. This is so because the device of Fig. 5-43(e) has the same CLOCKing characteristics exhibited by the MS D-FF in Fig. 5-32(a), a FLIP-FLOP which has no error-catching problem. So at the expense of a little additional hardware, an MS JK-FF can be produced that will have the less stringent input requirements of the MS D-FF.

Unusual FLIP-FLOPs Designed from JK-FFs. As a final example, we will design a special SET/TOGGLE (ST) FLIP-FLOP by using a JK-FF as the MEMORY element. The operation table given in Fig. 5-44(a) characterizes the ST-FF. From it we construct the state diagram shown in Fig. 5-44(b). This is easily accomplished by recognizing that the $0 \rightarrow 1$ transition is produced by both SET and TOGGLE operations giving a branching condition of $\overline{S}\overline{T} + \overline{S}T + S\overline{T} = \overline{S} + \overline{T}$. The $0 \rightarrow 0$ HOLD RESET branching condition (ST) follows from the $\sum = 1$ rule. Similarly, the $1 \rightarrow 0$ transition can occur only by a TOGGLE operation giving $\overline{S}T$ as the branching condition, and so on.

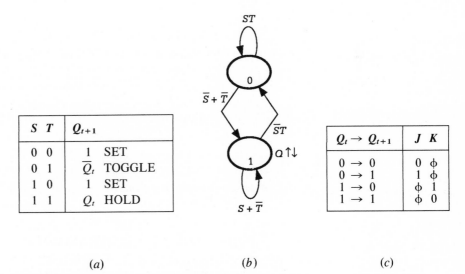

S T	Q_{t+1}	
0 0	1	SET
0 1	$\overline{Q}_t$	TOGGLE
1 0	1	SET
1 1	Q_t	HOLD

$Q_t \rightarrow Q_{t+1}$	J K
$0 \rightarrow 0$	0 ϕ
$0 \rightarrow 1$	1 ϕ
$1 \rightarrow 0$	ϕ 1
$1 \rightarrow 1$	ϕ 0

(a)   (b)   (c)

Fig. 5-44. *Design of a special SET/TOGGLE FLIP-FLOP. (a) Operation table. (b) State diagram. (c) State transition table for the JK-FF.*

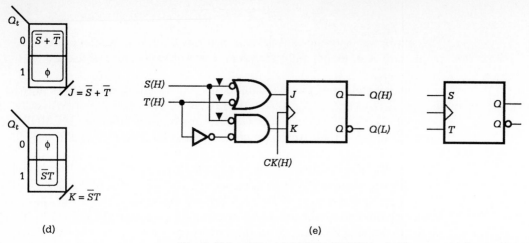

Q_t

| 0 | $\overline{S} + \overline{T}$ |
| 1 | ϕ |

$J = \overline{S} + \overline{T}$

Q_t

| 0 | ϕ |
| 1 | $\overline{S}T$ |

$K = \overline{S}T$

(d)

S(H) —
T(H) —

J   Q — Q(H)

K   Q̄ — Q(L)

CK(H)

S   Q

T   Q̄

(e)

Fig. 5-44 (contd.). *(d) NEXT STATE logic K-maps. (e) Logic diagram and circuit symbol for the RET ST-FF.*

By ANDing each outgoing branching condition of the state diagram in Fig. 5-44(b) with the appropriate *J, K* input requirements given in the state transition table of Fig. 5-44(c), we obtain the NEXT STATE K-maps shown in Fig. 5-44(d). Implementation of the minimum cover derived from these K-maps yields the logic circuit presented in Fig. 5-44(e). Since we have assumed the use of a RET JK-FF as the MEMORY element, the resulting ST-FF is also an RET FF, as indicated by its circuit symbol in Fig. 5-44(e). However, any triggering mechanism (PT, MS, ET, or DL) could have been chosen.

5.6.7   Setup and Hold-Time Requirements

Several times in the previous discussions we have made mention of the fact that a FLIP-FLOP will operate reliably only if the data inputs remain stable at their proper logic levels just before, during, and just after the triggering edge of the CLOCK waveform. To put this on a more formal and quantitative basis we say that the data inputs must meet the *setup and hold-time* requirements established by CLOCK, the *sampling variable* for synchronous FSMs. Referring to Fig. 5-45, the reader should note the following definition:

> *Setup time (t_{su}) is the time interval immediately preceding the ACTIVE (or INACTIVE) transition point (t_{tr}) of the triggering edge of CLOCK during which all data inputs must remain stable at their proper logic levels to ensure that the intended transition will be initiated.*

In Fig. 5-45 we show that the setup time starts just before the CLOCK waveform begins its triggering ascent (or descent) and ends at the ACTIVE (or INACTIVE) transition point, t_{tr}, of the triggering edge. If the setup time requirement is not met by the data inputs, there is no guarantee that CLOCK will sample the input data as intended and failure is possible. For this reason

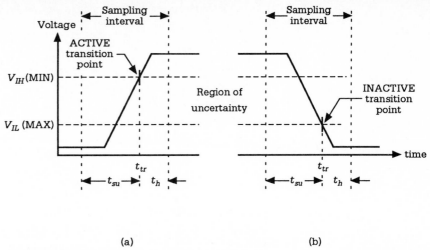

Fig. 5-45. *CLOCK voltage waveform showing setup and hold-time requirements for edge-triggered FLIP-FLOPS. (a) Rising edge-triggering. (b) Falling edge-triggering.*

the setup time is an important design parameter. Manufacturers will normally provide worst case setup time data for their FLIP-FLOPs.

The hold time is also an important design parameter. Again, referring to Fig. 5-45, we give the following definition:

> *The hold time (t_h) is the interval of time immediately following the ACTIVE (or INACTIVE) transition point (t_{tr}) during which the data inputs must retain their proper logic levels to ensure that the intended transition of the FSM will be successfully completed.*

Together the setup and hold times make up the *sampling interval* of the CLOCK waveform. That is,

$$\text{Sampling interval} = t_{su} + t_h \tag{5-7}$$

So, it can be stated that all data inputs must remain stable at their proper logic levels during the sampling period of CLOCK to ensure that the FLIP-FLOP will transition as intended.

The setup and hold-time requirements established by CLOCK are defined in Figs. 5-45(a) and (b) for RET-FFs and FET-FFS, respectively. Here, assuming positive logic, it is observed that the setup time always ends at the ACTIVE (or INACTIVE) transition point which is the upper (or lower) extremity of the region of uncertainty between the LOW and HIGH noise margins. Consequently, the setup time always includes the region of uncertainty and, as it turns out, is usually much larger than the rise or fall times for ordinary CLOCK waveforms.

The hold time begins at the ACTIVE (or INACTIVE) transition point and for ET-FFs ends following a period of time often much shorter than the setup time. For MS JK-FFs of the type shown in Fig. 5-34(a), the hold time

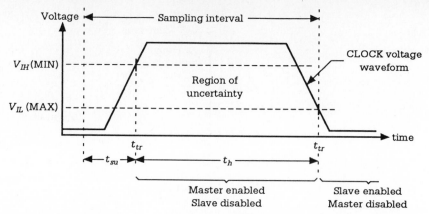

Fig. 5-46. *Setup and hold-time requirements for an MS JK-FF of the type shown in Figure 5-34(a).*

may be larger than the setup time as illustrated in Fig. 5-46. The hold time between the ACTIVE and INACTIVE transition points is necessary to ensure that the master CELL will not be irreversibly SET or RESET due to logic noise on the *J, K* lines while CLOCK is HIGH. Notice that between the ACTIVE and INACTIVE transition points the master CELL is ENABLED while the slave CELL is DISABLED and that the reverse is true following the INACTIVE transition point. In this figure account has not been taken of the path delay through the coupling INVERTER.

Awareness and proper use of the setup and hold-time requirements of FLIP-FLOPs is vital to good sequential machine design practice. A short discussion at the end of Sect. 5.8.4 explains why this is important. In Chapter 6 we will extend the concept of setup and hold-time requirements beyond FLIP-FLOPs to asynchronous FSMs of all types.

5.7 DESIGN OF SIMPLE FSMs WITH EDGE-TRIGGERED FLIP-FLOPs

We have designed or otherwise discussed pulse-triggered FLIP-FLOPs, master/slave FLIP-FLOPs, edge-triggered FLIP-FLOPs, and data-lockout FLIP-FLOPs. All these FLIP-FLOPs can function as the MEMORY for synchronous FSMs. However, some of these FLIP-FLOPs find specific application in modern design, while others are of little or no interest. The PT-FFs are the least important class of FLIP-FLOPs. Because of their stringent CLOCKing requirements, they are rarely used as memory elements. The PT D-FFs are of interest mainly as transparent D LATCHes with applications such as holding registers in microprocessor circuits and as special modules in asynchronous circuits. The MS-FFs have found extensive use as MEMORY elements in MSI, LSI, and VLSI applications primarily because their electronic simplicity and compactness make them economically attractive. Of course, care must be taken in using the MS-FFs to ensure that the strict requirements on setup and hold time are met as discussed in Sects. 5.6.4 and 5.6.7.

Where nearly ideal, high-speed sampling is required, and economic considerations are not a factor, ET-FFs may be the MEMORY elements of choice. The setup and hold-time requirements for these FLIP-FLOPs are the least stringent of all, and they possess none of the problems associated with either the PT- or MS-FFs discussed earlier. In this section we will undertake three relatively simple synchronous FSM designs by using ET-FFs. It is the purpose of this section to demonstrate by example the design methodology to be used, with emphasis on the procedure required to generate the NEXT STATE and output logic functions of an FSM. This will involve nothing new. Rather, it will be an extension of Sect. 5.6.6 applied to FSMs of two- and three-state variables.

5.7.1 Design of a Sequence Detector

Suppose it is required that we design a synchronous FSM that will produce a signal (say, Z) any time the sequence . . . 001 . . . appears in a continuous stream of data on a synchronized input line, X. The block diagram in Fig. 5-47(a) details the I/O specifications, where we have assumed rising edge triggering and an ACTIVE HIGH input and output.

The state diagram of Fig. 5-47(b) represents the Moore version of this sequence detector. The reader may verify that it satisfies the algorithm by following the sequence beginning with STATE ⓐ. Thus, if this FSM detects the order $\overline{X}$, $\overline{X}$, and X (meaning 0, 0, and 1), each on a successive rising edge of the CLOCK waveform, the transitions are ⓐ → ⓑ, ⓑ → ⓒ, and ⓒ → ⓓ, whereupon an output is issued in STATE ⓓ. On the other hand, if the sequence is broken at any stage, the FSM will not issue Z. For example, if X (logic 1) is detected in STATE ⓑ on the rising edge of CLOCK, the

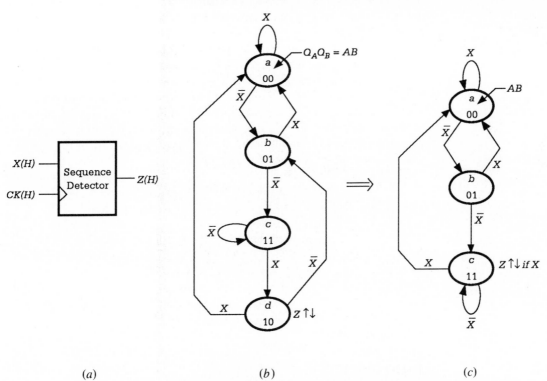

(a)   (b)   (c)

Fig. 5-47. *Representation of an FSM that will detect the sequence . . . 001 . . . (a) Block diagram. (b) and (c) Moore and Mealy versions of the state diagram.*

sequence is broken and the FSM will transit back to STATE $\textcircled{a}$. Or if the FSM detects $\overline{X}$ (logic 0) in STATE $\textcircled{c}$, the FSM will HOLD in STATE $\textcircled{c}$ until X is detected on the rising edge of CLOCK at which time the FSM will go to STATE $\textcircled{d}$ and issue Z. Once in STATE $\textcircled{d}$ the FSM will either transit to STATE $\textcircled{b}$ under input condition $\overline{X}$ or will transit to STATE $\textcircled{a}$ under X where the sequence will begin all over again.

The state diagram representing the Mealy version of the sequence detector is presented in Fig. 5-47(c). It is obtained from the Moore version in Fig. 5-47(b) by *merging* STATEs $\textcircled{c}$ and $\textcircled{d}$ to form STATE $\textcircled{c}$ with a conditional output. The reader can easily verify that the Mealy version also satisfies the algorithm for the sequence detector by again tracing the decision paths from STATE $\textcircled{a}$ through STATE $\textcircled{c}$.

Some explanation is required regarding the conditional output shown in STATE $\textcircled{c}$ of Fig. 5-47(c). The output Z is issued conditionally on X, which is an exiting (branching) condition. This is perfectly acceptable, however, since if X meets the setup and hold-time requirements of CLOCK, the FSM will also issue Z long enough to trigger any next-stage switching device for which Z is an input. In other words, a successful transition from STATE $\textcircled{c}$ must also result in the successful output of Z.

For the purposes of this discussion, let us design this sequence detector by using FET D-FFs as the MEMORY. The state transition table for the D-FF, reproduced in Fig. 5-48(a), is the means by which the MEMORY element is characterized. It is this information together with the sequential information contained in the state diagram that must be introduced into the design process. This we will do for both versions of this machine represented in Fig. 5-47.

In Fig. 5-48(b) are shown the NEXT STATE and output K-maps and minimum cover for the Moore version of the sequence detector. Since there are two state variables ($Q_A = A$ and $Q_b = B$), there must be two NEXT STATE K-maps, one for D_A (input to D-FF A) and one for D_B (input to D-FF B). The NEXT STATE K-maps are plotted by ANDing the input requirements indicated in the state transition table of Fig. 5-48(a) with the appropriate branching conditions read from the state diagram of Fig. 5-47(b), the same procedure that was followed in Sect. 5.6.6. For example, cell 0 (representing STATE $\textcircled{a}$) for bit A takes a $(0 \cdot X) + (0 \cdot \overline{X}) = 0$, while cell 1 (STATE $\textcircled{b}$) requires $(0 \cdot X) + (1 \cdot \overline{X}) = \overline{X}$, and cell 3 (STATE $\textcircled{c}$) is given a $(1 \cdot X) + (1 \cdot \overline{X}) = 1$, and so on. Or for bit B, cell 0 requires a $(0 \cdot X) + (1 \cdot \overline{X}) = \overline{X}$, while cell 2 (representing STATE $\textcircled{d}$) takes $(1 \cdot \overline{X}) + (0 \cdot X) = \overline{X}$, and so on. Remember that the cell coordinates of a NEXT STATE K-map represent the PRESENT STATE while the entry for each cell represents the NEXT STATE instruction for a particular state variable bit.

The output K-map in Fig. 5-48(b) is obtained directly from the state diagram in Fig. 5-47(b) by observing that an output exists only in STATE $\textcircled{d}$; hence the 1 in cell 2 of the Z K-map.

The same procedure is used for the design of the Mealy version of the sequence detector. By ANDing the input requirements indicated in the state transition table of Fig. 5-48(a) with the appropriate branching conditions taken from the state diagram in Fig. 5-47(c), there results the NEXT STATE K-maps and minimum cover given in Fig. 5-48(c). Notice that the unused STATE 10 is represented as a don't care in all K-maps, including that for

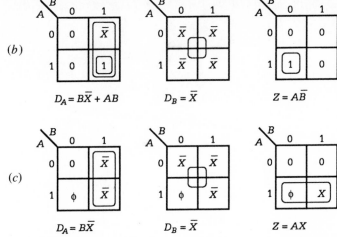

$Q_t \rightarrow Q_{t+1}$	D
$0 \rightarrow 0$	0
$0 \rightarrow 1$	1
$1 \rightarrow 0$	0
$1 \rightarrow 1$	1

(a)

Fig. 5-48. *Design of the sequence detector represented in Figure 5-47. (a) Characterization of the D-FF MEMORY. (b), (c) NEXT STATE K-maps, output K-maps, and minimum cover for the Moore and Mealy machines of Figures 5-47(b) and (c), respectively.*

the output Z, and that it is used to extract minimum cover as we have done. Obviously, the minimum cover for Mealy version is simpler than that for the Moore version, a consequence of the presence and use of the don't care in cell 2.

Since the Mealy version provides the simplest NEXT STATE logic, we will choose it to implement this FSM. Shown in Fig. 5-49 is the MIXED LOGIC circuit for the sequence detector designed as a Mealy machine by using the minimum cover given in Fig. 5-48(c). It can be seen that this machine satisfies the functional relationships given by Eqs. (5-2) in Sect. 5.3.2, which require that $Z = f(X,A,B)$. Remember that $A = Q_A$ and $B = Q_B$ and that these are the PRESENT STATE parameters since they are the outputs of the MEMORY section.

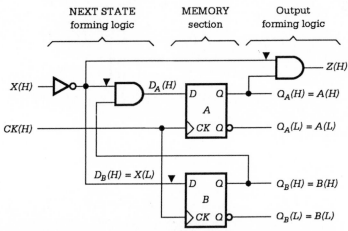

Fig. 5-49. *Logic circuit for the Mealy machine of Figure 5-47(c) as determined from the minimum cover given in Figure 5-48(c).*

5.7.2 Design of a Binary Up-counter

In Sect. 5.2 we discussed the concept of a logic state sequence produced by the FSM represented as a block diagram in Fig. 5-1. There the sequence was shown to be a repeating three-bit binary up count. We will now design this FSM by using as the MEMORY FLIP-FLOPs that yield the simplest NEXT STATE logic and that are triggered on the rising edge of the CLOCK waveform.

Shown in Figs. 5-50(a) and (b) are the block diagram and state diagram for the three-bit binary up-counter. A brief inspection of the state diagram, or the timing diagram of Fig. 5-1(b), reveals the TOGGLE character inherent in the binary number system. For this reason it is logical to assume that a T-FF would be a good choice for the MEMORY element, and it is. The state

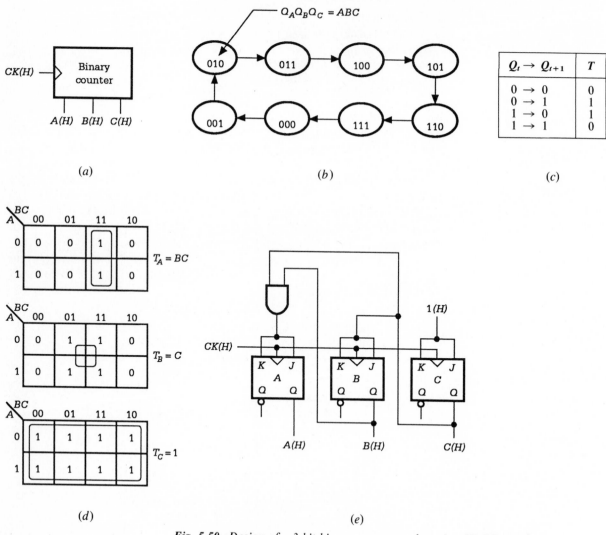

Fig. 5-50. *Design of a 3-bit binary up counter by using JK-FFs configured as T-FFs for the MEMORY (a) Block diagram. (b) State diagram. (c) Characterization of the MEMORY. (d) K-maps. (e) Logic circuit.*

transition table, used to characterize the MEMORY, is given in Fig. 5-50(c), and the resulting NEXT STATE K-maps and minimum cover are provided in Fig. 5-50(d). The K-maps are plotted by ANDing the input information in the T column of the state transition table with the appropriate unconditional branching conditions of the state diagram. Thus, one simply places a logic 1 in any cell for which the NEXT STATE is a bit TOGGLE operation $(0 \rightarrow 1$ or $1 \rightarrow 0)$.

The logic circuit for this binary counter is shown in Fig. 5-50(e). It is composed of three RET JK-FFs, which have been configured as T-FFs, and the minimum NEXT STATE logic taken from Fig. 5-50(d). The T-FFs, it will be recalled, are not usually available commercially and so must be obtained from other FFs, usually of the JK type. Observe that we have assigned $Q_A = A$, $Q_B = B$, and $Q_C = C$ as a means of simplifying the nomenclature for the state variables.

5.7.3 Design of a Binary/Gray Code Counter

For our third worked exercise let us design a two-bit binary/Gray code up-counter which has a count mode control X and four outputs B, C, RED, and GRN (for GREEN), as indicated in Fig. 5-51(a). The input X determines the count sequence according to the following:

$$\text{If } \overline{X}, \text{ then } BC = \ldots 00 \rightarrow 01 \rightarrow 10 \rightarrow 11 \rightarrow 00 \ldots \text{(binary)}$$

with output RED.

$$\text{If } X, \text{ then } BC = \ldots 00 \rightarrow 01 \rightarrow 11 \rightarrow 10 \rightarrow 00 \ldots \text{(Gray)}$$

with output GRN.

The count status outputs, B and C, are to be taken as two state variables. The output RED is to activate a red LED (light-emitting diode) any time the FSM is counting in binary, and the output GRN is to activate a green LED

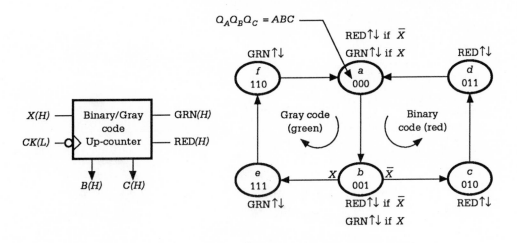

(a)   (b)

Fig. 5-51. *Representation of a 2-bit binary/Gray code up-counter. (a) Block diagram. (b) State diagram.*

any time the FSM counts in Gray code. The mode control X is changeable at any time except during the sampling interval given by Eq. (5-7), and the FSM is to be triggered on the falling edge of 1-Hz (1-cycle/sec) CLOCK waveform.

The state diagram that satisfies the requirements for this counter is shown in Fig. 5-51(b). The state variables $B = Q_B$ and $C = Q_C$ are chosen as the count outputs. A third state variable, $A = Q_A$, is used so that all states in the state diagram will be unique, as they must be. The choice of state variables to represent the count is, of course, arbitrary, but it is "logical" to assign the count to juxtaposed bits. From Fig. 5-51(b), it is seen that the right loop represents natural two-bit binary count under the input condition $\bar{X}$ while the left loop represents two-bit Gray code count under input condition X, both count sequences being on state variables B and C. Notice that this FSM has but one decision state, STATE $\textcircled{b}$, since for all other states branching is unconditional.

Let us first design this counter by using D-FFs. Shown in Fig. 5-52(a) is the state transition table for the D-FF which characterizes the MEMORY. By ANDing the input requirements of this table with the associated branching conditions taken from the state diagram of Fig. 5-51(b), we generate the NEXT STATE K-maps and minimum cover given in Fig. 5-52(b). Notice that don't cares are placed in cells 4 and 5 since STATEs 100 and 101 are unused. Minimum NEXT STATE logic cover is extracted from these maps by making use of these don't cares.

The K-maps for outputs GRN and RED are presented in Fig. 5-52(c). The entries for these K-maps are read directly from the state diagram. For example, cells 0 and 1 take the EVs representing the conditional outputs for STATEs 000 and 001, cells 4 and 5 represent don't care states, and the

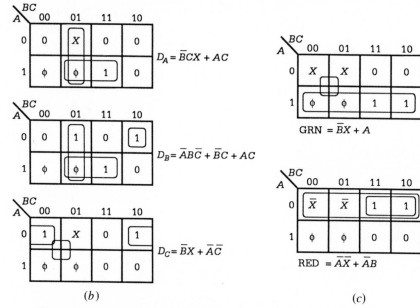

(a)

(b)

(c)

Fig. 5-52. *Design of the binary/Gray code up-counter with D-FFs. (a) Characterization of the MEMORY. (b), (c) NEXT STATE and output K-maps and minimum cover.*

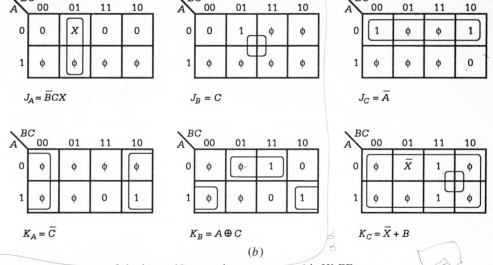

$Q_t \rightarrow Q_{t+1}$	J	K
$0 \rightarrow 0$	0	φ
$0 \rightarrow 1$	1	φ
$1 \rightarrow 0$	φ	1
$1 \rightarrow 1$	φ	0

(a)

$J_A = \bar{B}CX$

$J_B = C$

$J_C = \bar{A}$

$K_A = \bar{C}$

$K_B = A \oplus C$

$K_C = \bar{X} + B$

(b)

Fig. 5-53. Design of the binary/Gray code up-counter with JK-FFs.
(a) Characterization of the MEMORY. (b) NEXT STATE logic K-maps and minimum cover.

remainder of the cells take 1's or 0's depending on whether or not a given output is ACTIVE.

For comparison purposes we will now use JK-FFs as the MEMORY for this counter. Presented in Fig. 5-53(a) is the state transition table for the JK-FF MEMORY element reproduced here from Fig. 5-44(c) for the convenience of the reader. Following the procedure established in Sect. 5.6.6, we use this table together with the state diagram of Fig. 5-51(b) to plot the NEXT STATE logic K-maps shown in Fig. 5-53(b). Notice that don't cares are again placed in cells 4 and 5 of all K-maps since these cells represent unused states. The output K-maps for the JK-FF design remain the same as in Fig. 5-52(c).

The minimum cover for the NEXT STATE logic obtained from the JK-FF K-maps is much simpler than that extracted from the D-FF K-maps of Fig. 5-52(b). A gate/input tally of 3/7 for the *JK* K-maps compares to 9/21 for the *D* K-maps. This difference is due solely to the presence of the four don't cares in the state transition table for the JK-FF, don't cares which are not present in the state transition table for either the D-FF or the T-FF. From this information we can make the following statement:

> *With few exceptions the NEXT STATE logic for JK-FFs will be less complex than the NEXT STATE logic for either D-FFs or T-FFs.*

However, there is a trade-off. The use of JK-FFs requires twice as many NEXT STATE variables; hence twice as many K-maps must be plotted, as for either D-FFs or T-FFs. Thus, one must labor harder to achieve reduced hardware requirements. The design of the three-bit binary up-counter given in Sect. 5.7.2 is one of the few exceptions to this rule.

There is a method that will minimize the effort required to plot NEXT

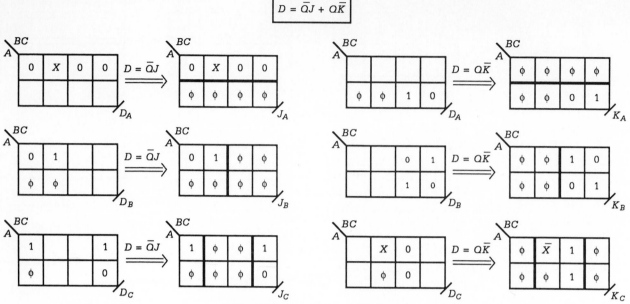

$$D = \bar{Q}J + Q\bar{K}$$

Fig. 5-54. *Conversion of D K-maps to JK K-maps by using Eq. (5-6).*

STATE *JK* K-maps. This method involves first plotting the *D* K-maps (or *T* K-maps) and then transforming them to *JK* K-maps. In Fig. 5-54 we demonstrate this method by using the results of the binary/Gray code counter design. The procedure is based on the transformation equation $D = \bar{Q}J + Q\bar{K}$ (Eq. (5-6) in Sect. 5.6.6) and is stated as follows:

> **D map → J map:** *Take all that is not in the state variable domain of the D K-map and transfer it to the J K-map (hence D = $\bar{Q}J$); then fill in the remaining cells with don't cares.*
>
> **D map → K map:** *Take all that is in the state variable domain of the D K-map and transfer it complemented to the K K-map (hence D = $Q\bar{K}$); then fill in the remaining cells with don't cares.*

The utility of this procedure lies in the fact that the *D* K-maps are easier to plot than the *JK* K-maps, and the transformation process that follows is easily executed.

Returning to the design example of this section, we implement the binary/Gray code up-counter by using the simpler *JK* NEXT STATE logic of Fig. 5-53(b) and the output logic of Fig. 5-52(c) to yield the results shown in Fig. 5-55. Here, we have chosen to use an assortment of gate-level logic to implement the NEXT STATE functions. Observe that by proper selection of the FLIP-FLOP outputs and external logic, the use of INVERTERs can be minimized.

The binary/Gray code up-counter just designed is a *self-correcting* counter. To understand what this means, consider that on power-up, this FSM may initialize into any one of the eight states, including either of the unused (don't care) states, 100 or 101. If the counter powers up into one of

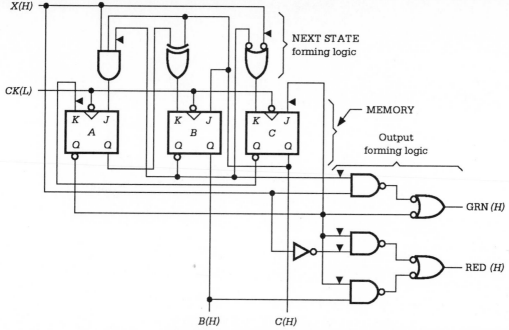

Fig. 5-55. *JK-FF implementation of the binary/Gray code up-counter of Figure 5-51.*

the six states of the state diagram in Fig. 5-52(b), it will function properly according to the logic level of X. However, if it should power up into STATE 100 or STATE 101, the FSM must be forced into the main routine (the state diagram) if it is to be self-correcting. This means that the NEXT STATE for either STATE 100 or STATE 101 must be a state existing in the state diagram. To determine if this is so, we introduce the present state values for each don't care state, in turn, into the logic expressions for D_A, D_B, and D_C given in Fig. 5-52(b). The result indicates that $100 \rightarrow 00X = (001$ or $000)$, and $101 \rightarrow 11X = (110$ or $111)$, all of which are part of the main routine. Thus, there are no extraneous subroutines into which the FSM can initialize, and so the counter is self-correcting.

5.8   DESIGN CONSIDERATIONS

There are a number of design intricacies and problem areas that we have purposely avoided in the previous sections. We did this because we wanted to focus the reader's attention on basic design concepts. Now, we move on to consider some of the design intricacies and problem areas that are certain to be encountered in any serious design endeavor. We believe that proper design techniques for the error-free operation of synchronous sequential machines must be given the highest priority.

5.8.1   Logic Optimization and State Code Assignment Rules

For the simple FSMs that have been considered so far, the state code assignments of the state diagrams were either provided by the authors without explanation [e.g., Figs. 5-36(a) and 5-47(b)] or fixed by the nature of the

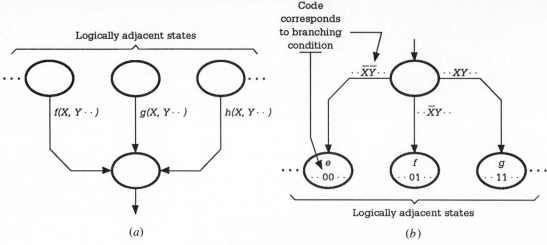

Logically adjacent states

$f(X, Y \cdots)$ $g(X, Y \cdots)$ $h(X, Y \cdots)$

(a)

Code corresponds to branching condition

$\cdots \overline{X}\overline{Y} \cdots$ $\cdots XY \cdots$

$\cdots \overline{X}Y \cdots$

e f g
$\cdots 00 \cdots$ $\cdots 01 \cdots$ $\cdots 11 \cdots$

Logically adjacent states

(b)

Fig. 5-56. *Logic adjacency rules for state code assignment. (a) "Into" rule. (b) "Out of" rule as it applies to multiple branching.*

FSM [e.g., Fig. 5-50(b)]. However, there are guidelines, or rules as we call them, that if applied can significantly reduce the complexity of the NEXT STATE logic functions for a given FSM. In Fig. 5-56 we illustrate the two logic adjacency "rules." The rules require that we make logically adjacent certain groups of states of a state diagram depending on the interbranching relationship between these states and others. The two rules (actually guidelines), named the "*into*" rule and the "*out of*" rule, are stated as follows:

"Into" Rule—The states of a state diagram that branch *into* a common state, as in Fig. 5-56(a), should be made logically adjacent provided that the branching conditions $f(X, Y, \ldots)$, $g(X, Y, \ldots)$, $h(X, Y, \ldots)$, and so on, for these states possess the commonality required for logically adjacent patterns.

"Out of" Rule—Those states that are the NEXT STATEs of a common state, as in Fig. 5-56(b), should be made logically adjacent in accordance with their input branching conditions.

The proper application of the "into" rule is illustrated in Fig. 5-57(a). Here, sufficient commonality exists between the branching conditions to yield the logically adjacent patterns shown in the D K-maps. However, by replacing the pivotal STATE 010 with 000, we break the unit distance coding and produce XOR-type EV patterns, as shown in Fig. 5-57(b). Of course, if a multilevel XOR optimization is the object of the design, then nonlogic adjacent state code assignments may be desirable.

Proper application of the "out of" rule is illustrated in Fig. 5-58(a) for the case of three state variables, four-way branching, and use of D-FFs as the MEMORY. Here, the entries in cell 4, for each of the three NEXT STATE variables, are seen to be simple subfunctions. But violation of the "out of" rule may lead to more complex subfunction entries as indicated in Fig. 5-58(b) and (c). Usually, the more complex the subfunction entries, the more complex will be the NEXT STATE logic functions. Notice that in

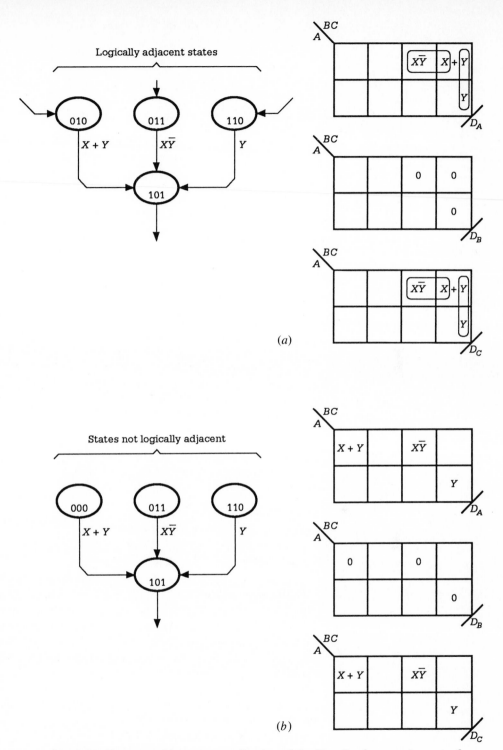

Fig. 5-57. *Proper and improper application of the "into" rule assuming use of D-FFs. (a) Proper application showing logically adjacent patterns in the NEXT STATE logic K-maps. (b) Improper application showing absence of logically adjacent patterns.*

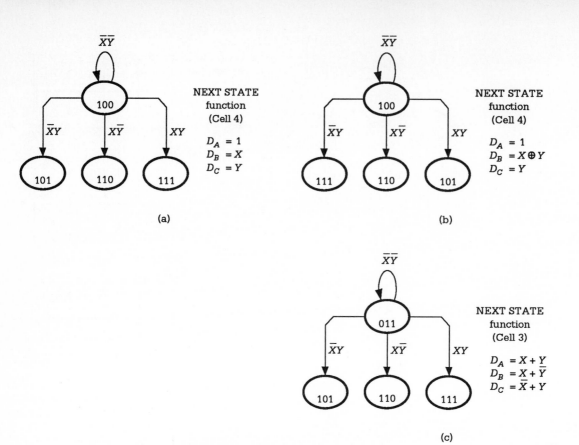

Fig. 5-58. *Proper and improper application of the "Out of" rule to four-way branching assuming the use of D-FFs. (a) Proper application leading to minimum subfunction entries. (b), (c) Improper application showing more complex subfunction entries.*

these examples the HOLD condition $\overline{X}\,\overline{Y}$ is treated as one of the four branching conditions, as it must be.

We emphasize that the "into" and "out of" rules are to be regarded as guidelines, not as hard, fast rules that must be followed at any cost. Their application to the state code assignment process for a state diagram can significantly reduce the complexity of the NEXT STATE logic functions. But this is *not* guaranteed. Furthermore, not all possible applications of these rules may be permitted within a given state diagram, particularly one having complex branching conditions. The state diagram of Fig. 5-47(b) is a simple example of this fact. The point is that the optimization process is not as straightforward as simply applying an adjacency rule or two. Nor can it be stated that a given rule favors one type of FLIP-FLOP over another type, though such rules have been attempted. The best advice that can be given to the reader at this time is to use these rules (guidelines) intelligently and cautiously on a trial-and-error basis, keeping in mind that their function is to *help* optimize (reduce) the NEXT STATE logic functions for a given synchronous FSM.

One further point needs to be made. In attempting to apply the logic adjacency rules, it may be necessary to increase the number of state variables which may, in turn, increase circuit complexity. In such cases, one must weigh the advantages gained by applying the logic adjacency rules with the disadvantages wrought by an increase in state variable count.

5.8.2 Output Race Glitches

Improper design of an FSM can lead to logic noise in the output, and this noise, in turn, can cause the premature triggering of a nonsynchronous next stage switching device to which the FSM is attached. So it may be important that FSMs be designed to output "clean" logic signals free of unwanted logic transients called *glitches*.

There are two main sources of output logic noise in an FSM. They are

1. Glitches produced by state variable race conditions.

2. Glitches produced by STATIC HAZARDs in the output logic.

In this and the following section we will consider these two sources of logic noise in turn, with emphasis placed on their removal by proper design methods.

Before beginning this task, it will be helpful to define the two types of glitches that are encountered as follows:

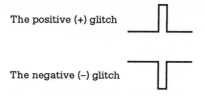

Such output glitches may, or may not, develop to an extent that they cross the switching threshold (transition point) of a next-stage device. However, if glitches are formed that cross the switching threshold, incorrect operation of an asynchronous next-stage device can result. Therefore, if we cannot guarantee that output glitches will not develop to this extent, we must eliminate them by one means or another.

Effect of Race Conditions. A *race condition* exists in an FSM any time a state-to-state transition involves a change of two or more state variables. Since no two state variable bits can change precisely at the same time, the FSM must cycle asynchronously along one of two (or more) *alternative paths* to reach the destination state, each alternative path being through a non-destination (race) state. If output discontinuities exist along alternative paths, *output race glitches* can occur. These glitches are also called *FUNCTION HAZARDs* and are discussed in Sect. 4.12 relative to combinational logic circuits.

Several examples of output race glitches produced by *race transitions* involving a change of two state variable bits are illustrated by the state diagram and K-map segments given in Fig. 5-59. Possibilities for race glitch production arise from the various combinations of conditional outputs, unconditional outputs, don't care states, and conditional and unconditional

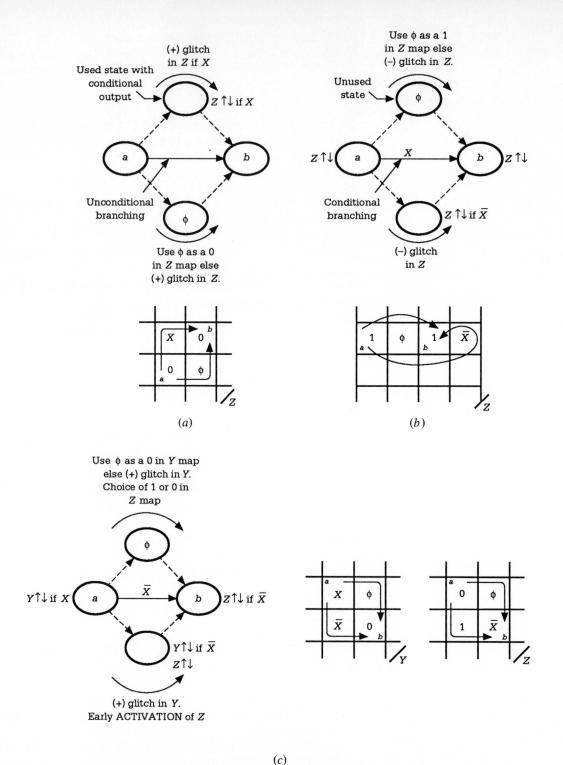

Fig. 5-59. (a) and (b) State diagram and K-map segments showing single output race glitch production for transitions involving changes in two state variables. (c) Race glitch analysis applied to an FSM having two output variables.

5 / Synchronous Sequential Machines

branching, as indicated in Figs. 5-59(a), (b), and (c). Notice that the presence of don't care states (unused states) may or may not produce output glitches depending on how they are used in the output K-map segments associated with the state diagram segments. The presence of a don't care in a K-map requires that it be assigned either a logic 1 or a logic 0 when looping out cover, the decision often being influenced by the presence of race conditions such as those described here.

Figure 5-59(c) deals with the issues of multiple outputs and output AC-TIVATION skew. With regard to multiple outputs, it is important to understand that a don't care state may be assigned different logic levels (1 or 0) in different output K-maps so as to minimize or eliminate race glitch production. Thus, the use of each don't care must be considered separately for each output.

Early or late ACTIVATION or DEACTIVATION of an output (called output ACTIVATION skew) is usually not considered to be a problem unless the timing constraints are so stringent that such ACTIVATION skews are prohibited. If such stringent timing constraints exist, problems other than ACTIVATION skew are likely to be far more serious and design limiting.

Additional explanation of Figs. 5-59(b) and (c) may be helpful to the reader. Both examples show conditional branching for the ⓐ→ⓑ transition. This means that if the branching condition meets the setup and hold-time requirements of CLOCK, the race condition will take place, and any outputs conditional on race states will be affected accordingly. Thus, the conditional output in Fig. 5-59(b) will not be ACTIVATED because it does not depend on the input branching condition required to initiate the race. On the other hand, two of the three conditional outputs in Fig. 5-59(c) will be ACTI-VATED since they both depend on the same branching condition that initiated the race, namely, $\overline{X}$. The conditional output Y in STATE ⓐ will not be activated any time after $\overline{X}$ goes ACTIVE and, hence, cannot be ACTI-VATED when the race condition is initiated.

The presence of one or more output race glitches may or may not pose a problem for the designer. If, for example, the outputs of an FSM drive devices that are not sensitive to logic noise, elimination of the output race glitches would not be necessary. In the discussion that follows, we will assume that the presence of output race glitches is undesirable and that they should be eliminated.

Elimination of Output Race Glitches. There are two methods by which output race glitches can be eliminated. They are as follows:

1. *Eliminate the race condition.* This can be accomplished by revising the state code assignment, by providing additional state variables, or by using *fly states*, all of which have the potential for increasing the complexity of the NEXT STATE logic.

   Revision of the state code assignment, as a means of removing the race condition, is sometimes difficult if not impossible to carry out. But it is an option which should be considered by the designer, particularly if it reduces the complexity of the NEXT STATE logic as well as eliminates the race condition. Providing additional state variables is almost always a viable option for removing the race condition, but it is nearly certain to increase the hardware requirements of the FSM. Remember that a MEMORY element (FLIP-FLOP) is required for each state variable.

A *fly state* is a one-way branching state which has been inserted between two states for the purpose of either creating a logically adjacent transition or adding a unit of time delay, or both. Any unused state can be used as a fly state, and on rare occasions a used state can serve as a fly state. The use of a fly state to create logic adjacency is illustrated in Fig. 5-60(a) for the case of a two-bit race condition. Notice that the don't care (unused) state becomes part of the state diagram with the removal of the race condition. The transition sequence of the FSM is

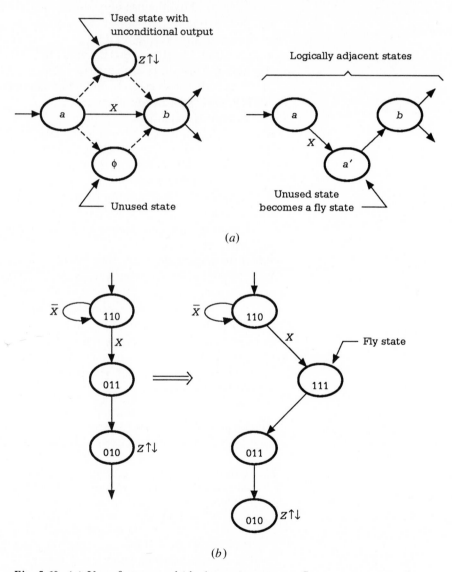

(a)

(b)

Fig. 5-60. *(a) Use of an unused (don't care) state as a fly state to remove the race condition causing the glitch. (b) A typical example of the use of a fly state to remove a race glitch.*

now $(a) \rightarrow (a') \rightarrow (b)$ with an added CLOCK time period. An example of the use of a fly state to remove a race glitch is given in Fig. 5-60(b). Here, we have taken the unused 111 STATE as the fly state to remove the alternative race path through the 010 output STATE.

2. *Filter Out the Race Glitch.* This approach requires that an AND gate or FLIP-FLOP be placed on any output line (of an FSM) on which a glitch can occur. The logic circuits provided in Fig. 5-61 illustrate how this can be accomplished. In Fig. 5-61(a) an AND gate is placed on the output $Z(H)$ of the RET FSM with an INVERTER providing the $CK(L)$ ENABLE input to the AND gate. However, this method has the inherent disadvantage that Z'(H) follows $CK(L) = \overline{CK}(H)$ for sustained $Z(H)$ signals. Hence, this method has limited use. A better method is shown in Fig. 5-61(b). Here, an FET D-FF is positioned on the output of the RET FSM. The FET D-FF functions as an effective filter for logic noise of any origin provided that the logic noise occurs as a result of a state transition immediately following the triggering edge of a CLOCK waveform. We assume that the CLOCK waveform is symmetric (or nearly symmetric) with an ACTIVE HIGH period greater than the pulse widths of the glitches to be filtered.

The timing diagram of Fig. 5-61(c) illustrates the results of applying the filter method of Fig. 5-61(b) to glitches produced by the race con-

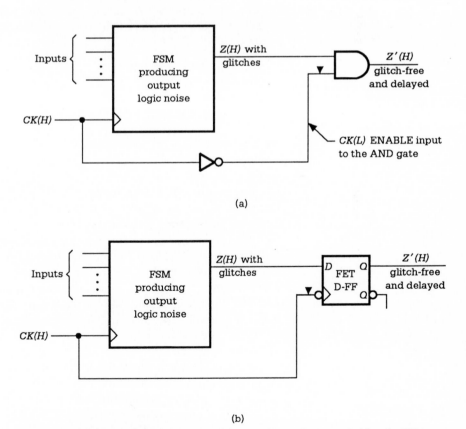

(a)

(b)

Fig. 5-61. *Methods for filtering out logic noise by producing a half-cycle CK delay of the output from an FSM. (a) Logic circuit showing the use of an AND gate and an INVERTER. (b) Logic circuit showing the use of an FET D-FF.*

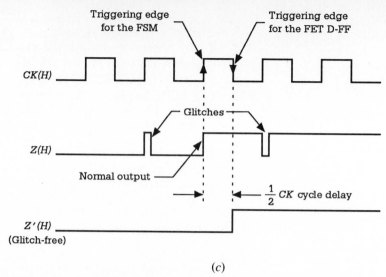

(c)

Fig. 5-61 (contd.). *(c) Timing diagram showing the results of applying the method of Figure 5-61(b) to logic noise of the type produced in Figure 5-59.*

ditions shown in Fig. 5-59. Race glitches produced in output $Z(H)$ never reach the output $Z'(H)$ because the logic noise is dissipated before the falling edge of CLOCK enables the signal $Z(H)$ to pass through the filtering device. The one-half CLOCK cycle delay of the output is the price that must be paid for this filtering action. Had we chosen to use an RET D-FF in place of the FET D-FF in Fig. 5-61, the filtering action would still take place but a full-CK-cycle delay in the output would result. In this case the output signal would be in phase with CLOCK rather than being antiphase to it as in Fig. 5-61.

An Example. As a practical example of race glitch analysis, consider the synchronous pulse width adjuster (PWA) shown in Fig. 5-62. Before beginning the analysis a short description of its operation is in order. It is the function of this FSM to generate a *clean* pulse P of width determined by the width of the input pulse X as indicated by the timing diagrams in Fig. 5-62(a). The intervals for which X is ACTIVE are to be one, two, or three CLOCK periods (T_{CK}) and are to be widely separated in time from each other by intervals of at least $5T_{CK}$ during which X must be INACTIVE. (It takes $4T_{CK}$ to complete the sequence.) The state diagram that meets the requirements of the PWA is shown in Fig. 5-62(b). Obviously, it is a Mealy machine of one input and one output. Notice that X implies logic 1 while $\overline{X}$ implies logic 0. Note also that the state codes have been assigned to satisfy both the "into" and "out of" rules, given in Sect. 5.8.1, for NEXT STATE optimization purposes.

In Fig. 5-63 we present the output race glitch analysis for the PWA. Shown are the three race conditions relative to states ⓑ, ⓒ, and ⓔ. Notice that two of the three possible race glitches are unavoidable should the FSM take the alternative paths that lead to race glitch production. The transition ⓑ → ⓒ through STATE ⓓ will unavoidably produce a $(+)$ glitch in P. A $(+)$ glitch due to the transition ⓑ → ⓒ through the don't care (unused) STATE 101 can be avoided by using the ϕ as a logic 0 in the P K-map.

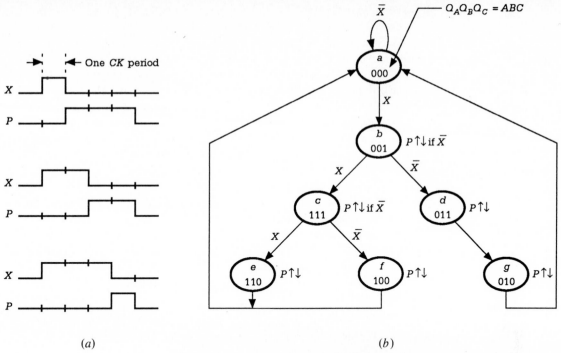

Fig. 5-62. *(a) Positive logic timing diagrams and (b) State diagram for a synchronous pulse width adjuster (PWA).*

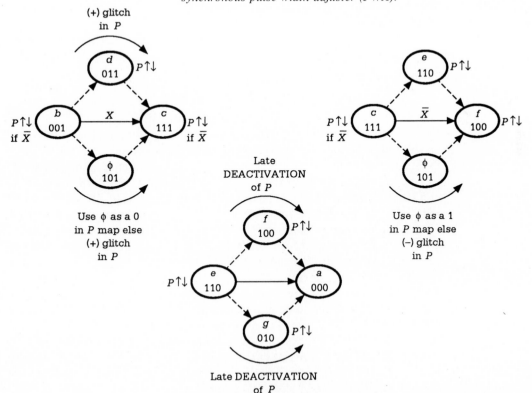

Fig. 5-63. *Race glitch analysis for the pulse width adjuster of Figure 5-62.*

However, the transition ⓒ→ⓕ through the φ-STATE 101 requires that the don't care be used as a logic 1 in the P K-map. Since we cannot have it both ways in the same output K-map, one of the race glitches through the φ-STATE will be unavoidable should the FSM take that alternative path.

To issue a clean output pulse interval, a requirement of the PWA, it is necessary that the output P be free of all glitches. One method of accomplishing this would be to filter out any glitches that may occur by placing an edge-triggered FLIP-FLOP on the output line as was done in Fig. 5-61(b). The trouble with this approach is that it delays the final output by one-half of a CLOCK cycle, a condition that may or may not be acceptable. There is a better way. By changing the state code assignment of STATE ⓒ in Fig. 5-62(b) to 101, the PWA becomes race glitch free. However, the "out of" rule is no longer satisfied relative to STATE ⓑ. A much better solution is to merge STATEs ⓒ and ⓓ and merge STATEs ⓔ, ⓕ, and ⓖ to form a four-state state diagram that is race glitch free and that conforms completely with the "out of" rule. It is left as an exercise for the reader to demonstrate this. Thus, the state diagram of Fig. 5-62(b) is unnecessarily complex, though it served our purpose in studying output race glitches.

5.8.3 STATIC HAZARDs in the Output Logic

In Sect. 4.12 the subject of HAZARDs (STATIC HAZARDs) was discussed as was appropriate for combinational logic circuits. What was said there with regard to HAZARDs is applicable here to the output logic of synchronous FSMs with *two* important qualifications. First, we must be conscious of the sequential details of the FSM in question:

> *A HAZARDous transition can exist in the output of a synchronous sequential machine if, and only if, that transition exists as part of the normal sequential operation of that machine.*

Thus, it seems clear that we must somehow incorporate the use of the state diagram into the STATIC HAZARD analysis procedure.

Second, we must be aware of the mixed-rail output characteristics of the FLIP-FLOPs that are used for the FSM's MEMORY. From the discussion in Sect. 4.12 and the timing characteristics for the BASIC CELLs given in Fig. 5-17, the conditions for SOP and POS HAZARD production can be established for output logic which is driven solely by the mixed-rail outputs (i.e., the state variables) from FLIP-FLOPs. These conditions are given in Fig. 5-64, and require that the mixed-rail outputs, $Q(H)$ and $Q(L)$, perform the same complementary function as do INVERTERs in purely combinational logic machines. Therefore, the conditions of Fig. 5-64 apply unambiguously to the output logic of a Moore machine, provided that that output logic has no INVERTERs. Purely combinational logic circuits, as in Sect. 4.12, require INVERTERs to perform the complementary operations. In such circuits SOP HAZARDs can be produced only by $1 \rightarrow 0$ input changes while POS HAZARDs can be produced only by $0 \rightarrow 1$ input changes, independent of input ACTIVATION LEVELs.

For a Moore machine, compatibility between the requirements of Fig.

Conditions for SOP HAZARD Production
(1) $1 \rightarrow 0$ change in Q when $Q(H)$ leads $Q(L)$. (2) $0 \rightarrow 1$ change in Q when $Q(L)$ leads $Q(H)$.

Conditions for POS HAZARD Production
(3) $0 \rightarrow 1$ change in Q when $Q(H)$ leads $Q(L)$. (4) $1 \rightarrow 0$ change in Q when $Q(L)$ leads $Q(H)$.

Fig. 5-64. *The four conditions for HAZARD production in output logic driven by the mixed-rail outputs of FLIP-FLOPs.*

5-64 and the character of the mixed-rail FLIP-FLOP outputs represented in Fig. 5-17 means that STATIC HAZARD production is possible. If there is no agreement between these figures, HAZARDs (as we define them) will not be produced. Accordingly, we draw the following conclusions relative to Moore machines:

FLIP-FLOPs whose mixed-rail outputs derive from NAND BASIC CELLs can produce HAZARDs only in POS output logic, but can do so on either a $0 \rightarrow 1$ or $1 \rightarrow 0$ change in Q.

FLIP-FLOPs whose mixed-rail outputs derive from NOR BASIC CELLs can produce HAZARDs only in SOP output logic, but can do so on either a $0 \rightarrow 1$ or $1 \rightarrow 0$ change in Q.

FLIP-FLOPs whose mixed-rail outputs derive from INVERTERs, such as in Fig. 5-32(c), can produce HAZARDs in SOP output logic on a $0 \rightarrow 1$ change in Q, or in POS output logic on a $1 \rightarrow 0$ change in Q.

Thus, the Q waveforms of Fig. 5-17(a) satisfy the conditions for POS HAZARD productions while those of Fig. 5-17(b) satisfy the conditions for SOP HAZARD production. Obviously, these conclusions are useful only if the FLIP-FLOP technology (NAND centered or NOR centered), is known to the designer. Furthermore, these conclusions are based on the assumption that output logic gates of the same logic family have approximately the same propagation delay and that the lead delays for these gates are uniform enough so that they can be disregarded. Asymmetric path delays associated with output logic gates or their leads may alter the conclusions given and may cause HAZARDs to be produced under different sets of conditions.

HAZARD production in the output logic of Mealy machines occurs under the same conditions as those established for the Moore machine except, in this case, the external inputs can form their own coupled terms. Thus, it is possible to have externally initiated HAZARDs produced in the output logic in addition to those produced by state variables. It must be remembered that in the output logic of Mealy machines, the complementary operations involving the external inputs must be performed by INVERTERs and that HAZARD analysis relative to these external inputs must proceed as discussed in Sect. 4.12.

Summary. STATIC HAZARDs are of concern only in the output forming logic of synchronous state machines. HAZARDs formed in the NEXT STATE forming logic are filtered by the MEMORY FLIP-FLOPs. STATIC HAZARDs produced in the output forming logic are of two types: those internally initiated by coupled state variables in either Moore or Mealy machines, and those externally initiated by external inputs in Mealy machines. Internally initiated HAZARDs can by eliminated by HAZARD cover or by hardware compatibility between FLIP-FLOPs and the output forming logic. Externally initiated HAZARDs can be eliminated by HAZARD cover. Any STATIC HAZARD can be eliminated by using a filtering stage on the FSM output.

There still remains the question of whether or not a HAZARD analysis is necessary. In fact, a complete HAZARD analysis need *not* be undertaken for any one of the following reasons:

- It is ascertained that the existance of HAZARDous transitions in output logic of an FSM poses no problem for the next stage.
- It is decided that filtering devices will be placed on the FSM's outputs to eliminate output race glitches or other logic noise.
- For a Moore machine it is known that the choice of output logic hardware in combination with the nature of the mixed-rail outputs of the FSM's FLIP-FLOPs cannot produce HAZARDs. Of course, this can be known by the designer only if the FLIP-FLOP technology (NAND or NOR) is known.

If none of the reasons given for the exemption of HAZARD analysis apply, a complete study of the HAZARD problem should be carried out on the FSM. To do this we recommend the following procedure:

1. Map each output logic function, extract minimum or reduced cover, and identify all coupled variables and associated coupled terms as discussed in Sect. 4.12.
2. Deduce the state-to-state transitions and branching requirements for each potential HAZARD from its coupled terms. This should be done in conjunction with the conditions for HAZARD production given in Fig. 5-64 and with the conclusions derived from them.
3. Compare the results of step 2 with the requirements of the state diagram. If agreement is found between the state diagram and a pair of coupled terms, determine the HAZARD cover and add it to the output expression to be implemented as discussed in Sect. 4.12. Disagreement between the state diagram and the requirements of a pair of coupled terms means that those coupled terms will not produce a HAZARD in the FSM under study and that HAZARD cover need not be added.

An Example. The presence of output race glitches aside, let us apply the foregoing procedure to the pulse width adjuster of Fig. 5-62. For the purposes of this discussion, let us assume the use of RET D-FFs each with mixed-rail outputs from a NOR BASIC CELL as in Fig. 5-17(b). Thus, SOP HAZARDs are possible while POS HAZARDs are not.

Presented in Fig. 5-65 is the SOP HAZARD analysis for this FSM. The minimum SOP cover, given in Fig. 5-65(a), reveals a HAZARD for either

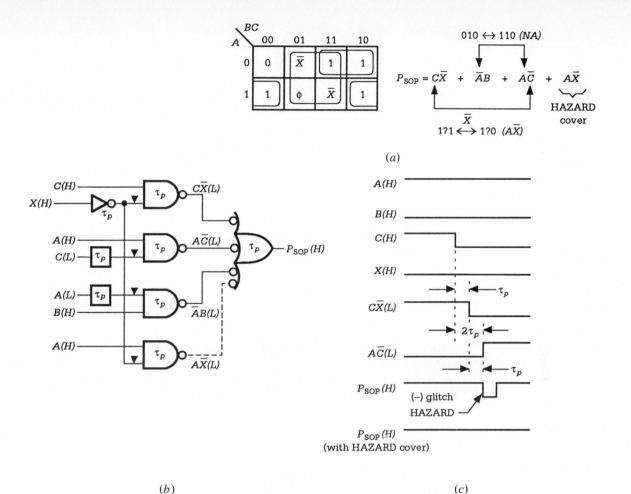

(a)

(b) (c)

Fig. 5-65. *HAZARD analysis for the pulse width adjuster of Figure 5-62.*
(a) K-map and minimum SOP cover for output P *showing an SOP HAZARD,*
produced by 111 → 100 race transition, and its cover. (b) Output logic with unit
path delays, τ_p. *(c) Timing diagram showing development of the (−) glitch SOP*
HAZARD.

the 111 → 101 → 100 or the 111 → 110 → 100 race path initiated by a 1̂ →
0 change in C. This was determined from the coupled terms $C\overline{X} + A\overline{C}$ which
require a change in C (see rectangular arrows) to initiate one of the two
transitions represented by 1?1 ↔ 1?0 under $\overline{X}$ branching conditions. Here,
the use of a question mark (?) represents the unspecified state variable B,
meaning Q_B. From an inspection of the state diagram, we see that only the
111 → 100 transition under $\overline{X}$ branching conditions satisfies *all* requirements
of the coupled terms. Since, by Fig. 5-64, SOP HAZARD production is
possible for a 1 → 0 change in C when $C(H)$ leads $C(L)$, and since this
condition is satisfied by the NOR BASIC CELL shown in Fig. 5-17(b) representing the outputs of FLIP-FLOP C, this is a valid HAZARDous transition. The other possible transition, 1 ? 1 → 1 ? 0, for a 0 → 1 change in
C under branching condition $\overline{X}$ does not exist in the state diagram, and so
is not a valid HAZARDous transition. Also, it is found that the coupled

terms $\overline{A}B + A\overline{C}$ are not applicable (NA) to HAZARD production since neither of the two unconditional transitions required by the coupled terms, and represented by 010 ↔ 110, exists in the state diagram of Fig. 5-62(b).

Once the PWA has stabilized into STATE 111, it will transit to STATE 100 if X is sampled LOW on the next triggering edge of CLOCK. It will make this transition by one of the two alternative race paths, shown in Fig. 5-63, depending on which FLIP-FLOP (B or C) changes first, but both involving a $1 \rightarrow 0$ change in C. In either case the HAZARD cover is the same and is found by ANDing the residues of the coupled terms. The results are then added to the expression to be implemented, as indicated in Fig. 5-65(a).

The development of the SOP HAZARD for the $101 \rightarrow 100$ race path is shown in the timing diagram of Fig. 5-65(c). Here, we assume the FLIP-FLOPs are NOR based. The output logic circuit is provided in Fig. 5-65(b) to help the reader account for the unit path delays, τ_p. These delays are assigned to each NAND gate and to the INVERTER. Since, by Fig. 5-17(b), $Q(H)$ leads $Q(L)$ for a $1 \rightarrow 0$ change in Q, boxed τ_p delays are also placed on all $Q(L)$ input lines to account for the mixed-rail timing character of the FLIP-FLOP outputs. Thus, the term $A\overline{C}(L)$ responds to a change in $C(H)$ after a $2\tau_p$ delay, one τ_p for the $C(L)$ output of FLIP-FLOP C (indicated by $\overline{C}$ in the term $A\overline{C}$) and one τ_p for the NAND gate. The term $C\overline{X}(L)$, on the other hand, responds to the change in $C(H)$ after only one path delay due to the NAND gate. When the two terms are ORed together, there results the SOP HAZARD shown in Fig. 5-65(c). It is important to note that the HAZARD cover, $A\overline{X}(L)$, is always logic 1 for either path of the $111 \rightarrow 100$ race transition and that adding it to the P_{SOP} minimum expression "ORs out" the HAZARD. Observe also that the term $\overline{A}B(L)$ is excluded from the timing diagram because it remains INACTIVE over the HAZARDous transition.

We know that STATIC HAZARDs will not be produced in POS output logic driven solely by edge-triggered FFs whose mixed-rail outputs are derived from NOR BASIC CELLs. Now, suppose we assume that the edge-triggered FFs are to be NAND based, that is, with mixed-rail outputs derived from NAND BASIC CELLs. An inspection of Figs. 5-17(a) and 5-64 indicates that the conditions are favorable for POS HAZARD production but not for SOP HAZARD production. Let us now consider the POS HAZARD problem.

The POS HAZARD analysis for the PWA is provided in Fig. 5-66 by assuming the use of ET FLIP-FLOPs with mixed-rail outputs from NAND BASIC CELLs. The minimum POS cover, shown in Fig. 5-66(a), is read in MAXTERM code and reveals a POS HAZARD for only the $001 \rightarrow 101 \rightarrow 111$ race path initiated by a $0 \rightarrow 1$ change in A (that is, $A \rightarrow \overline{A}$ in MAXTERM code) under branching condition X. This is determined from the coupled terms $(A + B + \overline{X})(\overline{A} + \overline{C} + \overline{X})$, which require a change in A, indicated by the rectangular arrows, to initiate one of the two race transitions represented by 001 ↔ 101 under branching condition X. Since we have read this information, in MAXTERM code, it is necessary for comparison purposes that we complement (at least in our minds) either the branching condition of the state diagram or that implied by the POS expression, not both. Thus, only the $001 \rightarrow 101$ race transition satisfies all requirements of the POS-coupled terms. The 001 ↔ 011 race transitions do not satisfy these requirements. Notice that the other possible transition, $001 \leftarrow 101$, does not exist in the state diagram and, therefore, cannot be HAZARDous.

Let us be clear on one point. The $001 \rightarrow 111$ transition will take place along one of two alternative race paths as shown in Fig. 5-63. For either race path the POS HAZARD cover is the same (though only one race path is HAZARDous) and is determined by ORing the residues of the coupled terms and then adding the result to the POS expression to be implemented, as indicated in Fig. 5-66(a).

The development of the POS HAZARD for the $001 \rightarrow 101$ race path is illustrated in Fig. 5-66(c) with the help of the logic circuit of Fig. 5-66(b). Since, by Fig. 5-17(a), $Q(H)$ leads $Q(L)$ for a $0 \rightarrow 1$ change in Q, boxed τ_p

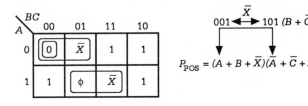

$$P_{POS} = (A + B + \bar{X})(\bar{A} + \bar{C} + \bar{X})(A + B + C) \cdot \underbrace{(B + \bar{C} + \bar{X})}_{\text{HAZARD cover}}$$

(a)

(b)   *(c)*

Fig. 5-66. *POS HAZARD analysis for the pulse width adjuster of Figure 5-62. (a) K-map and minimum POS cover for output P showing a POS HAZARD, produced by the $001 \rightarrow 101$ race path, and its cover. (b) Output logic with unit path delays, τ_p. (c) Timing diagram showing development of the $(+)$ glitch POS HAZARD and its elimination.*

delays are assigned to all $Q(L)$ input lines to account for the mixed-rail timing character of the NAND based FLIP-FLOPs. Unit τ_p delays are also assigned to each NOR gate and to the external input INVERTER. Thus, $A(H)$ leads $A(L)$, causing the term $(A + B + \overline{X})(L)$ to lead $(\overline{A} + \overline{C} + \overline{X})(L)$ and the result is the development of the POS HAZARD. Observe that the term $(B + \overline{C} + \overline{X})(L)$ is the HAZARD cover that removes the POS HAZARD. Also, note that the term $(A + B + C)(L)$ has been excluded from the analysis since it is always logic 1. In analyzing any POS timing diagram, we must remember that the waveforms for POS terms are ANDed together, not ORed as are the waveforms in SOP timing diagrams.

On the strength of the HAZARD analysis alone, it would seem that the best choice of hardware for the PWA (if that choice is permitted) is SOP output logic in combination with NAND-based FLIP-FLOPs. This eliminates the HAZARD problem while, at the same time, permits use of the simpler SOP output expression. Of course, if it is our plan to use an output filtering device to eliminate possible output race glitches, then any suitable hardware would suffice regardless of the existence or nonexistence of HA-ZARDs.

Further Discussion of the Output Glitch Problem. There still remain a few important points worth considering before departing from this section. If a designer cannot guarantee, by the methods outlined in this and the previous section, the absence of *all* output glitches (race or HAZARD), then placing filtering devices on all output lines offers one feasible solution. This is a common practice when dealing with highly complex synchronous sequential machines. The advantage of the "filter" approach is that it avoids the need to run race and HAZARD glitch analyses on the FSM. A disadvantage is that it adds hardware and time delay to the operation of the FSM. So if it is the intent to optimize the design by minimizing hardware and time delay, complete race and HAZARD glitch analyses should be carried out and corrective measures taken.

But there may be a much easier solution to the output glitch problem. If the FSM in question is to be the input stage to another synchronous FSM, no corrective action is necessary—*the next-stage MEMORY section will serve as the filtering device for any logic noise generated on the output logic of the previous stage*. It is worthwhile to remember this fact, especially when dealing with system-level designs. Also, remember that glitches in the output signals from an FSM are of concern only if they have a potential to cause problems. For example, if the outputs of an FSM are to drive light-emitting diodes, there is no need to remove the logic noise (assuming it exists) since the response of the human eye is too slow to pick up the glitches in the LEDs.

5.8.4 Asynchronous Inputs

The inputs to all FSMs considered so far were explicitly stated to be, or at least tacitly assumed to be, synchronous inputs. A synchronous input is one that is synchronized with CLOCK to the extent that it *cannot* change its logic level during a sampling interval (see Fig. 5-45). Any input that does not meet this requirement is said to be an *asynchronous input*, defined as follows:

> *An asynchronous input is one that can change logic levels at any time, particularly during the sampling interval established by the sampling variable.*

As was pointed out in Sect. 5.6.7, an input to a synchronous FSM must meet the setup and hold-time requirements established by CLOCK (the sampling variable) or proper transitions of the FSM cannot generally be guaranteed. Simply stated, a synchronous FSM may not function properly if asynchronous inputs are present. Remember that CLOCK is, by definition, an asynchronous input.

Rules Associated with Asynchronous Inputs. To reduce the probability for FSM malfunction, the following rules should be observed:

1. Branching Dependency Rule

> *Avoid branching dependency on more than one asynchronous input.*

The timing possibilities become too complex to predict the sequential behavior of an FSM when more than one asynchronous input controls the branching from a given state. Figure 5-67 demonstrates how a state diagram segment can be modified to satisfy the branching dependency rule. Since such modifications involve additional CLOCK periods (e.g., from STATE ⓒ to STATE ⓔ), care must be taken to make certain that

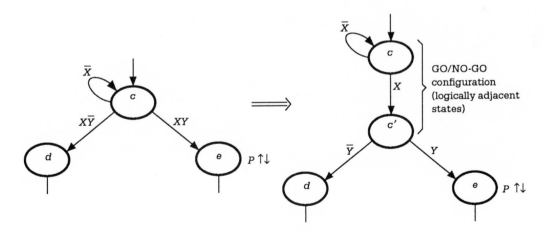

X,Y = Asynchronous inputs

Fig. 5-67. *Modification of a state diagram segment to eliminate state branching dependency on more than one asynchronous input.*

all timing requirements specified for the design are met. The GO/NO-GO configuration featured in Fig. 5-67 will be discussed shortly.

2. Conditional Output Rule

> *Do not attempt to generate an output conditional on an asynchronous input.*

An output that is conditional on an asynchronous input can, under certain conditions, be no more that an erroneous glitch. This is so because the asynchronous input can change at any time even before the output has been fully developed. Worse yet is the case where both branching and output from a state are conditional on an asynchronous input. Again, the timing possibilities become too complex to predict the behavior of the FSM. Figure 5-68 illustrates this and shows how expansion from a Mealy segment to a Moore segment can remove the output dependency on an asynchronous input. However, there still remains a problem due to the two-way branching from STATE ⓑ which depends on the asynchronous input, X. This problem will be discussed next in connection with the application of the "out of" rule as applied to asynchronous inputs. Notice the additional CLOCK period introduced by the expansion from a Mealy segment to a Moore segment in Fig. 5-68, which assumes that Z and W cannot be issued concurrently.

3. Restatement of the "Out of" Rule

> *The NEXT STATEs of a common state whose branching is conditional on an asynchronous input must be made logically adjacent.*

If the "out of" rule (as it applies to asynchronous inputs) is not obeyed, there is a small but nonzero probability that an *error transition* will occur in the FSM. Thus, observance of the revised "out of" rule is now a requirement, not just a guideline for logic optimization as it was for synchronous inputs in Sect. 5.8.1.

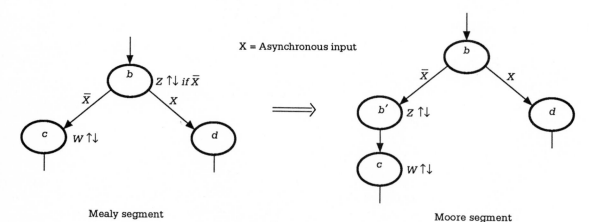

Fig. 5-68. *Modification of a state diagram segment to eliminate output generation dependency on an asynchronous input.*

The GO/NO-GO configuration shown in the modified state diagram segment of Fig. 5-67 provides a reliable means of dealing with an asynchronous input. The fact that STATEs $\bigcirc{c}$ and $\bigcirc{c'}$ are logically adjacent ensures that no error transitions are possible and that the transition $\bigcirc{c} \rightarrow \bigcirc{c'}$ can occur only if X is sampled ACTIVE. By making STATEs $\bigcirc{d}$ and $\bigcirc{e}$ logically adjacent also ensures that no error transitions are possible from STATE $\bigcirc{c'}$, but it cannot always be known which path will be taken by the FSM, to STATE $\bigcirc{d}$ or to STATE $\bigcirc{e}$, since the path decision must be made on the asynchronous input Y. However, the modified state diagram segment of Fig. 5-67 becomes completely acceptable if input Y is synchronized with CLOCK leaving X as the only asynchronous input other than CLOCK, the sampling variable. Thus, the GO/NO-GO configuration may be regarded as the only reliable means of handling asynchronous inputs short of synchronizing them.

Synchronizing the Input. A reliable approach to dealing with the problem of asynchronous inputs is to synchronize each asynchronous input with the CLOCK waveform before it is introduced to the NEXT STATE logic section of the FSM. This is accomplished simply by placing an edge-triggered D-FF on the line of an incoming asynchronous input. Shown in Fig. 5-69(a) is

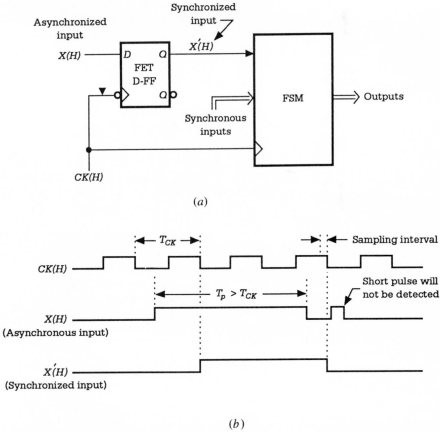

(a)

(b)

Fig. 5-69. *(a) Synchronization of an asynchronous input,* X. *(b) Timing diagram showing the results of applying a synchronizing cell to the asynchronous input.*

the synchronizing ET D-FF cell, and in Fig. 5-69(b) a timing diagram illustrating the results of its application to an asynchronous input $X(H)$. Notice that successive asynchronous $0 \to 1 \to 0$ pulses must be separated by at least one sampling period before they will appear as discrete synchronous pulses in the output of the synchronizing cell. More important, observe that this method is effective only if the individual asynchronous pulse widths are greater than the CLOCK period ($T_p > T_{CK}$). A "short" asynchronous pulse which cannot be detected and synchronized is shown in Fig. 5-69(b). The subject of short asynchronous pulses will soon be discussed.

In Fig. 5-69(a) an FET D-FF is used for an RET FSM so as to delay the synchronized signal $X'(H)$ no more than a half CLOCK cycle. To achieve the same result, an RET D-FF could be used for an FET FSM. Had we wished to delay the signal by as much as a full CLOCK cycle, we could have done so by using FLIP-FLOPs of the same triggering edge as the FSM. This has the advantage of bringing the input in phase with the system CLOCK (as opposed to being antiphase with CLOCK), but it is susceptible to CLOCK distribution problems as discussed in Sect. 5.8.5.

Stretching and Synchronizing the Input. There still remains the problem of "short" asynchronous input signals (pulses) that may be missed because their ACTIVE periods are too short to be sampled by the triggering edge of the CLOCK waveform. An example of a short pulse is shown in Fig. 5-69(b). To deal with short asynchronous pulses of this type, it is necessary that they be stretched so that they will be caught and treated as "long" asynchronous inputs by the FSM. One means of stretching short input signals is illustrated in Fig. 5-70(a). Here, a SET-dominant BASIC CELL, called a *catching cell,* is used to intercept and stretch the signal. All that is required to catch the short signal is that the signal exist long enought to SET the BASIC CELL. Since the stretched signal from the catching cell is still asynchronous, it must be dealt with within the FSM. The GO/NO-GO configuration, shown in Fig. 5-70(b), provides the best means of accomplishing this. Once the synchronized input X' has been sampled ACTIVE HIGH by CLOCK in STATE ⓠ and the transition is made to STATE ⓡ, an output

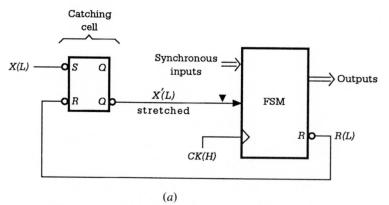

(a)

Fig. 5-70. *Stretching of a short input signal,* X. *(a) Use of a NAND BASIC CELL to catch the signal.*

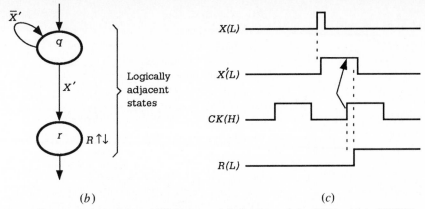

(b)   (c)

Fig. 5-70 (contd.). *(b) The GO/NO-GO configuration of the FSM which RESETs the catching cell. (c) Timing diagram showing the stretched signal X'(L) and path delay offsets for the FSM and catching cell.*

$R(L)$ is issued to RESET the catching cell so as to ready it for the next signal. Since a short input signal can be stretched up to a CLOCK period by the GO/NO-GO configuration of Fig. 5-70(b), it follows that such short signals must be spaced apart by at least one full CLOCK period, otherwise data could be lost. Of course, if it is intended that the signal X be stretched further, the output $R(L)$ can be issued in any state, including the final state of the sequence. But to do this would require proportionately greater spacing between short pulses if all data are to be detected. The timing diagram in Fig. 5-70(c) illustrates the stretching process just described. Notice that the offsets (dotted vertical lines) between $CK(H)$ and $R(L)$ and between $R(L)$ and $X'(L)$ represent path delays through the FSM and BASIC CELL, respectively.

Another means of dealing with short asynchronous input signals is to remove the GO/NO-GO configuration from inside the FSM and place it in combination with the catching cell on the input to the FSM. This arrangement is called a stretching/synchronizing cell and is illustrated by the use of FET D- and JK-FFs in Figs. 5-71(a) and (b). The obvious advantage of this arrangement is that it permits the short asynchronous input signal to be caught, stretched and synchronized prior to its introduction into the FSM. This eliminates the need to install a GO/NO-GO configuration in the state sequence (assuming one does not already exist) and it eliminates the need to generate

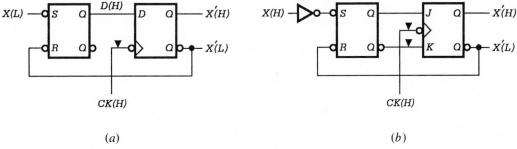

(a)   (b)

Fig. 5-71. *Stretching and synchronizing of short asynchronous inputs. (a) FET D-FF version and (b) FET JK-FF version of the stretching/synchronizing cell.*

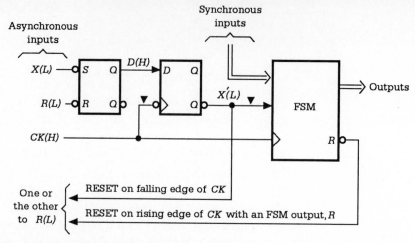

Fig. 5-72. *Logic circuit illustrating two options for RESETting the catching cell.*

an additional output $R(L)$, but at the cost of an additional FLIP-FLOP—a trade-off of sorts. The two alternatives for RESETting the catching cell are shown in Fig. 5-72.

The timing diagram in Fig. 5-73 illustrates the synchronizing and stretching of short, asynchronous input pulses by using the FET D-FF version given in Fig. 5-71(a). In this diagram the small offsets (or shifts) in the $D(H)$ waveform represent the path delay through the BASIC CELL. All other path delays are excluded for simplicity. Here, it can be seen that pulses P_1, P_3, and P_4 are each caught by SETting the BASIC CELL and then are synchronized and stretched to one CLOCK period as shown. Pulse P_2, however, is not caught because it occurred during the stretched period of P_1'. This leads to the following two *catching rules* for short pulses:

1. Short positive $(0 \rightarrow 1 \rightarrow 0)$ pulses must exist long enough to SET a BASIC CELL.

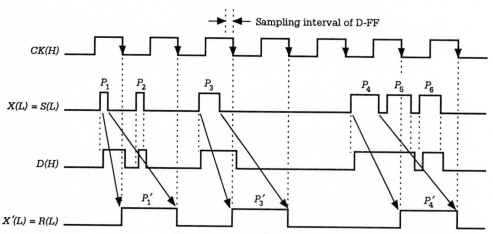

Fig. 5-73. *Timing diagram showing the stretching and synchronizing of short, widely spaced $0 \rightarrow 1 \rightarrow 0$ pulses.*

5 / Synchronous Sequential Machines

2. The outer edges of any two successive $0 \rightarrow 1 \rightarrow 0$ pulses (that is, the $0 \rightarrow 1$ edge of the first pulse and the $1 \rightarrow 0$ edge of the second) must be separated by at least two successive sampling intervals of CLOCK if the second pulse is to be caught, synchronized, and stretched.

Pulses P_2, P_5, and P_6 do not satisfy the requirements of the second catching rule and so are not caught, synchronized, and stretched. One obvious solution to this problem is to increase the frequency of the CLOCK waveform if it is possible and convenient to do so. If this cannot be done, the only other solution is to make certain that *all* data pulses arrive at intervals which satisfy the second catching rule. Failure to take either action will result in lost data.

The positive pulses P_4, P_5, and P_6 in Fig. 5-73 can be viewed as separating short negative $(1 \rightarrow 0 \rightarrow 1)$ pulses. By the second catching rule, it follows that negative pulses cannot be synchronized and stretched without using a LOGIC LEVEL CONVERTER (an INVERTER) on the asynchronous input to complement the logic waveform. Even then the pulses must be separated in agreement with the second catching rule. Notice that removal of the INVERTER in Fig. 5-71(b) accomplishes the logic level conversion.

Metastability and Synchronizer Circuits. Even with the use of a catching cell, there is still a very small but nonzero probability that a synchronizer ET FLIP-FLOP can be forced into a temporary metastable state that is either oscillatory or "hung up" somewhere between SET and RESET. This is possible in any FLIP-FLOP, including those in the MEMORY stage, if a data input to the FLIP-FLOP changes in close proximity to the triggering edge (i.e., during the sampling interval) of the CLOCK waveform. As a result, a "runt pulse" of the type shown in Fig. 6-19, could be produced and could, in turn, cause the FLIP-FLOP to enter the metastable state. Such a pulse may lack sufficient energy to resolve the FLIP-FLOP into either a SET or RESET condition.

Although there is only a very small probability that a synchronizer ET FLIP-FLOP could be forced into a metastable state, it can happen. The result could be the malfunction of the FSM to which the synchronizer FLIP-FLOP is attached. To greatly reduce the probability of such an occurrence, a second synchronizer FLIP-FLOP can be added as in Fig. 5-74(a), and for the case of $\Delta t_2 = \Delta t_1$ as in Fig. 5-74(b), both to be operated antiphase with the MEMORY FLIP-FLOPS. Here, it is assumed that if FF1 enters the metastable state on the rising edge of the CK waveform, there is an extremely high probability that it (FF1) will exit from that metastable state before the next rising edge of CK which must trigger FF2. Then, with probability near one, CK (in FF2) will sample a resolved condition from FF1 that is either a clean SET or clean RESET, but there is no way to predict which one.

If the metastable state in FF1 should last longer than one CLOCK period, there still exists the strong likelihood that FF2 will resolve into either a clean SET and clean RESET condition. Even so, it is possible to further reduce the probability that the metastable state will be passed on to the FSM. This can be done either by adding a third ET FLIP-FLOP to those in Fig. 5-74(a) or by increasing the CLOCK period. However, the reader must be advised that no scheme is possible that will reduce the probability to zero.

Because the metastable state may endure for a period of time that is unpredictable, concern about metastability is justifiable, and many studies of the metastable state have been made. The Annotated References at the end of this chapter cites work pertinent to this subect. The subjects of metastability and arbitration are covered in Sect. 6.9.1 of this book. Also, more information on these and related subjects can be gathered from sources cited in the Annotated References at the end of Chapter 6.

5.8.5   CLOCK SKEW

In synchronous sequential machines the triggering edge of the CLOCK waveform is assumed to reach each FLIP-FLOP of the MEMORY at approximately the same time. This has been assumed for all synchronous FSMs we have considered to this point. Sometimes, however, this does not happen due to the presence of asymmetric path delays in the CLOCK leads to the MEMORY devices. These delays can be caused by long wire leads, or by INVERTERs or DRIVERs placed in the CLOCK leads for fan-out purposes. When such delays become large enough to cause a shift in the triggering edge of one FLIP-FLOP relative to another, CLOCK SKEW is said to exist. CLOCK SKEW can become a serious problem if timing anomalies or incorrect data result.

Figure 5-74 illustrates one type of problem that can occur as a result of CLOCK SKEW. Here, input data $X(H)$ are supposed to pass serially through the two RET D-FFs on successive triggering edges of the $CK(H)$ waveform. The physical arrangement of the FLIP-FLOPs is given in Fig. 5-74(a) where path delays Δt_1 and Δt_2 are indicated on the CLOCK leads to FLIP-FLOPs 1 and 2, respectively. These delays may be present for a number of reasons some of which we have mentioned. However, their specific origin is unimportant to this discussion.

The proper operation of this FSM is illustrated in Fig. 5-74(b) for the case where the CLOCK lead delays are identical, $\Delta t_1 = \Delta t_2$. It is evident that FF2 picks up the output of FF1 on the following triggering edge of CLOCK such that the synchronized output of the second FLIP-FLOP goes ACTIVE at the same time that the synchronized output of the first FF goes INACTIVE. What we have just described is the operation of a two-bit serial-

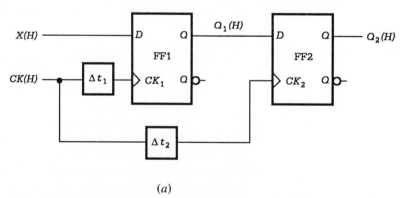

(a)

Fig. 5-74. *Illustration of a CLOCK SKEW problem in a synchronous FSM. (a) Logic circuit showing path delays in the CLOCK input leads.*

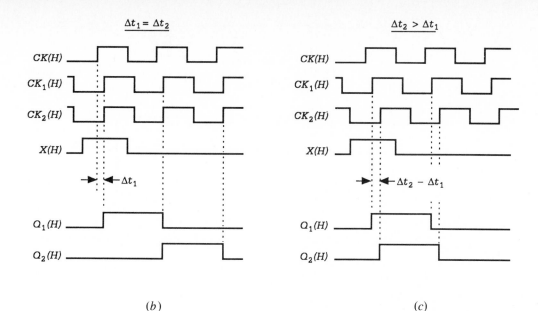

$$\Delta t_1 = \Delta t_2 \qquad\qquad \Delta t_2 > \Delta t_1$$

(b)   (c)

Fig. 5-74 (contd.). *(b) Timing diagram showing proper operation when $\Delta t_1 = \Delta t_2$. (c) Timing diagram showing a timing irregularity caused when $\Delta t_2 > \Delta t_1$.*

in/parallel-out (SIPO) SHIFT REGISTER. SHIFT REGISTERs will be the subject of a detailed discussion in Sect. 5.10.1.

When the CLOCK lead delays are unequal, as in Fig. 5-74(c), a CLOCK SKEW exists between the two FFs and a timing anomaly results. Now CK_2 of FF2 samples $Q_1(H)$ after a time equal to $\Delta t_2 - \Delta t_1$ rather than after a full CLOCK period as required for proper operation. Such timing anomalies can lead to unrecoverable errors in the operation of SHIFT REGISTERS and other devices.

Another type of CLOCK SKEW problem can occur when the data are presented in parallel, as shown in Fig. 5-75(a). This arrangement is that of a 2-bit parallel-in/parallel-out (PIPO) SHIFT REGISTER. Figure 5-75(b) illustrates the proper operation of this FSM when the data input is the same for both FLIP-FLOPs and when the CLOCK lead delays are symmetric (equal). In this case both FLIP-FLOP outputs are identical. However, when the CLOCK lead delays become asymmetric, data can be missed by one of the FLIP-FLOPs as shown in Fig. 5-75(c). This speaks in favor of the antiphase synchronization of an asynchronous input illustrated in Fig. 5-69. If a full-cycle delay is used in synchronizing the input, there is the risk of missed data. Clearly, the presence of a CLOCK SKEW complicates further the problem of asynchronous inputs.

The elimination of CLOCK SKEW in simple synchronous FSMs of the type just described is not usually a difficult task. Providing that the CLOCK SKEW is stable (that is, not time dependent), one simply balances the delays by using wire leads, INVERTER pairs, noninverting DRIVERs and the like. For high-frequency systems, transmission line delays on leads can be substantial, and this can cause the balancing procedure to become more difficult. In any case, the elimination of the CLOCK SKEW problems can be ensured only if all CLOCK lead delays are symmetric or very nearly so.

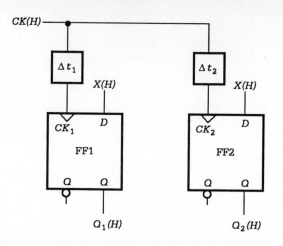

(a)

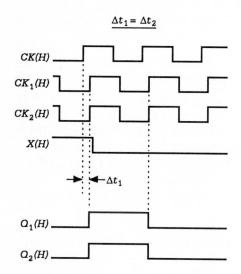

(b)

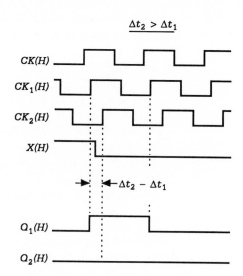

(c)

Fig. 5-75. *Sampling of incorrect data due to CLOCK SKEW. (a) Logic circuit showing CLOCK lead delays. (b) Timing diagram showing proper operation when $\Delta t_1 = \Delta t_2$. (c) Timing diagram showing sampling of incorrect data when $\Delta t_2 > \Delta t_1$.*

CLOCK SKEW problems are more difficult to diagnose and deal with in very complex systems where a SYSTEM CLOCK must drive synchronously the MEMORY elements of several independent devices. VLSI circuits are a good example. It is here where the use of master/slave FLIP-FLOPs can be helpful. Because MS FFs have a half-cycle output response to input change, they can eliminate the effects of CLOCK SKEWs of duration less than their sampling interval (see Fig. 5-46). Of course, this is not

5 / Synchronous Sequential Machines

the perfect solution, since MS FFs can themselves be a source of problems as was discussed in Sect. 5.6.4. The best advice that can be given to the designer of such systems is to "think symmetrically" when laying out a circuit. Try to avoid obvious sources of asymmetric path delays, particularly those associated with the SYSTEM CLOCK leads. Often, a conscious effort in this regard can save much time and expense.

5.8.6 Initialization and Reset of the FSM

An important part of the operation of any sequential machine is that it be *initialized* (on power-up) into a specific state or that it be *reset* into a specific state once it is in operation. If initialization and reset of FSMs were not possible, one can imagine the chaos that could result. Take, for example, the cruise control of an automobile. Failure of it to initialize or reset into a start-up state could be disastrous. Imagine not being able to initialize or reset the controller of one's computer. Equally important is the fact that no FSM should ever be designed such that it can initialize or reset into "hang" (unused) states or subroutines which are not part of the desired sequence. Whether the FSM is the controller for an elevator or traffic light system, or the controller for a robot or audio playback system, it should be obvious that initialization and reset capabilities are vitally important.

SANITY Circuits. What is needed is a signal that can be used to drive an FSM momentarily to a specific starting or reference state whenever it is necessary to do so—that is, during power-up to initialize the FSM or during operation to reset the FSM. Presented in Fig. 5-76(a) is a SANITY circuit as it might be used to power up or reset a three-bit MEMORY into the 001 STATE. It is called a SANITY circuit because it adds "sanity" to a situation that could otherwise be chaotic for the designer.

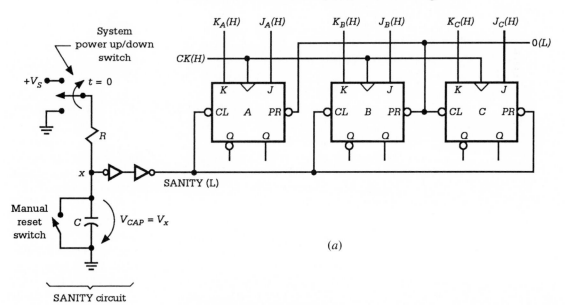

(a)

Fig. 5-76. *Use of a SANITY circuit to initialize or manually reset an FSM into the 001 STATE. (a) SANITY circuit and MEMORY section for the FSM.*

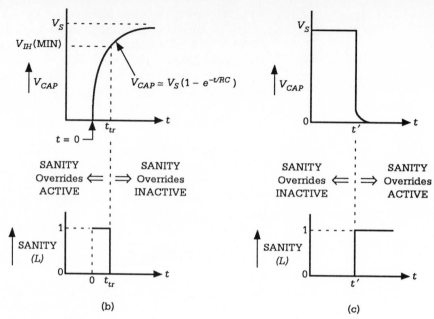

Fig. 5-76 (contd.). *(b) Initialization and (c) reset of the MEMORY showing the capacitor V-t characteristics and the resulting SANITY(L) logic signals.*

Figure 5-76(b) has been included to help the reader understand how the SANITY circuit operates. On system power-up, the SANITY circuit is connected to the supply voltage V_S, which charges the capacitor over time, and hence develops a voltage V_{CAP} across it as shown. At $t = 0$, when power is first supplied to the SANITY circuit, $V_x = V_{CAP} = 0$ and the node x and SANITY(L) output are both at the logic level $0(H) = 1(L)$. This logic level ACTIVATEs the SANITY asynchronous overrides of the MEMORY elements (the RET JK-FFs) driving them (the FLIP-FLOPs) momentarily into the 001 STATE. As the voltage V_{CAP} continues to rise, a point is reached when the ACTIVE transition point of the INVERTERs is crossed bringing the node x and the SANITY(L) output to the logic level $1(H) = 0(L)$, as indicated in Fig. 5-76(b). At this point and thereafter, all asynchronous overrides are DEACTIVATED and the MEMORY FLIP-FLOPs operate normally.

It is important for the reader to understand that the initialization process just described is made possible by energizing the SANITY circuit at the same time that the system devices are powered up. Thus, the SANITY switch and the system power switch are one and the same. Also, for a successful initialization, it is necessary that the time constant for the SANITY circuit, RC, be adjusted so that the voltage V_{CAP} reaches the ACTIVE transition point t_{tr}, shown in Fig. 5-76(b), after the system devices have equilibrated and after the CLOCK generating circuit has stabilized. In other words, the system devices and the CLOCK signal must be ready to operate normally the instant the SANITY asynchronous overrides are DEACTIVATED at time t_{tr}.

The time constant we refer to is simply the product of resistance R and capacitance C; that is,

$$\tau = RC \qquad (5\text{-}8)$$

where τ is the time constant in units of seconds (s) when R is in ohms (Ω) and capacitance is in farads (F). Thus, the unit equation for the time constant is (1 s) = (1 Ω)(1 F). Typical values of the time constant for SANITY circuits range from 10 ms to 100 ms, with values for R and C commonly in the range of 5 kΩ to 10 kΩ and 2 μF to 10 μF, respectively. Values for τ shorter than 10 ms may not permit the CLOCK generating circuit to stabilize or may not permit switch contact noise (discussed later) to damp out. Values larger than 100 ms may cause unnecessary delays in the initialization process and may even prevent the MEMORY FLIP-FLOPs from initializing simultaneously. As a "rule of thumb" the SANITY circuit will reach its steady-state level after a period of about 5τ following power-up, but the ACTIVE transition point of the threshold device will be reached well in advance of that.

Once the FSM is in operation it may be desirable to reset it into a specific state. If the reset and initialization states are the same, a single SANITY circuit will suffice. This is the case in Fig. 5-76(a). Here, closure of the manual switch across the capacitor resets the MEMORY as indicated in Fig. 5-76(c). Immediately following closure of the switch, the capacitor is discharged directly to ground, bringing the node x and SANITY(L) output to $0(H) = 1(L)$, abruptly ACTIVATING the SANITY asynchronous overrides and resetting the MEMORY.

If the SANITY circuit input to each asynchronous override is ANDed with an appropriate ENABLE input, an FSM can be initialized (or reset) into any state. To do this requires as many ENABLE inputs and INVERTERs as there are FLIP-FLOPs. It is left as an exercise for the reader to show how this is accomplished.

The SANITY circuit of Fig. 5-76(a) is adequate for many applications. However, there are other applications where it will not function properly. It suffers from fan-out capability, and it cannot be powered down and back up until a time about equal the time constant has elapsed. On power-down the capacitor must discharge through the resistor R, and this takes time determined by the time constant, RC. So, it is possible that the SANITY circuit will not reach a power-down condition before it is powered up again. Another problem is that a slow-input voltage change across the capacitor may cause a slow change in the output SANITY voltage, and this could fail to cause the SANITY asynchronous overrides of the MEMORY to be ACTIVATED simultaneously. The ACTIVE threshold voltages for the FFs will usually show some variation.

An improved SANITY circuit is shown in Fig. 5-77. It features an increased fan-out capability due to the parallel INVERTERs, a diode through which the capacitor can be discharged on power-down, and the use of a Schmitt trigger gate which causes the SANITY output voltage to change abruptly. The inverting Schmitt trigger functions logically in the same way as any INVERTER, but it has a higher $0 \rightarrow 1$ threshold voltage and a higher amplification factor than does an INVERTER. Consequently, the output voltage of the Schmitt trigger will change very rapidly even with a slow rise in the capacitor voltage V_{CAP}. Use of the Schmitt trigger in a SANITY circuit is not always necessary and is dependent on the type of logic family being used to implement the FSM as well as the value chosen for the time constant.

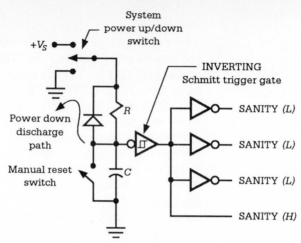

Fig. 5-77. *An improved SANITY circuit showing greater fan-out capability, faster capacitor discharge on power down, and sharper SANITY voltage transitions due to presence of the Schmitt trigger.*

Note that the SANITY circuit of Fig. 5-77 also provides a SANITY(H) output for devices (other than most FLIP-FLOPs) that have ACTIVE HIGH asynchronous PRESET and CLEAR overrides.

5.8.7 Debouncing Circuits

A common problem in digital system design is to provide human interface to the system. The use of push-button switches is a typical example. Asynchronous input signals from push-button switches often produce a phenomenon called "switch bounce" which derives from the mechanical structure of the switch. Multiple open/closed transitions may occur immediately following the depression or release of a button switch. Shown in Fig. 5-78(a) is a simple nondebounced single-pole/single-throw (SPST) mechanical switch and the multiple logic transitions, or contact noise, that may occur as a result of closing or opening the switch. Serious problems can result in an FSM if a high-frequency CLOCK catches the bounce signals. This is equivalent to the introduction of false data.

There are a number of circuits that can be used effectively to debounce a switch. The single-pole/single-throw switch of Fig. 5-78(a) can be debounced effectively by simply slowing down the rise (or fall) of voltage V_x following contact (or release) of the switch and by placing threshold devices such as a Schmitt trigger and an INVERTER on the output. Figure 5-78(b) shows how this can be accomplished. Here, following switch contact, the R-C component slows the voltage rise, as in Fig. 5-76(b), and a Schmitt trigger gate produces an abrupt change in the output voltage as its ACTIVE threshold is crossed. A similar behavior takes place for the voltage drop when the switch is opened. The result is the logic waveform shown in Fig. 5-78(b). Notice that the logic transitions are delayed slightly from the action of the switch. These delays are a result of the charging and discharging characteristics of the R-C circuit and the position of the ACTIVE and IN-ACTIVE transition points of the Schmitt trigger. The delays can be adjusted by altering the time constant, RC, but they must be long enough to wait out

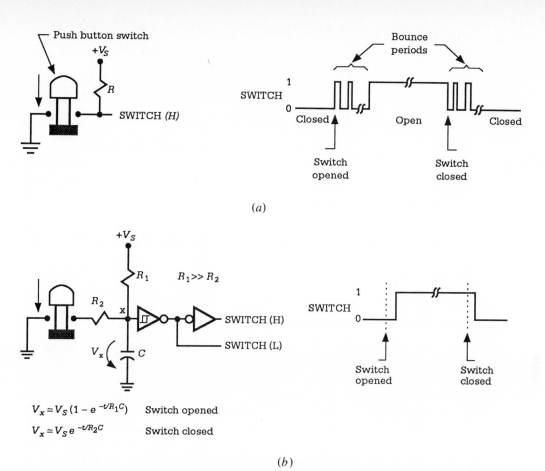

$$V_x \simeq V_S (1 - e^{-t/R_1 C}) \quad \text{Switch opened}$$

$$V_x \simeq V_S e^{-t/R_2 C} \quad \text{Switch closed}$$

(b)

Fig. 5-78. *Debouncing of a normally closed single-pole/single-throw mechanical switch. (a) A non-debounced switch and the resulting timing diagram showing contact logic noise. (b) A debouncing circuit for the switch in (a) and the resulting bounce-free timing diagram.*

the bounce periods shown in Fig. 5-78(b). Typical bounce periods are of the order of 10 ms or less so the *RC* time constant must be of that same magnitude or greater.

Use of a delay circuit like that in Fig. 5-78(b) is the only known method of debouncing a single-pole/single-throw switch. On the other hand, single-pole/double-throw (SPDT) switches can be debounced by using any one of a variety of different circuits. One circuit, shown in Fig. 5-79, performs a maximum speed debounce by making use of the SET and RESET action of a BASIC CELL. The manner in which it debounces a SPDT switch is remarkably simple. When the switch is thrown to the UP position the first contact bounce to achieve the $0(H) = 1(L)$ logic level on the $S(L)$ line SETs the CELL. All subsequent bounces are ignored. Similarly, when the switch is thrown to the DN position, the first bounce to reach the $0(H) = 1(L)$ logic level on the $R(L)$ line RESETs the CELL and all subsequent bounces are ignored. Observe that when the switch breaks with a given contact, any contact (release) noise that is produced can do nothing but HOLD the BASIC CELL in its present SET or RESET condition. Then when the switch makes

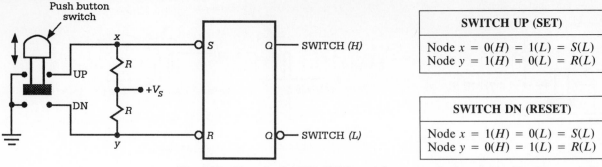

SWITCH UP (SET)
Node $x = 0(H) = 1(L) = S(L)$
Node $y = 1(H) = 0(L) = R(L)$

SWITCH DN (RESET)
Node $x = 1(H) = 0(L) = S(L)$
Node $y = 0(H) = 1(L) = R(L)$

Fig. 5-79. *Use of a BASIC CELL to debounce a single-pole/double-throw switch.*

contact with the next set of terminals, the CELL will alter its SET/RESET status, but only after a contact pulse has passed through a logic transition point.

5.9 DESIGN AND ANALYSIS OF MORE COMPLEX SYNCHRONOUS FSMs

In Sect. 5.7 we carried out the design of three simple FSMs, purposely avoiding the problem areas and design intricacies of Sect. 5.8. Now, it is necessary that we move on to more complex synchronous FSM designs where we can apply much of what we have learned in the previous sections. To assist in this endeavor we bring together the various parts of the design process into a more formal procedure. This will be followed by two examples illustrating some of the fine points of design. We conclude with a section dealing with the analysis of synchronous FSMs, roughly the reverse of the design process in one respect but similar in another.

5.9.1 A Design Procedure

For reference purposes in this section we present a seven-part design procedure. Considering the magnitude of the design problem, such a procedure must necessarily be complex often involving an interdependency of its parts. While not every design consideration has been included, the procedure is complete enough to serve as a guideline for successful FSM design.

Part I. Understand the Problem.

1. Develop a thorough understanding of the functional requirements and I/O specifications of the FSM to be designed.
2. Note any specific timing constraints that must be met.

Part II. Construct a State Diagram.

1. Choose a Moore or Mealy model and construct a fully documented state diagram that meets the requirements of the algorithm and timing constraints of the FSM. Use flowcharts and timing diagrams if necessary.
2. If asynchronous inputs are present, make certain that the branching dependency and conditional output rules given in Sect. 5.8.4 are obeyed.

Part III. Make State Code Assignments.

1. If optimization of the NEXT STATE logic is a requirement of the design, make state code assignments consistent with the logic adjacency rules of Sect. 5.8.1 wherever possible. In doing so, be aware of the asynchronous input problem.

2. If asynchronous inputs are present, decide at this point whether or not they are to be synchronized. If they are to be synchronized, proceed to Part IV. If not then make certain that the "out of" rule is satisfied by means of the GO/NO-GO configuration for each asynchronous input. This may require alteration of the state diagram and a reassignment of the state code.

Part IV. Obtain the Ouput Functions.

1. Choose the NEXT STATE and output logic hardware and MEMORY devices to be used and then determine the output functions. Knowing how the output functions are to be implemented and the character of the FLIP-FLOPs that will drive this implementation can influence the design strategy as discussed in Sect. 5.8.3. For ROM (or DECODER) implementation of the output functions, the required CANONICAL data can be obtained directly from the state diagram (see Part V). Otherwise map the output logic and extract minimum or reduced cover.

2. Determine whether or not logic noise in the outputs of the FSM will be a problem (see Sect. 5.8.3). If logic noise does not pose a problem for the next stage, proceed to Part V. If such noise will be a problem and if output race glitches and/or HAZARDs exist, then corrective action must be taken:

   a. If output race glitches are present, plan to filter the outputs (Sect. 5.8.2) or else eliminate the race conditions causing the glitches by using fly states or by returning to Part III for reassignment of the state codes.

   b. If output HAZARDs exist, plan to filter the outputs (if filtering has not already been planned due to race glitches) or use HAZARD cover to eliminate the HAZARD glitches. If neither of these options is acceptable, return to Step III for reassignment of the state codes. Note that this last option is by far the most difficult to carry out. Also, remember that HAZARDs may or may not be possible in the output logic of synchronous FSMs depending on the type of output logic chosen and the FLIP-FLOP technology used (see Sect. 5.8.3).

Part V. Obtain the NEXT STATE Logic Functions.

1. Determine the NEXT STATE logic functions. For ROM or DECODER implementation the required CANONICAL data can be obtained from the state diagram at the same time that the output data are obtained in Part IV, step 1. Otherwise, map the NEXT STATE logic by using the appropriate state transition table to characterize the FLIP-FLOP MEMORY element and then loop out minimum or reduced cover.

Part VI. Select the Initialization, Reset, and Debouncing Hardware.

1. If applicable select the hardware necessary for initialization and reset of the FSM and for switch debouncing. Refer to Sects. 5.8.6 and 5.8.7. Make certain that *all* timing constraints are met.

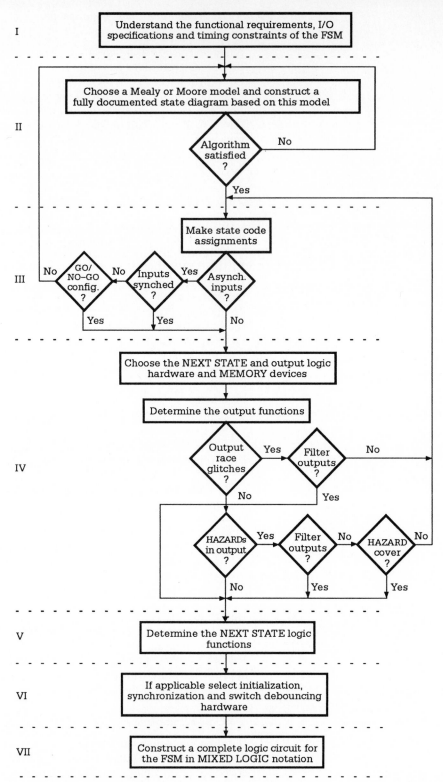

Fig. 5-80. *An abridged flow chart of the design procedure for synchronous FSMs showing seven major parts.*

Part VII. Construct the Logic Diagram.

1. Construct a complete logic circuit of the FSM in MIXED LOGIC notation and make any necessary comments for future reference.

The seven-part design procedure can be altered somewhat to satisfy the needs of a particular design project. For example, a relatively simple FSM design will often allow Parts III and IV to be carried out as a single step. Decisions regarding asynchronous inputs or output logic noise may not be required for some designs, or there may be no need to select initialization, reset, and switch debouncing hardware. But significant deviation from this procedure, particularly for relatively complex FSM designs, can result in extra effort, additional costs, or even the failure of the design project. Finally, it is always a good idea to test the FSM under expected operating conditions before it is put into use. In cases where real time testing is not convenient or permissible, computer-aided logic circuit simulations should be made.

Presented in Fig. 5-80 is an abridged flowchart for the design procedure just given. The purpose of the flowchart is to help the reader gain a better perspective of the design process and to provide a quick reference guide. A brief inspection of the flowchart indicates that some detail has been omitted. Even so, it emphasizes those aspects of the design procedure that we believe are the most important. The design examples given next will follow the intent of this flowchart even though specific mention may not always be made of each step.

5.9.2   Design of a One- to Three-Pulse Generator

As our first example let us design a pulse generator that will issue one, two, or three clean, discrete pulses depending on the setting of two switches, SW1 and SW0. A general description of this device is provided in Fig. 5-81. The requirements of this FSM are that the pulses be of ACTIVE duration equal to the ACTIVE duration of the CLOCK, that the two switches be set

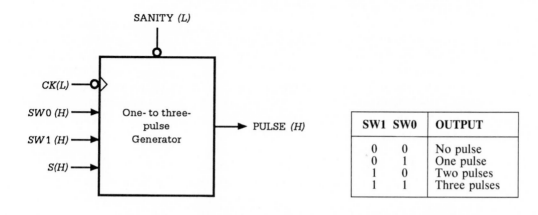

SW1	SW0	OUTPUT
0	0	No pulse
0	1	One pulse
1	0	Two pulses
1	1	Three pulses

(a)   *(b)*

Fig. 5-81. *Description of the one- to three-pulse generator. (a) Block diagram. (b) Operation table.*

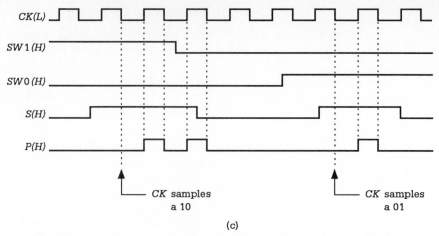

Fig. 5-81 (contd.). (c) Timing diagram showing the generation of pulses.

well in advance of the START (S) command, and that the switches remain stable at their settings for at least one CLOCK period following the START command. It is further required that the pulse generator be initialized into a nonoutput state, and that the START signal be returned to the INACTIVE condition following a pulse generating sequence and before initiating another pulse sequence. Such an FSM could be used to assist in the testing and debugging of another FSM.

The inputs S, $SW1$, and $SW0$ are asynchronous inputs. However, switch inputs $SW1$ and $SW0$ are of long duration compared to the CLOCK period and are set in advance of S, so they need not be synchronized or stretched. The START signal, assumed to be ACTIVATED by a push-button switch, should be debounced and either operated in a GO/NO-GO branching configuration or synchronized, as described in Sects. 5.8.4 and 5.8.7.

The block diagram of Fig. 5-81(a) reveals the ACTIVATION levels for the various inputs and the output of the one- to three-pulse generator. Notice that the CLOCK signal is ACTIVE LOW, meaning that the FSM is to be triggered on the falling edge of the CLOCK waveform. This is necessary if one requirement of the pulse generator is to be met, namely, that the ACTIVE duration of a pulse be equal to the ACTIVE duration of the CLOCK waveform. Had we chosen to use rising edge triggering, the ACTIVE widths of the pulses would be narrower than the ACTIVE duration of CLOCK by the amount of time equal to the propagation delay of the MEMORY.

The operation table in Fig. 5-81(b) shows the four available options which range from no pulse generation to the generation of three discrete pulses depending on the prior settings of switches $SW1$ and $SW0$. Figure 5-81(c) illustrates the generation of first two discrete pulses, then one discrete pulse for which the switch settings are $(SW1, SW0) = (1, 0)$ and $(0, 1)$, respectively. Here, we assume that the asynchronous input, S, will be debounced and synchronized to the rising edge of the CLOCK waveform. For this purpose an RET synchronizing cell must be used.

The flowchart that satisfies the algorithm and timing requirements for the one- to three-pulse generator is given in Fig. 5-82. The reader should use this flowchart in connection with the operation table of Fig. 5-81(b) to

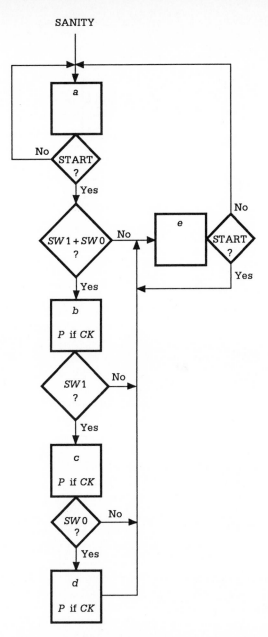

Fig. 5-82. *Flow chart for the one- to three-pulse generator.*

think through the operation of this FSM. As we indicated in Sect. 5.4.2, the flowchart should be viewed as a "thinking tool" whose primary function is to assist in the construction of the state diagram for an FSM. The sequential behavior of the one- to three-pulse generator seems to be complex enough to warrant use of the flowchart for this purpose.

Shown in Fig. 5-83(a) is the state diagram for the one- to three-pulse generator based on the flowchart in Fig. 5-82. Notice how much more vividly the sequential behavior is represented by the state diagram than by the flow chart. There are five states that require the use of at least three state variables which we name $Q_A Q_B Q_C = ABC$. The reader should verify that the outgoing

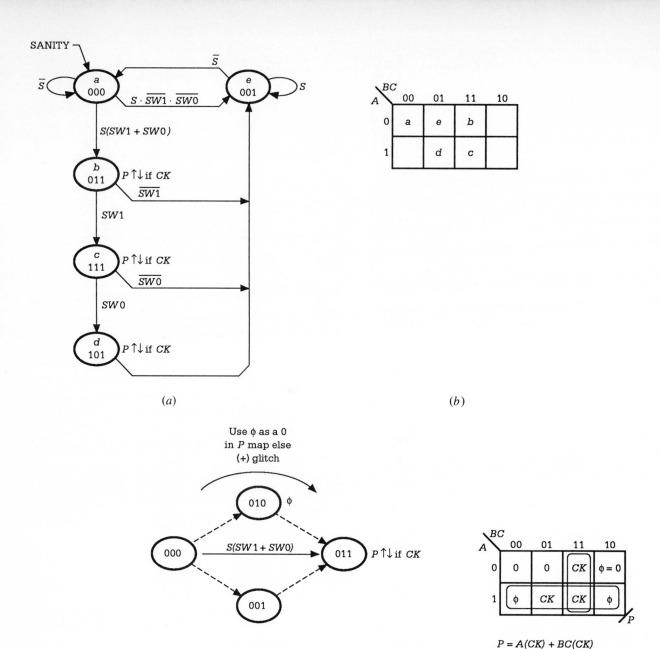

(a)

(b)

(c)

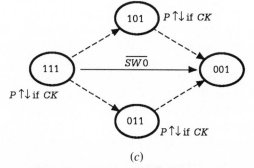

$P = A(CK) + BC(CK)$

Fig. 5-83. *Design of the one- to three-pulse generator. (a) State diagram. (b) State assignment map showing a high degree of compliance with the logic adjacency rules. (c) Output race glitch analysis, K-map and reduced cover showing no race glitches or HAZARDs.*

branching conditions for each state satisfy both the $\sum = 1$ rule and mutually exclusive requirement set forth in Sect. 5.4.1.

We must now make state code assignments for the five states as required in Part III of Fig. 5-80. Let us do so with the goal of achieving an output free of race glitches and HAZARDs. Also, let us continue to assume that the asynchronous input, S, is debounced and synchronized to the rising edge of the CLOCK waveform. Shown in Figs. 5-83(a) and (b) is one set of state code assignments that meet the requirements of our goal. Notice that compliance with either the "into" or "out of" rule is not possible for most of the states.

Moving on to Part IV of Fig. 5-80, we choose the NEXT STATE (NS) and output logic hardware and the MEMORY devices. Knowing at this time how the output logic section of an FSM is to be implemented and knowing the type of FLIP-FLOPs to be used forces us to deal with hardware-related problems that may affect our design strategy. For the present design, let us choose an FPLA implementation of the NS and output logic sections. So as to minimize the output requirements of the FPLA, we will choose to use FET D-FFs (as opposed to using JK FFs) for the MEMORY section. We will decide later if it is necessary to employ a particular FLIP-FLOP technology.

In Fig. 5-83(c) we carry out both race glitch and HAZARD analyses. Remembering that the FSM samples data on the falling edge of the CLOCK waveform (hence all conditional outputs are INACTIVE), we see that no race glitches can occur in the output if the don't care in cell 2 is taken as a logic 0 in the K-map for P. An inspection of the resulting reduced cover for P reveals no coupled terms; hence no gate HAZARDs are present regardless of the FLIP-FLOP technology that is chosen. The output logic is glitch free as shown.

The K-maps and minimum cover for the NEXT STATE logic functions are given in Fig. 5-84. Observe that full use is made of the don't cares that arise from the unused states, an option which is always available for the NEXT STATE logic but which is not always available for the output logic

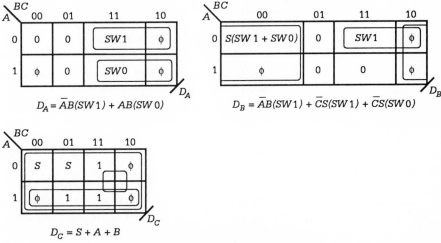

Fig. 5-84. *NEXT STATE K-maps and minimum cover.*

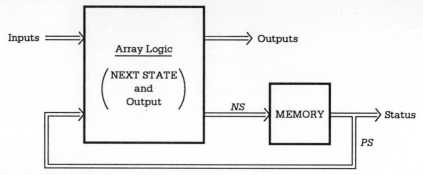

Fig. 5-85. *Model of Figure 5-4 as modified for array logic implementation of the NEXT STATE and output logic sections.*

due to possible race glitch formation. The NS logic cover together with that for the output function provide the p-term data which must be programmed into the FPLA.

Presented in Fig. 5-85 is the general model for any sequential machine as modified to permit implementation of the NEXT STATE and output logic sections with array logic (ROMs or PLAs). It is sometimes referred to as Huffman's model—after D. A. Huffman who, like Moore and Mealy, was a pioneer in the development of sequential machine design. Obviously, this model applies to either a Mealy or Moore machine since output dependency on the input is not specified. Such information is contained in the state diagram and in the program for the logic array. Input synchronizing cells and output filtering devices can be added as needed to any FSM modeled by Fig. 5-85.

To implement the NEXT STATE and output logic of this FSM with a PLA-type device, it is helpful to construct a p-term table, as discussed in Sect. 4.11.2. This has been done in Fig. 5-86(a), where we assume the use of an $8 \times 16 \times 4$ FPLA. There are seven inputs, nine p-terms, and four outputs that must be programmed. Since CLOCK is an ACTIVE LOW input, it is marked with an asterisk ($*$) as a reminder that each 1 or 0 appearing in that column must be complemented in the program for the PLA (see Sect. 4.11.4). The abridged symbolic representation of the programmed FPLA is shown in Fig. 5-86(b). Notice that the $CK(L)$ input has been accounted for in the programming of the FPLA.

The logic circuit for the one- to three-pulse generator is given in Fig. 5-87. It should be compared with the model of Fig. 5-83. Observe that the SANITY(L) input is connected to the $CL(L)$ override of each D-FF, meaning that on power-up, the FSM will initialize into STATE 000. No filtering device is necessary on the output $P(H)$ since its signals are glitch free.

Presented in Fig. 5-88 is the debouncing/synchronizing circuit for the pulse generator. It consists of a JK synchronizing cell connected in series with the debouncing circuit of Fig. 5-79. When switch S is pushed down, the UP terminal is connected to ground and the synchronizing cell is SET. Then when released, spring action connects the DN terminal to ground causing the cell to be RESET. It is assumed that the switch is stable in a given terminal contact position for a period much longer than the CLOCK period.

| p-terms | Inputs | | | | | | | Outputs | | | |

p-terms	A	B	C	S	SW1	SW0	CK*	D_A	D_B	D_C	P
$\overline{A}B$(SW1)	0	1	–	–	1	–	–	1	1	0	0
AB(SW0)	1	1	–	–	–	1	–	1	0	0	0
$\overline{C}S$(SW1)	–	–	0	1	1	–	–	0	1	0	0
$\overline{C}S$(SW0)	–	–	0	1	–	1	–	0	1	0	0
A	1	–	–	–	–	–	–	0	0	1	0
B	–	1	–	–	–	–	–	0	0	1	0
S	–	–	–	1	–	–	–	0	0	1	0
ACK	1	–	–	–	–	–	0	0	0	0	1
BC(CK)	–	1	1	–	–	–	0	0	0	0	1

* Indicates a complemented ACTIVE LOW input.

(a)

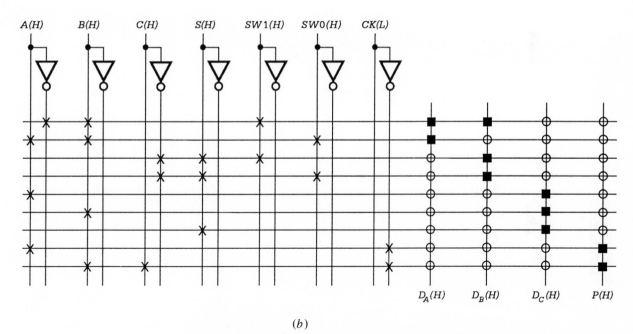

$D_A(H)$ $D_B(H)$ $D_C(H)$ $P(H)$

(b)

Fig. 5-86. *FPLA implementation of the one- to three-pulse generator in Fig. 5-83. (a) p-term table showing nine p-terms. (b) An abridged symbolic representation of an 8 × 16 × 4 FPLA programmed to provide the NEXT STATE and output logic.*

We have just completed the design of a synchronous FSM that can generate one to three clean pulses. We could have designed an FSM that would generate any number of such pulses. The system-level design of a multipulse generator is featured in Problem 5.69 at the end of this chapter.

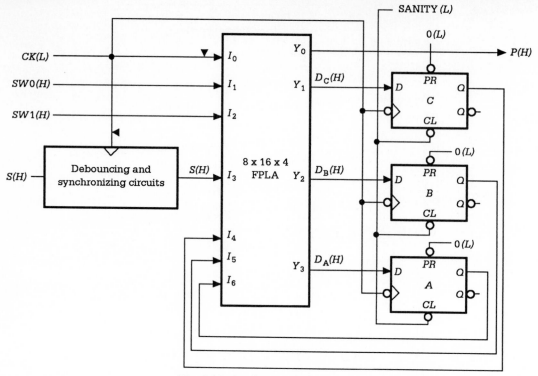

Fig. 5-87. *Implementation of the one- to three-pulse generator with FET D-FFs and an 8 × 16 × 4 FPLA.*

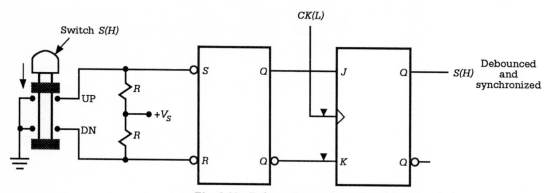

Fig. 5-88. *Debouncing and synchronizing circuit for asynchronous input S(H).*

5.9.3 Design of a Digital Combination Lock

For our second example we will design a circuit that can be used to unlock the access to something (say, a door to a room or automobile) by pushing buttons in some specific sequence. Such a device is usually called a digital combination lock (DCL). The block diagram in Fig. 5-89(a) indicates that two push-button switches, S1 and S2, control the operation of the DCL so as to issue signals LOCK and OPEN according to Fig. 5-89(b). Here, we have assumed that each switch is a mechanical single-pole/single-throw

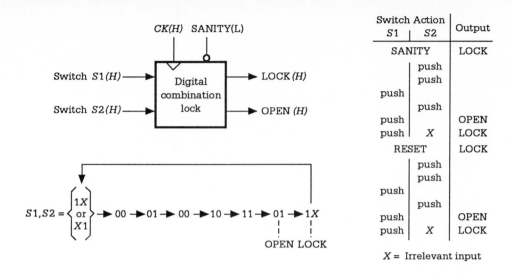

Switch Action S1	S2	Output
SANITY		LOCK
	push	
	push	
push		
	push	
push		OPEN
push	X	LOCK
RESET		LOCK
	push	
	push	
push		
	push	
push		OPEN
push	X	LOCK

X = Irrelevant input

(a)   *(b)*

Fig. 5-89. *Description of a digital combination lock. (a) Block diagram and logical switch sequence. (b) Operating sequence showing switch action and outputs.*

switch that changes the logic level of the signal each time it is pushed. Thus, an ON/OFF TOGGLE of a given switch is accomplished by pressing that switch twice. Since the switch signals are mechanically induced and are asynchronous, each must be debounced and synchronized. Also, we require that the DCL be initialized into the LOCK state and triggered on the rising edge of the CLOCK waveform.

It is the function of the DCL to issue the signal OPEN if, and only if, the sequence of five push-button operations, given in Fig. 5-89(b), is followed precisely. Deviation from this sequence at any point must require the FSM to return to the initialization state before another attempt can be made to complete the sequence. The output OPEN is issued upon successful completion of the switching sequence, and then return to the LOCK state occurs if switch S1 becomes ACTIVE. The flowchart in Fig. 5-90 presents the details of these sequential events. The reader should study and compare this flowchart with the operation table of Fig. 5-89(b).

The design of the DCL continues with the state diagram shown in Fig. 5-91(a). It is obvious from the state diagram for the DCL that the Moore model has been chosen for the implementation. Based on the flowchart of Fig. 5-90, it features five states with two- and three-way branching, two output states, and a state code assignment that involves the use of three state variables and unavoidable output race glitches. The three-way branching conditions require that the two switch inputs be synchronized so as to comply with the "branching dependency rule" relative to asynchronous inputs stated in Sect. 5.8.4. Notice that it is not possible to comply with the "out of" logic adjacency rule given in Sect. 5.8.1 since branching to the initialization (LOCK) state is required by all states. Limited compliance with the "into" rule can be seen.

We choose to use RET D-FFs and to implement the NEXT STATE and

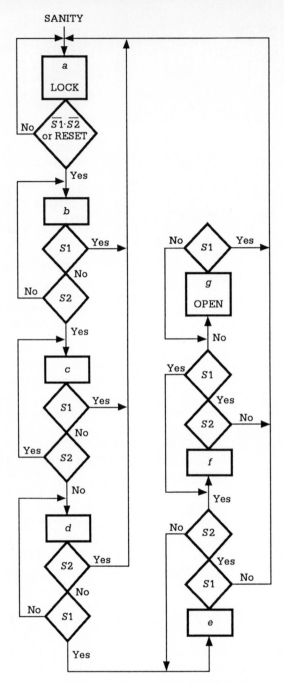

Fig. 5-90. *Flow chart for the digital combination lock described in Figure 5-89.*

output forming logic sections with a PROM. Recall from Sect. 4.11.1 that ROM implementation requires the use of CANONICAL data, which, for FSM design, are best obtained directly from the state diagram. The collapsed program table for ROM implementation of the DCL is presented in Fig. 5-91(c). It is constructed from the state diagram by considering the NEXT STATE branching requirements and outputs from each state taken one at a

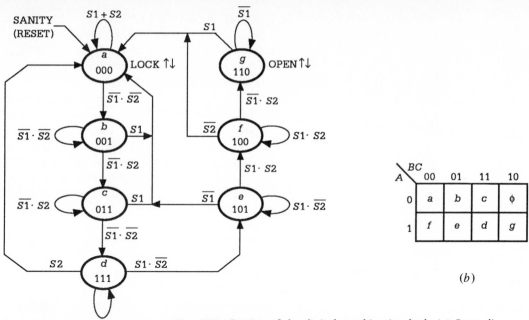

Fig. 5-91. *Design of the digital combination lock. (a) State diagram based on the flow chart of Figure 5-88. (b) State assignment map. (c) Collapsed program table for ROM implementation.*

(a)

State	PRESENT STATE			FSM Inputs		NEXT STATE			FSM Outputs	
Symbol	*A*	*B*	*C*	*S1*	*S2*	D_A	D_B	D_C	LOCK	OPEN
	0	0	0	0	0	0	0	1	1	0
a	0	0	0	1	0	0	0	0	1	0
	0	0	0	X*	1	0	0	0	1	0
	0	0	1	0	0	0	0	1	0	0
b	0	0	1	0	1	0	1	1	0	0
	0	0	1	1	X	0	0	0	0	0
	0	1	1	0	0	1	1	1	0	0
c	0	1	1	0	1	0	1	1	0	0
	0	1	1	1	X	0	0	0	0	0
	1	1	1	0	0	1	1	1	0	0
d	1	1	1	X	1	0	0	0	0	0
	1	1	1	1	0	1	0	1	0	0
	1	0	1	0	X	0	0	0	0	0
e	1	0	1	1	0	1	0	1	0	0
	1	0	1	1	1	1	0	0	0	0
	1	0	0	X	0	0	0	0	0	0
f	1	0	0	0	1	1	1	0	0	0
	1	0	0	1	1	1	0	0	0	0
g	1	1	0	0	X	1	1	0	0	1
	1	1	0	1	X	0	0	0	0	1

* Irrelevant input (X = 0 or 1).

(c)

time. For example, STATE ⓒ is the PRESENT STATE 011 of the FSM at some point in time. The NEXT STATE for STATE ⓒ is STATE 111 if $\overline{S1} \cdot \overline{S2}$ ($S1$, $S2 = 0$, 0) is sampled by CLOCK. But the NEXT STATE is STATE 011 (its own NEXT STATE) if the branching condition $\overline{S1} \cdot S2$ ($S1$, $S2 = 0$, 1) is sampled. Or, finally, the NEXT STATE for STATE ⓒ is STATE 000 if $S1$ ($S1 = 1$) is sampled independent of $S2$. In this table, as in several tables of Chapter 4, the symbol X appearing in the input column signifies an irrelevant input ($X = 0$ or 1). Notice that the program table indicates two outputs, the output LOCK in STATE 000 and the output OPEN in STATE 110, both represented by a logic 1 in the appropriate column of the table. An abridged symbolic representation of the programmed ROM would take on the appearance of Fig. 4-82. Such a representation for the DCL has not been presented in a figure due to the large number of MINTERMs involved.

Since we have acknowledged that output race glitches are unavoidable, we will use filtering devices, such as in Fig. 5-61, on the output lines. This action eliminates the need to run a race glitch analysis on this FSM. STATIC HAZARDs are not possible in the output logic of the DCL.

The circuit diagram for the ROM implementation of the DCL is given in Fig. 5-92. It is based on the model of Fig. 5-85 as was the previous FSM design. Here, use is made of a $2^5 \times 8$ ROM, that is, one of five inputs and eight outputs. Three of the ROM's eight outputs form the NEXT STATE address to RET D-FFs C, B, and A. Another two are used for the outputs, LOCK(H) and OPEN(H), which are filtered by FET D-FFs as shown. The

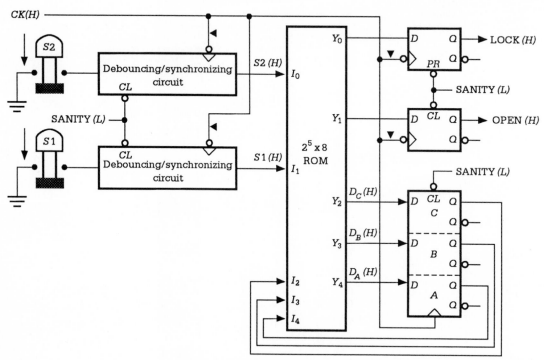

Fig. 5-92. *Circuit diagram for the ROM implementation of the digital combination lock.*

outputs of the RET D-FFs provide the PRESENT STATE feedback and FSM status.

The two switch inputs, $S1(H)$ and $S2(H)$, are introduced to the ROM from the debounce/synchronizing circuits shown in Fig. 5-93(a). Each circuit is composed of three parts: a debouncing circuit, a TOGGLE section, and a synchronizing cell. When the button of the normally closed switch is pushed, the switch signal becomes ACTIVE after a period of time about equal to R_1C. Once ACTIVE the logic level of the signal is reversed by the action of the RET JK-FF connected in the TOGGLE mode. From there the signal is synchronized by a FET D-FF and presented to the NS logic of the ROM.

All FLIP-FLOPs used in the design of the DCL, with the exception of the LOCK D-FF, can be asynchronously CLEARed by the SANITY(L) input from the SANITY circuit shown in Fig. 5-93(b). This is important for

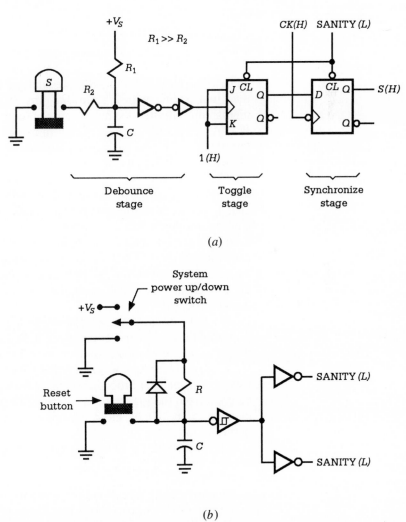

(a)

(b)

Fig. 5-93. Input conditioning and initialization circuits for the digital combination lock. (a) Debounce and synchronizing circuit. (b) Initialization/reset circuit.

the initialization and reset operations. On power-up the FSM is momentarily initialized into STATE ⓐ, the output LOCK is issued ACTIVE, and the output OPEN goes INACTIVE. At approximately the same time the *JK* and D-FFs of the debounce/synchronizing circuits are CLEARed, forcing switch signals $S1(H)$ and $S2(H)$ INACTIVE. This, in turn, causes the FSM to transit to and stabilize in STATE ⓑ where it must await the sequence of Fig. 5-89(b) to begin. Then by pushing button S2 the FSM will transit ⓑ → ⓒ, while a second push of S2 forces the transition ⓒ → ⓓ, and so on. The FSM can be RESET at any time by pressing down the RESET button [shown in Fig. 5-93(b)] and then releasing it to go up by spring action. The RESET condition has the same effect as the power-up SANITY operation, but without the need to power down the entire system. Because of the number of FLIP-FLOPs that must receive the SANITY(*L*) signal, it is necessary that the SANITY circuit has the fan-out capability shown in Fig. 5-93(b).

5.9.4 Analysis of Synchronous FSMs

The purpose of analyzing an FSM is to determine its sequential behavior. The procedure for FSM analysis is roughly the reverse of the procedure for FSM design given in Fig. 5-80. Thus, one begins with a logic circuit and ends with a state diagram. But the analysis process may involve much more than just ending with a state diagram. It may include timing analyses of inputs and outputs. Asynchronous inputs effects, output race glitches, and HAZARDs should be part of any thorough analysis of an FSM. Even user interfacing problems such as initializing and reset of the FSM, or debouncing of switch input signals should not be excluded.

The analysis of an FSM, beginning with the logic circuit, might be undertaken for a variety of reasons. For example, it is appropriate to apply an analysis procedure to a postdesign verification task. Or one might need to analyze another's work to understand it well enough to incorporate it into a new project. Finally, the application of an analysis procedure might be necessary simply to debug an FSM that does not function properly. But whatever the reason for its application—verification, understanding, or debugging of an FSM—it is clear that analysis plays an important role in the design process.

We have already considered the timing analyses of inputs and outputs as part of the design procedure given in Sect. 5.9.1. There, it will be recalled, we included the problems of asynchronous inputs, output race glitches, and HAZARDs. Now it is necessary that we focus our attention mainly on that part of the analysis procedure that deals with the construction of the state diagram from the logic circuit. It must always be remembered that the state diagram is the single most important means of describing the sequential nature of an FSM.

The procedure for obtaining the state diagram for an FSM from its logic circuit can be expressed in *five* easy steps:

1. Given the logic circuit for the FSM to be analyzed, examine it carefully for any obvious problems it may have, note the number and character of its FLIP-FLOPs and outputs (Mealy or Moore), and make note of any input conditioning and initialization requirements it may have.

2. Carefully read the NEXT STATE and output logic from the logic circuit.

3. Map the NEXT STATE and output logic expressions in K-maps appropriate for the FFs used.

4. From the K-maps construct the PRESENT STATE/NEXT STATE (PS/NS) table as demonstrated by the examples that follow. While this can be accomplished by using the K-maps for any MEMORY element (FF), it is most easily accomplished by using D K-maps. This is so because the NEXT STATE information for each bit is easily read from the single SET-oriented D K-maps. Therefore, it is recommended that JK and T K-maps be converted to D K-maps for this purpose.

5. Use the PS/NS TABLE together with the output K-maps to construct a fully documented state diagram for the FSM as detailed next.

An Example. To illustrate the analysis procedure, consider the logic circuit in Fig. 5-94(a). An inspection of this circuit reveals that it represents a Mealy machine having three state variables, one external input (assumed to be synchronous), and one output, and that it is initialized into the 010 STATE. Reading the NEXT STATE and output logic gives the results shown in Fig.

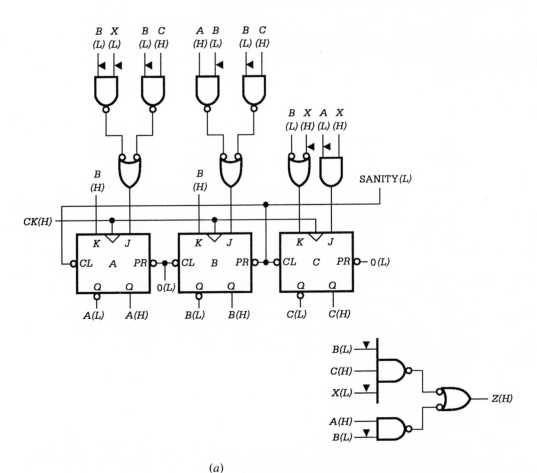

(a)

Fig. 5-94. *An FSM that is to be analyzed. (a) Logic circuit.*

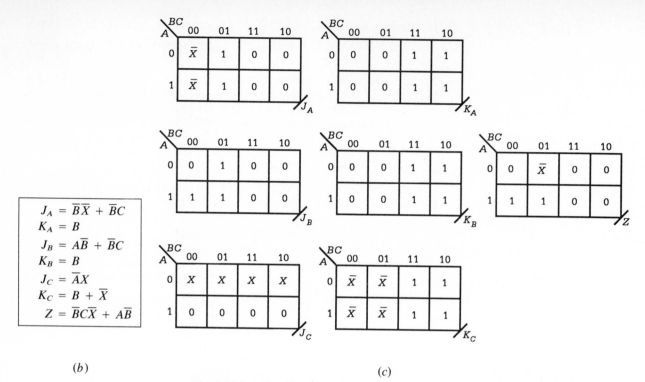

$$J_A = \overline{B}\overline{X} + \overline{B}C$$
$$K_A = B$$
$$J_B = A\overline{B} + \overline{B}C$$
$$K_B = B$$
$$J_C = \overline{A}X$$
$$K_C = B + \overline{X}$$
$$Z = \overline{B}C\overline{X} + A\overline{B}$$

(b)

(c)

Fig. 5-94 (contd.). *(b) Logic expressions and (c) K-maps for the NEXT STATE and output logic.*

5-94(b). These expressions are then mapped as in Fig. 5-94(c) ready for the next step which is to construct the PS/NS table. Notice the absence of the familiar don't cares in the J, K K-maps. This is expected and of no concern since all don't care information is necessarily lost once the expressions are reduced from CANONICAL form and implemented by a logic circuit.

As we indicated earlier in the analysis procedure, the PS/NS table is most easily constructed from D K-maps. Shown in Fig. 5-95(a) are the D K-maps which are obtained from the J, K K-maps by the conversion process illustrated in Fig. 5-54. The process of constructing the PS/NS table from the D K-maps involves taking the bit NEXT STATE data from each PRESENT STATE cell of the K-map and representing it in tabular form as in Fig. 5-95(b). For example, cell 0 of the K-maps represents PS 000. Since the NS entries are $D_A D_B D_C = \overline{X}0X$ for this state, one reads 100 for input condition $\overline{X}$ and 001 for input condition X. Cell 1, representing PS 001, has NS entries $D_A D_B D_C = 11X$, meaning 110 for $\overline{X}$ and 111 for X. Or for cell 2 (PS 010), the NS entries are $00X$, which gives 000 for input condition $\overline{X}$ and 001 for X. This process continues until all states (cells of the K-maps) are accounted for in the table. Observe that PRESENT STATEs (cells) for which the NS data entries are 1's or 0's only (no EVs) represent unconditional PS → NS transitions.

The state diagram, exclusive of outputs, is easily constructed from the PS/NS table by reading the branching requirements for each PS → NS transition. The output information is then added by reading the output K-map. This has been done for the FSM of Fig. 5-94 and the results are shown in

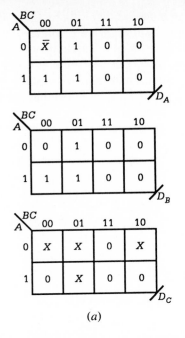

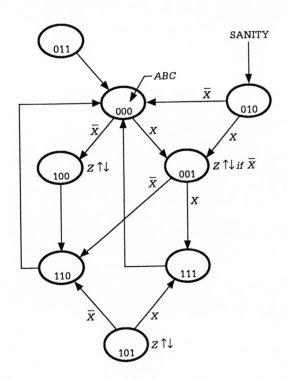

Fig. 5-95. (a) *Converted D K-maps and* (b) *PS/NS table for the FSM of Figure 5-94.* (c) *State diagram constructed from the PS/NS table of* (b).

Fig. 5-95(c). We recommend that initially the states be layed out in a somewhat random fashion and then later arranged into a more orderly array if desirable. Notice that a single NS assignment in the PS/NS table (for a given PS) signifies an unconditional transition and that an EV term in the output map represents a conditional (Mealy) output. Also, note that we recognize initialization into the 010 state by indicating a SANITY input to that state.

Now that the construction of the state diagram has been completed for the FSM, the analysis procedure can, and should be, extended to include any unusual characteristics of the FSM. This includes an analysis of the effects of asynchronous inputs, and the analyses of output race glitches and HAZARDs. First, it is clear that there is no entrance into either STATE 011 or STATE 101 (an output state), so these states are not part of any "useful" sequence. Second, the external input X must not be an asynchronous input since proper application of the "out of" rule in a GO/NO-GO configuration is lacking, and there is an output in STATE 001 that is conditional on this input. Finally, numerous output race glitches can occur during four of the six possible transitions, so output filtering devices should be employed. There are no STATIC HAZARDs possible. On the whole, this would seem to be a rather poorly designed FSM.

A More Complex Example. As our second example, let us analyze the multiple-input FSM whose NEXT STATE and output logic are given in Fig. 5-96(a), information that is assumed to have been obtained from some logic circuit not shown. Here, it is clear that A and B are the state variables and that D, N, S, W, X, and Y are the six inputs. A single output Z exists and is conditional on input N, which makes this FSM a Mealy machine.

The analysis continues with the plotting of the NEXT STATE and output K-maps shown in Fig. 5-96(b). The J, K K-maps are then converted to D K-maps, in Fig. 5-97(a), and the PS/NS table is constructed as in Fig. 5-97(b).

The PS/NS table of Fig. 5-97(b) is a greatly compressed table considering

$$J_A = BX\overline{Y}$$
$$K_A = \overline{B}\,\overline{N}D$$
$$J_B = \overline{A}S + AN$$
$$K_B = \overline{A}Y + A\overline{W}$$
$$Z = ABN$$

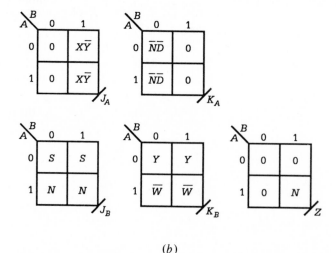

(a)

(b)

Fig. 5-96. *An FSM that is to be analyzed. (a) NEXT STATE and output logic as read from a logic circuit. (b) NEXT STATE and output K-maps.*

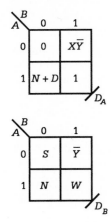

PS	Inputs	NS	PS	Inputs	NS
00	$\overline{S}$	00	11	$\overline{W}$	10
	S	01		W	11
01	$\overline{X}\,\overline{Y}$	01	10	$\overline{N}\overline{D}$	00
	$\overline{X}Y$	00		$\overline{N}D$	10
	$X\overline{Y}$	11		$N\overline{D}$	11
	XY	00		ND	11

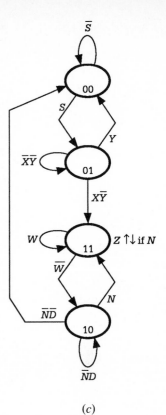

(a)	(b)	(c)

Fig. 5-97. (a) Converted D K-maps, (b) PS/NS table and (c) State diagram for the FSM represented in Figure 5-96.

that it represents $2^6 = 64$ possible NS entries due to the six inputs. So it is important to note that only the variables represented in the K-map cells of a given PRESENT STATE are expressed in CANONICAL form in the PS/NS table. Thus, if there is a PS with 3 input variables represented, then there would be 8 possible branching conditions (hence 8 possible NS entries) for that PRESENT STATE. Or if a PS has 4 input variables, 16 possible NS entries exist, and so on.

The final step is the construction of the state diagram shown in Fig. 5-97(c). As in the previous example, this is accomplished by reading from the PS/NS table the input branching conditions for each PS → NS transition. Notice that both the sum rule [Eq. (5-6)] and the mutually exclusive condition are automatically satisfied for each state. This is so because all possible combination of inputs for a given PS are represented in the input column of the PS/NS table. Finally, a brief inspection of the state diagram reveals that this FSM has no obvious problems.

5.10 MODULE AND BIT-SLICE DEVICES

In previous sections synchronous FSMs having various levels of complexity were studied as part of the process of developing the material in this chapter. Initially, we used the BASIC CELL as the MEMORY device for the design of FLIP-FLOPs. Then, by using FLIP-FLOPs as MEMORY devices, other more complex sequential machines were designed. Now, we will consider the design of some common but important FSMs which are packaged as devices with distinguishing names, names such as SHIFT REGISTERs and COUNTERs. These devices constitute two important classes of FSMs which are functionally different, and which have, generally, different utility. However, we will demonstrate later in this section and in Sect. 5.11 that both types of devices can be used as the MEMORY elements in the design of larger more complex FSMs. It is important for the reader to keep this last point in mind as we progress through the development of this material.

5.10.1 Registers

A shift register is a synchronous memory system consisting of an array of FLIP-FLOPs and the appropriate combinational logic all connected in such a way that binary information introduced into it can be shifted to the right or to the left depending on its design. For example, a SHIFT REGISTER could be designed to receive and transfer serial information bit by bit with each triggering edge of the CLOCK waveform. As a packaged device this could be called a serial-in/serial-out, or SISO, register. Another type of SHIFT REGISTER might be designed to receive parallel information and transfer it serially bit by bit with CLOCK, hence a parallel-in/serial-out, or PISO, register. A third possibility is the design of a SHIFT REGISTER that receives serial data and transfers it out in parallel, hence a serial-in/parallel-out, or SIPO, register. The acronyms SISO, SIPO, and PISO are commonly used to describe the operation modes of SHIFT REGISTERs, but they provide no indication of shift direction.

SHIFT REGISTERs are found in a wide variety of digital systems. In addition to permitting the temporary storage of binary data, they are useful for parallel-to-serial and serial-to-parallel conversion at the transmitting and receiving ends of a data communication line that delivers data serially. In this case, incoming parallel data must first be *serialized* by PISO register action. Then at the receiving end, it is *deserialized* by a SIPO register. In either case, the conversion process takes place bit by bit with each triggering edge of the CLOCK waveform.

Other applications of SHIFT REGISTERs include their use as MEMORY devices in complex FSM design and their use in the performance of arithmetic operations. In the latter case, the reader may recall from Sect. 4.9.2 that each bit shifted left or shifted right is equivalent to dividing or multiplying a binary number by 2, respectively. The serialization and deserialization capability of a shift register is also useful in serial or parallel arithmetic operations as is demonstrated by Problem 5.70 at the end of this chapter.

Registers that do not shift binary bits but simply hold (i.e., store) bit information and transfer it on the triggering edge of CLOCK are called *holding* or *storage* registers. The packaged devices are also known as parallel-in/parallel-out, or PIPO, registers. Their only function is to provide temporary storage of data. The PIPO operation is useful for filtering out logic noise in the output signals of FSMs or to provide ordered delivery of parallel data.

For reference purposes, we list the four register operation modes as follows:

PIPO, parallel-in/parallel-out
PISO, parallel-in/serial-out
SIPO, serial-in/parallel-out
SISO, serial-in/serial-out

A bidirectional (shift left or right) SHIFT REGISTER that is capable of operating in any of the four preceding modes, and that can store and output the same set of data indefinitely over any number of CLOCK periods, is called a *universal* SHIFT REGISTER. We will design this general type of SHIFT REGISTER after we have considered the simpler types.

Holding (PIPO) Registers. The PIPO register is the simplest of all registers since it consists of nothing more than an array of D-FFs each triggered on the same triggering edge of the CLOCK waveform. Shown in Figs. 5-98(a) and (b) are the state diagram and NEXT STATE K-map for a 1-bit-slice PIPO register. The subscript J refers to the Jth cell. Presented in Figs. 5-98(c) and (d) are the circuit diagram and symbol for a 4-bit PIPO register. Much larger units can be obtained by combining 4-bit-slice PIPOs. Four-bit and larger bit-sliced units are commercially available.

SHIFT REGISTERs. Only slightly more complex than the PIPO register is a unidirectional SHIFT REGISTER that can operate in any one of the four register modes, but that must shift or parallel load with each triggering edge of the CLOCK. In Fig. 5-99(a) we give the operation table and definitions for a 1-bit-slice parallel loadable right SHIFT REGISTER of this type. It has PIPO, PISO, SIPO, and SISO capability. Because it can shift as well as parallel load, a mode select input (S) must be included in the design. Thus, for $S = 0$ the NEXT STATE address for the Jth cell is the output Q_{J+1} from the next MSB stage to the left of the Jth stage. If $S = 1$, the device will perform the PIPO operation (no shift) with each triggering edge of CLOCK. Note that the direction "right" shift is used only for reference purposes since the device can be flipped about for left shift orientation if desired. But it is incapable of bidirectional shifting, that is, shifting either right or left on command.

Shown in Fig. 5-99(b) is the state diagram for the one-bit-slice SHIFT REGISTER as derived directly from the operation table given in Fig. 5-99(a). Since the Jth cell of the device must issue a Q_J when in either mode, and since there is no data HOLD condition explicitly provided in the operation table, the $0 \rightarrow 1$ and $1 \rightarrow 1$ branching conditions must be identical as indicated. The $1 \rightarrow 0$ and $0 \rightarrow 0$ branching conditions follow from the sum ($\sum = 1$) rule given by Eq. (5-3).

The NEXT STATE K-map for the Jth cell of this SHIFT REGISTER is shown in Fig. 5-99(c) together with the minimum cover. Implementation of the NEXT STATE function can be accomplished at the gate level or with modular devices. In Fig. 5-99(d) we take the latter approach by using a 2-to-1 MUX. Notice that the NEXT STATE is Q_{J+1} (a right shift) if $S = 0$, or it is P_J (a parallel load) if $S = 1$.

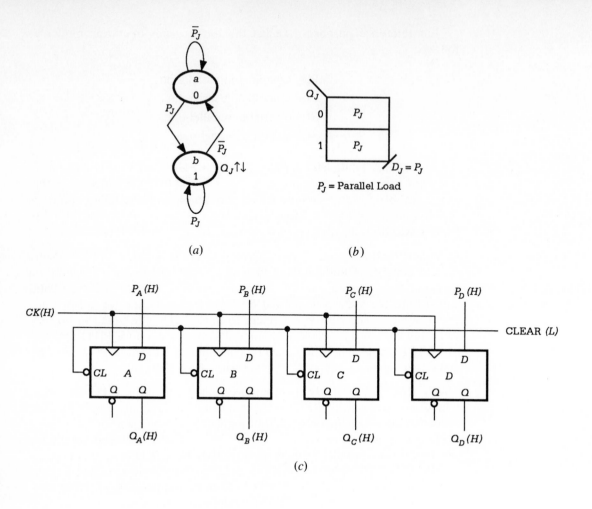

(a)

(b)

(c)

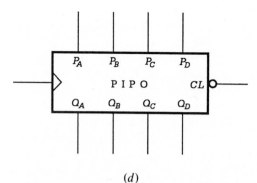

(d)

Fig. 5-98. *Design of a holding (PIPO) REGISTER. (a) State diagram and (b) K-map for the Jth cell. (c) and (d) Circuit diagram and symbol for a 4-bit holding REGISTER.*

The one-bit-slice SHIFT REGISTER in Fig. 5-99(d) can be cascaded to form an *n*-bit register. The logic circuit and circuit symbol for a four-bit-slice unit are given in Figs. 5-100(a) and (b). Notice the parallel connections for the CLOCK and asynchronous CLEAR inputs.

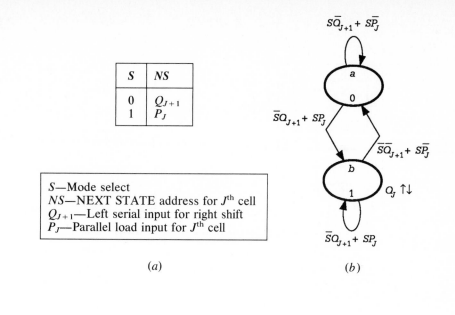

S	NS
0	Q_{J+1}
1	P_J

S—Mode select
NS—NEXT STATE address for J^{th} cell
Q_{J+1}—Left serial input for right shift
P_J—Parallel load input for J^{th} cell

(a)

(b)

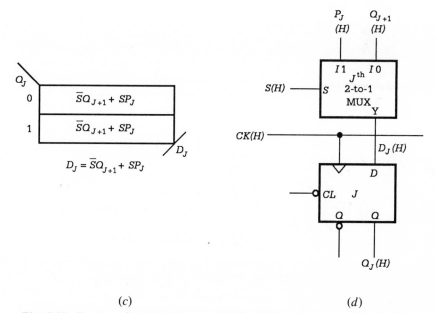

(c)

(d)

Fig. 5-99. *Design of a 1-bit slice parallel loadable right SHIFT REGISTER with PIPO, PISO, SIPO, and SISO capability. (a) Operation table and definitions. (b) State diagram and (c) NEXT STATE K-map and minimum cover assuming the use of D-FFs. (d) MUX implementation of the Jth cell.*

A timing diagram in Fig. 5-100(c) demonstrates the operation of the SHIFT REGISTER of Fig. 5-100(a). Included is a parallel load of 0101 (S = 1) followed by a series of shift zero right ($R0$) operations, a serial input ($L = 1$ when $S = 0$) also followed by a series of shift zero right operations, and finally an asynchronous CLEAR. Notice that this device is not capable of maintaining a constant output over any number of CLOCK periods independent of the serial and parallel inputs. That is, this SHIFT REGISTER

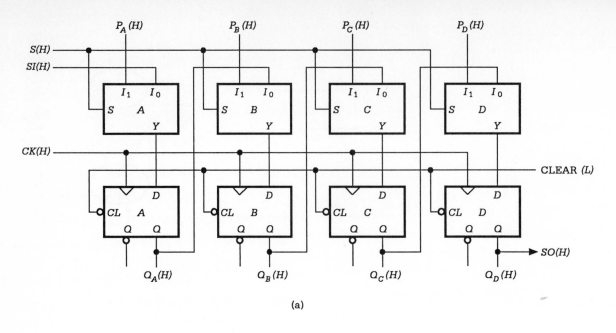

(a)

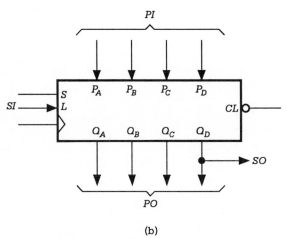

(b)

Fig. 5-100. *A 4-bit, right shift, parallel loadable REGISTER of the type shown in Figure 5-99. (a) Logic circuit. (b) Block diagram symbol.*

must either parallel load or shift right a 1 or 0 with each triggering edge of the CLOCK signal since a HOLD feature has not been designed into it. The following design of a more general type of SHIFT REGISTER will include this capability.

THE UNIVERSAL SHIFT REGISTER. Let us now design a SHIFT REGISTER that can parallel load, shift right or left a logic 1 or 0, or store the same binary word over any number of CLOCK periods. Such a device is generally known as a UNIVERSAL SHIFT REGISTER—it can perform all the normal register operations. Shown in Fig. 5-101(a) is the operation table for the Jth cell of this device. It is capable of maintaining a constant

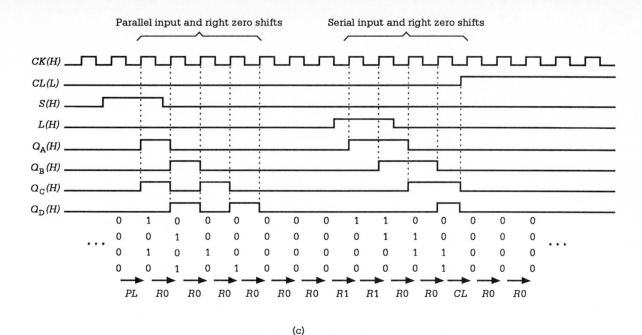

(c)

Fig. 5-100 (contd.). (c) *Timing diagram showing a parallel input of 1010 and a left serial input both with subsequent right zero shifts.*

output (the hold condition) when the mode select inputs are $S_1 S_0 = 00$. Or, if the mode select inputs are $S_1 S_0 = 01$, the operation is a right shift of 1 or 0 with a NEXT STATE of Q_{J+1}. However, if the mode select inputs are $S_1 S_0 = 10$, a left shift of 1 or 0 occurs with a NEXT STATE of Q_{J-1}. Note that for reference purposes we write the output word . . . Q_{J+1}, Q_J, Q_{J-1} . . . in the order of decreasing positional weight from left to right. Thus, the $J + 1$ cell is to the left of the Jth cell while the $J - 1$ cell is to the right. Finally, when the mode select inputs are $S_1 S_0 = 11$, the register action is a parallel load with a NEXT STATE of P_J.

The state diagram for a 1-bit-slice UNIVERSAL SHIFT REGISTER is presented in Fig. 5-101(b). It is obtained directly from the operation table by deducing the $0 \rightarrow 1$ and $1 \rightarrow 1$ branching conditions as shown, the only difference being that Q_J is 0 for the former and 1 for the latter. The remaining two branching conditions ($1 \rightarrow 0$ and $0 \rightarrow 0$) are obtained by applying the $\Sigma = 1$ rule as was done in Fig. 5-99(b).

The reader may find it curious that we are able to represent a relatively complex FSM, such as a SHIFT REGISTER, by a two-state state diagram. This is possible because each cell of the register performs a specific number of independent operations and involves the use of a single FLIP-FLOP together with the appropriate NEXT STATE forming logic. In this sense the state diagram representation of a UNIVERSAL SHIFT REGISTER is no different from, say, that of Fig. 5-43(a) which was used to design a JK-FF from a D-FF. We can generalize this concept by stating that any FSM, no matter how complex, can be represented by a two-state state diagram provided that a specific number of independent operations can be carried out by a single FLIP-FLOP together with the appropriate NS forming logic.

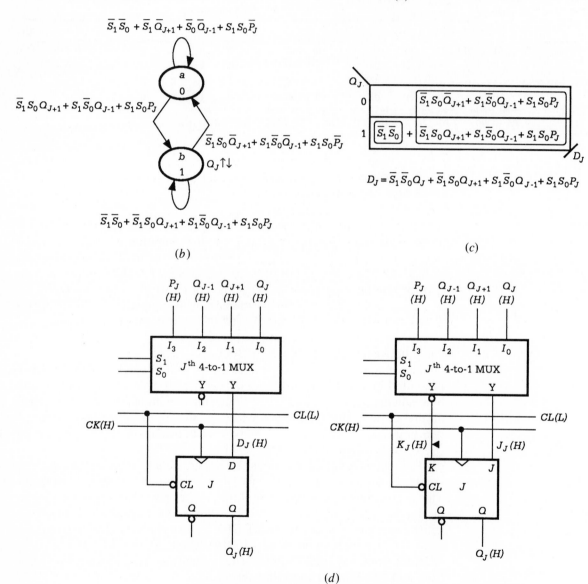

Mode Select			
S_1	S_0	NS	*Action*
0	0	Q_J	Hold
0	1	Q_{J+1}	Shift Right
1	0	Q_{J-1}	Shift Left
1	1	P_J	Parallel Load

(a)

$$\overline{S}_1\overline{S}_0 + \overline{S}_1\overline{Q}_{J+1} + \overline{S}_0\overline{Q}_{J-1} + S_1S_0\overline{P}_J$$

$$\overline{S}_1S_0Q_{J+1} + S_1\overline{S}_0Q_{J-1} + S_1S_0P_J$$

$$\overline{S}_1S_0\overline{Q}_{J+1} + S_1\overline{S}_0\overline{Q}_{J-1} + S_1S_0\overline{P}_J$$

$$\overline{S}_1\overline{S}_0 + \overline{S}_1S_0Q_{J+1} + S_1\overline{S}_0Q_{J-1} + S_1S_0P_J$$

(b)

$$D_J = \overline{S}_1\overline{S}_0Q_J + \overline{S}_1S_0Q_{J+1} + S_1\overline{S}_0Q_{J-1} + S_1S_0P_J$$

(c)

(d)

Fig. 5-101. *Design of a 1-bit slice UNIVERSAL SHIFT REGISTER. (a) Operation table. (b) State diagram. (c) K-map and minimum cover assuming D FLIP-FLOPS. (d) MUX implementation with D or JK FLIP-FLOPs.*

From the state diagram of Fig. 5-101(b) we can construct the first-order K-map for the 1-bit-slice UNIVERSAL SHIFT REGISTER. This is done in Fig. 5-101(c) where we have assumed the use of D FLIP-FLOPs. The minimum NEXT STATE logic cover for the Jth cell is obtained as indicated and is implemented with D or JK FLIP-FLOPs in Fig. 5-101(d). Here, we have chosen to use 4-to-1 MUXs for the NEXT STATE forming logic, D_J. Notice that MUX inputs I_3, I_2, I_1, and I_0 correspond to mode select inputs $S_1 S_0$, $S_1 \overline{S}_0$, $\overline{S}_1 S_0$, and $\overline{S}_1 \overline{S}_0$, respectively.

A 4-bit-slice UNIVERSAL SHIFT REGISTER is configured in Fig. 5-102(a) and is given the circuit symbol shown in Fig. 5-102(b), where two such units have been cascaded to produce an 8-bit-slice UNIVERSAL SHIFT REGISTER. Notice that both 4-bit units must be operated with the same mode select lines and with the same CLOCK and CLEAR lines. UNIVERSAL SHIFT REGISTERS of any size can be obtained by cascading (bit-slicing) units in this manner.

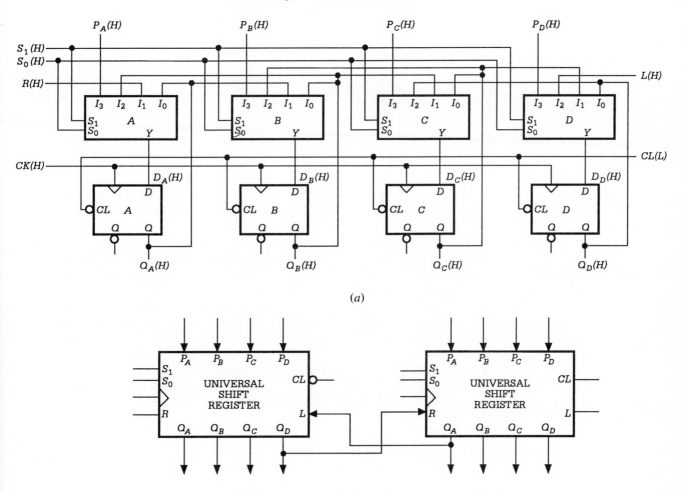

(a)

(b)

Fig. 5-102. A 4-bit-slice UNIVERSAL SHIFT REGISTER. (a) Logic circuit. (b) Block diagram symbol showing cascaded units.

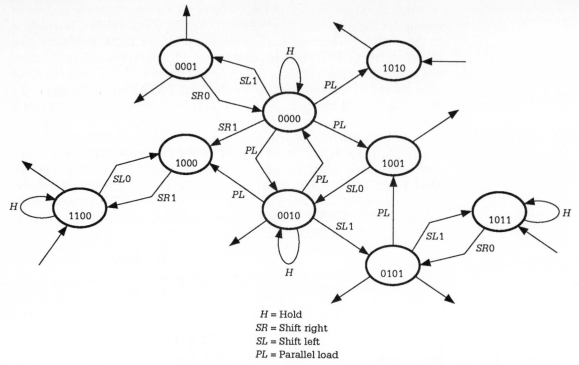

Fig. 5-103. *Portion of the state diagram for a 4-bit UNIVERSAL SHIFT REGISTER showing a few of the 256 possible state-to-state transitions and the register action required assuming that a shift has priority over a parallel load.*

The complete state diagram for a 4-bit universal shift register has 256 state-to-state transitions. This is so because each of the 16 states can branch to every other state and to itself. A small portion of such a state diagram is shown in Fig. 5-103 together with the register action required for each transition, assuming that a shift operation has priority over a parallel load. All 256 state-to-state transitions can be carried out by parallel load operations as in a 4-bit PIPO register.

The registers we have considered so far have been implemented with edge triggered FFs, mainly RET D-FFs. This need not be a requirement. In fact, there are commercial SHIFT REGISTER chips which are configured with MS JK-FFs. If such SHIFT REGISTERs are used, the designer must be aware that the error-catching feature, discussed in Sect. 5.6.4, poses a potential problem. If the *J* and *K* inputs are not maintained at their proper logic levels during the ACTIVE period of CLOCK, an irreversible SET or RESET of the master CELL can occur, and this can cause erroneous data to be CLOCKed to the register output. For this reason, it is often a more prudent design practice to use SHIFT REGISTER chips which employ edge triggered FLIP-FLOPs of either the D or JK type.

5.10.2 Counters

Counters form a class of FSMs for which each state code assignment is taken to be a number in a count sequence. Most simple counters are degenerate Moore machines which obey the basic model of Fig. 5-2 since their only outputs are the state variables. Binary counters are of this type and are classified as *modulus-N* counters, where N is the number of states of the sequence. Such binary counters are also classified as *divide-by-N* counters since the clock input frequency is divided by N if taken from the MSB output of the counter. Thus, the binary up-counter in Fig. 5-50 is a modulus-8 (or MOD-8) counter, but it is also a divide-by-8 counter since the waveform frequency taken from the A output of the counter is one-eighth that of the CLOCK frequency.

The state sequence of a counter need not conform to a regular binary count, up or down. Counters can be designed to count in any of the codes defined in Sect. 4.7.1 and in any direction. The counter that is designed in Sect. 5.7.3 is an example. We can even broaden the definition of the counter to include pseudorandom count sequences; a design presented at the end of this section is exemplary. Unconditional or conditional (Mealy) outputs may also be added at any point in a count sequence. When this is done the counter loses its simple character and emerges as a more complex machine, but it may still be regarded as a counter, as was done in Fig. 5-51.

The counters considered in this section and in previous sections are classified as *synchronous* counters—their FLIP-FLOPs are triggered by CLOCK simultaneously. Counters whose FLIP-FLOPs are each triggered by the output of the next LSB FLIP-FLOP are called *ripple* counters or *asynchronous* counters. These counters are considered only in Appendix 5.1 to this chapter.

A Bidirectional Gray Code Counter. We begin with the design of a 3-bit bidirectional Gray code counter featured in Fig. 5-104. The state diagram, shown in Fig. 5-104(a), indicates that when the direction control X is ACTIVE the count is up, and when it is INACTIVE the count is down. The NEXT STATE K-maps are plotted in Fig. 5-104(b), assuming the use of D-FFs, and a gate-minimum cover is extracted. Here, use is made of the XOR patterns discussed in Sect. 3.12.2. Observe that the term $A \oplus X$ appears in both the D_B and D_C expressions, thereby permitting the gate/input tally to be optimized at 9/18.

The use of JK-FFs instead of D-FFs should permit an even smaller gate/input tally as was pointed out in Sect. 5.7.3. Shown in Fig. 5-104(c) are the J, K K-maps which have been converted from the D K-maps by the procedure illustrated in Fig. 5-54. Now the gate-minimum cover yields a gate/input tally of 7/14, a result of the fact that the terms $B \oplus X$ and $A \oplus X$ can be used twice and three times, respectively. This is essentially a simple application of the multiple-output optimization algorithm discussed in Sect. 3.7.

The logic circuit for the 3-bit bidirectional Gray code counter is constructed in Fig. 5-104(d) where use is made of the optimum cover given in Fig. 5-104(c). Notice that this counter has provision for initialization and reset into STATE 000. Initialization into any other state would require the use of FLIP-FLOPs with both asynchronous PRESET and CLEAR overrides.

A Cascadable NBCD Up-counter. Our next design is that of a natural BCD (NBCD) up-counter. The NBCD sequence is very similar to the natural

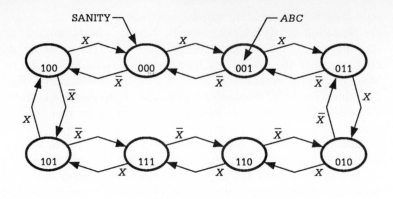

(a)

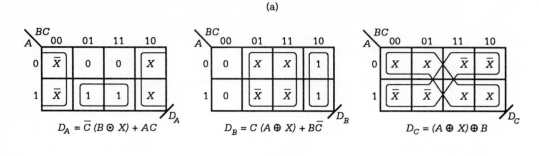

$$D_A = \overline{C}\,(B \odot X) + AC$$

$$D_B = C\,(A \oplus X) + B\overline{C}$$

$$D_C = (A \oplus X) \oplus B$$

(b)

Fig. 5-104. *Design of a 3-bit up/down Gray code counter. (a) State diagram. (b) NEXT STATE D K-maps and gate minimum cover.*

binary sequence except that the sequence is restarted after it reaches the value 1001. In decimal notation, the sequence is from 0 to 9 after which it starts over. A review of BCD codes is provided in Sect. 4.7.1.

In order that the NBCD counter be useful for a range of weighted digits (. . . 100, 10, 1, 0, 0.1, 0.01 . . .), it is necessary to build into the counter a bit-slice capability. That is, the counter must have ENABLE and CAR-RYOUT features so that the units can be cascaded to produce an *n*-digit NBCD counter. Shown in Fig. 5-105(a) is the state diagram for a 4-bit slice NBCD up-counter. It can be seen that it has the required ten states (0 through 9), that it is controlled by an ENABLE signal (*EN*) and that it generates a CARRYOUT (*CO*) when the 1001 state is reached. The purpose of the *CO* signal is to ACTIVATE the *EN* input of the next 4-bit stage. Thus, the *CO* of the least significant digit (LSD) counter module to the extreme right must be connected to the *EN* input of the next LSD counter module to its left, and so on. The *EN* input can be thought of as a CARRY input from the previous stage.

Note that there is a possible race glitch in *CO* when the FSM transitions from STATE 0111 to STATE 1000—there is a total of 24 alternative race paths for this transition, two of which pass through STATE 1001. However, this possible race glitch is of no concern since the glitch will have disappeared long before the next triggering edge of the CLOCK waveform, and hence could not prematurily activate the next stage.

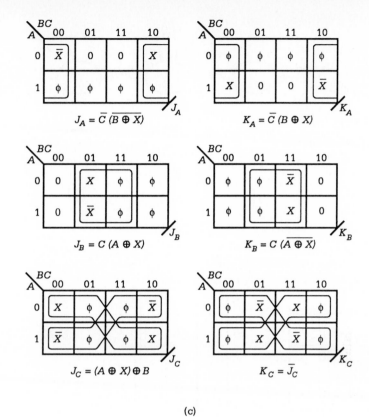

$$J_A = \overline{C}\,(\overline{B \oplus X})$$

$$K_A = \overline{C}\,(B \oplus X)$$

$$J_B = C\,(A \oplus X)$$

$$K_B = C\,(\overline{A \oplus X})$$

$$J_C = (A \oplus X) \oplus B$$

$$K_C = \overline{J_C}$$

(c)

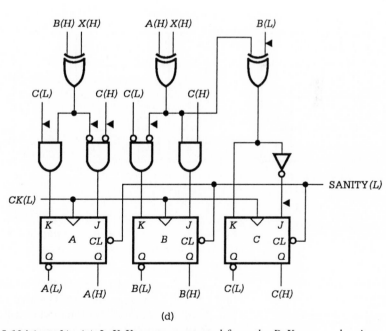

(d)

Fig. 5-104 (contd.). *(c) J, K K-maps, converted from the D K-maps, showing minimum cover. (d) Implementation with RET JK-FFs.*

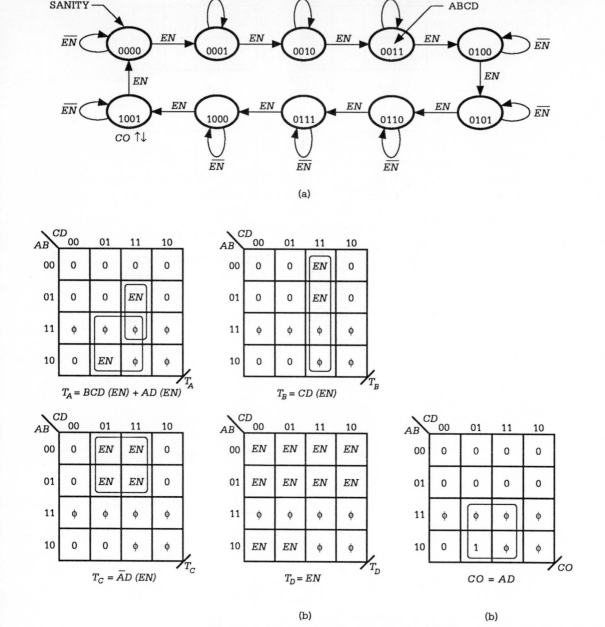

Fig. 5-105. *Design of a 4-bit cascadable NBCD up counter. (a) State diagram.*
(b) NEXT STATE T K-maps and minimum cover.

Experience from the 3-bit binary counter designed in Sect. 5.7.2 suggests that we should design the NBCD up-counter by using T-FFs. This we do in Fig. 5-105(b), where minimum cover is shown for each of the NEXT STATE *T* variables and for the single output, *CO*. Since there are six states left unused, there are six don't cares placed in cells 10 through 15, and these

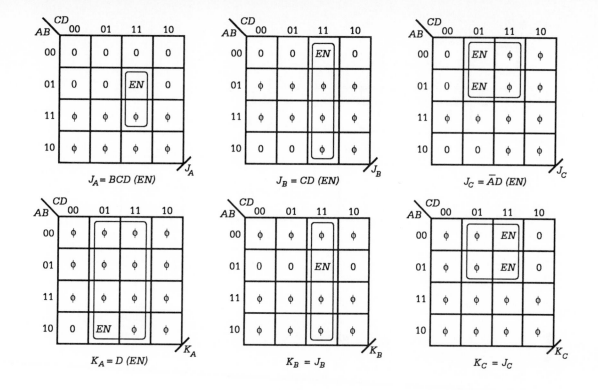

$$J_D = K_D = EN \quad \text{by inspection}$$

(c)

Fig. 5-105 (contd.). (c) J, K K-maps, converted from T K-maps, showing minimum cover.

are used in looping out minimum cover as shown. The gate/input tally for the NEXT STATE forming logic is 5/15.

Experience also tells us that a JK-FF implementation will likely lead to more optimum results than for the T-FF implementation. This is the case as indicated in Fig. 5-105(c). The J, K K-maps have been converted from the T K-maps of Fig. 5-105(b) by applying the expression $T = \overline{Q}J + QK$, a process similar to that for converting D K-maps to J, K K-maps. The gate/input tally for the NEXT STATE logic now stands at 4/12, a significant reduction over the T-FF value of 5/15.

Shown in Fig. 5-106(a) is the JK-FF implementation of the 4-bit-slice NBCD up-counter. A SANITY(L) input permits the counter to be initialized into STATE 0000 on power up or at any time during its operation if used as a RESET(L) input.

An n-digit NBCD counter is produced by bit-slicing the 4-bit module as shown in Fig. 5-106(b). The various digits, beginning with the least significant digit and ending with the most significant digit (MSD), take on whatever values are required. For example, a three-digit number might be interpreted as 37.5, meaning that the three modules represent the 10, 1, and 0.1 digits

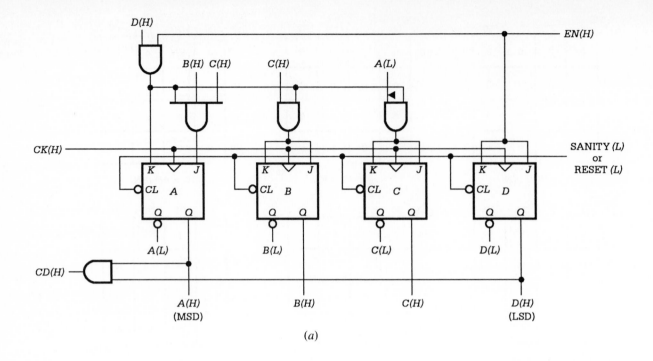

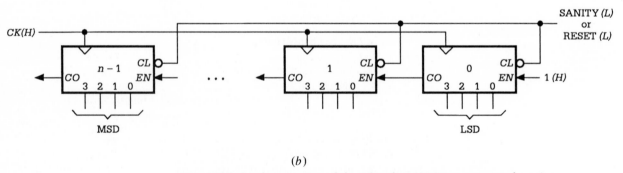

Fig. 5-106. *Implementation of the 4-bit-slice NBCD up counter by using RET JK-FFs. (b) Cascaded units for an* n-*digit BCD counter.*

from MSD to LSD, respectively. Attached to the seven-segment display logic shown in Fig. 4-50, and with the appropriate LED readout devices, the *n*-digit NBCD counter would have a useful human interface. Note that the counters can also be bit-sliced by CLOCKing each stage off of the *CO* output of the next LSB stage with the final LSB stage being driven by *CK*(H). In this case we set *EN* = 1 for all stages.

A Parallel Loadable Bidirectional Binary Counter. Let us now design a binary counter that can count up or down, that can be parallel loaded, and that can hold a given count indefinitely. The operation table and definitions for a 1-bit-slice counter of this type are presented in Fig. 5-107(a). The load (*LD*) and count enable (*EN*) inputs are the mode control inputs which determine whether the counter will hold, count, or parallel load as indicated in the table. A count function (*CNT*) represents an up- or down-count depending

LD	EN	NS	Action
0	0	Q_J	Hold
0	1	CNT	Count
1	X	P_J	Parallel Load

$$X = \text{Irrelevant Input}$$
$$LD = \text{Load}$$
$$EN = \text{Enable}$$

(a)

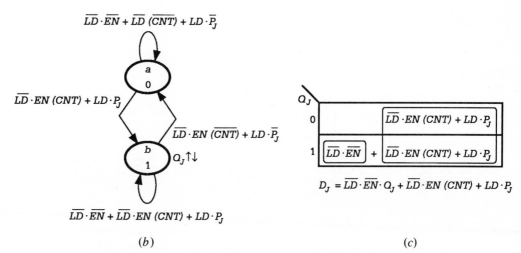

(b)　　　　　　　　　　　　　　　　　　　(c)

Fig. 5-107. *Design of a 1-bit slice parallel loadable up/down counter.*
(a) Operation table and definitions, and (b) State diagram for the Jth cell.
(c) K-map and minimum cover for the Jth cell assuming the use of
D FLIP-FLOPs.

on the count direction $D/\overline{U}$, and bit-slice capability is provided by the CAR-RYIN input (CI) and CARRY output features, as discussed later.

The state diagram for the Jth cell of this counter is constructed directly from the operation table and is shown in Fig. 5-107(b). Here, we recognize that the $0 \rightarrow 1$ branching condition must be $\overline{LD} \cdot EN(CNT) + LD \cdot P_J$ since $Q_J = 0$ and that the $1 \rightarrow 1$ branching condition is $\overline{LD} \cdot \overline{EN} + \overline{LD} \cdot EN(CNT) + LD \cdot P_J$ since $Q_J = 1$. The remaining two branching conditions are easily found by applying the $\sum = 1$ rule.

The first-order K-map representation of the 1-bit-slice parallel loadable up/down-counter is plotted from the state diagram in Fig. 5-107(c) by assuming the use of D-FFs. The minimum cover for the NEXT STATE function D_J is looped out as shown. A careful examination of the operation table of Fig. 5-107(a) reveals that this result could have been read directly from the operation table, a shortcut worth noting.

Now that the NEXT STATE forming logic for the Jth cell of this counter has been determined, it is necessary to select the hardware to be used for its implementation. As in the case of the universal shift register, the NEXT STATE logic function for this counter suggests the use of a 4-to-1 MUX. Shown in Fig. 5-108(a) is the second-order EV K-map plotted to accommodate the ACTIVE LOW LD input to a 4-to-1 MUX. This is accomplished

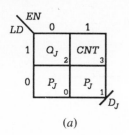

$D/\overline{U}$	Q_J	CI	CNT	CO	
0	0	0	0	0	
0	0	1	1	0	HALF ADDER
0	1	0	1	0	(Up count)
0	1	1	0	1	
1	0	0	0	0	
1	0	1	1	1	HALF SUBTRACTOR
1	1	0	1	0	(Down count)
1	1	1	0	0	

(a) (b)

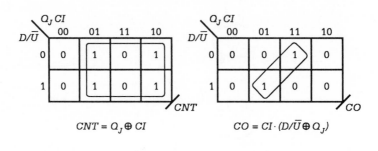

$$CNT = Q_J \oplus CI$$

$$CO = CI \cdot (D/\overline{U} \oplus Q_J)$$

(c)

Fig. 5-108. *Design of the 1-bit-slice parallel loadable up/down counter defined in Figure 5-107. (a) NEXT STATE K-map for ACTIVE LOW LD. (b) Truth table representing the increment and decrement operations for the Jth module of the counter. (c) K-maps for CNT (SUM or DIFFERENCE) and CO (CARRYOUT or BORROWOUT) showing gate-minimum cover.*

by complementing the *LD* (vertical) axis and by renumbering the cells accordingly, as discussed in Sect. 4.8.3.

A more complex problem centers on the means by which the count function *CNT* is to be implemented. Since increment- and decrement-type operations are involved, it is reasonable that a modular ADDER/SUBTRACTOR approach be used. However, for our purposes we will design the increment/decrement capability by using discrete gate logic. Shown in Fig. 5-108(b) is the truth table for HALF ADDER and HALF SUBTRACTOR operations of the Jth cell. A count direction input, $D/\overline{U}$, determines whether the count is up ($D/\overline{U} = 0$) or down ($D/\overline{U} = 1$). For the increment operation the variables *CI*, *CNT*, and *CO* are interpreted as CARRYIN, SUM (*S*) and CARRYOUT, whereas for the decrement operation these variables take on the meaning of BORROWIN (*BI*), DIFFERENCE (*D*), and BORROWOUT (*BO*), respectively.

The K-maps for *CNT* and *CO* are given in Fig. 5-108(c) together with the gate-minimum cover for each. Combining the information contained in Figs. 5-108(a) and (c) leads to the circuits for the 1-bit slice counter shown in Figs. 5-108(d) and (e). Observe that the external gate logic of Fig. 5-108(d) represents a HALF ADDER when $D/\overline{U}$ is 0 or a HALF SUBTRACTOR when $D/\overline{U}$ is 1 and that increment and decrement (by one) operations can result only when $CI(H) = 1(H)$. Also, note that the XOR gate to which

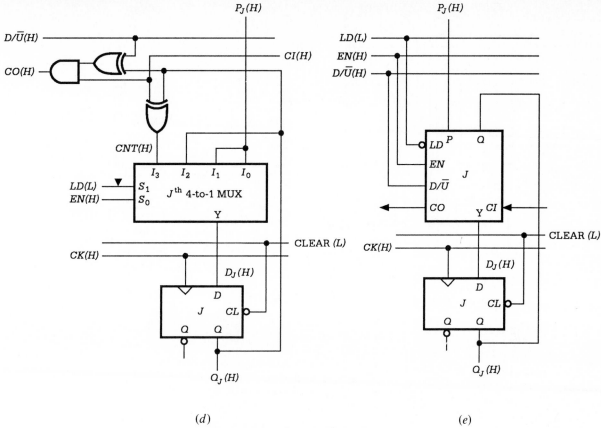

(d)   (e)

Fig. 5-108 (contd.). (d) MUX implementation of the Jth module with discrete gate logic used for the arithmetic operations. (e) A composite representation of (d).

$D/\overline{U}$ is connected acts as a controlled INVERTER, a fact that is readily deduced by comparing Figs. 4-6 and 4-15(e).

The logic circuit and block diagram symbol for a 4-bit-slice parallel loadable bidirectional counter are given in Figs. 5-109(a) and (b). Bit slicing is accomplished by adding similar units to the MSB side, thus connecting the CO of one unit to the CI of the other in ripple-carry fashion. Notice that the CI input of the LSB stage must be held at $1(H)$ so as to permit the increment and decrement operations.

To summarize, this counter has the capability of incrementing, decrementing, or holding a given parallel load input of four bits, or replacing that parallel load with another. The LD and EN inputs select the mode while D/U determines the count direction as stated earlier. The ripple-carry operation is similar to that for the parallel adder shown in Fig. 4-10.

The state diagram for the 4-bit parallel loadable up/down-counter involves 256 possible state-to-state transitions. This is so because each state of the 16 states can, in turn, branch to all other states and to itself in the parallel load mode. Presented in Fig. 5-110 is a portion of the state diagram showing some of the possible transitions and the counter action required. Here, the counter operations shown are hold (H), increment (I), decrement

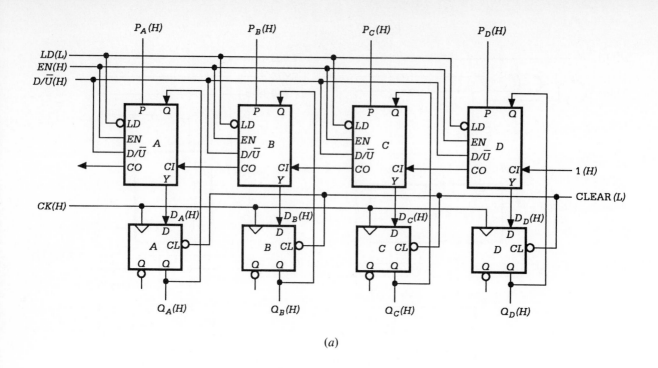

(a)

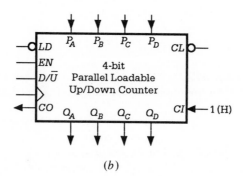

(b)

Fig. 5-109. *A 4-bit-slice parallel loadable up/down counter based on Figures 5-107 and 5-108. (a) Logic circuit. (b) Block diagram symbol.*

(D), and parallel load (PL). It is assumed that an increment or decrement has priority over a parallel load.

Design of a Pseudorandom Sequence Generator. To cap off our treatment of counters, let us design a 4-bit pseudorandom sequence generator first by using the parallel loadable up/down-counter of Fig. 5-109 and then by using the universal SHIFT REGISTER represented in Fig. 5-102(a). The object in doing this is to demonstrate the versatility of these two devices and to lay the groundwork for the use of such devices in the design of programmable controllers considered in the next section.

Figure 5-111(a) gives the state diagram for the 4-bit pseudorandom sequence generator we wish to design. The counter action for each state-to-state transition is indicated by a PL, an I, or a D. With this information, we plot the parallel input (P) K-maps, and the LOAD (LD), ENABLE (EN),

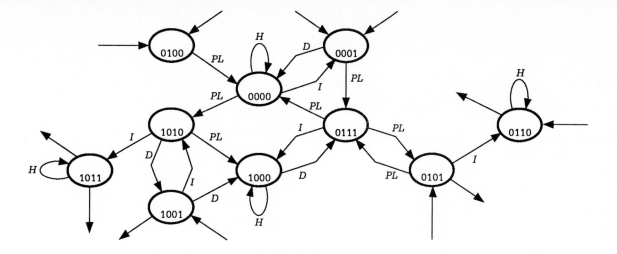

H = Hold
I = Increment (up 1)
D = Decrement (down 1)
PL = Parallel load

Fig. 5-110. *Portion of the state diagram for a 4-bit parallel loadable up/down counter showing a few of the 256 possible state-to-state transitions and the counter action required assuming that an increment or decrement has priority over a parallel load.*

and count direction $(D/\overline{U})$ K-maps, as shown in Figs. 5-111(b) and (c). It can be seen that don't cares are placed in the cells representing unused states and are used to loop out minimum cover. Certain cells are left blank so as to focus attention on what is happening with respect to the other cells. The blank cells represent don't care (ϕ) entries.

Some explanation of the K-map entries for Figs. 5-111(b) and (c) seems necessary. The parallel input K-maps of Fig. 5-111(b) are plotted with the understanding that D-FFs are used in the construction of each counter module. Thus, each state whose outgoing branching represents a parallel load (*PL*) operation has a NEXT STATE address which is the state code for the next state, a consequence of using the state transition table for the D-FF. For example, the NEXT STATE address for the 0101 STATE is 0010, and so 0, 0, 1, and 0 are placed in cells 5 of the maps for P_3, P_2, P_1, and P_0, respectively. Similarly, 0, 1, 1, and 0 are placed in cells 13 of these four K-maps, and so on.

The reader may wish to take note of the fact that when mapping an unconditional branching condition from a given state for D-FF implementation, *the NEXT STATE address for that state is always the state code assignment of the next state.* Use of this "shortcut" procedure can save time and minimize error. Remember, however, that it is the state transition table for the D-FF which characterizes the D-FF memory element and which is responsible for this shortcut procedure. Therefore, this procedure must be used only for D FLIP-FLOP implementation and then, only for unconditional branching conditions.

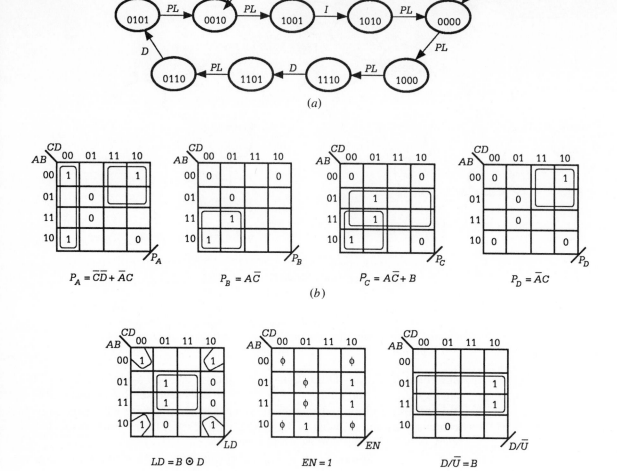

Fig. 5-111. *Design of 4-bit pseudo-random sequence generator by using a parallel loadable up/down counter. (a) State diagram showing counter action. (b) Parallel input data K-maps and minimum cover. (c) LOAD, ENABLE, and count direction K-maps shown with optimum cover.*

The LD, EN, and D/$\overline{\text{U}}$ K-maps of Fig. 5-111(c) are plotted by using the counter information given in Fig. 5-107(a) together with the state diagram of Fig. 5-111(a). Thus, the mode control parameters take on values (LD, EN) = (0, 1) for a count and (LD, EN) = (1, X) for a parallel load, where X represents an irrelevant input. For example, a logic 1 is placed in cells 0, 2, 5, 8, 10, and 13 of the LD K-map because these cells represent parallel load states. But don't cares are placed in these same cells of the EN K-map as required by the irrelevant input X in the operation table of Fig. 5-107(a). Logic 0's are placed in cells 6, 9, and 14 of the LD K-map since these cells represent count states. There are no 0's in the EN K-map because there are no hold conditions in the state diagram. Finally, logic 1 entries are placed in cells 6 and 14 of the D/$\overline{\text{U}}$ K-map to account for a decrement operation in STATEs 0110 and 1110, and a logic 0 is placed in cell 9 of this map because

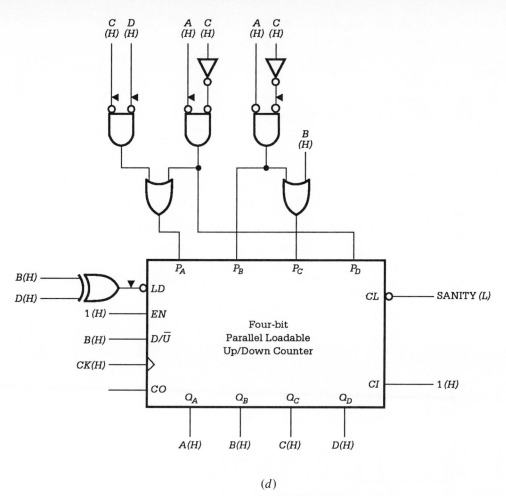

(d)

Fig. 5-111 (contd.). (d) Implementation with logic gates.

of the increment required in STATE 1001. Again, all blank cells represent φ entries and are left blank so as to focus attention on activity in the remainder of the cells.

Choosing optimum cover for the K-maps of Figs. 5-111(b) and (c) is easily accomplished by following the multiple-output optimization procedure discussed in Sect. 3.7. In this case, optimum cover is achieved simply by looping out the LD function as an XOR-type pattern. The results are implemented in Fig. 5-111(d). Notice that both the *EN* and *CI* inputs are held at 1(*H*) while the CLEAR (*CL*) input is connected to a SANITY(*L*) circuit so that the FSM can be initialized into STATE 0000. Note also that initializing the random sequence counter into any of the remaining eight states is not an option for this particular design. If this is a requirement for this counter, discrete D or JK FLIP-FLOPs with asynchronous PRESET and CLEAR overrides would have to replace the parallel loadable up/down-counter as the MEMORY.

As our final example, we design the same pseudorandom sequence counter by using the UNIVERSAL SHIFT REGISTER of Fig. 5-102(a) as

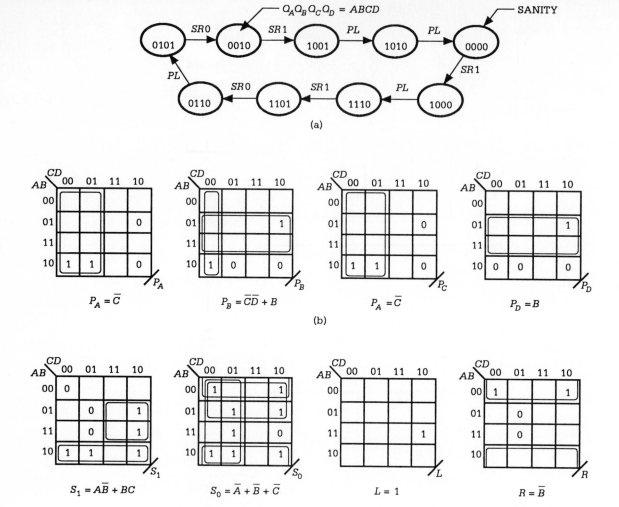

Fig. 5-112. *Design of a pseudo-random sequence generator by using a UNIVERSAL SHIFT REGISTER. (a) State diagram showing register action. (b) Parallel data input K-maps. (c) Mode control K-maps. (d) Serial data input K-maps.*

the MEMORY device. The general approach to this design is similar to that for the parallel loadable up/down-counter design just presented. Now, however, we must characterize the action of the register, rather than the up/down-counter, as the MEMORY. This is done in Fig. 5-112(a) where the register action is indicated for each state-to-state transition of the sequence by assuming that a shift (right or left) has priority over a parallel load. As was true in Fig. 5-103, the register actions, represented by SR0, SL0, SR1, and SL1, signify a shift right or left of 0 or 1, and *PL* represents a parallel load. No hold conditions exist.

The parallel input K-maps for the register implementation are presented in Fig. 5-112(b). They are plotted following the same procedure as was used to plot those of Fig. 5-111(b). Here again, the use of D-FFs in the imple-

5 / Synchronous Sequential Machines

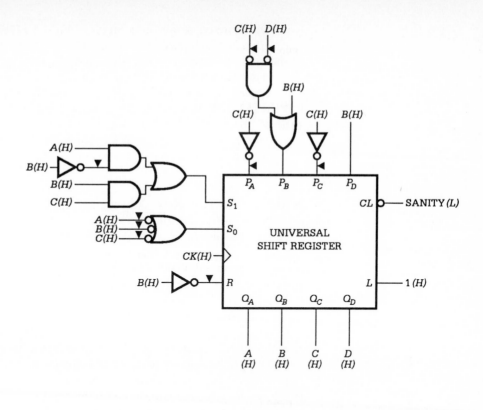

(e)

Fig. 5-112 (contd.). (a) Implementation by using discrete logic gates.

mentation of the UNIVERSAL SHIFT REGISTER requires that the NEXT STATE address for each parallel load (*PL*) state be the state code for the next state. Thus, 1's or 0's are entered in cells 6, 8, 9, and 10 since these are the parallel load states. As before, all blank cells represent don't care entries.

The K-maps for mode control inputs S_1 and S_0 are constructed by using the information in the operation table of Fig. 5-101(a) together with the state diagram of Fig. 5-112(a). These maps are given in Fig. 5-112(c). A shift right requires that $(S_1, S_0) = (0, 1)$; a shift left requires $(S_1, S_0) = (1, 0)$; and a parallel load requires $(S_1, S_0) = (1, 1)$. Thus, a 1 and 0 must be placed in cells 14 of the S_1 and S_0 K-maps, respectively, since STATE 1110 is a left shift state. Similarly, cell 5 (representing the right shift STATE 0101) requires that a 0 and 1 be placed in the respective S_1 and S_0 K-maps. Or for cell 9, representing the parallel load state 1001, 1's are placed in cells 9 of the mode control K-maps, and so on.

The serial left and serial right K-maps (the L and R maps) in Fig. 5-112(d) are plotted by entering a 0 or 1 into the appropriate map depending on whether a 0 or 1 shift is indicated for a particular state-to-state transition. For example, STATE 1110 requires a SL1 for the transition to STATE 1101, so a 1 is placed in cell 14 of the L K-map. Similarly, STATE 0101 requires an SR0 for the transition to 0010. Therefore, a 0 is placed in cell 4 of the R K-map.

Optimum cover for the K-maps of Figs. 5-112(b), (c), and (d) is achieved by adjusting the cover of the S_0 function so as to accommodate the four shared PIs from the P_A, P_B, P_C, and R functions. Observe that the L function is 1 since there are no SL0 operations in the state diagram.

Implementation of the pseudorandom sequence counter with the UNIVERSAL SHIFT REGISTER is shown in Fig. 5-112(e) and is observed to be of about the same complexity as that for the up/down-counter implementation in Fig. 5-111(d). Again, the asynchronous CLEAR input must be connected to the SANITY(L) circuit so that the counter can be properly initialized.

Self-correcting Counters. Any sequential machine that can be initialized into a state appropriate for its proper operation without the aid of a SANITY circuit, is said to be *self-correcting*. While some simple counters can be operated as self-correcting FSMs, most other FSMs cannot be and require the use of an initializing (SANITY) circuit. Rarely should attempts be made to make Moore and Mealy FSMs self-correcting. The use of SANITY circuit initialization is generally the established rule for such machines.

A counter is self-correcting if all don't care states lead into the main routine (state diagram) and is not self-correcting if there exists at least one extraneous subroutine, consisting of at least one don't care (unused) state, into which the counter could enter following power-up or reset. To determine whether or not a counter is self-correcting, it is only necessary to introduce the bits for each missing state into the NEXT STATE functions so as to find the NEXT STATEs for all missing states. If the FSM is found not to be self-correcting, it can be made so by selectively avoiding the use of those don't cares in the K-maps that correspond to states in the extraneous routine(s). Alternatively, the FSM can be designed so that all nonessential states be directed into one or more states of the required routine thereby eliminating all don't care states.

5.11 ALTERNATIVE APPROACHES TO FSM DESIGN

The "conventional" design of FSMs involves the use of individual FLIP-FLOPs for the MEMORY and the use of discrete gate logic for the NEXT STATE and output sections. Up to this point the emphasis has been mainly on this conventional approach to design. However, in Sects. 5.9.2 and 5.9.3 we deviated somewhat from this approach by introducing Huffman's model and the use of PLAs and ROMs to implement the NS and output forming logic. Then in Sect. 5.10.2 we deviated even further from conventional design by introducing the use of a parallel loadable up/down-counter and a UNIVERSAL SHIFT REGISTER as the MEMORY element for the design of a random sequence counter. So the idea of approaching FSM design from different perspectives is not new.

In this section we will continue to develop alternative means of designing synchronous FSMs. For comparison purposes all alternative designs will be based on the same state machine algorithm.

The state machine which is used throughout this section is represented by the state diagram in Fig. 5-113. It is observed to be a Mealy machine of

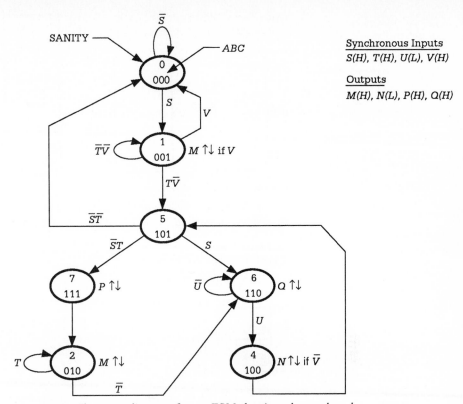

Fig. 5-113. *The state diagram for an FSM that is to be used to demonstrate various FSM design alternatives.*

seven states with one-, two-, and three-way branching. It has four synchronous inputs, one ACTIVE LOW, and four outputs also with one ACTIVE LOW. The state identifier symbols are the Arabic numerals corresponding to the state code assignments.

5.11.1 The Conventional Approach Revisited

The now-familiar Mealy model for conventional FSM design is given in Fig. 5-114. It includes input conditioning circuits (synchronizing and debouncing circuits, etc.) and an output holding register should they be needed. As the reader will recall, the three main sections of the model are the NEXT STATE forming logic, the MEMORY, and the output forming logic. For the conventional design we will use discrete gate logic for the NS and output logic sections and individual FLIP-FLOPs for the memory.

Shown in Fig. 5-115(a) are the K-maps and near optimum cover for the NS forming logic, assuming the use of D-FFs. The output K-maps and minimum cover are given in Fig. 5-115(b). Optimum NS logic cover would have resulted if the 1's in cells 4 and 6 of the D_A K-map had been looped out individually to form shared PIs with those of cells 4 and 6 of the D_C and Q K-maps. But little would have been gained in doing so.

The conventional implementation of this FSM is presented in Fig. 5-115(c). Here, we notice that the NS forming logic is fairly complex and that a SANITY(L) circuit is connected to the asynchronous CLEAR override of

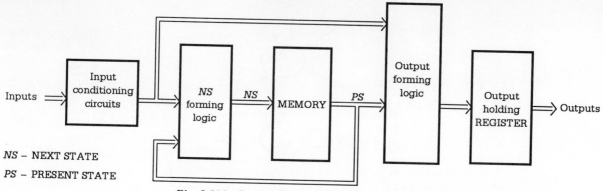

Fig. 5-114. *Conventional architecture for FSM design including input conditioning circuits and an output holding REGISTER.*

Inputs ⟹ Input conditioning circuits → NS forming logic → *NS* → MEMORY → *PS* → Output forming logic → Output holding REGISTER ⟹ Outputs

NS – NEXT STATE

PS – PRESENT STATE

Fig. 5-115. *Conventional design of the FSM in Figure 5-113. (a) Near optimum cover for the NEXT STATE forming logic assuming the use of D-FFs. (b) Output K-maps and minimum cover.*

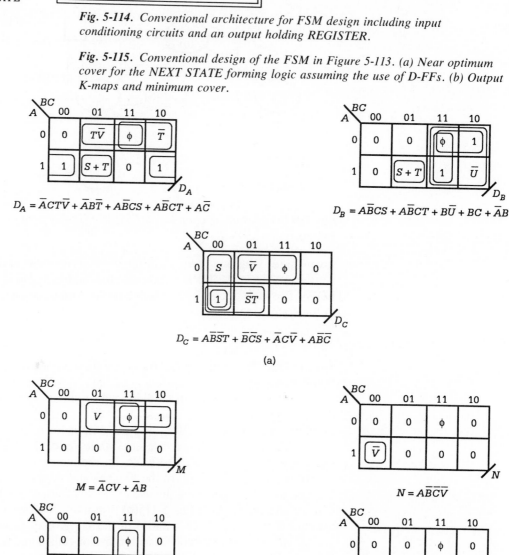

$D_A = \overline{A}CT\overline{V} + \overline{A}B\overline{T} + A\overline{B}CS + A\overline{B}CT + A\overline{C}$

$D_B = A\overline{B}CS + A\overline{B}CT + B\overline{U} + BC + \overline{A}B$

$D_C = A\overline{B}\,\overline{S}T + \overline{B}\,\overline{C}S + \overline{A}C\overline{V} + A\overline{B}\,\overline{C}$

(a)

$M = \overline{A}CV + \overline{A}B$

$N = A\overline{B}\,\overline{C}V$

$P = BC$

$Q = AB\overline{C}$

(b)

456

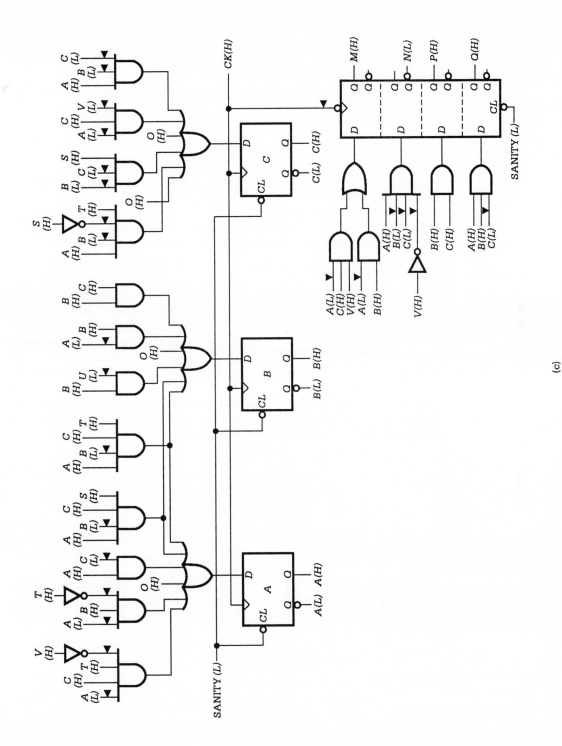

Fig. 5-115 (contd.). (c) Implementation by using discrete gate logic.

457

each FLIP-FLOP so that the FSM can be initialized (or reset) into STATE 000 on power-up, as required by the state diagram.

The output forming logic shown in Fig. 5-115(c) is connected to an output holding register, which is CLOCKed antiphase to the FLIP-FLOPs. The reason for using the output holding register is to filter out any output race glitches that may be produced by transition from STATE ⑤ to STATE ⓪ or from STATE ⑦ to STATE ②. No STATIC HAZARDs are possible as the output logic indicates. Notice that the asynchronous CLEAR of the output holding register is also connected to the SANITY(L) circuit for initialization and reset purposes.

5.11.2 Architecture Centered on a STATE DECODER

The deceptively complex architecture for an FSM designed around a STATE DECODER is represented by the block diagram in Fig. 5-116. Use is made of a STATE DECODER to reduce the complexity of the gate logic in the NS and output sections as indicated in the block diagram. But there is another important reason for using a STATE DECODER. It happens that this architecture is one of the simplest to use due to the fact that the outputs of the STATE DECODER are the state codes assigned to the states in the state diagram. As a result, *all the information needed to implement an FSM can be read directly from the state diagram without the need for K-maps.* While this architecture may be far from optimal, it is popular with designers because of the ease and simplicity with which an FSM can be implemented.

Shown in Fig. 5-117 are the NS K-maps reproduced from Fig. 5-115(a). Adjacent to these maps are written the NS logic expressions (for D-FFs) which have been read directly from the state diagram and which are of a form that can be directly used for the architecture of Fig. 5-116. A comparison of these expressions with their corresponding K-maps verifies their validity. Similarly, the composite output K-map in Fig. 5-117(b) verifies the

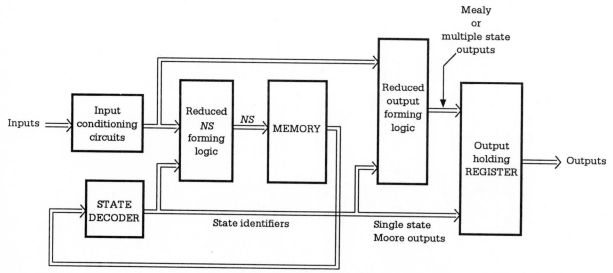

Fig. 5-116. *Architecture for FSMs designed around a STATE DECODER.*

validity of the output logic expressions (adjacent to the K-map) which have also been read directly from the state diagram.

Use of the information in Figs. 5-116, 5-117(a) and 5-117(b) results in the logic circuit presented in Fig. 5-117(c). Here, in comparison to conventional design, we observe a significant reduction in the number of gates required for the NS forming logic section, but at the expense of using a STATE DECODER. More important, the fact that this logic circuit has been constructed directly from the state diagram makes this a design alternative worth remembering. For purposes of testing protoboard circuits, such design alternatives can save both time and costs.

5.11.3 The Direct- and Indirect-Addressed MUX Approaches

The use of MUXs to implement the NS forming logic of an FSM can greatly reduce gate logic requirements. Furthermore, if certain requirements are met, the MUX approach permits the NS logic to be read directly from the

(a)

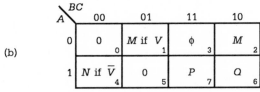

$D_A =$ (STATE 1) $T\overline{V}$ + (STATE 2) $\overline{T}$ + STATE 4 + (STATE 5) $(S + T)$
  + STATE 6

$D_B =$ STATE 2 + (STATE 5) $(S + T)$ + (STATE 6) $\overline{U}$ + STATE 7

$D_C =$ (STATE 0) S + (STATE 1) $\overline{V}$ + STATE 4 + (STATE 5) $\overline{S}T$

(b)

$M =$ (STATE 1) V + STATE 2
$N =$ (STATE 4) $\overline{V}$
$P =$ STATE 7
$Q =$ STATE 6

Fig. 5-117. *Design of the FSM in Figure 5-113 by using the architecture of Figure 5-116. (a) NEXT STATE D K-maps and DECODER cover. (b) Composite output K-map and DECODER cover.*

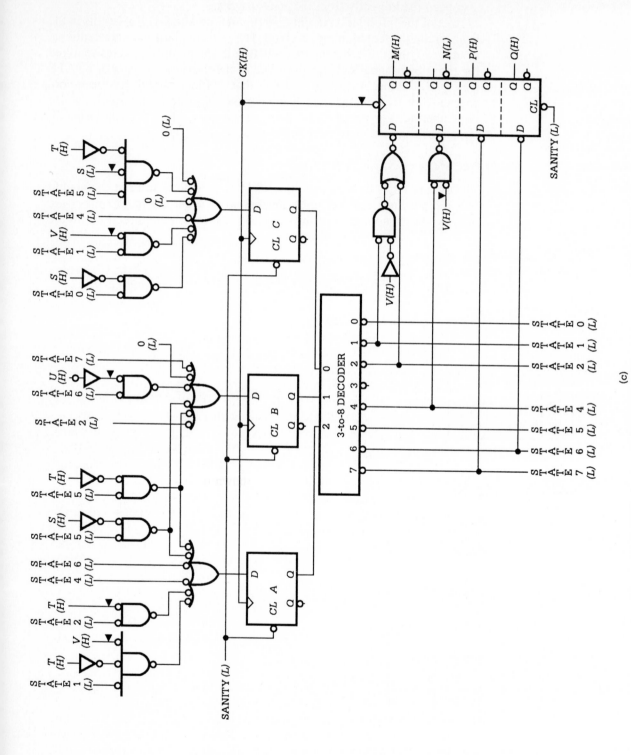

Fig. 5-117 (contd.). (c) *DECODER implementation of Figure 5-113 showing reduced NEXT STATE and output forming logic compared to Figure 5-115 (c).*

(c)

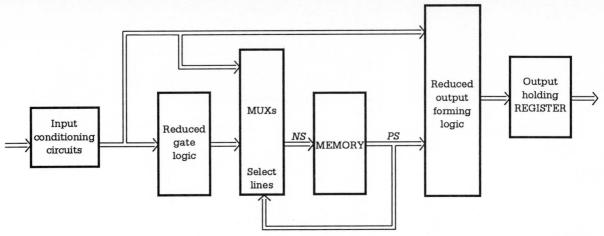

Fig. 5-118. *Architecture for FSMs designed around direct addressed MUXs.*

state diagram as in the previous example. The combination of these two advantages yields an approach which is one of the most powerful and versatile methods available for FSM design. In this section we will examine the use of MUXs of unrestricted size, and then we will consider the problem posed by MUX size limitation.

The architecture for the *direct MUX approach* to FSM design is represented by the block diagram in Fig. 5-118. This approach is characterized by MUX select inputs which are directly connected to the PS (feedback) variables. Thus, a three-state variable machine would require 8-to-1 MUXs, while a four-state variable FSM would need 16-to-1 MUXs, and so on. Obviously, the MUX size increases rapidly with numbers of state variables, which, in turn, is a measure of the size (number of states) of the machine. In most other respects this architecture is the same as that of Fig. 5-114.

Implementation of the FSM directly from the state diagram of Fig. 5-113 is easily accomplished with the results shown by the logic diagram in Fig. 5-119(a). The numbering of the MUX inputs corresponds to the state code assignments in the state diagram, and the logic for each input is read from the state diagram just as in Fig. 5-117(a). The present state (PS) feedback variables (i.e., the Q outputs of the FLIP-FLOPs) directly select the NS logic from the MUXs as indicated in the logic diagram.

The 3-to-8 decoder appearing in Fig. 5-119(a) serves two purposes. It reduces the gate logic requirement and permits the output forming logic to be implemented from information taken directly from the state diagram. In effect, the logic diagram of Fig. 5-119(a) conforms to a model which is a combination of the FSM architectures in Figs. 5-116 and 5-118—a very useful hybrid form.

Dealing with MUX Size Limitations. Implementation of very large synchronous sequential machines may require MUX sizes beyond what is available. If this is the case, the MUX size can be reduced by K-map compression but at the cost of increased external gate logic. Figures 5-119(b) and (c) illustrate this. Here, a map compression of one order reduces the MUX size from 8-to-1 to 4-to-1. However, the gate tally also increases from 3 to 11.

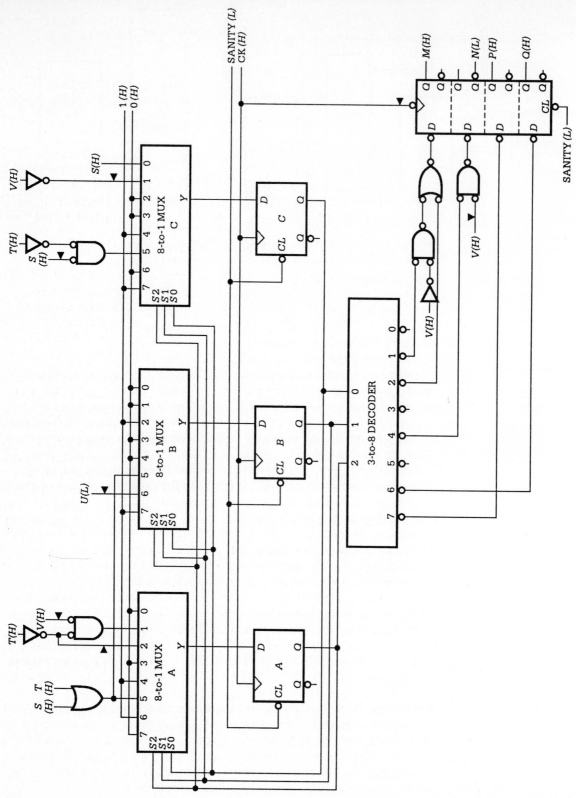

Fig. 5-119. (a) Direct address MUX implementation of the FSM represented in Figure 5-113.

(a)

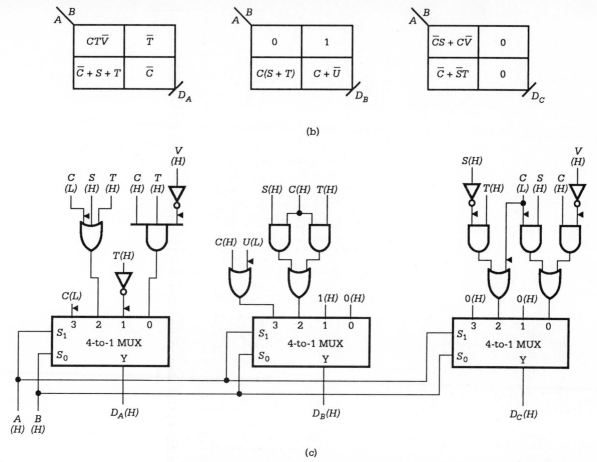

Fig. 5-119 (contd.). *(b) K-map compression required for MUX size reduction. (c) Implementation of the NS forming logic with 4-to-1 MUXs showing the resulting increase in external gate logic.*

There is often a better method of reducing MUX size that does not depend on K-map compression. This method is called the *indirect-addressed MUX approach* and is represented by the block diagram in Fig. 5-120. The addition of a MUX address converter permits MUX size reduction without K-map compression, but only if the number of unique MUX inputs can be reduced by factors of 2^n, where n is an integer.

In Fig. 5-121 the indirect-addressed MUX approach is applied to D_B because it is the only one of the three NS variables whose K-map has four (or fewer) unique entries. Hence, it is the only NS variable to which this method can be applied. The four unique K-map entries for D_B, appearing in Fig. 5-121(a), are arbitrarily assigned to the inputs of a 4-to-1 MUX as indicated in Fig. 5-121(b). By using the MUX of Fig. 5-121(b) together with the K-map of Fig. 5-121(a), we construct the truth table for the MUX select lines given in Fig. 5-121(c). How one implements the logic for the MUX select inputs, S_1 and S_0, is a matter of choice. For our design we accomplish this by using the 3-to-8 decoder and two NAND gates shown in Fig. 5-121(d).

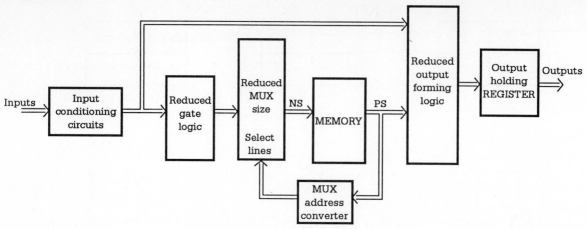

Fig. 5-120. *Architecture for FSMs designed around indirect addressed MUXs.*

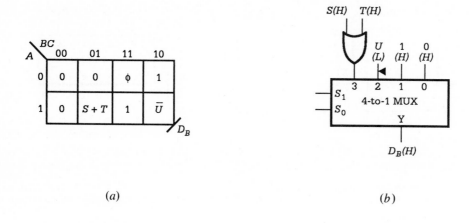

(a)

(b)

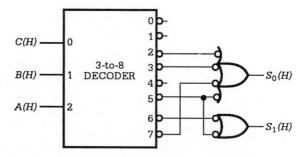

(c)

(d)

Fig. 5-121. *Size reduction of the B MUX in Figure 5-119 to conform to the indirect address MUX architecture of Figure 5-120. (a) NEXT STATE K-map for D-FF B. (b) Input requirements for the 4-to-1 B MUX. (c) MUX address table. (d) DECODER implementation of the MUX address converter logic.*

We are already familiar with the use of direct-addressed ROMs and PLAs to implement the NS and output forming logic of relatively simple FSMs, as in Sect. 5.9. Now, we will extend the application of these array logic devices to more complex FSM designs which include the presence of MIXED LOGIC inputs and the problem of array logic size limitations. Presented in Fig. 5-122 is the model of Huffman which we associate with array logic implementation (see Fig. 5-85). This FSM architecture could be called the *direct-addressed array logic approach*, meaning that the PS feedback lines from the MEMORY are connected directly to the array logic device(s).

We begin by designing the FSM of Fig. 5-113 with an $8 \times 20 \times 8$ FPLA. The *p*-term table for programming the FPLA is given in Fig. 5-123(a). We note that there are seven inputs, one of which is ACTIVE LOW (see asterisk), and seven outputs also with one output ACTIVE LOW. The manner in which the ACTIVE LOW input and ACTIVE LOW output are dealt with is discussed shortly. The 15 *p*-terms listed in the table of Fig. 5-123(a) are taken from the near-optimum cover given in Figs. 5-115(a) and (b). Recall from Sect. 4.11.2 and in particular from Fig. 4-85(b) that the dash in the input column of a *p*-term table signifies a nonconnection.

An asterisk ($*$) is placed on the U and N column headings in Fig. 5-123(a) to signify that these represent an ACTIVE LOW input and ACTIVE LOW output, respectively, and that some action is necessary to accommodate them. The action we take to accommodate the ACTIVE LOW input, U, is to complement any 1's and 0's appearing in that column. However, this is not an option for an ACTIVE LOW output from a PLA. Since we plan to use an output holding register to filter out any output race glitches that may occur, we will generate $N(L)$ from the register. For a review of the use of array logic devices with MIXED LOGIC inputs and outputs refer to Sect. 4.11.4.

Unlike the *p*-term table for programming a PLA, the ROM program table must be CANONICAL. The collapsed ROM program table for the FSM of Fig. 5-113 is given in Fig. 5-123(b). It represents $2^7 = 128$ MINTERMs since there are seven inputs. The use of the irrelevant input, X (representing 1 or 0), permits the table to be shortened to 15 rows instead of 128—hence, the name collapsed programmed table. Note that don't cares are *never* placed in an output column of either table in Fig. 5-123.

Both the ROM and PLA program tables in Fig. 5-123 signify the presence of the ACTIVE LOW input, U, and ACTIVE LOW output, N, in exactly

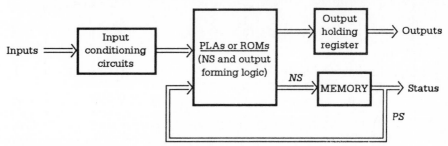

Fig. 5-122. *Architecture for FSMs designed around PLAs and direct addressed ROMs.*

	Inputs								Outputs						
p-Term	I_6 A	I_5 B	I_4 C	I_3 S	I_2 T	I_1 U*	I_0 V		Y_6 D_A	Y_5 D_B	Y_4 D_C	Y_3 M	Y_2 N*	Y_1 P	Y_0 Q
$\overline{A}CT\overline{V}$	0	–†	1	–	1	–	0		1	0	0	0	0	0	0
$\overline{A}B\overline{T}$	0	1	–	–	0	–	–		1	0	0	0	0	0	0
$A\overline{B}CS$	1	0	1	1	–	–	–		1	1	0	0	0	0	0
$A\overline{B}CT$	1	0	1	–	1	–	–		1	1	0	0	0	0	0
$A\overline{C}$	1	–	0	–	–	–	–		1	0	0	0	0	0	0
$B\overline{U}$	–	1	–	–	–	1	–		0	1	0	0	0	0	0
BC	–	1	1	–	–	–	–		0	1	1	0	0	1	0
$\overline{A}B$	0	1	–	–	–	–	–		0	1	0	1	0	0	0
$AC\overline{S}T$	1	–	1	0	1	–	–		0	0	1	0	0	0	0
$\overline{B}\overline{C}S$	–	0	0	1	–	–	–		0	0	1	0	0	0	0
$\overline{A}C\overline{V}$	0	–	1	–	–	–	0		0	0	1	0	0	0	0
$A\overline{B}\overline{C}$	1	0	0	–	–	–	–		0	0	1	0	0	0	0
$\overline{A}CV$	0	–	1	–	–	–	1		0	0	0	1	0	0	0
$A\overline{B}\overline{C}\overline{V}$	1	0	0	–	–	–	0		0	0	0	0	1	0	0
$AB\overline{C}$	1	1	0	–	–	–	–		0	0	0	0	0	0	1

(a)

* An ACTIVE LOW input or output.
† A nonconnection.

Fig. 5-123. *Design of the FSM in Figure 5-113 by using direct-addressed array logic and the model of Figure 5-122. (a) p-term table for implementation with an 8 × 20 × 8 FPLA, constructed from Figures 5-115(a) and (b).*

ROM Inputs								ROM Outputs						
I_6 A	I_5 B	I_4 C	I_3 S	I_2 T	I_1 U*	I_0 V		Y_6 D_A	Y_5 D_B	Y_4 D_C	Y_3 M	Y_2 N*	Y_1 P	Y_0 Q
(PS)								(NS)						
0	0	0	0	X†	X	X		0	0	0	0	0	0	0
0	0	0	1	X	X	X		0	0	1	0	0	0	0
0	0	1	X	0	X	0		0	0	1	0	0	0	0
0	0	1	X	X	X	1		0	0	0	1	0	0	0
0	0	1	X	1	X	0		1	0	1	0	0	0	0
1	0	1	0	0	X	X		0	0	0	0	0	0	0
1	0	1	0	1	X	X		1	1	1	0	0	0	0
1	0	1	1	X	X	X		1	1	0	0	0	0	0
1	1	0	X	X	1	X		1	1	0	0	0	0	1
1	1	0	X	X	0	X		1	0	0	0	0	0	1
1	1	1	X	X	X	X		0	1	0	0	0	1	0
0	1	0	X	0	X	X		1	1	0	1	0	0	0
0	1	0	X	1	X	X		0	1	0	1	0	0	0
1	0	0	X	X	X	0		1	0	1	0	1	0	0
1	0	0	X	X	X	1		1	0	1	0	0	0	0

(b)

* An ACTIVE LOW input or output.
† An irrelevant input.

Fig. 5-123 (contd.). *(b) Collapsed program table for implementation with a 2^7 × 8 ROM constructed directly from the state diagram in Figure 5-113.*

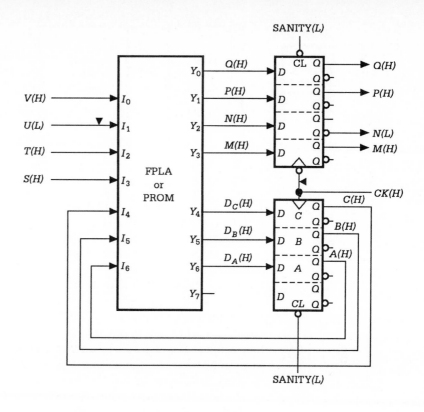

(c)

Fig. 5-123 (contd.). *(c) Logic circuit showing the use of an 8 × 20 × 8 FPLA or a $2^7 × 8$ PROM programmed according to Figure 5-123(a) or (b), respectively.*

the same manner. However, for the ROM program table there exists an alternative means of dealing with an ACTIVE LOW output that is not available to PLA *p*-term tables. The single 1 and 0's appearing in the *N* column of Fig. 5-123(b) can be complemented to generate the ACTIVE LOW logic level of this output. But if this is done, then *N(L)* must be taken from the $\bar{Q}(H)$ output of the holding register—a complementary I/O relationship that must be imposed when a *D(H)* register input is presented ACTIVE LOW.

The array logic (PLA or ROM) implementation of the FSM in Fig. 5-113 is represented by the logic circuit in Fig. 5-123(c). This circuit appears remarkably simple and uncomplicated because the details of the NS and output forming logic sections are contained in the programmed array logic device. Again this is called the direct-addressed PLA or direct-addressed ROM approach as opposed to the indirect ROM address architecture discussed next.

Dealing with ROM Size Limitations. There may come a point in the design of a complex FSM when the ROM size requirement of the design exceeds the ROM size that is available. When this happens the designer must alter the design strategy so as to meet the hardware limitations. One solution to this problem is to reduce the ROM size requirements by using the *indirect-addressed ROM approach* illustrated by the block diagram of Fig. 5-124.

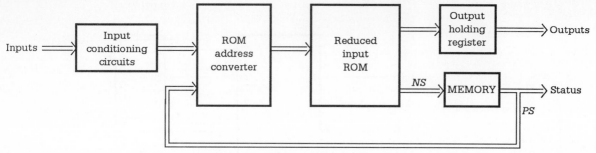

Fig. 5-124. Architecture for FSMs designed around indirect addressed ROMs.

Here, a ROM address converter permits a significant reduction in the number of required ROM inputs. The converter is commonly formed from MUXs and discrete gate logic.

The ROM program table presented in Fig. 5-125(a) is a reproduction of that in Fig. 5-123(b) except for the last four columns (I_3, I_2, I_1, and I_0) which form the indirect ROM address. The "I-WORDs" in these four columns are obtained somewhat arbitrarily by assigning to them a unique set of logic values for each unique set of the 7-bit ROM output WORDs appearing in columns Y_6 through Y_0 of the program table. Thus, rows 1 and 6 form the first unique set and are assigned the I-WORD 0000. Then rows 2 and 3 form the second unique set and are assigned the I-WORD 0001 while rows 5 and 15 combine to give the I-WORD address 0011. The remaining sets of ROM output WORDs are assigned I-WORDs in decimal order up through 1011.

The next step is to extract the indirect ROM address logic from the I-WORDs in Fig. 5-125(a). This is accomplished by constructing the third-order K-maps shown in Fig. 5-125(b), where use is made of the external input data in Fig. 5-125(a). For example, the entries for cell 0 are obtained from rows 1 and 2, which require that a logic 0 be placed in cell 0 for I_3, I_2, and I_1 and an S in cell 0 for I_0. Similarly, rows 3, 4, and 5 require that a logic 0 be placed in cell 1 of the I_3 and I_2 K-maps, but that $V + T\overline{V} = V + T$ be entered in cell 1 of the I_1 K-map and a $\overline{T}V + T\overline{V} = \overline{V}$ in cell 1 of the I_0 K-map, and so on. The final result is the logic circuit in Fig. 5-125(c), where use has been made of four 8-to-1 MUXs and external gates to generate the logic for the indirect ROM address. What we have accomplished here is to reduce the ROM size from seven inputs to four, but at the expense of the additional MUX and gate logic.

An alternative means of dealing with the problem of inadequate ROM size is to increase ROM input capability by using the multiplexed scheme illustrated in Fig. 4-91. Increased ROM output capability can be achieved by applying the scheme shown in Fig. 4-92. As to which of the foregoing alternatives is best suited to a given design is a matter of choice and, of course, the availability of the hardware.

5.11.5 Counters and SHIFT REGISTERS as MEMORY Elements

We now focus our attention on the MEMORY section of the FSM so as to help diversify design strategy. The use of an up/down-counter or a UNIVERSAL SHIFT REGISTER as the MEMORY element for the design of simple FSMs has been demonstrated in Sect. 5.10.2. All that remains, then,

Row	PS			External Inputs				ROM Outputs							ROM Address			
	A	B	C	S	T	U*	V	Y_6 D_A	Y_5 D_B	Y_4 D_C	Y_3 M	Y_2 N^*	Y_1 P	Y_0 Q	I_3	I_2	I_1	I_0
1	0	0	0	0	X†	X	X	0	0	0	0	0	0	0	0	0	0	0
2	0	0	0	1	X	X	X	0	0	1	0	0	0	0	0	0	0	1
3	0	0	1	X	0	X	0	0	0	1	0	0	0	0	0	0	1	0
4	0	0	1	X	X	X	1	1	0	0	1	0	0	0	0	0	1	1
5	0	0	1	X	1	X	0	0	0	1	0	0	0	0	0	0	0	1
6	1	0	1	0	0	X	X	1	0	0	0	0	0	0	0	0	0	0
7	1	0	1	0	1	X	X	1	1	0	0	0	0	0	0	1	0	0
8	1	0	1	1	X	1	X	1	1	1	0	0	0	0	0	1	1	1
9	1	1	1	X	X	1	X	1	1	0	0	0	0	1	0	1	1	0
10	1	1	1	X	X	0	X	0	0	0	0	0	1	0	0	0	0	1
11	1	1	1	X	0	X	X	1	1	1	1	0	0	0	0	0	0	0
12	0	1	0	X	1	X	X	0	0	0	1	0	0	0	1	0	1	1
13	0	1	0	X	X	X	0	1	1	1	0	1	0	0	1	0	1	0
14	1	0	0	X	X	X	1	0	0	1	0	0	0	0	1	0	0	0
15	1	0	0	X	X	X	X	1	0	1	0	0	0	0	0	0	1	1

*An ACTIVE LOW input or output.
†An irrelevant input.

(a)

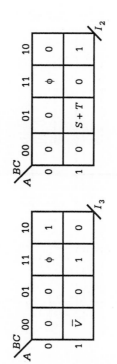

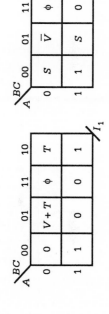

(b)

Fig. 5-125. Design of the FSM in Fig. 5-113 by using the indirect-addressed ROM architecture of Figure 5-124. (a) ROM program table with ROM address converter specifications. (b) ROM address K-maps read from the table given in (a).

469

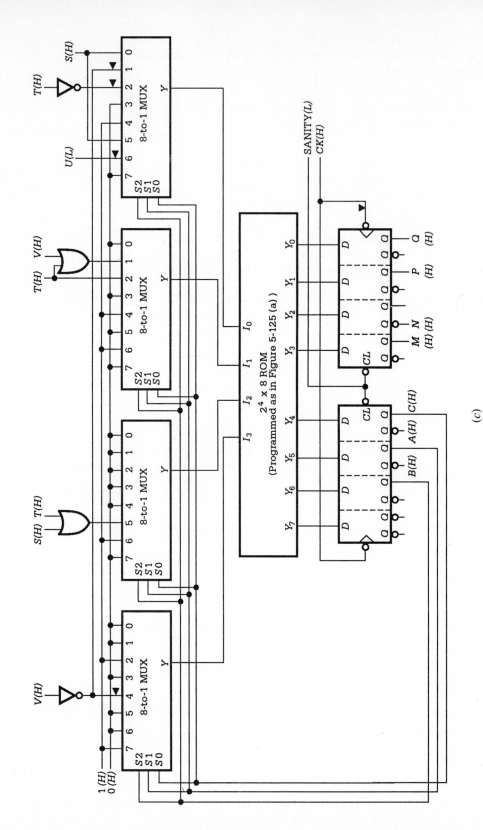

Fig. 5-125 (*contd.*). (c) *Logic circuit showing implementation with a $2^4 \times 8$ ROM.*

(c)

is to use these same devices as memory elements for more complex FSMs such as that in Fig. 5-113. This we do next. However, to make these designs more interesting we will also include some of the alternative approaches for NS and output logic implementation that were developed in previous sections.

The UNIVERSAL SHIFT REGISTER as the MEMORY Element. Shown in Fig. 5-126(a) is the state diagram of Fig. 5-113 but with an altered state code assignment that is more amenable for use with a UNIVERSAL SHIFT REGISTER MEMORY element. The branching legend reads: Branching Condition (Register Action). For example, the transition from STATE ⓪ to STATE ① requires $S(SL1)$, meaning that input S must be ACTIVE at the same time that the register shifts left a logic 1. Similarly, the transition from STATE ③ to STATE ⓪ requires $\bar{S} \cdot \bar{T}(PL)$, meaning that S and T must both be INACTIVE at the same time that the register is parallel loaded.

Used as the MEMORY element for the FSM in Fig. 5-113, the UNIVERSAL SHIFT REGISTER requires four sets of K-maps. These are pro-

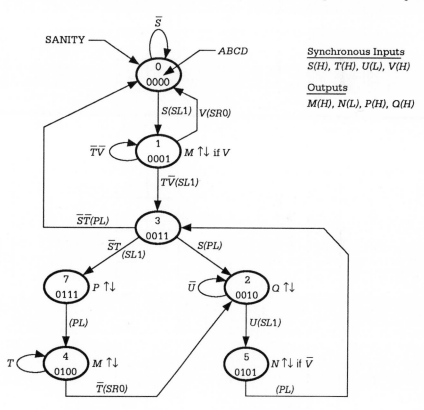

Branching legend: Branching Condition (Register Action)

(a)

Fig. 5-126. *A design of the FSM in Figure 5-113 centered on a UNIVERSAL SHIFT REGISTER. (a) State diagram showing new state code assignments and register action.*

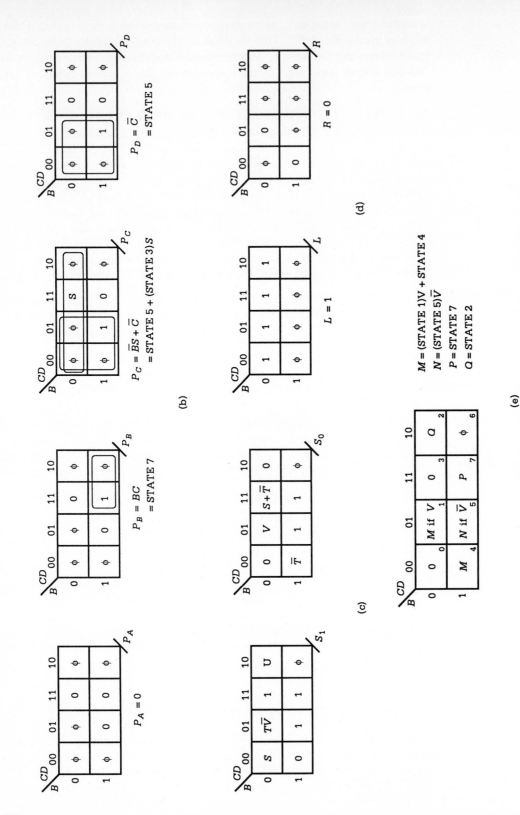

Fig. 5-126 (contd.). (b) Parallel load K-maps and minimum cover. (c) Mode select input K-maps. (d) Left and right serial input K-maps and minimum cover. (e) Composite output K-map and decoder cover.

472

vided in Figs. 5-126(b), (c), and (d). The four parallel load K-maps in Fig. 5-126(b) are given specific entries in cells 3, 5, and 7 since these cells represent the states for which there is a parallel load action required by the register. Keeping in mind that the UNIVERSAL SHIFT REGISTER of Fig. 5-101 is composed of D-FFs, we see that the four entries in each of the cells 3, 5, and 7 become 00SO, 0011, and 0100, respectively, as read across the P_A, P_B, P_C, and P_D K-maps. The remainder of the cells in these maps take don't cares since they represent non-PL states.

The mode select K-maps of Fig. 5-126(c) must take specific entries for all cells which represent used states. Thus, cell 6, representing the only unused STATE (0110), is assigned a don't care. The remaining seven cell entries are determined from the state diagram of Fig. 5-126(a) in connection with the operation table of Fig. 5-101(a). For example, the entries in cell 0 of the S_1 and S_0 K-maps are S and 0, respectively, since the register action is a shift left (10) operation conditional on S. Likewise, entries for cell 1 of the S_1 and S_0 K-maps are $T \cdot \overline{V}$, for a conditional shift left (10) operation, and V, for a conditional shift right (01) operation, respectively. Or the entries for cell 3 are 1 for the S_1 K-map since branching is either SL or PL (they must sum to 1 for the S_1 bit), and $S + \overline{S} \cdot \overline{T} = S + \overline{T}$ for the S_0 K-map since S and $\overline{S} \cdot \overline{T}$ are conditional PL branching conditions. It must be remembered that $(S_1, S_0) = (1, 1)$ for parallel loading.

Figure 5-126(d) shows the left (L) and right (R) serial input K-maps. These maps take logic 1 or logic 0 entries for shift operations depending on whether a 1 or a 0 is shifted left or right. For example, a 1 and ϕ are entered into cell 0 of the L and R K-maps, respectively, because only a SL1 operation exists—it does not matter what logic value is given to cell 0 of the R map. Similarly, cell 1 of these K-maps take 1 and 0, respectively, since both SL1 and SR0 operations exist for the 0001 STATE, and so on. Notice that cells 5 and 7, representing states whose branching conditions are strictly non-shifting, must take don't cares.

The composite output K-map for the FSM in Fig. 5-126(a) is given in Fig. 5-126(e). Also shown (to the right of this map) is the output logic required for the use of a STATE DECODER, logic which is read either directly from the state diagram or from the composite K-map. This information should be compared with that of Fig. 5-117(b).

A logic circuit can now be constructed from the information contained in Fig. 5-126. This is done in Fig. 5-127 where the design has been centered on the UNIVERSAL SHIFT REGISTER represented in Fig. 5-102(a). Notice that the logic for the mode controls (S_1 and S_0) is controlled by the PS variables via the mode select inputs to the 8-to-1 MUXs and that the output logic differs from that in Fig. 5-117(c) to the extent that use is made of different state decoder outputs.

The Parallel Loadable Up/Down-Counter as the MEMORY Element. We complete this portion of the design alternatives by using a parallel loadable up/down-counter as the MEMORY element for the design of the FSM represented in Fig. 5-113. The procedure to be used is basically the same as that used in the previous example where a UNIVERSAL SHIFT REGISTER served as the MEMORY. We begin with the state diagram in Fig. 5-128(a)

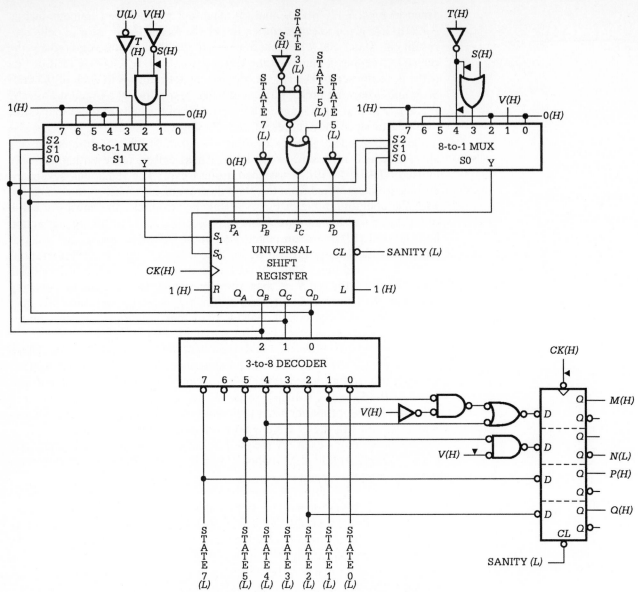

Fig. 5-127. *Implementation of the FSM in Figure 5-113 centered on a UNIVERSAL SHIFT REGISTER and a decoder.*

which has been recoded to agree more closely with the characteristics of the counter. Of course, the branching conditions remain the same, but the action required of the MEMORY device is now an increment (I), a decrement (D), or a parallel load (PL) as indicated in parentheses next to each branching condition.

The parallel load K-maps, shown in Fig. 5-128(b), are plotted following the same procedure as those in Fig. 5-126(b). However, in this case there is but a single state (STATE 2) for which parellel load action is required for outgoing branching. Thus, the entries for cell 2 are 0, $S + T$, $\overline{S} \cdot T$, and 0 respectively, for P_A, P_B, P_C, and P_D, as required for D-FFs. The remaining cells take don't cares since they represent "count action" states.

The LD and EN K-maps shown in Fig. 5-128(c) constitute the mode control K-maps for the counter. They are plotted by using the operation table of Fig. 5-107(a) together with the state diagram of Fig. 5-128(a). The value of each entry (0, 1, or entered variable) is based solely on whether the outgoing branching requires a count action or a parallel load action, and whether the action is conditional or unconditional. For example, the entries in cell 1 of the LD and EN K-maps are 0 and $T \cdot \overline{V} + V = T + V$, respectively, since count action is conditional for both transitions from STATE ① and since the operation table requires that $(LD, EN) = (0, 1)$. The count direction $(D/\overline{U})$ K-map in Fig. 5-128(c) takes cell values which are determined from the truth table in Fig. 5-108(b) and from the state diagram. Thus, only a decrement action (down-count) contributes a cell entry other than logic 0. Finally, the composite output K-map is presented in Fig. 5-128(d) together with the output logic appropriate for the use of a STATE DECODER.

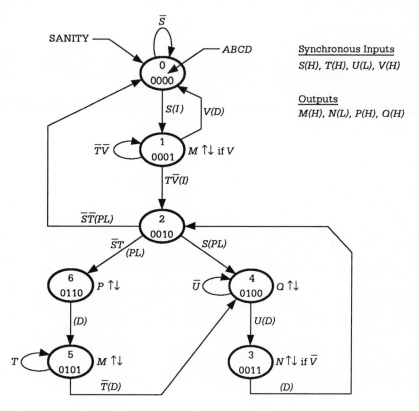

Branching legend: Branching Condition (Counter Action)

(a)

Fig. 5-128. *A design of the FSM of Figure 5-113 centered on a parallel loadable bidirectional counter. (a) State diagram showing new state code assignments and counter action.*

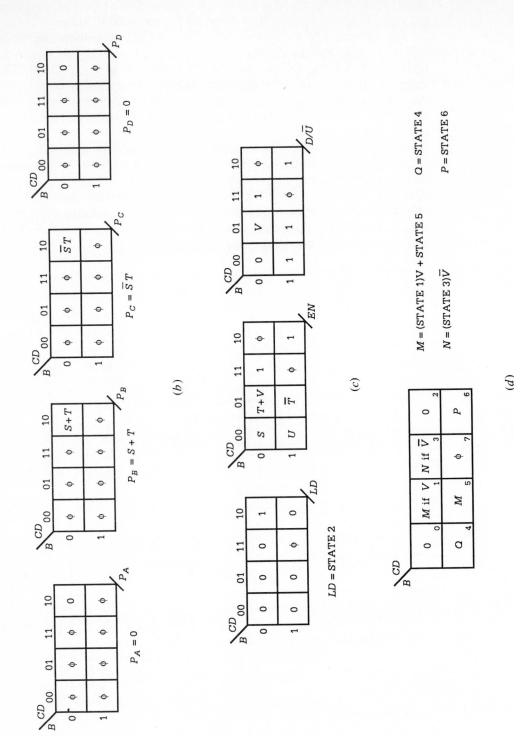

(b)

(c)

M = (STATE 1)V + STATE 5 Q = STATE 4

N = (STATE 3)$\bar{V}$ P = STATE 6

(d)

Fig. 5-128 (cond.). *(b) Parallel load K-maps and minimum cover. (c) Load, enable and count direction K-maps. (d) Composite output K-map and decoder cover.*

476

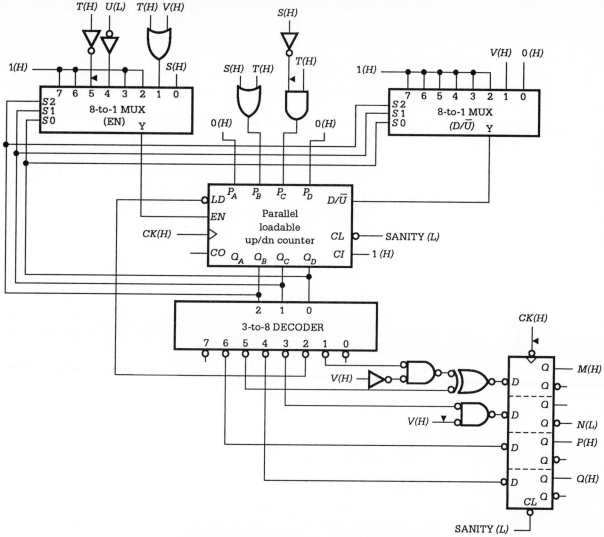

Fig. 5-129. *Implementation of the FSM in Figure 5-128(a) centered on a parallel loadable bidirectional counter and a decoder.*

Shown in Fig. 5-129 is the logic circuit for the design centered on the parallel loadable up/down-counter of Fig. 5-109. Here, we see that LD(*L*) is taken from the STATE 2(*L*) output of the DECODER as permitted by the LD K-map in Fig. 5-128(c). It is also observed that the PS (feedback) variables are the select lines for the MUXs which generate the *EN* and *D*/$\overline{U}$ signals to the counter. Notice that the output logic differs from that in Figs. 5-119(a) and 5-127 only because different STATE DECODER outputs are used.

5.11.6 Programmable FSM Architecture

The final step we take in the progression of alternative FSM designs is the inclusion of programmable FSM architecture. In the context of this chapter, the term "programmable" implies the use of array logic to generate any or all of the address signals for a counter, SHIFT REGISTER, or any other such device that serves as part of the MEMORY for an FSM. It is the nature of the programmable FSM architecture to have MEMORY sections more complex, but also more versatile, than any of those described in previous sections. In this sense our presentation here represents a fundamental departure from the more traditional approaches to FSM design we have considered so far.

Let us suppose we are asked to write down a list of features we consider to be the most desirable in any programmable FSM design methodology. Such a wish list would likely include a simple means of modifying the NS and output logic, and a modular approach that permits bit-slice capability to form larger FSMs, hence, built-in versatility. Other attractive features might include the ability to implement the FSM directly from a state diagram or from a listing of instructions, and the use of off-the-shelf MSI devices such as PROMs, MUXs, and DECODERs. To one extent or another, all these features are present in the design methodology described in this section.

Counters/MUXs as the MEMORY. We begin with a programmable FSM design centered on a modular 4-bit parallel loadable counter and MUXs as the MEMORY. Figure 5-130 illustrates the details of this architecture. Here, we observe that ROMs store the NEXT STATE and output logic program and issue signals that form the LD MUX address (ADR), EN MUX ADR, $D/\overline{U}$ MUX ADR, and PL ADR of the MEMORY. In addition, the ROMs also generate the FSM outputs (conditional and unconditional) without the aid of any additional logic.

It is our view that the programmable FSM architecture of Fig. 5-130 conforms to Huffman's model given in Fig. 5-122. This means that the counter, together with its associated MUXs, constitute the MEMORY of the state machine. This is obviously a more complex MEMORY than has been considered previously.

The state diagram, as modified to conform to the architecture of Fig. 5-130, is given in Fig. 5-131(a). We notice immediately that STATEs ② and ③ constitute what was originally a single state in Fig. 5-128(a). This modification is necessary so as to limit the state diagram to two-way outbranching. In fact there are two important restrictions associated with the use of the programmable architecture of Fig. 5-130. They are

1. Excluding the hold condition, branching from a given state is limited to two-way branching.
2. The counter actions required for two-way outbranching from a given state cannot both be a count (I or D) or both be a parallel load.

The foregoing two restrictions can be relaxed somewhat if the architecture of Fig. 5-130 is modified to include the placement of MUXs on the PL ADR inputs to the counter. For example, 2-to-1 MUXs on these inputs permit two-way PL branching from a given state while 4-to-1 MUXs permit three-way PL branching, and so on. This means increasing the ROM size to accommodate the added inputs and the extra address outputs.

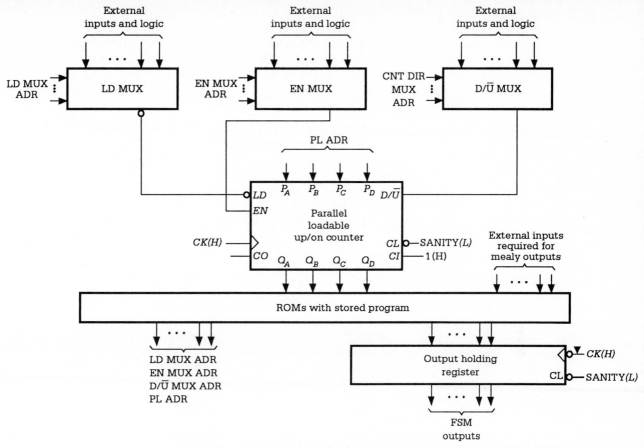

Fig 5-130. *Programmable FSM architecture centered on a modular 4-bit counter and MUXs as the memory.*

From an inspection of the state diagram in Fig. 5-131(a), it is evident that the original algorithm expressed by Fig. 5-128(a) is preserved only if $\overline{S} \cdot \overline{T}$ is assumed not to be a possible branching condition from STATE ③. The $\overline{S} \cdot \overline{T}$ branching condition is not defined for STATE ③ and, therefore, the SUM rule [Eq. (5-3)] is not obeyed. The branching constraint relative to STATE ③ and the additional CLOCK cycle required by the presence of the extra state can be viewed as the price that must be paid to use the programmable architecture of Fig. 5-130 for this particular FSM. The use of this architecture for other FSMs might not require such constraints.

It is possible to construct a logic circuit directly from the state diagram of Fig. 5-131(a) as we have done in previous examples. However, this task is facilitated by constructing the *PS/NS*/output instruction table shown in Fig. 5-131(b). Here, the format used is

---

Branching Condition (Counter ACTION) Destination — (Output)

This format can be viewed as an elementary high-level assembly-type language.

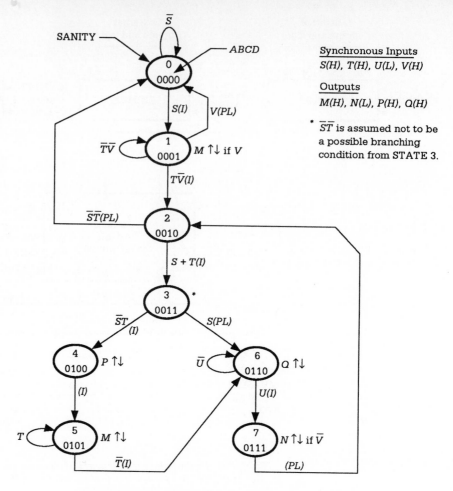

Branching legend: Branching Condition (Counter Action)

(a)

PS	NS and Output Instructions
0000	S (I) 1 $-$ (0)
0001	$T\overline{V}$ (I) 2, V (PL) 0 $-$ (M if V)
0010	$S+T$ (I) 3, $\overline{S}\overline{T}$ (PL) 0 $-$ (0)
0011	$\overline{S}T$ (I) 4, S (PL) 6 $-$ (0)
0100	(I) 5 $-$ (P)
0101	$\overline{T}$ (I) 6 $-$ (M)
0110	U (I) 7 $-$ (Q)
0111	(PL) 2 $-$ (N if $\overline{V}$)

Instruction Format:
Branching Condition (ACTION) Destination $-$ (Output)

(b)

Fig. 5-131. *Design of the FSM in Figure 5-128(a) by using the architecture of Figure 5-130. (a) Modified state diagram for up count, parallel load, and two-way branching. (b) The PS/NS/output instruction table for (a).*

480

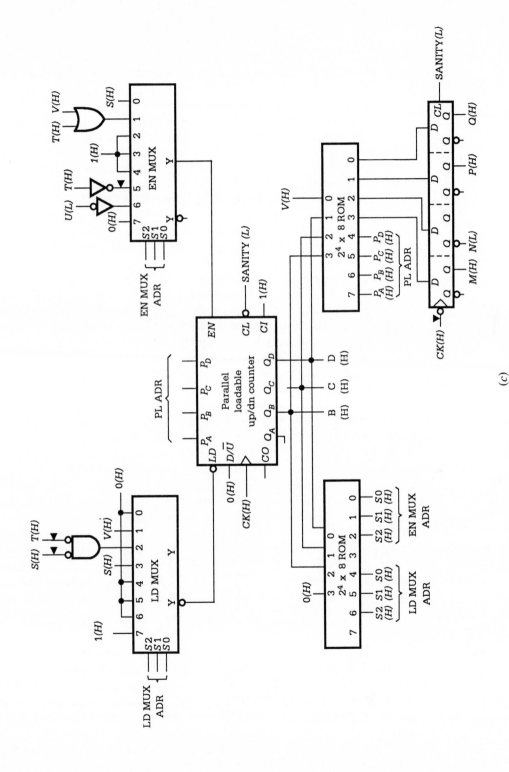

Fig. 5-131 (contd.). (c) Implementation by using two $2^4 \times 8$ ROMs and the necessary external NS and output logic.

(c)

An inspection of the instruction table in Fig. 5-131(b) reveals that it contains *all* of the sequential information of the state diagram. Indeed, this table provides a concentrated but simple means of representing the sequential behavior of an FSM. Of course, it is required that use be made of a well-defined MEMORY device such as a binary up/down parallel loadable counter or a UNIVERSAL SHIFT REGISTER. But most any sequential device, even a microprocessor, could be used as the MEMORY element as long as its operations are well defined.

The logic circuit for the FSM represented by Fig. 5-131(a) is shown in Fig. 5-131(c). While the external input logic to the LD and EN MUXs can be arbitrarily positioned, it is best positioned according to the PS/NS/output instruction table of Fig. 5-131(b) and always in consideration of the operation table for the counter given in Fig. 5-107(a). For example, the LD and EN MUX input logic required for branching from STATE ① is given by $LD = T \cdot \overline{V}(0, 1) + V(1, X) = V$, and $EN = (T + V)$, where the logic 0 and 1's and the irrelevant input X (appearing in parentheses) derive from the operation table of Fig. 5-107(a). (Note that $T \cdot \overline{V}$ could replace $T + V$ as the logic for input 1 of the EN MUX.) In similar fashion, the LD and EN MUX input logic required for branching from STATE ② is found to be $LD = (S + T)(0, 1) + \overline{S} \cdot \overline{T}(1, X) = \overline{S} \cdot \overline{T}$, and $EN = 1$. Or the input logic to terminal 3 of the MUXs is $LD = \overline{S} \cdot T(0, 1) + S(1, X) = S$ since $\overline{S} \cdot \overline{T}$ is not specified, and $EN = 1$, and so on. Observe that the $D/\overline{U}(H)$ input to the counter is held at $0(H)$, a consequence of the fact that no decrement count action exists in the state diagram of Fig. 5-131(a).

The logic circuit of Fig. 5-131(c) is not complete until the two ROMs have been programmed to generate the LD and EN MUX addresses, the PL address, and the outputs. This is done in Fig. 5-132. Figure 5-132(a) indicates the bit format to be used, and Fig. 5-132(b) is the collapsed program table in which X is used as an irrelevant input. A single external input (V) has been included with the ROM inputs since it is needed to generate the two Mealy outputs. Two ROMs are used instead of one ROM so as to maintain reasonable numbers of ROM outputs.

The program table in Fig. 5-132(b) differs from that of Fig. 5-123(b) in two important respects. First, don't cares are given to the PL address of any state whose branchings do not require PL counter action. STATEs ④, ⑤, and ⑥ in Fig. 5-132(b) are examples. Second, the address WORDs appearing in the LD and EN MUX ADR columns of Fig. 5-132(b) are determined by the positions of the LD and EN MUX input logic as well as by the state diagram (or PS/NS/output instruction table), a consequence of the programmable feature of this architecture.

Registers/MUXs as the MEMORY. Figure 5-133 shows the programmable FSM architecture which is centered on a MEMORY consisting of the UNIVERSAL SHIFT REGISTER and its associated MUXs. Except for the obvious differences in the MEMORY section, this architecture is the same as that in Fig. 5-130. Both architectures conform to Huffman's model given in Fig. 5-122.

One version of the state diagram in Fig. 5-113 which is suited to the architecture of Fig. 5-133 is presented in Fig. 5-134(a). Like that in Fig. 5-

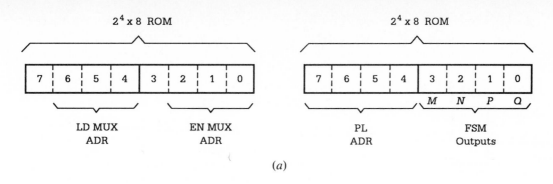

	ROM Inputs				ROM Outputs			
	PS							**FSM Outputs**
State Identifier	**B C D**	**V**	**LD MUX ADR**	**EN MUX ADR**	**PL ADR**	**M N* P Q**
0	0 0 0	X†	φ 0 0 0	φ 0 0 0	φ φ φ φ	0 0 0 0
1	0 0 1	0	φ 0 0 1	φ 0 0 1	0 0 0 0	0 0 0 0
1	0 0 1	1	φ 0 0 1	φ 0 0 1	0 0 0 0	1 0 0 0
2	0 1 0	X	φ 0 1 0	φ 0 1 0	0 0 0 0	0 0 0 0
3	0 1 1	X	φ 0 1 1	φ 0 1 1	0 1 1 0	0 0 0 0
4	1 0 0	X	φ 0 0 0	φ 1 0 0	φ φ φ φ	0 0 1 0
5	1 0 1	X	φ 0 0 0	φ 1 0 1	φ φ φ φ	1 0 0 0
6	1 1 0	X	φ 0 0 0	φ 1 1 0	φ φ φ φ	0 0 0 1
7	1 1 1	0	φ 1 1 1	φ 1 1 1	0 0 1 0	0 1 0 0
7	1 1 1	1	φ 1 1 1	φ 1 1 1	0 0 1 0	0 0 0 0

* An ACTIVE LOW output.
† An irrelevant input.

(b)

Fig. 5-132. *(a) Bit format for programmed instructions from the ROMs.*
(b) Collapsed program table for the ROMs.

131(a), it has been expanded from seven to eight states to meet the two-way outbranching limit established earlier. But also as before, this expansion is not without its price. The original algorithm can be preserved only if $\overline{S} \cdot \overline{T}$ is disallowed as a possible branching condition from STATE ⑦, acknowledging that the SUM rule is not obeyed for this state. Furthermore, an additional CLOCK period has been introduced by the presence of the extra state. Notice that the sum rule is obeyed for the combination of STATEs ③ and ⑦.

The state code assignments given to the state diagram in Fig. 5-134(a) favor a unidirectional SHIFT REGISTER. This is done intentionally to reduce the complexity of the MEMORY section. Also, it is to be noted that the shifting operations must involve all four state variable bits. In contrast, the counter actions indicated in Fig. 5-131(a) need involve only three of the four state variable bits. It is important that the reader understand this difference in MEMORY devices.

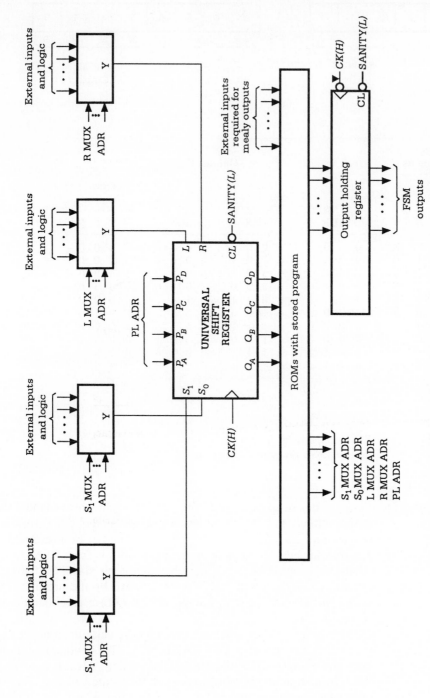

Fig. 5-133. Programmable FSM architecture centered on a 4-bit UNIVERSAL SHIFT REGISTER and MUXs as the memory.

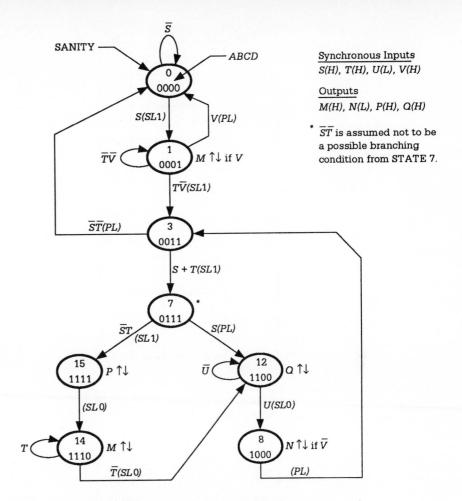

Branching legend: Branching Condition (Register Action)

(a)

State Identifier	PS	NS and Output Instructions
0	0000	S (SL1) 1 − (0)
1	0001	TV (SL1) 3, V (PL) 0 − (M if V)
3	0011	$S+T$ (SL1) 7, $\overline{S}\,\overline{T}$ (PL) 0 − (0)
7	0111	$\overline{S}T$ (SL1) 15, S (PL) 12 − (0)
15	1111	(SL0) 14 − (P)
14	1110	$\overline{T}$ (SL0) 12 − (M)
12	1100	U (SL0) 8 − (Q)
8	1000	(PL) 3 − (N if $\overline{V}$)

Instruction Format:
Branching Condition (ACTION) Destination − (Output)

Fig. 5-134. Design of the FSM in Figure 5-126(a) by using the architecture of Figure 5-133. (a) Modified state diagram for left shift, parallel load, and two-way branching only. (b) The PS/NS/output instruction table for (a).

(b)

485

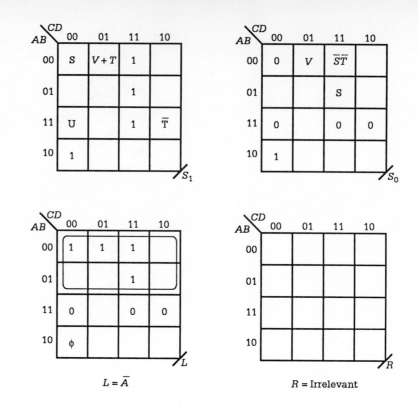

(c)

Fig. 5-134 (contd.). *(c) K-maps for mode control and serial input parameters.*

The PS/NS/output instruction table given in Fig. 5-134(b) provides the same information contained in the state diagram of Fig. 5-134(a), but in the form of a high-level assembly-type language. This table can be useful in constructing a logic circuit for this FSM. Also, the K-maps for the mode control and serial input parameters, provided in Fig. 5-134(c), can assist the reader in making the transition from the instruction table to the logic circuit in Fig. 5-134(d). The value of restricting the shift operation to left shift only is evident from an inspection of the serial input K-maps. Also, notice that the MUX input logic shown in Fig. 5-134(d) derives directly from the S_1 and S_0 K-maps in which the blank cells represent don't cares.

The logic circuit in Fig. 5-134(d) is constructed from the PS/NS/output instruction table and the K-maps given in Figs. 5-134(b) and (c). A comparison of this circuit with that in Fig. 5-131(c) reveals one important difference in design philosophy brought about by the character of the MEMORY that is used. In Fig. 5-134(d) the Mealy (conditional) outputs are generated by logic external to the ROM, whereas in Fig. 5-131(c) all output logic is contained within the ROM. If we use a $2^5 \times 8$ ROM on the output side of the register-based design, $V(H)$ could be included as a ROM input and all output logic could again be absorbed by the ROM.

Completion of the logic circuit in Fig. 5-134(d) is accomplished by programming the two ROMs as in Fig. 5-135. The bit format for programmed

5 / Synchronous Sequential Machines

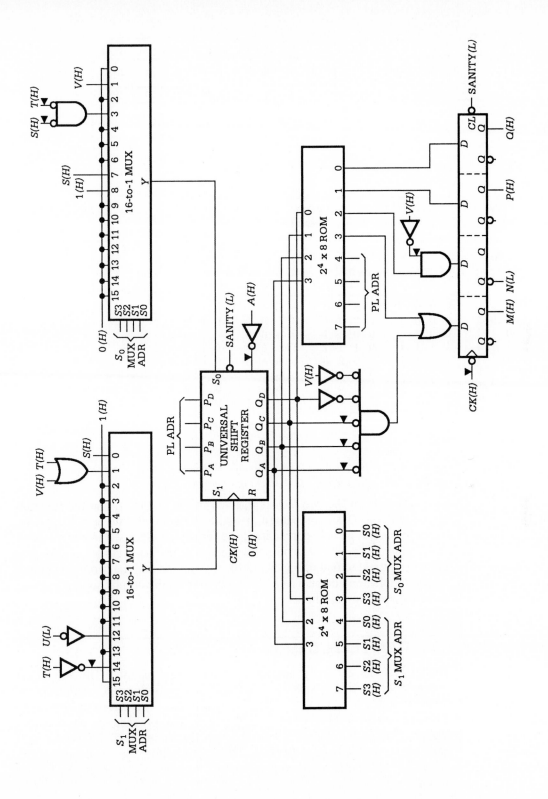

Fig. 5-134 (cond.). *(d) Implementation by using two $2^4 \times 8$ ROMs and the necessary external NS and output logic.*

(d)

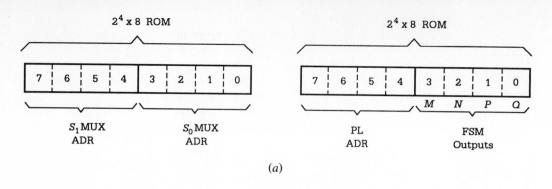

(a)

State Identifier	ROM Inputs (PS)				ROM Outputs S_1 MUX ADR				S_0 MUX ADR				PL ADR				FSM Outputs			
	A	B	C	D													M	N*	P	Q
0	0	0	0	0	0	0	0	0	0	0	0	0	φ	φ	φ	φ	0	0	0	0
1	0	0	0	1	0	0	0	1	0	0	0	1	0	0	0	0	0	0	0	0
3	0	0	1	1	0	0	1	1	0	0	1	1	0	0	0	0	0	0	0	0
7	0	1	1	1	1	1	1	1	0	1	1	1	1	1	0	0	0	0	0	0
15	1	1	1	1	1	1	1	1	0	0	0	0	φ	φ	φ	φ	0	0	1	0
14	1	1	1	0	1	1	1	0	0	0	0	0	φ	φ	φ	φ	1	0	0	0
12	1	1	0	0	1	1	0	0	0	0	0	0	φ	φ	φ	φ	0	0	0	1
8	1	0	0	0	1	1	1	1	1	0	0	0	0	0	1	1	0	1	0	0

* An ACTIVE LOW output.

(b)

Fig. 5-135. (a) Bit format for programmed instructions from the ROMs. (b) Program table for the ROM showing only used states.

instructions from the two ROMs is given in Fig. 5-135(a), and the program table for the ROMs is presented in Fig. 5-135(b). Only the eight states relevant to the design are represented, the remaining eight states being "don't care" states. Observe that no conditional outputs are included in this program in contrast to that of Fig. 5-132(b). Also notice that the N output column is not complemented since $N(L)$ is obtained from the output holding register as in the previous examples.

5.12 INTRODUCTION TO SYSTEM-LEVEL DESIGN

Digital systems generally operate in a sequential manner. The sequential process begins with a specific state and proceeds through a sequence of steps or states in a manner which is algorithmic. The algorithm is usually iterative, meaning that part or all of the sequential process can be repeated, thereby forcing the digital system through the same set of states over and over again, if commanded to do so. The word "state," as used here, refers to the logical status of all components of the digital system at some point in time.

To achieve good engineering designs the designer should have three qualities: experience, "artistic skill," and design intuition. These qualities, in addition to knowledge of the field, are needed at all stages of the design process. This is especially true in complex design problems where more than one workable solution is evident. It is not the purpose of this section to instill in the reader such qualities—this is not possible. Rather, we will provide a useful approach or method for attacking complex system-level design problems. We will also provide a relatively simple example of a system-level design to illustrate the use of this method.

5.12.1   The CONTROLLER and Data Path

One very common view of a digital system is the use of an FSM as a CONTROLLER for a set of component parts which comprise the *controlled system* called the *data path*. This view is expressed in Fig. 5-136, where all input and output conditioning logic has been omitted so as to focus attention on the main features of this architecture. Here, it is understood that the data path devices generally consist of a mixture of both sequential and combinational logic machines. Typical among these are registers, counters, ALUs, ROMs, PLAs, decoders, MUXs, shifters, comparators, analog-to-digital (A/D) converters, and the like. The architecture represented by Fig. 5-136 is one we emphasize in this text.

The CONTROLLER for a digital system is an FSM, perhaps one like that in Fig. 5-113. But the CONTROLLER is also the "brains" of the system. It is the function of the CONTROLLER to coordinate precisely the operation of the various components of the data path so as to perform the specific task(s) required of the system. Thus, the CONTROLLER must issue instructions (control signals) to the data path based on external input and feedback information from the data path. A configuration such as this, where the outputs on one unit are the inputs to another, and vice versa, is called a *handshake interface*. While Fig. 5-136 depicts a general handshake interface between the CONTROLLER unit and the data path unit (DPU), feedback from the DPU is not a requirement for all systems but is common in most. Note that both the CONTROLLER and the data path devices may receive signals from and issue signals to the outside world.

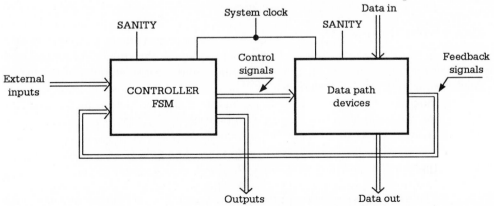

Fig. 5-136. Controller/data-path architecture for digital system design.

The design of the CONTROLLER requires that a fully documented state diagram be constructed. Construction of the state diagram can be facilitated by using a flowchart as a thinking tool. If the algorithm is complex, several stages of flowcharting may be necessary to approach or achieve an optimal design. The construction of timing diagrams might also be necessary to gain a correct perspective of the system requirements. Also, the combined use of a detailed flowchart and timing diagrams may be necessary to break the system down into manageable parts, parts that can be implemented with available hardware. The procedure for FSM (CONTROLLER) design is provided in Sect. 5.9.1.

5.12.2 The Functional Partition

Designing a complex digital system requires a "divide-and-conquer" approach. The system must be divided into subsystems that, in turn, must be broken down into well-defined parts that can be implemented with available hardware. The detailed block diagram that conveys this information is appropriately called the *functional partition* of the system. Thus, the functional partition contains a block representation of the CONTROLLER, *all* the peripheral devices that constitute the data path unit, the connections between the CONTROLLER and DPU devices, and all inputs from and outputs to the outside world. In other words, the functional partition contains all the information needed for "hookup" and operation of the system, given the details of the CONTROLLER design which is treated as a separate problem of the design process.

The functional partition and a detailed flowchart for the CONTROLLER of a digital system are usually interdependent and must be developed together. For a complex digital system this development process may require a "first-cut" representation of the functional partition and flowchart followed by their detailed representation. The use of timing diagrams may be a necessary part of these developments. Certainly, the artistic skill, intuition, and experience factors mentioned earlier play an important role here—a good designer has these qualities.

We emphasize the fact that there may be more than one good design for a given digital system. This is particularly true for complex systems. Whether or not a given design is judged best, good, or poor will usually depend on the environment in which the system must operate. For example, suppose one is to design the stepping motor control systems specifically for a robot. These control systems are necessary to keep the various mechanical parts from moving in a "jerky" fashion and to lessen the chance for mechanical malfunction. A complete design would include considerations of the mass, time, and distance relationships. The example that follows is a simple stepping motor control system which has no dedicated purpose. However, as the reader studies this design, it should be evident that there exists other designs which are possible but not necessarily better. Mass, time, and distance constraints are not considered in our example since a specific use for the stepping motor is not indicated.

The Stepping Motor Control System as an Example. Stepping motors are useful in systems where there is space only for a small motor to drive a relatively massive part. Linear angular accelerations and decelerations of the motor can prevent slippage, chattering, or jerky motion which could lead to mechanical failure. Stepping motors are used in computer printers to position the print heads, in robots to operate mechanical parts in some manner, in fluid control systems for accurate adjustment of fluid control valves, in the wire-wrap processing of circuit boards, and in a variety of other applications too numerous to mention.

The overall operational characteristics for the stepping motor control system are provided in Fig. 5-137. In Fig. 5-137(a) are shown the angular-velocity/time requirements of the motor. The GO command causes a linear angular acceleration of the motor while a HALT command causes a linear angular deceleration. A HOLD command causes the motor to maintain the angular velocity that is reached at the time the HOLD command is given. The motor must operate between zero speed and a maximum angular velocity that is set by the number of steps in the speed/time characteristic.

The physical picture for the overall system is presented in Fig. 5-137(b). Here, the control system receives one of three (nonoverlapping) asynchro-

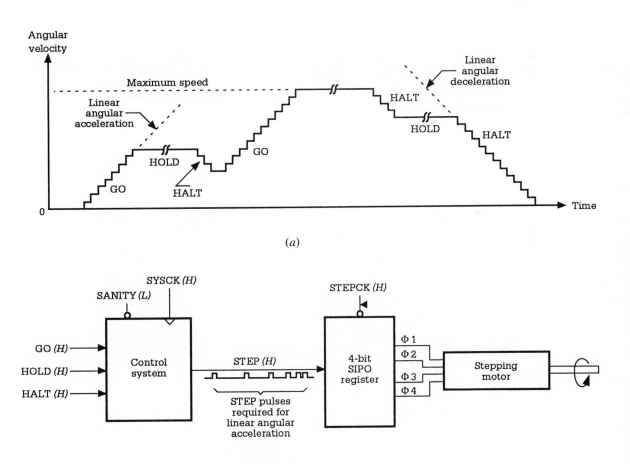

Fig. 5-137. Overall operational characteristic of the stepping motor control system. (a) Speed-time requirements. (b) Physical picture.

5.12 / Introduction to System-Level Design

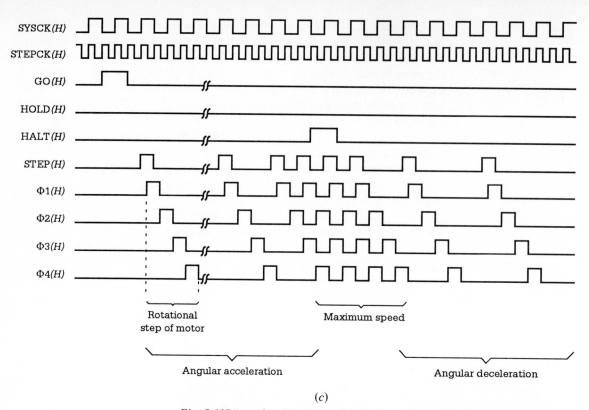

SYSCK(H)

STEPCK(H)

GO(H)

HOLD(H)

HALT(H)

STEP(H)

Φ1(H)

Φ2(H)

Φ3(H)

Φ4(H)

Rotational
step of motor

Maximum speed

Angular acceleration

Angular deceleration

(c)

Fig. 5-137 (contd.). (c) Acceptable timing relationships between synchronized external inputs and STEP pulse signals to the stepping motor.

nous input signals, GO, HOLD, or HALT and issues a series of STEP pulses in response to that input signal. In the physical picture, a GO signal is implied resulting in the STEP pulse series required to cause a linear angular acceleration of the motor. Each STEP pulse is received by the SIPO register, which, in turn, delivers a set of four phase pulses (Φ1, Φ2, Φ3, Φ4) to the power transistors of the stepping motor causing the motor to rotate by a certain amount. The SIPO register is triggered off of a STEPCK waveform which is synchronized with SYSCK, the waveform used to trigger the control system.

An acceptable timing relationship between external inputs, STEP pulse, and phase pulse signals to the stepping motor is given in Fig. 5-137(c). Each STEP pulse width is specified to be one period of the STEPCK waveform, which runs at exactly twice the frequency of the SYSCK waveform. When a STEP pulse is received by the register it is shifted from the LSB stage to the MSB stage in the register on each falling edge of the STEPCK pulse. Thus, a set of four time-shifted pulses is generated by each STEP pulse as illustrated in Fig. 5-137(c). The maximum rotational velocity is set by the frequency of the STEPCK waveform which is assumed to be low enough to match the inertial character of the motor. The SYSCK waveform can be generated from the STEPCK waveform by using a divide-by-two counter. Such a counter results if the $Q(L)$ output of an RET D FLIP-FLOP is connected to its input $D(H)$, while STEPCK is the CLOCK input and SYSCK is the $Q(H)$ output.

The functional partition for the stepping motor control system is shown in Fig. 5-138. Synchronous, nonoverlapping inputs GO, HOLD, and HALT are presented to the CONTROLLER from input conditioning circuits. The data path devices consist of the right SHIFT REGISTER of Fig. 5-100 operated as a SIPO register, a special synchronous up-counter whose design is covered by Problem 5.51 at the end of this chapter, and a special pulse-mode up/down-counter whose design is covered by Problem 6.42 in Chapter 6. Both counters have asynchronous parallel load capability on the LD command. This means that the asynchronous PRESET and CLEAR overrides on the FLIP-FLOPs of the counter are used to drive it into a given state and then the counting resumes or begins from that state. The two counters differ mainly in the way they are triggered. The up-counter triggers off of the falling edge of the SYSCK waveform; the up/down-counter triggers off of the falling edge of the UP (DECDLY) or DN (INCDLY) input pulse as

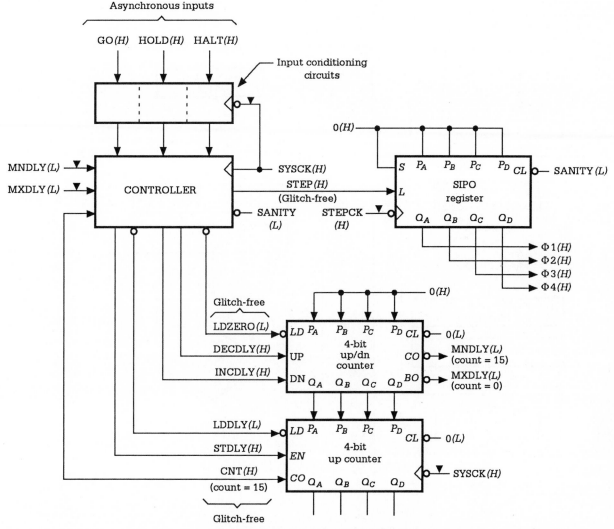

Fig. 5-138. *Functional partition for the stepping motor control system.*

deduced from the design examples in Sect. 6.8.3 in Chapter 6 (e.g. see Fig. 6-54). CARRYOUT (*CO*) or BORROWOUT (*BO*) signals are issued from the up/down-counter at and during the end count, 15 or 0, respectively. A *CO* (CNT) signal is issued by the up-counter on the falling edge of the SYSCK waveform at and during its end count of 15 and is sensed by the CONTROLLER on the next rising edge of SYSCK.

The detailed flowchart presented in Fig. 5-139 is to be understood in conjunction with the functional partition in Fig. 5-138. The basic algorithm involved is as follows: After a STEP pulse has been issued by the CONTROLLER (at some point in time), the up/down-counter is incremented (by 1), held at its present value, or decremented (by 1) depending, respectively, on the previous input command GO, HOLD, or HALT. The new count of the up/down-counter is the setup value which is loaded immediately into the

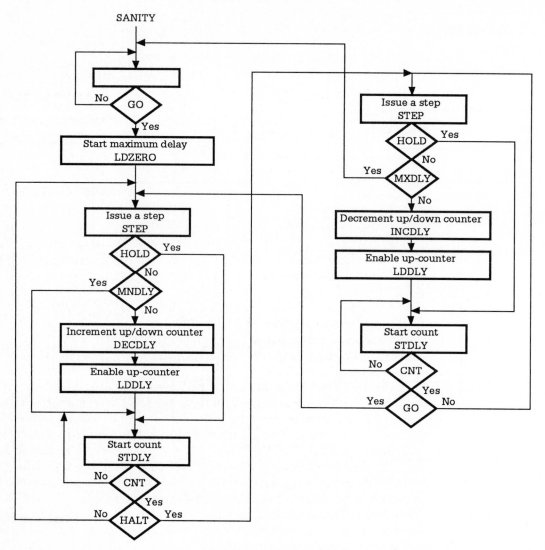

Fig. 5-139. First detailed flow chart for the stepping motor CONTROLLER.

up-counter by the LDDLY (LD) signal from the CONTROLLER. Then, on the STDLY (EN) signal from the CONTROLLER, the up-counter counts with SYSCK to 1111 (15_{10}) from the setup value after which a CNT signal is returned to the CONTROLLER indicating that the full count of 15 has been reached. Another STEP pulse is issued by the CONTROLLER and the above process is repeated. The count of the up-counter from the setup value to 15 is the delay time btween STEP pulses which is variable from 15 (setup value 0) to 0 (setup value 15). When the up/down-counter reaches its full setup count of 15, a MNDLY (CARRYOUT = CO) signal is returned to the CONTROLLER, thereby indicating that a minimum delay (maximum speed) has been reached. Similarly, when the up/down-counter reaches its lowest count (0), a MXDLY (BORROWOUT = BO) signal is returned to the CONTROLLER, which ends the process (stops the motor). The issuance of a DECDLY or INCDLY signal, required to increment of decrement the up/down-counter, can be avoided at any time during the process by singularly activating the HOLD input. Thus, the delay time between STEP pulses can be held constant following a HOLD pulse, and will remain so until a GO or HALT pulse is received by the CONTROLLER. Notice that after initialization, a GO input signal starts the process with a LDZERO output to the up/down-counter to establish a maximum delay time count (0 to 15) of the up-counter.

Shown in Fig. 5-140(a) is the final draft of the flowchart for the stepping motor CONTROLLER. This flowchart is a Mealy version of that in Fig. 5-

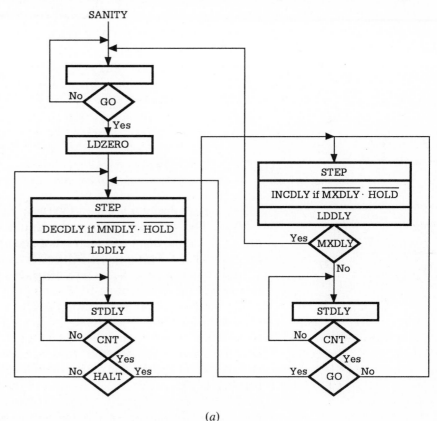

Fig. 5-140. *Final draft of sequential specifications for the stepping motor CONTROLLER. (a) Flow chart.*

(a)

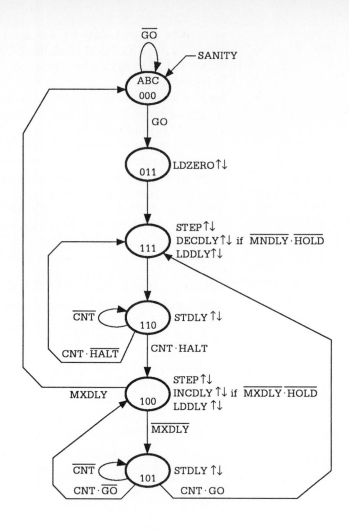

(b)

Fig. 5-140 (contd.). *(b) State diagram based on the flow chart in (a).*

139. The state diagram in Fig. 5-140(b) corresponds to the flowchart in Fig. 5-140(a). The state code assignment shown yields no output race glitches provided that the don't care states, 001 and 010, are not used for minimizing the output logic. The output LDZERO is an exception to this requirement.

As we indicated in Sect. 5.11, there are several architectures that can be used for the design of an FSM such as the stepping motor CONTROLLER. Because of the relatively large number of inputs and outputs, we choose an architecture centered around a PLA-type device. The NEXT STATE and output K-maps and the minimum cover best suited to this purpose are given in Fig. 5-141(a). Notice that the output forming logic is free of STATIC HAZARDs and output race glitches as required by the functional partition in Fig. 5-138.

The *p*-term table based on these results is given in Fig. 5-141(b). Two inputs and two outputs of this table have been marked with asterisks to

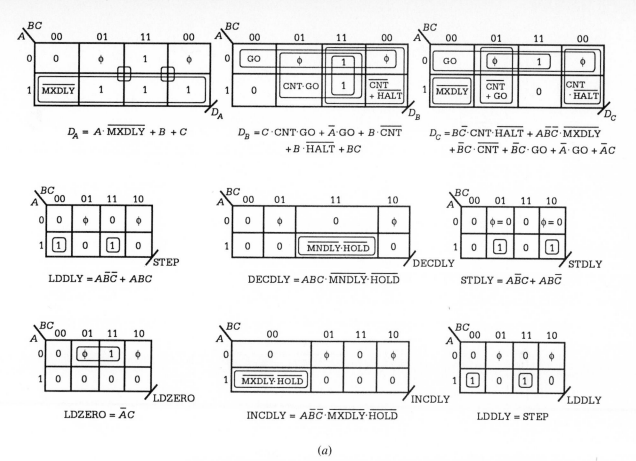

$$D_A = A \cdot \overline{\text{MXDLY}} + B + C$$

$$D_B = C \cdot \text{CNT} \cdot \text{GO} + \overline{A} \cdot \text{GO} + B \cdot \overline{\text{CNT}}$$
$$+ B \cdot \overline{\text{HALT}} + BC$$

$$D_C = B\overline{C} \cdot \text{CNT} \cdot \overline{\text{HALT}} + A\overline{B}\overline{C} \cdot \overline{\text{MXDLY}}$$
$$+ \overline{B}C \cdot \overline{\text{CNT}} + \overline{B}C \cdot \text{GO} + \overline{A} \cdot \text{GO} + \overline{A}C$$

$$\text{LDDLY} = A\overline{B}\overline{C} + ABC$$

$$\text{DECDLY} = ABC \cdot \overline{\text{MNDLY}} \cdot \overline{\text{HOLD}}$$

$$\text{STDLY} = A\overline{B}C + AB\overline{C}$$

$$\text{LDZERO} = \overline{A}C$$

$$\text{INCDLY} = A\overline{B}\overline{C} \cdot \overline{\text{MXDLY}} \cdot \overline{\text{HOLD}}$$

$$\text{LDDLY} = \text{STEP}$$

(a)

Fig. 5-141. *Design of the stepping motor CONTROLLER. (a) NEXT STATE and output K-maps and minimum cover for glitch-free output generation.*

indicate that they are ACTIVE LOW. ACTIVE LOW inputs can be accommodated by either complementing their columns in the p-term table or by using INVERTERs on their input lines, but not both. ACTIVE LOW outputs from a PLA-type device require the use of INVERTERs. Notice that there are 9 inputs, 8 outputs, and 20 p-terms indicated in the p-term table. Thus, the minimum size PLA required is $9 \times 20 \times 8$. Use of a ROM for this implementation would be an inefficient application of an array logic device, since only a small fraction of the 512 MINTERM capability would be utilized. For a review of array logic devices and their uses, the reader is referred to Sect. 4.11.

An overview of the stepping motor CONTROLLER architecture we use is shown in Fig. 5-142. Here, we have included the input conditioning circuits which catch, stretch, and synchronize each of the three asynchronous inputs as discussed in Sect. 5.8.4. The stretching operation is needed in the event the input signals are of duration less than one period of the SYSCK waveform. Also shown in Fig. 5-142 is the use of an AND gate to generate a STEP pulse which is ACTIVE only during the ACTIVE period of the SYSCK waveform. This satisfies the requirements of the timing diagram in Fig. 5-137(c).

INPUTS / **OUTPUTS**

p-terms	A I_8	B I_7	C I_6	MNDLY* I_5	MXDLY* I_4	CNT I_3	GO I_2	HALT I_1	HOLD I_0	STEP D_A O_7	D_B O_6	D_C O_5	LDDLY† O_4	LDZERO† O_3	DECDLY O_2	INCDLY O_1	STDLY O_0
$A \cdot \overline{MAXDLY}$	1	–	–	–	0	–	–	–	–	1	0	0	0	0	0	0	0
B	–	1	–	–	–	–	–	–	–	1	0	0	0	0	0	0	0
C	–	–	1	–	–	–	–	–	–	1	0	0	0	0	0	0	0
$C \cdot CNT \cdot GO$	–	–	1	–	–	1	1	–	–	0	1	0	0	0	0	0	0
$\overline{A} \cdot GO$	0	–	–	–	–	–	1	–	–	0	1	1	0	0	0	0	0
$B \cdot \overline{CNT}$	–	1	–	–	–	0	–	–	–	0	1	1	0	0	0	0	0
$B \cdot \overline{HALT}$	–	1	–	–	–	–	–	0	–	0	1	0	0	0	0	0	0
BC	–	1	1	–	–	–	–	–	–	0	0	1	0	0	0	0	0
$\overline{B}\,\overline{C} \cdot CNT \cdot \overline{HALT}$	–	0	0	–	–	1	–	0	–	0	0	1	0	0	0	0	0
$\overline{A}\,\overline{B}\,\overline{C} \cdot \overline{MXDLY}$	0	0	0	–	0	–	–	–	–	0	0	1	0	0	0	0	0
$\overline{B}C \cdot \overline{CNT}$	–	0	1	–	–	0	–	–	–	0	0	1	0	0	0	0	0
$\overline{B}C \cdot GO$	–	0	1	–	–	–	1	–	–	0	0	1	0	0	0	0	0
$\overline{A} \cdot GO$	0	–	–	–	–	–	1	–	–	0	0	0	0	0	0	0	0
$\overline{A}\,\overline{C}$	0	–	0	–	–	–	–	–	–	0	0	0	1	1	0	0	0
$\overline{A}\,\overline{B}\,\overline{C}$	0	0	0	–	–	–	–	–	–	0	0	0	1	0	0	0	0
ABC	1	1	1	–	–	–	–	–	–	0	0	0	1	0	0	0	0
$AB\overline{C} \cdot \overline{MNDLY} \cdot \overline{HOLD}$	1	1	0	0	–	–	–	–	0	0	0	0	0	0	1	0	0
$A\overline{B}\,\overline{C} \cdot \overline{MXDLY} \cdot \overline{HOLD}$	1	0	0	–	0	–	–	–	0	0	0	0	0	0	0	1	0
$A\overline{B}\,\overline{C}$	1	0	1	–	–	–	–	–	–	0	0	0	0	0	0	0	0
$AB\overline{C}$	1	1	0	–	–	–	–	–	–	0	0	0	0	0	0	0	1

* ACTIVE LOW inputs—complement columns or use INVERTERs.

† ACTIVE LOW outputs—use INVERTERs.

(b)

Fig. 5-141 (contd.). (b) *p-term table for PLA implementation.*

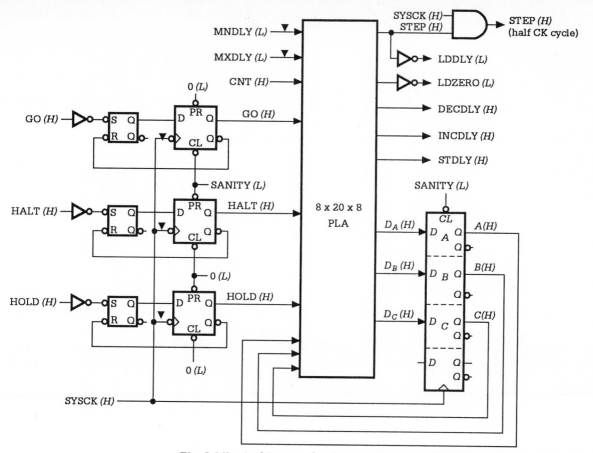

Fig. 5-142. *Architecture for the stepping motor CONTROLLER centered on a PLA and showing input conditioning circuitry.*

Many more useful examples of system-level design could be offered in this section, and each could be used to illustrate specific facets of the design process. This, however, is not practical given the space limitation of a text. As an alternative, we provide three relatively simple system-level design problems (Problems 5.69, 5.70, and 5.71) at the end of this chapter and two (Problems 6.50 and 6.51) at the end of Chapter 6. Also, to help the reader in the decision-making processes of these problems, we give a few suggestions regarding hardware, input conditioning, and so on. Again, we emphasize that these problems are open-ended in the sense that they have no single best solution. Consequently, the reader's design skills and engineering intuition can be exercised.

ANNOTATED REFERENCES

An excellent reference on the subject of synchronous sequential machines is the text by Fletcher. Chapters 7, 8, and 9 are of particular value. The notation and coverage of these three chapters closely parallel those used in the present text.

Fletcher, W. I., *An Engineering Approach to Digital Design*, Prentice Hall, Englewood Cliffs, N.J., 1980.

A good coverage of synchronous FSM design is provided in Chapters 6, 7, 8, and 9 of the book by Comer. The approach to design is roughly the same as used in the present text.

Comer, D. J., *Digital Logic and State Machine Design*, 2nd ed., Holt, Rinehart and Winston, New York, 1990.

The book by Ercegovac and Lang approaches the subject of synchronous sequential machine design from a very different point of view. These authors develop an algorithmic, high-level description of a sequential system and then use modular elements to synthesize it.

Ercegovac, M. D., and T. Lang, *Digital Systems and Hardware/Firmware Algorithms*, John Wiley, New York, 1985.

A conference paper by Chaney et al. draws attention to the possible effects of metastability in synchronizer circuits. Also, papers by Stroll and Veedrick provide a means by which the probabilities of metastability and the failure rate of synchronizers can be calculated.

Chaney, T. J., S. M. Ornstein, and W. M. Littlefield, "Beware the Synchronizer," Dig. COMPCON, San Francisco, Sept. 1972, pp. 317–319.

Stoll, P. A., "How to Avoid Synchronization Problems," VLSI Design, Nov.–Dec. 1982, pp. 56–59.

Veedrick, H. J. M., "The Behavior of Flip-Flops Used as Synchronizers and Prediction of their Failure Rate," IEEE Journal of Solid State Circuits, Vol. SC-15, No. 2 (Apr. 1980), pp. 169–176.

The software tools required to simulate FSMs are briefly reviewed in Chapter 7 of the text by Unger. Useful references are pvovided at the end of that chapter.

Unger, S. H., *The Essence of Logic Circuits*, Prentice Hall, Englewood Cliffs, New Jersey, 1989.

References for software tools used to minimize Boolean functions are provided in the Annotated References at the end of Chapter 3.

A partial listing of data manuals and handbooks for semicustom TTL and CMOS MSI and LSI logic design is provided in the Annotated References at the end of Chapter 4.

PROBLEMS

5.1 In Fig. P5-1 is the logic circuit for the SET-dominant GATED BASIC CELL. Complete the following timing diagram for this device. Neglect gate path delays. (Hint: Operate this machine as a BASIC CELL with an enable, G.)

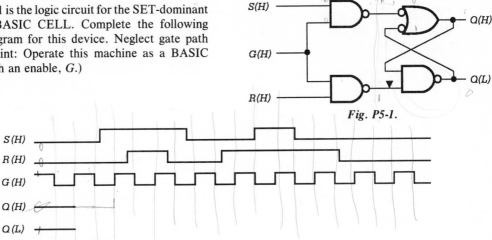

Fig. P5-1.

5.2 (a) Construct the state diagram for the SET-dominant GATED BASIC CELL of Problem 5.1. (Hint: Use Fig. 5-11(c) then AND the $0 \rightarrow 1$ and $1 \rightarrow 0$ branching conditions with G, and apply the SUM rule.)

(b) Use this state diagram together with a SET-dominant BASIC CELL as the memory to redesign the GATED BASIC CELL. What is the obvious difference between this design and that of Problem 5.1? (Hint: Follow the procedure indicated in Sect. 5.6.3.)

(c) Determine the output responses, $Q(H)$ and $Q(L)$, for the design in (b) by using the same input waveforms as in Problem 5.1. Compare these results with those of Problem 5.1 and explain any differences that are observed.

5.3 A PPT N-type FLIP-FLOP has the following operation table:

N	Q_{t+1}
0	$\overline{Q}_t$
1	1

(a) Construct the state diagram and state transition table for this FLIP-FLOP.

(b) Design the logic circuit for a positive pulse-triggered (PPT) N-FF by using a RESET-dominant BASIC CELL and AND/INV logic. Assume that inputs are $N(L)$ and $CK(H)$.

(c) Indicate what (if any) problems may arise if this FLIP-FLOP is not triggered by very narrow CK pulses.

5.4 Repeat the steps of Problem 5.3 for a PPT LN-type FLIP-FLOP defined by the following operation table. Use a SET-dominant BASIC CELL and NAND/INV logic in the design of this FLIP-FLOP. Assume that all inputs arrive ACTIVE HIGH.

L	N	Q_{t+1}
0	0	1
0	1	Q_t
1	0	$\overline{Q}_t$
1	1	0

5.5 Repeat the steps of Problem 5.3 for a negative pulse-triggered (NPT) GM-type FLIP-FLOP defined by the following operation table. Use a RESET-dominant BASIC CELL and AND/OR/INV logic to design this FLIP-FLOP. Assume that all inputs arrive ACTIVE HIGH.

G	M	Q_{t+1}
0	0	1
0	1	1
1	0	0
1	1	$\overline{Q}_t$

5.6 Repeat the steps of Problem 5.3 for an NPT SPOOF-FLOP (SF) FLIP-FLOP that operates according to the following operation table. Use a SET-dominant BASIC CELL and the NS forming logic of your choice for a gate-minimum implementation of this FLIP-FLOP. Assume that all inputs arrive from negative logic sources.

S	F	Q_{t+1}
0	0	1
0	1	$\overline{Q_t}$
1	0	0
1	1	1

5.7 Construct the four-state state diagrams representing the following edge-triggered FLIP-FLOPs:

(a) RET LN-FF (see Problem 5.4).

(b) FET GM-FF (see Problem 5.5).

(c) FET N-FF (see Problem 5.3).

(d) RET SF-FF (see Problem 5.6).

5.8 Find the gate-minimum logic circuit required for the following FLIP-FLOP conversions. Assume that all data inputs arrive ACTIVE HIGH.

(a) Convert an RET D-FF to an RET T-FF.

(b) Convert an RET JK-FF to an RET LN-FF (see Problem 5.4).

(c) Convert an FET GM-FF to an RET D-FF (see Problem 5.5).

(d) Convert a PPT D-FF to an NPT N-FF (see Problem 5.3).

(e) Convert an MS JK-FF to an MS SF-FF (see Problem 5.6).

5.9 **(a)** Convert an RET D FLIP-FLOP to a SET-dominant RET SR FLIP-FLOP. Assume that S and R arrive ACTIVE HIGH.

(b) Explain the difference between this FLIP-FLOP and the GATED BASIC CELL of Problem 5.1.

5.10 The function in the following K-map represents the SET-dominant form of the BASIC CELL. Implement the function Q_{t+1} by using only one 2-to-1 MUX and one AND gate (nothing else). Consider that the MUX has both an ACTIVE HIGH and an

ACTIVE LOW output. Take S and R ACTIVE LOW. (Hint: Compress the K-map into a first-order K-map with R and Q_t as entered variables.)

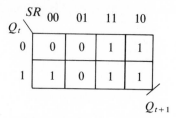

5.11 **(a)** Draw the state diagram which describes the operation of a JK FLIP-FLOP that has an ACTIVE LOW *synchronous* CLEAR input—one that takes effect only on the triggering edge of the CLOCK signal.

(b) Implement the FSM in (a) with an RET D-FF. Assume that both data inputs arrive ACTIVE HIGH. Use NAND logic only.

5.12 **(a)** Without using additional gates, make the appropriate connections to provide the GATED BASIC CELL in Fig. P5-1 with ACTIVE LOW asynchronous PRESET [$PR(L)$] and CLEAR [$CL(L)$] overrides.

(b) Use the resulting MIXED LOGIC timing diagram in Problem 5.1 to demonstrate the effect of $PR(L)$ and $CL(L)$ on the operation of this FSM. Alternate the ACTIVE periods of $PR(L)$ and $CL(L)$ with every other INACTIVE period of $G(H)$ beginning with $PR(L)$ on the first INACTIVE period of $G(H)$.

5.13 Provide ACTIVE LOW asynchronous PRESET and CLEAR overrides for the master/slave D FLIP-FLOP in Fig. 5-32(a). In doing so, consider the nature of the TRANSPARENT D LATCHes and the fact that they are constructed of NAND gates.

5.14 Provide ACTIVE LOW asynchronous PRESET and CLEAR overrides for the master/slave JK FLIP-FLOP in Fig. 5-34(a). Consider the error-catching characteristic of the MS JK-FF.

5.15 Given the circuit in Fig. P5-2, complete the following timing diagram and determine the logic function of f.

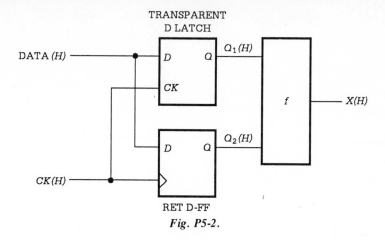

TRANSPARENT
D LATCH

Fig. P5-2.

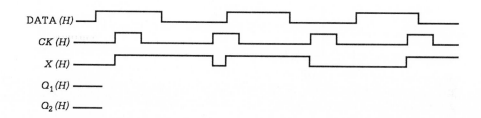

5.16 The MS D FLIP-FLOP was presented in Fig. 5-32(a) without the aid of state diagrams.

(a) Construct the state diagrams for the MS D-FF. (There are two.)

(b) As verification of their validity, use the state diagrams in (a) to design the MS D-FF. Use SET-dominant BASIC CELLs as the MEMORY elements.

5.17 The MS JK FLIP-FLOP in Fig. 5-34(a) can be thought of as two GATED BASIC CELLs which are interconnected by a handshake configuration—that is, a configuration where the outputs of one FSM are the inputs to the other and vice versa.

(a) Treating the MS JK-FF as a handshake configured device, construct the state diagrams that represent this FSM.

(b) As verification of their validity, use the state diagrams in (a) to design the MS JK-FF. Use SET-dominant BASIC CELLs as the MEMORY elements.

5.18 Practical semiconductor static RAM circuits are composed of simple, efficient FLIP-FLOP cells designed on the transistor level. However, it is possible to represent a static RAM cell at the logic level. Shown in Fig. P5-3 is the logic circuit and two block symbols for a static RAM cell. The circuit has

a data input D, an enable input EN, a read/write control $R/\overline{W}$, and a single output, O.

(a) Construct the state diagram for the static RAM cell. (Hint: Consider the similarity between the static RAM cell and the TRANSPARENT D LATCH of Figs. 5-23 and 5-24.)

(b) Demonstrate the operation of the static RAM cell by using a MIXED LOGIC timing diagram. Include all features of the cell.

5.19 Use the symbol in Fig. P5-3(b) to design a *linear* decoding scheme for a 4-bit RAM MEMORY that can store up to four words of one bit each. Plan to OR the four outputs. Assume that data are supplied to the four cells by a single data line and that a 2-to-4 line DECODER with ACTIVE HIGH outputs is used to enable individually the four cells.

5.20 Shown in Fig. P5-3(c) is the symbol for a static RAM cell that has two enable inputs, EX and EY, which replace the single enable input EN in Fig. P5-3(a).

(a) Use the symbol in Fig. P5-3(c) to design a *coincident* decoding scheme for a 16-bit (4 × 4) RAM MEMORY that will store up to 16 words, each of 1 bit. To accomplish this, use one 2-to-4 DECODER to control the EX enable inputs and a second 2-to-4 DECODER to control the EY enable inputs to the 16 RAM cells in the 4

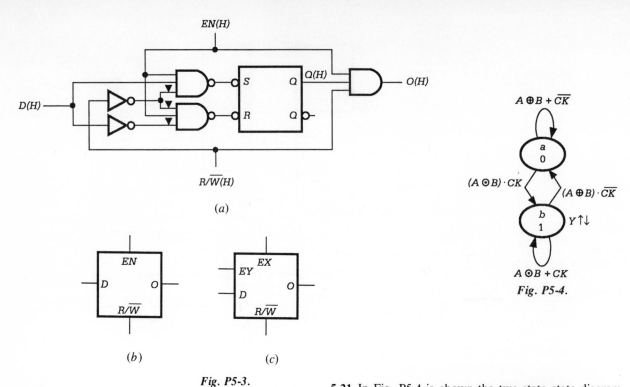

EN(H)

D(H)

Q(H)

O(H)

R/$\overline{W}$(H)

(a)

$$A \oplus B + \overline{CK}$$

a
0

$(A \odot B) \cdot CK$ $(A \oplus B) \cdot \overline{CK}$

b
1   Y↑↓

$A \odot B + CK$

Fig. P5-4.

EN

D   O

R/$\overline{W}$

(b)

EX
EY

D   O

R/$\overline{W}$

(c)

Fig. P5-3.

× 4 MEMORY array. Thus, the intersection of the two enables gives the location of a single MEMORY bit. Plan to OR the 16 outputs. Assume that data are supplied to all cells by a single data line and that the two DECODERs are designed with ACTIVE HIGH outputs.

(b) Indicate the 0100, 0111, and 1010 MEMORY locations and indicate how they are addressed.

(c) Discuss the prospect of using tristate outputs for the MEMORY cells and the possibility of a wired-OR output for the 16-bit MEMORY. Indicate how the logic circuit of Fig. P5-3(a) must be altered to accommodate a tristate output involving the two enable inputs. (See Fig. 4-90).

5.21 In Fig. P5-4 is shown the two-state state diagram for a CLOCKed DATA-CHECK MODULE.

  (a) Design the DATA-CHECK MODULE by using a SET-dominant BASIC CELL and the NEXT STATE logic required for a gate-minimum design. Assume that the inputs arrive ACTIVE HIGH and are synchronous with CLOCK and that the output is ACTIVE HIGH.

  (b) Complete the following MIXED LOGIC timing diagram for the DATA-CHECK MODULE and describe its operation. Take into account the gate delays for the A and B inputs relative to CLOCK.

5.22 Use D FLIP-FLOPs and NAND logic to design a logic circuit for the FSM represented by the state diagram in Fig. P5-5. Assume that all inputs and outputs are ACTIVE HIGH and that triggering occurs on the rising edge of the CLOCK waveform.

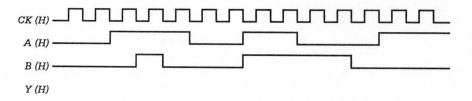

CK (H)

A (H)

B (H)

Y (H)

Timing diagram for Problem 5.21.

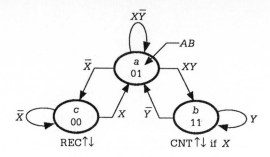

Fig. P5-5.

$X + \bar Y$

5.23 Construct a state diagram for an FSM that will sample (with CLOCK) a continuous stream of synchronized data on an input line X. The FSM is to issue an output Z any time the nonoverlapping sequence . . . 0101 . . . occurs. Make a glitch-free state assignment by using only two state variables. Verify your results. (See Sect. 5.8.2)

5.24 Design an optimum logic circuit for the FSM described in Problem 5.23 by using the following memory elements:

(a) RET D-FFs.

(b) FET JK-FFs.

(c) RET GM-FFs (see Problem 5.5).

Assume that X and Z are both ACTIVE HIGH.

5.25 Construct a four-state state diagram for an FSM that samples (with CLOCK) a continuous stream of synchronized data on an input line X. The FSM is to issue an output SEQDET any time the sequence . . . 0110 . . . occurs. Consider that the sequence may be overlapping as, for example, . . . 01101101 . . . . Make a glitch-free state code assignment by using only two state variables. (See Sect. 5.8.2)

5.26 Design a logic circuit for the FSM of Problem 5.25 by using the following memory elements and minimum external logic:

(a) FET D-FFs.

(b) MS JK-FFs.

(c) FLIP-FLOPs defined by the operation table:

L	Q_{t+1}
0	$\bar Q_t$
1	Q_t

Let X and SEQDET be ACTIVE HIGH.

5.27 Redesign the one- to three-pulse generator of Fig. 5-83(a) by relocating the SW1 and SW0 branching conditions as conditional outputs. In doing so, eliminate all return branching paths except that from the final state of the sequence. Use FET JK-FFs as D-FFs for the MEMORY elements and obtain an optimum design free of output race glitches and HAZARDs. Discuss the timing differences between this design and that of Fig. 5-83(a). Initialize this FSM properly.

5.28 **(a)** The state diagram in Fig. P5-6 describes the operation of a one-input/two-output synchronous FSM. Given the state assignment shown, make a complete output race glitch analysis on this FSM. Show with state diagram segments the origin of each possible glitch and indicate whether it is a (+) or (−) glitch.

(b) Use FET JK-FFs as T-FFs for the MEMORY to design an optimum glitch-free circuit for this FSM. Do not alter the state diagram. Assume that the input is synchronous and ACTIVE HIGH and that the outputs are ACTIVE HIGH. Filter the outputs if necessary.

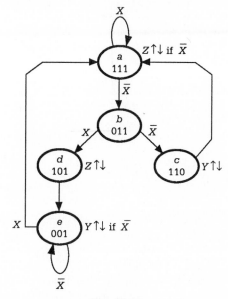

Fig. P5-6.

5.29 An FSM is to be designed that has the state diagram shown in Fig. P5-7.

(a) Given the state assignment shown, run a complete output race glitch analysis on this FSM. Show with state diagram segments the origin of each possible glitch and indicate whether it is a (+) or (−) glitch. Remember to include don't care states in your analysis. Is a HAZARD analysis necessary? Explain.

(b) Use RET D-FFs as the MEMORY and a ROM for the NEXT STATE and output logic to design a glitch-free logic circuit for this FSM. Follow the procedure in Sect. 5.9.3. Include an output filtering stage if it is necessary. Assume that the inputs are synchronous and that they arrive ACTIVE HIGH. Take the outputs ACTIVE LOW. Construct the program table for the ROM and a block diagram for the system.

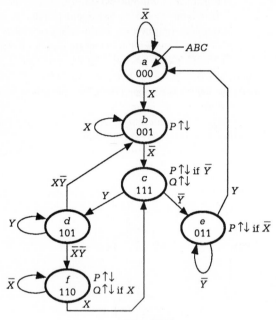

Fig. P5-7.

5.30 Presented in Fig. P5-8 is the state diagram for an FSM having two inputs (S and T) and one output Y.

(a) Analyze this FSM for output race glitches.

(b) Run a HAZARD analysis on this FSM. Assume that this FSM is to be implemented with FLIP-FLOPs whose outputs derive from SET-dominant (NAND-based) BASIC CELLs. Consider both SOP and POS output forming logic and use the gate/input tally as the basis for choosing an optimum design. Do not alter the state diagram.

(c) In consideration of the results in parts (a) and (b), design an optimum glitch-free logic circuit for this FSM. Use FET JK FLIP-FLOPs as the MEMORY. Assume that the inputs are asynchronous and that they arrive ACTIVE HIGH. Take the output ACTIVE HIGH.

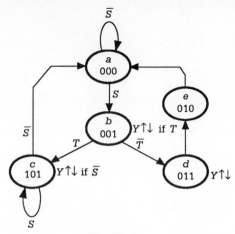

Fig. P5-8.

5.31 An FSM is to be designed that has the state diagram shown in Fig. P5-9.

(a) Run complete output race glitch and HAZARD analyses on this FSM and determine the requirements for glitch-free outputs. Assume that the output forming logic is driven by FLIP-FLOPs whose mixed-rail outputs derive from RESET-dominant (NOR-based) BASIC CELLS.

(b) Use RET D FLIP-FLOPs and a $5 \times 13 \times 5$ FPLA to design an optimum glitch-free logic circuit for this FSM. Filter the output signals only if necessary. Do not alter the state diagram. Initialize this FSM into the 000 STATE and assume that input X is asynchronous, that input Y is synchronous, and that both inputs arrive ACTIVE LOW. Let the outputs both be AC-

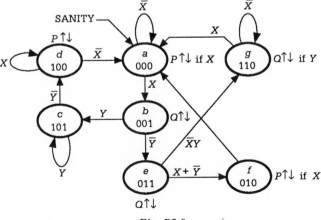

Fig. P5-9.

TIVE HIGH. Construct the p-term table and block diagram for this FSM.

5.32 In Fig. P5-10 is shown a possible synchronizing scheme for an asynchronous input X to an FSM. Observe that the synchronizing D FLIP-FLOP is CLOCKed in phase with the FSM. Use a timing diagram to compare this synchronizing scheme with that for the antiphase scheme of Fig. 5-71(a). Discuss any potential problems that are unique to the in-phase scheme.

5.33 Shown in Fig. P5-11 is the state diagram representing a serial NBCD-to-XS3 converter. A synchronous NBCD waveform is presented on the X input, and a synchronous XS3 waveform is issued on the Z output. Note that all output signals are issued on an exiting condition, and that the NBCD code arrives serially LSB first.

(a) Use RET D FLIP-FLOPs for an optimum design of this converter. Assume that the input X is synchronized to the falling edge of the CLOCK waveform. Initialize the FSM into the 000

STATE, and assume that X and Z are both AC-TIVE HIGH.

(b) Determine if glitches (of any kind) are present in the output signal, Z. If glitches are present, then take the necessary steps to eliminate them. Otherwise, do nothing. In any case, do not alter the state diagram. Complete the logic diagram for part a.

(c) Construct a timing diagram for NBCD input waveforms equivalent to decimal 2 followed by decimal 7. Thus, include logic waveforms for $X(H)$, $CK(H)$, $A(H)$, $B(H)$, $C(H)$, and $Z(H)$. Use a CLOCK waveform with a 50% duty cycle. Explain the difference in ACTIVE durations of the input and output pulses.

5.34 (a) Analyze the FSM given in Fig. P5-12 on page 508 by constructing a fully documented state diagram for it.

(b) Explain the consequences of having no initializing capability for this FSM.

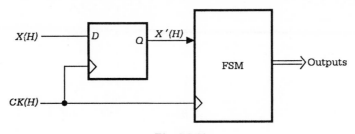

Fig. P5-10.

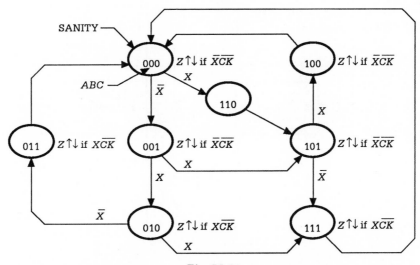

Fig. P5-11.

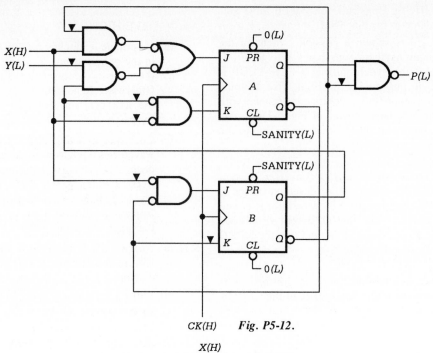

CK(H)   **Fig. P5-12.**

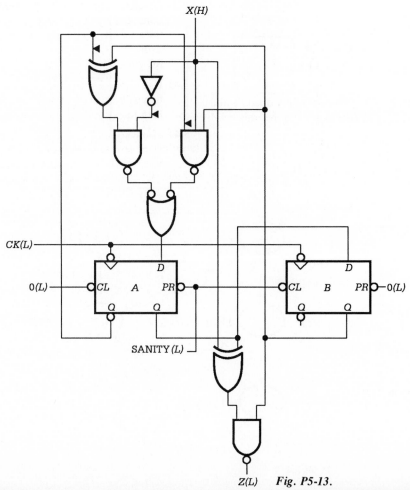

Z(L)   **Fig. P5-13.**

5.35 The following expressions are read from the logic circuit of an FSM implemented with T FLIP-FLOPs, A and B. The FSM has two inputs, X and Y, and one output, Z. Construct a three-state state diagram for this FSM. Merge any redundant states that may be present. (Hint: It may be helpful to convert the T K-maps to D K-maps.)

$$T_A = \overline{AB}\overline{X}\overline{Y} + AXY$$

$$T_B = \overline{A}BXY + \overline{B}\overline{Y} + A\overline{B}$$

$$Z = A\overline{X} + \overline{A}B$$

5.36 Carry out an analysis of the circuit in Fig. P5-13 on page 508 and end with a fully documented state diagram, including the SANITY input. Indicate any problems that this circuit might have.

5.37 The following expressions are read from the logic circuit of an FSM that is implemented with JK FLIP-FLOPs. The FSM has three inputs, X, Y, and Z, and one output, LOAD.

$$J_A = \overline{Y}Z + B\overline{Y} \qquad K_A = \overline{X} + \overline{B}$$

$$J_B = \overline{A}X\overline{Z} + AX \qquad K_B = A\overline{X}\overline{Z}$$

$$\text{LOAD} = \overline{AB}\overline{X}Y$$

From this information construct a fully documented state diagram. Indicate any problems that this FSM may have.

5.38 The following NEXT STATE and output logic expressions are read from a logic circuit for a seven-input/two-output synchronous FSM implemented with three FLIP-FLOPs.

$$D_A = \overline{ABC}\overline{W} + A\overline{C}\overline{N}$$

$$D_B = \overline{A}CX\overline{Y} + \overline{A}B + B\overline{C}$$

$$D_C = \overline{AB}\overline{C}S + \overline{AB}C\overline{Y} + BCH + ABN$$

$$\qquad + ABD + ABC$$

$$Z = \overline{B}CSY + ABC$$

$$K = A\overline{C}N$$

In the foregoing expressions, A, B, and C are the state variables, and Z and K are the outputs.

(a) Construct the state diagram for this FSM.

(b) Analyze this FSM for output race glitches.

5.39 Analyze the logic circuit in Fig. P5-14 and end with a fully documented state diagram. Note that the MEMORY elements are RET L FLIP-FLOPs which are defined by the operation table given in Problem 5.26. Indicate any problems this FSM may have. (Hint: It may be helpful to convert the L K-maps to D K-maps.)

5.40 (a) Collapse the state diagram shown in Fig. P5-15 into a state diagram of three states. It is required that outputs CLRCRY and PSCRY accompany the select outputs S_1, $S_0 = 1$, 1 and that each of the two sets of select outputs shown be assigned to separate states. Assume that any output race glitches that occur after the three-state process is complete have no effect on proper operation of the FSM.

(b) Design two logic circuits for the three-state FSM in (a), one with D-FFs and the other with JK-FFs. Compare the relative complexity of each.

5.41 Derive the expressions for V_x in Fig. 5-78(b). Assume that $R_1 \gg R_2$ and that the switch is opened (or closed) at $t = 0$ only after steady-state conditions are reached. State any simplifying assumptions that are made relative to the Schmitt trigger. (Note: This exercise involves solving a first-order RC circuit.)

5.42 Use FET D FLIP-FLOPs to design a binary counter that will count through states 1, 0, 4, 6, 7, 3, 1, ... Plan to initialize this counter into the 001 STATE.

5.43 Use RET JK FLIP-FLOPs to design the counter of Problem 5.42 as an optimum self-correcting counter. Thus, take the steps necessary to ensure that all unused states lead into the required state sequence but use only a minimum logic to accomplish this. Are SANITY circuit connections now necessary? Explain. (Hint: See the last parts of Sects. 5.7.3. and 5.10.2.)

5.44 Use RET LN type FLIP-FLOPs (defined in Problem 5.4) to design an optimum binary counter that will count through states 6, 3, 4, 2, 1, 6, ... Plan to initialize into STATE 110.

5.45 Use RET JK FLIP-FLOPs to design an optimum 5-bit counter that will count up in binary. Use only fourth-order EV K-maps to accomplish this.

5.46 A counter is to be designed that will count the following sequences in three-bit code:

Sequence I: . . . $0 \rightarrow 1 \rightarrow 3 \rightarrow 2 \rightarrow 0$ . . .   If x

Sequence II: . . . $7 \leftarrow 6 \leftarrow 4 \leftarrow 5 \leftarrow 7$ . . .   If $\overline{x}$

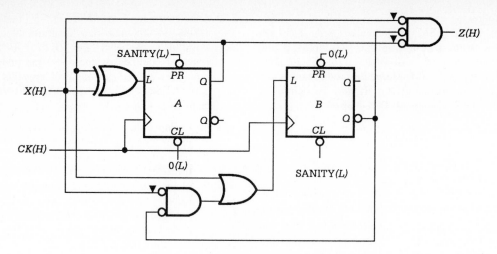

Fig. P5-14.

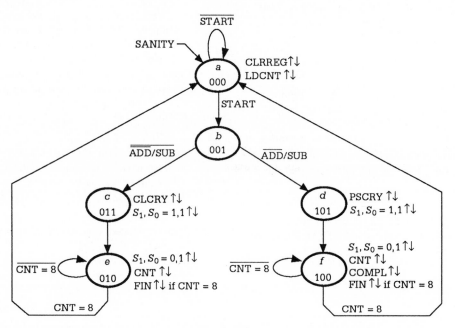

Fig. P5-15.

It is required that the counter change sequence at any time beginning with the complement of the state in the previous sequence. For example, if $x \rightarrow \bar{x}$ while in STATE 2 of sequence I, then sequence II will begin with STATE 5; that is, $010 \rightarrow 101$, and so on.

Implement the counter with RET D or T FLIP-FLOPs, whichever yield the simplest NEXT

STATE forming logic. Assume that x is an asynchronous input and that it arrives ACTIVE HIGH.

5.47 Make use of two 4-to-1 MUXs and RET D FLIP-FLOPs to design an optimum four-state FSM that will operate as a 2-bit up/down binary/Gray code counter with the following mode control and output requirements:

X	Y	Count	Outputs
0	0	UP Gray	UPGRY
0	1	DN Gray	DNGRY
1	0	UP binary	UPBIN
1	1	DN binary	DNBIN

It is required that the mode control inputs, X and Y, be permitted to change only on the falling edge of CLOCK in any state and that the counter counts immediately as required by the input changes. Assume that all inputs and outputs are ACTIVE HIGH. Use a 2-to-4 DECODER for the outputs.

5.48 Use four JK FLIP-FLOPs to design a SHIFT REGISTER that will generate the following continuous pulse train. Run a missing state analysis on this FSM and decide whether or not is necessary to initialize it with a SANITY circuit. (Hint: See the last parts of Sects. 5.7.3 and 5.10.2.)

5.49 Use three RET JK FLIP-FLOPs to design a counter that will generate the following continuous pulse trains:
Run a missing state analysis to determine if it is necessary to initialize this FSM by using a SANITY circuit. (Hint: See the last parts of Sects. 5.7.3 and 5.10.2.)

5.50 A counter is to be designed that will drive the following seven-segment display directly from its seven state variables, that is, from the FLIP-FLOP outputs. The counter must have a count-up control and bit-slice capability so that it can drive a display of seven-segment Arabic numerals greater than nine.

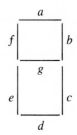

(a) Construct the state diagram for this counter.

(b) Assuming the use of D FLIP-FLOPs, map the state diagram directly into fourth-order EV K-maps and loop out minimum or near minimum cover. (Hint: There are many don't cares that can be used.)

(c) Use an $8 \times 32 \times 8$ FPLA to implement the NEXT STATE and output forming logic. Assume that the inputs and outputs are ACTIVE HIGH. Thus, construct the p-term table and block diagram for this FSM.

(d) Interface this counter with the code converter in Fig. 4-50 so that an NBCD input yields the appropriate Arabic numeral display.

Pulse train for Problem 5.48

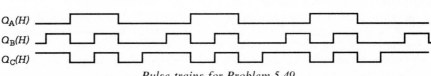

Pulse trains for Problem 5.49

5.51 (a) Design an optimum 4-bit, FET, synchronous, binary up-counter with a count ENABLE and asynchronous parallel load capability. This counter is described in Sect. 5.12 and has the block symbol shown in Fig. 5-138. It is the function of this counter to count up with CLOCK if enabled (EN ACTIVE). But when the LD command becomes ACTIVE, the counter must parallel load immediately via its FLIP-FLOPs' asynchronous PRESET and CLEAR overrides and then count up from that parallel load when LD is DEACTIVATED and when EN is ACTIVE. The counter must have an ACTIVE HIGH CARRYOUT (CO) output for cascading purposes and must have an ACTIVE LOW asynchronous CLEAR for initialization and reset purposes. (Hint: Consider using T-FFs as in Fig. 5-50. Use a collapsed truth table having inputs CL, LD, and PL to obtain the external logic for the asynchronous overrides of each FF.)

(b) Alter the design of part (a) so that each FLIP-FLOP of the counter can initialize either a logic 1 or a logic 0 and can be parallel loaded asynchronously with the $LD(L)$ command any time during counter operation. (Hint: To obtain the external logic, use a collapsed truth table having inputs PS, CL, LD, and PL.)

5.52 Design a ripple up-counter that will count through states ... 0, 1, 2, 3, 4, 0, ... and trigger on the rise of the pulse with outputs from the $Q(H)$ leads. (See Appendix 5.1 for help.)

5.53 Design a ripple down-counter that will count through states ... 6, 3, 2, 1, 6, ... and trigger on the falling edge of the pulse with outputs from the $Q(H)$ leads. Use JK-FFs and initialize this counter into the 001 state. (See Appendix 5.1 for details.)

5.54 Design a serial odd parity detector that will issue a positive pulse Z any time a series string of three nonoverlapping input intervals contains an odd number of positive pulses. Assume that X arrives ACTIVE HIGH and is synchronous with CLOCK. Optimize the design by making the most effective use of the "out of" and "into" rules for state code assignment. Plan to use FET JK FLIP-FLOPs and to initialize into an appropriate state. Let the output be issued ACTIVE LOW.

5.55 In Fig. P5-16 is shown a five-state state diagram with a shift-oriented state assignment. Implement this FSM with a 4-bit UNIVERSAL SHIFT REGISTER described in Figs. 5-101 and 5-102. Assume that X arrives ACTIVE HIGH.

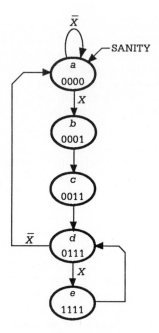

Fig. P5-16.

5.56 For the ROM/MUX implementation of an FSM shown in Fig. 5-125(c), find the logic required to reduce MUX size from 8-to-1 to 4-to-1. Take into account the fact that U is an ACTIVE LOW input.

5.57 Shown in Fig. P5-17 is the state diagram for an FSM having six inputs and four outputs.

(a) With RET D FLIP-FLOPs as the MEMORY, design a glitch-free logic circuit for this FSM that is centered on a STATE DECODER as in Sect. 5.11.2. Assume that $E + G + L = 1$ for STATE 3 and that all inputs and outputs are ACTIVE HIGH.

(b) Alter the design of part (a) so as to use a direct-address MUX approach as in Sect. 5.11.3.

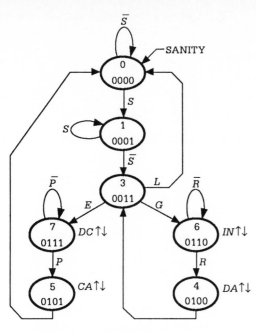

Fig. P5-17.

5.58 A direct-address array logic architecture, as described in Sect. 5.11.4, is to be used to design the FSM in Fig. P5-17. Assume that $E + G + L = 1$ and that all inputs and outputs are ACTIVE HIGH. Use D FLIP-FLOPs as the MEMORY.

   (a) Construct the collapsed program table for a direct-address ROM architecture.

   (b) Construct the program table for a direct-address PLA architecture.

5.59 Without altering the state code assignment, use the state diagram in Fig. P5-17 to design a logic circuit centered on the UNIVERSAL SHIFT REGISTER and a state DECODER as in Fig. 5-127. Assume that $E + G + L = 1$ for STATE 3 and that all inputs and outputs are ACTIVE HIGH.

5.60 Use the parallel loadable up/down-counter, as in Fig. 5-129, to design the FSM of Fig. P5-17. Alter the state code assignments in the state diagram to make the most efficient use of the counter. Assume that $E + G + L = 1$ and that all inputs and outputs are ACTIVE HIGH.

5.61 Shown in Fig. P5-18 is the state diagram for an FSM having five inputs and four outputs. Without altering the state code assignment, design this FSM centered on the UNIVERSAL SHIFT REGISTER and a STATE DECODER as in Fig. 5-127. Assume that inputs U and W arrive ACTIVE LOW and that inputs X, Y, and Z are ACTIVE HIGH. Take output

Q ACTIVE LOW and let outputs P, R, and S be issued ACTIVE HIGH. Also, assume that all inputs arrive synchronized with CLOCK.

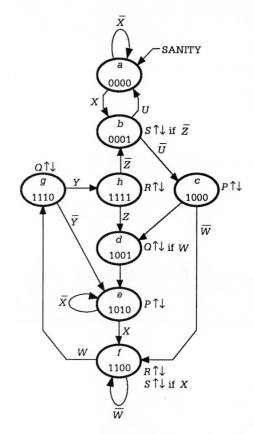

Fig. P5-18.

5.62 Repeat Problem 5.61 for a design centered on a parallel loadable up/down-counter and a STATE DECODER as in Fig. 5-129. Do not alter the state code assignments.

5.63 Shown in Fig. P5-19 is an FSM which has been implemented with a parallel loadable up-counter. Analyze this FSM by constructing its state diagram(s). Indicate any problems this FSM may have.

5.64 Shown in Fig. P5-20 is an FSM which has been implemented by using a UNIVERSAL SHIFT REGISTER. Analyze this FSM by constructing its state diagram(s). Indicate any problems this FSM may have.

5.65 Repeat Problem 5.61 for a design based on the programmable architecture indicated in Fig. 5-130. Do not alter the state code assignments in the state diagram.

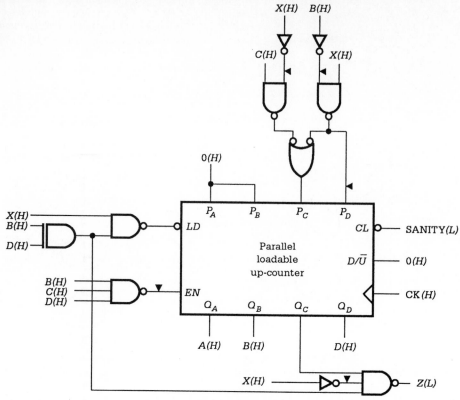

Fig. P5-19.

5.66 Repeat Problem 5.61 for a design based on the programmable architecture shown in Fig. 5-133. Do not alter the state code assignments in the state diagram.

5.67 An FSM is to be designed that will issue an output Z according to the timing diagram in Fig. P5-21. In words, if CLOCK samples S ACTIVE with X and Y INACTIVE, then Z is issued on Y following XY or on X following $\overline{X}Y$ provided that all these events are spaced one CLOCK period apart. If these conditions are not met, then Z will not be issued, and the FSM must wait for S to be sampled INACTIVE before it can return to the initial state and start the process over again. The output Z is issued for one CLOCK period only after which the FSM must return to the initial state.

(a) Construct the state diagram for this FSM and make a 3-bit state code assignment that is free of output race glitches. Plan to initialize into the 000 STATE.

(b) Use an architecture of your choice to design a logic circuit for this FSM. Assume that S, X and Y are ACTIVE HIGH inputs that are synchro-

nized to CLOCK and that the output Z is ACTIVE LOW. Design for a glitch-free output without using a filtering stage.

5.68 A traffic light controller is to be designed that will operate a traffic light at the intersection of a main highway and a less frequently used farm road. Traffic sensors are placed on both the highway and the farm road to indicate when traffic is present. If no traffic is sensed on the farm road, traffic on the highway is allowed to flow. But when a vehicle activates the sensor on the farm road, the signal will switch immediately if the traffic sensor on the highway is not active. Otherwise, the vehicle on the farm road must wait for 30 seconds or until the highway is clear, whichever occurs first. Once the farm road vehicle has gone, the system must permit traffic to resume on the highway.

In designing the controller, it is assumed that two interval timers are available, one for the 30-second interval and the other for a 5-second yellow light interval. The exact nature of these timers is not important (at this point) except that they each accept an input to signal the start of the interval and that they return an output to indicate when the spec-

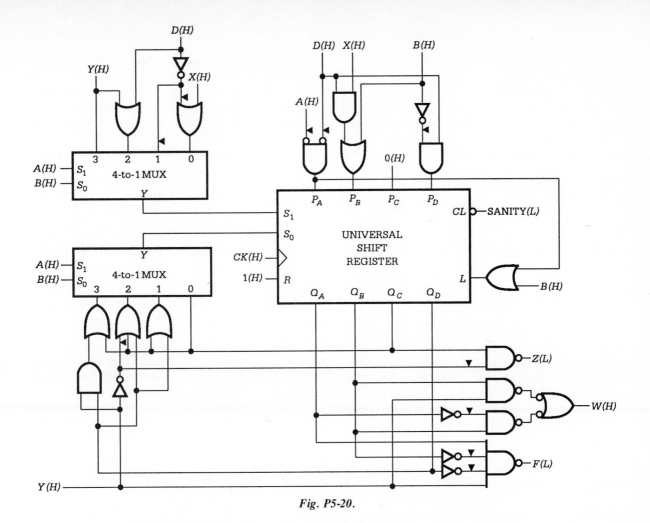

Fig. P5-20.

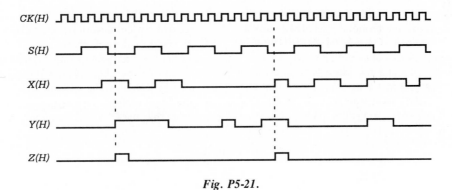

Fig. P5-21.

ified time has elapsed. Upon receiving the input signal, the timer resets and begins timing. At the end of the specified time, the output signal is ACTIVATED and remains ACTIVE until the input signal is DEACTIVATED.

(a) Construct a suitable state diagram for the traffic light controller and make any reasonable state code assignment applicable for an architecture centered on a UNIVERSAL SHIFT REGISTER. Use the architecture featured in Fig. 5-127 for a glitch-free design of this FSM. Assume that all inputs and outputs are ACTIVE HIGH. Use the following abbreviations and take F and H as asynchronous inputs.

Inputs: Farm = F; Highway = H; 30 second = 30; 5 second = 5.

Outputs: Farm, red = FR; farm, green = FG; farm, yellow = FY; Highway, red = HR; highway, green = HG; highway, yellow = HY; start 30 sec = S30; start 5 sec = S5.

(b) Repeat part (a) for a design of this FSM centered on a parallel loadable up/down counter as in Fig. 5-129. Thus, make a state code assignment that is best suited to this architecture.

(c) Repeat part (a) for a design centered on a PLA architecture as in Fig. 5-122. Use RET D-FFs as the MEMORY memory elements and plan to minimize the number of FLIP-FLOPs required for this design by keeping the MSB of the state codes INACTIVE.

5.69 Design at the system level an optimum or near optimum multipulse generator that will issue, on the PULSE output, 0 to 99 clean (glitch-free), evenly spaced pulses which have an ACTIVE duration the same as that of the system CLOCK (SYSCK). To do this, design and interconnect two NBCD down-counters each with ACTIVE LOW BORROWOUT (BO) capability. Also, consider that the START signal, required to initiate the pulse generation process, is an asynchronous input. Assume that the pulse count settings are made prior to the START signal and that these settings are loaded asynchronously into the counters (via the FLIP-FLOP asynchronous PR and CL overrides) on a LOAD(L) command from the CONTROLLER to the LD(L) inputs of the counters. Let the counters be rising edge triggered on their EN(H) input. Take care to initialize the system properly and debounce the START switch input. Use a CNT(H) command from the CONTROLLER to initiate the pulse generation process and an END(L) signal from the counters to end it. (Note: The use of flowcharts and timing diagrams may be necessary to gain a complete understanding of the system operation. Also, it will be necessary to connect two counters in series by connecting the BO(L) output of one to the EN(H) of the other. See Problem 5.51 for a counter design similar to that required in this problem but with EN = CK.)

5.70 Design at the system level an optimum or near optimum 8-bit parallel-to-serial ADDER/SUBTRACTOR system. To accomplish this, consider the use of a single FULL ADDER, a D FLIP-FLOP with PRESET and CLEAR overrides, a controlled INVERTER, and a binary counter to ADD or SUBTRACT serially (in 2's complement) two 8-bit numbers which are stored in separate 8-bit UNIVERSAL SHIFT REGISTERs. Assume that the system begins operation on receipt of the asynchronous command START, and that the results are presented serially. Upon completion of the 8-bit addition or subtraction process the system is to issue the finish signal FIN. Take care to initialize the system properly and to debounce any switch inputs. (Hint: Refer to Problem 5.40 and Fig. P5-15 and make any appropriate changes.)

5.71 Complete the system-level design of the traffic light control system described in Problem 5.68. Use the architecture described in part c of Problem 5.68. Pay particular attention to timer requirements as they pertain to the choices of counters and clock frequency. Initialize the system properly and deal with any asynchronous inputs that may be present. Assume that the CLOCK frequency is 13.1 kHz.

APPENDIX 5.1

Ripple Counters

Ripple counters are characterized by the use of T FLIP-FLOPs which are series triggered. Each T FLIP-FLOP is triggered off the output from the previous one except for the LSB FLIP-FLOP, which is triggered off of the CLOCK signal. For this reason, ripple counters are often called "asynchronous" counters.

Shown in Fig. A5-1 are the state diagrams, logic circuit, and timing diagram for an n-bit ripple down-counter—one that counts down in binary. Note that a ripple counter is also a divide-by-2^n counter. Thus, the first stage (T-FF$_0$) reduces the CLOCK frequency to $f_{CK}/2$, the second stage (T-FF$_1$) to $f_{CK}/4$, and so on, and the last stage reduces the CLOCK frequency to $f_{CK}/2^n$, where n is the number of FLIP-FLOPs in the counter. This points to one of the main uses of ripple counters—frequency division. Such counters can be very useful in applications where specific time delays must be built into the operation of a particular device. However, all ripple counters suffer from cumulative noise generation due to the series (ripple) triggering. This fact may limit the usefulness of these counters.

The ripple counter in Fig. A5-1 is designed to count down in binary, that is, to count through states . . . $2^n - 1$, $2^n - 2$, . . . , 7, 6, 5, 4, 3, 2, 1, 0, $2^n - 1$, ... A ripple counter can be designed to count up or down according to the relation

$$UP = (RET) \odot (Q_{CK}) \oplus (Q_o)$$

Thus, the counter will count up (UP = 1) if the T-FFs are triggered on the falling edge (RET = 0) off of the $Q(H)$ outputs ($Q_{CK} = 1$) when the count is read from the $Q(H)$ outputs ($Q_o = 1$). Further change in any one of the

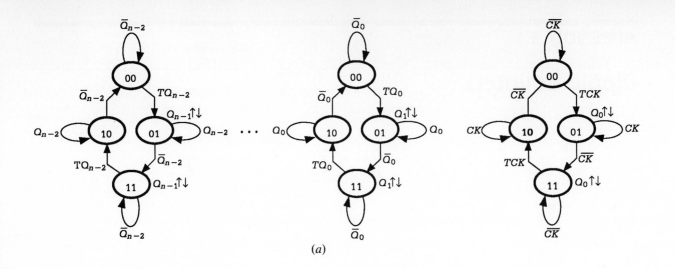

(a)

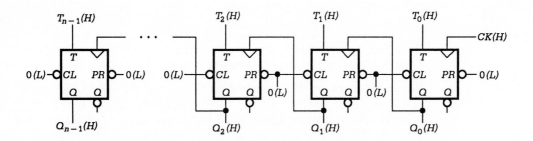

(b)

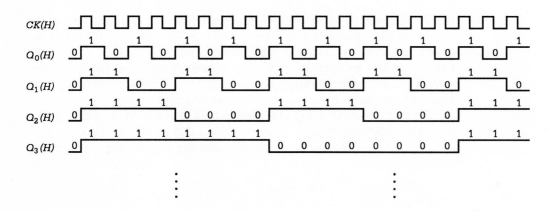

(c)

Fig. A5-1.

three parameters (RET, Q_{CK}, or Q_o) will cause the counter to count down (UP = 0). For example, if RET = 0, Q_{CK} = 1, and Q_o = 0, the ripple counter will count down in binary.

Ripple counters can also be designed to count through fewer states than 2^n by connecting external logic to the appropriate PRESET and CLEAR overrides. As an example, suppose it is required that a ripple counter count through states . . . 0, 1, 2, 3, 4, 0, ... In this case, $CL_2(L) = CL_0(L) = AC(L)$ while all the other asynchronous overrides are maintained at $0(L)$. These connections are necessary since the counter must ripple through STATEs 5 → 6 → 7 in going from STATE 4 to STATE 0. Thus, by applying the generic form of the state transition table for the basic cell to the 101 → 000 transition, the R K-maps yield the minimum cover $A \cdot C$ when STATEs 6 and 7 are taken as don't care states. All other S and R K-maps yield logic 0.

Initialization of ripple counters can be accomplished by using SANITY circuits in a manner similar to that used for synchronous counters. However, there is one major difference that must be dealt with. Ripple counters whose state sequences are other than a 2^n state binary up- or down-count must have external logic connected to the PS and CL overrides of the FLIP-FLOPs so as to direct the count, as indicated. For these counters, it is necessary to OR the SANITY line with the external logic before connecting it to the appropriate PR or CL override terminal.

CHAPTER 6

Asynchronous Sequential Machines

6.1   INTRODUCTION

In Chapter 5 the emphasis was directed toward synchronous sequential machine design. There we developed a rather thorough understanding of the concepts necessary for the meaningful design of these machines. Now it is necessary to move on to another important type of machine—the asynchronous FSM (finite state machine). In Fig. 6-1 we provide an overview of the various types of digital machines. Observe that combinational logic machines can be either synchronous or asynchronous depending on how they are used, although by themselves they must be considered as asynchronous machines.

A major aim of this chapter is, of course, to develop a working level understanding of asynchronous finite state machines and their design. But the mission of this chapter is really much broader than that. In the course of the various discussions, the reader will develop a better understanding of those concepts involved in synchronous machine design. In fact, an understanding of asynchronous sequential machines is required before synchronous sequential machines can be fully understood.

So why have we waited until now to present asynchronous sequential machine concepts and design methodologies? The answer lies in the fact that the treatment of asynchronous FSMs requires a certain level of maturity which can be obtained only through experience with the simpler synchronous FSMs. The study of asynchronous sequential machines forces us to deal with the complexities of sequential machines in greater depth.

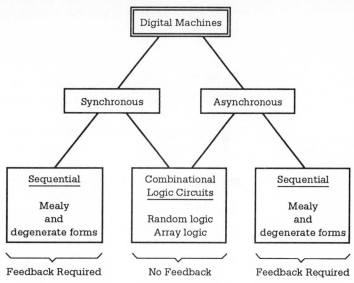

Fig. 6-1. *Breakdown of the various classes of digital machines.*

6.1.1 Features of Asynchronous FSMs

As we shall soon learn, all sequential machines have certain characteristics in common. However, there are some features owned more or less exclusively by asynchronous FSMs. They are

- The presence of MEMORY in the absence of the familiar CLOCKed FLIP-FLOP.
- The appearance of the asynchronous machine as a combinational logic circuit with feedback.

Other more subtle features characterize asynchronous FSMs. These features include the possible existence of certain undesirable effects such as logic noise due to HAZARDs in both the NEXT STATE (NS) and output forming logic sections, CRITICAL RACES (races that can produce error transitions), and ENDLESS CYCLES (oscillations). As we know, logic noise can also be generated in the output forming logic of synchronous sequential machines. However, we also know that such noise can be easily eliminated by using output holding registers. The action of the CLOCK in the memory stage and in the output holding register "filters out" any undesirable logic noise that may exist in the NEXT STATE and output logic sections, respectively, and ENDLESS CYCLES and CRITICAL RACES are eliminated by the SYSTEM CLOCK control of the (NS) data which must pass through the MEMORY elements (FLIP-FLOPs). In an asynchronous system, CLOCK if present, cannot play the role of filtering out logic noise and coordinating the passage of the NS data. Consequently, the benefits of CLOCK, which we take for granted in synchronous machines, no longer exist in asynchronous FSMs. The reality is that HAZARDs, CRITICAL RACES, and ENDLESS CYCLEs can occur in asynchronous FSMs and, if present, can cause these machines to malfunction. A detailed study of these timing defects and the actions required to eliminate them constitute a major portion of this chapter.

6.1.2 Need for Asynchronous FSMs

It is perhaps natural to believe that the data processing in and passage through a sequential machine must be regulated by some periodic sampling function, the SYSTEM CLOCK. This, of course, is a requirement of the synchronous sequential machine. But we never questioned the absence of a CLOCK in the combinational logic circuits of Chapter 4, yet these circuits are asynchronous machines of a type—those without feedback (i.e., non-sequential). Why then the concern about the need for a CLOCK to regulate synchronous sequential operations? And when is it advantageous, if ever, to perform sequential operations asynchronously? The complete answers to these questions will be forthcoming but only after we have considered most of the contents of this chapter. For now let it suffice to say that it may be desirable or even necessary to use asynchronous designs for the following reasons:

1. The speed requirements of the system may exceed the capability of the synchronous machine. Properly designed, a synchronous FSM can only approach (not equal) the speed of a properly designed asynchronous FSM performing the same sequential operation(s). There are exceptions to this rule.
2. Use of a SYSTEM CLOCK to synchronize a given sequential machine may not be possible or desirable. CLOCK distribution problems (CLOCK SKEW) may seriously limit the use of synchronous designs, particularly in complex digital systems operated at high frequencies.
3. Since FLIP-FLOPs and CLOCK oscillator circuits are absent, an asynchronous design may occupy less real estate on an IC chip than an equivalent synchronous design. However, this statement requires some qualification and will be discussed at a later point in this chapter.

Clearly, there is potential for use of asynchronous machines. In fact, it is predictable that designers *will* become more familiar with this type of machine, that asynchronous design techniques *will* improve, and that asynchronous FSM methods *will* play an important role in the design of future superhigh-speed microprocessors and computers. It is the judgment of many digital designers that synchronous IC system design (in general) has been pushed to its practical limit and that new approaches to digital design must be developed if future expectations are to be realized.

6.2 THE LUMPED PATH DELAY MODELS FOR ASYNCHRONOUS FSMs

In synchronous FSMs the MEMORY stage is formed by using FLIP-FLOPs. But if asynchronous FSMs are characterized by the absence of such devices as FLIP-FLOPs, how does memory manifest itself in these machines? The answer to this question lies in the fact that data transport through an FSM is not instantaneous. Propagation time delays are an inherent part of any circuit, and it is these path delays that constitute the memory stage of an asynchronous FSM. Recall that this is precisely the basis for the heuristic development of the BASIC CELL presented in Sect. 5.5.1. It is this heuristic development that provides the basis for the generalized and more formal treatment which follows.

Consider the models for the asynchronous Mealy (general) and Moore sequential machines shown in Fig. 6-2. These models, which we call *lumped path delay* (LPD) models, are applicable to FSMs said to operate in the *fundamental mode*. Operation in the fundamental mode assumes that no

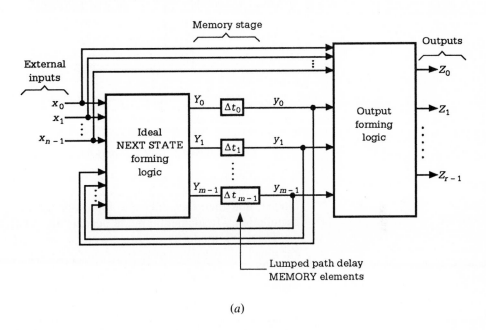

(a)

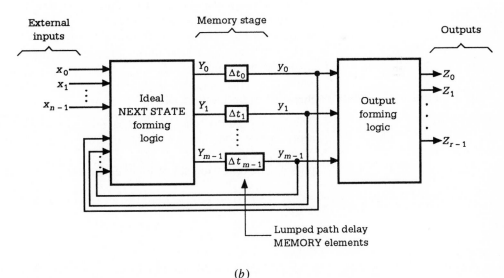

(b)

Fig. 6-2. *The lumped path delay (LPD) model for asynchronous sequential machines. (a) The Mealy (general) finite state machine. (b) Moore (degenerate Mealy) machine.*

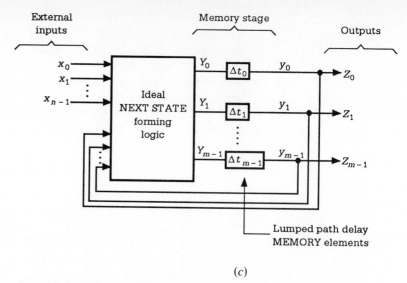

External inputs | Memory stage | Outputs

x_0
x_1
$\vdots$
x_{n-1}

Ideal NEXT STATE forming logic

Y_0 — Δt_0 — y_0 — Z_0
Y_1 — Δt_1 — y_1 — Z_1
$\vdots$
Y_{m-1} — Δt_{m-1} — y_{m-1} — Z_{m-1}

Lumped path delay MEMORY elements

(c)

Fig. 6-2 (contd.). *(c) Most degenerate Mealy model with feedback.*

external input to an FSM makes a change until all internal signals are stabilized, and then that only one input changes at a time. Because of these assumptions, fundamental mode designs are relatively simple to carry out.

The LPD model is characterized by the fact that the NEXT STATE forming logic is treated as ideal (free of path delays) and that the propagation time delays are separated out into a minimum number of distinct lumped memory elements, $\Delta t_0, \Delta t_1, \Delta t_2, \ldots, \Delta t_{m-1}$, such that one delay element appears in each gate/feedback loop. It is these lumped MEMORY elements taken in toto that constitute the MEMORY stage for the asynchronous FSMs. Thus, since data passage through an asynchronous FSM cannot be instantaneous, there is data storage however brief it might be. Note that the models in Fig. 6-2 have traditionally been labeled "fundamental mode models" even though the term "fundamental mode" applies strictly to an operating mode not a model.

6.2.1 Functional Relationships and the Stability Criteria

The Y and y parameters given in Fig. 6-2 are commonly used for LPD models representing asynchronous FSMs operated in the fundamental mode. These parameters are functionally related to each other and to the inputs and outputs by the following set of logic equations written in subscript notation:

$$Y_j(t) = y_j(t + \Delta t) \qquad\qquad j = 0, 1, \ldots, m - 1$$

$$Y_k(t) = Y_k[x_i(t), y_j(t + \Delta t)] \qquad k = 0, 1, \ldots, m - 1 \qquad (6\text{-}1)$$

and

$$Z_l(t) = Z_l[x_i(t), y_j(t)] \qquad\qquad l = 0, 1, \ldots, r - 1$$

where, for each k and l, $i = 0, 1, \ldots, n - 1$ and $j = 0, 1, \ldots, m - 1$. Here, the four parameters are defined by

$$x_i = x_{n-1}, \ldots, x_1, x_0 = \text{input state}$$

$$Y_k = Y_{m-1}, \ldots, Y_1, Y_0 = \text{NEXT STATE}$$

$$y_j = y_{m-1}, \ldots, y_1, y_0 = \text{PRESENT STATE}$$

and

$$Z_1 = Z_{r-1}, \ldots, Z_1, Z_0 = \text{output state}$$

the components of which have been arranged in positionally weighted form so as to represent binary WORDs. Notice that the inputs, x_i, to an FSM may themselves be multivariable functions, implying that one FSM may be controlled by another FSM.

Inspection of Eqs. (6-1), or the LPD models of Fig. 6-2, indicates that corresponding NEXT STATE and PRESENT STATE variables are separated in time by distinct lumped delay MEMORY elements, Δt_j. This leads directly to the important *stability criteria* stated as follows:

If the PRESENT STATE is logically equal to the NEXT STATE at any point in time, then

$$Y_j(t) = y_j(t) \quad or \quad Y_t = y_t \quad \textit{(for all j)} \qquad (6\text{-}2)$$

and the asynchronous FSM is stable in that state. Or if the PRESENT and NEXT STATES are not logically equal, then

$$Y_j(t) \neq y_j(t) \quad or \quad Y_t \neq y_t \quad \textit{(for any j)} \qquad (6\text{-}3)$$

and the asynchronous FSM is unstable in that state and will transit to another state.

The functional relations given by Eqs. (6-1) and the stability criteria expressed by Eqs. (6-2) and (6-3) define the LPD (fundamental mode) models given in Fig. 6-2. Here, the presence of a lumped MEMORY element for each gate/feedback loop ensures that all path delays within the NS forming logic are represented. A much less attractive alternative is the *distributed path delay model* which requires a MEMORY element for each gate and as many state variables. The LPD model has the decided advantage of simplicity—it requires a minimum of lumped MEMORY elements and hence a minimum number of state variables. Use of the distributed path delay model would be prohibitively difficult for all but the simplest of FSMs. Problem 6.24 at the end of this chapter serves as a simple example.

6.2.2 The State Transition Table for the LPD Model

A state transition table for the LPD model can be derived from the stability criteria given by Eqs. (6-2) and (6-3). The results are shown in Fig. 6-3 where we have labeled as stable the condition for which $y_t = Y_t$, and as unstable the condition for which $y_t \neq Y_t$. Here, $y_t \rightarrow y_{t+1}$ represents a transition from the PRESENT STATE to the NEXT STATE, implying that $y_{t+1} =$

Y_t	$y_t \rightarrow y_{t+1}$	
0	$0 \rightarrow 0$	Stable
0	$1 \rightarrow 0$	Unstable
1	$0 \rightarrow 1$	Unstable
1	$1 \rightarrow 1$	Stable

$y_t \rightarrow y_{t+1}$	Y_t
$0 \rightarrow 0$	0
$0 \rightarrow 1$	1
$1 \rightarrow 0$	0
$1 \rightarrow 1$	1

      (a)   (b)

Fig. 6-3. *The state transition table for the LPD model (a) as derived from the stability criteria given by Eqs. (6-2) and (6-3) and (b) altered to the form familiar for FLIP-FLOPs.*

Y_t. Notice the similarity between the state transition table of Fig. 6-3(b) and that for the D FLIP-FLOP presented in Fig. 5-23(c). The state transition table for the LPD model is essential to the design of asynchronous FSMs operated in the fundamental mode and will be used extensively throughout this chapter.

6.3 THE STATE DIAGRAM, K-MAPS, AND NEXT STATE MAP

The sequential behavior of any FSM (synchronous or asynchronous) is revealed most effectively by a fully documented state diagram representing the algorithm for that machine. However, the state diagram itself does not indicate whether the machine is synchronous or asynchronous, nor does it suggest what model is to be used for the design. While the state diagram is a most essential part of the design process, it alone is not sufficient to carry out a complete design project. In this section we will focus mainly on the salient features of the state diagram and its graphical representations, the EV K-maps and NEXT STATE map, as they apply to asynchronous FSM design. In doing so we will be laying the groundwork for what will prove to be a powerful approach to asynchronous FSM design. Our plan of attack is to approach this subject from a general perspective, for reference purposes, followed immediately by a simple example which illustrates the important points.

The Fully Documented State Diagram. The essential features of the fully documented state diagram are illustrated in Fig. 6-4(a), which represents a portion of some algorithm for a general *n*-input/*m*-output FSM. To reinforce what was learned in Sect. 5.4, we draw the reader's attention to the following features of the state diagram in Fig. 6-4(a):

- The state code assignment and literal name are given within each enclosed state symbol.
- The branching paths of the FSM are given by arrows *out* of each state.
- The branching conditions for each outgoing path is a Boolean expression (SOP or POS) in terms of the external input variables, x_i.
- The output notation (if any) is placed to the side of the enclosed state symbol by using the familiar up/down arrow ($\uparrow \downarrow$) notation. Each output expression is read from an output K-map or from the state diagram, and

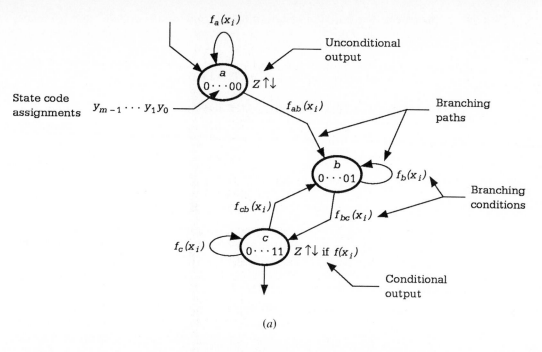

(a)

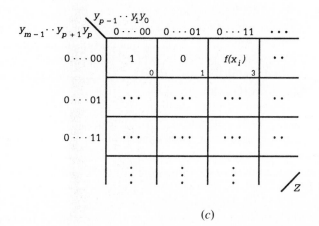

(b)

(c)

Fig. 6-4. Representation of a portion of some arbitrary algorithm for an asynchronous FSM. (a) The major features of a fully documented state diagram. (b) EV K-map for the kth NEXT STATE variable. (c) EV K-map for the output Z.

is generally a function of both the external input variables and feedback variables. In Fig. 6-4(a) both unconditional and conditional outputs are represented. The conditional output notation expressed by $Z \uparrow \downarrow$ if $f(x_i)$ indicates that the output Z will be issued only if the input conditions $f(x_i)$ are satisfied while the FSM is in STATE $\textcircled{c}$.

The SUM Rule and Mutually Exclusive Condition. In constructing a *fully documented* state diagram, it is essential that *all possible* outgoing branching conditions (BCs) from each state be accounted for. All possible BCs from any *n*-way branching j^{th} STATE are accounted for if, and only if,

$$\sum_{i=0}^{n-1} (BC)_{i \leftarrow j} = 1 \qquad (6\text{-}4)$$

in an SOP state diagram. In words, Eq. (6-4) requires that all n outgoing (as opposed to incoming) branching conditions from the jth STATE of an SOP state diagram Boolean SUM (OR) to logic 1. The reader will recall that Eq. (6-4) was previously expressed by Eq. (5-3) and was dubbed the "SUM rule." The dual of Eq. (6-4), which could be called the "product rule," would require that all outgoing branching conditions AND to logic 0 in a POS state diagram. However, this dual requirement would rarely (if ever) be applied since state diagrams are conventionally MINTERM code based as are K-maps.

There is yet another requirement of the outgoing branching conditions from a given state. With few exceptions all possible branching conditions from a given state [as expressed in Eq. (6-4)] must be *mutually exclusive* such that no possible branching condition can cause two different branching paths from that state. Therefore, all possible outgoing branching conditions from any *n*-way branching j^{th} STATE of an SOP state diagram are each mutually exclusive if, and only if,

$$(BC)_{i \leftarrow j} = \overline{\sum_{\substack{k=0 \\ k \neq i}}^{n-1} (BC)_{k \leftarrow j}} \qquad i = 0, 1, \ldots, n - 1 \qquad (6\text{-}5)$$

Equation (6-5) states that the branching condition from the jth STATE to the ith STATE must be the complement of the Boolean SUM of all remaining branching conditions from the jth STATE for *n*-way branching, and that this requirement must apply to each of the *n* branching conditions in turn. The mutually exclusive principle, which was previously stated in Sect. 5.4.1, is proved and applied in Appendix 6.1 to this chapter.

Equations (6-4) and (6-5) must usually be satisfied by each state in any fully documented state diagram. If unaccounted for branching conditions occur, the FSM could transit to some undefined state outside of the state diagram. Of course, if such unaccounted for branching conditions are not possible, then Eq. (6-4) need *not* be satisfied. Likewise, if an overlapping branching condition is not possible, then the mutually exclusive requirement of Eq. (6-5) need *not* be satisfied. There will be occasions later in this chapter when it will be desirable to violate the requirements of Eq. (6-4) or Eq. (6-5). When this happens we will draw the reader's attention to them.

EV K-maps. For design purposes the EV K-maps representing an asynchronous FSM are a necessary extension of the state diagram. They commit the designer to a specific model. When using the LPD model, the K-maps for the NEXT STATE variables are easily constructed by applying the state transition table of Fig. 6-3(b) to each state of the state diagram. One EV K-map is constructed for each NEXT STATE variable and takes the general form of the mth order K-map for the kth NS variable shown in Fig. 6-4(b). Notice that the PRESENT STATE variables form the coordinate axes of the K-map and that the cell entries are functions only of the external inputs, x_i. Once constructed, these EV K-maps are used to obtain the minimum cover needed for HAZARD analysis and optimum NEXT STATE forming logic of the FSM. So as to take full advantage of the simplicity that the EV method has to offer, it is important that the reader hold to the format of Fig. 6-4(b). That is, zero-order compression maps (with 1's and 0's) should rarely, if ever, be used for this purpose. The useful limit of K-map application is reached soon enough with the EV K-map method. Use of conventional mapping methods would make all but the simplest designs prohibitively difficult.

The output expressions for very simple FSMs can usually be read directly from the state diagrams. More complex FSMs require the use of an EV K-map for each output variable as we learned in Chapter 5. Figure 6-4(c) shows the general format which should be used, one similar to that used for the NEXT STATE variables.

The NEXT STATE map. The detailed sequential behavior of any FSM is elegantly contained in the state diagram for that machine. In fact, the state diagram, as treated in this text, is the single most important expression of the algorithm for the sequential machine. However, there are certain complexities unique to asynchronous FSMs which can be identified in state diagrams only after the designer has achieved a certain skill level in reading these diagrams. As a teaching aid in developing this skill level, we recommend a limited use of the NEXT STATE map, which is nothing more than a graphical representation of the state diagram. In Fig. 6-5(a) we give a set of hypothetical branching conditions for the state diagram segment of Fig. 6-4(a). The NEXT STATE map is constructed from these branching conditions by first transferring the branching data in Fig. 6-5(a) to the map in MINTERM code, and then by applying the K-map domain concept and stability criteria of Eqs. (6-2) and (6-3).

It should be helpful to summarize the important features of the NEXT STATE map. The features to note about the NEXT STATE map are

- The horizontal axis has coordinates of the n-bit external input code $x_{n-1} \ldots x_1 x_0$, the vertical axis has coordinates of the m-bit PRESENT STATE (feedback) code $y_{m-1} \ldots y_1 y_0$, and the cell entries represent the m-bit NEXT STATE code $Y_{m-1} \ldots Y_1 Y_0$.
- The NEXT STATE code, in effect, tells the FSM where to go next.
- Circled states are stable according to the stability criterion of Eq. (6-2) while those left uncircled are unstable according to Eq. (6-3).
- The curved horizontal arrows indicate FSM response due to changes in the external input variables.

$$f_a(x_i) = \bar{x}_{n-1} \ldots \bar{x}_1 x_0$$

$$f_{ab}(x_i) = \bar{x}_{n-1} \ldots \bar{x}_1 \bar{x}_0 + \bar{x}_{n-1} \ldots x_1 x_0$$

$$= \bar{x}_{n-1} \ldots (x_1 \odot x_0)$$

$$f_b(x_i) = \bar{x}_{n-1} \ldots \bar{x}_1 \bar{x}_0 + \bar{x}_{n-1} \ldots \bar{x}_1 x_0$$

$$= \bar{x}_{n-1} \ldots \bar{x}_1$$

$$f_{bc}(x_i) = \bar{x}_{n-1} \ldots x_1 x_0$$

$$f_c(x_i) = \bar{x}_{n-1} \ldots x_1 x_0$$

$$f_{cb}(x_i) = \bar{x}_{n-1} \ldots \bar{x}_1 x_0$$

(a)

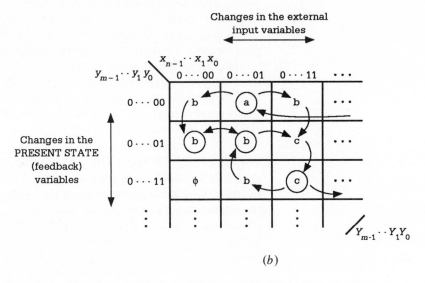

(b)

Fig. 6-5. *(a) Sample branching conditions and (b) NEXT STATE map representation for the generalized FSM of Figure 6-4.*

- The curved vertical arrows indicate FSM response due to changes in the feedback variables.
- The don't care (ϕ) shown in the 0 . . . 110 . . . 00 cell indicates that that NEXT STATE code is unused.

The NEXT STATE map of Fig. 6-5(b) yields information that could have been easily read from the state diagram whose branching conditions are expressed in Fig. 6-5(a): The FSM in STATE ⓐ will be in a holding condition (i.e., will branch onto itself) for input conditions $\bar{x}_{n-1} \ldots \bar{x}_1 x_0$; or it will transit to STATE ⓑ under one of the branching conditions $[\bar{x}_{n-1} \ldots \bar{x}_1 \bar{x}_0] + [\bar{x}_{n-1} \ldots x_1 x_0]$, where the first term leads the FSM to STATE ⓑ while the second sends the FSM on to STATE ⓒ. Similarly, the FSM in STATE ⓑ has a holding condition of $[\bar{x}_{n-1} \ldots \bar{x}_1 \bar{x}_0] + [\bar{x}_{n-1} \ldots \bar{x}_1 x_0] = \bar{x}_{n-1} \ldots \bar{x}_1$ and can branch to STATE ⓒ under input conditions $\bar{x}_{n-1} \ldots x_1 x_0$. The analysis of the sequential behavior of this hypothetical FSM continues in this manner until the branching conditions out of each state have been completely defined to meet the requirements of Eq. (6-4).

Again we emphasize that the NEXT STATE map is to be regarded as a *learning tool* and not as an end in itself. All the information needed to study the sequential behavior of a given FSM is contained in the state diagram. In fact, the state diagram provides a better perspective of the FSM's algorithm than does the NEXT STATE map. Moreover, as the complexity of the algorithm increases, a point is rapidly reached where the construction of a NEXT STATE map becomes impractical.

A Simple Example. We summarize this and the previous section by using the LPD model to design and analyze a simple asynchronous FSM. Consider the state diagram for a two-input/two-output asynchronous FSM given in Fig. 6-6(a). Notice that Eqs. (6-4) and (6-5) are satisfied for each of the states and that the 10 STATE is a don't care state since it is not used. Also, note that this FSM has an unconditional output at STATE $\textcircled{b}$ and a conditional output at STATE $\textcircled{c}$. The K-maps for the NEXT STATE variables are constructed by applying the state transition table of Fig. 6-3(b) to each state with the results shown in Fig. 6-6(b). The output K-map is given in Fig. 6-6(c). Observe that in all K-maps a don't care is placed in cell 2 (the unused state) and that it used to extract minimum cover for the NEXT STATE functions.

Now, we construct the NEXT STATE map from the state diagram as shown in Fig. 6-6(d). To do this we make use of MINTERM code, the domain concept for K-maps, and the stability criteria given by Eqs. (6-2) and (6-3). Beginning with STATE $\textcircled{a}$ it is seen from the state diagram that the holding

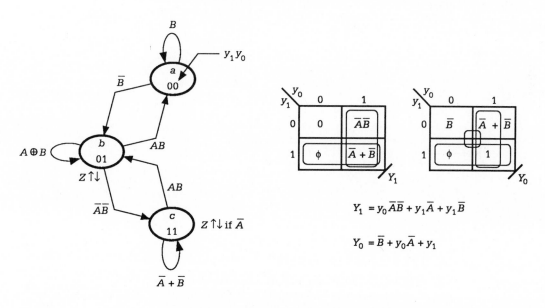

$$Y_1 = y_0\overline{A}\overline{B} + y_1\overline{A} + y_1\overline{B}$$

$$Y_0 = \overline{B} + y_0\overline{A} + y_1$$

(a)   (b)

Fig. 6-6. *A simple asynchronous FSM used to illustrate the use of (a) the state diagram, and (b) the K-maps for the NEXT STATE variables,* Y_1 *and* Y_0.

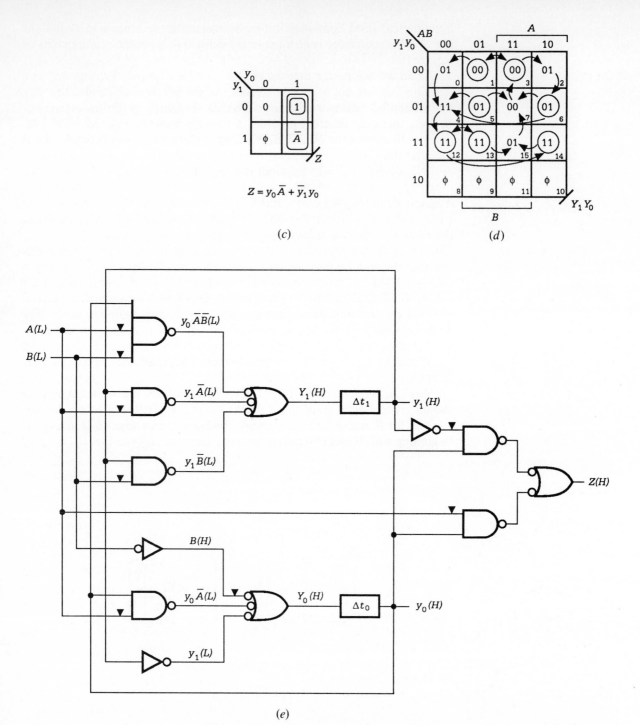

$$Z = y_0 \overline{A} + \overline{y}_1 y_0$$

(c)

(d)

(e)

Fig. 6-6 (contd.). *Illustrating the use of (c) the output K-map, and (d) the NEXT STATE map. (e) Circuit diagram for ACTIVE LOW inputs showing the position of the two lumped path delay MEMORY elements.*

6 / Asynchronous Sequential Machines

condition for the FSM in this state is B, so we place a circled NEXT STATE code 00 in the B domain of the $y_1y_0 = 00$ row. Remember that a circled entry indicates that the FSM is stable ($y_t = Y_t$) in that particular cell. Next, we see that the FSM can transit from the ⓪⓪ STATE to the ⓪① STATE under one of two possible branching conditions, $\overline{A}\overline{B}$ OR $A\overline{B} = \overline{B}$. This means that an uncircled 01 must be placed in each cell of domain $\overline{B}$, that is, in cells 0 and 2, as shown in the $y_1y_0 = 00$ row. Next, it is observed from the state diagram that the FSM in STATE ⓑ = 01 has a holding condition $A \oplus B$, requiring that a circled 01 be placed in cells of the $\overline{A}B + A\overline{B}$ domains (cells 5 and 6). Then, since the FSM will transit from STATE ⓪① to STATE ⓪⓪ under branching condition AB, a 00 (uncircled) must be placed in cell 7 so as to tell the FSM that the NEXT STATE is ⓪⓪. Similarly, the FSM will transit from STATE ⓪① to STATE ①① under branching condition $\overline{A}\overline{B}$ so a 11 is placed in cell 4, thereby forcing the FSM into STATE ①① of cell 12. Finally, in STATE ⓒ = ①① the holding condition is $\overline{A} + \overline{B}$ requiring that a circled 11 be placed in the $\overline{A} + \overline{B}$ domains. Exit from STATE ①① is possible from cells 13 or 14 but only for an input condition AB which forces the FSM to *cycle* from STATE ①① to STATE ⓪⓪ via STATE ⓪① (cell 15). Notice that, since the 10 STATE is left unused, don't cares are placed in cells 8 through 11.

Perhaps the reader may have already noticed that the NEXT STATE map is nothing more than a combination of the Y_1 and Y_0 K-maps of Fig. 6-6(b) represented in CANONICAL (conventional) form. Thus, use of either the state diagram or the NEXT STATE K-maps provides an alternative means of constructing the NEXT STATE map.

The term *cycle* is used to indicate a series of unstable transitions (that is, changes in the state variables) under the same input conditions. A cycle was just demonstrated by using the NEXT STATE map in Fig. 6-6(b). Another cycle can be seen in this NEXT STATE map when the FSM cycles from STATE ⓪⓪ (cell 1) to STATE ①① (cell 12) via STATE 01 (cell 0) under input conditions $\overline{A}\overline{B}$.

As can be seen, the NEXT STATE map does indeed describe the sequential detail of the FSM. However, the state diagram does the same thing in a less complicated, more elegant way. Examples are the two cycle paths just described in the discussion of the NEXT STATE map. These cycles are easily identified in the state diagram of Fig. 6-6(a) as transitions ⓒ→ b → ⓐ under branching conditions AB and the reverse transitions under branching conditions $\overline{A}\overline{B}$. It is important to remember that a fully documented state diagram contains all of the information needed to describe the sequential character of the FSM in question and that the NEXT STATE map is no more (or less) than a graphical realization of the state diagram. We will continue limited use of the NEXT STATE map as a learning tool but our real interest lies in developing the full potential of the state diagram.

The LPD circuit diagram for this FSM, shown in Fig. 6-6(e), is constructed from the NS and output logic expressions given in Figs. 6-6(b) and (c). A minimum of two lumped MEMORY elements is needed to represent all path delays in the NS forming logic. Thus, Δt_1 represents all delays associated with the y_1 feedback variable while Δt_0 represents all delays associated with the y_0 feedback variable. LPD logic circuits, such as the one in Fig. 6-6(e), are usually drawn without the fictitious Δt elements.

6.4 DESIGN AND ANALYSIS OF THE BASIC CELL

Chapter 5 and specifically Sect. 5.5 developed the concept of MEMORY through time delay, and heuristically explained the means by which a BASIC CELL controls MEMORY through SET and RESET commands. At that time the state diagrams, logic circuits, and state transition tables for the NAND- and NOR-centered BASIC CELLs were also developed from these heuristic arguments. A heuristic treatment of the BASIC CELL was necessary in Chapter 5 because the BASIC CELL is an asynchronous machine which requires asynchronous methods for its formal development. Now the reader has the background needed for a more formal treatment of this subject matter.

In this section we will design and analyze the BASIC CELL by using asynchronous methods. In the process of doing so we will provide additional insight into the character of the BASIC CELL, develop the first-cut procedures needed for the design and analysis of more complex asynchronous FSMs, and arrive at some fundamental conclusions regarding asynchronous FSMs generally. First, however, we will design a simple asynchronous FSM which has characteristics similar but not identical to that of a BASIC CELL.

6.4.1 Design of a Pseudo-BASIC CELL

Our design begins with the state diagram for the simple asynchronous FSM given in Fig. 6-7(a). Notice that the requirements of Eq. (6-4) are met for this FSM. We use the LPD state transition table of Fig. 6-3(b) to obtain the EV K-map for the NEXT STATE variable Y, the minimum cover, and the LPD circuit diagram, as shown in Figs. 6-7(b) and (c). The PRESENT and NEXT STATE parameters (y and Y) take the place of Q_t and Q_{t+1}, respectively, which were used in Sect. 5.5.1. Remember that the LPD model requires that the PRESENT STATE variable y be separated in time from the NEXT STATE variable Y by the lumped path delay element Δt.

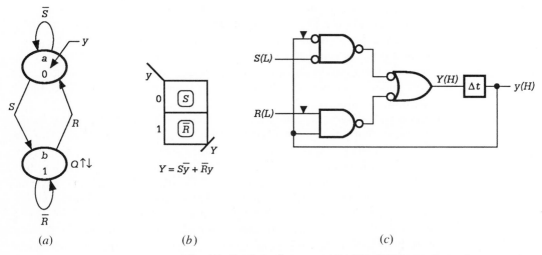

$$Y = S\bar{y} + \bar{R}y$$

(a)   (b)   (c)

Fig. 6-7. *Design of a pseudo BASIC CELL. (a) State diagram. (b) EV K-map for the NEXT STATE variable Y. (c) LPD circuit diagram.*

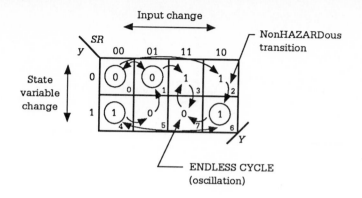

(d)

Fig. 6-7 (contd.). *(d) NEXT STATE map.*

From the treatment given in Sect. 5.5 it is evident that the circuit of Fig. 6-7(c) is not the one expected for the BASIC CELL. In fact, it does not meet the requirements of the BASIC CELL. An inspection of the state diagram in Fig. 6-7(a) reveals that this FSM goes into an oscillation for input conditions $(S, R) = (1, 1)$. To verify this, use will be made of the NEXT STATE map shown in Fig. 6-7(d). The reader will recall from Sect. 6.3 that the NEXT STATE map is constructed with the external input code as the horizontal axis and the PRESENT STATE code as the vertical axis. The cell entries are the NEXT STATE code. Next, we apply the stability criteria of Eqs. (6-2) and (6-3) and circle the stable states while leaving uncircled those states that are unstable. Thus, in a NEXT STATE map, input induced transitions of the FSM occurring between stable (circled) states will remain at the same PRESENT and NEXT STATE logic levels. However, if the external input changes cause the FSM to transit to an unstable state, a y variable transition must occur (vertical arrows). Now, we see that an input-induced transition to either cell 3 or cell 7 results in an ENDLESS CYCLE (oscillation) between the two unstable states—both cells 3 and 7 represent unstable states. So it is clear that this machine does *not* satisfy the requirements of the BASIC CELL, as mentioned earlier. It is, therefore, a pseudo-BASIC CELL of a type.

6.4.2 The SET-Dominant BASIC CELL

If the state diagram of Fig. 6-7(a) had been a valid description of the BASIC CELL, the analysis we just completed should have yielded either the NAND- or NOR-centered circuit shown in Fig. 5-18. But this did not happen. Why? To understand why, we must analyze the NAND-centered BASIC CELL given in Fig. 5-18(a). Reconstructed in Fig. 6-8(a) is the LPD circuit diagram for this FSM which is the same as that in Fig. 5-11(a). The circuit is read and yields the functional relationship between the PRESENT and NEXT STATE parameters given by $Y = S + \overline{R}y$. Then, by using this information, we construct the EV K-map and the resulting state diagram shown in Figs. 6-8(b) and (c). Now it becomes apparent that the state diagrams of Figs. 6-

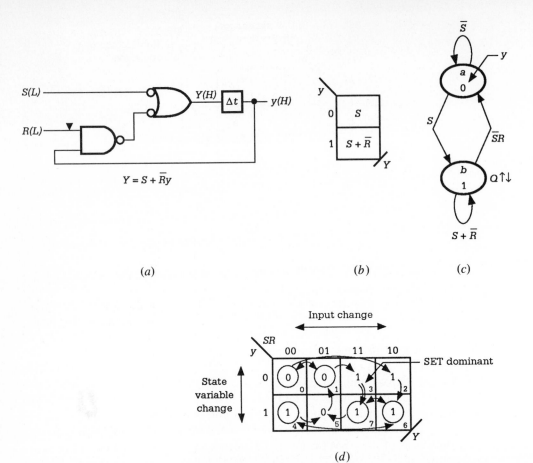

$$Y = S + \bar{R}y$$

(a)

(b)

(c)

(d)

Fig. 6-8. *Analysis of the NAND-centered BASIC CELL. (a) LPD circuit diagram. (b) K-map for the circuit in (a). (c) State diagram derived from the K-map in (b). (d) NEXT STATE map for the SET-dominant BASIC CELL.*

8(c) and 6-7(a) are not the same and, therefore, do not represent the same machine. This is not surprising since the circuits are not the same.

The analysis of the BASIC CELL is continued by constructing and analyzing the NEXT STATE map for this machine. This is done in Fig. 6-8(d) where the results show that the oscillation problem revealed in Fig. 6-7 no longer exists. That is, once the machine enters cell 3 the output must change, forcing the machine into cell 7 where it is stable. This is a SET operation as indicated by the double arrow. Therefore, we conclude, as we did in Sect. 5.5.1, that the NAND-centered BASIC CELL in Fig. 6-8 is SET dominant. The SET-dominant character of this CELL is also revealed by the branching conditions of the state diagram. Thus, if both S and R take on logic 1 values, the only permissible transitions are to STATE ①, where the FSM remains (stabilizes) until an input condition $\bar{S}R$ forces branching to STATE ⓪.

Under input conditions $(S, R) = (1, 1)$ the SET-dominant BASIC CELL is caused to reside stably in cell 7 shown in Fig. 6-8(a). In this state the BASIC CELL loses its mixed-rail output character as indicated in Fig. 6-9.

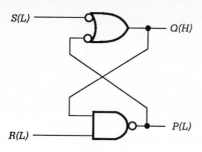

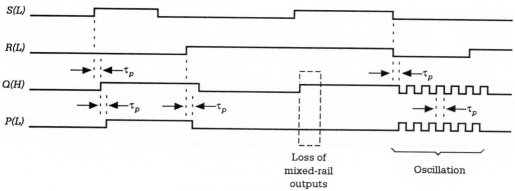

Fig. 6-9. *Timing diagram for the SET-dominant BASIC CELL showing loss of the mixed-rail outputs for the* (S, R) = (1, 1) *condition, and the oscillatory behavior that results when S and R change 1 → 0 simultaneously.*

Here, the BASIC CELL is represented in the familiar "cross-coupled" form of Fig. 5-18(a), and the outputs are given as $Q(H)$ and $P(L)$ to emphasize the fact that the mixed-rail output character is not always preserved.

If the inputs $S(L)$ and $R(L)$ undergo a simultaneous $1 \to 0$ (equivalent to LV → HV) change, the BASIC CELL may become metastable and either "hang up" in a state that is neither SET nor RESET or oscillate. This condition is illustrated in Fig. 6-9, which represents a logic (ideal) simulation of the BASIC CELL. The oscillation occurs because the identical cross-coupled NAND gates drive each other to simultaneously transit in opposite MIXED LOGIC directions after each unit path delay τ_p. Under ideal conditions, oscillatory behavior of this type is predictable and indicative of a possible metastable condition. However, an actual physical test of a BASIC CELL will most likely not yield these same results since metastability is a low-probability condition. But it can occur! (The subject of metastability is discussed in greater detail in Sect. 6.9.1.) It is because of the loss of mixed-rail output character and the possibility of metastable behavior that the $(S, R) = (1, 1)$ condition is normally avoided in BASIC CELLs used in FSM design. Recall that the $(S, R) = (1, 1)$ condition does not exist in the generic form of the state transition table given in Fig. 5-15(c). Thus, use of this state transition table for FSM design permits us to design FSMs which are relatively free of the problems associated with the $(S, R = 1, 1)$ condition.

6.4.3 The RESET-Dominant BASIC CELL

The state diagram for the SET-dominant BASIC CELL is characterized by the branching conditions S and $\overline{S}R$ for state transitions $\textcircled{a} \rightarrow \textcircled{b}$ and $\textcircled{b} \rightarrow a$, respectively, as shown in Fig. 6-8(c). Thus, the branching conditions for the RESET-dominant BASIC CELL must be $S\overline{R}$ and R, respectively, for the same state transitions. In Fig. 6-10(a) we give the state diagram for the RESET-dominant BASIC CELL, which is identical to that given in Fig. 5-13(c). Again, we note that Eq. (6-4) is satisfied for all (both) states, a requirement of any fully documented state diagram with unrestricted branching conditions.

The EV K-map and its POS cover, and the NOR-centered circuit diagram in LPD notation are presented in Figs. 6-10(b) and (c). The NEXT STATE map is given in Fig. 6-11. The RESET-dominant character of this circuit is evident from both the state diagram and the NEXT STATE map for the $(S,$

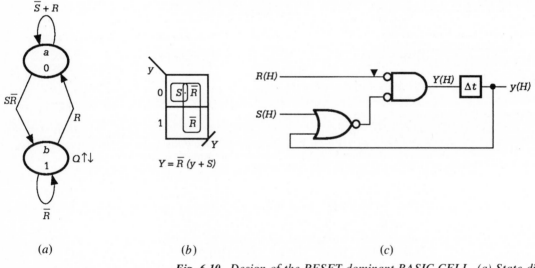

$$Y = \overline{R}\,(y + S)$$

(a)   (b)   (c)

Fig. 6-10. *Design of the RESET-dominant BASIC CELL. (a) State diagram. (b) NEXT STATE K-map. (c) LPD circuit diagram.*

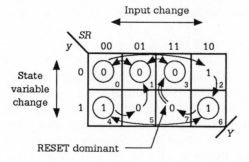

Fig. 6-11. *NEXT STATE map for the RESET-dominant BASIC CELL.*

R) = (1, 1) input condition. For example, the double arrow in the NEXT STATE map indicates that once the machine inters cell 7, it must transit to cell 3 and remain there until different input conditions force branching, a RESET operation. Then from this result we conclude, as we did in Sect. 5.5.1, that the NOR-centered BASIC CELL in Fig. 6-9 is RESET dominant. Furthermore, we see that the SET-dominant CELL is an SOP logic form whereas the RESET-dominant CELL is a POS logic form.

It is left to the reader to demonstrate with a timing diagram that an ideal RESET-dominant BASIC CELL behaves in a manner similar to that of Fig. 6-9.

The state transition tables for the SET and RESET dominant BASIC CELLs are derived in Sect. 5.5.1 and are presented in Figs. 5-15(a) and (b) together with the generic form in Fig. 5-15(c). The character and implications of these state transition tables is discussed at some length in Sect. 5.5.1 and will not be repeated here.

6.5 DESIGN CONSIDERATIONS

The preceding section is intended to be only a brief introduction to asynchronous FSM design and analysis. Much more must be known regarding the complexities of asynchronous sequential machines before meaningful designs are possible. It is the goal of this section to provide a detailed discussion of these complexities. In fact, by section's end it is hoped that the reader will have gained a deeper insight not only into the design of asynchronous FSMs but also into the design of synchronous FSMs. We regard the subject matter of this section as essential to the development of good design practices for asynchronous sequential machines.

In Section 6.1 we indicated that unwanted effects such as ENDLESS CYCLEs (oscillations), CRITICAL RACEs, and HAZARDs can exist in asynchronous FSMs, and in Fig. 6-7 we demonstrated the existence of an ENDLESS CYCLE in a simple two-input/one-output asynchronous FSM. Now it is necessary that we study these problems in sufficient detail so that they can be identified and eliminated if necessary.

6.5.1 Cycles and ENDLESS CYCLEs

The transition of an asynchronous FSM from one stable state to another stable state through one or more unstable states is called a *cycle*, as was stated earlier. When an asynchronous FSM enters a cycle that ends in an unstable state of that cycle, an ENDLESS CYCLE or *oscillation* results. While cycles are necessary to the proper operation of an asynchronous FSM, ENDLESS CYCLEs must always be avoided.

An ENDLESS CYCLE exists in an asynchronous FSM if $f_{cb}(x_i)$ and $f_{bc}(x_i)$ in Fig. 6-4(a) possess a common branching condition given by $f_{bc} \cdot f_{cb}$. A simple example is shown in Fig. 6-12(a). As can be seen, this FSM has a cycle from STATE ⓒ to STATE ⓐ under input conditions $A\bar{B}$. But it also has an ENDLESS CYCLE between STATEs ⓑ and ⓒ. The ENDLESS CYCLE exists because the branching condition $\bar{A}\bar{B}$ is common to both $f_{bc} = A\odot B$ and $f_{cb} = \bar{B}$, the to-and-from branching conditions between STATEs

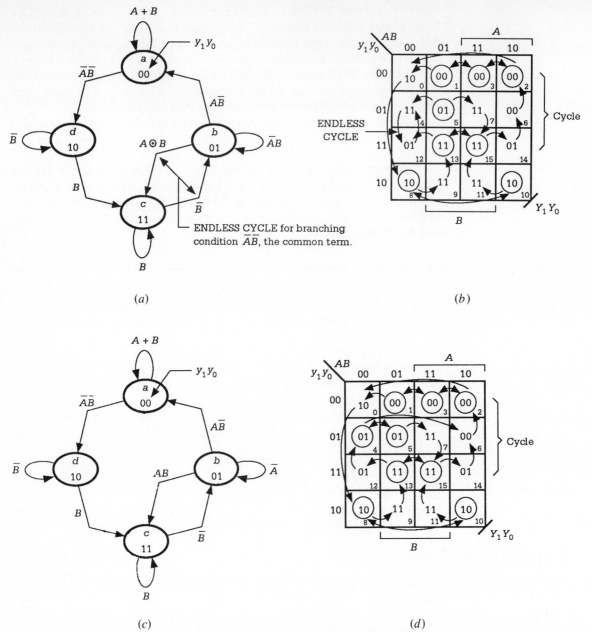

(a)

(b)

(c)

(d)

Fig. 6-12. *An asynchronous FSM illustrating a cycle and an ENDLESS CYCLE (oscillation). (a) State diagram. (b) NEXT STATE map. (c) State diagram corrected for the ENDLESS CYCLE. (d) NEXT STATE map showing alteration of cell 4.*

ⓑ and ⓒ. Thus, when the FSM receives the input condition $\overline{A}\,\overline{B}$ while in either STATE ⓑ or STATE ⓒ, it will go into oscillation between these states and will continue to oscillate until the input condition is changed.

The presence of the cycle and ENDLESS CYCLE is also clearly shown by the NEXT STATE map of Fig. 6-12(b), which has been constructed directly from the state diagram. Notice that the FSM can enter the END-LESS CYCLE from either STATE ⓑ or STATE ⓒ under input condition $\overline{A}\,\overline{B}$, but that the exit route (from the ENDLESS CYCLE) may be uncertain. That is, under branching condition $\overline{A}B$ the machine will transit either to STATE ⓑ or to STATE ⓒ depending on where the FSM resides at the instant the input condition $\overline{A}B$ is sensed by the NEXT STATE logic of the FSM. In fact, the FSM's only access to a holding condition in STATE ⓑ is by way of the ENDLESS CYCLE. For input condition $A\overline{B}$ the FSM will transit to stable STATE ⓐ in any case. The reader should verify that this sequential detail is easily determined from an inspection of the state diagram.

Now, let us suppose that the FSM of Fig. 6-12(a) should not have an ENDLESS CYCLE, but rather should transit to STATE ⓑ and hold in this state under branching conditions $\overline{A}\,\overline{B}$ according to some hypothetical algorithm. Removing the undesirable branching condition $\overline{A}\,\overline{B}$ for the ⓑ → ⓒ transition and "attaching" it to the holding condition for STATE ⓑ, now $\overline{A}B + \overline{A}\,\overline{B} = \overline{A}$, results in the corrected state diagram and NEXT STATE map shown in Figs. 6-12(c) and (d). Observe that the ENDLESS CYCLE has been eliminated by the reassignment of the $\overline{A}\,\overline{B}$ branching condition. This amounts to changing the NEXT STATE code of cell 4 in Fig. 6-12(d) from 11 to 01.

6.5.2 Race Conditions

The subjects of race conditions and their possible effects in synchronous FSMs was discussed in Sect. 5.8.2. There, attention was focused on the output glitches that could result from race conditions. Output race glitches can also occur in asynchronous FSMs. The output race glitch analysis that was discussed in Sect. 5.8.2 can be applied to asynchronous sequential machines to determine if the output will contain such logic noise. In the case of asynchronous FSMs, however, other more serious problems can result from race conditions. The nature of these problems and their identification and elimination will be explored at some length in this section.

The set of alternative cycle paths that lead to the same state is called a *race*. This we learned in Sect. 5.8.2. In an asynchronous FSM a race results when the FSM undergoes a transition to a NEXT STATE that differs from the PRESENT STATE by two or more bits. There are $N!$ alternative race paths for a race condition involving the change of N bits. Since no two feedback variables can change precisely at the same time, one variable will always change before another, even though the time span between the two events may be extremely small. Thus, the alternative race path taken by the FSM will depend on which feedback variable changes first to meet the stability criterion of Eq. (6-2), and this is not usually predictable.

Shown in Fig. 6-13 is a race condition for the ⓟ → ⓠ transition under branching condition f_{PQ} in a generalized state diagram segment. Let us assume that only two state variables must change for the transition (. . . 10. . to . . . 01. .) and that the race path to STATE ⓠ is through either STATE ⓡ or STATE ⓢ, depending on which of the two state variable bits involved changes first. The final residence of the FSM under branching condition f_{PQ} is predictable from the state diagram provided that a term in f_{PQ} is not also contained in either f_R or f_S, and provided that race STATE R and S have

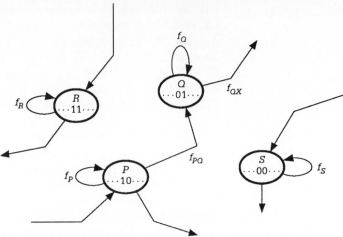

Fig. 6-13. *Generalized state diagram segment showing a race condition for the transition from STATE* $\textcircled{P}$ *to STATE* $\textcircled{Q}$.

branching paths to the destination STATE Q (not shown). Thus, if f_{PQ} is contained only in the holding condition f_Q for STATE Q, and in f_{RQ} and f_{SQ}, the FSM will reside stably in STATE Q following a cycle path through either STATE S or STATE R. Note that the possibility of output race glitches should also be kept in mind. For example, if STATEs S and R are output states but STATEs P and Q are not, output race glitches will be produced.

A race is illustrated in the state diagram of Fig. 6-14(a). Here, it is seen that under branching conditions $\overline{A}B$ the FSM will transit from STATE $\textcircled{b}$ to STATE $\textcircled{c}$ by way of the race condition which is established when the NEXT STATE code must change from 01 to 10 one bit at a time. Consequently, if y_1 changes before y_0 the FSM will transit directly to STATE $\textcircled{c}$ and hold there, but if y_0 should change before y_1, the FSM must transit to STATE $\textcircled{c}$ via STATE $\textcircled{a}$, which takes longer than by the other path. In other words, while in STATE $\textcircled{b}$, a branching condition $\overline{A}B$ will force the FSM into either STATE $\textcircled{a}$ or STATE $\textcircled{c}$. If forced into STATE $\textcircled{c}$, the FSM will reside there stably since $\overline{A}B$ is contained in the holding condition for STATE $\textcircled{c}$ which is $\overline{A} + B$. But if forced into STATE $\textcircled{a}$ with a branching condition $\overline{A}B$, the FSM must cycle $a \rightarrow d \rightarrow \textcircled{c}$ since neither the holding condition for STATE $\textcircled{a}$ nor the holding condition for STATE $\textcircled{d}$ contain $\overline{A}B$.

The race condition just described is also shown in the NEXT STATE map of Fig. 6-14(b). It can be identified by the oppositely oriented arrows in cell 5. But it should be observed that the race condition is more prominently displayed by the state diagram than by the NEXT STATE map. Also, it should seem clear that the EV K-maps or the minimum cover derived from them [see Fig. 6-14(c)] offer little, if any, help in identifying the race condition.

A race condition may or may not be a desirable feature in an FSM. If the alternative path delays can be tolerated, races can at times be introduced deliberately to achieve a more minimum NEXT STATE logic cover. However, if the presence of a race produces a more complex NEXT STATE logic circuitry, or if the race leads to possible output race glitches, or if the

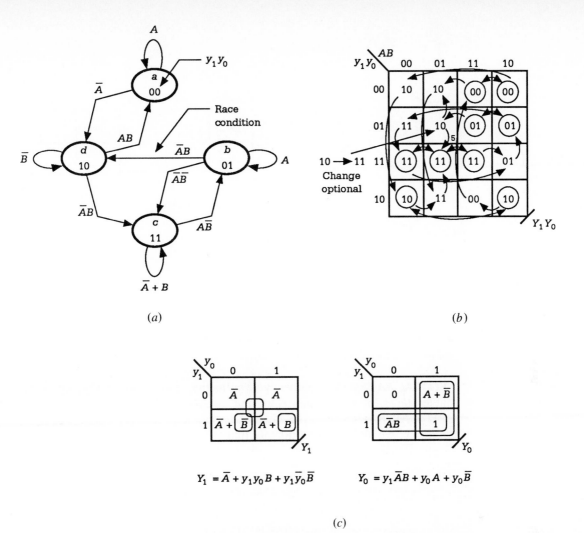

$$Y_1 = \overline{A} + y_1 y_0 B + y_1 \overline{y_0} \overline{B}$$

$$Y_0 = y_1 \overline{A} B + y_0 A + y_0 \overline{B}$$

(c)

Fig. 6-14. *A four-state asynchronous FSM illustrating a race condition. (a) State diagram representation. (b) NEXT STATE map representation. (c) EV K-maps for the NEXT STATE variables and minimum cover.*

asymmetry of the path delays raises questions of reliability, the race should be removed. In Fig. 6-14(d) we demonstrate the removal of a race by using the FSM represented in Fig. 6-14(a). Observe that the branching condition $\overline{A}B$, which previously lead to the race, has been reassigned to the ⓑ → ⓒ transition, which now has the branching condition $\overline{A}B + \overline{A}\,\overline{B} = \overline{A}$. This is equivalent to changing the NEXT STATE code of cell 5 in Fig. 6-14(b) from 10 to 11. Removal of the race condition also alters the NEXT STATE logic cover. By comparing Fig. 6-14(c) with Fig. 6-14(e), we see that the cover for Y_0 has been significantly reduced, which is justification enough for the removal of the race condition. Note that any concerns about the existence of race conditions in an FSM can be eliminated simply by making all state-to-state transitions logically adjacent, a common practice in asynchronous FSM design.

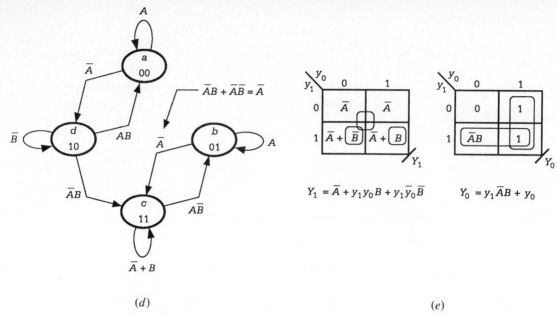

$$\overline{A}B + \overline{A}\overline{B} = \overline{A}$$

$$Y_1 = \overline{A} + y_1 y_0 B + y_1 \overline{y}_0 \overline{B}$$

$$Y_0 = y_1 \overline{A}B + y_0$$

(d)

(e)

Fig. 6-14 (contd.). *(d) State diagram showing removal of the race condition. (e) Revised K-maps and minimum cover.*

CRITICAL RACEs. Alternative cycle paths that lead to two or more different stable states, or to a stable state and an ENDLESS CYCLE, are called CRITICAL RACEs. CRITICAL RACEs can lead to error transitions. Accordingly, we state unconditionally that

> *CRITICAL RACEs must never be permitted to exist in the design of any asynchronous sequential machine.*

The presence of even one CRITICAL RACE will most likely lead to the malfunction of the FSM.

Figure 6-13 shows that a CRITICAL RACE is possible if any term in the branching condition f_{PQ} is also contained in any two (or all) of the holding conditions f_Q, f_R, and f_S for STATEs $\textcircled{Q}$,$\textcircled{R}$, and $\textcircled{S}$, or their exiting conditions (e.g., f_{QX} from STATE $\textcircled{Q}$), or any combination of these that may occur. The exiting condition from any STATE $\textcircled{Q}$, $\textcircled{R}$, or $\textcircled{S}$ may lead to another state with holding condition containing f_{PQ}, or, if the design is really wrought with problems, it may lead to an oscillation. Any of these possibilities constitutes an "illegal" asynchronous design and must be avoided at all cost. Note that the condition for a CRITICAL RACE in Fig. 6-13 results for any set of equal and nonzero Boolean products involving the branching condition f_{PQ} and two or more holding or exiting conditions. Examples are $f_{PQ} \cdot f_Q = f_{PQ} \cdot f_R \neq 0$ or $f_{PQ} \cdot f_R = f_{PQ} \cdot f_S \neq 0$.

A CRITICAL RACE is illustrated in Fig. 6-15(a), again for a simple two-input asynchronous FSM. It can be seen that the transition from STATE $\textcircled{c}$ to STATE $\textcircled{a}$ under branching condition $A\overline{B}$ requires that both state variables must change. However, since they cannot change simultaneously, the FSM

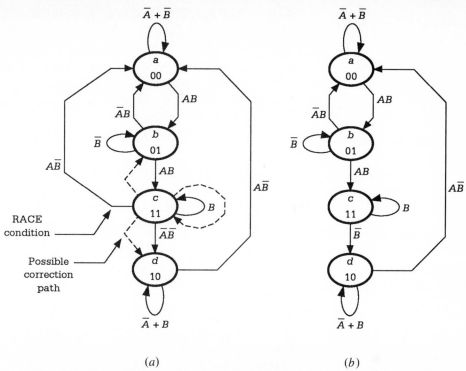

(a)	(b)

Fig. 6-15. *State diagrams showing (a) a CRITICAL RACE and three possible correction paths (dashed arrows), and (b) the CRITICAL RACE removed by using correction path $\textcircled{c}$ to $\textcircled{d}$.*

will finally reside in STATE $\textcircled{b}$ if y_1 changes before y_0, or will reside in STATE $\textcircled{a}$ via STATE $\textcircled{d}$ if y_0 changes before y_1. Notice that the CRITICAL RACE is easily identified in the state diagram since $A\overline{B}$, the branching condition for the race, is contained in the holding conditions for both STATE $\textcircled{a}$ and $\textcircled{b}$. This means that when the FSM attempts to transit from STATE $\textcircled{c}$ to STATE $\textcircled{a}$, under input conditions $A\overline{B}$, it will finally reside stably in either STATE $\textcircled{a}$ or $\textcircled{b}$ until the input conditions are again changed. But it is not predictable which STATE ($\textcircled{a}$ or $\textcircled{b}$) will be favored by the race.

Since the sequential behavior of the FSM in Fig. 6-15(a) is not predictable, steps must be taken to remove the CRITICAL RACE if this FSM is to be useful. Shown by the dashed arrows in Fig. 6-15(a) are three possible correction paths for the race branching condition $A\overline{B}$, any one of which, if used, would effectively remove the CRITICAL RACE. Let us assume that the original algorithm expressed in Fig. 6-15(a) must be retained following removal of the CRITICAL RACE. In this case the correction path would be from STATE $\textcircled{c}$ to STATE $\textcircled{d}$ with a new branching condition $\overline{A}\,\overline{B} + A\overline{B} = \overline{B}$ as indicated in Fig. 6-15(b). Now, if the branching condition $A\overline{B}$ is given while the FSM is in STATE $\textcircled{c}$, the FSM will cycle $\textcircled{c} \rightarrow d \rightarrow \textcircled{a}$ suffering only an additional path delay in route to STATE $\textcircled{a}$. Had the holding condition for STATE $\textcircled{c}$ been chosen as the correction path, the new holding condition would be $B + A\overline{B} = A + B$. Or, if the correction path were taken to be $\textcircled{c}$ to $\textcircled{b}$, the branching condition for the $\textcircled{c}$ to $\textcircled{b}$ transition would be $A\overline{B}$. In any case, it is important for the reader to understand that if the state diagram in Fig. 6-15(a) had been constructed correctly in the first place, there would be no need to make a correction. One should always bear in mind the following:

> *A CRITICAL RACE can be easily identified in a state diagram of any complexity simply by identifying the race condition that leads to two or more stable states—states whose holding conditions contain the branching condition for the race transition.*

Obviously, CRITICAL RACEs *cannot* exist in a state diagram whose state-to-state transitions are all logically adjacent (unit distance coded).

At this point in our development the reader should have the impression that most of the complexities unique to asynchronous FSMs can be recognized and dealt with effectively by use of the state diagram. This is true! Moreover, we will continue to develop skills in using the state diagram so as to make full use of its enormous design potential.

6.5.3   STATIC HAZARDs

Nearly all of Section 4.12 was devoted to the subject of STATIC HAZARDs in combinational logic circuits. There, the HAZARDs were externally initiated in nonsequential circuits. That is, external input changes were shown to produce HAZARDous transitions in asynchronous circuits without feedback. What was presented in Sect. 4.12 is also applicable to asynchronous FSMs except that now we must take into account the sequential nature of the machine and the fact that the HAZARDs may be internally initiated as well as externally initiated.

In Sect. 5.8.3 we discussed the subject of STATIC HAZARDs in the output forming logic of synchronous sequential machines. In that section we discovered that HAZARDs could be produced in the output logic of both Mealy and Moore machines under certain conditions.

We have already mentioned in Sect. 4.12 that two other types of HAZARDs are possible in asynchronous circuits. We named FUNCTION HAZARDs and ESSENTIAL HAZARDs as examples. FUNCTION HAZARDs were discussed briefly in Sect. 4.12 and more extensively in Sect. 5.8.2 under the guise of output race glitches. ESSENTIAL HAZARDs will be the subject of a detailed discussion later in this chapter. A fourth type of HAZARD, called a DYNAMIC HAZARD, is also possible in combinational logic circuits. DYNAMIC HAZARDs, which appear as multiple glitches in nonsteady-state outputs, are usually the result of poor logic design and should not exist in circuits designed by following the methodologies supported in this text.

In this section, with the help of the familiar state diagram and optimum cover derived from the K-maps, we will discuss simple and reliable methods for identifying and eliminating STATIC HAZARDs in asynchronous FSMs. The treatment provided in this section will lead quite naturally into the subject of ESSENTIAL HAZARDs which are the most complex of the HAZARD-type timing defects.

Before we begin our discussions it should be helpful to review some of the terminology which was used previously in Sects. 4.12 and 5.8.3 and which will be used again in this section.

HAZARD. An actual or potential, transient or permanent timing error in a logic circuit that results from an explicitly located asymmetric path delay associated with a gate and/or its input leads.

Gate Path Delay. The propagation delay time associated with a gate and the input leads to which it is connected. For simplicity we will assume that such path delay is concentrated at the output of the gate. In IC devices path delays result primarily from distributive capacitance and inductance in both the leads and gate transistors. The faster the switching speeds of the switching devices the more important become the lead contributions to the total path delay. Usually, it is not possible to separate out the contributions from leads and switching devices.

Coupled Variable. Any variable appearing complemented in one term and uncomplemented in another term of the same expression. Examples of SOP and POS coupled variables are given in the NEXT STATE equations

$$Y_0 = y_0 A \overline{B} + \overline{B} C + \overline{y}_0 \overline{A} C + y_1 y_0 \overline{C} + y_1 \overline{B}$$

and

$$Y_1 = (y_1 + \overline{A} + C)(y_1 + B)(\overline{y}_1 + \overline{y}_0 + A)(y_0 + \overline{A} + B)$$

where y_1, y_0, A, and C are the coupled variables in these expressions.

Coupled Term. Any term of a pair of terms that share a *single* couple variable. In the preceding expressions only the pairs of terms $\overline{B} C$, $y_1 y_0 \overline{C}$ in Y_0, and $(y_1 + B)$, $(\overline{y}_1 + \overline{y}_0 + A)$ in Y_1 are coupled terms.

Residue. That part of a coupled term that remains after the coupled variable has been removed.

HAZARD Cover. The ANDed residues of the SOP coupled terms representing a HAZARDous transition, or the ORed residues of the POS coupled terms representing a HAZARDous transition.

Combinational HAZARD. A HAZARD that can be identified and eliminated by means of combinational logic. STATIC HAZARDs are combinational HAZARDs whereas ESSENTIAL HAZARDs are not.

STATIC HAZARD Identification and Elimination. STATIC HAZARDs are produced by explicitly located asymmetric path delays due to INVERTERs. They can be produced by a $1 \rightarrow 0$ change in an SOP coupled variable or a $0 \rightarrow 1$ change in a POS coupled variable provided that the change represents a valid transition—that is, provided that the change agrees with the state diagram. A coupled variable requires the use of an INVERTER, which in turn produces the asymmetric path delay responsible for the HAZARD. HAZARDs of this type produce positive $(+)$ or negative $(-)$ glitches in otherwise steady-state outputs of FSMs.

STATIC HAZARDs in asynchronous FSMs are easily identified and eliminated by following a four-step procedure:

1. From the minimum or reduced cover expressions for each NEXT STATE variable, Y_i, look for the coupled variables and coupled terms.

2. Use the state diagram together with the information provided by the coupled terms involved to verify that each coupled variable change is associated with a valid STATE transition. Make certain that the coupled terms satisfy the branching condition requirements for the transition because, if they do not, a valid HAZARDous transition is not indicated.

3. Determine the HAZARD cover for each HAZARDous transition by ANDing the residues of coupled SOP terms, or ORing the residues of coupled POS terms. Remember that STATIC HAZARDs cannot be produced if two or more coupled variables are present in or implied by the coupled terms (see Sect. 4.12 for a discussion of FUNCTION HAZARDs). Thus, internally initiated HAZARDs (those initiated by a change in state variable) must take place under constant external input conditions. Similarly, externally initiated STATIC HAZARDs (those caused by a change in an external input condition) must take place under conditions that the initial and final states are the same.

4. Check to see whether or not each HAZARD cover is already contained in the appropriate NEXT STATE variable expression. If not, choose for each the minimum cover permitted by the EV K-maps and then add that cover to the appropriate Y_i expression to be implemented.

A Simple STATIC HAZARD Analysis. Consider the four-state asynchronous FSM in Fig. 6-16(a) which has a STATIC SOP HAZARD in Y_0 internally initiated by a change in y_1 when the machine transits from STATE $\copyright$ to STATE $\textcircled{b}$. This HAZARD is clearly revealed by the minimum cover for Y_0

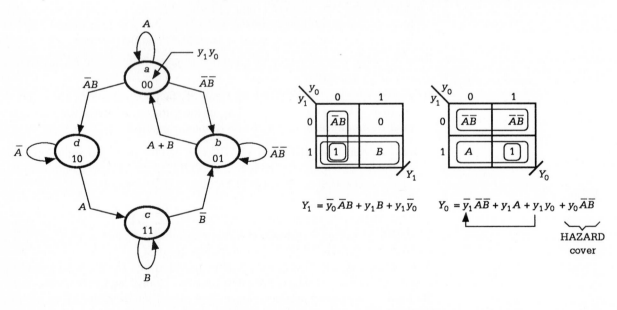

$$Y_1 = \bar{y}_0 \bar{A}B + y_1 B + y_1 \bar{y}_0$$

$$Y_0 = \bar{y}_1 \bar{A}\bar{B} + y_1 A + y_1 y_0 + y_0 \bar{A}\bar{B}$$

HAZARD cover

(a) (b)

Fig. 6-16. *An asynchronous FSM with a HAZARDous transition from STATE $\textcircled{11}$ to STATE $\textcircled{01}$. (a) State diagram. (b) K-maps and minimum SOP and HAZARD cover showing the coupled variables and the coupled terms.*

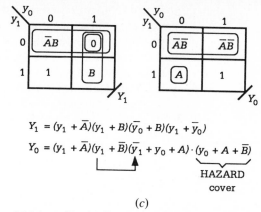

$$Y_1 = (y_1 + \overline{A})(y_1 + B)(\overline{y}_0 + B)(y_1 + \overline{y}_0)$$

$$Y_0 = (y_1 + \overline{A})(y_1 + \overline{B})(\overline{y}_1 + y_0 + A) \cdot (y_0 + A + \overline{B})$$

HAZARD
cover

(c)

Fig. 6-16 (contd.). *(c) K-maps and minimum POS cover.*

(see arrow) given in Fig. 6-16(b). Notice that the coupled variable change is $y_1 \rightarrow \overline{y}_1$ for an SOP HAZARD, the coupled terms are $y_1 y_0$ and $\overline{y}_1 \overline{A} B$, and the STATE change must be ⑪ to ⑪, a requirement of the coupled terms. An inspection of the state diagram indicates that the ⓒ→ⓑ transition under branching condition $\overline{B}$ satisfies the requirements of the coupled terms, namely, that $\overline{A}B$ be contained in $\overline{B}$ and that the STATE change be ⑪ → ⑪). The ANDed residues of the coupled terms yield the SOP HAZARD cover $(y_0) \cdot (\overline{A}B) = y_0 \overline{A}B$. Removal of the HAZARD is accomplished by adding this HAZARD cover to the minimum cover for Y_0, as shown in Fig. 6-16(b). Notice that the terms $\overline{y}_1 \overline{A}B$ and $y_1 A$ do not form a STATIC HAZARD since they contain more than one coupled variable. Furthermore, their ANDed residues yields $\overline{A}B \cdot A = 0$ for coupled variable y_1 and $\overline{y}_1 \overline{B} \cdot y_1 = 0$ for coupled variable A. Thus, no cover exists, so none is required.

To determine whether or not a STATIC POS HAZARD is possible, we must inspect the optimal POS cover given in Fig. 6-16(c). The $0 \rightarrow 1$ coupled variable change in the expression for Y_0 is $y_1 \rightarrow \overline{y}_1$ (see arrow), the coupled terms are $(y_1 + \overline{B})$ and $(\overline{y}_1 + y_0 + A)$, and the STATE change must be ⑪ → ⑪ as required by the coupled terms (read in MAXTERM code). A check of the POS state diagram [mentally produced by complementing the branching conditions of the SOP state diagram in Fig. 6-16(a)] indicates that the ⑪ → ⑪ transition under branching condition $(A + \overline{B})$ meets the requirements of the coupled terms $(y_1 + \overline{B})$ and $(\overline{y}_1 + y_0 + A)$. Therefore, a STATIC POS HAZARD is produced in this FSM. ORing the residues of the coupled terms yields the HAZARD cover shown in Fig. 6-16(c). Notice that coupled terms $(y_1 + \overline{A})$ and $(\overline{y}_1 + y_0 + A)$ cannot produce a POS STATIC HAZARD simply because they contain two coupled variables, y_1 and A. Also, their ORed residues are logic 1 indicating that no HAZARD cover exists.

The presence of the SOP HAZARDous transition in the FSM of Fig. 6-16 can be verified with a timing diagram. Presented in Fig. 6-17(a) is the MIXED LOGIC diagram, and in Fig. 6-17(b) is the MIXED LOGIC timing diagram for the ⑪ → ⑪ transition of this machine. To simplify the timing diagram we have purposely excluded the terms $\overline{y}_0 \overline{A}B$, $y_1 \overline{y}_0$, and $y_1 A$ since they are logic 0 for this transition ($y_0 = 1$ and $A = 0$). Furthermore, we have assigned the same unit path delay to each gate and INVERTER, forgetting for the moment that the logic circuit of Fig. 6-17(a) is represented in LPD notation. Thus, the duration of the ($-$) glitch HAZARD indicated in the y_1 output is one unit of path delay. Also shown in the timing diagram of Fig. 6-17(b) is the HAZARD cover $y_0\overline{A}\,\overline{B}$, which was obtained following the procedure given earlier. ORed with the minimum cover for Y_0, the $y_0\overline{A}\,\overline{B}$ term removes the STATIC HAZARD in y_0 as can be deduced in Fig. 6-17(b).

The STATIC HAZARD shown in Fig. 6-17(b) is nondisruptive in the sense that the FSM finally resides in the proper ⑴ STATE immediately following the brief improper transition to STATE ⑽. However, there is a short delay in achieving stability in the ⑴ STATE, and this could be highly disruptive to any next stage FSM to which y_0 is attached if the HAZARD is sufficiently well developed. Also, if it is required that STATE ⑽ issue an output signal, an output glitch will result which could be disruptive depending, of course, on how that output signal is used.

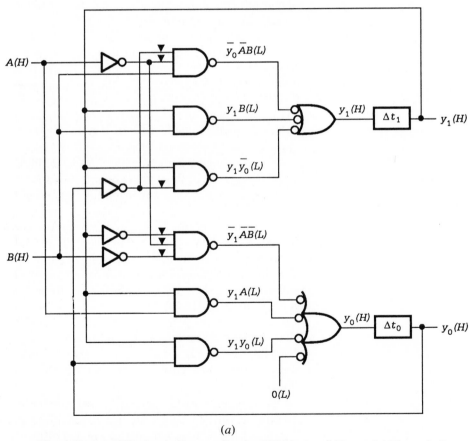

(a)

Fig. 6-17. (a) LPD circuit diagram for the FSM logic of Figure 6-16(b) excluding HAZARD cover $y_0\overline{A}\overline{B}$.

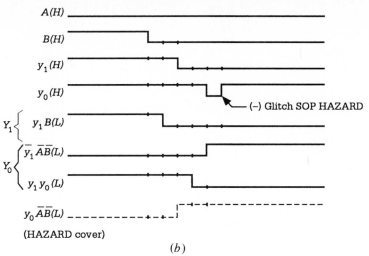

Fig. 6-17 (contd.). *(b) MIXED LOGIC timing diagram for the FSM of Figure 6-16 showing a $(-)$ glitch SOP HAZARD initiated by a $1 \to 0$ change in y_1, and showing the HAZARD cover (dashed waveform) required for the removal of the HAZARD.*

A More Complex STATIC HAZARD Analysis. Now let us move on to a somewhat more complex FSM. Given in Fig. 6-18(a) is the state diagram for a four-state asynchronous machine which will be shown to have two STATIC SOP HAZARDs: an externally initiated HAZARD when the FSM is in STATE ⓑ and an internally initiated HAZARD when the FSM undergoes a transition from ⓒ to ⓑ. We follow the procedure for HAZARD analysis given previously by first mapping the state diagram. Then we extract minimum cover as shown in Fig. 6-18(b). The coupled variables and coupled terms are identified by arrows on the two state variable expressions. To assist us in the analysis of FSMs containing multiple sets of coupled terms, it is helpful to construct a table. This is done in Fig. 6-18(c), where four suspect HAZARDous transitions are indicated. Observe that the coupled terms for the ⑪ → ⑩ transition and for the first ⑪ → ⓪① transition do not lead to HAZARD formation and are labeled NA (not applicable) because they do not satisfy the requirements of the respective branching conditions indicated in the state diagram. This leaves but two valid STATIC HAZARDs, one for the ⓪① → ⓪① transition and the other for the second ⑪ → ⓪① transition shown. These HAZARDs are removed by adding the terms (ANDed residues) $\bar{y}_1 y_0 \bar{A}$ and $y_0 AB$ to the minimum cover for Y_0 as indicated in Fig. 6-18(b).

Notice that the coupled terms, $\bar{y}_1 y_0 B$ and $\bar{y}_1 \bar{A}\bar{B}$, represent an externally initiated HAZARD under conditions $\bar{A} \cdot (B \to \bar{B})$ for which the initial and final states must be the same (STATE ⓪① in this case). Recall that STATIC HAZARDs cannot be formed if two or more coupled variables are present in or implied by the coupled terms.

We have just learned that not all pairs of coupled terms in the NS logic expressions for an asynchronous FSM are necessarily HAZARDous. Therefore, if it is the intent of the designer to produce a minimum circuit design, a detailed HAZARD analysis of the type just discussed is required. Such an analysis is not difficult to carry out. Nor does the analysis require much

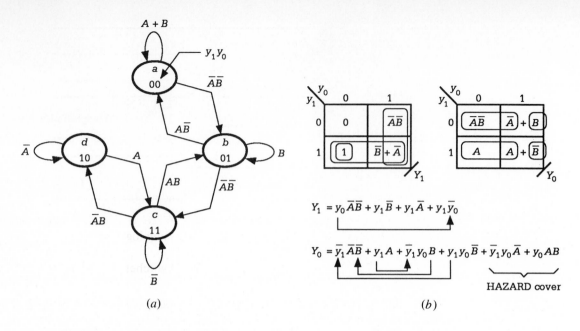

(a) (b)

NEXT STATE Variable	Coupled Variable Change	Coupled Terms	State Transition and Branching Condition*	HAZARD Cover
Y_1	$y_0 \rightarrow \bar{y}_0$	$y_0\bar{A}\bar{B},\ y_1\bar{y}_0$	$(11) \xrightarrow{\bar{A}\bar{B}} (10)$	NA†
Y_0	$B \rightarrow \bar{B}$	$\bar{y}_1 y_0 B,\ \bar{y}_1\bar{A}\bar{B}$	$(01) \xrightarrow{\bar{A}} (01)$	$\bar{y}_1 y_0 \bar{A}$
Y_0	$y_1 \rightarrow \bar{y}_1$	$y_1 y_0 \bar{B},\ \bar{y}_1\bar{A}\bar{B}$	$(11) \xrightarrow{\bar{A}\bar{B}} (01)$	NA
Y_0	$y_1 \rightarrow \bar{y}_1$	$y_1 A,\ \bar{y}_1 y_0 B$	$(11) \xrightarrow{AB} (01)$	$y_0 AB$

* Deduced from the coupled terms.
† Not applicable. The state transition does not occur under the input conditions required by the coupled terms.

(c)

Fig. 6-18. *A two-input/two-output asynchronous FSM with two STATIC HAZARDs. (a) State diagram. (b) EV K-maps, minimum SOP cover, and HAZARD cover. (c) HAZARD analysis table showing the presence of two STATIC HAZARDs.*

extra work on the part of the designer, since the EV K-maps and cover (minimum or not) for the NEXT STATE variables are already a requirement of the design process. On the other hand, if an optimum circuit design is not important, a "shotgun" approach to HAZARD elimination can be used. To do this one would simply include the redundant cover for *all* pairs of coupled terms in the NEXT STATE expressions. Of course, this could increase the hardware requirements considerably, a sacrifice that the designer may or may not be willing to make.

Are STATIC HAZARDs serious timing defects? Will their presence in an asynchronous FSM cause the FSM to malfunction? The answer to these

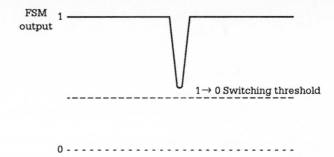

Fig. 6-19. Example of an underdeveloped negative glitch HAZARD in an FSM output signal which may not cause malfunction in the next stage.

questions requires some discussion. STATIC HAZARDs can be produced in both the NEXT STATE or output forming logic sections of an asynchronous FSM. HAZARDs that are produced in the NS forming logic may cause the FSM to malfunction if the HAZARD glitches develop to the extent that the switching thresholds are exceeded. This is in contrast to synchronous FSMs where the formation of HAZARDs in the NS forming logic is of no concern because of the filtering action of the FLIP-FLOPs. STATIC HAZARDs that are formed in the output forming logic of an asynchronous FSM may cause malfunction of a next-stage device, but again only if the HAZARD glitch is sufficiently well developed. Illustrated in Fig. 6-19 is an example of a negative glitch HAZARD that never fully develops and, therefore, may not cause problems in a next stage. The point is that, if it cannot be guaranteed that all HAZARD glitches formed in an asynchronous FSM will not cross the switching threshold, they must be considered as fully developed HAZARDs and dealt with accordingly.

Use of Exhaustive Decoder Devices. Combinational IC devices such as DE-CODERs, MUXs, and ROMs contain exhaustive decoder sections which are inherently noisy (logically)—any two logically adjacent CANONICAL terms are, by definition, coupled terms. So, the probability is high that at least one pair of coupled terms exists for a valid HAZARDous transition in an asynchronous FSM. However, underdeveloped STATIC HAZARDs of the type shown in Fig. 6-19 are quite common in such IC devices, and this fact makes it possible for exhaustive decoder devices to be used in some asynchronous circuits. Still, a wiser choice would be to avoid the use of all exhaustive decoder devices in asynchronous FSM design since there is always the possibility, if not a likelihood, that one or more HAZARDs will be produced and will be developed sufficiently to cause problems either in the FSM itself or in a next stage device that it controls. Remember that HAZARDs produced by commercial IC devices such as DECODERs, MUXs, and ROMs cannot be eliminated directly since there is no user access to their decoder sections.

6.5.4   ESSENTIAL HAZARDs

Elimination of all CRITICAL RACEs and STATIC HAZARDs from an asynchronous FSM operated in the fundamental mode does not ensure proper operation of the FSM. Noncombinational HAZARDs produced by explicitly located asymmetric path delays in gates and/or leads can cause the FSM to malfunction. These HAZARDs, called ESSENTIAL HAZARDs (or E-HAZARDs), are multiple-order steady-state HAZARDs in the sense that they involve the change of two or more state variables in otherwise steady-state output signals. By this definition STATIC HAZARDs are first-order combinational HAZARDs because they involve the change of only one state variable. Note that the term "essential" does not imply "indispensible" or "necessary" but, rather, refers to the fundamental mode of FSM operation that is required for the possible existence of an E-HAZARD. Without exception, E-HAZARDs cannot be eliminated by adding redundant cover as can STATIC HAZARDs.

The requirements for ESSENTIAL HAZARD formation are as follows:

1. The asynchronous FSM must be operated in the fundamental mode.
2. There must be at least two state (feedback) variables and at least one external input.
3. There must be at least two paths of propagation of an external input to the last state variable to change (the first invariant state variable): one path directly to that state variable and another path indirectly to the same state variable by way of one or more different state variables.
4. An asymmetric path delay must be explicitly positioned in the direct path of an input to the last state variable to change such that a critical race is established at the race gate between that input variable and one or more state variables via the indirect path. If the race is won by the state variable(s) an ESSENTIAL HAZARD is formed. Explicitly located asymmetric path delays of sufficient duration may ensure that the race is won by the state variable(s) and, hence, that the E-HAZARD is always formed.

The requirements for ESSENTIAL HAZARD formation are illustrated by the block diagrams in Fig. 6-20, where the fictitious lumped MEMORY elements have been purposely omitted for the sake of simplicity. In Fig. 6-20(a) we give the requirements for second-order E-HAZARD formation which involves the change of two state variables in response to a change in input X: when an "unplanned" delay in the t_1 (X) path to y_a exceeds the combined delays in paths t_2 and t_3, hence

$$t_1 > t_2 + t_3$$

there results a successive change in y_b and y_a (in that order) before the NS forming logic can completely respond to the change in X which initiated the first transition. This unintended process is the ESSENTIAL HAZARD that causes the FSM to malfunction. In the event that the condition $t_1 = t_2 + t_3$ should exist, a critical race is established between input X and the feedback variable y_b, and the outcome is uncertain. Here, y_a is the second state variable to change and y_b is the first to change.

The E-HAZARD shown in Fig. 6-20(a) can be removed by placing a counteracting delay (dashed box) in the y_b feedback path so that

$$t_1 < t_2 + t_3 + \Delta t$$

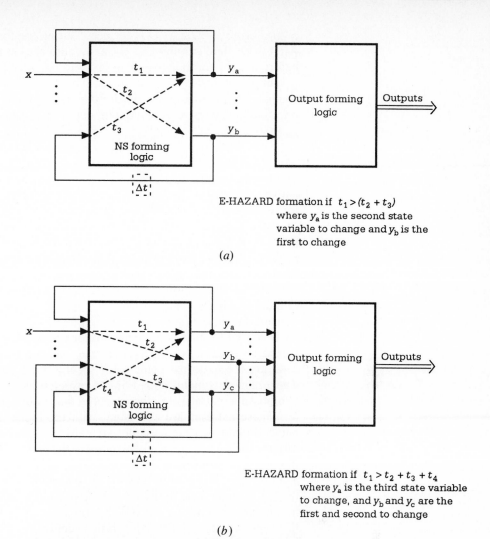

E-HAZARD formation if $t_1 > (t_2 + t_3)$
where y_a is the second state
variable to change and y_b is the
first to change

(a)

E-HAZARD formation if $t_1 > t_2 + t_3 + t_4$
where y_a is the third state variable
to change, and y_b and y_c are the
first and second to change

(b)

Fig. 6-20. *Requirements for ESSENTIAL HAZARD formation. (a) Second-order E-HAZARDs. (b) Third-order E-HAZARDs.*

The added delay removes or reverses the asymmetry in path delay that caused the HAZARD, *but it also slows down circuit response*. Note that the counteracting delay, Δt, can also be placed on either the t_2 or t_3 internal path, but to do so may cause the formation of other E-HAZARDs. For this reason, it is best to use an external feedback path for this purpose. The nature of the electronic circuits which produce counteracting delays of this type are discussed briefly in Sect. 6.8.3.

Shown in Fig. 6-20(b) are the requirements for third-order E-HAZARD formation involving the change of three state variables in response to a change in the input X. When the delay in the t_1 path to y_a is of sufficient duration to cause the condition

$$t_1 > t_2 + t_3 + t_4$$

a third-order E-HAZARD is formed. During the formation of this E-HAZARD three state variables y_b, y_c, and y_a are caused to change (in that order) before the NS forming logic can respond completely to the change in input X that initiated the first transition. Thus, y_a is the third state variable to change. Elimination of the third-order E-HAZARD is accomplished by appropriately placing a counteracting delay Δt (dashed box) such that

$$t_1 < t_2 + t_3 + t_4 + \Delta t$$

The counteracting delay may be placed on either or both of the feedback paths, y_b or y_c. It is best not to use any of the internal paths for this purpose for reasons given earlier.

If we compare the path delay requirements for second- and third-order E-HAZARD formation, we conclude that the greater the order of a potential E-HAZARD, the more difficult it becomes to meet the requirements for its formation. Second-order E-HAZARDs are much more likely to cause FSM malfunction than third-order E-HAZARDs, and fourth-order E-HAZARD formation is highly unlikely.

Certain sequential patterns are required before an ESSENTIAL HAZARD can be formed in an asynchronous FSM. These patterns are best revealed by using a NEXT STATE map or state diagram. Shown in Fig. 6-21(a) are the patterns required for second-order E-HAZARD formation in a generalized asynchronous FSM as revealed in a NEXT STATE map. The two possible patterns for the case $k \neq b$ are illustrated in Figs. 6-21(b) and (c). These patterns are consistent with the block diagram of Fig. 6-20(a), namely, that two state variables are caused to change in turn before the NS forming logic can completely respond to the change $\overline{X_1} \rightarrow X_1$, which initiated the first transition. If $k = b$ in Fig. 6-21(a), then the FSM will eventually and correctly stabilize in STATE $\textcircled{b}$ regardless of any delayed response of the NS forming logic to the change $\overline{X_1} \rightarrow X_1$. So, if $k = b$, the FSM transits to the correct state (STATE $\textcircled{b}$) but by a roundabout path. We will discuss this type of timing defect later in this section.

The E-HAZARD patterns revealed in the NEXT STATE maps of Fig. 6-21 are also revealed in the fully documented state diagram segment shown in Fig. 6-22. This is not surprising since the NEXT STATE map is nothing more than a graphical representation of the state diagram. Let us follow the sequence of events that lead to E-HAZARD formation by using the state diagram segment in Fig. 6-22. Properly operating, the FSM should transit from $\textcircled{a}$ to $\textcircled{b}$ under branching condition . . . $x_1 x_0$ when $\overline{x}_1 \rightarrow x_1$ and should hold in STATE $\textcircled{b}$ under the condition . . . $x_1 x_0$. The transition $\textcircled{a} \rightarrow \textcircled{b}$ amounts to a change of $0 \rightarrow 1$ in the y_0 state variable. However, if it happens that the complete response to $\overline{x}_1 \rightarrow x_1$ is sufficiently delayed in the x_1 path to y_1 of the NEXT STATE (NS) forming logic following the initiation of the $\textcircled{a} \rightarrow \textcircled{b}$ transition, the FSM will continue to sense . . . $\overline{x}_1 x_0$ rather than . . . $x_1 x_0$ and will undergo another transition from STATE $\textcircled{b}$ to STATE $\textcircled{c}$. This second transition occurs because y_0 (the first state variable to change) wins the race with x_1, at the race gate thereby forcing Y_1 to change from 0 to 1. Thus, the delayed response of the NS forming logic to $\overline{x}_1 \rightarrow x_1$ has caused the FSM to sequence incorrectly $\textcircled{a} \rightarrow b \rightarrow \textcircled{c}$.

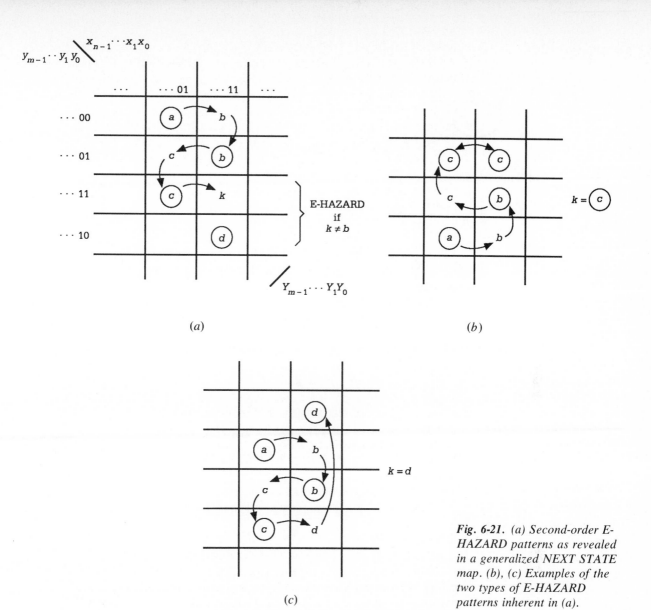

Fig. 6-21. (a) Second-order E-HAZARD patterns as revealed in a generalized NEXT STATE map. (b), (c) Examples of the two types of E-HAZARD patterns inherent in (a).

What happens next as a result of the complete response of the NS forming logic to the $\bar{x}_1 \rightarrow x_1$ change determines what type of E-HAZARD is actually formed. If the FSM either holds at STATE c or transits on to and holds at STATE d in final response to the $\bar{x}_1 \rightarrow x_1$ change, an E-HAZARD is formed. This is a malfunction of the FSM since the FSM incorrectly transits either from a to c or from a to d via the E-HAZARD path. In either case, the requirements for E-HAZARD formation are shown in Fig. 6-22 to be that the input condition I_{bc} be contained in branching condition f_a, that the input condition I_{ab} be contained in branching condition f_b, that the input condition I_{ab} not be contained in branching condition f_{cb}, and that there be at least one invariant state variable per transition in the E-HAZARD path. Here, it follows that a given transition can take place only if an input condition, I

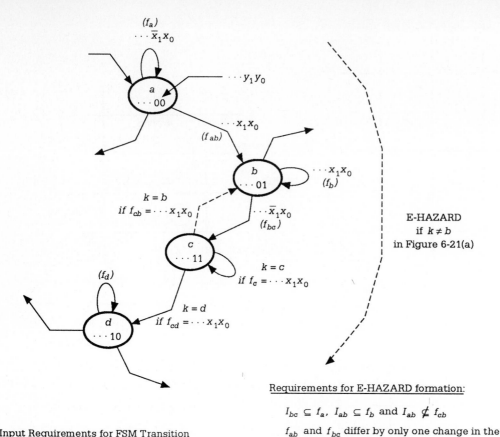

(f_a)
$\cdots \bar{x}_1 x_0$

$\cdots y_1 y_0$

a
$\cdots 00$

$\cdots x_1 x_0$
(f_{ab})

b
$\cdots 01$

$\cdots x_1 x_0$
(f_b)

$k = b$
if $f_{cb} = \cdots x_1 x_0$

$\cdots \bar{x}_1 x_0$
(f_{bc})

E-HAZARD
if $k \neq b$
in Figure 6-21(a)

c
$\cdots 11$

$k = c$
if $f_c = \cdots x_1 x_0$

(f_d)

d
$\cdots 10$

$k = d$
if $f_{cd} = \cdots x_1 x_0$

Requirements for E-HAZARD formation:

$I_{bc} \subseteq f_a,\ I_{ab} \subseteq f_b$ and $I_{ab} \not\subseteq f_{cb}$

f_{ab} and f_{bc} differ by only one change in the sampling variable

Requirements for d-trio formation:

$I_{bc} \subseteq f_a,\ I_{ab} \subseteq f_b$ and $I_{ab} \subseteq f_{cb}$

f_{ab} and f_{bc} differ by only one change in the sampling variable

Input Requirements for FSM Transition

$a \longrightarrow b$	$I_{ab} \subseteq f_{ab}$
$b \longrightarrow c$	$I_{bc} \subseteq f_{bc}$
$c \longrightarrow d$	$I_{cb} \subseteq f_{cb}$

etc.

Fig. 6-22. *General requirements for second-order E-HAZARD and d-trio formation shown by using a state diagram segment equivalent to the NEXT STATE map in Figure 6-21(a).*

(e.g., I_{ab}), is contained in the branching condition, f (e.g., f_{ab}), for that transition, as indicated in Fig. 6-22. Mathematically, the three requirements for E-HAZARD formation are represented as $I_{bc} \subseteq f_a$, $I_{ab} \subseteq f_b$, and $I_{ab} \not\subseteq f_{cb}$. An additional requirement is that branching conditions f_{ab} and f_{bc} differ logically by only a single change in the sampling variable.

There is a third possible transition path that the FSM can take once it has reached STATE $\textcircled{c}$ following the initial response to $\bar{x}_1 \rightarrow x_1$ described earlier. In the case of $k = b$ in Fig. 6-21(a), the FSM will transit back to STATE $\textcircled{b}$, which is the correct final state. This third alternative path is shown in Fig. 6-22 by the dashed arrow from $\textcircled{c}$ to $\textcircled{b}$. A pattern such as $\textcircled{a} \rightarrow b \rightarrow \textcircled{c} \rightarrow \textcircled{b}$ is called a *d-trio* (delay-trio) HAZARD, meaning that an asymmetric path delay causes three state changes before the FSM finally resides stably in the correct state. Therefore, we conclude that a d-trio HAZARD is a timing defect that causes two unintended additional changes in state variables before the FSM finally resides stably in the proper state. The requirements for d-trio HAZARD formation are shown in Fig. 6-22 to be the same as that for second-order E-HAZARD formation except that now the requirement is $I_{ab} \subseteq f_{cb}$ so as to provide a path back to STATE (b). Higher-order HAZARDs of this type are also possible.

Summary. A second-order E-HAZARD or d-trio is formed if an asymmetric path delay of sufficient duration is placed on the direct input path to the second state variable to change (first invariant state variable), providing

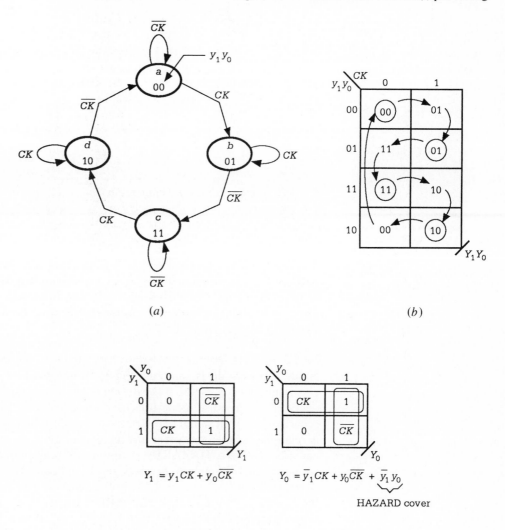

(a)

(b)

(c)

Fig. 6-23. *Design of an asynchronous toggle circuit. (a) State diagram. (b) NEXT STATE map. (c) K-maps and minimum SOP cover including STATIC HAZARD cover.*

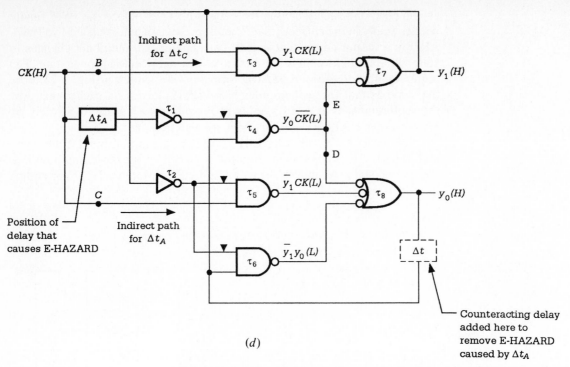

Fig. 6-23 (contd.). (d) Circuit diagram in distributed path delay form showing gate delays and an unintended path delay Δt at position A that will cause an E-HAZARD to form if $(\Delta t + \tau_1) > (\tau_5 + \tau_8)$. Also shown is the counteracting delay (dashed box) required to eliminate the HAZARD.

that the requirements given in Fig. 6-22 are met. The second-order E-HAZARD or d-trio is eliminated by placing a counteracting delay on the feedback line corresponding to the first state variable to change—the indirect path. The counteracting delay must be equal to or greater than the net delay which caused the E-HAZARD or d-trio.

An ESSENTIAL HAZARD Analysis. In Fig. 6-23 we design the TOGGLE circuit. This circuit is sometimes called a nongated RET T-FF if an output Q is issued, and is useful in the design of asynchronous sequential machines operated in the pulse mode (discussed in Sect. 6.8). The TOGGLE circuit is particularly well suited to our purposes at this time because its architecture is relatively simple and yet it has the capability of producing both STATIC and ESSENTIAL HAZARDs. Skipping, for the moment, to Fig. 6-23(c), we notice that the minimum cover for Y_0 indicates that the STATIC HAZARD occurs in y_0 while in STATE ⑪ and that it is externally initiated by the change $CK \rightarrow \overline{CK}$. The HAZARD cover is $\overline{y}_1 y_0$ as determined by the procedure presented earlier.

Shown in Fig. 6-23(a) is the state diagram for the TOGGLE circuit and in Fig. 6-23(b) its NEXT STATE map representation. When we compare the patterns in these figures with those in Figs. 6-21 and 6-22, we see that the TOGGLE circuit contains potential E-HAZARDs of the type produced when $k = d$. The presence of E-HAZARDs can be demonstrated by placing an explicitly located asymmetric path delays in the logic diagram. Presented in Fig. 6-23(d) is the logic diagram for the TOGGLE circuit which is obtained from the NEXT STATE logic expressions in Fig. 6-23(c). In this circuit the LPD notation is dropped, and each gate is assigned a path delay τ_i (representing the gate and its inputs) to permit the various paths to be easily tracked. This logic circuit is said to be represented in *distributed* path form. Now, if an adjustable delay Δt is placed at position A in the direct path to y_1 and if $(\Delta t_A + \tau_1) > (\tau_5 + \tau_8)$, the second-order E-HAZARD is formed. In this case the CRITICAL RACE (to race gate 4) between $\overline{CK}$ and y_0 is won by y_0 via the indirect path through gates 5 and 8. The result is a change in the second state variable y_1 caused by a change in y_0. Notice that the output of race gate 4 ($y_0\overline{CK}$) agrees logically with the transit requirements for the second y-variable change represented by $01 \rightarrow$ ⑪ in Fig. 6-23(a). Note also that the STATIC HAZARD cover, $\overline{y}_1 y_0$, has been included but that it plays no role in the formation or elimination of this E-HAZARD, a fact that may not be true in other FSMs.

The development of the E-HAZARD in the TOGGLE circuit is shown in the ideal timing diagrams of Fig. 6-24. Here, a unit path delay τ_p is assigned to each gate and to each INVERTER. In Fig. 6-24(a) the asymmetric path delay at position A is assigned the value $\Delta t_A = 4\tau_p$. This delay causes the FSM to transit incorrectly

$$\text{⑩} \rightarrow 01 \rightarrow \text{⑪} \rightarrow \text{⑩}$$

for input change $\overline{CK} \rightarrow CK$. Properly operating, the FSM should transit ⑩ $\rightarrow$ ⑪ for that input change, but instead it malfunctions as indicated.

If we assign to each gate in Fig. 6-23(d) a unit path delay τ_p, as we have just done, the condition $(\Delta t_A + \tau_1) > (\tau_5 + \tau_8)$ for E-HAZARD formation becomes $\Delta \tau_A > \tau_p$. This suggests that when Δt_A is little more than a gate path delay, the E-HAZARD will form. This is demonstrated by the timing diagram in Fig. 6-24(b), where we have set $\Delta t_A = 2\tau_p$. A close inspection of the timing diagram reveals that, as a result of a CK change from 0 to 1, the FSM transits through the sequence

$$\text{⑩} \rightarrow 01 \rightarrow 11 \rightarrow 01 \rightarrow \text{oscillation}$$

ending in an oscillation. The FSM should transit ⑩ $\rightarrow$ ⑪ for the input change $\overline{CK} \rightarrow CK$. But because of the small delay $2\tau_p$ at position A, y_0 wins the race against CK at gate 4 but only by τ_p. This causes y_1 to change twice in succession following the initial change in y_0 and forces the FSM back into the original cycle and, hence, into oscillation.

Typically, E-HAZARDs cause oscillatory behavior of the type described above under ideal conditions when the asymmetric path delay causing the HAZARD has a duration slightly greater than the minimum value required to produce the HAZARD. Thus, such oscillatory behavior indicates the onset of E-HAZARD formation but only for logic simulations involving identical gates. For the TOGGLE circuit of Fig. 6-23(d) oscillatory behavior occurs when Δt_A is in the range of τ_p to $2\tau_p$. For $\Delta t_A > 3\tau_p$ the E-HAZARD manifests itself as in Fig. 6-24(a).

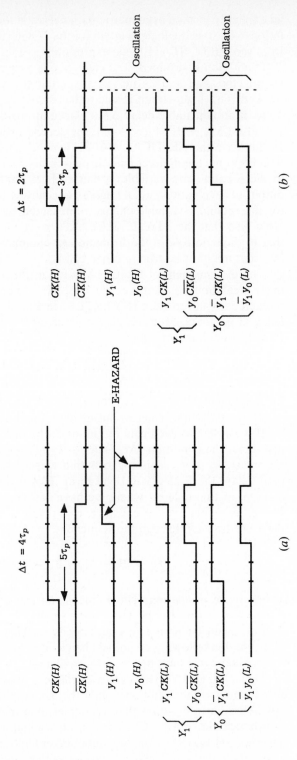

Fig. 6-24. *Ideal MIXED LOGIC timing diagrams for the toggle circuit of Figure 6-23(d) showing the effect of the E-HAZARD produced by an asymmetric path delay Δt located at position A. (a) Case of Δt = 4τ_p. (b) Case of Δt = 2τ_p showing the oscillatory behavior that results from the E-HAZARD. Note: For these timing diagrams each gate and each INVERTER is assigned a unit path delay of τ_p.*

Removal of the E-HAZARD caused by Δt_A in the TOGGLE circuit of Fig. 6-23(d) is accomplished by placing on the y_0 feedback line (dashed box) a counteracting delay Δt which satisfies the condition $(\Delta t_A + \tau_1) < (\tau_5 + \tau_8 + \Delta t)$ but which also slows down circuit response. Placement of this delay on the y_1 feedback line would not satisfy the requirement for eliminating the E-HAZARD.

E-HAZARD production also occurs when an asymmetric path delay Δt is placed at position C or at position D in Fig. 6-23(d). We leave it as an exercise for the reader to determine the path delay requirements for E-HAZARD formation at each of these positions. An asymmetric path delay Δt placed at position B produces a special STATIC HAZARD under the condition that $(\Delta t_B + \tau_3) > (\tau_1 + \tau_4)$. However, this is not an E-HAZARD, as we have defined it, since it involves a race to gate 7 between the sampling variable CK and its complement $\overline{CK}$. Note that this special STATIC HAZARD can form only if the ANDed residues of the coupled p-terms (to the race gate) is not logic 0. An E-HAZARD cannot be formed by placing a delay of any magnitude of position E. Why not?

A More Complex ESSENTIAL HAZARD Analysis. Presented in Fig. 6-25 are the state diagram and NEXT STATE map for an FSM which possesses two timing defects, an E-HAZARD and a d-trio. The formation paths of these two HAZARDs are revealed on the state diagram by dashed arrows and in the NEXT STATE map by solid arrows. Here, it is seen that the E-HAZARD pattern is the same as that shown in Figs. 6-21 and 6-22 for the case where $k = \textcircled{c}$ and that the d-trio HAZARD pattern agrees with that for which $k = b$. Notice that the requirements necessary for E-HAZARD formation, namely, $I_{ba} \subseteq f_c$, $I_{cb} \subseteq f_b$, and $I_{cb} \not\subseteq f_{ab}$, are met if $I_{cb} = AB$ and that the requirements for d-trio formation, $I_{dc} \subseteq f_a$, $I_{ad} \subseteq f_d$, and $I_{ad} \subseteq f_{cd}$, are met if $I_{ad} = \overline{A}\overline{B}$. Notice also that the E-HAZARD and d-trio paths involve a change in a single input variable and that each state-to-state transition involves one invariant state variable.

Let us trace the sequence of events leading to the formation of the two timing defects by using the state diagram in Fig. 6-25(a). Consider that the FSM initially resides (stably) in STATE $\textcircled{11}$ under holding conditions $\overline{A}B$. When the input change $\overline{A} \rightarrow A$ is sensed, the transition $\textcircled{11} \rightarrow 01$ is initiated and a $1 \rightarrow 0$ change occurs in y_1. Now, if the complete response of the NS forming logic to the change $\overline{A} \rightarrow A$ is delayed in the A path to y_0, the NS logic still senses $\overline{A}B$ after y_1 has changed $1 \rightarrow 0$. Because the FSM still senses $\overline{A}B$ instead of AB, it will transit $01 \rightarrow \textcircled{00}$ with a $1 \rightarrow 0$ change in y_0. The FSM will then reside stably in STATE $\textcircled{00}$ even after the NS forming logic completely responds to the input change $\overline{A} \rightarrow A$. In short, the FSM should transit $\textcircled{c} \rightarrow \textcircled{b}$ under branching conditions AB, but because of an unintended, explicitly located, path delay in the direct path of A to y_0, the FSM malfunctions by erroneously undergoing a transit $\textcircled{c} \rightarrow b \rightarrow \textcircled{a}$. The reader may verify that the NEXT STATE map in Fig. 6-25(b) provides this same information but in a somewhat less vivid manner.

Next we trace the formation path of the d-trio HAZARD as depicted in the state diagram of Fig. 6-25(a). Begin with the FSM in STATE $\textcircled{00}$ under holding conditions $\overline{A}\overline{B}$. When the input change $B \rightarrow \overline{B}$ is sensed by the FSM,

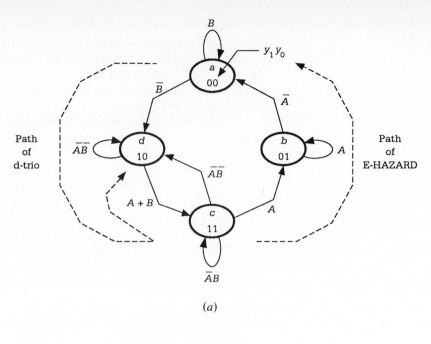

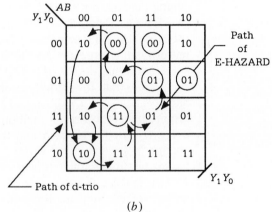

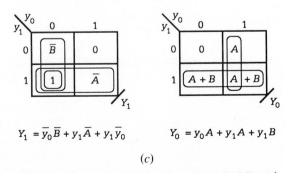

$$Y_1 = \bar{y}_0\bar{B} + y_1\bar{A} + y_1\bar{y}_0 \qquad Y_0 = y_0 A + y_1 A + y_1 B$$

(c)

Fig. 6-25. *An FSM showing the presence of an E-HAZARD and a d-trio. (a) State diagram representation. (b) NEXT STATE map representation. (c) NEXT STATE K-maps and minimum cover showing no STATIC HAZARDs.*

the $\overline{00} \rightarrow 10$ transition is initiated causing a $0 \rightarrow 1$ change in y_1. If an unintended, explicitly located delay exists in the direct B path to y_0, the second state variable to change, the NS forming logic continues to sense input conditions $\overline{A}B$ instead of $\overline{A}\,\overline{B}$, thereby causing the FSM to transit 10 $\rightarrow \overline{11}$. Finally, in complete response to the $B \rightarrow \overline{B}$ change, the FSM senses $\overline{A}\,\overline{B}$ and transits 11 $\rightarrow \overline{10}$, completing the d-trio HAZARD path. Again, one notes that the same information is provided by the NEXT STATE map of Fig. 6-25(b), but less vividly. Note that this d-trio path is nondisruptive to FSM operation, provided that this FSM is not used to drive another.

No HAZARD analysis of an FSM is complete without checking for the presence of STATIC HAZARDs. This is done in Fig. 6-25(c) where we show the K-maps and minimum cover for the NEXT STATE variables, Y_1 and Y_0. The NEXT STATE logic expressions indicate that no STATIC HAZARDs are possible since no coupled terms exist in either expression.

The details of E-HAZARD and d-trio formation are best displayed by the use of timing diagrams, and this we will do. First, however, a logic circuit must be constructed so that the HAZARD-producing delays can be located

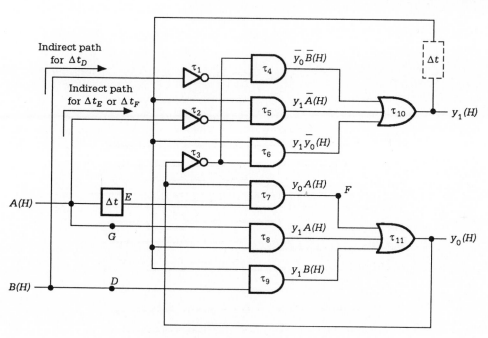

E-HAZARD if $(\Delta t_E + \tau_7) > [\tau_2 + \tau_5 + \tau_{10} + (\tau_8 \text{ or } \tau_9)]$

E-HAZARD if $(\tau_7 + \Delta t_F) > [\tau_2 + \tau_5 + \tau_{10} + (\tau_8 \text{ or } \tau_9)]$

d-trio if $\Delta t_D > (t_1 + t_4 + \tau_{10})$

(a)

Fig. 6-26. *Formation of the E-HAZARD and d-trio revealed in the FSM of Figure 6-25. (a) Circuit diagram showing position (E or F) of the asymmetric path delay Δt that will produce the E-HAZARD, the position (D) of an asymmetric path delay Δt that will cause the d-trio to form, and the position of the counteracting delay (dashed box) that will eliminate these timing defects.*

6.5 / Design Considerations

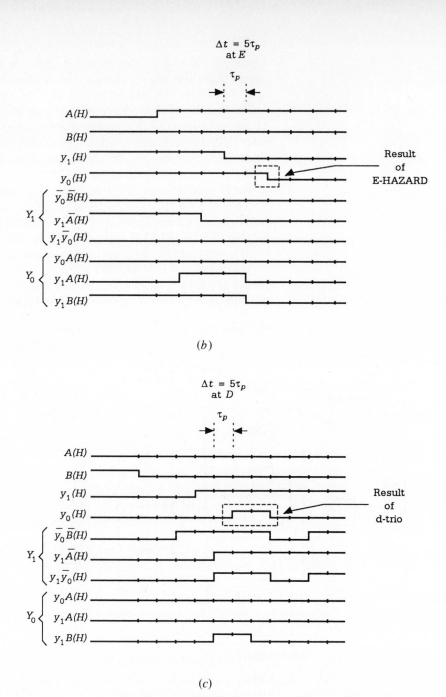

(b)

(c)

Fig. 6-26 (contd.). *(b) Timing diagram showing the formation of the E-HAZARD.*
(c) Timing diagram showing formation of the d-trio.

and delay paths can be traced. Shown in Fig. 6-26(a) is the logic circuit
which has been constructed by applying the LPD model to the NEXT
STATE logic given in Fig. 6-25(c). Here, the LPD notation has again been
dropped and the circuit is represented in distributed path delay notation by
assigning discrete delays to each gate.

Before we construct a timing diagram, it will be helpful to establish the requirements for E-HAZARD formation and elimination as deduced from the logic circuit in Fig. 6-26(a). The position of the unintended delay Δt is shown in Fig. 6-26(a) to be at position E for E-HAZARD formation and at D for d-trio formation. These positions (E and D) were ascertained by tracing the direct and indirect paths of the inputs to the second y-variable to change. For example, position E is in the direct path of the sampling variable A to the second state variable to change, y_0, via gates 7 and 11, where gate 11 is the race gate. The indirect path of the input A to race gate 11 is by way of gates 2, 5, and 10, feedback variable y_1 (the first y-variable to change) and gate 8 or 9. Thus, input A and feedback variable y_1 are engaged in a critical race to gate 11 by way of the direct and indirect paths. If Δt_E is sufficiently large, then $(\Delta t_E + \tau_7) > (\tau_2 + \tau_5 + \tau_{10} + \tau_9)$ and y_1 wins the race causing the formation of the E-HAZARD. A counteracting delay Δt which is placed in the y_1 feedback path [dashed box in Fig. 6-26(a)] and which meets the condition $(\Delta t_E + \tau_7) < (\tau_2 + \tau_5 + \tau_{10} + \tau_9 + \Delta t)$ eliminates the E-HAZARD. Notice that the input $y_0 A$ to race gate 11 agrees logically with the state variable and branching condition requirements for the first state variable change [$\textcircled{11} \rightarrow 01$ in Fig. 6-25(a)].

Now, it can be seen that the timing diagram in Fig. 6-26(b) verifies the formation of the E-HAZARD under the condition that $\Delta t = 5\tau_p$ at position E when all gates are assigned a unit path delay, τ_p. The change in input conditions $\overline{A}B \rightarrow AB$ should result in the transition $\textcircled{11} \rightarrow \textcircled{01}$. But because of the presence of the delay $\Delta t_E = 5\tau_p$, the formation of the E-HAZARD causes the FSM to transit incorrectly $\textcircled{11} \rightarrow 01 \rightarrow \textcircled{00}$. The E-HAZARD forms because the unintended delay Δt keeps the terms $y_0 A(H)$ and $y_1 A(H)$ from going ACTIVE as they normally would following the change $\overline{A} \rightarrow A$ shown in the timing diagram.

From the logic diagram in Fig. 6-26(a), we can deduce the requirements for d-trio HAZARD formation just as we did for E-HAZARD formation. The presence of an unintended delay at position D is in the direct path of sampling variable B to state variable y_0 via gates 9 and 11. The indirect path of input B to gate 9 is by way of gates 1, 4, and 10, and feedback variable y_1. Thus, a critical race to gate 9 exists between input B and feedback variable y_1. If Δt_D is sufficiently large such that $\Delta t_D > (\tau_1 + \tau_4 + \tau_{10})$, then y_1 wins the race and the d-trio is formed. A counteracting delay Δt in the feedback path (dashed box) which meets the condition $\Delta t_D < (\tau_1 + \tau_4 + \tau_{10} + \Delta t)$ removes the d-trio.

The formation of the d-trio HAZARD is verified by the timing diagram in Fig. 6-26(c) for the case $\Delta t_D = 5\tau_p$, where, as before, all gates are assigned a unit path delay, τ_p. The d-trio is initiated by the change in input conditions $\overline{A}\,\overline{B} \rightarrow \overline{A}B$, causing the FSM to transit incorrectly $\textcircled{00} \rightarrow 10 \rightarrow \textcircled{11} \rightarrow \textcircled{10}$ when it should have transited $\textcircled{00} \rightarrow \textcircled{10}$. The two additional transitions amount to the production of a two-path delay glitch in the y_0 output signal and, hence, constitute a HAZARD which cannot be eliminated by combinational means.

Not all explicitly located asymmetric path delays lead to E-HAZARD or d-trio formation when the delays are present in the direct path of a sampling variable to a state variable. A typical example is the placement of a delay of any magnitude at position G in Fig. 6-26(a). No E-HAZARD or d-trio is possible in this case because the output of gate 8 ($y_1 A$) does not satisfy the transit requirements for first or second y-variable change deduced from the state diagram in Fig. 6-25(a). For E-HAZARD or d-trio production to occur the output of gate 8 must be compatible with either $\overline{y}_1 \overline{A}$ or $y_1 \overline{A} B$ if gate 8 is the race gate, or compatible with $y_o A$ or $\overline{y}_0 \overline{B}$ if gate 11 is the race gate. However, the logic circuit shows the output of gate 8 to be $y_1 A$. It is left as an exercise for the reader to verify that an E-HAZARD is formed by an asymmetric path delay Δt placed at position F in Fig. 6-26(a) if the requirement $(\tau_7 + \Delta t_F) > (\tau_2 + \tau_5 + \tau_{10} + \tau_9)$ is satisfied.

Summary of SOP Race-Gate Requirements for E-HAZARD and D-trio Formation.

1. A race gate (ANDing or ORing) must be in the *direct path* of the sampling variable to the second y-variable to change (first invariant y-variable), and the asymmetric path delay required to produce the E-HAZARD or d-trio must be present in that direct path. For example, the AND race gate (gate 9) in Fig. 6-26(a) is in the direct path of B to y_0 (the second y-variable to change), and the delay Δt must be present at position D.
2. An ANDing race gate must generate a p-term that is logically compatible with the transit requirements for the *second* y-variable change. For example, $y_1 B$ from AND gate 9 in Fig. 6-26(a) must be logically compatible with $y_1 \overline{A} B$, the requirement for the $10 \rightarrow \textcircled{11}$ transition in the d-trio path shown in Fig. 6-25(a). That is, $y_1 \overline{A} B$ must be contained in the NS expression for the second y-variable to change (Y_0), which it is, as indicated in Fig. 6-25(c).
3. An ORing race gate must have a direct-path p-term input that is logically compatible with the transit requirements for the *first* y-variable change. For example, the $y_0 A$ input to the OR race gate (gate 11) is logically compatible with $y_0 A B$—the requirement for the $\textcircled{11} \rightarrow 01$ transition which initiates the E-HAZARD process. Thus, $y_0 A B$ must be contained in the NS expression for the second y-variable to change (Y_0), which it is, as shown in Fig. 6-25(c).
4. The indirect path for any race gate (ANDing or ORing) must be logically compatible with *all* nonsampling external input variables and must permit the first y-variable change. For example, the indirect path to OR race gate 11 for E-HAZARD formation in Fig. 6-26(a) must be compatible with nonsampling input B, which it is, and must permit a $1 \rightarrow 0$ change in y_1. Note that on rare occasions the inclusion of STATIC HAZARD cover can provide an indirect path to a race gate.
5. An asymmetric path delay of sufficient magnitude placed in the direct path to the race gate causes the race (between direct and indirect paths to the race gate) to be won by the indirect path. The first y-variable to change causes the second y-variable to change, resulting in the formation of a second-order E-HAZARD or d-trio.

The race gate requirements for E-HAZARD and d-trio formation given apply only to SOP logic. A similar set of requirements can be developed for POS logic. The task of doing this is left to the reader as an end-of-chapter problem (Problem 6.19).

A special STATIC HAZARD can result when two coupled p-terms to an ORing race gate contain the sampling variable as the only coupled variable. The delay required to produce the special STATIC HAZARD must be present in the INVERTER-less path of the sampling variable to the ORing race gate and must be larger than the delay of the INVERTER in the remaining path of the sampling variable to the same ORing race gate. The HAZARD is eliminated by adding to the Y expression involved the HAZARD cover (ANDed residues) for the coupled p-terms. In a similar manner, a special STATIC HAZARD can be formed by two coupled s-term inputs to an ANDing race gate in POS logic. Such a HAZARD can be eliminated by adding the ORed residues of the s-terms to the Y expression containing the s-terms.

An example of a special STATIC HAZARD results if a delay $\Delta t_B > \tau_1$ is present at position B in Fig. 6-23(d). In this case the coupled p-terms are those from NAND gates 3 and 4, and the race (to NAND gate 7) is won by p-term $y_0 CK$, thereby causing the FSM to malfunction. The HAZARD is eliminated by adding the term $y_1 y_0$ to the Y_1 expression. While this eliminates the STATIC HAZARD for any value of delay at position B, it allows the formation of an E-HAZARD under the condition that $(\Delta t_B + \tau_3) > (\tau_1 + \tau_4 + \tau_8 + \tau_{HC})$, where τ_{HC} is the path delay associated with the HAZARD cover $y_1 y_0$. In this case the race gate is gate 7 and the path is ⑪ → 10 → ⑩⓪ → ⑩①. Note that the requirements established for E-HAZARD formation are met.

Perspective on HAZARDs. Almost all multiple state asynchronous FSMs designed to operate in the fundamental mode will have the potential to form at least one E-HAZARD or at least one d-trio HAZARD. Fortunately, the requirements for E-HAZARD and d-trio HAZARD formation are not always easily met in the NEXT STATE forming logic of an asynchronous FSM, and even when they are met there is often enough inertial delays associated with the feedback paths to prevent their formation and the malfunction that may result. Still, an active E-HAZARD or d-trio represents one of the most serious types of timing defects known to exist in asynchronous FSMs designed to operate in the fundamental mode. For this reason, designers of asynchronous FSMs must be aware of the potential for these defects to cause FSM malfunction and take reasonable precautions to ensure that this does not happen. The precautions include avoidance of excessive delays on certain leads or in certain gates (including INVERTERs) and, if necessary, the placement of counteracting (preventive) delays on specific feedback lines. If preventive delays are placed in feedback lines, it must be understood that such delays will slow down circuit response. In such cases the designer must weigh the increased probability for successful FSM operation due to the preventive delays with the resulting reduction in circuit response. Thus, massive delays on feedback lines may be totally unacceptable for some high-speed FSM applications. If so, restricted augmentation of the minimum path delay requirements for E-HAZARD formation (without insertion of feedback delays) is possible by DeMorgan transformation of the race gate. But this can be done if an INVERTER is required for the sampling variable in the race gate term. The transformation $y_0 \overline{CK} \to \overline{y_0} + \overline{CK}$ in Fig. 6-23(d) is an example.

It must be remembered that an active ESSENTIAL HAZARD in an

asynchronous FSM cannot be eliminated by the combinational logic methods used to eliminate STATIC HAZARDs. Also, while it is possible (but not probable) that an asynchronous FSM can operate properly with STATIC HAZARDs present, the presence of even one active E-HAZARD will most assuredly cause the FSM to malfunction. Therefore, it is strongly recommended that designers of asynchronous FSMs eliminate all STATIC HAZARDs from their designs and take the necessary precautions to minimize the probability for E-HAZARD formation. The effort to accomplish these aims will not be minimal to be sure, but the end reward of successful FSM operation will more than justify the means of achieving it.

6.5.5 Setup- and Hold-Time Requirements

The essence of the LPD model presented in Sect. 6.2 is the lumped MEMORY element that separates (in time) the corresponding NEXT STATE and PRESENT STATE variables. It was from this fundamental character of the model that the stability criteria [Eqs. (6-2) and (6-3)] were established. There, as will be recalled, it was stated that an asynchronous FSM is considered stable in the jth STATE of some sequence if, and only if, $Y_j(t) = y_j(t)$ for all j. So what requirements must be met by the FSM to satisfy the stability criteria contained in Eq. (6-2)? To answer this question properly, it is necessary to look at the operation of an asynchronous FSM at the STATE level. For this purpose we define (actually redefine from Sect. 5.6.7) the following *timing parameters*:

Sampling Variable. A specific input variable whose passage through its ACTIVE (or INACTIVE) transition point is required to initiate a state-to-state transition. Thus, the sampling variable, also called the *triggering variable,* is the last input variable permitted to change its logic level in the initiation of a state-to-state transition.

Setup Time (t_{su}). The time interval immediately preceding the ACTIVE (or INACTIVE) transition point of the sampling variable during which *all* inputs must be maintained at their proper logic levels.

Hold Time (t_h). The time interval immediately following the ACTIVE (or INACTIVE) transition point of the sampling variable during which *all* inputs must be maintained at their proper logic levels to ensure successful completion of the transition.

Intimately connected with these three timing parameters is the asynchronous input condition. It is appropriate at this time that we restate from Sect. 5.8.4 the definition of asynchronous input:

Asynchronous Input. Any external input to an FSM that can change logic levels at any time particularly during the setup and hold times established by the sampling variable. By definition, the sampling variable for any given STATE is an asynchronous input.

In the discussions that follow, use will be made of the timing parameters to develop a better understanding of the state transition process in any FSM, synchronous or asynchronous. In addition, the reader will gain an appreciation for the heavy restrictions which must be placed on the use of asynchronous inputs in such FSMs.

The timing parameters just defined are illustrated in Fig. 6-27 by the rising edge of the voltage waveform for a sampling variable, x_0, and a transition from STATE $\textcircled{p}$ to STATE $\textcircled{q}$. Here, it is seen that prior to the beginning of the setup time (t_{su}) the external inputs, exclusive of the sampling variable, may change logic levels, but the FSM still retains its present STATE $\textcircled{p}$. During the setup period all external inputs must remain logically stable including the sampling variable which is approaching its triggering (ACTIVE transition) threshold t_{tr}. At the triggering threshold the sampling variable logically samples the logic status of the variables involved in the branching operation. An additional period of time, the hold time t_h, is required for the effects of the sampling variable change to settle in, after which the FSM will reside in a stable STATE $\textcircled{p}$ if $Y_j(t) = y_j(t)$ for all j, or will transit (cycle) on to another state if $Y_j(t) \neq y_j(t)$ for any j. Illustration of the setup- and hold-time requirements relative to the falling edge of the sampling variable waveform follows in a similar manner. Note that, in general, the setup and hold times will vary from STATE to STATE, and that their values for an ACTIVE transition of the sampling variable will usually not be the same as those for an INACTIVE transition.

Each state of an asynchronous FSM will generally have its own sampling variable and its own unique setup and hold-time requirements established by the sampling variable. Therefore, an asynchronous FSM can correctly transit from one stable state to another only if the setup- and hold-time requirements established for that transition are met by the appropriate input variables. In Figs. 6-28(a), (b), and (c) are shown three branching configurations commonly found in asynchronous FSM design. The first, in Fig. 6-

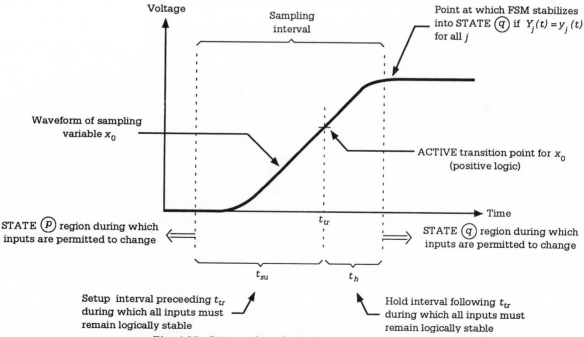

Fig. 6-27. *Rising edge of voltage waveform for sampling variable x_0 showing the setup- and hold-time requirements established by x_0 for a transition $\textcircled{p}$ to $\textcircled{q}$*

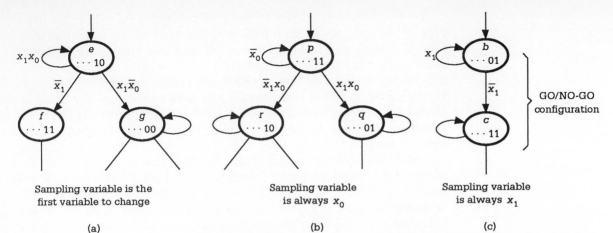

Sampling variable is the
first variable to change

(a)

Sampling variable
is always x_0

(b)

Sampling variable
is always x_1

(c)

Fig. 6-28. Examples of three common branching configurations showing the sampling variable involved for each. (a), (b) Two two-way out-branching configurations where setup- and hold-time requirements of the sampling variable must be met. (c) The GO/NO-GO configuration for which branching is independent of setup- and hold-time considerations.

28(a), features two-way outbranching from a STATE e where the sampling variable is the first variable (x_1 or x_0) to change. If input x_1 is the first to change, then x_0 must meet the setup- and hold-time requirements of x_1 so that a predictable transition will take place. Similarly, if x_0 is the first variable to change, then input x_1 must meet the setup- and hold-time requirements of x_0 to ensure a predictable transition. Thus, in either case, failure of the nonsampling variable to meet the setup- and hold-time requirements established by the sampling variable could result in an unpredictable transition which is unacceptable FSM behavior.

The two-way outbranching configuration in Fig. 6-27(b) differs from that in Fig. 6-28(a) in that the sampling variable is always x_0. So x_1 must meet the setup- and hold-time requirements of x_0 if a predictable transition is to result.

The GO/NO-GO configuration shown in Fig. 6-28(c) deserves special attention because it is the only configuration whose outbranching does not depend on setup- and hold-time considerations. The reader may recall that this GO/NO-GO configuration was cited in Sect. 5.8.4 as being the only reliable means of dealing with asynchronous inputs. And so it remains! The GO/NO-GO configuration in Fig. 6-28(c) differs from that in Fig. 5-67 only to the extent that in the latter case, CLOCK is the implied sampling variable. In Fig. 6-28(c), x_1 is the only input controlling the branching from STATE b, though there may be numerous other inputs associated with an FSM having this configuration in its state diagram. This GO/NO-GO configuration cannot fail for any change in input x_1 that crosses the switching threshold.

Naturally, we would prefer to design all asynchronous FSMs by using the GO/NO-GO in Fig. 6-28(c) since it does not depend on setup- and hold-time considerations and, accordingly, cannot fail. Unfortunately many algorithms require the use of two (rarely more than three) outbranching decisions as in Figs. 6-28(a) and (b), and multiple-branching configurations such as these place severe constraints on the permissible intervals of input change.

To reinforce what has just been learned, consider the branching configuration in Fig. 6-28(b). Let us suppose that at some point in time an FSM resides in STATE $\textcircled{p}$ under holding condition $\bar{x}_0$, and that a transition to STATE $\textcircled{q}$ is required. In Fig. 6-29 we illustrate the requirements for a successful transition $\textcircled{p} \rightarrow \textcircled{q}$ under branching conditions $x_1 x_0$. Notice that the sampling interval must lie within the proper stable region (in this case the HV region) of the input variable waveform before a successful transition $\textcircled{p} \rightarrow \textcircled{q}$ can be assured. This, of course, is precisely what is expressed in Fig. 6-27.

The problems posed by multiple-branching conditions, as in Figs. 6-28(a) and (b), are not trivial and cannot be ignored if meaningful and reliable FSM designs are to result. Let us be clear on one point which emerges from the previous discussion:

> *The sampling variable must be considered the only asynchronous input controlling the branching from any given STATE of a sequential FSM.*

This implies the following:

> *No two (or more) external inputs controlling the branching in an FSM should be allowed to change in close proximity to each other.*

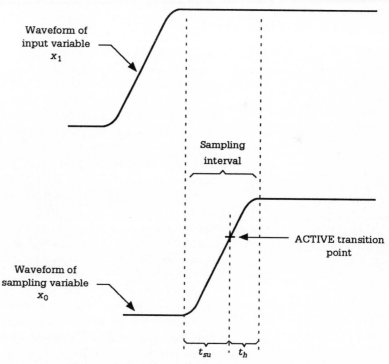

Fig. 6-29. *An input voltage waveform that meets the setup- and hold-time requirements established by the sampling variable,* x_0, *in Figure 6-28(b).*

If this restriction is not observed, stabilized entry of the FSM into a given state cannot be assured. Remember that all inputs in an asynchronous FSM are, by definition, asynchronous unless they are constrained by some means to work "synchronously" with the input designated as the sampling variable.

For an FSM operated synchronously, CLOCK is the designated sampling variable for each state of the machine; hence CLOCK is the universal sampling variable in such a machine. All other external inputs controlling the branching from a particular state must necessarily meet the setup- and hold-time requirements for that state imposed by CLOCK if the FSM is to operate reliably. Failure to meet this requirement can result in unpredictable behavior as discussed at the end of Sect. 5.8.4.

In the general case of a multiple-branching asynchronous FSM, there is no universal sampling variable for the machine. Each state may have its own unique sampling variable. Therefore, to deal with multiple-branching conditions in asynchronous FSMs, the designer really has but one choice:

> *Constrain all external inputs to meet the setup- and hold-time requirements imposed on each state by the sampling variable for that state.*

This means that all (other) input variables in an FSM must be *synchronized* with the sampling variable to the extent that their logic levels cannot be altered during the setup and hold times established by the sampling variable. Thus, the inputs can change their logic levels but only after the FSM has stabilized according to the stability criteria $Y_j(t) = y_j(t)$ for all j. What we have just stated is the fundamental requirement for branching in any sequential machine. Again we state that failure to meet this requirement could lead to unpredictable behavior and malfunction.

How the setup- and hold-time requirements are met for any given state in a multiple-branching asynchronous FSM is determined by the designer of the FSM in consideration of what the FSM must do and the nature of the external world that must provide the FSM's input signals. A simple example would be the case of an asynchronous FSM whose multiple branching from any given state is controlled by push-button (manual) switch variables. In this case the setup- and hold-time requirements for branching from such a state could be satisfied by pushing the input buttons one at a time, making certain that the sampling variable button for that state is pushed last of all. In effect, this synchronizes all (other) inputs to the sampling variable—they are forced to maintain their proper logic levels during the setup and hold times imposed by the sampling variable. But there could be several states in an FSM that have multiple-branching conditions, and each of these states could have its own unique sampling variable. This would require the designer to identify the sampling variable of each such state so that the switch buttons could be pushed in the precise order required for reliable branching.

Section 6.9 discusses other alternatives in dealing with the setup- and hold-time requirements imposed by the sampling variable. There, the hand-shake interface is reintroduced as a means of controlling the timing between external inputs. Section 6.9 also discusses in depth the phenomenon called metastability.

The arguments just presented point to the importance of maintaining strict control over all inputs to an asynchronous (or synchronous) FSM. It is not enough just to design an FSM and hope that it will work when it is put into operation.

6.5.6 Rules for State Diagram Construction

In previous sections of this chapter we have stated or at least implied some of the rules regarding proper construction of the state diagram for asynchronous FSM design. For example, we have already alluded to the need for logically adjacent states so as to avoid possible CRITICAL RACEs, and we have addressed the problem of multivariable branching conditions. Now, it is necessary that we place these rules on a somewhat more formal basis and add to them a few other rules and guidelines which follow quite naturally from the previous sections.

STATE Code Assignment Rule (Logic Adjacency Rule)

> *With few exceptions, all state-to-state transitions must be logically adjacent, that is, unit-distance coded.*

Failure to observe the *logic adjacency rule* for an asynchronous FSM may cause CRITICAL RACEs and possible malfunction of the FSM. One exception is the creation of an intentional non–CRITICAL RACE condition for the expressed purpose of gaining a more optimum design as discussed in Sect. 6.5.2.

When the logic adjacency rule cannot be satisfied for a particular state diagram, extra "fly states" may be added as illustrated in Fig. 6-30. However, it must be remembered that the maximum number of states permitted for m state variables is

$$N_{\max} = 2^m \tag{6-6}$$

If the addition of extra states is not desirable or possible, additional state variables can be added. If neither of these options is desirable, the race condition can be eliminated by following the procedure described in Sect. 6.5.2. In any case, it is essential that *no* CRITICAL RACE be left in the design of the FSM.

Branching Condition Rule

> *State branching dependency on more than one asynchronous input must be avoided.*

Remember that the sampling variable must be the only asynchronous variable controlling branching from a particular state. All other external inputs in a multivariable branching condition must be constrained to meet the setup- and hold-time requirements established by the sampling variable—there are

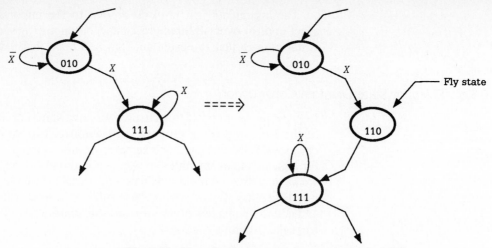

Fig. 6-30. Use of a fly state to prevent a race condition.

no exceptions. How this can be accomplished is discussed in Sect. 6.5.5 and later in Sect. 6.9. If the branching condition rule is not obeyed, dependable operation of the asynchronous FSM cannot be assured.

Output Rules

1. Do not attempt to generate an output on a fly state.
2. Do not attempt to generate an output conditional on a branching (exiting) condition.

The reason for these rules should seem obvious. An output on a fly state or an output conditional of a branching condition will be nothing more than a glitch, possibly too short in duration to trigger the digital device(s) receiving the output signal.

Examples of how *not* to generate outputs in an asynchronous FSM are given in Fig. 6-31. Notice that the unconditional output $P \uparrow \downarrow$ in Fig. 6-31(a) is in violation of output rule 1 and that the conditional output in Fig. 6-31(b) violates output rule 2. Recall that output generation schemes such as those in Fig. 6-31 are proper for synchronous FSMs since CLOCK is understood to be the sampling variable.

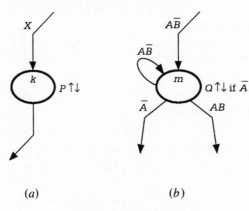

(a) (b)

Fig. 6-31. Examples of improper generation of outputs. (a) Unconditional output on a fly state [violation of Rule (1)]. (b) Output generation conditional on the exiting condition [violation of Rule (2)].

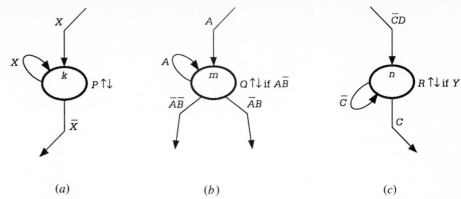

Fig. 6-32. Examples of proper generation of outputs. (a) Unconditional output. (b) and (c) Conditional outputs.

Examples of *proper* generation of unconditional and conditional outputs in an asynchronous FSM are given in Fig. 6-32. It seems clear that the unconditional output in Fig. 6-32(a) is sustained to the extent that input X is held ACTIVE, whereas the unconditional output of Fig. 6-31(a) can be sustained no longer than a path delay. Similarly, the output Q of Fig. 6-32(b) will be sustained for as long as $A\bar{B}$ is ACTIVE. In Fig. 6-32(c) is shown the case of an output, R, that is conditional on an input condition not part of the branching requirements for that STATE. These output assignments together with their many variations are commonly used in asynchronous design.

Expansion and Merge Rules. The LPD model discussed in Sect. 6.2 is valid for the general Mealy FSM as well as for its degenerate forms, the Moore machines, represented in Fig. 6-2. It is desirable at this time to show how the designer can change the state diagram for an asynchronous FSM from a Moore form to that of a Mealy machine or vice versa, or simply merge or expand the state diagram to alter the number of states. There are really only two rules to remember:

Merging or expansion of states in a state diagram should be performed under the following conditions:

1. If the change yields a less complex NEXT STATE or output forming logic design *without* unacceptably altering the basic algorithm or timing constraints of the FSM.
2. If it is necessary to correct for timing defects or to alter the path delays of the circuit so as to meet timing specifications.

Successful merging or expansion of the state diagram requires not only a thorough understanding of the FSM itself but also an understanding of the functional environment within which the machine must operate. Following any such merge or expansion the designer must determine whether or not any timing problems have been introduced by the state diagram alteration (e.g., setup and hold-time problems, HAZARDs, or CRITICAL RACEs).

Examples of typical Mealy $\leftrightarrow$ Moore expansions and merges are presented in Fig. 6-33. Observe that the basic algorithm has been preserved in

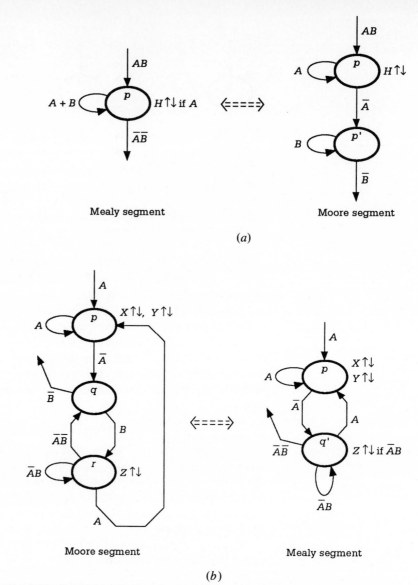

Fig. 6-33. Examples of state diagram expansion and contraction (merge). (a) Simple Mealy ↔ Moore conversion. (b) More complex Mealy ↔ Moore conversion.

each conversion but that no account has been taken of any timing constraints which might have been in effect. The addition or removal of states changes the timing conditions of the FSM since each state-to-state transition represents a certain amount of delay. For example, in the Moore segment of Fig. 6-33(b), STATE q is used as a fly state presumably for purposes of maintaining logically adjacent transitions. Its merger with STATE r produces the Mealy segment shown which changes the timing conditions of the FSM. Note that in either segment the output Z is permitted to be ACTIVE only for input conditions $\overline{A}B$. Also, note that all outgoing branching conditions for each state satisfy Eqs. (6-4) and (6-5).

Section 5.8.6 dealt with the subject of initialization and reset of synchronous FSMs. There, we described the use of a SANITY circuit to PRESET or CLEAR a FLIP-FLOP. Of course, FLIP-FLOPs are not present in asynchronous FSMs that are designed to operate in the fundamental mode. But SANITY circuits, of the type shown in Fig. 5-77, are still required to initialize these FSMs as described in the paragraphs that follow.

To initialize (or reset) an asynchronous FSM, it is necessary that the outputs of the SANITY circuit [SANITY(H) or SANITY(L)] be connected to the appropriate gates so as to drive the present state variables (hence the FSM) into the required initial state. With reference to the conjugate gate forms shown in Fig. 2-28, we set out the MIXED LOGIC connections necessary to initialize the gates of which all digital machines are composed.

To Initialize Logic 0:

Connect SANITY(L) to ACTIVE HIGH input ANDing gate forms, NAND and AND.

Connect SANITY(H) to ACTIVE LOW input ANDing gate forms, NOR and OR.

To Initialize Logic 1:

Connect SANITY(L) to ACTIVE LOW input ORing gate forms, NAND and AND.

Connect SANITY(H) to ACTIVE HIGH input ORing gate forms, NOR and OR.

Note that these connection requirements are satisfied inside FLIP-FLOPs as, for example, the JK-FF shown in Fig. 5-30(b).

In making the connections indicated, care should be exercised to plan for the fan-in requirements of the gates and the fan-out capabilities of the SANITY circuit(s). Related to these remarks is the admonition to analyze the logic circuit carefully before making the SANITY connections. In most cases the goal should be to initialize only those gates that are essential to the initialization process. The JK-FF in Fig. 5-30 is a simple case in point. On the other hand, implementation of an asynchronous FSM with array logic, such as PLAs or PALs, may require the initialization of entire AND or OR arrays.

6.6   DESIGN AND ANALYSIS OF ASYNCHRONOUS FSMs

Up to this point we have applied the lumped path delay (LPD) model to the design of simple asynchronous FSMs and have discussed many factors which must be considered before successful designs can be achieved. Now it is desirable to gather this information together into procedural form followed by a few simple but useful design examples.

The design procedure that follows is presented in *five* parts for future reference purposes. We emphasize that the procedure we present is to be considered as a guide only, not a rigorous routine that must necessarily be followed. Deviations from this procedure are expected depending on the complexity of the FSM involved.

Part I. Understand the Problem

1. Develop a thorough understanding of the functional requirements and I/O specifications of the asynchronous FSM to be designed.
2. Take note of any specific timing constraints that must be met.

Part II. Construct the State Diagram

1. Construct a fully documented state diagram which is free of ENDLESS CYCLEs and which is in agreement with the algorithm and timing constraints for the FSM. In particular, make certain that all inputs controlling the branching from each STATE meet the setup- and hold-time requirements of that STATE. If necessary, use timing diagrams to test for branching reliability. Use the GO/NO-GO configuration whenever possible, even at the expense of using extra states or state variables.
2. If desirable make a preliminary judgment as to the type of outputs (conditional or unconditional) required and merge or expand the state diagram accordingly. Repeat step 1; then go to Part III.

Part III. Make State Code Assignments

1. Unit-distance code all State-to-state transitions in the state diagram so that they are logically adjacent; then go to Part IV.
2. If unit-distance coding is not possible, do one of the following:
   a. Add fly states, or merge or expand the state diagram as needed to establish logic adjacency and repeat Part II.
   b. Check for CRITICAL RACEs. If present, remove them (see Sect. 6.5.2) and repeat Part II; then go to Part IV.
   c. Go to Part IV if only nonCRITICAL RACE conditions exist, if the alternative paths of the race cause no timing problems or output glitches, and if the race conditions contribute to a more optimum circuit. Otherwise, remove all race conditions as in 2(b), repeat Part II, and go to Part IV.

Part IV. Obtain the NEXT STATE and Output Functions

1. Use the state diagram and the LPD model to map each NEXT STATE variable (Y_i) and each output by taking the feedback variables (y's) as the map axes.
2. Loop out minimum or reduced cover for each NEXT STATE variable and output.
3. Check the minimum (or reduced) cover for STATIC HAZARDs and potential E-HAZARDS and d-trios by following the procedures given in Sects. 6.5.3 and 6.5.4. Eliminate the STATIC HAZARDs and make note of all potential E-HAZARDS and d-trios for future reference, or take preventive action, or both. Now go to Part V.

Part V. Construct the Logic Diagram

1. Construct the circuit in MIXED LOGIC notation by taking into account the I/O specifications given in Part I.
2. If possible, implement and test the circuit under actual operating conditions. The use of a computer-aided simulation is also an acceptable testing procedure.

The five-part design procedure first given is flexible and may be altered somewhat to meet the needs of a particular design project. However, significant deviation from the intent of any part of it may result in an improper design and malfunction of the FSM. To provide a better perspective of the intent of each part of the design procedure, we offer the abridged flowchart shown in Fig. 6-34. Although some detail has been omitted, it represents those features we wish to emphasize.

6.6.2   Design Examples

We have covered the basic design procedure, and it is now desirable to illustrate its use with a few useful examples. In the examples that follow we will design a commonly used FLIP-FLOP and three devices of our own choosing, including two that were designed as synchronous FSMs in Chapter 5. An attempt will be made to cover the essential features of the design methodology emphasized in this chapter, but without specific mention of the various steps.

The Edge-Triggered D FLIP-FLOP. Let us assume that this device is to trigger on the rising edge of the CLOCK (CK) and that the DATA (D) input must always be constrained to meet the setup and hold-times established by CLOCK, the sampling variable. We may further assume that both inputs, CK and D, arrive from positive logic sources and that the device is to have two outputs $Q(H)$ and $Q(L)$.

Presented in Fig. 6-35(a) is the four-state state diagram for this device which we reproduce here from Fig. 5-36(a). Notice that there are no END-LESS CYCLEs. Also, there are no race conditions since all state-to-state transitions are logically adjacent (i.e., are unit distance coded).

In Figs. 6-35(d) and (c) are shown the K-maps and minimum cover for the NEXT STATE variables, Y_1 and Y_0, and the output, Q. An inspection of the Y expressions indicates that neither has a STATIC HAZARD. The coupled terms $y_1\overline{D}CK$ and $y_0\overline{CK}$ in Y_1 do not produce a STATIC HAZARD in STATE ⑪ on a $CK \rightarrow \overline{CK}$ change, and the HAZARD produced in STATE ⑪ and associated with the coupled terms $\overline{y}_1DCK$ and $y_0\overline{CK}$ in Y_0 is covered by $\overline{y}_1y_0$.

There is considerable potential for E-HAZARD and d-trio formation as can be seen from a brief inspection of the state diagram in Fig. 6-35(a). If certain unintended asymmetric path delays exist in the circuit, E-HAZARD formation will occur along path ⓑ$\rightarrow c \rightarrow$ⓓ$\rightarrow$ⓐ under branching condition $\overline{DCK}$ from STATE ⓑ, or along path ⓓ$\rightarrow a \rightarrow$ⓑ$\rightarrow$ⓒ under branching condition $D\overline{CK}$ from STATE ⓓ. And d-trio formation will occur along path ⓐ$\rightarrow b \rightarrow$ⓒ$\rightarrow$ⓑ, or along path ⓒ$\rightarrow d \rightarrow$ⓐ$\rightarrow$ⓓ. Notice that the d-trio paths cause no glitches in the output Q and, therefore, amount only to a

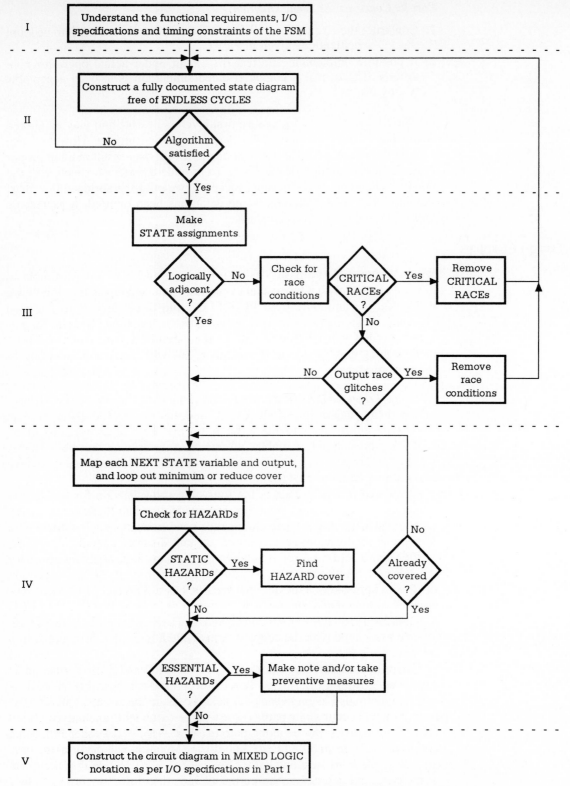

Fig. 6-34. *An abridged flow chart of the design procedure for asynchronous FSMs showing five major parts.*

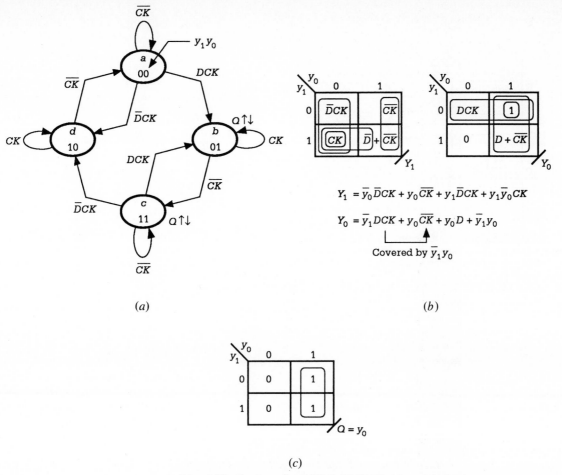

$$Y_1 = \bar{y}_0 \overline{DCK} + y_0 \overline{CK} + y_1 \overline{DCK} + y_1 \bar{y}_0 CK$$

$$Y_0 = \bar{y}_1 DCK + y_0 \overline{CK} + y_0 D + \bar{y}_1 y_0$$

Covered by $\bar{y}_1 y_0$

(a)   *(b)*

(c)

Fig. 6-35. *Design of a RET D FLIP-FLOP. (a) State diagram. (c) K-maps and minimum cover for the NEXT STATE variables. (c) Output K-map.*

small delay in achieving state stability. The E-HAZARD paths, on the other hand, are very disruptive since they cause either an unintended SET or RESET of the FLIP-FLOP. Both E-HAZARD paths are of the type for which $k = d$ in Fig. 6-22.

This brings us to Figs. 6-35(d) and (e), where we have constructed the logic circuit diagram based on the LPD model and have provided the circuit symbol for this FLIP-FLOP. For comparison purposes later, the gate/input tally is 9/26, which, as usual, excludes INVERTERs.

Our D FLIP-FLOP in Fig. 6-35(d) has been designed with a mixed-rail output by connecting an INVERTER to the output line $y_0(H)$. For this reason we have given this FLIP-FLOP the same circuit symbol as that given to the RET D-FF in Fig. 5-38(c). However, it must be understood that the IN-VERTER form of the mixed-rail output does not take the place of the unique mixed-rail outputs produced from a BASIC CELL, the outputs that are common to most FLIP-FLOPs. In Sect. 6.7.2 we will discuss methods of design that will provide this unique mixed-rail output on FLIP-FLOPs and, at the same time, yield simpler but not necessarily less trouble-prone designs.

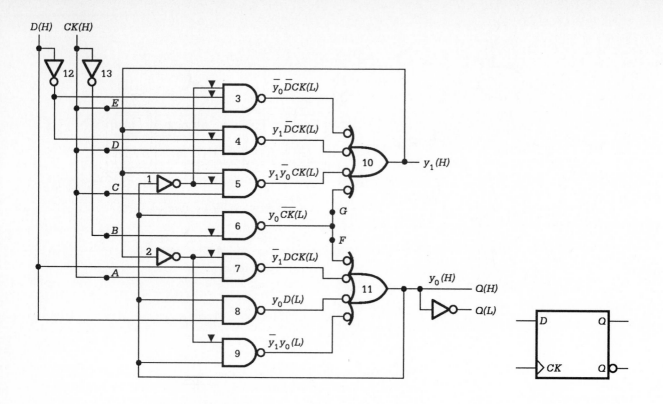

(d) (e)

Fig. 6-35 (contd.). *(d) Circuit diagram derived from the LPD model. (e) Circuit symbol.*

Special-Purpose Edge-Triggered FLIP-FLOP. For this example we will design a special SET/TOGGLE (ST) edge-triggered FLIP-FLOP which has the circuit symbol given in Fig. 6-36(a) and which operates according to the table given in Fig. 6-36(b), reproduced from Fig. 5-44(a). It is assumed that CLOCK samples the data on its falling edge (hence FET) and that inputs S and T must be constrained to meet the setup and hold-times established by CLOCK. The polarization levels required for the inputs and output are implied by the circuit symbol.

In Fig. 6-36(c) is shown the state diagram for this ST FLIP-FLOP. It is derived directly from the operation table given in Fig. 6-36(b). For example, the SET transition from STATE ⓐ to STATE ⓑ is determined from the combination of the SET and TOGGLE conditions, or $(\overline{S}\,\overline{T} + \overline{S}T + S\overline{T}) \cdot \overline{CK} = (\overline{S} + \overline{T}) \cdot \overline{CK}$, and the RESET transition is just the TOGGLE condition, $ST\overline{CK}$, all triggered on the falling edge of the CLOCK, $\overline{CK}$. The remainder of the branching conditions are determined from the application of Eqs. (6-4) and (6-5). Observe that the completed state diagram has no ENDLESS CYCLEs or CRITICAL RACEs.

The K-maps and minimum cover given in Fig. 6-36(d) reveal one STATIC HAZARD in Y_1 with HAZARD cover as indicated. Also, there is significant potential for E-HAZARD and d-trio production as can be seen from the state diagram. If the appropriate delays exist in the circuit, E-HAZARD formation

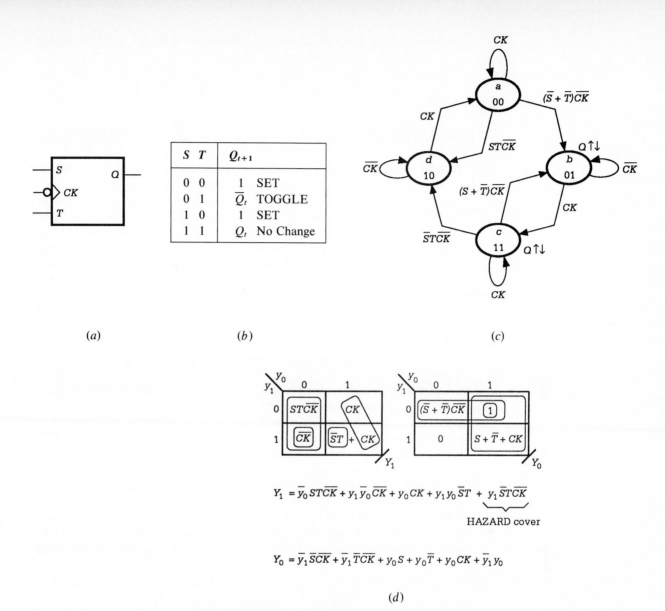

Fig. 6-36. *Design of a special SET/TOGGLE (ST) FET-FLIP-FLOP. (a) Circuit symbol. (b) Operation table. (c) State diagram. (d) K-maps and minimum cover showing HAZARD COVER.*

can occur along path $(a) \rightarrow b \rightarrow (c) \rightarrow (d)$ under branching condition $\overline{STCK}$ from STATE (a), along path $(b) \rightarrow c \rightarrow (d) \rightarrow (a)$ under branching condition $\overline{STCK}$ from STATE (b), or along path $(d) \rightarrow a \rightarrow (b) \rightarrow (c)$ under branching condition $(\overline{S} + \overline{T})CK$ from STATE (d). The first E-HAZARD path would cause a (+) glitch in Q, the second path would produce an unintended RESET of the FLIP-FLOP, and the last path would lead to an unintended SET condition. All four E-HAZARD paths are of the type for which $k = d$ in Fig. 6-22. A d-trio can occur along path $(a) \rightarrow b \rightarrow (c) \rightarrow (b)$ but only under branching condition $\overline{TCK}$. This d-trio causes no problem in the output signal but does cause a delay in achieving state stability. Remember that should an E-HAZARD or d-trio become a problem, it could be eliminated by loading the appropriate y-variable feedback line with a delay at least equal to the net delay that caused the HAZARD, assuming that the delay magnitude is known.

One- to Three-Pulse Generator. We leave the subject of FLIP-FLOPs to design an asynchronous FSM that was previously designed in Sect. 5.9.2 as a synchronous machine. This is done so as to illustrate the differences that exist between synchronous and asynchronous FSM design. In Fig. 6-37 we provide the block diagram and operation table for the one- to three-pulse generator which were given previously in Fig. 5-81. In comparing the two figures, it is evident that oscillator (OSC in the asynchronous design) has replaced CLOCK in the synchronous design. It is the function of this machine to generate a single set of one, two, or three clean pulses depending on the prior settings of the two debounced control switches, SW_1 and SW_0. The settings of the control switches must be made and settled in prior to the use of the START button S, which is also debounced. We will assume that the oscillator operates at a frequency which does not exceed the limitations established by the worst case gate path delays inherent in the machine and that S is of long duration compared to the OSC (sampling variable) period.

The state diagram for an asynchronous FSM that meets the requirements of the one- to three-pulse generator is presented in Fig. 6-38(a). A brief inspection of this state diagram indicates that it is a Mealy machine (it has conditional outputs) of three state variables and it has no ENDLESS CYCLEs or CRITICAL RACEs. Notice that we have arbitrarily separated the pulses by one OSC period, the minimum separation. Greater separation can be obtained in multiples of OSC by adding more states. Note also that this state diagram is formed exclusively of GO/NO-GO branching configurations.

Shown in Fig. 6-38(b) are the K-maps and minimum cover for the three NEXT STATE variables. The expression for Y_1 reveals one STATIC HAZARD which occurs in STATE 011 on the $OSC \rightarrow \overline{OSC}$ change, as indicated

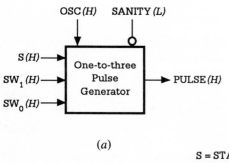

SW_1	SW_0	Output
0	0	No output
0	1	One pulse
1	0	Two pulses
1	1	Three pulses

(a)

(b)

S = START
SW = Switch
OSC = Oscillator

Fig. 6-37. *I/O specifications for the one- to three-pulse generator. (a) Circuit symbol. (b) Operation table.*

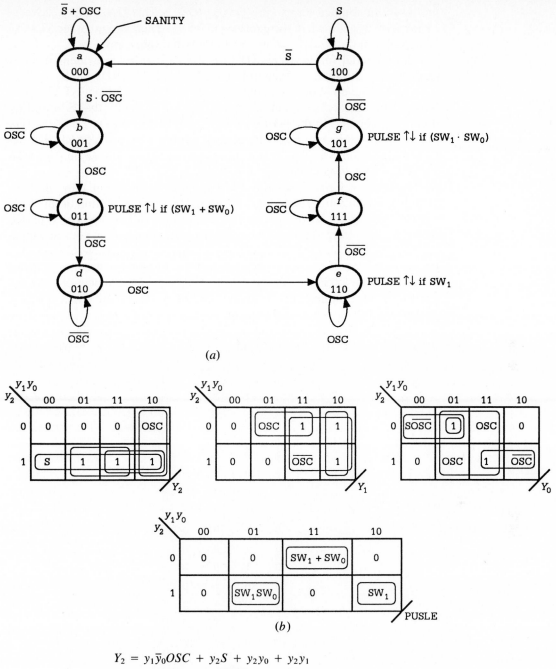

(a)

$Y_2 = y_1\bar{y}_0 OSC + y_2 S + y_2 y_0 + y_2 y_1$

$Y_1 = \bar{y}_2 y_0 OSC + y_1 \overline{OSC} + y_1 \bar{y}_0 + \underbrace{\bar{y}_2 y_1}$

HAZARD cover

$Y_0 = \bar{y}_2 \bar{y}_1 S\overline{OSC} + y_0 OSC + y_2 y_1 \overline{OSC} + \bar{y}_2 \bar{y}_1 y_0$

$PULSE = \bar{y}_2 y_1 y_0 SW_1 + \bar{y}_2 y_1 y_0 SW_0 + y_2 y_1 \bar{y}_0 SW_1 + y_2 \bar{y}_1 y_0 SW_1 SW_0$

Fig. 6-38. *Design of a one- to three-pulse generator. (a) State diagram. (b) K-maps and minimum cover for HAZARD-free operation.*

by the set of coupled terms $(\bar{y}_2 y_0 OSC,\ y_1\overline{OSC})$. This HAZARD is covered by the term $\bar{y}_2 y_1$ shown in the expression for Y_1. There are no other STATIC HAZARDs in the Y expressions or the output expression.

The potential for E-HAZARD formation exists as is evident from an inspection of the state diagram in Fig. 6-38(a). E-HAZARDs can be initiated from any of seven states. The initiation of E-HAZARD from STATE $\textcircled{g}$ is not possible under the conditions required for S, namely, that S be of long duration compared to the period for OSC. All E-HAZARDs are of the type for which $k = d$ in Fig. 6-22 with one exception. The E-HAZARD that can be initiated from STATE $\textcircled{f}$ is of the type for which $k = \textcircled{c}$ in Fig. 6-22. There are no d-trios possible in this FSM.

The two-level circuit for this FSM requires a gate/input tally of 20/64, including the output forming logic portion. The implementation of this FSM is best suited to an array logic device such as an FPLA. Presented in Fig. 6-39(a) is the p-term table suitable for programming a PLA and in Fig. 6-39(b) is given the abridged symbolic representation for an $8 \times 16 \times 4$ FPLA (or FAPLA) which has been programmed to function as the one- to three-pulse generator of Fig. 6-38. Here, the feedback line connections are not shown but are implied. For a review of FPLAs and their applications to design, the reader is referred to Sect. 4.11.3.

The SANITY(L) circuit, discussed in Sect. 5.8.6, is applied to the operation of the one- to three-pulse generator as indicated in Fig. 6-39(b). Here,

p-term	y_2	y_1	y_0	OSC	S	SW_1	SW_0	y_2	y_1	y_0	Pulse
	Inputs							Outputs			
$y_1\bar{y}_0 OSC$	—	1	0	1	—	—	—	1	0	0	0
$y_2 S$	1	—	—	—	1	—	—	1	0	0	0
$y_2 y_0$	1	—	1	—	—	—	—	1	0	0	0
$y_2 y_1$	1	1	—	—	—	—	—	1	0	0	0
$\bar{y}_2 y_0 OSC$	0	—	1	1	—	—	—	0	1	0	0
$y_1\overline{OSC}$	—	1	—	0	—	—	—	0	1	0	0
$y_1\bar{y}_0$	—	1	0	—	—	—	—	0	1	0	0
$\bar{y}_2 y_1$	0	1	—	—	—	—	—	0	1	0	0
$\bar{y}_2\bar{y}_1 S\ \overline{OSC}$	0	0	—	0	1	—	—	0	0	1	0
$y_0 OSC$	—	—	1	1	—	—	—	0	0	1	0
$y_2 y_1\overline{OSC}$	1	1	—	0	—	—	—	0	0	1	0
$\bar{y}_2\bar{y}_1 y_0$	0	0	1	—	—	—	—	0	0	1	0
$\bar{y}_2 y_1 y_0 SW_1$	0	1	1	—	—	1	—	0	0	0	1
$\bar{y}_2 y_1 y_0 SW_0$	0	1	1	—	—	—	1	0	0	0	1
$y_2 y_1\bar{y}_0 SW_1$	1	1	0	—	—	1	—	0	0	0	1
$y_2\bar{y}_1 y_0 SW_1 SW_0$	1	0	1	—	—	1	1	0	0	0	1

(a)

Fig. 6-39. *PLA implementation of the one- to three-pulse generator of Fig. 6-38. (a) p-term table.*

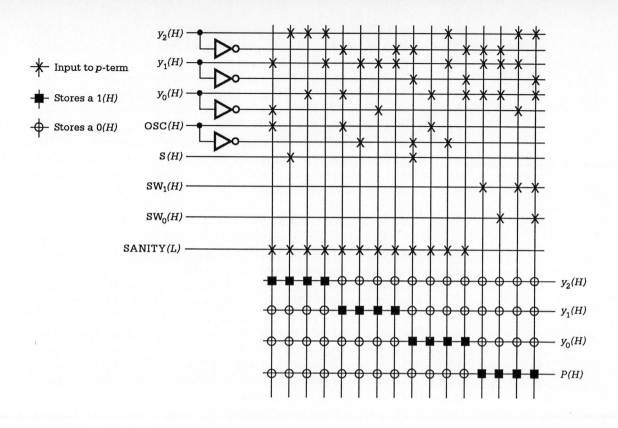

(b)

Fig. 6-39 (contd.). *(b) Abridged symbolic representation of an 8 × 16 × 4 FPLA (or FPAL) programmed to function as the one- to three-pulse generator.*

the FSM is driven into STATE 000 on power-up by connecting SANITY(*L*) to each *p*-term line that is associated with a *y* variable as indicated by the 12 connections (*X*'s). The only provisos are that the START button S not be pushed until after the machine has settled into the 000 STATE, a period of about $\tau = RC$, which is the time constant given by Eq. (5-8) for the SANITY circuits discussed in Sect. 5.8.6, and that both $S(H)$ and SAN-ITY(*L*) input signals be of long duration compared to the period for OSC. These conditions avoid the complications that might result from having branching dependency on more than one asynchronous input (the branching condition rule). Equally acceptable would be to power-up the one- to three-pulse generator into the 100 STATE, ⓗ. This would be accomplished by disconnecting the SANITY(*L*) input to the four $y_2(H)$ *p*-term lines in Fig. 6-39(b) and adding in their place a single *s*-term (vertical) line connecting SANITY(*H*) to $y_2(H)$ with the understanding that the provisos given above still apply. A SANITY(*H*) signal should be taken from the SANITY(*L*) circuit by using an odd number of INVERTERs, preferably three for buffering purposes. In effect, a SANITY(*L*) input disables (CLEARs) an AND stage having ACTIVE HIGH inputs, while a SANITY(*H*) input enables (PRESETs) an OR stage having ACTIVE HIGH inputs. The polarization levels of the SANITY circuits used must be reversed for the case of ACTIVE LOW inputs to the AND or OR stages. Refer to Sect. 6.5.7 for a review of this subject.

Use of a PROM to implement the one- to three-pulse generator would not be a good choice for two reasons. First, fully developed STATIC HAZARDs produced in a ROM device cannot be eliminated by combinational logic methods simply because one does not have access to the exhaustive decoder section of the ROM. So, if used in an asynchronous FSM, PROMs will produce logic noise and may cause the FSM to malfunction. Second, an eight-input PROM generates 256 MINTERMs of which only a small fraction are actually used, an inefficient use of the device.

Digital Combination Lock/Sequence Detector. As a final example for this section, let us design a two-input/two-output FSM that is to function as a digital combination lock, or sequence detector, according to the I/O specifications given in Fig. 6-40. Like the example just discussed, this machine has been previously designed as a synchronous FSM (in Sect. 5.9.3) and is redesigned here as an asynchronous FSM for comparison purposes. It is a requirement that the entire sequence provided in Fig. 6-40(b) be followed without deviation before the FSM will issue the output OPEN. The inputs A and B are treated the same as switches S1 and S2 in Fig. 5-89. It is assumed that a SANITY(L) circuit will be used to start up the FSM in the initial LOCK STATE and that each switch action will be debounced and widely separated in time from the other. The latter assumption is necessary to ensure that the setup- and hold-time requirements are met for each state. The complement of the switch variable may be taken as the OFF condition.

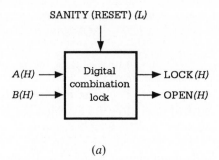

(a)

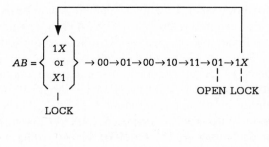

(b)

Fig. 6-40. I/O specifications for a digital combinational lock. (a) Block diagram symbol. (b) Input sequence leading to the issuance of the output OPEN.

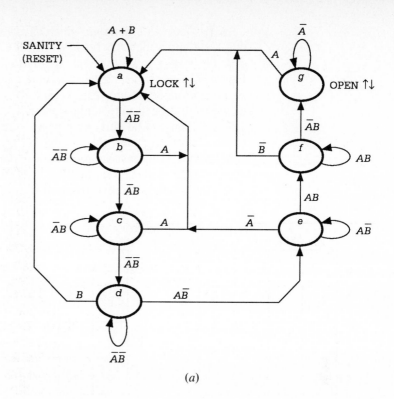

(a)

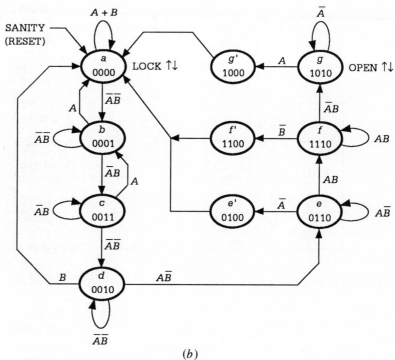

(b)

Fig. 6-41. *Design of a digital combination lock. (a) State diagram as derived from Figure 5-91(a) but without fly states and state code assignments. (b) State diagram with fly states and state code assignments.*

6.6 / Design and Analysis of Asynchronous FSMs

591

A first-cut state diagram which satisfies the sequential requirements of this FSM and which has no ENDLESS CYCLEs is constructed in Fig. 6-41(a). It is seen to be the same as that used for the synchronous design in Fig. 5-91(a). However, it has one major drawback as the state diagram for an asynchronous design. An excessively large number of STATE variables would have to be used to achieve logically adjacent state-to-state transitions. This problem can be solved by using unused states as fly states to branch back to STATE $\textcircled{a}$, but this would still require the use of more states than is necessary. The best solution is to make use of states in the original sequence as fly states, provided that these states are *not* output states, and then "fill in" elsewhere with unused states. This is done in Fig. 6-41(b) where

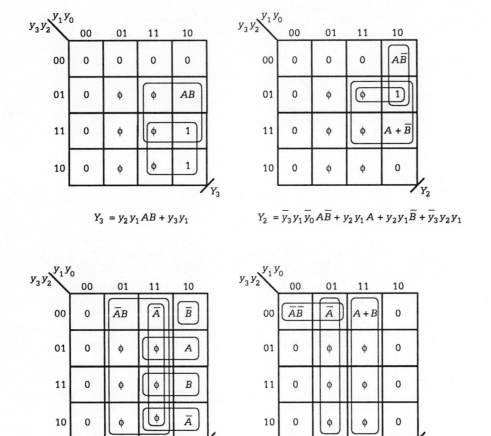

$$Y_3 = y_2 y_1 AB + y_3 y_1$$

$$Y_2 = \bar{y}_3 y_1 \bar{y}_0 A\bar{B} + y_2 y_1 A + y_2 y_1 \bar{B} + \bar{y}_3 y_2 y_1$$

$$Y_1 = \bar{y}_3 \bar{y}_2 y_1 \bar{y}_0 \bar{B} + \bar{y}_3 y_2 y_1 A + y_3 y_2 y_1 B + \bar{y}_3 \bar{y}_2 y_1 \bar{A}$$

$$+ y_0 \bar{A}B + y_1 y_0 \bar{A} + \underbrace{\bar{y}_3 \bar{y}_2 y_1 \bar{A}\bar{B} + y_3 y_1 \bar{A}B}_{\text{HAZARD cover}}$$

$$Y_0 = \bar{y}_3 \bar{y}_2 y_1 \bar{A}\bar{B} + \bar{y}_1 y_0 \bar{A} + y_1 y_0 A + y_1 y_0 B$$

(c)

Fig. 6-41 (contd.). (c) K-maps and minimum cover for the NEXT STATE variables.

NEXT STATE Variable	Coupled Variable Change	Coupled Terms	State Transition and Branching Condition*	HAZARD Cover
Y_1	$y_2 \rightarrow \bar{y}_2$	$\bar{y}_3 y_2 y_1 A$, $\bar{y}_3 \bar{y}_2 y_1 \bar{y}_0 B$	$0110 \xrightarrow{A\bar{B}} 0010$	NA†
Y_1	$y_3 \rightarrow \bar{y}_3$	$y_3 \bar{y}_2 y_1 \bar{A}$, $\bar{y}_3 \bar{y}_2 y_1 \bar{y}_0 B$	$1010 \xrightarrow{\bar{A}\bar{B}} 0010$	NA
Y_1	$y_0 \rightarrow \bar{y}_0$	$y_1 y_0 \bar{A}$, $\bar{y}_3 \bar{y}_2 y_1 \bar{y}_0 B$	$0011 \xrightarrow{\bar{A}\bar{B}} 0010$	$\bar{y}_3 \bar{y}_2 y_1 \bar{A}\,\bar{B}$
Y_1	$y_3 \rightarrow \bar{y}_3$	$y_3 y_2 y_1 B$, $\bar{y}_3 y_2 y_1 A$	$1110 \xrightarrow{AB} 0110$	NA
Y_1	$A \rightarrow \bar{A}$	$\bar{y}_3 y_2 y_1 A$, $y_0 \bar{A} B$	$0111 \xrightarrow{B} 0111$	NA
Y_1	$A \rightarrow \bar{A}$	$\bar{y}_3 y_2 y_1 A$, $y_1 y_0 \bar{A}$	$0111 \longrightarrow 0111$	NA
Y_1	$y_2 \rightarrow \bar{y}_2$	$y_3 y_2 y_1 B$, $y_3 \bar{y}_2 y_1 \bar{A}$	$1110 \xrightarrow{\bar{A}\bar{B}} 1010$	$y_3 y_1 \bar{A} B$
Y_0	$y_1 \rightarrow \bar{y}_1$	$y_1 y_0 B$, $\bar{y}_1 y_0 \bar{A}$	$0011 \xrightarrow{\bar{A}\bar{B}} 0001$	NA

* Deduced from the coupled terms.
† Not applicable. The state transition does not occur under the input conditions required by the coupled terms, or does not occur at all.

(d)

Fig. 6-41 (contd.). *(d) STATIC HAZARD analysis.*

ten states have been given state code assignments (by using four state variables) which makes the FSM free of CRITICAL RACEs and free of output race glitches.

The K-maps and minimum cover for the state diagram of Fig. 6-41(b) are given in Fig. 6-41(c). A STATIC HAZARD analysis of the NEXT STATE variable expressions is provided in Fig. 6-41(d). The analysis indicates that of the several sets of coupled variables that exist only two sets yield valid STATIC HAZARDs. Both HAZARDs exist in the Y_1 expression and are internally initiated. The minimum HAZARD cover for both is indicated in Figs. 6-41(c) and (d). Note that no externally initiated HAZARD is possible since the FSM never undergoes a transition through the 0111 STATE. It is left as an exercise for the reader to verify that there are six E-HAZARD paths and one d-trio path in this digital combination lock.

The outputs, OPEN and LOCK, are generated by the terms $Y_3 \bar{Y}_2 Y_1$ and $\bar{Y}_3 \bar{Y}_2 \bar{Y}_1 \bar{Y}_0$, respectively, as read from the state diagram. These together with the NEXT STATE forming logic make a total gate/input tally of 24/91 (excluding INVERTERs) assuming no limits are placed on the number of gate inputs that are permitted.

The digital combination lock (or sequence detector) we have just designed is, in reality, a controller since the outputs, OPEN and LOCK, must produce action in some other part of a system as, for example a solenoid or a data bank, and so on. Thus, this example could be regarded as an introduction to system-level design. However, there is much more to system-level design than has been offered here. Section 6.7 takes yet another step in that direction by developing the concept of the NESTED machine.

A brief inspection of the flowchart in Fig. 6-34 indicates that analysis is an integral part of the design process. No design is complete without checking the FSM for the various defects that may cause it to malfunction, or that may cause the NEXT STATE, to which it is attached, to malfunction. Still, there are times when a designer must determine the sequential nature of a logic circuit for an asynchronous FSM so as to determine its function or to explain why it does not operate properly. In any case, the approach is roughly the reverse of the design procedure given in Fig. 6-34 to the extent that a state diagram is derived from the logic circuit. The analysis may then be continued by using that state diagram together with all other information to troubleshoot the FSM for any possible problems it may have. On the whole, the analysis procedure is quite similar to that given in Sect. 5.9.4 for synchronous FSMs.

Procedure for Asynchronous FSM Analysis. The procedure for asynchronous FSM analysis is presented in the following six steps:

1. Given the circuit for an asynchronous FSM in either SOP or POS form and assuming that it has been designed to operate in the fundamental mode, reconstruct the circuit in MIXED LOGIC LPD form (if this has not already been done) so as to identify the NEXT STATE and PRESENT STATE variables.
2. Read the LPD circuit to obtain the NEXT STATE and output logic expressions.
3. Map the NEXT STATE and output logic expressions in EV K-maps that have as their coordinate axes the PRESENT STATE variables.
4. Construct the PRESENT STATE/NEXT STATE (PS/NS) TABLE (in CANONICAL SOP form) directly from the EV K-maps. Inclusion of the output data in the PS/NS TABLE is necessary only when the output forming logic is complex.
5. Construct the fully documented state diagram from the PS/NS TABLE.
6. Analyze the state diagram for possible ENDLESS CYCLEs (oscillations), CRITICAL RACEs, output race glitches, STATIC HAZARDs, and ESSENTIAL HAZARDs (including d-trios). If the FSM contains either an oscillation or CRITICAL RACE or both, the usual procedure is to break off the analysis at this point and eliminate these defects by following some known algorithm for the FSM's operation. It makes little sense to continue the analysis of an FSM that contains timing defects that will cause it to malfunction. However, once the FSM is free of oscillations and CRITICAL RACEs, the analysis should continue.

An Example. As an example, Fig. 6-42(a) shows a two-input/one-output logic circuit for an asynchronous FSM which is to be analyzed. From its appearance we judge that it should be analyzed by using the LPD model, and so we reconstruct it as in Fig. 6-42(b), where the PRESENT and NEXT STATE variables are easily identified. The next step is to read the LPD circuit for the NEXT STATE and output logic expressions and then plot the corresponding K-maps. These results are presented in Fig. 6-42(c).

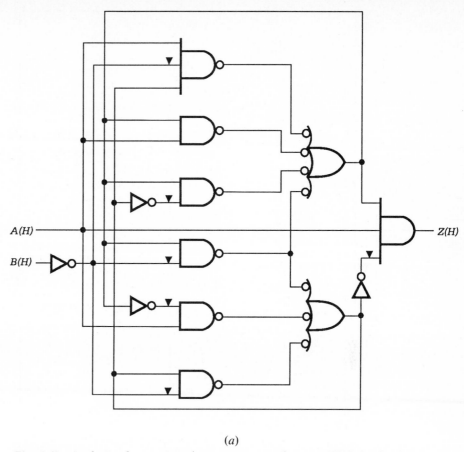

(a)

Fig. 6-42. *Analysis of a two-input/one-output asynchronous FSM. (a) Logic circuit.*

The final steps are to construct the PS/NS TABLE from the K-maps and then use these K-maps to derive the state diagram. This is done and the results are given in Figs. 6-42(d) and (e). Since the single output is a conditional output, the FSM is identified as a Mealy machine.

The defect analysis begins with the state diagram where it is apparent that the FSM has no ENDLESS CYCLEs. However, it does have a CRITICAL RACE. In attempting to transit $\widehat{c} \rightarrow \widehat{a}$ the FSM will stabilize into either STATE $\widehat{a}$ or STATE $\widehat{d}$ since the holding conditions for both these states contain the $\widehat{c} \rightarrow \widehat{a}$ branching condition $\overline{A}B$. The FSM can end up in STATE $\widehat{a}$ because the branching condition from $\widehat{b}$ to $\widehat{a}$ is $\overline{A}B$. Assuming that the algorithm on which this FSM is based requires a transiton from $\widehat{c}$ $\rightarrow \widehat{a}$, we can remove the CRITICAL RACE by using STATE $\widehat{b}$ as a fly state [dashed arrow in Fig. 6-42(e)]. This causes no change in the Y_1 expression given in Fig. 6-42(c) but adds a fourth term to the Y_0 expression which now reads $Y_0 = \overline{y}_1 A + y_0 \overline{B} + y_1 \overline{B} + y_1 y_0 \overline{A}$. Having accomplished this change, we continue with the analysis of the FSM.

An inspection of the NEXT STATE logic expressions, following removal of the CRITICAL RACE, reveals the presence of no STATIC HAZARDs. For example, the coupled terms $(y_0 A\overline{B}, y_1 \overline{y}_0)$ in the expression for Y_1 shown

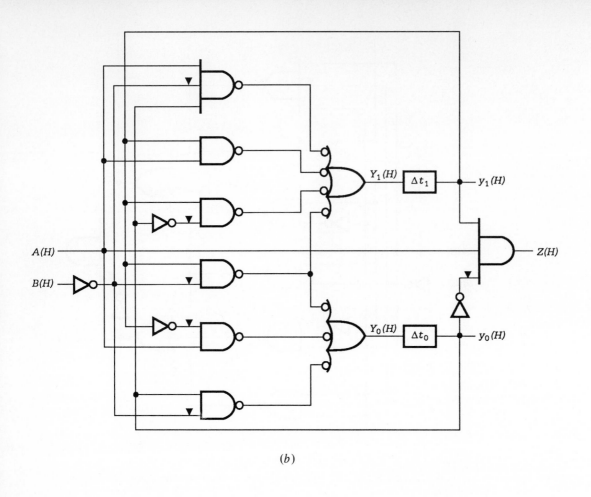

(b)

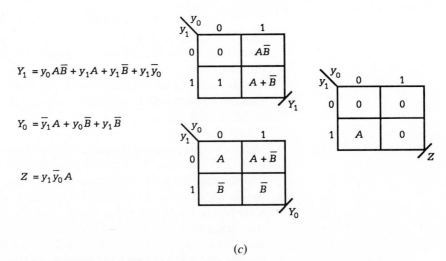

$$Y_1 = y_0 A\overline{B} + y_1 A + y_1 \overline{B} + y_1 \overline{y_0}$$

$$Y_0 = \overline{y_1} A + y_0 \overline{B} + y_1 \overline{B}$$

$$Z = y_1 \overline{y_0} A$$

(c)

Fig. 6-42 (contd.). *(b) LPD logic circuit showing the PRESENT and NEXT STATE logic variables. (c) NEXT STATE and output logic expressions and derived K-maps.*

PRESENT STATE (PS)	External inputs	NEXT STATE (NS)	PRESENT STATE (PS)	External inputs	NEXT STATE (NS)
00	$\overline{A}$	00	10	$\overline{B}$	11
	A	01		B	10
01	$\overline{A}\overline{B}$	01	11	$\overline{A}\overline{B}$	11
	$\overline{A}B$	00		$\overline{A}B$	00
	$A\overline{B}$	11		$A\overline{B}$	11
	AB	01		AB	10

(d)

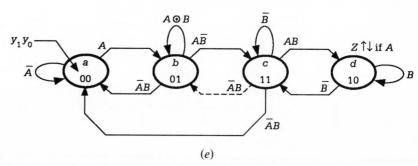

(e)

Fig. 6-42 (contd.). *(d) PRESENT STATE/NEXT STATE table. (e) Derived state diagram showing correction path (dashed arrow) for the CRITICAL RACE path* ⓒ → ⓐ.

in Fig. 6-42(c) indicate that a state transition from ⑪ → ⑩ would have to occur under branching conditions $A\overline{B}$. But according to the state diagram this transition can take place only for branching conditions AB. Thus, there can be no STATIC HAZARD in Y_1. Similarly, the coupled terms $(y_1\overline{B}, \overline{y}_1 A)$ in the Y_0 expression require a ⑪ → ⑴ transition under branching condition $A\overline{B}$, and this does not occur. Therefore, no STATIC HAZARD exists in Y_0. Remember that for any internally initiated HAZARDous transition (i.e., one initiated by a state variable) the external input conditions must remain constant.

The potential for E-HAZARD and d-trio formation exists as is evident from an inspection of the corrected state diagram. The d-trio path is seen to be ⓑ → c → ⓓ → ⓒ. If this timing defect were to be activated by an unintended delay in the B path to the y_0 state variable, a critical error would result since the FSM would issue the conditional output Z while momentarily in STATE ⓓ. This type of error differs from an output race glitch in that the duration of the output Z is dependent on the magnitude of the unintended delay that caused the d-trio. E-HAZARD formation is also possible in the corrected FSM along the path ⓓ → c → b → ⓐ under branching condition $\overline{A}\,\overline{B}$. This E-HAZARD is of the type for which k = ⓒ in Fig. 6-22 (STATE ⓑ serves as a fly state) and would result if an unintended delay occurred in the B path to y_1. Thus, if these defects exist, they could be eliminated by placing a counteracting delays of sufficient duration on both the y_1 and y_0 feedback lines.

6.7 THE NESTED MACHINE APPROACH TO ASYNCHRONOUS FSM DESIGN

The single most distinguishing feature of the LPD model is its lumped MEMORY element. The lumped MEMORY element not only serves as the MEMORY stage of the FSM, but it also can be regarded as the most rudimentary form of asynchronous machine. If we view the MEMORY element in this way, we are lead quite naturally to the concept of the NESTED asynchronous machine. This follows since now the lumped MEMORY element can be replaced with any one of a variety of other asynchronous machines which would serve as the MEMORY element. Use of the NESTED machine concept can permit the design of certain complex asynchronous FSMs which might otherwise not be practical or even possible. The idea here is that the design of a complex asynchronous FSM can be accomplished in parts, one part driving another, and so on, a concept common to digital design. Furthermore, the application of the NESTED machine concept can often lead to FSM designs that are less prone to HAZARD formation. In this section we will explore some of the simpler forms of the NESTED machine models and provide a few simple examples of their application.

6.7.1 The NESTED Machine Models

The most general NESTED machine model for an asynchronous FSM is presented in Fig. 6-43(a). It is seen to be nothing more than a Mealy model with a NESTED asynchronous machine as its MEMORY stage. The simplest NESTED asynchronous machine, other than the most degenerate case of the lumped MEMORY element, is the BASIC CELL. So, if we replace the NESTED asynchronous machine MEMORY stage of Fig. 6-43(a) with a bank of BASIC CELLs, we obtain the general NESTED CELL model of an asynchronous Mealy machine shown in Fig. 6-43(b). In this model the NEXT STATE forming logic is the SET/RESET control logic for the bank of NESTED BASIC CELLS, which have now become the MEMORY elements. It is interesting to note that the only difference between synchronous and asynchronous models resides in the nature of MEMORY element used. Thus, if we replace the bank of asynchronous BASIC CELLS, in Fig. 6-43(b), with FLIP-FLOPs regulated by CLOCK, we end up with the general model for synchronous FSMs in Fig. 5-4. This means that the reader should already have some familiarity with the NESTED machine concept from the experiences of Chapter 5.

A series of degenerate forms can be derived from the general NESTED CELL model of Fig. 6-43(b) just as was done for the LPD models in Fig. 6-2. One of the simplest degenerate forms is that shown in Fig. 6-43(c). Here, the SET/RESET logic is the controller for a single NESTED CELL, and no output forming logic exists. This model, as we shall soon learn, is of particular value in designing FLIP-FLOP type asynchronous machines. It has the advantage of supplying a mixed-rail output while, at the same time, offering a means of reducing hardware requirements and, in many cases,

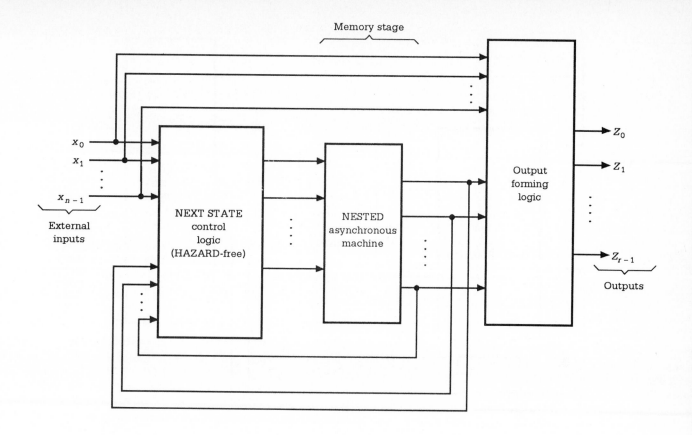

(a)

Fig. 6-43. *NESTED machine models for asynchronous FSMs. (a) General Mealy model.*

reducing the potential for FSM malfunction. In fact, without making specific mention of it, the NESTED CELL model was applied to the design of pulse-triggered and edge-triggered FLIP-FLOPs in Sects. 5.6.3 and 5.6.5. This is evidence of the fine line that separates synchronous and asynchronous machine design, a matter that will be discussed further in Sect. 6.8.

6.7.2 Design Examples

In the design examples given in this section, use will be made of both the general and degenerate forms of the NESTED CELL model illustrated in Fig. 6-43. We will apply the NESTED CELL approach to the redesign of the familiar rising edge-triggered (RET) D FLIP-FLOP, to the design of the edge-triggered JK and *data-lockout* (DL) FLIP-FLOPs, and to the redesign of the one- to three-pulse generator considered in the previous section.

Redesign of the RET D FLIP-FLOP. The state diagram of Fig. 6-35(a) can be collapsed to form a three-state FSM controller for a BASIC CELL, as shown in Fig. 6-44(a). The output Q is replaced by outputs SET and RESET, which become the inputs to the BASIC CELL. Notice that these outputs

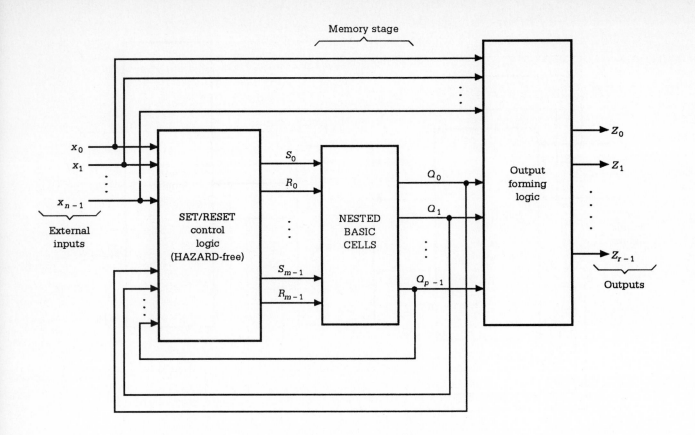

(b)

(c)

Fig. 6-43 (contd.). (b) NESTED CELL version of the Mealy model. (c) A simple degenerate form of the NESTED CELL model.

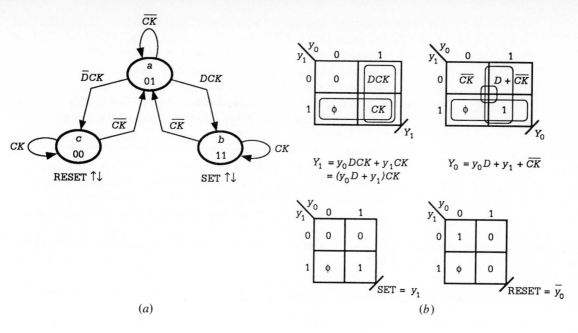

$Y_1 = y_0 DCK + y_1 CK$
$\quad = (y_0 D + y_1)CK$

$Y_0 = y_0 D + y_1 + \overline{CK}$

SET $= y_1$

RESET $= \overline{y_0}$

(a)

(b)

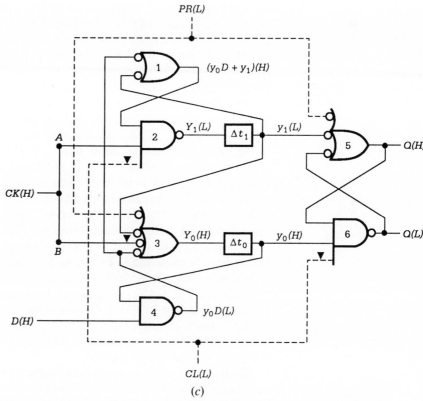

(c)

Fig. 6-44. *Design of the RET D FLIP-FLOP by using the NESTED CELL approach. (a) State diagram for the SET/RESET controller. (b) K-maps and minimum cover for the NEXT STATE and output logic. (c) LPD logic circuit showing connections for PRESET and CLEAR overrides (dashed leads).*

are properly located according to the algorithm for the D FLIP-FLOP. That is, the device must SET on the rising edge of CLOCK if D is ACTIVE or must RESET on the rising edge of CLOCK if D is NOT ACTIVE.

The state diagram in Fig. 6-44(a) represents an asynchronous controller since it provides the control logic for a BASIC CELL. The controller can be designed by using either the LPD model or the NESTED CELL model, hence a second application of the NESTED CELL model. In the latter case, the output CELL would be driven by two BASIC CELLs of the controller together with any NEXT STATE logic that is needed. We choose the LPD model design shown in Figs. 6-44(b) and (c) since it yields the more optimum design. Notice that the NEXT STATE variable expressions reveal no STATIC HAZARDs. Though not obvious from a first inspection of these expressions, there are two potential d-trios present that will be discussed shortly. The circuit diagram in Fig. 6-44(c) requires only six NAND gates for its implementation and forms the basis for the design of commercial RET D FLIP-FLOPs. We have included the fictitious lumped MEMORY delay notation as an aid to reading the circuit.

Additional PRESET and CLEAR overrides can be added as indicated in the circuit diagram of Fig. 6-44(c). Notice that it is necessary to connect the $PR(L)$ override to both NAND gate 3 and NAND gate 5. Then if $PR(L) = 1(L)$, the RET D-FF is SET since $Q(H) = 1(H)$ and $y_0 = 1(H)$ causing $Q(L) = 1(L)$. The D-FF is CLEARed by the $CL(L)$ input to NAND gates 2 and 6. Now, if $CL(L) = 1(L)$, the D-FF is RESET since $Q(L) = 0(L)$ and all three inputs to NAND gate 5 are $0(L)$, yielding $Q(H) = 0(H)$. These connections should be compared with those in Fig. 5-30(b) for the pulse triggered JK-FF.

The gate/input tally for the D FLIP-FLOP of Fig. 6-44 is 6/13 if the PRESET and CLEAR overrides are excluded. This may be compared with a gate/input tally of 9/26 for the design of the RET D FLIP-FLOP represented in Fig. 6-35, a substantial difference.

As mentioned earlier, the RET D-FF in Fig. 6-44 has the potential for d-trio production. A brief inspection of the state diagram in Fig. 6-44(a) indicates that a d-trio can be formed beginning in STATE ⓒ under branching conditions $\overline{DCK}$. Since y_1 is the second state variable to change, an unintended delay of sufficient duration in the CLOCK path to y_1 will result in the formation of the d-trio ⓒ → a → ⓑ → ⓐ. This represents an unrecoverable error because the D-FF has suffered an unintended SET the duration of which is delay dependent. Use of the state diagram in Fig. 6-44(a) with the circuit in Fig. 6-44(c) indicates that the minimum requirement for d-trio formation is $\Delta t_A > \tau_3 + \tau_4 + \tau_1$ when the delay Δt_A appears at position A in Fig. 5-44(c). If active, this d-trio can be eliminated by placing in the y_0 path to NAND gate 4 a counteracting delay equal to or greater than the delay Δt_A that caused the d-trio.

Similarly, a d-trio can be formed beginning in STATE ⓑ under branching conditions $\overline{D}CK$. Since y_0 is the second state variable to change in this case, an unintended delay of sufficient size in the CLOCK path to y_0 will result in the d-trio path ⓑ → a → ⓒ → ⓐ. This represents an unintended RESET of the FLIP-FLOP and is, therefore, a critical error. If we use both the state diagram and logic circuit in Fig. 6-44, we find that the minimum requirement for formation of this d-trio is $\Delta t_B > \tau_2$ when the unintended delay is at position B in Fig. 6-44(c). In round numbers, this means that this RET D-FF can be caused to malfunction if an unintended delay of little more than a gate path delay occurs in the CLOCK path to gate 3. If active, this d-trio can be removed by placing a counteracting delay of at least Δt_B duration in the y_1 feedback path to NAND gate 3.

The development of the d-trio for delay position B in Fig. 6-44(c) is illustrated in the timing diagram of Fig. 6-45. Here, we have taken the delay Δt_B to be $3\tau_p$ and have assigned a unit path delay τ_p to each gate. Notice that since $D = 0$ for this timing sequence, terms y_0DCK and y_0D are both logic 0 and are not represented in the timing diagram—they make no contribution. A close inspection of the timing diagram indicates that as the delay Δt_B shrinks from $3\tau_p$ to τ_p the d-trio disappears but without passing through an oscillation stage.

As we pointed out in Sect. 6.6.2, the mixed-rail outputs that typify the NESTED CELL designs of FLIP-FLOPs are of a unique type. There is a certain degree of symmetry associated with the mixed-rail outputs $Q(H)$ and $Q(L)$ in response to the inputs SET and RESET (see the timing diagrams in Fig. 5-17). This symmetry is not shared with the mixed-rail outputs that are generated from the $Q(H)$ line and an INVERTER split from it to produce $Q(L)$, as is required in the LPD circuit of Fig. 6-35(d).

An important feature of the edge-triggered D FLIP-FLOP is that it possesses a "data-lockout" feature. Once the data signal has been stored in the FLIP-FLOP, further changes in D will have no effect on the output response until the next sampling interval of CLOCK. Thus, no improper transitions are possible for defect-free operation. Of course, D may change erratically prior to the initiation of t_{su} (see Fig. 6-27), but the output response will depend only on the logic level of D held steady during the sampling interval of CLOCK.

The RET JK FLIP-FLOP. This design example follows closely that of the RET D FLIP-FLOP in Figs. 5-36 and 5-38. Here, we will apply the NESTED CELL model of Fig. 5-37 to the design of the rising edge-triggered (RET) JK FLIP-FLOP. It is our goal to obtain a near optimum design for this FLIP-FLOP and then analyze it for potential timing defects.

Presented in Fig. 6-46(a) is the four-state state diagram that correctly represents the RET JK-FF and that will be used for its design. A three-state

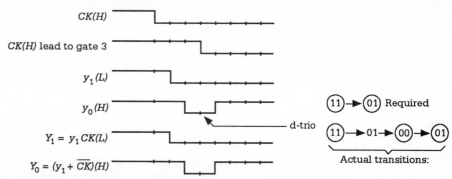

Fig. 6-45. *Timing diagram for the RET D FLIP-FLOP of Figure 6-44 showing the development of a d-trio due to a $3\tau_p$ asymmetric path delay on the CK path to y_0(H).*

state diagram for this FLIP-FLOP can be constructed, one similar to that of Fig. 6-44(a), by introducing Eq. (5-6) into the state diagram in Fig. 6-44(a). This results in a design that is the RET version of the JK-FF shown in Fig. 5-43(d). Verification of this is left as an exercise for the reader.

Shown in Figs. 6-46(b) and (c) are the K-maps, minimum cover, and logic circuit for the NESTED CELL design of the RET JK-FF. Notice that the NEXT STATE logic for the two BASIC CELLs is free of STATIC HAZARDs. The logic circuit in Fig. 6-46(c) represents a near optimum design since the ORed terms for S_A and R_A have been merged with the input to BASIC CELL A (gates 5 and 6).

The RET JK-FF in Fig. 6-46 has the potential to malfunction as a result of formation of either one of four E-HAZARDs. The state diagram in Fig. 6-46(a) indicates that one E-HAZARD path is $\widehat{00} \to 01 \to \widehat{11} \to \widehat{10}$ under branching condition $JK \cdot (\overline{CK} \to CK)$ when an asymmetric path delay exists in the direct path to BASIC CELL A at position E_1 [see Fig. 6-46(c)]. The indirect path of the CLOCK signal is through gates 7, 8, and 9. Thus, the minimum path delay required to form this E-HAZARD is $\Delta t_{E1} > \tau_7 + \tau_8$

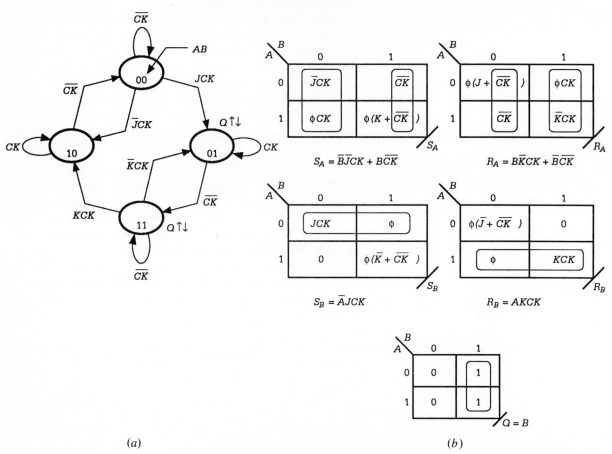

(a)

(b)

Fig. 6-46. *Design of the RET JK FLIP-FLOP by using the NESTED CELL model. (a) State diagram. (b) K-maps and minimum cover for the NEXT STATE and output forming logic of the BASIC CELLS.*

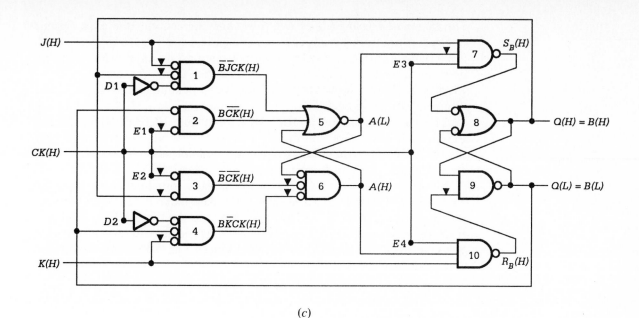

Fig. 6-46 (contd.). *(c) Logic circuit.*

$+ \tau_9$ in terms of the distributed gate delays. Note that the critical race is to gate 2 and that the output of gate 2 satisfies the transit requirements for the second y variable change as indicated by the state diagram in Fig. 6-46(a). This E-HAZARD can be eliminated (if active) or otherwise prevented by placing a counteracting delay equal to or greater than Δt_{E1} in the $B(L)$ feedback path.

A second E-HAZARD path is deduced from the state diagram of Fig. 6-46(a) to be $\textcircled{11} \rightarrow 10 \rightarrow \textcircled{00} \rightarrow \textcircled{01}$ under branching condition $JK \cdot (\overline{CK} \rightarrow CK)$ when an asymmetric path delay $\Delta t_{E2} > (\tau_{10} + \tau_9 + \tau_8)$ is present in the direct path to BASIC CELL A at position E_2. The indirect path is through gates 10, 9, and 8 and the critical race is to gate 3. Notice that the output of gate 3 is $\overline{B}\,\overline{CK}$, which agrees with the transit requirements for the second y variable change as deduced from Fig. 6-45(a). Elimination (or prevention) of this E-HAZARD results if a counteracting delay, equal to or greater than the net delay causing the HAZARD, is placed in the $B(H)$ feedback path. E-HAZARD and d-trio formation can also occur when asymmetric path delays are placed at positions E3, E4, D1, and D2. The analyses of these potential defects are left as an exercise for the reader.

The operation of the edge-triggered JK-FF is charcrerized by a data-lockout feature as are all edge-triggered FLIP-FLOPs. Once data have been stored in the FLIP-FLOP on the triggering edge of the CLOCK waveform, further changes in the J and K data signals will have no effect until the next sampling interval of CLOCK. Remember, it is the logic levels of J and K, held steady throughout the sampling interval of CLOCK, that are stored and that determine the output response.

If it is necessary to add asynchronous PRESET and CLEAR overrides to the JK-FF of Fig. 6-46(c), it is only necessary that use be made of gates 7, 8, 9, and 10 in a manner similar to that shown in Fig. 5-30(b).

The JK Data-Lockout FLIP-FLOP. For some applications there are special timing constraints that require the use of a FLIP-FLOP that is edge triggered, has a half-CLOCK-cycle delay in output response to an input change, but does not have the error-catching mode feature of the master/slave FLIP-FLOP. Such a device is known as the Data-Lockout FLIP-FLOP. It was discussed briefly in Sect. 5.6.5.

The state diagram for the RET JK data-lockout FLIP-FLOP is given in Fig. 6-47(a). The "ears" on the six-state state diagram are responsible for the half-CLOCK-cycle delay. Notice that it is not possible to collapse this state diagram to one of four states without losing this delay capability.

Shown in Figs. 6-47(b) and (c) are the K-maps, minimum cover, and logic circuit for a NESTED CELL design of the data-lockout FLIP-FLOP. Note that the simple NEXT STATE forming logic (to the two BASIC CELLs) is HAZARD-free and that it is obtained by using the generic form of the state transition table for the BASIC CELL given in Fig. 5-15(c). By count, the gate/input tally for this design is 13/34.

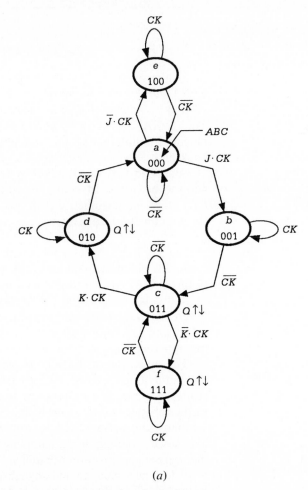

(a)

Fig. 6-47. *Design of a RET JK data LOCKOUT FLIP-FLOP by using the NESTED CELL model. (a) State Diagram.*

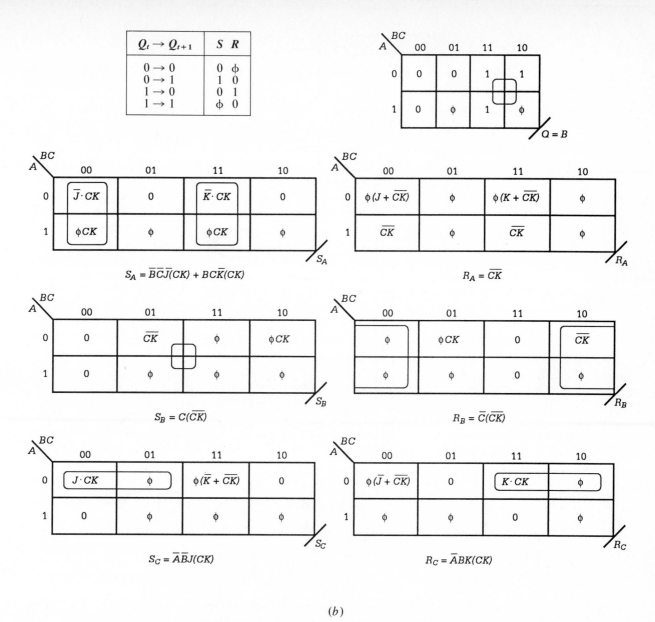

$Q_t \rightarrow Q_{t+1}$	S	R
$0 \rightarrow 0$	0	ϕ
$0 \rightarrow 1$	1	0
$1 \rightarrow 0$	0	1
$1 \rightarrow 1$	ϕ	0

$S_A = \overline{B}\overline{C}\overline{J}(CK) + BC\overline{K}(CK)$

$R_A = \overline{CK}$

$S_B = C(\overline{CK})$

$R_B = \overline{C}(\overline{CK})$

$S_C = \overline{A}\overline{B}J(CK)$

$R_C = \overline{A}BK(CK)$

(b)

Fig. 6-47 (contd.). (b) K-maps and minimum cover for the output Q and the SET and RESET parameters S and R.

Redesign of the One- to Three-Pulse Generator. As our last example, we will apply the general NESTED CELL model of Fig. 6-43(b) to a redesign of the one- to three-pulse generator considered earlier in Sect. 6.6.2. To do this, use will be made of the state diagram in Fig. 6-38(a) except that now it will be mapped by applying the state transition table for the BASIC CELL as was done in the design of the data-lockout FLIP-FLOP above.

Presented in Fig. 6-48(a) are the K-maps and minimum STATIC-HAZARD-free cover for the SET/RESET logic and output logic. A partial NAND-

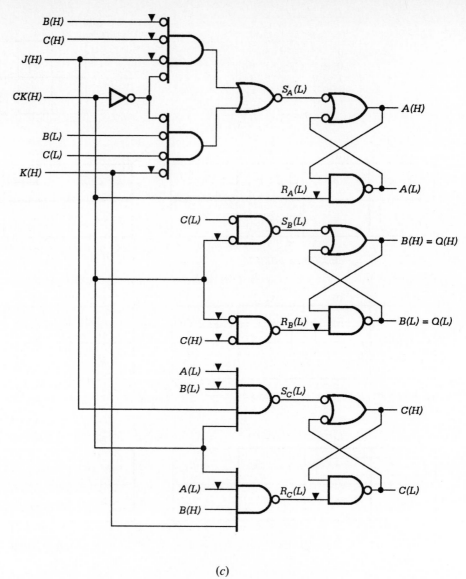

(c)

Fig. 6-47 (contd.). *(c) Logic circuit.*

centered implementation of this FSM is given in Fig. 6-48(b) where, for simplicity, we have omitted the SET/RESET logic.

Summary. The NESTED machine models should be regarded as a generalization of the LPD models. Or put another way, the LPD models can be thought of as the most degenerate forms of the NESTED machine concept. Although this argument suggests that there is but one model (the NESTED machine model) from which all others are derived, we shall continue to distinguish between the NESTED CELL model and the LPD model for reference purposes. In any case, asynchronous FSMs that are designed with

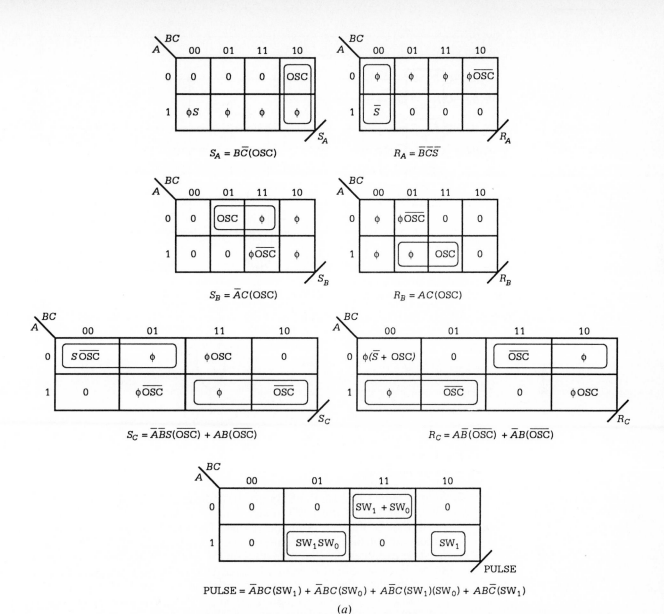

$$S_A = B\overline{C}(\text{OSC})$$

$$R_A = \overline{B}\,\overline{C}\,\overline{S}$$

$$S_B = \overline{A}C(\text{OSC})$$

$$R_B = AC(\text{OSC})$$

$$S_C = \overline{A}\,\overline{B}S(\overline{\text{OSC}}) + AB(\overline{\text{OSC}})$$

$$R_C = A\overline{B}(\overline{\text{OSC}}) + \overline{A}B(\overline{\text{OSC}})$$

$$\text{PULSE} = \overline{A}\,\overline{B}C(\text{SW}_1) + \overline{A}BC(\text{SW}_0) + A\overline{B}C(\text{SW}_1)(\text{SW}_0) + AB\overline{C}(\text{SW}_1)$$

(a)

Fig. 6-48. *Design of the one- to three-pulse generator by using the NESTED CELL model of Figure 6-38(b). (a) K-maps and minimum cover for the SET/RESET logic and output forming logic.*

these models are operated in the fundamental mode. This is in contrast to those that operate in the PULSE mode, the subject of the next section.

There are specific design applications for which one or another of the two models (LPD or NESTED CELL) should be considered preferential. The following gives some of these applications together with the recommended model:

1. FLIP-FLOPs are best designed by using the NESTED CELL model. The advantages are mixed-rail outputs, reduced circuit complexity, and reduced potential for STATIC HAZARD production.

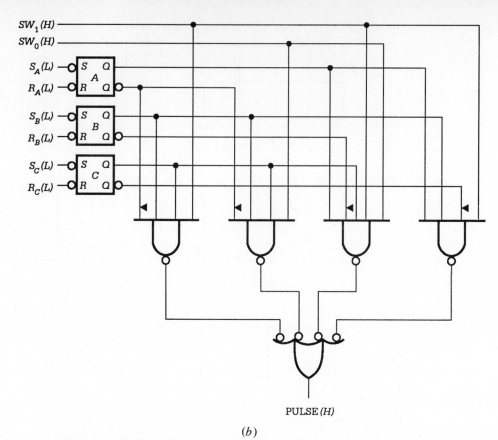

(b)

Fig. 6-48 (contd.). (b) Circuit diagram showing only the NESTED CELLs and output forming logic.

2. Simple nonFLIP-FLOP FSMs are best designed with the LPD model. The advantage of the LPD model is its simplicity of application and a result that is often an optimum design. The disadvantage is that the resulting circuits are often prone to have STATIC HAZARDs. Thus, care must be exercised to eliminate all STATIC HAZARDs during the design stage. For the case of potential E-HAZARD or d-trio formation, the designer should make note of their potential for causing malfunction of the FSM and either provide a plan for their removal, should they become a problem, or take preventive measures during the design stage.

3. In the case of complex asynchronous FSMs, the designer should consider using the NESTED machine models. This permits the FSM to be divided into more tractable parts and also tends to minimize STATIC HAZARD formation. If the FSM requires a large number of state variables, computer aided methods may be required.

The foregoing information is offered only as a guide. Certainly, there will be "gray area" cases and others that go counter to these recommendations. Usually, if the designer has a thorough understanding of the operation and I/O specifications of the FSM to be designed, an appropriate model will emerge quite naturally. If not, it may be necessary to design the FSM by using more than one model so that the "best" design can be chosen. We recommend this approach whenever it is reasonable to do so. Better designs can never emerge unless designers are willing to make the extra effort to explore new approaches to a given design problem.

6.8 THE PULSE MODE APPROACH TO ASYNCHRONOUS FSM DESIGN

Asynchronous FSMs which are designed to operate with nonoverlapping pulsed inputs and which use unclocked FLIP-FLOPs as MEMORY elements are called *PULSE mode sequential machines*. The PULSE mode approach offers a simple and reliable means of designing CLOCK-independent FSMs but at the price of greatly restricted input signal conditions. The previous sections in this chapter dealt exclusively with asynchronous FSMs that were designed to operate in the fundamental mode, a mode characterized, in part, by overlapping input signals and the potential to form certain types of timing defects such as ENDLESS CYCLEs, CRITICAL RACEs, and ESSENTIAL HAZARDs.

It is the goal of this section to present the PULSE mode approach as an alternative means of designing certain types of asynchronous FSMs. We will show that the PULSE mode approach has certain features in common with synchronous FSM design and avoids many of the problems associated with the fundamental mode. However, these apparent advantages are offset by the severe restrictions placed on the input signals. In fact, it is for this reason that we have chosen to defer treatment of the PULSE mode approach until this time.

6.8.1 PULSE Mode Models and System Requirements

The generalized model for PULSE mode FSM design is shown in Fig. 6-49. It is unique in the sense that its MEMORY stage is composed of un-CLOCKED (nongated) T FLIP-FLOPs. However, as we indicate shortly, other types of MEMORY elements may be used but only if certain conditions are met. Degenerate forms of this model follow as in Figs. 5-2 and 5-3.

The inputs to PULSE mode FSMs must be nonoverlapping pulses that are at least minimally separated such that the leading edge of any one pulse lies outside of the sampling interval for the trailing edge of any previous pulse. Examples of such pulses are shown in Fig. 6-50. Here, the positive pulses are shown to have ACTIVE durations (pulse widths) with no upper bound but with a lower limit sufficient to trigger the FLIP-FLOP MEMORY

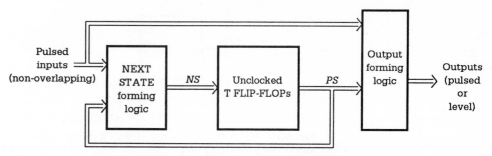

Fig. 6-49. General (Mealy) model for PULSE mode FSM design.

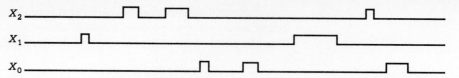

Fig. 6-50. *Examples of nonoverlapping and at least minimally separated positive pulses having ACTIVE durations with no upper bound.*

elements. "Runt" pulses, those that are not fully developed, must not be permitted since their effect on the FLIP-FLOPs is unpredictable. The complement of the PULSE trains shown in Fig. 6-50 are examples of negative pulses having no upper bound on their INACTIVE durations. It should be understood that the "at least minimally separated" restriction placed on these input pulses is equivalent to the requirement that inputs to a fundamental mode circuit be permitted to change only when the stability criteria $Y_i(t) = y_i(t)$ are satisfied. In fact, proper operation of any FSM (synchronous or asynchronous) can be ensured only if all MEMORY elements of the FSM achieve stability prior to the change of an input logic level.

Choice of MEMORY Elements. The choice of MEMORY elements for PULSE mode FSM design is quite limited. Positive $(0 \rightarrow 1 \rightarrow 0)$ pulses of unrestricted ACTIVE duration require the use FET T-FFs, while negative $(1 \rightarrow 0 \rightarrow 1)$ pulses of unrestricted INACTIVE duration require RET T-FFs. The various options are summarized in Fig. 6-51 where we have included the use of edge-triggered JK-FFs operated as T-FFs.

If further restrictions are placed on the inputs to PULSE mode circuits, use of other MEMORY elements becomes possible. Figures 6-52(a), (b), and (c) show examples of alternative MEMORY elements that must be operated

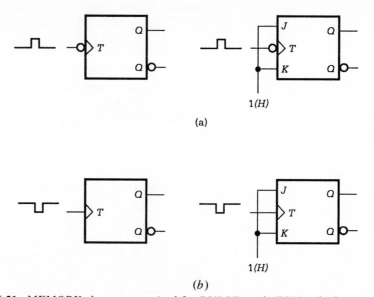

Fig. 6-51. *MEMORY elements required for PULSE mode FSMs which receive pulses of unrestricted pulse width. (a) Nongated FET T-FFs for positive pulses. (b) Nongated RET T-FFs for negative pulses.*

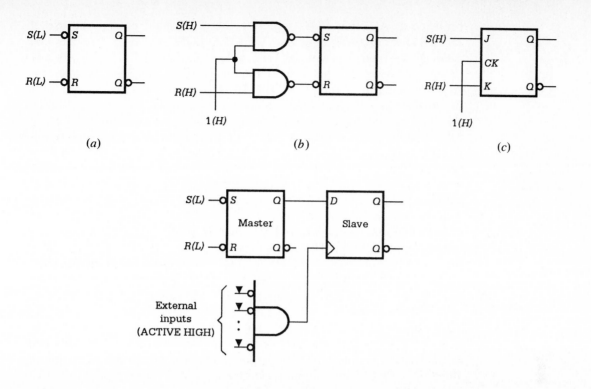

Fig. 6-52. *Examples of alternative MEMORY elements for PULSE mode FSMs which receive positive pulses of restricted ACTIVE duration. (a) BASIC CELL. (b) GATED SR LATCH operated as BASIC CELL. (c) Pulse triggered JK-FF operated as a BASIC CELL. (d) Example of data-controlled-triggering.*

as BASIC CELLs, and must receive inputs having ACTIVE durations that are limited to the delays positioned on the Q feedback lines. By our limiting the pulse widths to the feedback delays, we ensure that only one MEMORY element will be activated in response to an input change. Also, if we map with the generic form of the state transition table for the BASIC CELL [(Fig. 5-15(c)], we eliminate the possibility that the S and R inputs to these MEMORY elements can go ACTIVE at the same time. Thus, under these conditions the pulse-triggered JK-FF in Fig. 6-52(c) becomes a suitable memory element for a PULSE mode FSM, but only if it is operated as a BASIC CELL.

Under certain conditions CLOCKed FLIP-FLOPs can be used as components of MEMORY elements for PULSE mode FSMs. The requirement is simply that FLIP-FLOP triggering be controlled by the external inputs to the FSM. The MEMORY element shown in Fig. 6-52(d) is an example of one that is suitable for use in PULSE mode designs. It must be operated in a master/slave fashion so that the master and slave sections are never triggered on the same rising (or falling) edge of a data pulse. The presence of the NOR gate shown in Fig. 6-52(d) helps to accomplish this for ACTIVE HIGH inputs. An external input will SET or RESET the BASIC CELL (via the NEXT STATE logic) on the rising edge of its pulse but will trigger the RET D-FF (slave section) only on the falling edge of the same pulse. The presence of ACTIVE LOW inputs to the NOR gate requires the use of INVERTERs or, if all inputs are ACTIVE LOW, an AND gate could be used. An OR gate replaces the NOR gate if an FET D-FF is used. In any case, it is easy to see that a restriction is placed on the minimum tolerable ACTIVE duration of the pulse. The ACTIVE period of a given pulse must be greater than the combined minimum path delay through the NEXT STATE logic and master section (neglecting the presence of the NOR gate); otherwise the pulse will not be picked up by the slave section and data will be lost. This is quite the opposite problem from that of the BASIC CELL–type elements in Figs. 6-52(a), (b), and (c) where an upper bound must be placed on the ACTIVE duration of each pulse. The reader should note that the master CELL of Fig. 6-52(d) can be replaced by any of the MEMORY elements in Fig. 6-51.

It is important to understand that the MEMORY elements in Fig. 6-51 require that the NEXT STATE logic be designed for T FLIP-FLOPs, while those of Fig. 6-52 require a NEXT STATE logic designed for the BASIC CELL. All MEMORY elements require that the pulses be nonoverlapping but only the MEMORY elements in Fig. 6-52 require pulses of restricted ACTIVE duration.

Presented in Fig. 6-53 is a model which is applicable to the use of the alternative MEMORY elements of the type shown in Figs. 6-52(a), (b), and (c) and which is appropriate for array or discrete logic implementation of the NEXT STATE and output logic functions. This model differs from that of Fig. 5-85 only to the extent that delays of duration Δt are placed on the feedback lines. As will be demonstrated later, these delays are necessary to ensure that only one MEMORY element can be activated in response to an input pulse.

6.8.2 PULSE Mode FSM Characteristics and Design Procedure

Like synchronous FSMs and unlike fundamental mode FSMs, properly designed and operated PULSE mode machines cannot have ENDLESS CYCLEs, CRITICAL RACEs, or ESSENTIAL HAZARDs. Aside from the unique problems associated with the use of BASIC CELLS in PULSE mode

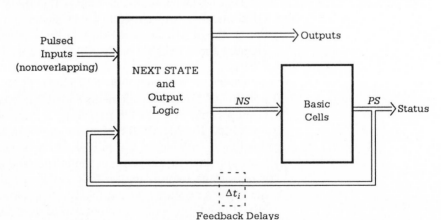

Fig. 6-53. General model for PULSE mode circuits which use BASIC CELLs as the MEMORY elements.

design, the only timing defects common to PULSE mode FSMs are output race glitches and STATIC HAZARDs in the output forming logic. So, while PULSE mode circuits are operated asynchronously, they can have timing defects no different from those of synchronous FSMs. Moreover, PULSE mode FSMs whose inputs are nonoverlapping and at least minimally separated cannot have asynchronous input problems since the inputs are, in effect, synchronized to each other. Compared to fundamental mode FSMs, PULSE mode FSMs are much simpler to design with fewer potential problems. However, the attractive features of the PULSE mode approach are offset by input restrictions that greatly limit PULSE mode FSM application.

Design Procedure. Not surprisingly, the design procedure for PULSE mode FSMs follows that of synchronous FSMs, but with some significant differences, as outlined here:

1. Construct a state diagram by using single variable outgoing branching conditions from each state. There can be as many outgoing branchings from a given state as there are inputs, and *all* branching variables will either be uncomplemented or complemented but never mixed. Variables not used as outgoing branching conditions from a given state are assumed to act as holding conditions for that state. Thus, state holding conditions need not be specified unless it is desirable for emphasis. Remember that the sum rule [Eq. (6-4)] cannot be applied to PULSE mode FSMs since only one pulse can be ACTIVE at any given time.

   Logically adjacent state-to-state transitions are not necessary except to eliminate output race glitches. Since CLOCK-triggered output holding registers cannot be used to filter out logic noise in PULSE mode designs, output race glitches and STATIC HAZARDs in the output forming logic must not be permitted to exist if they can cause a problem. Fly states cannot normally be used in PULSE mode circuit design. Why not?

2. Select a suitable MEMORY element of the type shown in Figs. 6-51 or 6-52. At this point it is necessary to know the nature of the input pulses. Pulses of unlimited pulse width require FLIP-FLOPs of the type shown in Fig. 6-51. If the pulse widths fall within certain limits, MEMORY elements of the type given in Fig. 6-52 can be used. In the latter case, delays at least as long as the longest possible input pulse width must be placed on the feedback lines in accordance with Fig. 6-53. Use of the MEMORY element in Fig. 6-52(d) is the exception.

3. Map the NEXT STATE and output logic functions from the state diagram just as is done in synchronous FSM design. If T-FFs of the type shown in Fig. 6-51 are chosen as the MEMORY elements, then use must be made of the state transition table for T-FFs given in Fig. 5-26(c). However, if the choice is a MEMORY element of the type presented in Fig. 6-52, use must be made of the generic form of the state transition table for the BASIC CELL given in Fig. 5-15(c). In this latter case, either delays must be positioned on all feedback lines or the input pulse widths must be limited to less than the shortest path delays through the circuit. Again, use of the MEMORY element in Fig. 6-52(d) is the exception.

4. Loop out minimum or reduced cover for the NEXT STATE and output logic functions by following the same procedure as for synchronous FSMs. When minimizing output functions, care should be taken not to create output race glitches involved with don't care states. See Sect. 5.8.2

for a detailed discussion of this matter. Also, since STATIC HAZARDs can be formed in the output forming logic, it may be necessary to analyze the output function for these defects, particularly if their presence can cause a problem in the next stage. Remember that while the inputs must always be pulsed signals, the FLIP-FLOP outputs may be either level (hence overlapping) or pulsed signals.

6.8.3 Design Examples

The following design examples illustrate the application of the PULSE mode approach to asynchronous sequential machine design. An attempt is made to cover most of the salient features of this method but leaving some to discovery and end-of-chapter problems.

Design of a Simple PULSE Mode Sequence Recognizer. Our first example is the design of a PULSE mode circuit having two inputs, X and Y, and one output, Z, as indicated by the block diagram in Fig. 6-54(a). It is the function of this FSM to issue the output Z coincidentally with the last input pulse in the sequence . . . Y-X-X . . . whenever this sequence occurs. Hence, the output is to be a pulsed output as opposed to a *level* output which is independent of input pulse width. We assume that the signals are positive pulses of unrestricted ACTIVE duration.

The state diagram appropriate for this sequence recognizer is presented in Fig. 6-54(b). Notice that the sequence . . . Y-X-X . . . is easily recognized and that all inputs are accounted for by single variable branchings from each state. However, it is not necessary to indicate the holding conditions (as we

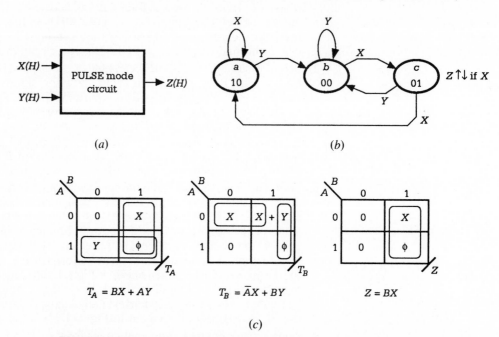

Fig. 6-54. *Design of a simple sequence recognizer by using the PULSE mode approach. (a) Block diagram. (b) State diagram applicable to PULSE mode design. (c) NEXT STATE and output K-maps and minimum cover.*

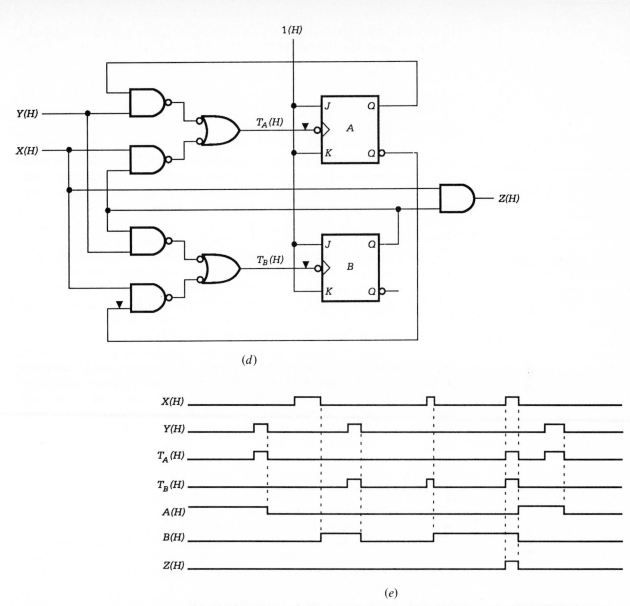

Fig. 6-54 (contd.). (d) Logic circuit using FET JK-FFs as the MEMORY elements. (e) Timing diagram (excluding gate delays) showing the sequence beginning in STATE 10.

have done) since for the PULSE mode such conditions are understood. The conditional output in STATE ⓒ is necessary since it is required that Z be issued coincidentally with the second X pulse following a Y pulse, and since FET T-FFs must be used as the MEMORY elements. If this is not obvious to the reader at this time, it will be when we consider the resulting timing diagram for the logic circuit.

The NEXT STATE and output logic K-maps and minimum cover are presented in Fig. 6-54(c). The NEXT STATE K-maps are plotted by using the state transition table for T-FFs given in Fig. 5-26(c). There is no need to examine the T expressions for STATIC HAZARDs since these defects cannot be generated in the NEXT STATE forming logic of PULSE mode FSMs which are properly designed and operated.

Shown in Fig. 6-54(d) is the PULSE mode logic circuit for this simple sequence recognizer. It can be seen that the MEMORY elements are FET JK-FFs operated as T-FFs and that the NEXT STATE logic is connected to the $CK(L)$ inputs of these FLIP-FLOPs as required by Fig. 6-51(a).

A more detailed account of the operation of this PULSE mode FSM is best provided by a timing diagram. Figure 6-54(e) shows the MIXED LOGIC timing diagram for this FSM beginning with STATE ⑩. For simplicity, we have neglected gate path delays. The pulses are of variable widths and are nonoverlapping with leading edges which are required to exist outside of the sampling interval for the trailing edge of any previous pulse. That the output pulse Z is issued coincidentally with the second X pulse following a Y pulse is verified by the timing diagram and logic circuit of Figs. 6-54(e) and (d), respectively. Since $Z = BX$, B must be ACTIVE when the second X pulse goes ACTIVE if Z and X are to have the same ACTIVE period. Had we placed Z as an unconditional output associated with STATE ⓐ in Fig. 6-54(b), then Z (i.e., $Z = A$) would not be a pulsed output coincident with X, but instead would be a level output coincident with A.

For contrast, we will now implement the sequence recognizer of Fig. 6-54(b) by using BASIC CELLs as the MEMORY elements and by applying the model shown in Fig. 6-53. This represents a significant departure from the previous design in that two FLIP-FLOP input signals are needed for each state variable, and an upper bound must be placed on the ACTIVE duration of the input pulses. Also, it is now necessary to assign the conditional output Z to STATE ⑩ for reasons that will be made clear in the discussion that follows.

In Fig. 6-55(a) are given the NEXT STATE and output K-maps and minimum cover for the BASIC CELL design of the sequence recognizer, assuming nonoverlapping input pulses which are at least minimally separated. The resulting logic circuit is presented in Fig. 6-55(b). Notice that delays of duration Δt are placed on all feedback lines. This is necessary so that the input pulse widths are limited to durations of Δt—not to the best case gate path delays for the transitions, which is very short. Also indicated in the logic circuit is the fact that the BASIC CELLs can be replaced by ACTIVE LOW input pulse-triggered (PT) JK-FFs operated as BASIC CELLs. This substitution is made possible by designing the NEXT STATE forming logic with the generic form of the state transition table for the BASIC CELL [Fig. 5-15(c)] which precludes the $(S, R) = (1, 1)$ condition. If the $(S, R) = (1, 1)$ condition were permitted to occur, the PT JK-FFs would go into oscillation. For further discussion on this matter, the reader is referred to Sect. 5.6.3 and Fig. 5-29(b).

The reassignment of the conditional output Z to STATE ⑩ in the state diagram of Fig. 6-54(b) and the need for delays Δt on the feedback lines in Fig. 6-55(b) are best explained by using the timing diagram shown in Fig. 6-55(c) where, for simplicity, we have neglected all gate path delays. Let us begin in STATE ⑩⓪ and follow the sequence of events leading to the output Z. For simplicity, use is made of the gate numbering system in Fig. 6-55(b) together with positive logic values as read from the MIXED LOGIC timing

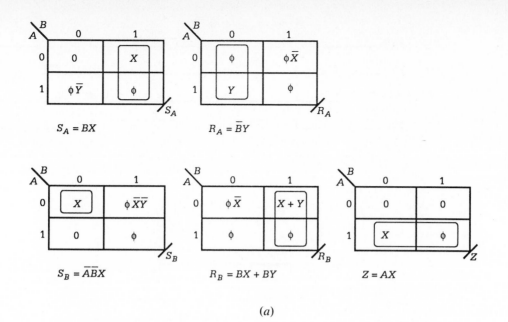

$S_A = BX$ $R_A = \bar{B}Y$

$S_B = \bar{A}\bar{B}X$ $R_B = BX + BY$ $Z = AX$

(a)

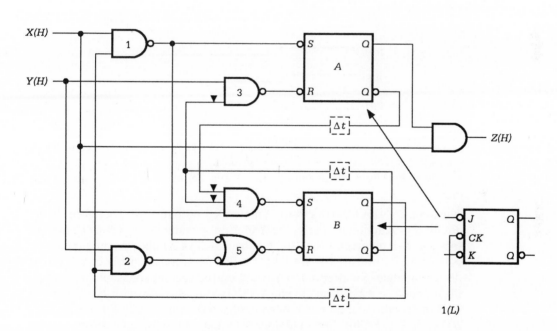

(b)

Fig. 6-55. *PULSE mode design of the sequence recognizer in Figure 6-54(b) by using BASIC CELLs as the MEMORY elements. (a) NEXT STATE and output K-maps. (b) Logic circuit showing delays Δt in the feedback lines.*

diagram. Beginning with X and Y both at logic 0, the outputs of all gates are 0. Then when X goes $0 \rightarrow 1$, gate 4 goes $0 \rightarrow 1$ and SETs BASIC CELL B. If the original X pulse is still ACTIVE (logic 1) when the signal from BASIC CELL B reaches gate 1, then BASIC CELL A will SET, causing an error

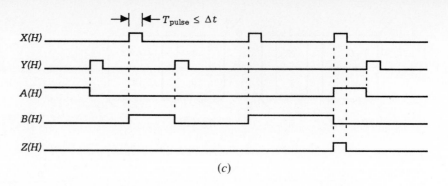

(c)

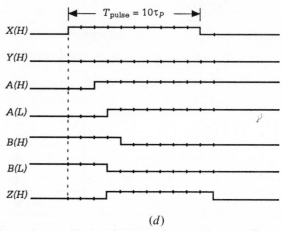

(d)

Fig. 6-55 (contd.). (c) *Timing diagram showing the overall operation of the pulse mode circuit.* (d) *Timing diagram for the 01 → 10 transition showing the development of the output pulse, Z, in gate path delay units* τ_p.

transition to STATE ⑪ and the unintended output of Z. Therefore, if the circuit is to function properly, the X or Y pulse width (T_pulse) must be limited to a duration no greater than the feedback delays Δt plus the minimum gate path delay for transitions. We take $T_\mathrm{pulse} \le \Delta t$ as a reasonably conservative design practice. So after a Y pulse and on the second consecutive X pulse, the output $Z = AX$ appears to go ACTIVE coincidentally with X as required. Had the conditional output Z been associated with STATE ⑪ giving $Z = BX$, as in Fig. 6-54(b), it would appear that an output of very short duration (a glitch) would be issued. The only other alternative is to place an unconditional output Z on STATE ⑩ to give $Z = A$, a level output issued coincidentally with A.

That the output Z in Fig. 6-54(d) cannot be issued precisely coincidentally with X is shown by the detailed timing diagram for the ⑪ → ⑩ transition in Fig. 6-55(d). This timing diagram is constructed under the assumption that all gates (including those of the BASIC CELLs) have unit path delays τ_p and that $\Delta t = 10\tau_p$. Notice that the Z pulse is ACTIVE for an $8\tau_p$ period but is delayed relative to X by $3\tau_p$. Thus, as Δt approaches $2\tau_p$, the Z pulse width approaches 0. Notice that if the conditional output Z is assigned to

STATE $\textcircled{01}$, giving $Z = BX$, there results an output glitch of duration τ_p that is independent of the pulse width.

Delay Circuits. Much was said in Sect. 6.5.3 about the use delays in feedback lines as a means of preventing E-HAZARD formation. Now, use of the PULSE mode model, shown in Fig. 6-53, requires the presence of such delays in the feedback lines. So, it is reasonable that the reader be given some idea of how delay circuits are constructed even though such topics generally fall outside the scope of this text. Shown in Fig. 6-56(a) is the circuit for a nearly ideal inertial delay element of the type that could be used

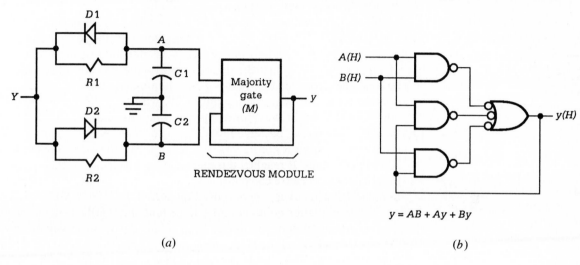

$$y = AB + Ay + By$$

(a)   (b)

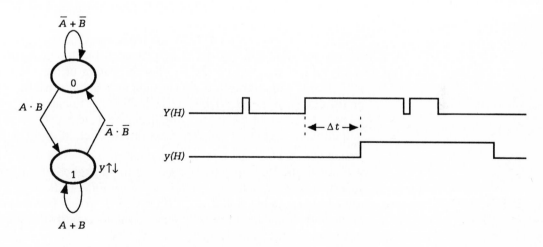

(c)   (d)

Fig. 6-56. *An inertial delay element. (a) Circuit made up of resistors R, diodes D, capacitances C, and a majority gate M configured as RENDEZVOUS MODULE. (b) Logic diagram and majority function for the RENDEZVOUS MODULE. (c) STATE DIAGRAM for the RENDEZVOUS MODULE. (d) Timing diagram showing development of the inertial delay. Δt, and the filtering action of the R-C circuit.*

in the logic diagram of Fig. 6-55(b) or that could be used as a counteracting delay in logic circuits of the type shown in Figs. 6-23(d) and 6-26(a). The delay circuit consists of resistors ($R1 = R2$), capacitors ($C1 = C2$), diodes ($D1 = D2$), and a *majority gate, M,* configured as an FSM called a *RENDEZVOUS MODULE* or *MULLER C* (concurrency) *MODULE*. The logic circuit for the RENDEZVOUS MODULE and the majority function derived from it are given in Fig. 6-56(b). By inspection, the majority function, $y = AB + Ay + By$, is logic 1 iff a majority (in this case two) of the inputs are logic 1. Thus, the output of the RENDEZVOUS MODULE can go to logic 1 only if A and B both become logic 1, and will go to logic 0 only if A and B both become logic 0. The sequential character of the RENDEZVOUS MODULE is completely specified by its state diagram in Fig. 6-56(c).

The RENDEZVOUS MODULE together with the R-C circuit in Fig. 6-56(a) ensures the creation of the delay Δt indicated in the timing diagram of Fig. 6-56(d). Although the detailed analysis of this circuit is complex and beyond the scope of this text—it is a nonlinear second-order circuit—its operation can be understood qualitatively with little difficulty. On the rising edge of the Y input waveform the RC time constant at node B is smaller than that at node A because diode $D2$ is turned ON (with a low resistance compared to either $R1$ or $R2$) while diode $D1$ is OFF (with a high resistance). As a result, node B reaches the high-voltage threshold of the majority gate before node A because capacitor $C2$ can charge up through the low resistance of diode $D2$. However, the output of the RENDEZVOUS MODULE cannot go HIGH until after nodes A and B have both gone HIGH, hence the delay. The reverse is true for the falling edge of the Y waveform where now diode $D1$ is ON and $D2$ is OFF, resulting in a smaller time constant at node A than at node B. Thus, node A reaches low voltage before node B since the capacitor $C1$ can discharge through the low resistance of diode $D1$. Then since the output of the RENDEZVOUS MODULE cannot go LOW until both nodes A and B are LOW, a time delay results.

The magnitude of the delay Δt produced by the circuit in Fig. 6-56(a) can be adjusted somewhat by altering the values of the R's and C's in the R-C circuit, and a threshold device such as a Schmitt trigger can be placed on the output of the RENDEZVOUS MODULE to minimize waveform distortion. Notice that the narrow input pulses have no effect on the delayed output pulse, a result of the low pass filtering action of the R-C circuit.

Design of a PULSE Mode Digital Combination Lock. As a second example we will design the PULSE mode version of the digital combination lock which was previously designed as a synchronous FSM in Sect. 5.9.3 and as an asynchronous fundamental mode FSM in Sect. 6.6.2. Because of the nature of the pulsed inputs, our present design will necessarily deviate significantly from the two previous designs.

The pulse sequence for the digital combination lock is given in Fig. 6-57(a). This sequence of events (pulses) matches that of Fig. 5-89, except in the latter case the input pulses can be overlapping. In the present design, inputs X and Y are assumed to be pulsed signals from push-button switches which are interlocked in such a manner that only one switch can be activated at any one time. As indicated by the pulse sequence, the output OPEN is issued coincidentally with the X pulse after which a second X pulse initializes

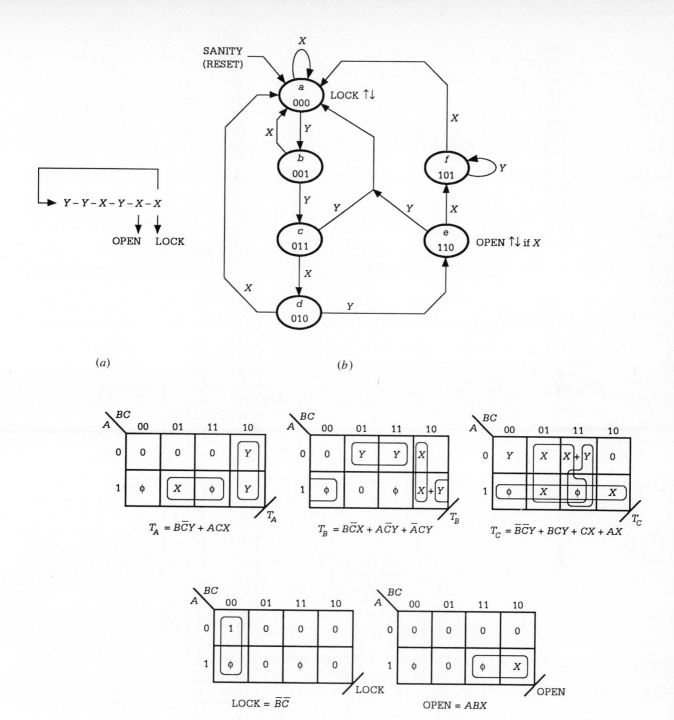

(a)

(b)

$$T_A = B\bar{C}Y + ACX$$

$$T_B = B\bar{C}X + A\bar{C}Y + \bar{A}CY$$

$$T_C = \bar{B}\bar{C}Y + BCY + CX + AX$$

$$LOCK = \bar{B}\bar{C}$$

$$OPEN = ABX$$

(c)

Fig. 6-57. *PULSE mode design of a digital combinational lock. (a) Input pulse sequence. (b) State diagram. (c) NEXT STATE and output K-maps and minimum cover.*

| p-Term | Inputs | | | | | Outputs | | | | |
	A	B	C	X	Y	T_A	T_B	T_C	LOCK	OPEN
$B\overline{C}Y$	–	1	0	–	1	1	0	0	0	0
ACX	1	–	1	1	–	1	0	0	0	0
$B\overline{C}X$	–	1	0	1	–	0	1	0	0	0
$A\overline{C}Y$	1	–	0	–	1	0	1	0	0	0
$\overline{A}CY$	0	–	1	–	1	0	1	0	0	0
$\overline{B}\,\overline{C}Y$	–	0	0	–	1	0	0	1	0	0
BCY	–	1	1	–	1	0	0	1	0	0
CX	–	–	1	1	–	0	0	1	0	0
AX	1	–	–	1	–	0	0	1	0	0
$\overline{B}\,\overline{C}$	–	0	0	–	–	0	0	0	1	0
ABX	1	1	–	1	–	0	0	0	0	1

(d)

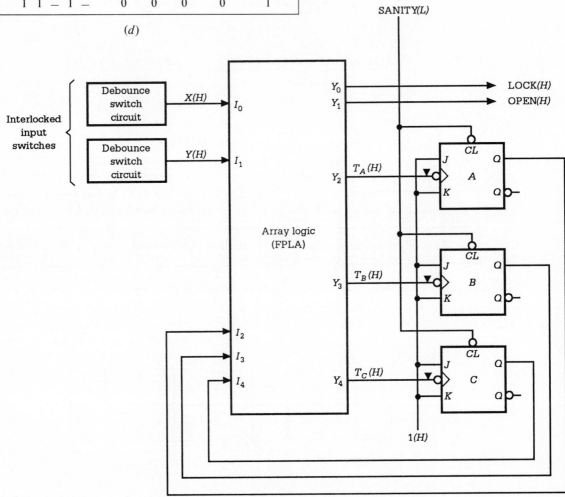

(e)

Fig. 6-57 (contd.). *(d) p-term table. (e) Implementation with a 6 × 14 × 6 FPLA.*

the FSM into the LOCK state. Any deviation from this sequence must also result in resetting the FSM into the LOCK state.

The state diagram that satisfies the requirements for the PULSE mode design of the digital combination lock is shown in Fig. 6-57(b). The pulse sequence is easily identified, and both inputs are accounted for by the single variable branching conditions from each state. Notice that there are three race conditions, none of which leads to an output race glitch. Also, observe that the position of the conditional output OPEN implies the use of falling edge-triggered FLIP-FLOPs as the MEMORY elements. We assume that the ACTIVE durations of the inputs have no upper bound and select FET T-FFs for the MEMORY. Next we map out the NEXT STATE logic expressions together with the output functions as shown in Fig. 6-57(c).

All that remains at this point is to decide how the digital combination lock is to be implemented and what to use for the input conditioning circuits. We choose to use a PLA for the NEXT STATE and output forming logic. Presented in Figs. 6-57(d) and (e) are the p-term program table and logic circuit for PLA implementation of this PULSE mode FSM. We see that there are 5 inputs (not counting the SANITY input), 11 p-terms, and 5 outputs. Therefore a $5 \times 12 \times 5$ PLA or larger will suffice. Note that PULSE mode FSMs are initialized by a SANITY circuit in the same manner as are synchronous FSMs, that is, by means of the PRESET and CLEAR overrides associated with the FLIP-FLOPs.

The input conditioning circuits for switches X and Y amount to nothing more than debouncing circuits of the type shown in Fig. 5-88, but without the synchronizing stage. To make certain that the pulse signals from these switches are always nonoverlapping and at least minimally separated, the switches should be mechanically interlocked (as indicated) so that only one switch can be activated at any one time.

Design of a Candy Bar Vending Machine Controller. As our final design example, we will design a PULSE mode FSM that can control the operation of a candy bar vending machine. The candy bars each cost 40 cents and are dispensed automatically by the machine after correct change has been deposited. The machine accepts nickels, dimes, and quarters only. It consists of a controller (of our design), an electromechanically operated coin changer for nickel return, an accumulator register with counter and comparator to keep account of the coin exchange, and an electromechanically operated candy bar dispensing mechanism. In this section we will concern ourselves only with the design of the controller. The system-level design of this vending control system is the subject of Problem 6.50 at the end of this chapter.

The denomination of each coin is evaluated as it passes through the receiver of the machine (via a coin slot) and the accumulator is automatically updated. After each coin has left the receiver, the controller receives from the accumulator register, counter, and comparator one of three signals: $<$ 40 cents, $>$ 40 cents, or $=$ 40 cents. If $<$ 40 cents (underpayment), the machine awaits the deposit of another coin. If $>$ 40 cents (overpayment), the machine returns to the customer one nickel at a time by means of the coin changer until correct change remains. Or if $=$ 40 cents, the candy is dispensed (dropped) to the customer, the accumulator is cleared, and the controller is returned to its initialization state ready for the next customer.

A state diagram which satisfies the requirements of the candy bar vending machine just described is given in Fig. 6-58(a) together with definitions of the abbreviations that are used. Here again, one notices the simplicity of the state diagram for the PULSE mode design; each branching condition is a single uncomplemented input variable and no holding conditions need to be indicated. Of course, all inputs must be nonoverlapping pulses that are at least minimally separated. However, no special provisions are needed to generate these pulses since this vending machine cannot operate otherwise. Note that it is not necessary to indicate the holding conditions on the various states because all input variables not used for outbranching from a given state are implied holding conditions for that state.

The conditional outputs DC, DA and CA in Fig. 6-58(a) are pulsed outputs as are all conditional outputs in PULSE mode designs; the unconditional output RN is, of course, a level output. Thus, the FSM will be designed as a Mealy machine with mixed outputs (pulsed and level) and with FET T-FFs which are assumed to be produced from FET JK-FFs.

The K-maps and minimum cover for the NEXT STATE and output forming logic are given in Fig. 6-58(b). Here, full use is made of the don't cares arising from the three unused states. This is permissible for the output logic since there are no race conditions, hence no chance for output race glitch formation. The implementation of this PULSE mode FSM is left to the reader as an exercise.

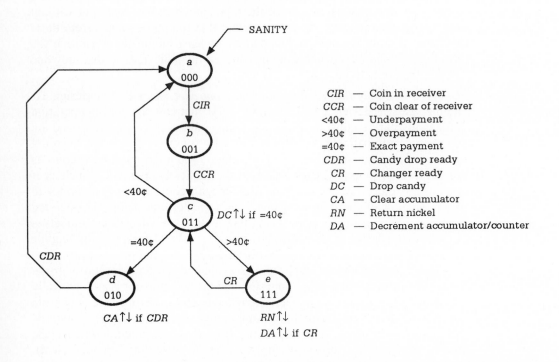

CIR	— Coin in receiver
CCR	— Coin clear of receiver
<40¢	— Underpayment
>40¢	— Overpayment
=40¢	— Exact payment
CDR	— Candy drop ready
CR	— Changer ready
DC	— Drop candy
CA	— Clear accumulator
RN	— Return nickel
DA	— Decrement accumulator/counter

(a)

Fig. 6-58. *PULSE mode design of a candy bar vending machine. (a) Fully documented state diagram and definitions of abbreviations.*

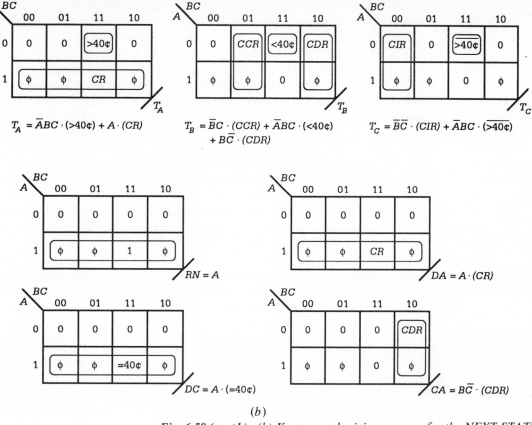

$$T_A = \bar{A}BC \cdot (>40\text{¢}) + A \cdot (CR)$$

$$T_B = \bar{B}C \cdot (CCR) + \bar{A}BC \cdot (<40\text{¢}) + B\bar{C} \cdot (CDR)$$

$$T_C = \bar{B}\bar{C} \cdot (CIR) + \bar{A}BC \cdot (\overline{>40\text{¢}})$$

$$RN = A$$

$$DA = A \cdot (CR)$$

$$DC = A \cdot (=40\text{¢})$$

$$CA = B\bar{C} \cdot (CDR)$$

(b)

Fig. 6-58 (contd.). (b) K-maps and minimum cover for the NEXT STATE and output logic.

6.8.4 Analysis of PULSE Mode FSMs

The procedure used to analyze a PULSE mode circuit is the reverse of that used for design. One begins with a logic circuit, reads the NEXT STATE and output logic expressions, and maps the results. Then the K-maps are used to construct the PRESENT STATE/NEXT STATE (PS/NS) table, which, in turn, is used to construct the state diagram. A similar procedure was followed in Sects. 5.9.4 and 6.6.3 for synchronous FSMs and asynchronous (fundamental mode) FSMs, respectively. However, the example that follows demonstrates that PULSE mode FSM analysis has its own unique characteristics which set it apart from the analyses discussed previously.

An Example. Consider the PULSE mode logic circuit presented in Fig. 6-59(a). It is easily identified as a PULSE mode circuit because of the manner in which the FLIP-FLOPs are used. It is also seen to be a Mealy machine since the output is a function of input Y. The NEXT STATE and output logic expressions are read from the logic circuit and are provided in the figure. These expressions are mapped and the results are shown in Fig. 6-59(b). The PS/NS table can be constructed from these K-maps, but it is usually worth the effort to convert the T K-maps to D K-maps, as we have done in Fig. 6-59(c), and then construct the PS/NS table from the D K-maps. Either way, the results for the PS/NS table are the same and are given in Fig. 6-59(d). Certain of the entries in the PS/NS table have been lined out because they pertain to input conditions which are not applicable to PULSE mode circuits. Of those entries that remain, only single uncomplemented input variables are relevant. Thus, the $(01) \rightarrow (11)$ transition occurs under branching condition Y while the $(11) \rightarrow (10)$ transition takes place under branching condition X, and so on. The result is the state diagram in Fig. 6-59(e) representing a four-state Mealy machine. Again we note that an indication of the holding condition for STATE (00) is not essential to the construction of the fully documented state diagram for this FSM. Such holding conditions are implied.

No FSM analysis is complete without making an assessment of the timing defects it may have. However, in properly designed and operated PULSE mode FSMs, ENDLESS CYCLEs and CRITICAL RACEs are not possible, and STATIC HAZARDs can exist only in the output forming logic. This FSM has none of these defects. Moreover, an inspection of the state diagram in Fig. 6-59(e) indicates that no output race glitches exist. Therefore, this FSM will operate properly, provided that the input pulses are nonoverlapping and at least minimally separated. Remember that the use of the nongated

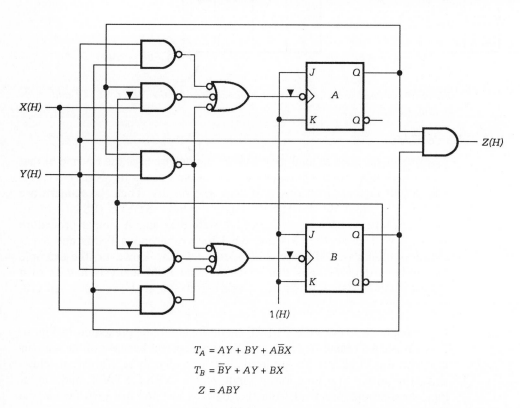

$$T_A = AY + BY + A\bar{B}X$$

$$T_B = \bar{B}Y + AY + BX$$

$$Z = ABY$$

(a)

Fig. 6-59. *Analysis of a PULSE mode FSM. (a) Logic circuit and NEXT STATE and output logic expressions as read from the circuit.*

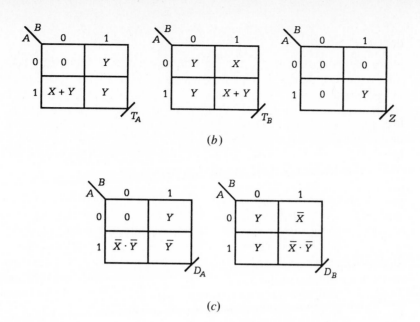

(b)

(c)

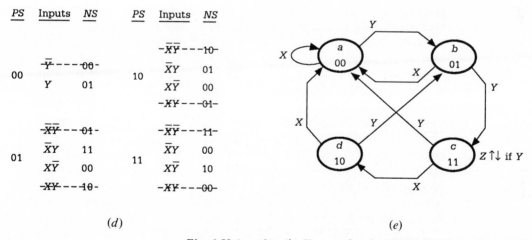

(d)

(e)

Fig. 6-59 (contd.). *(b) K-maps for the NEXT STATE and output logic. (c) D-FF K-maps converted from the T-FF K-maps in (b). (d) The PS/NS table obtained from the K-maps. (e) State diagram derived from the PS/NS table.*

FET T-FFs as the MEMORY elements permits the ACTIVE duration of the positive pulses to be unrestricted.

The algorithm for this FSM is easily read from the state diagram in Fig. 6-59(e). It may be stated as follows:

Output Z is issued coincidentally with any three consecutive Y pulses following any number of X pulses or following no X pulse.

In symbolic notation and beginning in STATE ⓪⓪, this agorithm can be expressed as the pulse/output sequence

$$\ldots X\text{-}X \ldots Y\text{-}Y\text{-}Y^*\text{-}Y\text{-}Y\text{-}Y^*\text{-}X\text{-}Y\text{-}Y\text{-}X\text{-}X\text{-}Y\text{-}Y\text{-}Y^*\text{-}X\text{-}Y \ldots$$

where the asterisk (*) indicates an output condition. Thus, this FSM is a pulse sequence detector that recognizes an uninterrupted series of three Y pulses whenever they occur. It is left as an exercise for the reader to show that the state diagram of Fig. 6-59(e) can be collapsed to one having three states.

6.9   MODULAR ELEMENTS AND THE HANDSHAKE INTERFACE

Modular elements are devices that have been designed to perform a task that is frequently required. Their existence saves the designer the trouble of redesigning the device each time its performance is needed. FLIP-FLOPs, registers, and counters may be considered as simple examples of modular elements that are designed to be operated in a synchronous environment. However, in this section we focus our attention on the design of modular elements which will operate in an asynchronous system-level environment. This involves the use of the *handshake interface* in which the outputs of one FSM are the inputs to another, and vice versa.

Asynchronous finite state machines that are designed to operate in the fundamental mode (by using the LPD or NESTED CELL models) have their problems and limitations, as we have pointed out at various times in previous sections of this chapter. Recall that FSMs operated in the fundamental mode require that no more than one input controlling the branching from a given state be permitted to change logic levels at any given time. Also, fundamental mode FSMs will malfunction if ENDLESS CYCLEs, CRITICAL RACEs, or active ESSENTIAL HAZARDs are present. Even the presence of a single STATIC HAZARD in the NEXT STATE forming logic of a fundamental mode FSM can cause it to malfunction.

PULSE mode FSM design is offered in Sect. 6.8 as a means of overcoming some of the problems and limitations inherent in fundamental mode machines—but at the expense of even more severe input constraints. It will be recalled that the inputs to PULSE mode FSMs must be nonoverlapping pulses which are at least minimally separated. When this requirement is met, the PULSE mode FSM will operate free of ENDLESS CYCLEs, CRITICAL RACEs, STATIC HAZARDs (in the NEXT STATE forming logic), and E-HAZARDs and d-trios. However, as a result of the input constraints, PULSE mode FSM designs have very limited applications.

Ideally, we would prefer to design asynchronous FSMs which have all the desirable characteristics of the fundamental mode and PULSE mode models but with none of their problems and limitations. While these idealized design goals can never be fully realized, we can come close. In this section we will consider design strategies that make use of the handshake interface in a manner that avoids some of the problems and limitations which characterize fundamental mode and PULSE mode FSM designs. We will demonstrate, however, that these design strategies are not without their price tag, the price being increased hardware requirements and increased time delays over those which result from fundamental and PULSE mode designs.

In this section we present two important modular elements: the ARBI-

TER MODULE and the D-FLOP MODULE. Several others exist and are presented as problems at the end of this chapter.

6.9.1 The ARBITER MODULE

The ARBITER MODULE is used to control access to a protected system by means of the handshake interface illustrated in Fig. 6-60. In effect the ARBITER MODULE permits only one of two contending input requests to have access to a protected system at any given time even if both requests are made simultaneously. If both requests are ACTIVE, access is granted to the first to go ACTIVE—hence, first-in, first-out. The second request signal will be granted access only after the DONE signal for the first has been received from the protected system. Single-access requests are granted without contention, but, if both go ACTIVE at precisely the same time, access will be granted to one, and only one, on the basis of some electronic imbalance in the logic circuit. Any metastable condition that exists in the ARBITER MODULE must be resolved before access can be granted to one or the other of the input requests.

Shown in Fig. 6-61(a) are the state diagrams that satisfy the requirements of a two-request ARBITER MODULE. An inspection of these state diagrams reveals that the ARBITER MODULE consists mainly of four TRANSPARENT D-LATCHes (pulse-triggered D-FFs) and a SET-dominant BASIC CELL. The inputs to the BASIC CELL are $Q_x \odot$ DONEX $= S$ and $Q_Y \odot$ DONEY $= R$, and the output, Q, of the BASIC CELL is the triggering signal to the TRANSPARENT D-LATCHes in antiphase fashion as shown. Inputs REQX and REQY are obviously the data inputs to the LATCHes. Thus, a full handshake configuration exists, since the outputs of the REQX and REQY TRANSPARENT D-LATCHes are also inputs to the BASIC CELL (via its NEXT STATE forming logic), and vice versa. Note that a DONEX or DONEY input signal to the ARBITER MODULE is issued by the protected system following successful receipt of the respective output, GRANTX or GRANTY. Consequently, the "DONE" signals are a necessary part of the handshake interface.

Since the ARBITER MODULE appears to consist of three simple, well-known FSMs, its logic circuit can be constructed directly from the state diagrams without the assistance of the familiar K-map step. This is done in Fig. 6-61(b), where two XOR gates provide the NEXT STATE forming logic

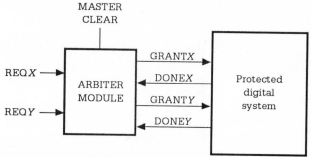

Fig. 6-60. *Block diagram of the ARBITER MODULE and its handshake interface with a protected digital system.*

for the special BASIC CELL. The two REQ D-LATCHes may be thought of as an input register, the two GRANT D-LATCHes as the output register, and the special BASIC CELL (discussed later) as the MEMORY. Alternatively, the input and output D-LATCHes could be regarded as the NEXT STATE and output logic FSMs, respectively, that form a special version of the NESTED machine model in Fig. 6-43. Note that we have arbitrarily chosen the REQ and DONE inputs and the GRANT outputs to be ACTIVE HIGH, but alteration of their ACTIVATION LEVELs can be made if appropriate changes are also made in the hardware. Note also that we have included in Fig. 6-61(b) the MASTER CLEAR (MCL) connections required to initialize (or reset) the ARBITER MODULE into its standby state. The two NAND gates together with the asynchronous CLEAR inputs to the TRANSPARENT D-LATCHes provide this feature, as explained shortly. The block circuit symbol for the ARBITER MODULE is given in Fig. 6-61(c).

To help the reader understand the operation of the ARBITER MODULE, we have constructed a MIXED LOGIC I/O table for the BASIC CELL in

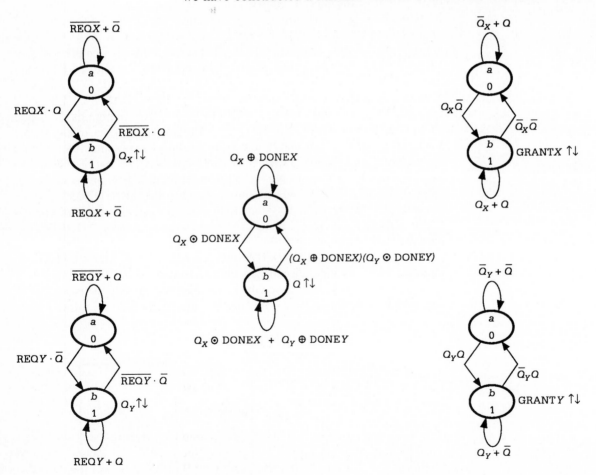

(a)

Fig. 6-61. *Design of a two-request ARBITER MODULE. (a) State diagrams.*

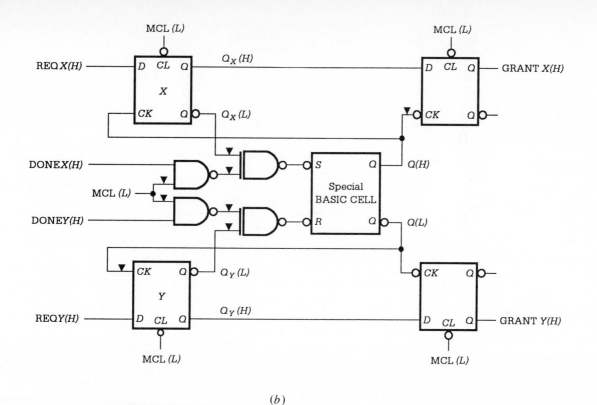

(b)

ARBITER MODULE

REQX

DONEX

DONEY

REQY

GRANTX

GRANTY

CL

(c)

Fig. 6-61 (contd.). *(b) Logic circuit derived from the state diagram in (a) showing MASTER CLEAR connections. (c) Block circuit symbol. (d) MIXED LOGIC I/O table for the BASIC CELL of the ARBITER MODULES.*

Fig. 6-61(d). The ARBITER MODULE is initialized (or reset) by applying a 1(L) to the MCL(L) inputs, which forces the MODULE into the standby condition (4) of the I/O table. In condition (4) under normal operation the two REQ D-LATCHes are transparent and the two output D-LATCHes are in the RESET state. Thus, in condition (4) both GRANTX and GRANTY are DEACTIVATED due to the $1(H) = 0(L)$ outputs of the BASIC CELL.

Input Condition	$Q_X \odot DONEX(L)$ $[S(L)]$	$Q_Y \odot DONEY(L)$ $[R(L)]$	$Q_{t+1}(H)$	$Q_{t+1}(L)$	
(1)—Hold	$0(L)$	$0(L)$	$Q_t(H)$	$Q_t(L)$	Hold
(2)—GRANTX	$0(L)$	$1(L)$	$0(H)$	$0(L)$	RESET
(3)—GRANTY	$1(L)$	$0(L)$	$1(H)$	$1(L)$	SET
(4)—Standby	$1(L)$	$1(L)$	$1(H)$	$0(L)$	———

(d)

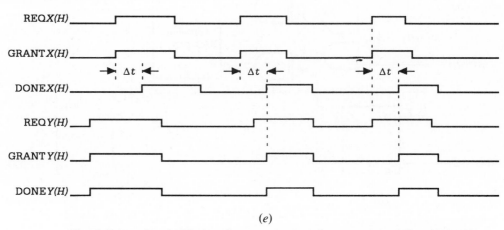

(e)

Fig. 6-61 (contd.). *(e) Timing diagram showing the operating of the ARBITER MODULE when the GRANTY output is connected to the DONEY input, and the GRANTX output is connected to the DONEX input via a large delay* Δt.

Let us assume that the MCL(L) inputs are all maintained at $0(L)$ so that the ARBITER MODULE can perform normally. Thus, when all inputs to the ARBITER MODULE are INACTIVE, the module is in its standby state. Then, if only REQX goes ACTIVE, the BASIC CELL is driven into condition (2) which ACTIVATEs the GRANTX D-LATCH and holds the X input D-LATCH in its SET state. When the DONEX signal is received from the protected system, the BASIC CELL returns to standby condition (4) and awaits the next request for access. Similarly, if REQY is the only ACTIVE request, the BASIC CELL is forced into condition (3) and GRANTY is ACTIVATED. This holds the Y input D-LATCH in its SET state until a DONEY signal is received after which the BASIC CELL is returned to the standby state.

Now, if both REQX and REQY go ACTIVE, one immediately after the other, the BASIC CELL will reside in condition (2) or (3) and the module will issue either a GRANTX or a GRANTY signal depending on which request (X or Y) is received first. If REQX is first, then a GRANTX signal is issued immediately and GRANTY is delayed until a DONEX signal is received from the protected system. When the DONEX signal is received, the BASIC CELL is forced into condition (3), via condition (1), permitting GRANTY to go ACTIVE. Similarly, if REQY is the first to go ACTIVE, GTANTY is issued immediately and GRANTX is delayed until DONEY is received from the protected system. When the DONEY signal is received, the BASIC CELL is forced into condition (2), via condition (1), thereby causing GRANTX to go ACTIVE. By the same reasoning, we note that GRANTX and GRANTY will go INACTIVE in the order that REQX and REQY go INACTIVE but only after the respective DONE signals are received from the protected system. Thus, in general, any change in the GRANTX and GRANTY signals will occur in the order that REQX and REQY change but only following receipt of the respective DONE signals by the MODULE.

The operation of the ARBITER MODULE just described is summarized by the timing diagram in Fig. 6-61(e). Here, the GRANTY output is connected to the DONEY input and the GRANTX output is connected to the DONEX input both by way of a large delay Δt representing a response delay from the protected system.

Metastability and Arbitration. Once the ARBITER MODULE has been placed in the standby mode, simultaneous $0 \rightarrow 1$ changes in the inputs REQX(H) and REQY(H) (hence, simultaneous $1 \rightarrow 0$ changes in the $S(L)$ and $R(L)$ inputs to the BASIC CELL) may, on rare occasions, cause the MODULE to behave improperly or unpredictably. To understand why this may happen and to understand the steps that can be taken to minimize the problems that may result, it is necessary to discuss the operation of the BASIC CELL as it pertains to the operation of the ARBITER MODULE. Remember that the ARBITER MODULE is the first device we have dealt with that requires a BASIC CELL to be operated in the $(S, R) = (1, 1)$ condition, the standby condition.

It may be recalled from Fig. 6-9 that when the S and R inputs to a BASIC CELL undergo simultaneous $1 \rightarrow 0$ changes, the BASIC CELL breaks into a logic oscillation, one that continues until one (or both) of the inputs changes $0 \rightarrow 1$. This logic oscillation occurs because the cross-coupled NAND gates are identical in every respect and because they drive each other in a manner that causes the output $Q(H)$ to alternate between ACTIVE and INACTIVE states, exactly antiphase to output $P(L)$. Such behavior in a BASIC CELL is called *logic instability*. It may be regarded as the "ideal" logic response of the BASIC CELL to simultaneous $1 \rightarrow 0$ changes in the logic inputs S and R.

The behavior of real NAND gates in a BASIC CELL differs from that just described. Real NAND gates are not identical. Their gain and their noise margins are not identical, and their $0 \rightarrow 1$ and $1 \rightarrow 0$ switching thresholds and propagation delays normally differ somewhat from gate to gate. Furthermore, the input voltage signals have finite rise and fall times. As a result of these factors, there is a very small but nonzero probability that a real BASIC CELL will enter a "temporary" condition called *metastability* following simultaneous LV to HV changes (equivalent to $1 \rightarrow 0$ changes logically) in the inputs, or following an input "runt" pulse, one that barely reaches the switching threshold. In the metastable condition, the BASIC CELL will either reside in an unresolved state somewhere between SET and RESET (but in neither) or it will oscillate. Experiments have shown that the time the BASIC CELL may spend in the metastable state usually amounts to few tens of gate delays, but, potentially, it could be much longer. BASIC CELLs composed of NAND (or NOR) gates with relatively large propagation time to rise time ratios (τ_p/τ_{LH}) tend to oscillate in the metastable state while BASIC CELLs with gates having τ_p/τ_{LH} ratios less than unity tend to favor unresolved behavior in the metastable state. In either case, the BASIC CELL will eventually emerge from the metastable state and reside stably in a SET or RESET condition, an outcome that is not usually predictable. Experimental work on this subject is cited in the Annotated References at the end of this chapter.

Thus, an ARBITER MODULE may be caused to operate improperly or unpredictably if, on rare occasions, its BASIC CELL is forced into a temporary condition of metastability by simultaneous LV to HV changes in the REQX and REQY inputs or by a runt pulse in one of these input signals. Unfortunately, the metastable condition cannot be prevented from occurring. Metastability is a statistical phenomenon that is fundamental to feedback systems such as cross-coupled NAND gates and cannot be eliminated by any electronic "fix it" scheme. The "fix it" scheme will have its own small but nonzero probability of going metastable. However, the oscillatory behavior resulting from the metastable state can be prevented, and the *metastable exit time* (the statistical time interval between entrance into and exit from the metastable state) can be reduced if not minimized. These measures can be accomplished in a CMOS BASIC CELL by adjusting certain switching thresholds of the cross-coupled NAND gates higher than normal and by adding INVERTERs (to the BASIC CELL outputs) that have switching thresholds lower than normal. If done properly, the result is a *special BASIC CELL* (BASIC CELL plus INVERTERs) as in Fig. 6-61(b) that will prevent oscillation and possibly reduce the metastable exit time should the BASIC CELL become metastable. Oscillation, resulting from the metastable condition, is prevented because the NAND gates can no longer drive each other in an antiphase manner. Also, to compensate for the threshold adjustments in both the NAND gates and the INVERTERs, gain elements may need to be installed on the INVERTER outputs. The details of how all this can be accomplished and the effect that such alterations may have on such factors as the gain bandwidth product and propagation delays, are complex and are beyond the scope of this text. However, a good treatment of this subject is cited in the Annotated References at the end of this chapter.

Once accomplished, the ARBITER MODULE will operate properly under any set of input conditions. However, on rare occasions, arbitration decisions may be delayed for potentially long periods of time due to the metastable "hangup" condition. Even so, upon emerging from such a metastable state, whenever that occurs, the ARBITER will make a clean decision and grant a request for access to one, and only one, contending request. The outcome of this decision may not be predictable.

The phenomenon of metastability is fundamental to the task of arbitration and cannot be eliminated by altering the design of the ARBITER. The only sure way to minimize the probability of a metastable hangup is to minimize competition for access to the protected system. Obviously, the greedier the contending requests are for access the greater will be the probability that the ARBITER will go metastable at some point in time.

A natural side effect of altering the thresholds of the NAND gates in a BASIC CELL is the effect this alteration has on the $0 \rightarrow 1$ and $1 \rightarrow 0$ propagation delays. The effect on the propagation delays of the BASIC

CELL cannot itself account for the overall behavior of the special BASIC CELL in the ARBITER MODULE just described. However, it can be demonstrated logically that a proper choice of edge propagation delays can, under certain conditions, eliminate the oscillatory behavior exhibited in Fig. 6-9. We leave this task as an exercise for the reader.

Application of the ARBITER MODULE. One way to overcome the severe input restrictions placed on PULSE mode designs is to find a means of producing nonoverlapping pulses from asynchronous input signals which can be overlapping. The use of the ARBITER MODULE provides such a means as demonstrated in the paragraphs that follow.

Shown in Fig. 6-62(a) is a PULSE mode FSM driven by an ARBITER MODULE and the logic (two AND gates) necessary to produce nonoverlapping pulses from two asynchronous inputs X and Y. The PULSE mode FSM is the same as that in Fig. 6-54. Nonoverlapping pulses to the PULSE mode FSM are assured by connecting the GRANT outputs of the ARBITER MODULE to their respective DONE inputs, and by connecting the mixed-rail outputs from the ARBITER MODULE to the AND gates as shown.

The operation of the PULSE mode FSM is illustrated by the timing diagram in Fig. 6-62(b). Note that overlapping input pulses X and Y yield nonoverlapping pulses GX and GY even when the ACTIVE periods of the input pulses are coincidental. This is so because of the action of the ARBITER MODULE.

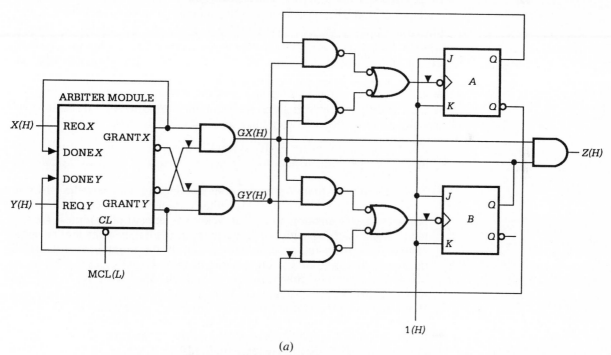

(a)

Fig. 6-62. *Application of the ARBITER MODULE to the simple PULSE mode sequence recognizer of Figure 6-54. (a) Logic circuit showing the ARBITER MODULE connections necessary to produce non-overlapping pulses from two asynchronous input signals.*

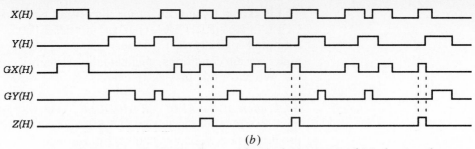

(b)

Fig. 6-62 (contd.). (b) Timing diagram for the logic circuit of (a) showing the occurance of an output pulse Z(H) any time the pulse sequence . . . Y − X − X . . . occurs.

A more likely use of the ARBITER MODULE is to arbitrate contending requests for access to a multiple-branching fundamental mode FSM. Since any two or more inputs controlling the branching to such an FSM cannot be permitted to change simultaneously, it is necessary to either constrain the input sources to meet the setup and hold-time requirements of each state's sampling variable or to interpose an input conditioning circuit that will accomplish this. The use of an ARBITER stage for this purpose is the subject of Problems 6.47 and 6.49 at the end of this chapter.

6.9.2 Externally Asynchronous/Internally Clocked Systems

In this section we describe digital logic architecture that invokes the advantages of both the synchronous and asynchronous design methodologies, but without many of their disadvantages. From the perspective of the external world such a system operates asynchronously but is internally clock driven following a specific handshake protocol. The term "handshake" refers to a logic configuration whereby the outputs of one FSM become the inputs to another, and vice versa.

A well-designed externally asynchronous/internally clocked (EAIC) system should have the following characteristics:

- Correct operation of an EAIC system is assured regardless of when the inputs change with respect to themselves or with respect to the internal CLOCK.
- The initiation of a state transition does not depend on a periodic CLOCK signal, as in a synchronous FSM; rather it depends on completion of the previous state transition and on an acknowledgment of readiness for the next. Thus, each transition event occurs immediately after an acknowledgment has been received indicating that all data inputs have satisfied the set-up and hold-time requirements of the internal CLOCK.
- The EAIC system is delay insensitive and will operate correctly without CLOCK distribution problems (CLOCK SKEW).
- When input signals are stored in the MEMORY elements of the EAIC system, they are, in effect, synchronized with respect to the internal CLOCK. Once stored, further data input is temporarily "locked out," and the stored information is retained in MEMORY until after the outputs have been updated with that information, a process similar to edge-triggered FLIP-FLOP operation.
- FSMs designed with the EAIC architecture can be initialized (or reset)

into any state, and logic noise in the outputs can be "filtered out" by using a standard output holding register triggered antiphase to the internal CLOCK.

- The NEXT STATE and output forming logic for the EAIC system can be implemented by using many of the same alternatives discussed in Sect. 5.11.

The features of the EAIC system represent a compromise between asynchronous fundamental mode and synchronous approaches to sequential machine design. All the advantages of a synchronous system design are realized, but without the possibility of CLOCK SKEW. A disadvantage of the EAIC system is that additional logic is required to produce the internal CLOCKing mechanism. As a result, the response time of the EAIC system will normally lie somewhere between equivalent synchronous and fundamental mode designs.

In Fig. 6-63 is shown the block diagram for one version of the EAIC architecture. The EAIC system consists of an input register, a MEMORY register, a RENDEZVOUS MODULE (RMOD) that generates the CLOCK signal, NEXT STATE and output forming logic, and an output holding register should it be necessary to filter the output signals. The input and MEMORY registers are composed of D-FLOP MODULEs that combine with the RMOD to produce the handshake interface required for EAIC system operation. The input and MEMORY registers receive identical CLOCK signals from the RMOD and CLOCK driver. Note that the purpose of the CLOCK driver is to boost the CLOCK signal for greater fan-out capability. For small systems the CLOCK driver can be omitted.

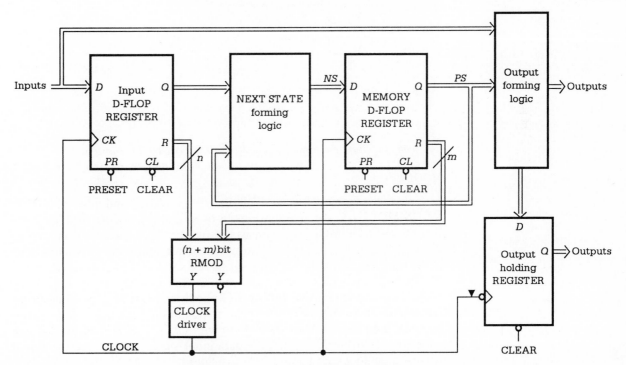

Fig. 6-63. *Block diagram for an externally asynchronous/internally clocked system.*

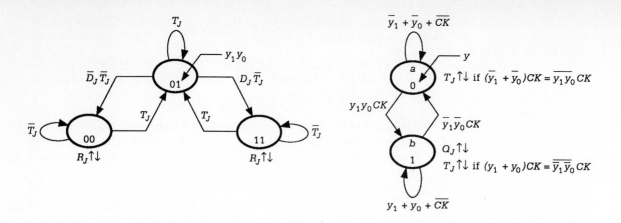

Resolver FSM   Output FSM

(a)

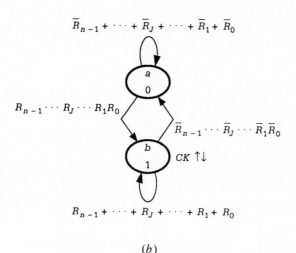

(b)

Fig. 6-64. *(a) State diagrams for the Jth D-FLOP MODULE showing its resolver and output FSM components. (b) State diagram for an* n-*bit RENDEZVOUS MODULE.*

The state diagrams for the Jth D-FLOP MODULE and *n*-bit RENDEZ-VOUS MODULE are given in Figs. 6-64(a) and (b). Figure 6-64(a) shows that the D-FLOP MODULE consists of two FSMs, a resolver FSM and an output FSM. The resolver FSM stores the input data D_J on the falling edge of the triggering variable T_J, and issues a request R_J (to the RENDEZVOUS MODULE) for a CLOCK signal when it (the resolver FSM) enters the RESET condition (STATE ⑩) or SET condition (STATE ⑪). The storage function of the resolver FSM is similar to that of the FET D-FF as is evident by comparing Figs. 6-64(a) and 6-44(a) with CLOCK complemented. Thus, once stored, further change in the data input signal cannot alter the stored value—a data-lockout feature.

When the requests for CLOCK, R, are received from all n D-FLOPs, the RENDEZVOUS MODULE in Fig. 6-64(b) issues a CLOCK signal. This CLOCK signal, together with the state variable outputs (y_1, y_0) of the resolver circuit, cause the output FSM in Fig. 6-64(b) to make a transition that depends on the resolved condition (SET or RESET) of the resolver FSM. In the SET condition (STATE ⑪) the output FSM undergoes a transition to STATE ① conditionally on $y_1 y_0$CK and issues an ACTIVE Q_J signal. In the RESET condition (STATE ⑩) the transition is to STATE ⓪ conditionally on $\bar{y}_1 \bar{y}_0$CK, and Q_J becomes INACTIVE. In either case, a tiggering signal T_J is issued by the output FSM, sending the resolver FSM back to its unresolved state (STATE ⑪) where request R_J becomes INACTIVE. When all requests for CLOCK, R, become INACTIVE, the RENDEZVOUS MODULE undergoes a transition to STATE ⓪ where CLOCK is INACTIVE. An INACTIVE CLOCK signal causes all triggering variables, T, to become INACTIVE, which, in turn, forces the storage of input data in all D-FLOPs and a repetition of the process just described.

Shown in Figs. 6-65(a) and (b) are the fundamental mode K-maps for the resolver and output FSMs of the D-FLOP, respectively, as obtained from Fig. 6-64. The cover used here is not minimum in every case but is necessary to help ensure that the EAIC system is delay insensitive. The system must operate correctly with any delays inserted in the CLOCK distribution leads.

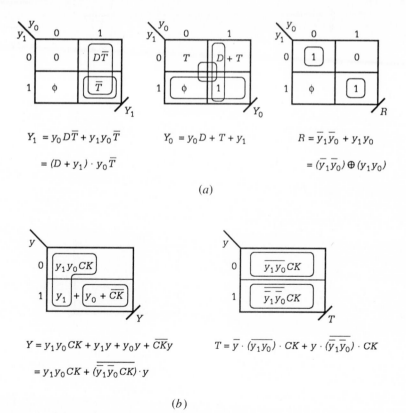

$$Y_1 = y_0 D\bar{T} + y_1 y_0 \bar{T}$$
$$= (D + y_1) \cdot y_0 \bar{T}$$

$$Y_0 = y_0 D + T + y_1$$

$$R = \bar{y}_1 \bar{y}_0 + y_1 y_0$$
$$= (\bar{y}_1 \bar{y}_0) \oplus (y_1 y_0)$$

(a)

$$Y = y_1 y_0 CK + y_1 y + y_0 y + \overline{CK}y$$
$$= y_1 y_0 CK + (\overline{y_1 y_0 CK}) \cdot y$$

$$T = \bar{y} \cdot (\overline{y_1 y_0}) \cdot CK + y \cdot (\overline{\bar{y}_1 \bar{y}_0}) \cdot CK$$

(b)

Fig. 6-65. *Design of the D-FLOP MODULE. (a) K-maps and minimum cover for the resolver FSM. (b) K-maps and minimum cover for the output FSM.*

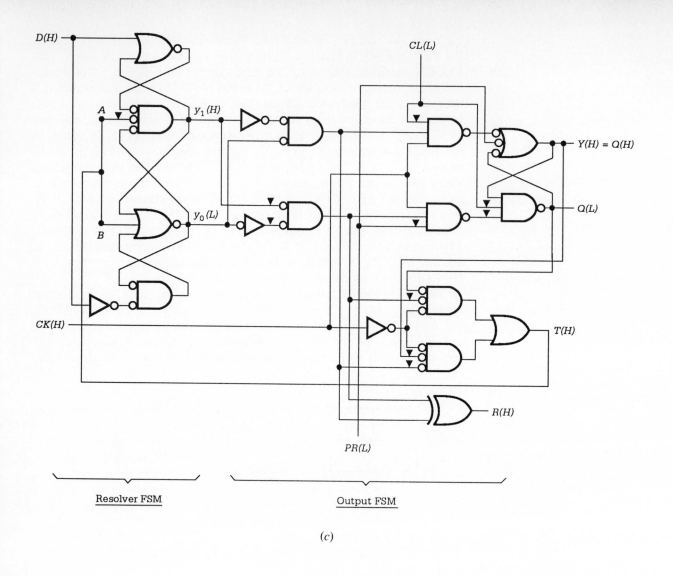

Resolver FSM

Output FSM

(c)

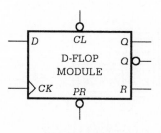

Block symbol

(d)

Fig. 6-65 (contd.). *(c) Logic circuit for the D-FLOP MODULE showing the ACTIVE LOW CLEAR and PRESET override connections and the resolver and output FSM sections. (d) Block circuit symbol.*

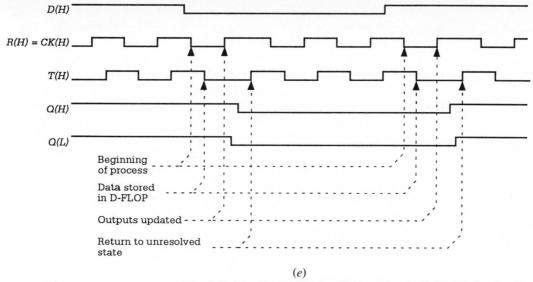

Beginning of process

Data stored in D-FLOP

Outputs updated

Return to unresolved state

(e)

Fig. 6-65 (contd.). *(e) Timing diagram for the D-FLOP showing its operation when request* R *is connected to the input* CK.

The logic for the request parameter, R, is taken as an XOR function to guarantee that R will go ACTIVE only when the resolver FSM is stable in one or the other of the two resolved states. However, an OR function would also suffice. The logic expressions for Y and T in Fig. 6-65(b) are also chosen to meet stability requirements. For example, the timing parameter, T, cannot go ACTIVE until the resolver FSM has stabilized into a resolved state and until the output has been updated.

The logic circuit for the D-FLOP is provided in Fig. 6-65(c) where the resolver and output sections are clearly indicated. The resolver FSM is recognized as being similar to that of Fig. 6-44(c) but with NOR logic and CLOCK ACTIVE LOW. Other differences stem from the choice of cover taken for Y_1. A somewhat different version of the D-FLOP (called the Q-FLOP) is presented in work on Q-modules cited in the Annotated References of this chapter.

The two NOR gates and two INVERTERs shown in Fig. 6-65(c) provide a metastable detection stage for the GATED BASIC CELL. The NOR gates and INVERTERs are designed with special switching thresholds so as to detect and resolve the metastable state should it occur. The details of how this is accomplished is beyond the scope of this book. However, the work on Q-modules, cited in the Annotated References at the end of this chapter, offers a simple cross coupled CMOS INVERTER design for the metastable detection stage.

Inherent in the EAIC system is the requirement for an upper bound to be placed on the propagation delay through the NEXT STATE forming logic, Δt_{NS}. An inspection of Fig. 6-65(c) indicates that correct operation of the EAIC system can be guaranteed only if $\Delta t_{NS} < (\Delta t_{T\text{-logic}} + \Delta t_{Resolver} + \Delta t_{RMOD})$, where the delay period in parentheses is approximately one half the CLOCK period. This requirement is easily met by NEXT STATE forming logic of two or three levels. Note that the $T(H)$ generating logic (T-logic) and the RMOD logic each constitute two levels of logic with respect to propagation delay. The resolver circuit logic (including the detection stage) constitutes approximately four levels of logic.

Asynchronous PRESET and CLEAR overrides can be added to the D-FLOP MODULEs so as to permit initialization (or reset) into any state of an EAIC system. The connections for these overrides are shown in Fig. 6-65(c). A block circuit symbol, presented in Fig. 6-65(d), provides a macroscopic view of all the D-FLOP terminals including the PRESET (PR) and CLEAR (CL) inputs.

In Fig. 6-65(e) is given the timing diagram for the D-FLOP MODULE illustrating its operation when request R is connected to CLOCK. This timing diagram is obtained from a computer simulation. It shows that the D-FLOP process begins with the falling edge of the $R = CK$ signal which forces the T variable LOW, thereby storing the data in the resolver FSM. The process continues with the updating of the output on the following rising edge of the $R = CK$ signal, and ends with a return of the resolver FSM to its unresolved state on the next rising edge of the T signal. Any delay could be placed in the R-to-CK connection resulting only in a slower performance of the D-FLOP.

An EAIC system of n total D-FLOPs must be operated by using a RENDEZVOUS MODULE (RMOD) having at least n inputs. The K-map, minimum cover, and logic circuit for an n-bit RMOD are presented in Fig. 6-66. This design is shown with ACTIVE HIGH and LOW outputs which provide the CLOCK signals for the EAIC system. If fewer than n inputs are needed for EAIC system operation, the unused inputs can be connected to those in use without affecting the operation of the RMOD.

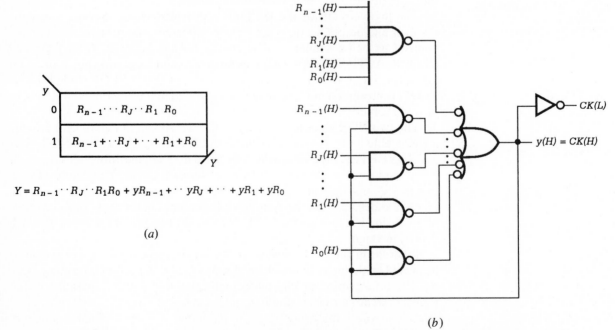

$$Y = R_{n-1} \cdots R_J \cdots R_1 R_0 + yR_{n-1} + \cdots yR_J + \cdots + yR_1 + yR_0$$

(a)

(b)

Fig. 6-66. *Design of the* n-*bit RENDEZVOUS MODULE of Figure 6-64(b). (a) K-map and minimum cover. (b) Logic circuit showing ACTIVE HIGH and ACTIVE LOW MODULE outputs.*

A Simple Example of an EAIC System. The application of the EAIC system approach to a simple sequence recognizer is given in Fig. 6-67. The logic circuit, shown in Fig. 6-67(a), consists of two input D-FLOP MODULEs (the input register), two MEMORY D-FLOP MODULEs (the MEMORY or output register), NEXT STATE and output forming logic, and a four-input RMOD. The LPD model design of this sequence recognizer is featured in Problem 6.21 at the end of this chapter.

The timing diagrams for the operation of the EAIC system in Fig. 6-67(a) are taken from a computer simulation and are presented in Figs. 6-67(b) and (c). The frequency of the $CK(H)$ signal in Fig. 6-67(b) is too high to be resolved. The timing diagram in Fig. 6-67(c) is a blow-up of the dashed box region in Fig. 6-67(b). It reveals the details of the EAIC system operation just before and just after a SEQACK(H) transition. Notice how the four R signal transitions must be "bundled" to produce a CLOCK signal transition. Thus, a given CLOCK transition can occur only after all inputs to the RMOD have previously undergone the same transition.

One alternative is to design the output FSM of the D-FLOP by using the

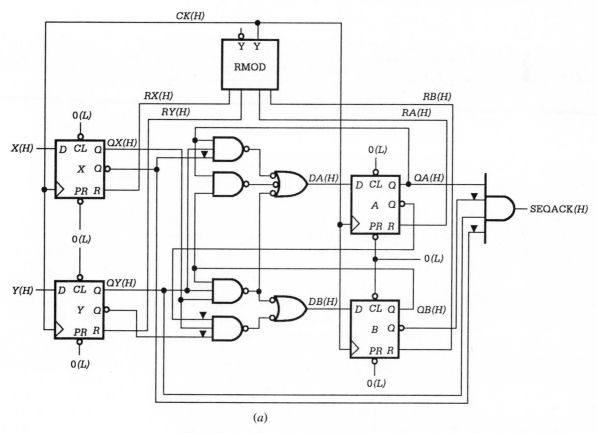

(a)

Fig. 6-67. *Application of the EAIC system to a simple sequence recognizer. (a) Logic circuit showing the input and MEMORY D-FLOPs, the NEXT STATE and output forming logic, and a four-input RENDEZVOUS MODULE.*

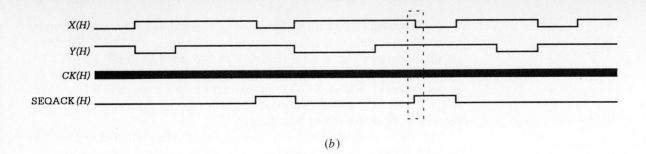

(b)

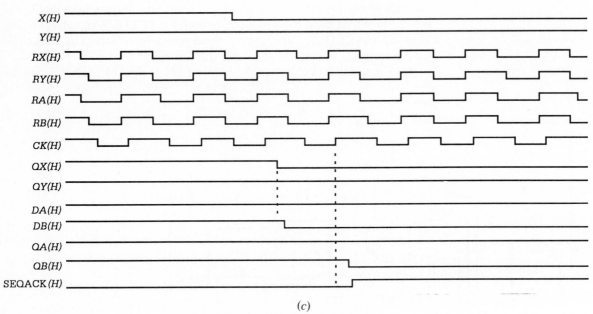

(c)

Fig. 6-67 (contd.). (b) Timing diagram for the sequence recognizer of (a) showing an output SEQACK any time in the sequence . . . 10–11–01 . . . occurs in the X, Y input signals. (c) Blow-up of the dashed box region in (b) showing details of system operation inclusive of the delays through the NEXT STATE and output forming logic.

LPD model in lieu of the NESTED CELL model used in Fig. 6-65. This alternative design should not alter the performance of the D-FLOP but will alter the character of the mixed-rail outputs $Q(H)$ and $Q(L)$.

Scope of EAIC System Application. Application of the EAIC system appears to be nearly unlimited since it can replace any of the synchronous system architecture described in Sect. 5-11. A D-FLOP MODULE can be converted to any X-FLOP MODULE (e.g., a JK-FLOP or T-FLOP MODULE) by using the conversion logic discussed in Sect. 5.6.6, and any D-FLOP register or counter can be designed for use in an EAIC system by following the procedures outlined in Sect. 5.10.

The book by Unger is a classic in the field of asynchronous sequential machines. Both fundamental mode and PULSE mode asynchronous machines are discussed. Chapter 4 provides an excellent coverage of CRITICAL RACES, combinational HAZARDs and essential HAZARDs. However, the treatment is theorem/lemma oriented and may not be easy reading for the beginner. Chapter 6 discusses asynchronous sequential circuits that return "done" signals. Original works by Muller, Unger, Huffman, McCluskey, and others are cited.

Unger, S. H., *Asynchronous Sequential Switching Circuits,* Wiley-Interscience, New York, 1969.

Asynchronous sequential machines designed to operate in the fundamental mode are discussed in the text by Fletcher. The notation used closely parallels that used by the present authors. Chapter 10 includes adequate treatment of asynchronous FSM design and analysis and includes discussions of timing defects such as ENDLESS CYCLEs, CRITICAL RACEs, and HAZARDs. Some useful examples are given and a bibliography is provided.

Fletcher, W. I., *An Engineering Approach to Digital Design,* Prentice Hall, Englewood Cliffs, N.J., 1980.

A condensed, relatively advanced, but readable treatment of asynchronous FSMs is given in Chapters 12 and 13 of Dietmeyer's book. The subjects of HAZARDs and state assignment techniques, among others, are covered. Important earlier works are cited.

Dietmeyer, D. L., *Logic Design of Digital Systems,* Allyn & Bacon, Boston, 1978.

A book by Nagle et al. and another by McCluskey cover the subjects of both fundamental mode and PULSE mode asynchronous sequential machines. The book by McCluskey emphasizes testable circuit design and adheres rigidly to the ANSI/IEEE Standard for logic symbols. Chapters 7, 8, and 9 of McCluskey's book provide adequate coverage of fundamental mode and PULSE mode asynchronous machines. Chapter 3 considers the subject of combinational HAZARDs in considerable detail. Chapter 8 of Nagle et al. covers the subjects of fundamental and PULSE mode finite state machines in a traditional manner. Bibliographies are included in both texts.

Nagle, H. T., Jr., B. D. Carroll, and J. D. Irwin, *An Introduction to Computer Logic,* Prentice Hall, Englewood Cliffs, N.J., 1975.

McCluskey, E. J., *Logic Design Principles with Emphasis on Testable Semicustom Circuits,* Prentice Hall, Englewood Cliffs, N.J., 1986.

The unpublished work of Sproull and Sutherland provides an excellent treatment of the MOS design of asynchronous FSM modules. A knowledge of petri nets is recommended. Volume II of their work has been of significant value to the present author in matters pertaining to asynchronous modules, and in particular the ARBITER MODULE. An extensive bibliography of the pertinent subject areas is provided.

Sproull, R. F., and I. E. Sutherland, *Asynchronous Systems,* Vols. I, II and III, Sutherland, Sproull and Associates, Inc., Pittsburgh, Penn., (unpublished manuscript, 1986).

An excellent article by Rosenberger et al. describes the CMOS design and analysis of Q-modules which are used in externally asynchronous, internally CLOCKed systems. A good bibliography provides the background for this work.

Rosenberger, F. U., C. E. Molnar, T. J. Chaney, and T. P. Fang, "Q-Modules: Internally Clocked Delay-Insensitive Modules," *IEEE Trans. Compt.,* Vol. 37, (Sept. 1988), pp. 1005–1018.

An article by Chaney and Molnar provides some interesting experimental observations on the metastable state in BASIC CELLs.

Chaney, T. J., and C. E. Molnar, "Anomalous Behavior of Synchronizer and Arbiter Circuits," *IEEE Trans. Compt.,* Vol. C-22 (Apr. 1973), pp. 421–422.

The software tools required to simulate FSMs are briefly reviewed in Chapter 7 of the more recent text by Unger. Useful references are given at the end of that chapter.

Unger, S. H., *The Essence of Logic Circuits,* Prentice-Hall, Englewood Cliffs, N.J., 1989.

References for software tools used to minimize Boolean function are provided in the Annotated References at the end of Chapter 3.

A partial listing of data manuals and handbooks for semicustom TTL and CMOS MSI and LSI logic design is provided in the Annotated References at the end of Chapter 4.

PROBLEMS

6.1 The logic circuit in Fig. P6-1 is that of a special MEMORY cell.

   (a) Determine its state diagram by following a procedure similar to that for the SET-dominant basic cell in Sect. 6.4.2.

   (b) Analyze this FSM for any problems or potential problems it may have.

   (c) Construct the state transition table and operation table for this special MEMORY cell.

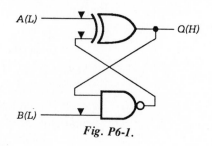

Fig. P6-1.

6.2 An asynchronous FSM is to be designed which has two inputs, A, and B, and one output, Z. It is the function of this machine to issue an output signal Z any time a change occurs in A or B such that their logic levels are not the same.

(a) Construct the Moore and Mealy forms of the state diagram for this machine.

(b) Use the LPD model and either form of the state diagram to design this FSM. End with a suitable logic circuit.

(c) What conclusion can be drawn from this exercise regarding combinational logic in general?

6.3 The state diagram of Fig. 6-6(a) is MINTERM code based as are all state diagrams and K-maps presented in this text. We know that a MINTERM code based state diagram must satisfy the requirements of Eqs. (6-4) and (6-5), unless, of course, certain restrictions are placed on the input sources.

(a) Convert the state diagram of Fig. 6-6(a) to one that is MAXTERM code based and verify that the *dual* of each equation, Eq. (6-4) and (6-5), is satisfied. Thus, from these results, write the generalized dual forms for Eqs. (6-4) and (6-5). (Hint: To convert a state diagram from MINTERM code to MAXTERM code form, or vice versa, complement all branching conditions.)

(b) Plot the NEXT STATE and output EV K-maps from the MAXTERM code based state diagram by using the LPD model, and loop out minimum POS cover. Compare these results with those of Fig. 6-6.

6.4 A two-input (A, B) and single output (Z) asynchronous FSM operates in the following manner:

Whenever $B = 1$, then $Z = 0$. But if $B = 0$ and $Z = 0$, a change in A causes Z to be $Z = 1$. The output Z cannot change to $Z = 0$ until B changes to $B = 1$.

Use the LPD model and optimum NOR logic to design this FSM for glitch-free operation and assume that the inputs arrive such as to meet the setup and hold-time requirements for each state. Assume that the inputs arrive from positive logic sources and that the output is ACTIVE LOW. (Hint: The state diagram should be one of three states.)

6.5 Two SET-dominant GATED (CLOCKED) BASIC CELLs were used in Fig. 5-34(a) to configure the MS JK-FF. By using the LPD model, design the SET-dominant GATED BASIC CELL. Begin with its state diagram and first-order EV K-map, and end with the logic circuit. Assume that the inputs S, R, and G (for GATE), arrive from positive logic sources and that the output $y = Q$ is ACTIVE HIGH.

6.6 Follow the same procedure as in Problem 6.5 to design the RESET-dominant form of the GATED BASIC CELL. Assume that the inputs S, R and G and the output Q are all ACTIVE HIGH. (Hint: Loop out the entered variables of the first-order K-map in MAXTERM code.)

6.7 The state diagram in Fig. P6-2, has been coded in *one-hot* code (see Fig. 4-42). The design of such an asynchronous FSM is said to be by the "one-hot" method.

(a) Are there any oscillations or CRITICAL RACEs associated with the FSM in Fig. P6-2? Explain.

(b) Are output race glitches possible in the FSM of Fig. P6-2? If so, under what conditions can they be avoided? Give a detailed answer by using state diagram segments as in Fig. 5-59.

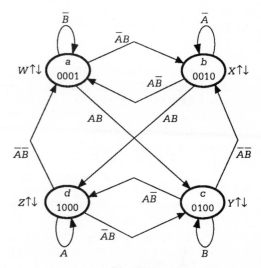

Fig. P6-2.

6.8 (a) Use the LPD model to find the optimum cover for the FSM in Fig. P6-2 of Problem 6.7. The design must be free of STATIC HAZARDs and output race glitches.

(b) What conclusion is reached regarding STATIC HAZARD production in this FSM?

(c) Are E-HAZARDs and d-trios possible in this FSM? Explain your answer.

(d) What are the advantages and disadvantages of the one-hot method applied to asynchronous FSM design?

6.9 (a) With the help of the information in Fig. 6-47(a), construct a fully documented state diagram for the rising edge triggered D data lockout FLIP-FLOP.

(b) Discuss how this data lockout FLIP-FLOP is similar to yet different from the D master slave FLIP-FLOP of Fig. 5-32(a).

6.10 Refer to Problem 6.9. Use the LPD model to carry out the necessary steps to obtain an optimal design (logic circuit) for the D data lockout FLIP-FLOP free of STATIC HAZARDs.

6.11 Refer to Problem 6.9. Identify four E-HAZARD paths that, if active, will cause the FSM to malfunction. Give the race gate and the initiating branching condition for each E-HAZARD. Also, identify two d-trio paths and explain why they do not have the potential to cause the FSM to malfunction.

6.12 The state diagram in Fig. 5-23(b) represents a positive pulse triggered (PPT) D FLIP-FLOP or TRANSPARENT D-LATCH. It can also be called a TRANS-HI MODULE. In Sect. 5.6.3 this FLIP-FLOP was designed by using the NESTED CELL model, though no mention of this was made at that time.

    (a) Find an optimum design for the PPT D-FF of Fig. 5-23(b) by using the LPD model. Check for STATIC HAZARDs, and end with an appropriate LPD logic circuit. Verify that the gate/input tally (excluding INVERTERs) is 4/9 with HAZARD cover included.

    (b) Construct a timing diagram for this FSM and verify that the TRANSPARENT D-LATCH will break into oscillation if STATIC HAZARD cover is not included. Demonstrate with the timing diagram that inclusion of the STATIC HAZARD cover removes the HAZARD.

6.13 The STATE DIAGRAM in Fig. 5-29(b) represents a positive pulse-triggered (PPT) JK FLIP-FLOP. In Sect. 5.6.3 this FLIP-FLOP was designed by using the NESTED CELL model of Fig. 6-43(c). Design the PPT JK-FF of Fig. 5-29(b) by using the LPD model. Check for STATIC HAZARDs and end with an optimum LPD circuit having mixed-rail outputs and $PR(L)$ and $CL(L)$ overrides. Compare this circuit with that of Fig. 5-30(b). Which if either is the preferred design? Explain.

6.14 The FSM of Fig. 6-16(a) has the potential to malfunction by the action of a single E-HAZARD.

    (a) Identify the E-HAZARD path, the branching condition under which it is initiated, and the race gate.

    (b) In terms of Fig. 6-21, classify this E-HAZARD as to its type.

    (c) Does a d-trio path exist? If so, give the path and identify the race gate and the initiating branching conditions.

6.15 Refer to Problem 6.14.

    (a) Use the logic circuit of Fig. 6-17(a) in conjunction with the results of Problem 6.14 and follow the procedure in Sect. 6.5.4 to determine the minimum path delay required to produce the E-HAZARD. To facilitate this analysis assign a unit path delay τ_p to each gate and to each IN-VERTER. Which gate is the race gate?

    (b) Verify the results of (a) by using a timing diagram of the type shown in Fig. 6-26(b). Place a counteracting delay in the appropriate feedback path and demonstrate removal of the E-HAZARD.

6.16 The state diagram is provided in Fig. P6-3 for a three state asynchronous FSM.

    (a) Analyze this FSM for possible ENDLESS CYCLEs, CRITICAL RACEs, STATIC HAZARDs and output race glitches. Base the STATIC HAZARD analysis on SOP cover obtained by using the LPD model. Use a timing diagram to demonstrate the existence of any STATIC HAZARD or output race glitch that may exist.

    (b) Analyze this FSM for possible POS STATIC HAZARDs. If present, use timing diagrams to demonstrate their existence and removal with the appropriate HAZARD cover.

    (c) Analyze this FSM for possible ESSENTIAL HAZARDs and d-trios. If any exist, determine the paths, race gates, branching conditions, and minimum delay requirements for their production. Assume that each gate and INVERTER has a unit path delay τ_p. Use a timing diagram to demonstrate the formation and removal of any E-HAZARD or d-trio that can be produced.

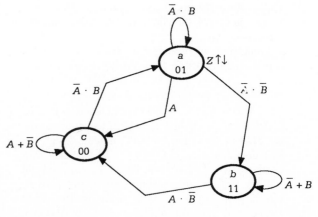

Fig. P6-3.

6.17 The FSM in Fig. 6-18(a) appears to have the potential to malfunction along either one of two E-HAZARD paths.

(a) Use the logic circuit, derived from the NEXT STATE expressions in Fig. 6-18(b), and the STATE DIAGRAM in Fig. 6-18(a) together with timing diagrams to explain why this FSM can or cannot be caused to malfunction by E-HAZARD action.

(b) Are any d-trio paths present? Explain.

6.18 The state diagram for the DATA CHECK MODULE is given in Fig. P5-4 at the end of chapter 5.

(a) Use the LPD model to design the DATA CHECK MODULE as an asynchronous FSM. Assume that the inputs and output are all ACTIVE HIGH.

(b) Repeat part (a) by using the NESTED CELL model for the design.

(c) Demonstrate the operation of the DATA CHECK MODULE with a MIXED LOGIC timing diagram for the design in part (b). Neglect propagation path delays.

6.19 The requirements for E-HAZARD and d-trio formation in SOP logic circuits is presented in Sect. 6.5.4.

(a) Summarize the requirements for E-HAZARD and d-trio formation applicable to POS logic circuits. Include race gate requirements in your answer.

(b) Use timing diagrams to test the validity of the requirements given in (a) on the POS logic circuit that results from Fig. 6-16.

6.20 A sequence recognizer is to be designed that receives two inputs, X and Y, and issues a glitch-free output SEQVALID any time the nonoverlapping pattern of input changes is . . . $XYXYYXYY$ . . . . The FSM is to monitor this sequence of changes beginning when both X and Y become INACTIVE following initialization, or following completion of the correct sequence, or following any incorrect sequence.

(a) Construct a state diagram for this FSM that is suitable for the LPD model and that is free of oscillations, CRITICAL RACEs, and output race glitches.

(b) Implement this asynchronous FSM by using the NESTED CELL model and a PLA-type device. Hence, construct a p-term table by using the state diagram from (a) and the appropriate K-map cover which must be HAZARD-free. Assume that the inputs arrive from positive logic sources which are never permitted to change at

the same time, and hence are never in violation of the setup- and hold-time requirements for any given state. Also, assume that the output SEQVALID is issued ACTIVE HIGH.

(c) Construct the abridged symbolic representation for the programmed PLA. Initialize by means of an asynchronous CLEAR override in RESET-dominant BASIC CELLs.

(d) Make a table showing all potential second order E-HAZARD and d-trio paths associated with this FSM. Include the branching conditions required to initiate each defect.

6.21 A sequence recognizer is to be designed which has two inputs, A and B, and a single output SEQACK. It is required that the output SEQACK go ACTIVE any time the sequence . . . 10-11-01 . . . is recognized.

(a) Construct a four state state diagram for this FSM that is suitable for the LPD model and that is free of CRITICAL RACEs and oscillations.

(b) Assume that the inputs arrive from positive logic sources which are never permitted to change at the same time, and hence are never in violation of the setup- and hold-time requirements for any given state. Assume also that the output is issued ACTIVE LOW. Use the LPD model and discrete logic for a defect-free, optimum design of this FSM. Compare the NEXT STATE logic of this design with that of Fig. 6-67(a).

(c) Identify any potential E-HAZARD, d-trio, or special delay HAZARD paths that may exist. Give the race gate and minimum path delay requirements for each. Verify the existence of each potential defect in a MIXED LOGIC timing diagram.

6.22 An LPD logic circuit yields the following expressions for the NEXT STATE logic:

$$Y_1 = \bar{y}_1 \bar{Y}_0 \bar{A} + y_0 AB + y_1 \bar{B} + y_1 y_0$$

$$Y_0 = y_1 \bar{A} B + y_0 B + \bar{y}_1 y_0 A + y_1 y_0 \bar{A}$$

$$Z = \bar{y}_1 y_0 A$$

(a) Construct the state diagram and NEXT STATE map for this FSM. (Hint: See Sect. 6.6.3.)

(b) Analyze this machine for any possible defects or potential problems it may have.

6.23 A badly designed (nonsense) asynchronous FSM is designed by using the NESTED CELL model discussed in Sect. 6.7. When this circuit is read, the following logic expressions are obtained:

$$S_A = \overline{B}\overline{Y} + B\overline{X}Y \qquad S_B = AX$$

$$R_A = BX + B\overline{Y} \qquad R_B = \overline{A}\overline{Y} + \overline{X}$$

$$Z = \overline{A}B$$

(a) Obtain the state diagram for this machine. (Hint: It may be helpful to transform the second-order K-maps to LPD form by using the transformation equation $Y = \overline{y}S + y\overline{R}$.)

(b) Analyze this FSM for any defects it may have.

6.24 The circuit in Fig. P6-4 is represented by the distributed path delay model for which a path delay, NEXT STATE variable and PRESENT STATE variable exist for each gate.

(a) Analyze this circuit by following the procedure outlined in Sect. 6.6.3, and verify that a four state state diagram results.

(b) Merge the appropriate states to form the two state state diagram that would be valid for the LPD model of the same FSM.

6.25 A special duty rising edge triggered N FLIP-FLOP is defined by the following operation table:

BASIC CELL of the special FET ST-FF in Fig. 6-36(c). (Hint: The SET and RESET branching conditions must include $\overline{Q}$ and Q, respectively, so as to accommodate the TOGGLE character of this FLIP-FLOP.)

(b) Verify with a timing diagram based on the STATE DIAGRAM in (a) that this ST-FF is semi-transparent and that it does not satisfy all the requirements of a true edge triggered FLIP-FLOP.

6.27 (a) Design the RET D-FF by using the NESTED CELL model applied to the controller for the BASIC CELL shown in Fig. 6-44(a). Compare the results with those of Fig. 6-44(c).

(b) The logic circuit in part (a) appears to have the potential to malfunction as a result of d-trio formation along either one of two paths. Use this logic circuit together with the state diagram in Fig. 6-44(a) to explain why d-trio formation is not possible by either path. (Hint: Consider the indirect path requirement for d-trio formation.)

(c) Verify the results of part b by using timing diagrams. Simplify the analysis by assigning a unit path delay, τ_p, to each gate and INVERTER.

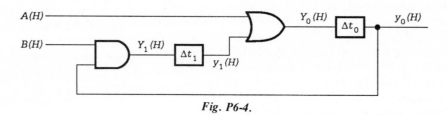

Fig. P6-4.

N	Q_{t+1}
0	$\overline{Q}_t$
1	1

The N FLIP-FLOP is to be designed by using exclusively the NESTED CELL model of Fig. 6-43(c). Construct the three-state state diagram for the controller of the BASIC CELL and apply the NESTED CELL model to the design of this controller. Assume that N and CLOCK both arrive from positive logic sources.

6.26 (a) Construct a three-state state diagram representing the SET/RESET controller for the

6.28 The fundamental mode logic circuit in Fig. P6-5 is called the TOGGLE MODULE. The TOGGLE MODULE is composed of two SET-dominant GATED BASIC CELLS (see Problem 6.5) and an INVERTER. It is the function of the TOGGLE MODULE to accept transitions on the T input and send them alternately to its X and Y outputs starting with the X output following a CLEAR (or RESET) override signal.

(a) Obtain the state diagram for the TOGGLE MODULE.

(b) Identify any possible E-HAZARD paths this FSM may have relative to delays placed at points A and B, and find the minimum path delay requirements for each. For simplicity, assign a

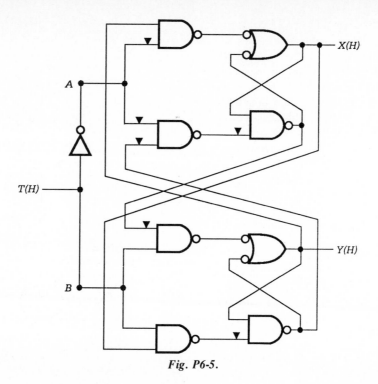

Fig. P6-5.

unit path delay τ_p to each gate and to the IN-VERTER.

(c) Show the connections required to equip the logic circuit in Fig. P6-5 with an asynchronous CLEAR override.

6.29 Replace the input circuit between nodes A and B in Fig. P6-5 with the new input circuit shown in Fig. P6-6. Construct a timing diagram for this new input circuit and determine the new minimum path delay requirements for E-HAZARD formation in terms of the delay Δt. Compare these results with those obtained in part (b) of Problem 6.28. Again assign the unit path delay τ_p to each gate.

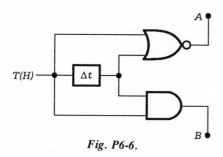

Fig. P6-6.

6.30 The state diagram in Fig. P6-7 represents the SE-LECTOR MODULE. It is the function of this module to steer transition input signals, c, to either the

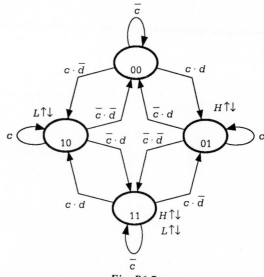

Fig. P6-7.

H or L output, depending on whether the input d is ACTIVE or INACTIVE as shown.

(a) Use the LPD model to design an optimal logic circuit for the SELECTOR MODULE that is free of STATIC HAZARDs. Assume that the inputs, c and d, arrive from positive logic sources and that the outputs, H and L, are AC-TIVE HIGH.

(b) Discuss whether or not E-HAZARDs or d-trios can cause this FSM to malfunction.

(c) Construct the NEXT STATE map for the SELECTOR MODULE by following the procedure given in Sect. 6.3.

6.31 Shown in Fig. 6-66 is the design of an n-input RENDEZVOUS MODULE (RMOD) that is based on the LPD model.

(a) Use Fig. 6-64(b) and the NESTED cell approach (see Sect. 6.7.1) to design a four-input RMOD.

(b) Comment on the advantages and disadvantages of this design over that of Fig. 6-66(b). Direct your comments mainly to the operation of the BASIC CELL and any timing problems that may occur.

6.32 Design a special MEMORY cell that will accept either of two SET signals, S_1 or S_0, or either of two RESET signals, R_1 or R_0, and that will not permit any $(S, R) = (1, 1)$ condition to cause a transition. Use the NESTED CELL model and minimum NAND logic to do this, and assume that the special MEMORY cell operates with ACTIVE LOW inputs. Equip this MEMORY cell with ACTIVE LOW asynchronous PRESET and CLEAR overrides and sketch its circuit (block) symbol.

6.33 (a) Analyze the logic circuit in Fig. P6-8 by constructing its state diagram. (Suggestion: Use the expression $Y = \bar{y}S + y\bar{R}$ to convert the S and R K-maps to LPD K-maps; then construct the PS/NS table.)

(b) Are E-HAZARDs and d-trios possible in this FSM? Explain.

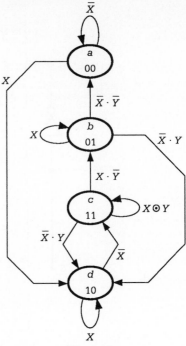

Fig. P6-9.

(b) Construct the NEXT STATE map from the state diagram following the example in Fig. 6-6 and indicate in this map the problems identified in part (a).

(c) Suggest a workable algorithm for this machine and then make the appropriate changes in the state diagram and NEXT STATE map.

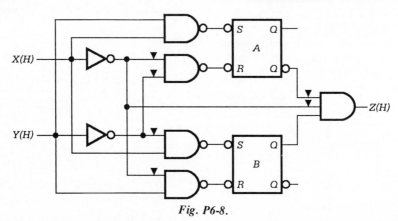

Fig. P6-8.

6.34 Shown in Fig. P6-9 is the state diagram for an FSM that has some inherent problems which will cause it to malfunction.

(a) Identify these problems and indicate how they would cause malfunction of this machine.

6.35 A PULSE mode digital combination lock (DCL) is to be designed for a vault. It is the function of the DCL to issue the signal OPNVAULT coincidentally with the last Y pulse in the positive PULSE sequence . . . X-X-Y-X-Y . . . , and then return immediately to the initialization state and reissue a LOCK signal. Note that this sequence cannot be overlapping.

(a) Use FET JK-FFs as the MEMORY to design the DCL. Assume that the inputs arrive as non-overlapping, at least minimally separated pulses from negative logic sources, and that the outputs are issued ACTIVE HIGH.

(b) Use MEMORY elements of the type shown in Fig. 6-52(d) to design the DCL. Assume that the nonoverlapping inputs and the outputs are all ACTIVE HIGH.

6.36 A fundamental mode circuit is to be designed that will detect the direction of rotation of a circular shaft as shown in Fig. P6-10. Two light beams are caused to fall incident on the end surface of the shaft half of which is reflecting and half nonreflecting. Two photocells, A and B, are located at the proper angle of reflection relative to the two beams so that whenever a beam strikes a reflecting surface the photocell receiving the reflected beam will generate a voltage signal. For the shaft position shown in Fig. P6-10, the logic input to the FSM is $AB = 01$.

It is the requirement of this asynchronous FSM that the output CCW be ACTIVE any time the shaft is rotating counterclockwise and INACTIVE any time it is rotating clockwise (CW). The output response to rotational direction must be as fast as possible.

Use the LPD model to design an optimum logic circuit for this FSM that is free of ENDLESS CYCLEs, CRITICAL RACEs, output race glitches, and STATIC HAZARDs. Discuss whether or not E-HAZARDs and d-trios are possible in this FSM. If they are possible, give the path and race gate requirements for each of these potential defects. Also, discuss the limitations of this design with respect to the sensitivity to shaft oscillations.

6.37 Design a dual–mode edge-triggered FLIP-FLOP that will operate as an RET D FLIP-FLOP if $P = 1$ or as an FET T FLIP-FLOP if $P = 0$. Use the NESTED CELL approach to design the three-state controller (for the BASIC CELL) similar to that of Fig. 6-44(a). Design for a minimum gate count assuming that all inputs arrive from positive logic sources.

6.38 The timing diagram shown in Fig. P6-11 represents the operation of a TRANS-LO MODULE. It is the function of this module to steer the transition input T to the output Z only when the data input D is ACTIVE LOW.

(a) Construct the state diagram for the TRANS-LO MODULE from its timing diagram.

(b) Use the LPD model and the results of (a) to design a HAZARD-free logic circuit for this module.

(c) Discuss any timing problems this FSM may

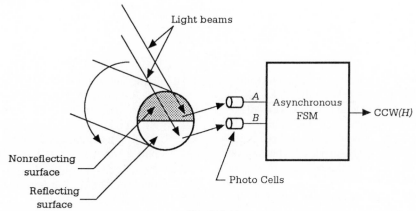

Light beams

A

Asynchronous FSM

B

CCW(H)

Nonreflecting surface

Reflecting surface

Photo Cells

Fig. P6-10.

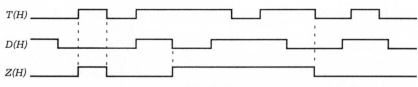

T(H)

D(H)

Z(H)

Fig. P6-11.

have. Assume that the gate delays may not be the same. Use timing diagrams to support your answers.

6.39 Shown in Fig. P6-12 is the logic circuit for the PULSE SYNCHRONIZER MODULE. It is the function of this module to issue one or a series of pulses synchronized with control inputs.

(a) Construct the state diagram for this module and describe its operation. Use a timing diagram to demonstrate the operation of this module.

(b) Discuss any inherent problems, or lack of them, that this module may have.

(c) If the potential exists for E-HAZARD and d-trio formation, give the initiating branching condition, race gate, and minimum path delay requirements for each.

CALL MODULE will steer that access request (either REQX or REQY) to its respective output, X or Y, provided that the "other" request line is INACTIVE at the time ACK is received. Thus, REQX → X if r is sent to and ACK is received from the system when REQY is INACTIVE. Similarly, REQY → Y if r is sent to and ACK is received from the system when REQX is INACTIVE. A second request can be granted access if ACK is ACTIVE when the first request is withdrawn.

(a) Construct the two state diagrams for this module. [Hint: One version of the CALL MODULE consists of two RENDEZVOUS MODULEs, of the type shown in Fig. 6-56(b), together with the appropriate NEXT STATE logic.]

(b) Construct the logic circuit for the CALL MOD-

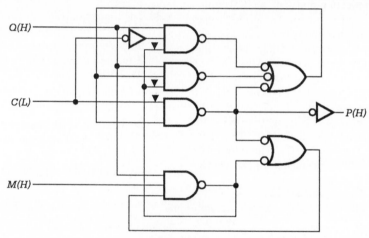

Fig. P6-12.

6.40 Shown in Fig. P6-13 is the block diagram illustrating the handshake between a CALL MODULE and a digital system. It is the function of the CALL MODULE to first issue a signal, r, to the system indicating that an access request signal has been made on one of two lines, REQX or REQY, but not on both. Then, if the system acknowledges receipt of the request by sending back a signal ACK to the CALL MODULE while the request is ACTIVE, the

ULE by using two RENDEZVOUS MODULEs, an XOR gate and two NOR gates. Assume that the request signals, REQX and REQY, arrive ACTIVE HIGH, and that the ACK input is ACTIVE LOW. Also, let all outputs be issued ACTIVE HIGH.

(c) Design the entire CALL MODULE as a single two-state FSM. Try both the LPD model and the NESTED CELL approaches. Explain in what way this design is functionally different from that of parts a and b.

6.41 Design the ARBICALL MODULE by combining the ARBITER MODULE with the CALL MODULE of Problem 6.40. Discuss how it might be applied to a protected system.

6.42 Use four FET JK FLIP-FLOPs and the necessary external logic to design a PULSE mode, up/down binary counter that has asynchronous parallel load

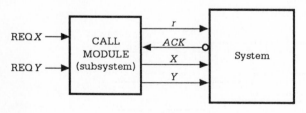

Fig. P6-13.

capability. This counter is described in Sect. 5.12 and has the block symbol shown in Fig. 5-138. It is the function of this counter to count up or down on the falling edge of either the UP or DN input pulse (never both at the same time). When the LD command becomes ACTIVE, the counter must parallel load immediately via its FLIP-FLOPS' asynchronous PRESET and CLEAR overrides, and then count up or down from that loaded value. The counter must be equipped with ACTIVE LOW CARRYOUT (CO) and BORROWOUT (BO) outputs for cascading purposes and must have an ACTIVE LOW asynchronous CLEAR. [Hint: The external logic required for the asynchronous overrides is best obtained from a truth table with inputs PR, CL, LD, and PL (for parallel load).]

6.43 Use the fundamental mode model to design an optimum or near optimum asynchronous controller for the multipulse generator described in Problem 5.69 of Chapter 5. The controller FSM must be free of all STATIC HAZARDs, output race glitches, and CRITICAL RACEs and must be initialized into an appropriate state. Use OSC as the name for the CLOCK input to the controller. Note that the output, LOAD, must be issued over a full OSC cycle and that outputs, PULSE and CNT, may be issued simultaneously. Try both the LPD model and NESTED CELL approach and compare the two results.

6.44 A PULSE mode asynchronous FSM is to be designed that functions as a controller for a security area where proprietary (or classified) documents are stored. It is required that when occupied the security area must be occupied by no more and no fewer than two people. Access to the security area is through an outer door (D1), along a narrow corridor and through an inner door (D2), as shown in Fig. P6-14. The corridor is equipped with two narrow (planar) light beams, X and Y, which are spaced 16 inches apart. The light beams cover the cross-sectional area of the corridor and fall incident on vertical photodetector cells. The inner door (normally locked) is unlocked and the outer door (normally unlocked) is locked when the second person passes the Y beam. The system permits one or both of the people in the access corridor to change their minds at any time and exit the corridor. However, any attempt by a third person to pass through the check beams, X and Y, will set off an alarm and lock both doors until the authorities arrive to reset the system.

The light beams (X and Y) are designed such that they will detect only body mass one at a time. Any attempt by two persons to pass as one in either direction will set off the alarm. (Note: This design may discriminate against unusually stout people.)

A red occupancy light is monitored on a remote control panel. The light is turned on when the second person passes the Y check beam on entering and is turned off when the second person passes the X check beam on exiting the corridor.

Construct a minimum-state state diagram for this FSM, and design for a HAZARD-free output forming logic. Plan to implement the NEXT STATE and output forming logic by using a PLA. Thus, construct a p-term table and block diagram for this FSM. Assume that a single pulse is sufficient to unlock (or lock) a door or turn on a light and maintain that condition until another pulse causes the reverse action. Also, assume that the FSM is initialized (or reset) into a STATE for which door D1 is unlocked and door D2 is locked.

6.45 Design the rotation detector of Problem 6.36 by using the EAIC system approach described in Sect. 6.9.2. Implement the NEXT STATE and output forming logic with discrete logic. Use an edge-triggered D-FF to filter out any logic noise that is produced by the output forming logic.

6.46 Design the digital combination lock in Fig. 5-91 by using the EAIC system described in Sect. 6.9.2. Implement the NEXT STATE and output forming logic by using a PLA and use an output holding register to filter any logic noise that may appear in the output signals. Thus, construct the p-term table for

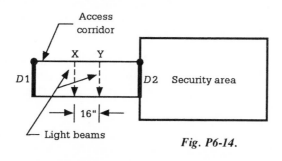

Mnemonics

UD1 Unlock door D1
LD1 Lock door D1
UD2 Unlock door D2
LD2 Lock door D2
LTON Light ON
LTOFF Light OFF

Fig. P6-14.

the NEXT STATE and output forming logic, and construct a block diagram for the system. Note that the input signals need only be debounced since synchronization of the inputs is an inherent feature of the EAIC system. Make certain that the EAIC system is properly initialized.

6.47 Configure the fundamental mode sequence recognizer, described in Problem 6.20, with an input conditioning circuit that will permit it to be operated with asynchronous inputs, X and Y. Discuss any assumptions that are made and any limitations that this system may have.

6.48 Configure the PULSE-mode digital combination lock of Problem 6.35 with an input conditioning circuit that will permit it to be operated with asynchronous (possibly overlapping) inputs X and Y. Discuss any assumptions that are made and any limitations that this system may have.

6.49 The input restrictions for fundamental mode and PULSE mode FSMs are specific. Fundamental mode machines must receive well developed, transition controlled inputs that never change simultaneously; and the inputs to PULSE mode FSMs must be nonoverlapping pulses. Thus, special input conditioning circuits must be used if these machines are to be operated by asynchronous inputs as suggested by Problems 6.47 and 6.48.

(a) Design an ARBITER unit for a four-input fundamental mode FSM that will ensure that no two, three or four inputs will change at the same time and that will protect the FSM from runt pulses. Make certain that the input signal changes are presented to the FSM in the order that they arrive at the conditioning circuit. Simultaneous input changes must be arbitrated by the input conditioning circuit on the basis of competing pairs of signals and some imbalance i.. the circuit, and must be presented to the FSM nonsimultaneously. Note that precise timing constraints need not be considered since this is a hypothetical problem. Verify the results with a timing diagram. (Hint: Use a combination of six two-input ARBITERs and four three-input RENDEZVOUS MODULEs.)

(b) Alter the circuit of part (a) so that the input conditioning circuit is suitable for driving a four-input PULSE mode FSM when the inputs are all asynchronous. To do this, the four asynchronous inputs must be converted to nonoverlapping pulses which are at least minimally separated. Verify the results with a timing diagram. (Hint: See Fig. 6-62(a).)

6.50 Carry out a system-level design of the PULSE mode candy bar vending machine of Fig. 6-58. To do this, follow the example given in Sect. 5.12. That is, construct a complete FUNCTIONAL PARTITION for the control system and plan to include the following data-path devices: a 4-bit adder, a PIPO (accumulator) register, a 4-bit down-counter with asynchronous parallel load, and a 4-bit comparator. Also, include box symbols for the coin changer (CC), the electromechanical candy bar dispenser (CBD), the electronically sensing coin receiver (CR), and a price change strapping unit ($PCSU$) to alter the price of the candy bars. Assume that the vending machine receives only nickels, dimes, and quarters which are encoded according to the number of nickels: 5 cents = 0001, 10 cents = 0010, and 25 cents = 0101.

The adder maintains a current account of all the change received, encoded as the number of nickels, and stores the result in the accumulator register. The register then presents that data to the counter where it is loaded to the comparator on the CCR (coin clear of receiver) command. The comparator evaluates < 40 cents, > 40 cents, or = 40 cents and sends the result to the controller. The action then is: wait for more change if < 40 cents, return a nickel (RN) if > 40 cents and decrement the accumulator/counter (DA) if the changer is ready (CR), or dispense a candy bar (DC) if = 40 cents and clear the accumulator register (CA) if the candy drop dispenser is ready (CDR). Note that the $PCSU$ input to the comparator is set at 40 cents encoded as 1000.

Plan to implement the CONTROLLER of Fig. 6-58 by using FET JK-FFs (as PULSE mode MEMORY elements) and suitable external logic. Remember to initialize the FSM into the proper state.

6.51 Complete the system-level design of the asynchronous multipulse generator described in Problems 6.43 and 5.69. Thus, construct the functional partition for this system and include the results of Problem 6.43. Plan to use counters of the type described in Problem 6.42 rather than those required in Problem 5.69. Demonstrate the operation of the asynchronous multipulse generator system with a MIXED LOGIC timing diagram for 12 pulses.

Proof of the Mutual Exclusive Branching Requirement Given by Eq. (6-5)

The requirements of Eq. (6-4) in Sect. 6.3 must be satisfied if all possible outgoing (as opposed to incoming) branching conditions are to be accounted for. However, meeting these requirements does not guarantee that all possible outgoing branching conditions from a given state are mutually exclusive of each other as required by Eq. (6.5). If a set of outbranching conditions (including the holding condition) from a given state are not mutually exclusive, then it is possible for the FSM to transit along two or more different branching paths under the same set of input conditions leaving the outcome uncertain. Clearly, this is unacceptable unless, of course, the mutually shared branching conditions never occur.

In this appendix we will prove that all possible outgoing branching conditions (BCs) from any n-way branching j^{th} STATE of an SOP state diagram are mutually exclusive iff

$$(BC)_{i \leftarrow j} = \overline{\sum_{\substack{k=0 \\ k \neq 1}}^{n-1} (BC)_{k \leftarrow j}} \qquad i = 0, 1, \ldots, n-1 \qquad (6\text{-}5)$$

In words, Eq. (6-5) states that the branching condition from the jth STATE to the ith STATE is the complement of the Boolean SUM (union) of all remaining branching conditions from that j^{th} STATE for n-way branching and that this requirement applies to each of the n outbranching conditions in turn.

For the general case, consider that there are N variables on which branching depends for the jth STATE of an FSM. Let there be 2^N STATEs to which the jth STATE can transit and, therefore, 2^N uniquely defined branching conditions controlling these transitions. For this general case Eq. (3-13) applies in the form

$$\sum_{i=0}^{2^N-1} (BC)_{i\leftarrow j} = 1$$

which is Eq. (6-4), where $(BC)_i$ now refers specifically to the ith MINTERM branching condition. Next, let us partition this expression in the form

$$\sum_{i=0}^{r} (BC)_{i\leftarrow j} + \sum_{\substack{k=r \\ k\neq i}}^{2^N-1} (BC)_{k\leftarrow j} = 1$$

which we are permitted to do in 2^N unique ways. Then by applying to this expression the OR law

$$f_i + \bar{f}_i = 1$$

which is applicable to any Boolean function, we arrive at the result

$$\sum_{i=0}^{r} (BC)_{i\leftarrow j} = \overline{\sum_{\substack{k=r \\ k\neq i}}^{2^N-1} (BC)_{k\leftarrow j}}$$

Finally, by taking $r = 0$, Eq. (6-5) results and the proof is complete.

A brief inspection of the branching conditions for the state diagram in Fig. 6-6(a) indicates that the requirements of Eq. (6-5) are satisfied. For example, by applying Eq. (6-5) to State $\textcircled{b}$ we obtain

$$f_{bc} = \overline{f_{bb} + f_{ba}}$$

meaning

$$\overline{A}\overline{B} = \overline{A \oplus B + AB}$$
$$= (A \odot B)(\overline{AB})$$
$$= (\overline{A}\overline{B} + AB)(\overline{A} + \overline{B})$$
$$= \overline{A}\overline{B}$$

Likewise, we see that $f_{cc} = \overline{f}_{cb}$ and $f_{ab} = \overline{f}_{aa}$, both of which are true by inspection. Thus, all possible branching conditions for this state diagram are uniquely defined. There are no undefined or overlapping branching conditions.

Now let us apply Eq. (6-5) to the state diagram segment of Fig. A6-1. Here, we can see that the requirements of Eq. (6-5) are not satisfied for STATE $\textcircled{p}$ since

$$f_{PM} \neq \overline{f_{PP} + f_{PN}}$$

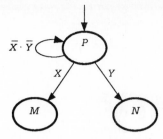

Fig. A6-1. *STATE DIAGRAM segment showing branching conditions from STATE P which satisfy the requirements of Eq. (6-4) but do not satisfy Eq. (6-5).*

that is,

$$X \neq \overline{\overline{X}\,\overline{Y} + Y} = X\overline{Y}$$

Similarly,

$$f_{PN} \neq \overline{f_{PP} + f_{PM}}$$

However, the requirements of Eq. (6-4) are satisfied for STATE $\widehat{P}$:

$$f_{PP} + f_{PM} + f_{PN} = 1$$

that is,

$$\overline{X}\,\overline{Y} + X + Y = 1$$

These results mean that all possible branching conditions from STATE $\widehat{P}$ are accounted for but are not mutually exclusive. The branching condition XY is shared by $f_{PM} = X$ and $f_{PN} = Y$. If XY should occur while in STATE $\widehat{P}$, the branching path for the FSM would be uncertain. As the reader will note, the branching conditions of Fig. A6-1 are proper for this state diagram segment if the input condition XY is never permitted to occur while the FSM is in STATE $\widehat{P}$.

The point to be made here is that Eqs. (6-4) and (6-5) are tests that the designer should use to determine whether or not all possible branching conditions from any given state are accounted for and, if they are, whether or not each branching condition is mutually exclusive of all others from that state. The requirements of these equations must be met for each state if all possible branching conditions can occur. If there exist branching conditions which can never occur, the requirements of these equations need not be enforced.

Glossary

Accumulator: a register used to store the results of an arithmetic operation.

ACTIVATE: to make ACTIVE.

ACTIVATION level: the physical state, HV or LV, designated to be the ACTIVE state for a signal.

ACTIVATION LEVEL INDICATOR: a symbol [(H) or (L)] that is attached to a signal name that establishes the physical state, HV or LV, that is to be the ACTIVE state for the signal.

ACTIVE: a descriptor that denotes an action condition and that implies logic 1.

Active device: any device that provides current (or voltage) gain.

ACTIVE HIGH: indicates a positive logic source.

ACTIVE LOW: indicates a negative logic source.

Active mode: the physical state of a BJT in which it can be used to amplify a signal.

ACTIVE state: the logic 1 state of a logic device.

ACTIVE transition point: the point in a voltage waveform where a digital device passes from the INACTIVE state to the ACTIVE state.

Addend: an operand to which the augend is added.

ADDER: a digital device that adds two binary numbers to give a SUM and a CARRY.

ADDER/SUBTRACTOR: a combinational logic device that can perform either the addition or subraction (by sign-compliment arithmetic) operations.

Adjacency laws: expressions that, when applied, lead to collapsed truth tables from which minimum functions can be read.

Adjacent cell: a K-map cell whose coordinates differ from that of another cell by only one bit.

Adjacent pattern: an XOR pattern involving a complemented function in one cell of a K-map and the uncomplemented form of the same function in an adjacent cell.

Algorithm: a step-by-step procedure for accomplishing a specific task.

Alternative race path: one of two or more transit paths an FSM can take during a race condition.

ALU: arithmetic and logic unit.

AND: an operator requiring that all inputs to an AND gate be ACTIVE before the output is ACTIVE.

AND function: the function that derives from the definition of AND.

AND gate: a physical device that performs the electrical analog of the AND function.

AND laws: a set of Boolean identities based on the AND function.

AND-OR-INVERT gate: a physical device that performs

662

the electrical analog of SOP with an ACTIVE LOW output or POS with an ACTIVE HIGH output.

AND plane: the ANDing stage of a PLD such as a ROM or PLA.

ARBICALL MODULE: a module that combines the ARBITER MODULE with the CALL MODULE to control access to a protected system.

ARBITER MODULE: a device that is designed to control access to a protected system by arbitration of contending signals.

Arithmetic and logic unit: a physical device that is capable of performing either arithmetic or logic operations.

Arithmetic SHIFTER: a SHIFTER that is capable of generating and preserving a sign bit.

Array logic: any of a variety of logic devices, such as ROMs or PLAs, that are composed of an AND array and/or an OR array.

Assert: ACTIVATE

Associative law: a law of Boolean algebra that states that the location of parentheses in a *p*-term or *s*-term does not matter.

Associative pattern: an XOR pattern in a K-map which allows a term of an XOR or EQV function to be looped out with the same term in an adjacent cell provided that the XOR or EQV connective is preserved in the process.

Asynchronous: clock independent—having no fixed time relationship.

Asynchronous input: an input that can change at any time, particularly during the sampling interval.

Asynchronous override: an input such as PRESET or CLEAR that, when ACTIVATED, interrupts the normal operation of a FLIP-FLOP.

Augend: an operand that is added to the addend in addition.

B: BORROW.

Base: radix. Also, one of three regions in a BJT.

BASIC CELL: a basic MEMORY cell usually composed of either cross-coupled NAND gates or cross-coupled NOR gates.

BCD: binary coded decimal.

BI: BORROWIN.

Bias: voltage ''influence'' on a circuit segment.

Bidirectional counter: a counter that can count up or down.

Binary: a number system of radix two; having two values or states.

Binary code: a combination of bits that represent alphanumeric or arithmetic information.

Binary coded decimal: any of several methods of representing decimal digits as binary numbers.

Binary WORD: a linear array of juxtaposed bits that represents a number or that conveys an item of information.

Bipolar junction transistor: An npn or pnp transistor.

Bit: a binary digit.

Bit-slice: partitioned into identical parts such that each part operates on one bit in a multibit word.

BJT: Bipolar junction transistor.

BO: BORROWOUT.

Boolean algebra: the mathematics of logic attributed to the mathematician George Boole (1815–1864).

Boolean product: AND or intersection operation.

Boolean sum: OR or union operation.

BORROWin: the borrow input to a subtractor.

BORROWout: the borrow output from a subtractor.

Boundary: the separation of logic domains in a K-map.

Branching condition: the input requirements necessary to cause a state-to-state transition in an FSM.

Branching path: a state-to-state transition path in a state diagram.

BUFFER: a DRIVER.

Bus: a collection of signal lines which operate together to transmit a group of related signals.

Byte: a group of 8 bits.

C: CARRY. Also, the collector terminal in a BJT.

CALL MODULE: a module designed to control access to a protected system by issuing a request for access to the system and then granting access after receiving acknowledgement of the request from the system.

CANONICAL: made up of terms that are either all MINTERMs or all MAXTERMs.

CANONICAL truth table: a truth table made up of 1's and 0's.

Capacitance, C: the constant of proportionality between total charge on a capacitor and the voltage across it, $Q = CV$, where C is given in farads [F].

Capacitor: a two terminal energy storing element for which the current through it is determined by the time-rate of change of voltage across it.

CARRY GENERATE: a function that is used in an LAC ADDER.

CARRYin: the carry input to a binary adder.

CARRY-LOOK-AHEAD: (see LOOK-AHEAD-CARRY)

CARRYout: the carry output from an ADDER.

CARRY PROPAGAGE: a function that is used in an LAC adder.

CARRY SAVE: a fast addition method for three or more binary numbers where the carries are saved and added to the final sum.

Cascade: to combine identical devices in series such that any one device drives another; to bit-slice.

Catching cell: A BASIC CELL that is used to intercept and stretch an input signal.

Cell: the intersection of all possible domains of a K-map.

Central processing unit: a processor that contains the necessary logic hardware to fetch and execute instructions.

CI: CARRYIN.

Circuit: a combination of elements (e.g., logic devices) that are connected together to perform a specific operation.

CK: CLOCK

CLEAR: an asynchronous input used in FLIP-FLOPs, registers, counters and other sequential devices, that, when activated, forces the internal state of the device to logic 0.

CLOCK: a regular source of pulses which control the timing operations of a synchronous sequential machine.

CLOCK SKEW: a phenomenon that is generally associated with high frequency CLOCK distribution problems in synchronous sequential systems.

CMOS: complementary configured MOSFET in which both NMOS and PMOS are used.

CO: CARRYOUT.

Code: a system of binary words used to represent decimal or alphanumeric information.

CODE CONVERTER: a device designed to convert one binary code to another.

Collapsed truth table: a truth table containing irrelevant inputs.

Collector: one of three regions in a BJT.

Combinational HAZARD: a HAZARD that is produced within a combinational logic circuit.

Combinational logic: a configuration of logic devices in which the outputs occur in direct, immediate response to the inputs without feedback.

Commutative law: the Boolean law that states that the order in which variables are represented in a p-term or s-term does not matter.

COMPARATOR: a combinational logic device that compares the values of two binary numbers and issues one of three outputs indicating their relative magnitudes.

Compatibility: a term used in this text to indicate agreement between the LOGIC LEVEL of an input signal to a logic symbol (e.g., a gate) and the input LOGIC LEVEL indicated for that symbol.

Complement: the value obtained by logically inverting the state of a binary digit; the relationship between numbers that allows numerical subtraction to be performed by an addition operation.

Complementary metal oxide semiconductor: a form of MOS that uses both p- and n-channel transistors (in pairs) to form logic gates.

Complementation: a condition that results from logic incompatibility; the MIXED LOGIC equivalent of the NOT operation.

Computer: a digital device that can be programmed to perform a variety of tasks (e.g., computations) at extremely high speed.

Conditional branching: state-to-state transitions that depend on the input status of the FSM.

Conditional output: an output that depends on one or more external inputs.

Conjugate gate forms: a pair of logic circuit symbols that derive from the same physical gate and that satisfy the DeMorgan relations.

Connective: a Boolean operator symbol (e.g., AND or XOR).

CONTROLLER: that part of a digital system that controls the data path.

Conventional K-map: a K-map whose entries are exclusively 1's and 0's.

Counter: a sequential logic circuit designed to count through a particular sequence of states.

Counteracting delay: a delay placed on an external feedback path to eliminate an E-HAZARD or d-trio.

Count sequence: a repeating sequence of binary numbers that appears on the outputs of a counter.

Coupled term: one of two terms containing only one coupled variable.

Coupled variable: a variable that appears complemented in one term of an expression (SOP or POS) and that also appears uncomplemented in another term of the same expression.

Cover: a reduced form of a function extracted from a K-map by the loop-out process.

CPU: central processing unit.

CRITICAL RACE: a race condition that can result in a transition to an erroneous state.

Current, I: the flow or transfer of charged matter (e.g., electrons) given in amperes [A].

Cutoff mode: the physical state of a BJT in which no significant collector current is permitted to flow.

Cycle: successive and uninterrupted state transitions in an asynchronous sequential machine.

Data bus: a parallel set of conductors which are capable of transmitting or receiving data between two parts of a system.

Data lockout: the property of a FLIP-FLOP that permits the data inputs to change immediately following a RESET or SET operation without affecting the FLIP-FLOP output.

Data lockout FLIP-FLOP: a one-bit memory device which has the combined properties of a master/slave FLIP-FLOP and an edge triggered FLIP-FLOP.

Data path: the part of a digital system that is controlled by the CONTROLLER.

DATA SELECTOR: a MULTIPLEXER.

DCL: digital combination lock.

DEACTIVATE: to make INACTIVE.

Debounce: to remove the noise that is produced by a mechanical switch.

Debouncing circuit: a circuit that is used to debounce a switch.

Decade: a quantity of 10.

DECODER: a combinational logic device that will ACTIVATE a particular MINTERM code output line determined by the binary input.

Decrement: reduction of a value by some amount (usually by 1).

Delay: the time elapsing between related events in process.

Delay circuit: a circuit whose purpose it is to delay a signal for a specified period of time.

DeMorgan relations: mixed logic expressions of DeMorgan's laws.

DeMorgan's laws: a property that states that the complement of the Boolean product of terms is equal to the Boolean sum of their complements; or that states that the complement of the Boolean sum of terms is the Boolean product of their complements.

DEMULTIPLEXER: a combinational logic device (usually a DECODER) in which a single input is selectively steered to one of a number of output lines.

Depletion mode: a normally ON NMOS that has a conducting n-type drain-to-source channel in the absence of a gate voltage but that looses its conducting state when the gate voltage reaches some negative value.

D FLIP-FLOP: a one-bit memory device whose output value is set to the D input value on the triggering edge of the clock signal.

D-FLOP MODULE: a MEMORY element that is used in an EAIC system and that has characteristics similar to that of a D FLIP-FLOP.

Diad: a grouping of two logically adjacent MINTERMs or MAXTERMs.

Diagonal pattern: an XOR pattern formed by identical EV subfunctions in any two diagonally located cells of a K-map whose coordinates differ by two bits.

Difference: the result of a subtraction operation.

Digit: a single symbol in a number system.

Digital: related to discrete quantities.

Digital combination lock: a sequence recognizer that can be used to unlock something.

Digital engineering design: the design and analysis of digital devices.

Digital signal: a logic waveform composed of discrete logic levels (e.g., a binary digital signal).

Diode: a two-terminal passive device consisting of a pn junction that permits significant current to flow only in one direction.

Diode-transistor logic: logic circuits consisting mainly of diodes and BJTs.

Direct address approach: an alternative approach to FSM design where PS feedback is direct to the NS logic (e.g., MUXs or PLDs).

Distributed path delays: a notation in which a path delay is assigned to each gate or INVERTER of a logic circuit.

Distributive law: The dual of the factoring law.

Divide-by-N counter: a binary counter of N states whose MSB output divides the CLOCK input frequency by N.

Dividend: the quantity that is being divided in a division operation.

DIVIDER: a combinational logic device that performs the binary division operation.

Divisor: the quantity that is divided into the dividend.

DMUX: DEMULTIPLEXER.

Domain: a range of logic influence or control.

Domain boundary: the vertical or horizontal line or edge of a K-map.

Don't care: a non-essential MINTERM or MAXTERM (denoted by the symbol ϕ) which can take either a logic 1 or logic 0 value without affecting the overall function to which it applies.

DPU: data processing unit.

Drain: one of three terminals of a MOSFET.

Driver: a one-input device whose output can drive substantially more inputs than a standard gate.

DTL: diode-transistor logic.

D-trio: a type of ESSENTIAL HAZARD that causes a fundamental mode machine to transit to the correct state via an unauthorized path.

Duality: a property of Boolean algebra that results when the AND and OR operators (or XOR and EQV operators) are interchanged simultaneously with the interchange of 1's and 0's.

Dual relations: two Boolean expressions that can be derived one from the other by duality.

Duty cycle: in a periodic waveform, the percentage of time the waveform is ACTIVE.

DYNAMIC HAZARD: a multiple-glitch HAZARD that results when multiple input-to-output paths are present in a combinational logic circuit.

EAIC system: externally asynchronous/internally clocked system.

ECL: emitter-coupled logic.

Edge-triggered FLIP-FLOP: a FLIP-FLOP that is triggered on either the rising edge or falling edge of the CLOCK waveform and that exhibits the data-lock-out feature.

E-HAZARD: ESSENTIAL HAZARD.

Electron: the majority carrier in an n-type conducting semiconductor.

Electronic switch: a voltage or current controlled switching device.

Emitter: one of three terminals of a BJT.

Emitter-coupled logic: a high-speed nonsaturating logic family.

EN: ENABLE.

Enable: a condition that permits a logic device to operate normally.

ENABLE: an input that is used to enable (or disable) a logic device.

ENCODER: a digital device that converts digital signals into coded form.

ENDLESS CYCLE: an oscillation that occurs in asynchronous FSMs.

Enhancement mode: a normally OFF NMOS that develops an n-channel drain-to-source conducting path (i.e., turns ON) with application of a gate voltage.

Entered variable: a variable entered in a K-map.

EPI: ESSENTIAL PRIME IMPLICANT.

EPROM: erasable programmable read-only memory.

EQUIVALENCE: the output of a two-input logic gate that is ACTIVE if, and only if, its inputs are logically equivalent (i.e., both ACTIVE or both INACTIVE).

EQV: EQUIVALENCE

EQV function: the function that derives from the definition of EQUIVALENCE.

EQV gate: a physical device that performs the electrical analog of the EQV function.

EQV laws: a set of Boolean identities based on the EQV function.

Erasable programmable read-only memory: a ROM that can be user programmed many times.

Error catching: a problem in a JK master/slave FLIP-FLOP where a 1 or 0 is caught in the master cell when CLOCK is ACTIVE and is issued to the output by the slave cell when CLOCK goes INACTIVE.

ESSENTIAL HAZARD: a sequential HAZARD that can occur as a result of an explicitly located delay in an asynchronous FSM that has at least three states and that is operated in the fundamental mode.

ESSENTIAL PRIME IMPLICANT: a PRIME IMPLICANT that must be used to achieve minimum cover.

EV: entered variable.

EV K-map: a K-map that contains EVs.

EV truth table: a truth table containing EVs.

Even parity: an even number of 1's (or 0's) depending on how even parity is defined.

EVM: entered variable K-map.

Excitation table: state transition table.

EXCLUSIVE OR: a two-variable function that is ACTIVE if only one of the two variables is ACTIVE.

Expansion of states: opposite of merging of states.

Extender: a circuit or gate that is designed to be connected to a digital device to increase its fan-in capability—also called an *expander*.

Factoring law: the Boolean law that permits a variable to be factored out of two or more *p*-terms that contain the variable in an sop expression.

Fall time: the period of time it takes a voltage (or current) signal to change from 90% to 10% of its high value.

Falling edge-triggered: activation of a device on the falling edge of the triggering (sampling) variable.

False data rejection: the feature of a code converter that indicates when unauthorized data has been issued to the converter.

Fan-in: the maximum number of inputs a gate may have.

Fan-out: the maximum number of equivalent gate inputs that a logic gate output can drive.

FDR: false data rejection.

Feedback path: a signal path from output to input of the same logic device.

FET: falling edge-triggered. Also, field effect transistor.

Field programmable logic array: one-time user programmable PLA.

FILL Bit: the bit of a combinational SHIFTER that receives the fill logic value in a shifting operation.

Finite state machine: a sequential machine that has a finite number of states into which it can reside stably.

FLIP-FLOP: a one-bit MEMORY element that exhibits sequential behavior controlled exclusively by a CLOCK input.

Flow chart: a chart that is made up of an interconnection of action and decision symbols for the purpose of representing the sequential nature of something.

Fly state: a state (in a state diagram) whose only purpose is to remove a race condition.

Forward bias: a voltage applied to a *p-n* junction diode in a direction as to cause the diode to conduct or turn ON.

FPLA: field programmable logic array.

Frequency, f: the number of waveform cycles per unit time in Hz or $[s^{-1}]$.

Frequency division: the reduction of frequency by a factor of f/n usually by means of a binary counter, where n is the number of states in the counter.

FSM: finite state machine, either synchronous or asynchronous.

FULL ADDER: a combinational logic device that adds two binary bits to a CARRYin bit and issues a SUM bit and a CARRYout bit.

FULL SUBTRACTOR: a combinational logic device that subtracts a subtrahend bit and a BORROWin bit from a minuend bit, and issues a DIFFERENCE bit and a BORROWout bit.

Fully documented state diagram: a state diagram that specifies all input branching conditions and output conditions, that satisfies the sum rule and mutually exclusive requirement, and that has been given a proper state code assignment.

Function: a Boolean expression representing a specific binary operation.

Functional partition: a diagram that gives the division of device responsibility in a digital system.

FUNCTION GENERATOR: a combinational logic device that generates logic functions usually by use of a MUX.

FUNCTION HAZARD: a hazard that is produced when two or more coupled variables change in near proximity to each other.

Fundamental mode: the operational condition for an asynchronous FSM in which no input change is permitted to occur until the FSM has stabilized following any previous input change.

Fusible link: an element in a PLD that can be "blown" to store a logic 1 or logic 0 depending on how the PLD is designed.

Gate: a physical device (circuit) that performs the electrical analog of a logic function. Also, one of three terminals of a MOSFET.

GATED BASIC CELL: a BASIC CELL that responds to its S and R input commands only on the triggering edge of a GATE or CLOCK signal.

Gate/input tally: the gate and input count associated with a given logic expression. The tally may or may not include INVERTERs.

Gate-minimum logic: logic requiring a minimum number of gates, and may include XOR and EQV gates in addition to two-level logic.

Gate path delay: the interval of time required for a gate output response to an input signal.

Glitch: a momentary unauthorized excursion in an otherwise steady state signal.

GO/NO-GO configuration: a solitary, conditional, out-branching path in a state diagram.

Gray code: a reflective unit distance code.

Ground: a reference voltage level usually taken to be zero volts.

HALF ADDER: a combinational logic device that adds two binary bits and issues a SUM bit and a CAR-RYout bit.

HALF SUBTRACTOR: a combinational logic device that subtracts one binary bit from another and issues a DIFFERENCE bit and a BORROWout bit.

Handshake interface: a configuration between two devices whereby the outputs of one device are the inputs to the other and vice versa.

Hang state: an isolated state into which an FSM can reside (on power up) but which is not part of the authorized routine.

HAZARD: a glitch or unauthorized transition that is caused by an asymmetric path delay in an IN-VERTER, gate or lead during a logic operation.

HAZARD cover: the redundant cover that removes a STATIC HAZARD.

Heuristic: by empirical means or by discovery.

HOLD condition: branching from a given state back into itself, or the input requirements necessary to effect such branching action.

Holding register: a PIPO register that is used to filter output signals.

Hold time: the interval of time immediately following the transition point during which the data inputs must remain logically stable to ensure that the intended transition of the FSM will be successfully completed.

Hole: the absence of a valence electron—the majority carrier in a p-type conducting semiconductor.

HV: high voltage.

Hybrid function: a function containing both SOP and POS terms.

IC: integrated circuit.

IMPLICANT: a term in a reduced or minimized expression.

INACTIVE: not ACTIVE and implying logic 0.

INACTIVE transition point: the point in a voltage waveform where a digital device passes from the ACTIVE state to the INACTIVE state.

Incompatibility: a condition where the input to a logic device and the input requirement of that device are of opposite ACTIVATION levels.

Incompletely specified function: a function which con-

tains nonessential MINTERMs or MAXTERMs (don't cares).

Increment: increase usually by one.

Indirect address approach: an alternative approach to FSM design where PS feedback to the NS logic is by way of a converter for the purpose of reducing MUX or PLD size.

Inertial delay element: a delay circuit based mainly on an RC component.

Initialize: to drive a logic circuit into a beginning state.

Input: a signal or line into a logic device that controls the operation of that device.

Integrated circuit: an electronic circuit that is usually constructed entirely on a single small semiconductor chip called a monolith.

Intersection: AND operation.

"Into" rule: one of two guidelines used to reduce the complexity of the NEXT STATE logic functions for an FSM.

Inversion: the inverting of a signal from HV to LV or vice versa.

INVERTER: a physical device that performs inversion.

Involution: double COMPLEMENTATION of a variable or function.

I/O: input/output

Irrelevant input: an input whose presence in a function is nonessential.

Island: a K-map entry that must be looped out of a single cell.

JK FLIP-FLOP: a type of FLIP-FLOP that can perform the SET, RESET, HOLD and TOGGLE operations.

Juxtapose: to place side by side.

Karnaugh map: a graphical representation of a logic function named after M. Karnaugh (1953).

Kirchhoff's current law: the algebraic sum of all currents into a circuit element or circuit section must be zero.

Kirchhoff's voltage law: the algebraic sum of all voltages around a closed loop must be zero.

K-map: Karnaugh map.

LAC: LOOK-AHEAD-CARRY

Large-scale integrated circuits: IC chips that contain 100 to 1,000 gates.

Latch: an alternative name for a BASIC CELL or FLIP-FLOP.

Least significant bit: the bit (usually at the extreme

right) of a binary WORD that has the lowest positional weight.

LED: light-emitting diode.

Logic: the computational capability of a digital device that is interpreted as either a logic 1 or logic 0.

Logic adjacency: two logic states whose state variables differ from each other by only one bit.

Logic circuit: a digital circuit that performs the electrical analog of some logic function or process.

Logic diagram: a digital circuit schematic consisting of an interconnection of logic symbols.

Logic family: a particular technology such as TTL or CMOS that is used in the production of ICs.

Logic instability: the inability of a logic circuit to maintain a stable logic condition. Also, an oscillatory conditon in an asynchronous FSM.

LOGIC LEVEL: logic status, that is, either positive logic or negative logic.

LOGIC LEVEL CONVERSION: the act of converting from positive logic to negative logic or vice versa.

Logic map: any of a variety of graphical representations of a logic function.

Logic noise: undesirable signal fluctuations produced within a logic circuit as a result of input changes.

Logic waveform: a rectangular waveform between ACTIVE and INACTIVE states.

LOOK-AHEAD-CARRY: the feature of a "fast" ADDER that anticipates the need for a carry and then generates and propagates it more directly than does a PARALLEL ADDER.

Loop-out: the action that identifies a prime implicant in a K-map.

Loop-out protocol: a minimization procedure whereby the largest 2^n group of logically adjacent MINTERMs or MAXTERMs are looped out in the order of increasing n ($n = 0, 1, 2, 3, \ldots$).

LPD: lumped path delay.

LSB: least significant bit.

LSD: least significant digit.

LSI: large scale integration.

Lumped path delay model: a model, applicable to FSMs that operate in the fundamental mode, that is characterized by a lumped MEMORY element for each state variable/feedback path.

LV: low voltage.

MAGNITUDE COMPARATOR: COMPARATOR.

Majority function: a function that becomes ACTIVE only when a majority of the inputs become ACTIVE, and becomes INACTIVE only when a majority of the inputs become INACTIVE.

Majority gate: a logic gate that yields a majority function.

Map: usually a Karnaugh map.

Map compression: a reduction in the order of a K-map.

MAP heading variable: the variable for which a conventional fourth-order K-map is constructed in minimizing a function of five or more variables.

MAP KEY: the order of K-map compression, hence 2^{N-n}, where N is the number of variables in the function to be mapped and n is the order of the K-map to be used.

Master/slave FLIP-FLOP: a FLIP-FLOP characterized by a master (input) stage and a slave (output) stage that are triggered by CLOCK antiphase to each other.

MAXTERM: a POS term that contains all the variables of the function.

MAXTERM code: a code in which complemented variables are assigned logic 1 and uncomplemented variables are assigned logic 0—the opposite of MINTERM code.

Mealy machine: an FSM that conforms to the Mealy model.

Mealy model: the general model for a sequential machine where the OUTPUT STATE depends on the INPUT STATE as well as the PRESENT STATE.

Mealy output: a conditional output.

Medium-scale integrated circuits: IC chips that contain say 13 to 99 gates according to one convention.

MEMORY: the ability of a digital device to store and retrieve binary WORDs on command.

MEMORY element: a device for storing and retrieving one bit of information on command. In asynchronous FSM terminology, a fictitious lumped path delay.

Merge: the concatenation of buses to form a larger bus.

Merging of states: in a state diagram, the act of combining states to produce a state diagram of fewer states.

Metal-oxide-semiconductor: the material constitution of an important logic family (MOS) used in IC construction.

Metastability: an unresolved state of an FSM which is neither a SET nor RESET or which is logically unstable.

Metastable exit time: the statistical time interval between entrance into and exit from the metastable state.

MEV: Map entered variable.

Minimization: the process of reducing a logic function to its simplest form.

Minimum cover: the optimally reduced representation of a logic expression.

MINTERM: a term in an SOP expression where all variables of the expression are represented in either complemented or uncomplemented form.

MINTERM code: a logic variable code in which complemented variables are assigned logic 0 while uncomplemented variables are assigned logic 1—the opposite of MAXTERM code.

Minuend: the operand from which the subtrahend is subtracted in a subtraction operation.

MIXED LOGIC: the combined use of the positive and negative logic systems.

Mixed-rail output: dual, logically equal outputs of a device (e.g., a FLIP-FLOP) where one output is issued ACTIVE HIGH while the other is issued ACTIVE LOW, but which are not issued simultaneously.

Mnemonic: a short single group of symbols (usually letters) that are used to convey a meaning.

Model: the means by which the major components and their interconnections are represented for a digital machine or system.

Module: a device that performs a specific function and that can be added to or removed from a system to alter the system's capability. A common example is a FULL ADDER.

Modulus-N counter: (see divide-by-N counter).

Monad: a MINTERM (or MAXTERM) that is not logically adjacent to any other MINTERM (or MAXTERM).

Moore machine: a sequential machine that conforms to the Moore model.

Moore model: a degenerate form of the Mealy (general) model in which the OUTPUT STATE depends only on the PRESENT STATE.

Moore output: an unconditonal output.

MOS: metal-oxide-semiconductor.

MOSFET: metal-oxide-semiconductor field effect transistor.

Most significant bit: the bit (usually the extreme left) of a binary WORD that has the highest positional weight.

MSB: most significant bit.

MSD: most significant digit.

MSI: medium scale integration.

MULLER C MODULE: a RENDEZVOUS MODULE

Multilevel logic minimization: minimization involving XOR-type patterns.

Multiple-output minimization: optimization of more than one output expression from the same logic device.

Multiplex: to select or gate (on a time-shared basis) data from two or more sources onto a single line or transmission path.

MULTIPLEXER: a device that multiplexes data.

Multiplicand: the number being multiplied by the multiplier.

Multiplier: the number being used to multiply the multiplicand.

MULTIPLIER: a combinational logic device that will multiply two binary numbers.

Mutually exclusive requirement: a requirement in state diagram construction that forbids overlapping branching conditions—that is, it forbids the use of a possible branching condition for more than one branching path.

MUX: MULTIPLEXER.

NAND-centered BASIC CELL: cross-coupled NAND gates forming a BASIC CELL.

NAND gate: a physical device that performs the electrical analog of the NOT AND function.

Natural binary code: a code for which the bits are positioned in a binary WORD according to their positional weight in polynomial notation.

Natural binary coded decimal: a four-bit, ten-WORD code that is weighted 8, 4, 2, 1 and that is used to represent decimal numbers.

NBCD: natural binary-coded-decimal.

n-channel: an n-type conducting region in a p-type substrate.

Negative logic: a logic system in which high voltage (HV) corresponds to logic 0 and low voltage (LV) corresponds to logic 1. The opposite of positive logic.

Negative pulse: a 1-0-1 pulse.

NESTED CELL: a BASIC CELL that is used as the MEMORY in an asynchronous FSM design.

Nested machine: any asynchronous machine that serves as the MEMORY in the design of a larger sequential machine.

NEXT STATE: a STATE that follows the PRESENT STATE in a sequence of states.

NEXT STATE forming logic: the logic hardware in a sequential machine whose purpose it is to generate the NEXT STATE function input to the MEMORY.

NEXT STATE function: the logic function that defines the NEXT STATE of an FSM given the PRESENT STATE.

NEXT STATE MAP: a composite K-map where the entries for each cell are the NEXT STATE functions for the PRESENT STATE represented by the co-ordinates of that cell.

NEXT STATE variable: the variable representing the NEXT STATE function.

NMOS: an *n*-channel MOSFET.

Noise immunity: the ability of a logic circuit to reject unwanted signals.

Noise margin: the maximum voltage fluctuation that can be tolerated in a digital signal without crossing the switching threshold of the switching device.

Non-restoring logic: logic that consists of passive switching devices such as diodes or pass transistors that cannot amplify but that can only dissipate power.

NOR-centered BASIC CELL: cross-coupled NOR gates forming a BASIC CELL.

NOR gate: a physical device that performs the electrical analog of the NOT OR function.

NOT function: an operation that is the logic equivalent of COMPLEMENTATION.

NOT laws: a set of Boolean identities based on the NOT function.

npn: refers to a BJT having a *p*-type semiconductor base and an *n*-type semiconductor collector and emitter.

NPT: negative pulse-triggered.

NS: NEXT STATE.

Octad: a grouping of eight logically adjacent MINTERMs or MAXTERMs.

Odd parity: an odd number of 1's or 0's depending on how odd parity is defined.

Offset pattern: an XOR pattern in a K-map in which identical subfunctions are located in two non-diagonal cells that differ in cell coordinates by two bits.

Ohm's law: voltage is linearly proportional to current, $V = RI$, where R is the constant of proportionality called the resistance (in ohms).

One-hot code: a non-weighted code in which there exists only one 1 in each WORD of the code.

One's complement: a system of binary arithmetic in which a negative number is represented by complementing each bit of its positive equivalent.

Operand: a number or quantity that is to be operated on.

Operation table: a table that defines the functionality of a FLIP-FLOP or some other device.

OPI: OPTIONAL PRIME IMPLEMENT.

OPTIONAL PRIME IMPLICANT: a PRIME IMPLICANT whose presence in a minimum function produces alternative minimum cover.

OR: an operator requiring that the output of an OR gate be ACTIVE if one or more of its inputs are ACTIVE.

Order: refers to the number of variables on the axes of a K-map.

OR function: a function that derives from the definition of OR.

OR gate: a physical device that performs the electrical analog of the OR function.

OR laws: a set of Boolean identities based on the OR function.

OR plane: the ORing stage of a PLD.

Outbranching: branching from a state exclusive of the HOLD branching condition.

"Out of" rule: one of two guidelines used to reduce the complexity of the NEXT STATE logic functions for an FSM.

Output: a concluding signal issued by a digital device.

Output forming logic: the logic hardware in a sequential machine whose purpose it is to generate the output signals.

Output holding register: a register, consisting of D FLIP-FLOPs, that is used to filter out output logic noise.

Output race glitch: a FUNCTION HAZARD that is produced by a race condition in a sequential machine.

Overflow error: a false magnitude or sign that results from a left shift in a SHIFTER when there are insufficient WORD bit positions at the SPILL end.

Packing density: the practical limit to which switches of the same logic family can be packed in an IC chip.

PAL: programmable array logic.

PALU: programmable arithmetic and logic unit.

PARALLEL ADDER: a cascaded array of FULL ADDERs where the CARRYout of a given FULL

ADDER is the CARRYin to the next most significant stage FULL ADDER.

Parallel Load: the simultaneous loading of data inputs to devices such as registers and counters.

Parity bit: a bit appended to a binary WORD to create or remove even or odd parity.

PARITY DETECTOR: a combinational logic device that will detect an even (or odd) number of 1's (or 0's) in a binary WORD.

PARITY GENERATOR: a combinational logic device that will append a logic 1 (or logic 0) to a binary WORD so as to generate an even (or odd) number of 1's (or 0's).

Passive device: any device that is incapable of producing voltage or current gain and, thus, only dissipates power.

Pass transistor switch: a MOS transistor switch that functions as a nonrestoring switching device and that does not invert a voltage signal.

p-channel: a p-type conducting region in an n-type substrate.

PDP: power-delay product.

Period: the time in seconds [s] between repeating portions of a waveform, hence, the inverse of the frequency.

Physical truth table: an I/O specification table based on a physically measurable quantity such as voltage.

PI: PRIME IMPLICANT.

Pipeline: a processing scheme where each task is allocated to specific hardware (joined in a line) and to a specific time slot.

PIPO: parallel-in/parallel-out operation mode of a register.

PISO: parallel-in/serial-out operaton mode of a register.

PLA: programmable logic array.

Planar format: a two-dimensional array of conventional fourth-order K-maps used to minimize functions of more than four variables.

PLD: programmable logic device.

PMOS: a p-channel MOSFET.

p-n junction diode: (see diode)

pnp: refers to a BJT having an n-type semiconductor base and a p-type semiconductor emitter and collector.

POLARIZED MNEMONIC: a contracted signal name onto which is attached an ACTIVATION LEVEL INDICATOR.

POS: product-of-sums.

POS HAZARD: a STATIC HAZARD that is produced by a 0-to-1 input change in POS logic.

Positional weighting: a system in which the weight of a bit in a binary WORD is determined by its polynomial representation.

Positive logic: the logic system in which HV corresponds to logic 1 and LV corresponds to logic 0.

Positive pulse: a 0-1-0 pulse.

Power, P: the product of voltage, V, and current, I, and is given in units of watts [W].

Power-delay product: the product of the average power dissipated by a logic device and its propagation delay time.

PPT: positive pulse-triggered.

PR or PRE: PRESET.

PRESENT STATE: the logic STATE of an FSM at a given instant.

PRESENT STATE/NEXT STATE table: a table that is produced from the NEXT STATE K-maps and that is used to construct a fully documented state diagram in an FSM analysis.

PRESET: an asynchronous input that is used in FLIP-FLOPS to set them to a logic 1 condition.

PRIME IMPLICANT: a group of 2^n adjacent MINTERMs or MAXTERMs which are sufficiently large that they cannot be combined with other 2^n groups in any way to produce terms of fewer variables.

PRIORITY ENCODER: a logic device that generates a coded output based on a set of prioritized data inputs.

Product-of-sums: in a Boolean expression, the ANDing of ORed terms.

Programmable array logic: any PLD that can be programmed only in the AND plane.

Programmable logic array: any PLD that can be programmed in both the AND and OR planes.

Programmable logic device, PLD: any two-level, combinational array logic device from the families of ROMs, PLAs, or PALs.

Programmable read-only memory: a once-only user programmable ROM.

PROM: programmable read-only memory.

Propagation delay: in a logic device, the time interval of an output response to an input signal.

PS: PRESENT STATE.

PS/NS: PRESENT STATE/NEXT STATE.

PT: pulse triggered.

P-term: a Boolean product term—one consisting only of ANDed variables.

P-term table: a table that consists of p-terms, inputs and outputs and that is used to program PLA type devices.

Pull-down resistor: a resistor that causes a signal on a line to remain at low voltage.

Pull-up resistor: a resistor that causes a signal on a line to remain at high voltage.

Pulse: an abrupt change from one level to another followed by an opposite abrupt change.

Pulse mode: an operational condition for an asynchronous FSM where the inputs are required to be nonoverlapping pulse signals.

Pulse-triggered: a triggering mechanism requiring a CLOCK pulse, often a very narrow CLOCK pulse.

Pulse width: the ACTIVE duration of a positive pulse or the INACTIVE duration of a negative pulse.

Quad: a grouping of four logically adjacent MINTERMs or MAXTERMs.

Quotient: the result of a division operation.

R: RESET.

Race condition: a condition in sequential circuits where a transition from one state to another involves two or more alternative paths.

Race gate: the gate to which an external input signal and a feedback variable are in race contention to determine whether or not an E-HAZARD or d-trio will be produced.

Radix: the number of unique symbols in a number system—same as the base of a number system.

RAM: random access memory.

Random access memory: a read/write memory system in which all memory locations can be accessed directly independent of other memory locations.

RC: resistance/capacitance or resistor/capacitor.

Read only memory: a PLD that can be programmed only in the OR plane.

Redundant cover: nonessential and nonoptional cover in a function representation.

REDUNDANT PRIME IMPLICANT: a PRIME IMPLICANT that yields redundant cover.

Reference matrix: a K-map that contains multiple-output entries. Also called a *composite output K-map*.

Reflective code: a code that has a reflection (mirror) plane midway through the code.

REGISTER: a digital device, configured with FLIP-FLOPs and other logic, that is capable of storing and shifting data on command.

Remainder: in division, the dividend minus the product of the divisor and the quotient.

RENDEZVOUS MODULE: a majority gate device whose output becomes ACTIVE when all external inputs become ACTIVE and becomes INACTIVE when all external inputs become INACTIVE.

RESET: a logic 0 condition or an input to a logic device that sets it to a logic 0 condition.

Residue: the part of term that remains when the coupled variable is removed.

Resistance, R: the voltage drop across a conducting element divided by current through the element, given in ohms.

Resistor-transistor logic: a logic family that consists of BJTs and resistors.

Restoring logic: logic consisting of switching devices such as BJTs and MOSFETs that can amplify.

Reverse bias: a voltage applied to a p-n junction diode in a direction as to minimize conduction across the junction.

Reverse saturation current: the current through a p-n junction diode under reverse bias.

Ripple carry: the process by which a PARALLEL ADDER transfers the carry from one FULL ADDER to another.

Ripple counter: a counter whose FLIP-FLOPs are each triggered by the output of the next LSB FLIP-FLOP.

Rise time: the period of time it takes a voltage (or current) signal to change from 10% to 90% of its high value.

Rising edge-triggered: activation of a logic device on the rising edge of the triggering (sampling) variable.

ROM: read-only memory.

Round-off error: the amount by which a magnitude is diminished due to an underflow or SPILL-OFF in a SHIFTER undergoing a right shift.

RPI: REDUNDANT PRIME IMPLICANT.

RTL: resistor-transistor logic.

S: SET. Also, the source terminal of a MOSFET.

Sampling interval: the interval that is the sum of the setup and hold times.

Sampling variable: the last variable to change in initiating a state-to-state transition in a sequential machine.

SANITY circuit: an resistor/capacitor (RC) type circuit that is used to initialize an FSM into a particular state.

Saturation mode: the physical state of a BJT in which collector current is permitted to flow.

Schmitt trigger: an electronic gate with hysteresis and high noise immunity that is used to "square up" pulses.

SELECTOR MODULE: a device whose function it is to steer one of two input signals to either one of two outputs depending on whether a specific input is ACTIVE or INACTIVE.

Self-correcting counter: a counter for which all states lead into the main sequence or routine.

Sequence detector (recognizer): a sequential machine that is designed to recognize a particular sequence of input signals.

Sequential machine: any digital machine with feedback paths whose operation is a function of both its past history and present input data.

SET: a logic 1 condition or an input to a logic device that sets it to a logic 1 condition.

Set-up time: the interval of time prior to the transition point during which all data inputs must remain stable at their proper logic level to ensure that the intended transition will be initiated.

SHIFT REGISTER: a register that is capable of shifting operations.

Shift: the movement of binary WORDS to the left or right in a shifter or shift register.

SHIFTER: a combinational logic device that will shift or rotate data asynchronously upon presentation.

Sign bit: a bit appended to a binary number (usually in the MSB position) for the purpose of indicating its sign.

Sign-complement arithmetic: 1's or 2's complement arithmetic.

Sign-magnitude representation: a means of identifying positive and negative binary numbers by a sign and magnitude.

SIPO: serial-in/parallel-out operation mode of a register.

SISO: serial-in/serial-out operation mode of a register.

Slice: that part of a circuit or device that can be cascaded to produce a larger circuit or device.

Small-scale integration: IC chips that, by one convention, contain up to 12 gates.

SOP: sum-of-products.

SOP HAZARD: a STATIC HAZARD that is produced by a 1-to-0 input change in SOP logic.

Source: one of three terminals of a MOSFET. The origin of a digital signal.

SPILL bit: the bit in a SHIFTER that is spilled off (lost) in a shifting operation.

Square wave: a rectangular waveform.

SSI: small scale integration.

Stability criteria: the requirements that determine if an asynchronous FSM, operated in the fundamental mode, is stable or unstable in a given state.

Stable state: any logic state of an asynchronous FSM that satisfies the stability criteria.

Stack format: a three-dimensional array of conventional fourth-order K-maps used for function minimization of more than four variables.

State: a unique set of logic values that characterize the logic status of a machine at some point in time.

State code assignment: unique code WORDs that are assigned to each state to identify that state for mapping purposes.

State diagram: the diagram or chart of an FSM that shows the state sequence, branching conditions, and output information necessary to describe its sequential behavior.

State transition table: a table, descriptive of the operation of a FLIP-FLOP, that is used to characterize that FLIP-FLOP as the MEMORY in an FSM design. Also, called an *excitation table*.

State variable: any variable whose logic value contributes to the logic status of a machine at any point in time. Any bit in the state code assignment of a state diagram.

STATIC HAZARD: an unwanted glitch in an otherwise steady state signal that is produced by an asymmetric path delay through an INVERTER or other gate.

Static RAM: a nonvolatile form of RAM—does not need periodic refreshing to hold its information.

Steering logic: logic based primarily on pass transistor switches.

S-term: a Boolean sum term—one containing only ORed variables.

Stretching stage: an input conditioning stage that catches a short input signal and stretches it.

Substrate: the supporting or foundation material in and on which a semiconductor device is constructed.

SUBTRACTOR: a digital device that subtracts one binary WORD from another to give a DIFFERENCE and BORROW.

Subtrahend: the operand being subtracted from the minuend in a subtraction operation.

Sum-of-products: in a Boolean expression, the ORing of ANDed terms.

Sum rule: a rule in state diagram construction that requires that all possible branching conditions be accounted for.

Switching speed: a device parameter that is related to its propagation delay time.

Synchronizer circuit: a logic circuit (usually a D FLIP-FLOP) that is used to synchronize an input signal with respect to CLOCK.

Synchronous machine: a sequential machine that is CLOCK driven.

System level design: a design that includes CONTROLLER and data path sections.

Tabular minimization: a minimization procedure that uses tables exclusively.

T FLIP-FLOP: a FLIP-FLOP that operates in either the TOGGLE or HOLD mode.

Time constant: the product of resistance and capacitance given in units of seconds [s]—a measure of the recovery time of an R-C circuit.

Timing diagram: a set of logic waveforms showing the time relationships between two or more logic signals.

Toggle: repeated but controlled transitions between any two states, for example, between the SET and RESET states.

Transfer characteristic: for a transistor switch, a plot of V_{out} versus V_{in}.

TRANS-HI MODULE: PPT TRANSPARENT D LATCH.

TRANS-LO MODULE: NPT TRANSPARENT D LATCH.

Transistor: a three-terminal switching device that exhibits current or voltage gain.

Transistor-transistor logic: a logic family in which bipolar junction transistors provide both logic decision and current gain.

Transition: in a digital machine, a change of one state or level to another.

Transmission gate: a pass transistor switch.

TRANSPARENT D LATCH: a two state D FLIP-FLOP in which the output, Q, tracks the input, D, when CLOCK is ACTIVE if TRANS-HI or INACTIVE if TRANS-LO. A PT D FLIP-FLOP.

Triggering threshold: the point beyond which a transition is said to take place.

Triggering variable: sampling variable.

Tristate bus: as used in this text, the wire-ORed output lines from a multiplexed scheme of PLDs having tristate ENABLEs.

TRISTATE DRIVER: an active logic device that operates in either a disconnect mode or an inverting (or noninverting) mode.

Truth table: a table that provides an output value for each possible input condition to a combinational logic device.

TTL: transistor-transistor (BJT) logic.

Two-level logic: logic consisting of only one ANDing and one ORing stage.

Two-phase CLOCKing: two synchronized CLOCK signals that have nonoverlapping ACTIVE or nonoverlapping INACTIVE waveforms.

Two's complement: one's complement plus one.

Unconditional branching: state-to-state transitions that can take place independent of the input status of the FSM.

Unconditional output: an output of an FSM that does not depend on an input signal.

Union: OR operation.

Unit distance code: a code in which only one bit is permitted to change between any two of its adjacent states.

Universal FLIP-FLOP: a JK FLIP-FLOP.

Universal gate: a NAND or NOR gate.

UNIVERSAL SHIFT REGISTER: a shift register capable of performing PIPO, PISO, SIPO, and SISO operations in addition to being capable of performing the HOLD condition.

Unstable state: any logic state in an asynchronous FSM that does not satisfy the stability criteria.

Unweighted code: a code that cannot be constructed by any mathematical weighting procedure.

VEM: variable entered map

Venn diagram: shaded circular diagrams that are used to represent the AND and OR operations between variables.

Very-large scale integrated circuits: IC chips that contain more than 1,000 gates.

Vietch diagram: a type of Karnaugh map.

VLSI: very-large-scale integrated circuits.

Voltage, V: the potential difference between two points, in units of volts [V]. Also, the work required to move a positive unit test charge against an electric field.

Voltage waveform: a physical waveform of voltage as a function of time in which rise and fall times exist.

Weighted code: a binary code in which the bit positions are weighted with different mathematically determined values.

Wired logic: an arrangement of logic circuits in which the outputs are physically connected to form an "implied" AND or OR function.

WORD: (see binary WORD)

XNOR: (see EQUIVALENCE)

XOR: EXCLUSIVE OR.

XOR function: the function that derives from the definition of EXCLUSIVE OR.

XOR gate: a physical device that performs the electrical analog of the XOR function.

XOR laws: a set of Boolean identities that are based on the XOR function.

XOR pattern: any of four possible K-map patterns that result in XOR type functions.

XS3 code: NBCD code plus three.

Zero banking: a feature of an NBCD-to-seven-segment conversion that blanks out the seven-segment display if all inputs are zero.

Index